Contents

2   NATURAL VIBRATION ABOUT EQUILIBRIUM   54

3   FORCED HARMONIC VIBRATION   116

Contents

Preface

Speed of rotating machines has changed very little from the time of the muscle powered machines of the stone age to the water and air powered ones of the classical and medieval times and finally to the Watts engine. Rotating speeds were, in general, below 1000 rpm at the end of the 19th Century. At that time, in the 1870's, Dr. Gustaf Patrik de Laval, a swedish engineer, invented the milk separator which had to work at 6000 to 10,000 rpm. De Laval's first units were horse-wheel or hand-driven with geared step-up of speed. He soon saw the need for a direct drive and the steam turbine was born. The day that de Laval presented his marine steam turbine to the World Columbian Exposition, opened in 1893 in Chicago by President Cleveland, marks the beginning of the era of high speeds. De Laval turbines worked to 42,000 rpm speeds, way above the critical speed. In a few years, there was a 40-fold increase in rotating speed which was equivalent to a 1600-fold increase in the unbalance forces. It was the time that the brilliant solutions in vibration theory developed by the great mathematicians, from Newton to Poincaré, would find problems to be applied to. Sound and vibration became a separate branch of physics and Lord Rayleigh's "Theory of Sound" appeared. Subsequently, mechanical vibration became an engineering discipline and W. Hort's "Technische Schwingungslehre" was published in 1910 by Julius Springer.

Most vibration topics taught to undergraduate students were developed in the time period between the work of Newton and that of Lord Rayleigh by mathematicians who had very little to do with applications. On the other hand, our century may be labeled the era of vibration applications, and they are too numerous to put in one book. Therefore, such a textbook should include the basic theory and some notable or pedagogically useful applications. The balance, in the sense introduced by Truesdell in his "Six Lectures on Natural Philosophy," is one of the manifestations of the taste of the authors who, having been practicing engineers for a lifetime, would not have anything to add to the Leonardo da Vinci's dictum: "O students, study mathematics, and do not build without foundations. . . ."

During the past 50 years, most engineering authors have omitted the original references, replacing them with recent textbooks or references. A practical reason was posed for this practice—that more recent references are easier to find for further study. This, of course, is not exactly true. It is much easier to locate in libraries Newton's *Principia mathematica* or Lagrange's *Mécanique analytique,* perhaps in English translations, in libraries than it is to find many recent references. Moreover, no student of modern poetry would consider his or her education complete without reading Homer or Shakespeare, although they may have very little in common with Howard Nemerov's poetry. Yet very few engineering students have read Newton or Euler, despite the fact that most of the material they learn was written by them. Finally, every engineer should take seriously the counsel of Leibnitz: "It is most useful to trace the sources of memorable inventions. . . . That is so because such knowledge helps not only the history of letters give each his own and encourage others to pursue like glory, but also, when method is disclosed by shining examples, the art of discovery graten." Of course, to revive the accurate referencing, nearly forgotten for so long and controversial, is not easy. Suggestions and comments by readers will be of great help.

In C. Truesdell words, "experiment is necessary to see for yourself if something you read is true but also to find something you have not already read." We have no control over the availability of laboratory facilities to readers while they study this book. However, SIMULAB is a vehicle that provides the reader with simulated experience aiming at making the computer screen an observation window into a laboratory where most of the principles described here can be demonstrated. This, of course, is not a substitute for laboratory experience. In its absence, however, it can provide the reader with an alternative route to reach the understanding of the subject. The reader can practice on his or her computer, viewed as a laboratory window, using the keyboard as a control console.

Desegregation of design in all engineering science courses has been advocated as an alternative to specific design courses in engineering education, in the tradition of schools such as the Ecole Polytechnique. In this sense, major emphasis is placed on design, not in the form of providing ample "design formulas" but using examples and design problems in most chapters which are true case studies and truly open ended, with the belief that this can contribute to the dissemination of design to the engineering curricula instead of its segregation into design courses. This offers the possibility for teaching a vibrations course as a design course.

In an earlier edition, this book, entitled *Vibration Engineering,* was one of the first in engineering to include a rather complete set of FORTRAN code for vibration analysis. We have decided not to include it here in printed form but in the form of electronically recorded software, not only in the interest of space but to allow for debugging and upgrading the code as new hardware and software become available. With the contemporary rate of change in computers, the average of 10 years between editions of a book is certainly a very long time. Thus source code is not included in the text, but the basic algorithms are written in pseudocode for easy implementation in the high-level language of the reader's preference.

In recent years, engineers have realized the importance of using vibration analyses to monitor and diagnose machinery conditions for more effective predictive maintainance. Chapters 13 and 14 were written as an introduction to this methodology.

Since this is a completely rewritten new edition, the authors will be obliged if the readers help in the identification of misprints, errors, and points of ambiguity in the text, the problems, or the software.

ACKNOWLEDGMENTS

The authors gratefully acknowledge the help provided by Professors A. Vakakis, University of Illinois–Urbana, A. Unal, Santa Clara University/Reliance Technical Services Inc., and H. Hamidzadeh, South Dakota State University, who read parts of the manuscript and offered valuable suggestions for improvement of the text.

Andrew D. Dimarogonas
Sam Haddad

Historical
Introduction

---

1 THE ORIGINS OF VIBRATION THEORY

The development of vibration theory, as a subdivision of mechanics, came as a natural result of the development of the basic sciences it draws from, mathematics and mechanics. These sciences were founded in the middle of the first millennium B.C. by the ancient Greek philosophers. Of course, people were using the underlying principles in their everyday life long before that, sometimes in a systematic way. For example, geometry and other branches of mathematics were used extensively during the second and third millennia B.C. in Mesopotamia and Egypt in problems such as land surveying. The rules they developed and used were generally empirical in nature and no attempt was made to deduct these rules from fundamental principles in a rigorous way.

The scientific method of dealing with nature started with the Ionian school of natural philosophy, whose leader was Thales of Miletos (640–546 B.C.), the first of the seven wise men of antiquity. He is perhaps better known for his legendary discovery of the electrical properties of yellow amber (electron) and introduction

*Notes for the Historical Introduction appear, as numbered, at the end of the chapter.

of the term *electricity*, the phenomenon observed when rubbing electron against a wool cloth. More important, Thales introduced the concept of the logical proof for abstract propositions [Hunt, 1978].[1] Thales, who also was a very successful entrepreneur, traveled extensively in Mesopotamia and Egypt and became familiar with the knowledge that the people there had of geometry and astronomy.

Pythagoras of Samos (ca. 570–497 B.C.), a close contemporary of Buddha, Confucius, and Lao-Tse, can be considered a student of the Ionian school. After traveling to Babylon and Egypt, and probably India, he moved from Samos to the Croton area of southern Italy and established the Pythagorean school, the first institution of higher education and scientific research. The primary contributions of the Pythagorean school were the development of the theory of numbers and the theory of music and harmony.

The term *vibration* has been used from the time of Aeschylos.[2] Differences in the pitch of sound have been understood since the development of music. Although it has been suggested that a musical instrument existed in 13,000 B.C.,[3] it is certain that an understanding of music and of consonance dates back to 3000 B.C. in China, where the philosopher Fohi wrote two monographs on the theory of music [Skudrzuk, 1954]. Pythagoras has quantified the theory of music and related it to his theory of numbers. Boethius [in Lindsay, 1972] reported a legendary incident involving Pythagoras.

> He was passing a metal workers' shop and heard the hammers when struck produce somehow a single concord from their diverse sounds. Surprised to find that which he had long been seeking he went into the shop and after long consideration concluded that it was the variation in the force of those using the hammers that produced the diversity of sounds. To verify this he had the men exchange hammers. But it turned out that the character of the sounds did not depend on the strength of the men but remained the same even after the hammers were exchanged. On noting this he weighed the hammers. Now there happened to be five hammers, and those two which gave the consonance of an octave (diapason) were found to weigh in the ratio of 2 to 1. He took that one which was double the other and found that its weight was four-thirds the weight of a hammer with which it gave the consonance of a fourth (diatessaron). Again he found that this same hammer was three-halves the weight of a hammer with which it gave the consonance of a fifth (diapente). Now the two hammers to which the aforesaid hammers had been shown to bear the ratio of 4 to 3 to 2, respectively, were found to bear to each other the ratio of 9 to 8. The fifth hammer was rejected, for it made no consonance with the others. On returning home Pythagoras tried to determine by various researches whether the whole theory of consonances could be explained by these proportions. Thus he attached equal weights to strings and judged their consonances by the ear. Again, he varied the procedure by doubling or halving the length of reeds and using the other proportions. In this way he achieved a very considerable degree of certainty.

> Often as a means of testing the proportions he would pour cyathi (bowls) of fixed weight into vessels, and with a bronze or iron rod, strike the vessels containing the various weights. He was overjoyed to find no reason to alter his conclusions. He then proceeded to examine the length and thickness of strings. In this way he discovered the principle of the monochord of which we shall speak hereafter. The monochord was called canon not merely from the wooden ruler by which we measure the length

of strings corresponding to a given tone, but because it forms for this type of investigation so definite and precise a standard that no inquirer can be deceived by dubious evidence.

This incidence is illustrated in Figure 1.

In this way, Pythagoras established a rational method of measuring sound frequencies, at least integer fractions and multiples of basic sounds of musical instruments. The standard frequency for calibration would be the perception of trained musicians, which were accurate enough for the needs of the time. Pythagoras founded not only the science of acoustics but also the theory of vibration, since the relation of sound and vibration was well known in his school.

Pythagoras conducted experiments with hammers, strings, pipes, and shells. He established the first vibration research laboratory (shown in Figure 2, based on a drawing by Boethius) the first known research laboratory [Dimarogonas, 1990]. Moreover, he invented the monochord,[4] a purely scientific instrument, to conduct experimental research into the vibrations of taut strings and to set a standard for vibration measurements (Figure 2).

Contrary to the traditional belief that Galileo Galilei was first to observe the isochronism of the pendulum, it probably was long known to ancient cultures, such

Figure 1 Pythagoras observing the hammer sounds in the smithy. (From Hugo Sprechshart. 1488. *Flores Musicae*.)

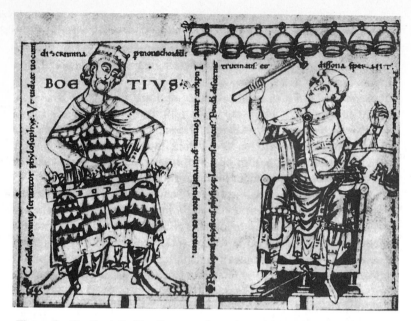

Figure 2 On the left, Boetius experimenting with the Monochord. On the right, Pythagoras performing the hammer experiments with bells in his laboratory. (From the Boethius manuscript at Cambridge.)

as those of the Greeks and Chinese. Legend suggests Daedalus (middle of the second millennium B.C.) as the inventor of the pendulum. It appears as a bob for spinning and leveling devices on vases of the sixth century B.C. Spinning was also known in India in 2500 B.C. The first indication of use of the pendulum as a timing device is mentioned by Aristophanes (450–388 B.C.). In his *Frogs* (*Ranae*), he made a direct reference: "the music should be balanced with an oscillator."[5] It can be assumed that the pendulum was known to the ancient peoples. In fact, the balance (which is a compound pendulum) was known in 4500 B.C., and according to the dimensions of the balances or drawings that have been found, they would oscillate at about 1 Hz, a practical frequency for a balance. It can further be assumed that isochronism was known. By this time, the mechanical clock was well known. Because the escapement mechanism, although known in some form to Chinese and Arabs, was not then available in a practical form, the pendulum was not widely used for time measurements.

The first monograph on acoustics, *On Acoustics*, was written by Aristoteles. He thus introduced the term, traditionally attributed to Sauver (1653–1716) [Lindsay, 1966].

Static equilibrium rules were known empirically in very early times, but a rigorous organization into seven postulates and seven propositions was first proposed by Archimedes (ca. 287–212) in his work *On the Equilibrium of Planes*.

In his *Mechanics*, Aristoteles[6] attempted earlier to formulate statics in the framework of the general laws of motion. He understood the vectorial character of forces and introduced the law of the parallelogram for force addition. To prove the static equilibrium equation $\Sigma F_i x_i = 0$, he introduced the principle of virtual

4   Historical Introduction

work $\Sigma F_i \delta_i = 0$. In *Physics*, Aristoteles remarks: "No one could say why a thing one's set in motion should stop anywhere; for why should it stop there rather than here? So that a thing will either be at rest or must be moved ad infinitum, unless something more powerful gets in its way." This can be considered a fair analogy to Newton's first law: "Every body continues in its state of rest, or in uniform motion in a right line unless it is compelled to change that state by forces imposed upon it." It should be noted, however, that Aristoteles does not pose this as an axiom. He offers a proof, as a *reductio ad absurdum*.

Acceleration was known and related to force. In a commentary on Aristoteles' *On Heavens (De Caelo)*, Simplicius stated that the force between two terrestrial bodies depends on their distance and mass. Further, he asserts that acceleration is proportional to force, for the same mass, in analogy to Newton's second law. Aristoteles himself asserts that "the force is proportional to the weight for traveling the same distance at the same time" (*Physics*); "the relative speeds of two bodies will be in inverse ratio to their respective sizes" (*On the Heavens*); "as to the cause of such acceleration . . . the bodies are endowed with greater force" (Simplicius, *Commentary on Physics*). Newton's second law is: "The change of motion is proportional to the motive force impressed and is made in the direction of the right line in which this force is impressed."

On the principle of action and reaction, Aristoteles states: "The agent is itself acted upon by that on which it acts," and he broadens the scope of this axiom to cover heat and fluid flow (*On the Heavens*). Newton's third law states: "To every action there is always opposed an equal reaction; or, the mutual actions of two bodies upon each other are always directed to contrary parts."

On the relation of motion to potential and kinetic energy, Alexander of Afrodisias, a commentator on Aristoteles, introduced the idea of the conservation of energy[7]: "For the thing which, by causing them to pass from heaviness in potentia to heaviness in actu, causes them to be in a different state from the one they were before, is the cause of natural motion."

The basic difference between Aristotelian and Newtonian dynamics does not lie in the equations that express the respective concepts. The basis for the former is the philosophical conception of the world's harmony, based on a teleological view of nature. The basis for the latter is an *a priori* conception of time and space.

The fact that for a linear system there are frequencies that can provide harmonic motion was long known to musicians, but it was first stated as a law of nature by Pythagoras. Moreover, he proved experimentally with his hammer experiments that natural frequencies are system properties and do not depend on the magnitude of the excitation. He proceeded experimentally to prove, as reported by Theon of Smyrna (second century A.D.) that:

1. The natural frequency of a string is inversely proportional to its length and diameter; it increases with increasing tension "with other proportions" (not specified). It is quite probable that Pythagoras was aware of the correct rule of the dependence of the natural frequency on the tension, since weight could be measured with accuracy.

2. The natural frequency of longitudinal vibration of columns is inversely proportional to the length of the columns.

3. The statement above regarding columns is also true of vessels. Pythagoras changed their natural frequencies by pouring water into them.

4. Pythagoras also tested disks, but no results were reported. There is a report in Platon's *Phaedon* (A.D. 108) that Hippasos (a student of Pythagoras said to have been killed for divulging secrets of the Pythagoreans) tested four bronze disks and found that their natural frequencies were inversely proportional to their thickness.

Vibration instruments were used in ancient Greece and China. Herodotos (ca. 484–425 B.C.) reported[8] on a vibration transducer, a shield coated with a thin sheet of bronze. By holding it against the ground and listening to the sound of the vibrating bronze plate, it was possible to detect in the sixth century B.C. the digging of underground tunnels in Barca, a North African town where Libya is today, when it was under Persian siege. One can recognize a mechanical vibration transducer-amplifier, very similar in principle to a reed tachometer. Similar systems were known in China in 370 B.C. [Needham, 1962].

For the same technical application, Vitruvius [first century B.C.] reports on the first use of a pendulum as another type of mechanical vibration transducer-amplifier. In 214 B.C. the architect Tryphon of Alexandria used hanged vases, which began to oscillate in response to underground shocks when the enemy, using iron tools, was digging tunnels under the city during the siege of the Illyrian town Appolonia by Philip V, the king of Macedonia. One can recognize here an understanding of the pendulum principle and the principle of resonance and sympathetic vibrations. The latter term was introduced not by Galileo but by Plotinus (3rd Century A.D.) *"The strings move sympathetically."*[2]

The term *balancing*, as with vibrations, dates back to Aeschylos. Use of the term *tuning* implies that the concept must have been known with regard to musical instruments in very early times. For machinery, tuning was performed in catapults to achieve the desired prestress for maximum strain energy storage[9]: "Thus, with tight wedging, catapults are tuned to the proper pitch by musical sense of hearing." Vitruvius reports that to help with theater acoustics, they would tune large vases to different notes and consonances and distribute them throughout the audience, using them effectively as mechanical amplifiers or, better, echo systems [Lindsay 1972].

China has long experienced deadly earthquakes that have had devastating economic, social, and political consequences. In about A.D. 132, the imperial government wanted a way to detect distant earthquakes so that they would be able to respond swiftly. The scientist and mathematician Chang Cheng invented an instrument, a 3-m-long pendulum (Figure 3) using falling balls, to record direction and perhaps magnitude. In fact, while Cheng was trying to convince the authorities of the utility of his invention [Needham, 1962], an incident occurred that convinced the public that the instrument had magic properties. It recorded an earthquake in the province of Lung-Hsi, 400 miles away, of such small magnitude that nobody had noticed it. Later, news of the earthquake came from the distant province. The natural frequency of the instrument was approximately 0.3 Hz, and apparently the motion recorded was of too low frequency to be felt by people.

Figure 3 Chinese seismograph, second century A.D. (Reproduced from Temple, R. 1989. *The Genius of China*. New York: Simon & Schuster.)

The legend of Jericho,[10] where the walls were demolished in resonance with musical instruments, may have been only symbolic. It makes clear, however, that engineers in Joshua's time knew well the phenomena associated with resonance and with sympathetic vibration.

In the ancient world, there was substantial progress in vibration theory and an extended understanding of the basic principles of natural frequency, vibration isolation, vibration measurements, and resonance and sympathetic vibrations. This body of knowledge had very limited use, however, due to the low level of production technology and machinery speeds. Moreover, many branches of mathematics were already extensively developed, but calculus and computational mathematics were at too early a stage to allow for analytical treatment of vibration.

2 THE EARLY MODERN ERA

The early modern era is marked by the works of Galileo and Newton. Moreover, it is marked by the early stages of mechanization and the industrial revolution. The utilization of chemical energy with its associated high power per unit machinery volume, introduced numerous vibration problems. This, together with the development of calculus and continuous mechanics, led to the rapid development of vibration theory by the mid-nineteenth century. At the time of Galileo and Newton, physics and mechanics were much more developed than is generally assumed. Their fundamental contribution is that they revived and redefined these sciences at times when progress in natural science was being demanded. There were several previous attempts to revive physics and mechanics in the first half of the second millennium A.D. However, the time was not yet appropriate.

Galileo (1564–1642) was born in Pisa to a noble Florentine family. He was educated in Greek, Latin, and logic in the monastery of Vallombrosa near Florence and continued at the University of Pisa, where he was to study medicine. He soon threw all his energy into studying the works of Euclid and Archimedes and became familiar with the work of Leonardo da Vinci in mechanics. During his stay in Pisa (1589–1592) he conducted his famous experiments with falling bodies, which were published in 1590 in the treatise *De Motu Gravium*. The principal conclusions were that:

1. All bodies fall from the same height at equal times.
2. The final velocities are proportional to the times.
3. The spaces fallen are proportional to the square of times.

These conclusions, being in some disagreement with Aristoteleian mechanics, did not win him friends. He did not obtain a degree and had to return to Florence and give private lessons to earn a living. In 1592 he was given a Chair at the University of Padua, and in 1610 he returned to Florence as "philosopher and mathematician extraordinary" to the Great Duke of Tuscany. There he published his work *Discorsi . . . ,*[11] in which he included his ideas on vibration and strength of materials.

Galileo's discussions on the isochronism of the pendulum and the study of resonance and forced vibration stirred new interest in vibration and acoustics. His death interrupted his work on the development of a pendulum clock, and his son continued the work. It was Huygens, however, who developed the pendulum clock, the first accurate device for time measurements. Moreover, the deviation from isochronism due to the pendulum's nonlinearity was noted and various designs were devised to yield an accurate pendulum clock.

While the originality of Galileo's ideas is controversial,[12] his works and life contributed substantially to a movement that brought together many men with scientific interests in experimental work. This movement culminated in the establishment in Naples of the Accademia Saccretorum Naturae in 1560 and the Academia dei Lincei in Rome in 1606, with Galileo as a member. After Galileo's death in 1642, the Academia del Cimento was organized in Florence, and in 1662 the Royal Society was begun in London. In 1966, the French Academy of Science and in 1770 the Berlin Academy of Science were organized. All these academies published their transactions, and these had a great influence on the development of science in the eighteenth and nineteenth centuries.

Isaac Newton was born on Christmas day in 1642, the day of Galileo's death. He enrolled at Trinity College to study chemistry, but early in his first year studied Euclid and later, other mathematicians. In 1665 he began thinking of the rate of change, or *fluxion*, of continuously varying quantities. He later became professor of mathematics at Cambridge and president of the Royal Society of London. His laws of motion were published in his *Philosophia Naturalis Principia Mathematica* in 1687, perhaps the most admired scientific treatise of all time. Although the laws of motion were already known in one form or other, it was the development of calculus by Newton and Leibnitz that made them applicable to the problems of physics and mechanics.

Beginning in 1686, Jakob Bernoulli attempted to relate Huygens' theory of a swinging body to Archimedes' law of the lever, that is, to rational mechanics; Newton's *Principia* offered the vehicle. Bernoulli revisited the problem in 1703, recognizing that for bodies in motion, the system of applied forces is equivalent to the system of reversed accelerations per unit mass. Between then and 1750, differential equations of equilibrium of motion for various collections of bodies were derived by Taylor, Johann and Daniel Bernoulli, Euler, Clairaut, and D'Alembert.[13]

As stated above, extensive experimental results were available for vibrating strings since the time of Pythagoras, and further results were obtained by Galileo and Marinus Mersenne (1588–1648), a Franciscan friar. The vibration modes and nodal points were observed by Joseph Sauveur (1653–1716), who also identified the *fundamental natural frequency* and *harmonic tones*. Daniel Bernoulli explained the experimental results by the *principle of superposition* of harmonics and introduced the idea of expressing any small oscillation as a sum of the independent simple harmonics,[14] each with its own frequency and amplitude. The problem of the vibrating string was first solved mathematically by Lagrange,[15] who considered it as a sequence of small masses.

The wave equation was introduced by D'Alembert in a memoir to the Berlin Academy in 1750.[16] In his memoir he also used it for longitudinal vibration of air columns in pipe organs. Experimental results for the same problem were obtained by Pythagoras. The solution of the string equation is due to Daniel Bernoulli, D'Alembert, and Euler, although Joseph Sauveur (1653–1716) and Brook Taylor (1685–1731) have earlier obtained approximate solutions.

Euler[17] obtained a differential equation for the lateral vibration of bars and determined the functions we now call *normal functions* and the equation we now call the *frequency equation* for beams with free, clamped, or simply supported ends. Chladni[18] investigated these vibration problems and the longitudinal and torsional vibration of bars. Furthermore, working on the problem of oscillations of ships, in 1838, Euler advanced the hypothesis that any body has three orthogonal axes through its center of mass, about each of which it oscillates freely in small motion, with an arbitrary amplitude for each. Moreover, in 1846 he discovered that permanent rotation about an axis is possible only if the products of inertia about the axis are zero, and shortly thereafter, that Newton's laws apply not only for a body as a whole but to every part of it. He then expressed the second law as the principle of linear momentum.

Euler and Jakob Bernoulli attempted to solve the problem of vibrating plates and shells analytically. Euler[19] considered an elastic membrane to consist of two systems of stretched strings perpendicular to each other and obtained the membrane differential equation. Jakob Bernoulli (1759–1789) obtained a differential equation of the vibrating plate, considering it as consisting of two systems of beams perpendicular to each other. Further, Euler proved for a variety of vibration problems that the principle of small oscillations and superposition were direct consequences of the law of linear momentum, and not independent axioms. He further developed an idea of Jakob Bernoulli regarding the principle of rotational momentum, which, together with the principle of linear momentum, constitute the basic axioms of mechanics.

Chladni further investigated plates and shells—bells in particular. His work stirred great interest in the subject. At the suggestion of Napoleon, who was impressed by a demonstration of Chladni's experiments, the French Academy proposed as the subject of a prize, the investigation of vibrating plates. Sophie Germain (1776–1831) won the prize after a lengthy procedure and controversy [Timoshenko, 1953]. She derived the correct differential equation, but there was concern regarding the rigor of the proof and the correctness of the boundary conditions. Further improvements were made by Poisson and Kirchhoff, but it was Navier[20] (1785–1863) who gave a rigorous theory of the bending vibration of plates. He further investigated the general equations of equilibrium and vibration of elastic solids. He formed an expression for the work done by all forces in a small relative displacement and obtained the differential equations by way of the calculus of variations.

Solution of the differential equations of motion for an elastic solid was treated by Poisson[21] and Clebsch[22] (1833–1872), who founded the general theory of vibrations. The theory of vibration of thin rods, in particular the theory of torsional vibration, was brought under the general equations of vibration of elastic solids by Poisson.[23]

3 MODERN TIMES

At the end of the nineteenth century, the theory of vibration was very extensively developed. At the same time, there was rapid progress in high-speed machinery building, in particular the development of locomotives and steam turbines. De Laval was experimenting with 30,000-rpm turbine rotors and the theory already developed found numerous applications.

The first systematic treatise on vibration was written by Lord Rayleigh [Rayleigh, 1894]. He formalized the idea of normal functions, as introduced by Daniel Bernoulli and Clebsch, and introduced the ideas of generalized forces and generalized coordinates. He further introduced systematically the energy and approximate methods in vibration analysis, without solving differential equations. This idea was developed further by W. Ritz.[24] Rayleigh introduced a correction to the lateral vibration of beams due to rotatory inertia, and Timoshenko[25] the correction due to shear deformation.

Whirling of shafts was anticipated by W. A. Rankine.[26] Extensive analytical investigations were done by Dunkerley and Reynolds,[27] and De Laval observed and resolved experimentally most rotor dynamics problems, experimenting with steam turbines in the last quarter of the nineteenth century. The whirling problem was solved by A. Föppl.[28] His analysis is sometimes inaccurately credited to Jeffcott[29] and the De Laval rotor is sometimes misnamed the "Jeffcott rotor."

In the 1920s the turbine industry designed machines to operate at substantially higher loads and at speeds above the lowest critical speed, and this introduced the modern-day rotor dynamics problems, which were treated by B. L. Newkirk[30] and A. T. Kimball.[31] Gyroscopic effects were introduced by A. Stodola.[32] The influence of fluid bearings was investigated by Stodola and further quantified by B. L. Newkirk and H. D. Taylor[33] and by A. Stodola.[34]

Vibration of shafts and beams of engineered shapes was first studied by Frahm,[35] in particular, torsional vibration of ship main shafts. Frahm's method was developed in tabular form by H. Holzer,[36] E. Guembel,[37] and M. Tolle.[38] For lateral vibration of bars, the HGT method was developed by F. van den Dungen[39] for lumped mass beams in Belgium and by Hohenemser and Prager[40] for continuous but discretized beams, although it is generally misnamed the *Myklestad*[41]–*Prohl*[42] *method*. This points up an interesting aspect of the development of vibration theory: In the 1920s and 1930s electric power was in great demand in the United States, due to the rapidly developing industry. Turbine manufacturers hired several engineers from Europe, notable among them Timoshenko, den Hartog, and Myklestad, who were all hired by Westinghouse. N. O. Myklestad, a Norwegian, attended the University in Copenhagen in the late 1920s and probably was familiar with van den Dungen's work. Prohl, who probably was unaware of van den Dungen's work, stated in his paper that the method was outlined for him by H. Poritsky, a General Electric mathematician who came from Russia. Prohl used van den Dungen's method for the computation of critical speeds of rotors, and his paper had a very substantial impact on turbomachinery design in the United States. The van den Dungen method was further developed in matrix form by W. Thomson[43] and was called the *transfer matrix method*.

Finite element methods are now used extensively. The basis of the idea of continuum discretization as it applies to beams involves the Holzer–Guembel–Tolle–van den Dungen method and the transfer matrix method. Its use for a general continuum was suggested by Courant.[44] It was developed for structures by M. J. Turner, R. W. Clough, H. H. Martin, and L. J. Topp.[45] As noted earlier, W. Ritz extended energy methods introduced by Lord Rayleigh,[46] and Galerkin introduced a generalization of this idea.[47]

The nonlinear behavior of the pendulum was observed in Huygens' time (1629–1695).[48] Systematic study of nonlinear systems was introduced by H. Poincaré,[49] G. Duffing,[50] and B. van der Pol.[51] Approximate methods were developed by Duffing, Linstedt,[52] and others, and the general treatment of stability problems was developed by Liapounov.[53]

The first systematic treatises on vibration were part of Lord Rayleigh's *Theory of Sound* or parts of mechanics textbooks, such as the classic book by August Föppl.[54] The first engineering vibration textbook was written in 1910 by Hort,[55] an engineer at Siemens-Schuckert in Berlin. This book had a profound effect on vibration teaching since it established vibration as a separate field of mechanics and as a course in mechanical and electrical engineering. Moreover, it became a standard reference, though very seldom referenced, for all subsequent vibration textbooks, and its outline has served as the standard for vibration textbooks up to the present day.

Hort's book was followed by a number of vibration textbooks in the 1920s.[56] Shortly thereafter, Timoshenko[57] and den Hartog,[58] both students of Prandtl at one time, transferred this practice to the United States. Their textbooks were not particularly original but had a profound influence on both teaching and research on the subject in the United States and were instrumental in making the United States the center of developments in the field of vibration research.

TABLE 1 VIBRATION CHRONOLOGY

People	Events
Fohi (3000 B.C.)	Consonances; book on music
Thales of Miletos (640–546 B.C.)	Scientific method
Pythagoras of Samos (ca. 570–497 B.C.)	Natural frequency; experimental physics; theory of numbers
	Vibration of the taut string
Herodotos (ca. 484–425 B.C.)	
Aristophanes (450–388 B.C.)	Oscillator
Platon (ca. 429–347 B.C.)	Sympathetic vibration
Aristoteles (384–322 B.C.)	Laws of motion; book on acoustics
Euclides (330–275 B.C.)	Pendulum use as vibration meter
Alexander of Aphrodisias (early third century B.C.)	Kinetic and potential energy
Archimedes (ca. 287–212 B.C.)	Laws of statics, hydrostatics
Vitruvius (first century B.C.)	Chinese seismograph
Boethius (A.D. 480–524)	
Leonardo da Vinci (1452–1519)	
Galileo Galilei (1564–1642)	Measurement of pendulum frequency
Marinus Mersenne (1588–1648)	
Christian Huygens (1629–1695)	Pendulum clock; nonlinearity
Isaac Newton (1642–1727)	Laws of motion
Gottfried Leibnitz (1646–1716)	Calculus
Joseph Sauveur (1653–1716)	Vibration harmonics
Brook Taylor (1685–1731)	
Daniel Bernoulli (1700–1782)	Wave equation
Leonhard Euler (1707–1783)	Principle of superposition
Jean le Rond D'Alembert (1717–1783)	
Joseph-Louis Lagrange (1736–1813)	Lagrange's equation
C. A. Coulomb (1736–1806)	Torsional vibration
E. F. F. Chladni (1756–1824)	
Jakob Bernoulli (1759–1789)	
Sophie Germain (1776–1831)	Vibration of plates
Simeon-Dennis Poisson (1781–1840)	
Claude Louis Marie Henri Navier (1785–1836)	Vibration of solids
Augustin Cauchy (1789–1857)	French Revolution
J. M. C. Duhamel (1797–1872)	
W. J. M. Rankine (1820–1872)	Critical speeds of shafts
A. Clebsch (1833–1872)	
John William Strutt, Lord Rayleigh (1842–1919)	Energy methods; first vibration treatise
Carl G. P. deLaval (1845–1913)	
Henri Poincaré (1854–1912)	Nonlinear vibration
Heinrich Hertz (1857–1894)	
A. Stodola (1859–1942)	Rotor-bearing dynamics

Historical Introduction

4 CONCLUSION

There were four distinct periods in the development of vibration engineering (see Table 1):

1. *Sixth Century BC–AD 1600*. In the ancient world, there was substantial progress in vibration theory and an extended understanding of the basic principles of natural frequency, vibration isolation, vibration measurements, and resonance and sympathetic vibrations, with very limited use in engineering.

2. *1600–1850*. The early modern era coincided with the early stages of mechanization and the industrial revolution. The utilization of chemical energy and the development of calculus and continuum mechanics led to the rapid development of vibration theory by the mid-nineteenth century.

3. *1850–1920*. At the end of the nineteenth century there was rapid progress in machinery building, in particular the development of locomotives and steam turbines. The theory already developed found numerous applications.

4. *1920s*. In the 1920s the turbine industry designed machines to operate at substantial loads and speeds above the lowest critical speed, and this introduced modern-day rotor dynamics and vibration analysis. Many of the earliest works of some great mechanicians of this century, such as Prandtl, A. Föppl, Stodola, von Mises, von Kármán, and Prager, during the period when most of the important vibration concepts were identified, dealt with vibration.

NOTES

[1]Names/dates in brackets refer to the reference lists at the ends of chapters. Notes are given primarily for clarification or for historical reasons.

[2]Liddel, W. G., and Scott, R. 1879. *Greek–English Lexicon*. Oxford: Clarendon Press.

[3]Sandars, N. K. 1968. *Prehistoric Art in Europe*. London: Harmondsworth.

[4]Theon of Smyrna (second century A.D.). *On Mathematical Matters Useful in Reading Platon*. Edited and translated into French by Jean Dupuis. 1892. Paris: Librairie Hachette.

[5]Aristophanes (450–388 B.C.). *Frogs*. Oxford: Clarendon Press, 1905.

[6]Cohen, M. R., and Drabkin, I. E. 1958. *A Source Book in Greek Science*. Cambridge, Mass.: Harvard University Press.

[7]Alexander of Aphrodisias (early third century B.C.). *Commentary on Aristotle's Metaphysics*. Translated in Sir Thomas L. Heath. 1932. *Greek Astronomy*. London: J.M. Dent.

[8]Herodotos. *Stories*. Translated by A. D. Godley. 1921. Vol. 2 (of four), No. 118 of the Loeb Classical Library. London: Heinemann; New York: G.P. Putnam's Sons.

[9]Gordon, J. E. 1978. *Structures*. New York: Plenum Press.

[10]Joshua 6:20.

[11]Galileo Galilei. 1638. *Discorsi e demonstrationi matematiche intorno a due nuove scienze attenenti alla mechanica e ai movimenti locali*. Translated by Crew, H., and de Salvio, A. 1914. New York: Macmillan.

[12]Truesdell, C. 1987. *Great Scientists of Old as Heretics in "The Scientific Method."* Charlottesville, Va.: University Press of Virginia.

[13]Truesdell, C. 1984. *An Idiot's Fugitive Essays on Science*. New York: Springer-Verlag.

[14]1750. *Reflections et éclaircissements sur les nouvelles vibrations des cordes exposés dans les Mémoires de l'Academie de 1747 et 1748*. Royal Academie of Berlin, p. 147.

[15]1759. *Recherches sur la nature et la propagation du son*. Miscellanea Taurinensia, t. I.

[16]*Recherches sur le courbe que forme une corde tendue mise en vibration*, Royal Academy of Berlin, 1747, p. 214.

[17]Euler, L. 1744. Aditamentum, De curvis elasticis. In *Methodus inveniendi lineas curvas maximi minimive proprietate gaudentes*. Lausanne. Bernoulli, D. 1751. De vibrationibus . . . laminarum elasticarum. *Comment. Acad. Sci. Imperialis Petropolitanae, 13*.

[18]Chladni, E. F. F. 1802. *Die Akoustik*. Leipzig.

[19]Euler, L. 1767. *Novi Comment. Acad. Sci. Imperialis Petropolitanae, 10*, 243.

[20]Navier, Claude L. M. H. May 1821. *Mem. Acad. Sci. Paris*.

[21]1829. Mémoire sur l'équilibre et le mouvement des corps élastiques, *Mem. Acad. Sci. Paris, 8*.

[22]1862. *Theorie der Elastizität fester Körper*. Leipzig.

[23]Poisson, 1829, loc. cit.

[24]1909. *Crelle's J., 85*.

[25]1916. *Philos. Mag., 41*: 744; *43*: 125.

[26]1869. On the centrifugal force of rotating shaft. *Engineer (London), 27*: 249.

[27]1883. On the whirling and vibration of shafts. *Philos. Trans.*, Ser. A, 279–359.

[28]1895. Das Problem der DeLaval'schen Turbinenwelle. *Civilingenieur, 61*: 333–342. Stodola, A. 1916. Neuere Beobachtungen über die kritischen Umlaufzahlen von Wellen. *Schweiz. Bauztg., 68*: 210–214.

[29]1919. *Philos. Mag.*, Mar., pp. 304–314.

[30]1924. Shaft whipping. *Gen. Electr. Rev., 27*(3): 169–178.

[31]1924. Internal friction theory of shaft whirling, *Gen. Electr. Rev., 27*(4): 244–251. 1925. Internal friction as a cause of shaft whirling. *Philos. Mag.*, Ser. 6, *49*: 724–727.

[32]1918. Neue kritische Drehzahlen als Folge der Kreiselwirkung der Läufräder. *Z. Gesamte Turbinenwes., 15*: 269–275.

[33]1925. Shaft whipping due to oil action in journal bearings. *Gen. Electr. Rev., 25*(8): 559–568.

[34]1925. Kritische Wellenstörung infolge der Nachgiebigkeit des Ölposters im Lager. *Schweiz. Bauztg., 85*: 265. Also, 1927. *Steam and Gas Turbines*. New York: McGraw-Hill, 1975, p. 15. [Gash, R., and Pfützner, H., 1975]

[35]1902. *Z. VDI*, p. 797.

[36]1907. *Schifbau, 8*: 823, 866, 904.

[37]1912. *Z. VDI, 56*: 1025.

[38]1921. *Regelung der Kraftmachinen*. Berlin.

[39]Van Den Dungen, M. F-H., 1928, *Les Problèmes Généreaux de la Technique des Vibrations*, Mem. Sc. Phys., L'Academie des Sciences, Paris, Gauthier-Villars.

[40]Hohenemser, K., Prager, W., 1933. Dynamik der Stabwerke. Berlin: Julius Springer.

[41]1944. A new method of calculating natural modes of uncoupled bending vibration of airplane wings and other types of beams. *J. Aeronaut. Sci.*, vol. 11 pp. 153–162.

[42]1945. A general method for calculating critical speeds of flexible rotors. *J. Appl. Mech., 12*: 142.

[43]1950. Matrix solution for the vibration of non-uniform beams. *J. Appl. Mech., 17*: 337–339. For an extensive discussion, see Pestel, E. C., and Leckie, F. A. 1963. *Matrix Methods of Elastomechanics*. New York: McGraw-Hill.

[44]1943. Variational methods for the solution of problems of equilibrium and vibrations. *Bull. Am. Math. Soc., 49*: 1–23.

[45]1956. Stiffness and deflection analysis of complex structures. *J. Aeronaut. Sci., 23*: 805–824.

[46]1909. *Ann. Phys., 28*: 737.

[47]Galerkin, B. G. 1915. Reihenentwicklungen für einige Fälle des Gleichgewichts von Platten und Balken. *Wjestn. Ing. Petrograd.*, *19*.

[48]Huygens, C. 1673. *Horologium oscillatorium*. Ostwald's Klassiker 192.

[49]1892. *Sur les courbes définies par une équation différentielle*, in *Oevres*. Paris: Gauthier-Villars.

[50]1918. *Erzwungene Schwingungen bei veränderlicher Eigenfrequenz*. Braunschweig: Vieweg.

[51]1927. Forced oscillations in a system with nonlinear resistance. *Philos. Mag.*

[52]1883. Differentialgleichungen der Störungstheorie. *Comment. Acad. Sci. Imperialis Petropolitanae, 31*.

[53]1907. *Ann. Fac. Sci. Toulouse, 2*: 203–469; originally published in Russian.

[54]Föppl, A. 1899. *Vorlesungen über der Technischen Mechanik*. Berlin: Julius Springer.

[55]1910. *Technische Schwingungslehre*. Berlin: Julius Springer.

[56](a) Föppl, O. 1923. *Grundzüge der Technischen Schwingungslehre*. Berlin: Julius Springer. (b) Eason, A. B. 1923. *The Prevention of Vibration and Noise*. London: Oxford Technical Publications. (c) Schneider, E. 1928. *Mathematische Schwingungslehre*. Berlin: Julius Springer. (d) van den Dungen, M. F.-H. 1926. *Cours de technique des vibrations*. Brussels. (e) Geiger, J. 1927. *Mechanische Schwingungen und ihre Messe*. Berlin: Julius Springer.

[57]1928. *Vibration Problems in Engineering*. New York: D. van Nostrand.

[58]1934. *Mechanical Vibration*. New York: McGraw-Hill.

REFERENCES AND FURTHER READING

ARISTOTELES. *On Acoustics*, Oxford Edition. Oxford: Clarendon Press.

BECK, T. 1900. *Beiträge zur Geschichte des Maschinenbaues*. Berlin: Julius Springer.

BOETHIUS (A.D. 480–524). Concerning the principles of music. In Lindsay, R. B. 1972. *Acoustics: Historical and Philosophical Development*. Stroudsburg, Pa.: Dowden, Hutchinson & Ross.

BOYER, C. B. 1968. *A History of Mathematics*. Princeton, N.J.: Princeton University Press.

DIMAROGONAS, A. D. 1990. The origins of vibration theory. *J. Sound Vibrat., 140*(2): 181–189.

DIMAROGONAS, A. D. 1978. *Lectures in History of Technology* (in Greek). 2 vols. Patras, Greece: Patras University Press.

HARTENBERG, R. S., and DENAVIT, J. 1964. *Kinematic Synthesis of Linkages*. New York: McGraw-Hill.

HUNT, F. V. 1978. *Origins in Acoustics*. New Haven, Conn.: Yale University Press.

LINDSAY, R. B. 1966. The story of acoustics. *J. Acoust. Soc. Am., 39*(4): 629–644.

LOVE, A. E. H. 1927. *A Treatise on the Mathematical Theory of Elasticity*. New York: Dover Publications.

NEEDHAM, J. 1962. *Science and Civilization in China*, Vol. 4, Part I. Cambridge: University Press.

RAYLEIGH, J. W. S. 1894. *Theory of Sound*. New York: Dover Publications, 1946.

SEIREG, A. 1969. *Mechanical Systems Analysis*. Scranton, Pa.: International Textbook Co.

SKUDRZUK, E. 1954. *Die Grundlagen der Akustik*. Vienna: Springer-Verlag.

SZABÓ, I. 1979. *Geschichte der mechanischen Prinzipien*. Stuttgart: Birkhäuser.

TIMOSHENKO, S. P. 1953. *History of Strength of Materials*. New York: McGraw-Hill.

VITRUVIUS (first century B.C.). De architectura. In Lindsay, R. B. 1972. *Acoustics: Historical and Philosophical Development*. Stroudsburg, Pa.: Dowden, Hutchinson & Ross.

1

Kinematics of Vibration

1.1 PERIODIC MOTION

Variations in physical phenomena that take place more or less regularly and repeat themselves in respect to time are described as *oscillations*. The 24-hour variation of daylight, the swinging of a pendulum, and the variation of the voltage in an electric alternator are typical examples of oscillations.

The name *mechanical vibration* has been used, in general, to describe small oscillations of mechanical systems and structures. By "small" we mean oscillations associated with displacements that are small compared with the dimensions of the oscillating subject. This distinction, of course, is not free of ambiguity and has not been widely accepted; nevertheless, we shall imply it throughout this book unless stated otherwise.

Mechanical vibration is sometimes an undesirable result of imperfections associated with the design, manufacturing, or operation of mechanical devices and systems, such as the vibration associated with the unbalanced masses of rotating

*Oscillation of the cabin scene from Charlie Chaplin's *The Gold Rush*.

or reciprocating machines. There are other instances when "perfect" machines or structures can vibrate mechanically due to external excitations associated with turbulent fluid flow, unsymmetric heating, earthquakes, and so on. These mechanical vibrations are called *forced vibration*.

There are instances when mechanical vibration is the result of a system property called *instability*, if there is a mechanism to supply energy to sustain the vibration. Vibrations of this type, termed *self-excited vibrations*, are among the most undesirable and least controllable.

Mechanical vibration is not always unwanted. In fact, vibration is useful in many engineering processes, such as mixing of dispersions and aggregates—concrete mixing, for example. Vibration is also used for the opposite purpose: to unmix or sort mixtures—stones from sand, for example. Further, vibrating devices are used to transport aggregates, such as granular materials, and to feed discrete components to manufacturing processes, such as in vibrating feeders. Vibration is not only often wanted but also pleasing, as in music for example, or utilized for medical purposes, such as in massage and in ultrasonic processes for diagnosis or treatment.

Mechanical vibration, both wanted and unwanted, is always associated with the fluctuation of mechanical loads, therefore with fluctuation of mechanical stresses and fatigue failure of mechanical components. Moreover, vibration can have other unwanted effects, such as the loosening of threaded connections, friction and wear, and damage of electronic and other delicate components. Finally, vibration can affect comfort, performance, and the health of people subjected to it, as in sickness due to ship (or high-rise building) oscillation. These facts make it imperative that engineers understand the vibration behavior of every mechanical component, machine, structure, and system.

To study vibration, the motion of a point or component as its position changes with time, it is expedient to express it in the form of a function of time. A function $x(t)$ which has the same value for values of t that differ by a constant T is called a *periodic function*:

$$x(t) = x(t + nT), \qquad n = 1, 2, 3, \dots \qquad (1.1)$$

If the function $x(t)$ describes an oscillation, the smallest quantity T for which equation (1.1) is satisfied is called the *period of the oscillation*. We said "smallest" because any multiple of the period T obviously satisfies equation (1.1). The reciprocal f of the period is called the *frequency of the oscillation* and is usually measured in hertz (Hz):

$$f = \frac{1}{T} \qquad (1.2)$$

The half-difference,

$$A = \tfrac{1}{2}(x_{max} - x_{min}) \qquad (1.3)$$

is called the *amplitude of the oscillation*.

To give an example of an oscillation, we consider the mechanism of Figure 1.1, called a *Scotch yoke*. A wheel of radius r rotates about a horizontal axis through its center O. A pin P is attached to the wheel and slides in a slot attached to a stem, which can reciprocate in a horizontal guide. We notice that if the wheel

Portrait of Hertz. (Courtesy of the Library of Congress.)

HEINRICH RUDOLF HERTZ (1857–1894)

Hertz, a German physicist and professor at the Karlsruhe Institute of Technology and the University of Bonn, contributed to the theory of electromagnetic waves and the theory of elastic contacts. He discovered the photoelectric effect and confirmed experimentally Maxwell's theory of electromagnetism. The unit of frequency (hertz) was given his name.

---

rotates with an angular velocity ω, the tip Q of the stem will be displaced from a middle position by

$$x = r \cos \theta = r \cos \omega t \tag{1.4}$$

if we start measuring the time when the pin P is at the rightmost point on the x-axis. If the recording paper moves with velocity $V = \omega r$, the function $x = r \cos \omega t$ will be recorded as the well-known sinusoidal curve. The cyclic motion of the pin P is mapped onto the paper as a sinusoidal motion, and this type of motion is called *harmonic*. By references to this mechanism, one can always visualize the reciprocity between cyclic and harmonic motion. The quantity ω is the angular velocity of the cyclic motion and is also defined as the *circular frequency* of the harmonic motion. It is obvious that the harmonic motion $x = r \cos \omega t$ is periodic with amplitude r, frequency $f = \omega/2\pi$, and period $T = 2\pi/\omega$.

The velocity and acceleration for the horizontal harmonic motion of point Q are determined by finding the time derivatives of the displacement $x(t)$, namely $v(t) = \dot{x}(t) = -\omega r \sin \omega t$ and $a(t) = \ddot{x}(t) = -\omega^2 r \cos \omega t$, respectively (Figure 1.2).

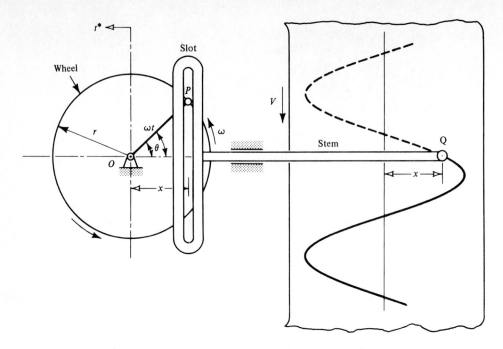

Figure 1.1 Scotch yoke: a harmonic motion mechanism.

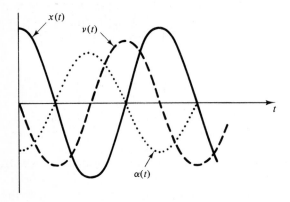

Figure 1.2 Displacement, velocity, and acceleration of harmonic motion.

The periodic motion of Figure 1.3a is not purely harmonic in the plane (x, t), but one can draw a coordinate system (x^*, t^*) on which the motion is harmonic. In this system, the motion is expressed by the equation

$$x^* = X \sin \omega t^*, \qquad \omega = \frac{2\pi}{T} \tag{1.5}$$

while in the system (x, t) the same motion is expressed by the equation

$$x = x_0 + X \sin \omega(t - t_0) \tag{1.6}$$

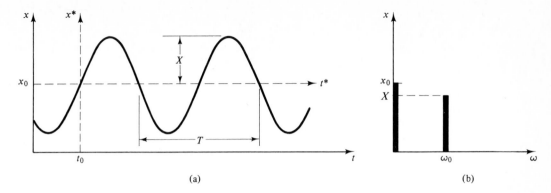

Figure 1.3 Time- and frequency-domain description of a harmonic motion.

Any motion given by the equation

$$x = x_0 + a \sin \omega t + b \cos \omega t \tag{1.7}$$

can be represented by a harmonic motion

$$x - x_0 = X \sin \omega(t + t_0) = X \cos\left[\omega(t + t_0) - \frac{\pi}{2}\right] \tag{1.8}$$

where $a > b$, $X = (a^2 + b^2)^{1/2}$, $t_0 = 1/\omega$ arctan b/a. Any oscillation of the type shown in Figure 1.3a is called *harmonic oscillation*[1] and can be expressed either in the form $X \cos(\omega t + \theta)$ or $X \sin(\omega t + \theta)$.

It can be seen that harmonic motion can be described completely by only two scalar quantities: amplitude X and frequency ω. This can be graphically represented in an amplitude versus frequency diagram. In this diagram the harmonic motion is represented by one point with coordinates ω, X (Figure 1.3b), while in the displacement versus time diagram (Figure 1.3a), to represent the same motion one has to specify a curve, that is, an infinite number of points. Figure 1.3a and b represent the same function in the *time domain* and the *frequency domain*, respectively. The expression of the information included in Figure 1.3a, in the form shown in Figure 1.3b, is called *transformation*. One can transform from one representation of a harmonic function to the other in both directions.

It is apparent that some information is lost in this transformation—namely the beginning of time t_0—but this usually does not have an engineering significance. However, for superimposed vibration it is important.

Two harmonic oscillations are called *synchronous* if they have the same frequency (or angular velocity ω). For example, [3 cos 10t, 4 cos 10t], [3 cos 10t, -5 sin 10t], [3 cos 10t, 2 cos(10t $-$ 0.5)] are couples of synchronous oscillations. We note that two synchronous oscillations do not necessarily have maximum values at the same time. For the last example used, 3 cos 10t has maximum value 3 at

---

[1]Harmonic functions were known to the ancients. The first tables of trigonometric functions known were those of Hipparchus of Nicaea (ca. 180–125 B.C.). The name *sine* appears to be due to mistranslation: the original Hindu term *jiba* (chord) was confused with the Arab word *jaib* (inlet) in the English translation from Arabic and was translated *sinus*, the Latin word for "inlet," in 1150 by Robert of Chester.

$t = 0, 2\pi/10, 4\pi/10, \ldots$. The second oscillation appears to be delayed by a time $0.5/10$. We call this the *time lag*. We observe also that the constant 0.5 has the dimension of angle and it represents the product *angular velocity $\times$ time lag*. This is called the *phase angle*. It is apparent that the concept of phase angle applies only to two synchronous oscillations. The harmonic motion is a particular case of periodic motion. In general, the periodic motion will be something of the type shown in Figure 1.4.

To express the relative measured values of the amplitude, velocity, and acceleration, the technical unit *decibel* is used, defined as $dB = 20 \log_{10}(z/z_0)$, where z is the quantity under consideration and z_0 a reference value. Some reference values in current use are:

$$\text{Velocity:} \quad v_0 = 10^{-8} \text{ m/s}$$

$$\text{Acceleration:} \quad a_0 = 1 \text{ } \mu g = 9.81 \times 10^{-6} \text{ m/s}^2$$

For example, 20 dB means a value 10 times the reference value, 40 dB means 100 times the reference value, and so on. For the relative measurement of the frequency, the *octave* is used: If two frequencies have ratio 2:1, they differ by one octave.

☐ **Example 1.1**

Find the amplitudes of each of the following two harmonic synchronous oscillations and the phase angle between them:

$$x_1(t) = 3 \cos 20t - 4 \sin 20t$$

$$x_2(t) = 1.5 \sin(20t - 30°) - 2 \cos 20t$$

Solution   We will rewrite the functions describing the vibration in the form of single harmonic functions $X \sin(\omega t - \phi)$ and then we will compute the phase angle between them. The two vibrations have the same frequency, 20 rad/s. We write the two oscillations in the form (1.8)

$$x_1(t) = (3^2 + 4^2)^{1/2} \sin 20(t + t_1) = 5 \sin 20(t + t_1)$$

$$t_1 = \frac{1}{20} \arctan \frac{3}{-4} = 0.125 \text{ s}$$

$$x_2(t) = 1.5 \sin 20t \cos 30° - 1.5 \cos 20t \sin 30° - 2 \cos 20t$$

$$= (-2 - 1.5 \sin 30°) \cos 20t + 1.5 \cos 30° \sin 20t$$

$$= -2.75 \cos 20t + 1.3 \sin 20t = (2.75^2 + 1.3^2)^{1/2} \sin 20(t + t_2)$$

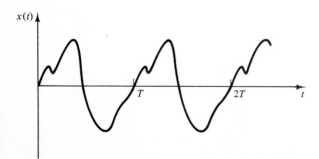

$x(t)$

T

$2T$

t

Figure 1.4   Periodic motion.

$$= 3.04 \sin 20(t + t_2)$$

$$t_2 = \frac{1}{20} \arctan \frac{-2.75}{1.3} = \frac{1}{20} \times (-1.13) = -0.056 \text{ s}$$

$$\theta = 20(t_2 - t_1) = 20 \times 0.181 = 3.62 \text{ rad}$$

Therefore, $X_1 = 5$, $X_2 = 3.04$, and $\theta = 207°$.

(This example demonstrates the equivalent single harmonic function to a sum of harmonic functions of the same frequency and the determination of the phase angle between them.)

$\square$

1.2 VECTOR AND COMPLEX NUMBER REPRESENTATION

A convenient way of representing harmonic oscillations is with complex numbers. Figure 1.5a shows part of the mechanism of Figure 1.1. If x represents the real axis and y the imaginary axis, the pin P can be considered as the tip of a vector **OP**, which can be represented by a complex number

$$z = x + iy \tag{1.9}$$

From the theory of complex numbers we know that for $r = |\mathbf{OP}|$,

$$z = re^{i\theta} = re^{i\omega t} \tag{1.10}$$

because if we start measuring the time when the point P crosses the x-axis, $\theta = \omega t$. The displacement $x = r \cos \omega t$ equals the real part of the complex number z, because a complex number can be written as

$$z = re^{i\omega t} = r(\cos \omega t + i \sin \omega t) \tag{1.11}$$

Therefore,

$$x = \text{Re}(z) = r \cos \omega t \tag{1.12}$$

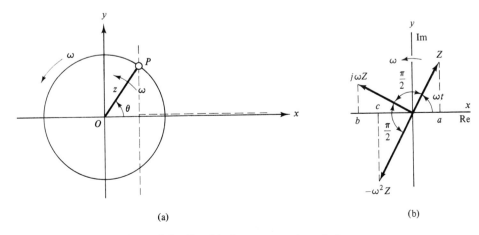

(a)   (b)

Figure 1.5 Graphical representation of phasor.

Therefore, any harmonic motion corresponds to a rotation of a vector of constant length or a complex number of constant magnitude r with a constant angular velocity ω. At time t, the value of x is the projection of the rotating vector on the x-axis or the real part of the rotating complex number. The complex number $re^{i\omega t}$ carries amplitude and phase angle information and it is called *phasor*, or *phasor transform* of the harmonic motion.

Phasor representation of harmonic vibration is convenient and can be handled by graphical and computer methods. Velocity $v = i\omega r e^{i\omega t} = i\omega z$ and acceleration $a = -\omega^2 r e^{i\omega t} = -\omega^2 z$ are also phasors and can be represented by vectors in the same plane with z (Figure 1.5b). The phase angle $\pi/2$ of the velocity and π of the acceleration in respect to the direction of the displacement is due to multiplication by i and -1, respectively.

In Section 1.1 we have seen that two synchronous harmonic oscillations can be added: in particular, the amplitude of the resulting oscillation is $X^2 = X_1^2 + X_2^2$. If the two harmonic oscillations of amplitudes X_1 and X_2 have different frequencies, ω_1 and ω_2, respectively, the resulting oscillation will be periodic only when the ratio of the two natural frequencies is a rational number, that is, the ratio of two integers.

The harmonic oscillation $x = x_0 + r \cos(\omega t + \theta)$ can also be represented by the x-component of a vector of length r rotating with an angular velocity ω, emanating from the point $x = x_0$ of the x-axis, having an angle θ with it (Figure 1.6a). Vector properties apply here, such as vector addition (Figure 1.6b). If the vectors represent asynchronous vibration, vector addition applies, but at any value of time the resulting vector has different magnitude because the two vectors rotate with different velocities. This is illustrated in an interesting phenomenon that occurs when two harmonic motions of the same amplitude but of slightly different frequencies are imposed on a vibrating body (Figure 1.7).

Let $x_1 = X \cos \omega t$, $x_2 = X \cos(\omega + \delta\omega)t$. The motion of the body, then, is the superposition of the two vibrations:

$$X = x_1 + x_2 = X \cos \omega t + X \cos(\omega + \delta\omega)t \qquad (1.13)$$
$$= X[\cos \omega t + \cos(\omega + \delta\omega)t]$$

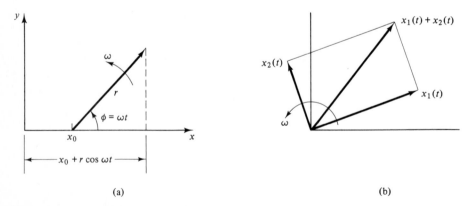

(a)   (b)

Figure 1.6 Vector representation of the harmonic motion.

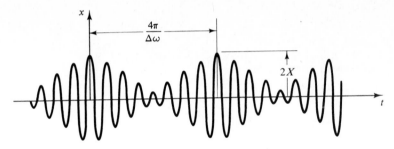

Figure 1.7 Beating phenomenon.

From trigonometry (see Appendix III), $\cos x + \cos y = 2 \cos[(x + y)/2] \cos [(x - y)/2]$. Then

$$x = X[2 \cos(\omega t + \omega t + \delta \omega t)/2)] \cos\left(\frac{-\delta\omega}{2}\right)t$$

$$= 2X \cos\left(\frac{\delta\omega}{2}\right)t \cos\left(\omega + \frac{\delta\omega}{2}\right)t$$

(1.14)

The amplitude of x is seen to fluctuate between $-2X$ and $2X$ according to the $2X \cos(\delta\omega/2)t$ term, while the general motion of x is a cosine wave of angular velocity equal to $(\omega + \delta\omega/2)$. This special pattern of motion is known as the *beating phenomenon*.[2] Whenever the amplitude reaches a maximum, there is said to be a beat. The beat frequency is equal to

$$f_b = \frac{\delta\omega/2}{2\pi} = \frac{\delta\omega}{4\pi} \qquad \text{cycles/s}$$

(1.15)

and the period $T = 1/f_b = 4\pi/\delta\omega$ seconds.

1.3 THE SIMPLE PENDULUM

Most of the vibrations of mechanical systems may be represented by a simple harmonic motion. Many others, although of a different type, may also be approximated by a simple harmonic motion, provided that their amplitude remains small. Consider, for example, a simple pendulum, also called a *mathematical pendulum*,[3] consisting of a material point, a small bob of mass m attached to a cord of length l, which may oscillate in a vertical plane (Figure 1.8a). For example, this pendulum could represent a weight hanging from a crane hook. At a given time t, the cord forms an angle θ with the vertical. The forces acting on the bob are its weight W and the force T exerted by the cord (Figure 1.8c). The bob has an acceleration

[2]Lord Rayleigh. 1894. *Theory of Sound*. New York: Dover Publications, 1946. The earliest known measurement of frequency was by counting the beats of the unknown vibration with another vibration of known frequency.

[3]Galileo Galilei, *Discorsi e dimostrationi* . . . , Ostwald's Klassiker 11, 24, 25; and Huygens, C. 1673. *Horologium oscillatorium*. Paris.

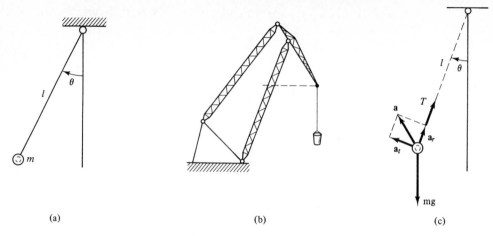

Figure 1.8 Simple pendulum.

with radial component a_r and tangential component a_t; the cord has angular velocity $\partial\theta/\partial t = \omega$ and angular acceleration $\partial^2\theta/\partial t^2 = \alpha$.

Resolving the momentum vector $m\mathbf{a}$ into tangential and normal components, with $m\mathbf{a}_t$ directed to the right (i.e., in the direction corresponding to increasing values of θ) and observing that $a_t = \alpha l = l\ddot\theta$, we write, by virtue of Newton's second law of motion (discussed in more detail in Chapter 2),

$$\sum F_t = m\mathbf{a}_t, \qquad -W \sin\theta = ml\ddot\theta$$

Noting that $W = mg$ and dividing through by ml, we obtain

$$\ddot\theta + \frac{g}{l}\sin\theta = 0 \tag{1.16}$$

For oscillations of small amplitude, we may replace $\sin\theta$ by θ, expressed in radians. At this point one has to know how small is "small," in other words, how large an error is introduced with this simplification. This can be done in three ways:

1. Analytically, as explained in Chapter 12
2. Numerically, as also explained in Chapter 12
3. Experimentally, as done in SIMULAB Experiment 1.1

Assuming that the amplitude is small in the foregoing sense,

$$\ddot\theta + \omega^2\theta = 0 \tag{1.17}$$

where $\omega^2 = g/l$. This is a simple linear ordinary differential equation with constant coefficients and the reader probably knows that it is satisfied by the harmonic functions $\cos\omega t$ and $\sin\omega t$. To exemplify this, the solution of this equation will be sought[4] in a general polynomial function of time $\theta(t) = c_0 + c_1 t + c_2 t^2 + c_3 t^3 +$

---

[4]Morse, P. M. 1948. *Vibration and Sound.* New York: McGraw-Hill.

$\cdots$, where $c_0, c_1, c_2, \ldots$ are yet undetermined constants. Substituting in equation (1.17) and grouping the powers of t yields

$$(2!\, c_2 + \omega c_0^2) + (3!\, c_3 + \omega^2 c_1)t + (4!\, c_4 + \omega^2 c_2)t^2 + (5!\, c_5 + \omega^2 c_3)t^3 + \cdots = 0$$

Portrait of Galileo. (Engraving of 1624 by Ottavio Leoni.)

GALILEO GALILEI (1564–1642)

Mathematician and founder of modern physics. Professor at the universities of Pisa and Padua. Renowned for his epoch-making contributions to astronomy, physics, and philosophy. Pioneer in experimental physics and systematic user of scientific instrumentation, he observed the motion of the pendulum and concluded that the period of vibration is constant and depends on the length and not on the mass of the pendulum. In his *Discourses Concerning Two New Sciences*, 1638, there is a remarkable discussion of the vibration of bodies. The originality of his discoveries was seriously disputed, but their influence on Western science has been profound.

This is true for any value of t only if all quantities in parentheses are zero. This gives equations to express all c's in terms of c_1 and c_0. If we do this, $\theta(t) = c_0[(1 - (\omega t)^2/2! + (\omega t)^4/4! - \cdots] + c_1[(\omega t)/1! - (\omega t)^3/3! + \cdots)]$. The functions in brackets are Taylor expansions about zero of $\cos \omega t$ and $\sin \omega t$,[5] respectively. Therefore, the general solution of equation 1.17[6] is

$$(t) = c_0 \cos \omega t + c_1 \sin \omega t \tag{1.18}$$

where c_0 and c_1 are yet undetermined constants. Therefore, the pendulum executes harmonic oscillation with circular frequency ω of the oscillation equal to $(g/l)^{1/2}$. The period of the small oscillations of a pendulum of length l is $T = 2\pi/\omega = 2\pi(l/g)^{1/2}$ and does not depend on the amplitude. This property was called *isochronism*.[7]

SIMULAB Experiment 1.1

Using the SIMULAB program PENDULUM, investigate the validity of the assumption of small amplitudes, for a pendulum with $l = 1$ m. More specifically, measure the period of vibration for different amplitudes of oscillation of the pendulum and observe its dependence on the amplitude. Plot the period of vibration for a range of amplitudes $\pi/10 < \theta_0 < \pi/2$ in increments of $\pi/20$ and comment on the isochronism of the pendulum.

Solution   No attempt will be made at this point to explain the dependence of the frequency of oscillation upon its amplitude. That will be done in Chapter 12. The linearization hypothesis will be tested only by the simulated experiment. We shall make the pendulum oscillate at different amplitudes and measure the period of oscillation. We will then compare it with that predicted by the linear theory.

We place the SIMULAB diskette in drive A and type

```
A> PENDULUM
```

On the upper right of the lab screen, there is a menu. Selection is performed by the keypad direction keys. For most menu items, there is a submenu at the bottom of the page. Select a menu and submenu. Hit + to increase, − to decrease, ENTER to invoke *quit*. We set the length to 1 m and the width of the rod to zero. We set the angle increment to $\pi/20 = 0.15708$ rad. We measure the time T (10 periods for better accuracy) that it takes the pendulum to return to the initial position, selected each time to be $i\pi/20$, $i = 1, 2, \ldots, 10$. The result is plotted in Figure SE1.1. One can observe that for oscillation angles below 30°, the deviation is insignificant for many engineering applications. For some applications requiring very high accuracy such as in time measurements, the acceptable range might be much more narrow.

---

[5]The series representation for the sin and cos functions was discovered by Newton and independently by Leibnitz in 1676.

[6]The solution for equation (1.17) was found by Daniel Bernoulli in 1739 and Euler developed the general theory of ordinary linear differential equations, usually encountered in vibration analysis (*Institutiones calculi integralis*, St. Petersburg, 1768–1770).

[7]Usually attributed to Galileo Galilei (*Dialogues Concerning Two New Sciences*, 1638; available as a Dover paperback) but known to ancient Greeks and Chinese.

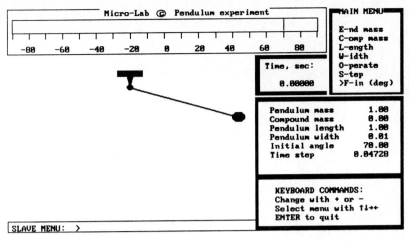

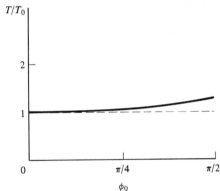

Figure SE1.1

1.4 HARMONIC ANALYSIS

The harmonic functions $g_n(\omega t) = \cos n\omega t$ or $\sin n\omega t$ satisfy the equation

$$\frac{1}{\pi} \int_0^T g_m(\omega t) g_n(\omega t) \, dt = \begin{cases} 0 & \text{for } m \neq n & \text{(a)} \\ 1 & \text{for } m = n \neq 0 & \text{(b)} \end{cases} \qquad (1.19)$$

as one can easily show, if T is the period $2\pi/\omega$. Functions that satisfy equation (1.19a) are called *orthogonal* and if, in addition, they satisfy (1.19b), they are called *orthonormal*.

The *Fourier theorem*[8] states that any function $g(t)$ that is periodic over the interval $0 \angle t \angle T$ can be represented with arbitrary accuracy by the series

$$g(t) = \frac{a_0}{2} + \sum_{n=1}^{\infty} \left(a_n \cos \frac{2\pi nt}{T} + b_n \sin \frac{2\pi nt}{T} \right) \qquad (1.20)$$

---

[8]Fourier (*Théorie analytique de la chaleur*, 1822) introduced the idea, which was subsequently developed in its present form by Daniel Bernoulli.

Portrait of Fourier. (Courtesy of the Library of Congress.)

JEAN BAPTISTE JOSEPH FOURIER (1768–1830)

French mathematician, professor at the Ecole Polytechnique and secretary of the Paris Academy of Sciences. Best known for his contribution to the theory of heat flow, in *Théorie analytique de la chaleur*, published in 1822, which included the ideas that led to the development of Fourier analysis.

---

called the *Fourier series*, where the constants a_n and b_n are called *Fourier coefficients*.[9] If we multiply equation (1.20) by cos $n\omega t$ or sin $n\omega t$ and integrate from 0 to T, using the orthogonality relations (1.19), we can prove that

$$a_0 = \frac{2}{T} \int_0^T g(t) \, dt$$

$$a_n = \frac{2}{T} \int_0^T g(t) \cos n\omega t \, dt, \qquad b_n = \frac{2}{T} \int_0^T g(t) \sin n\omega t \, dt \tag{1.21}$$

This procedure is called the *harmonic analysis* of periodic functions. Given any periodic function $g(t)$ with period T, one has to perform the integrations (1.21), insert the constants a_n and b_n in equation (1.20), and truncate. The latter means that one can take only a finite number of terms in the series. There is no rule that

---

[9]Sufficient conditions for a periodic function $g(t)$ to be expressible in terms of a converging Fourier series are called the *Dirichlet conditions* (1828); that is, the function is single valued, has a finite number of discontinuities, maxima and minima, and a finite integral of its absolute value over a period (Dirichlet, L., *Werke*, Berlin, 1889). Periodic functions encountered in mechanical vibrations generally satisfy these conditions.

Kinematics of Vibration   Chap. 1

gives general guidelines as to the number of terms needed to approximate the function $g(t)$ to a certain degree of desired accuracy. This depends on the function $g(t)$ itself.

The circular frequency $\omega = 2\pi/T$ is called *fundamental* and the function $g(t)$ is considered as sum of harmonic functions that have frequencies equal to the fundamental and its integer multiples. The latter functions are called *harmonics*.

For very simple forms of the function $g(t)$, the integrations (1.21) can be carried out directly. If this is not convenient or possible, or if the function is measured experimentally and since the integrals are proper, one should not have any difficulty integrating numerically[10] with a simple rule such as the trapezoidal rule, to name just one example.

If experimentally determined, the function $g(t)$ is given by way of a number of points $x_1, x_2, \ldots, x_N$, and at times $t_1, t_2, \ldots, t_N$, respectively. This is the case, for example, when $g(t)$ is the vibration of a device of some sort, measured by way of a vibration measuring instrument. If the times $t_1, t_2, \ldots, t_N$ are not arbitrary, as often happens, but are taken at $N + 1$ equidistant points over the period T, the coefficients a_n and b_n can be calculated numerically, replacing with summations the integrals in equations (1.21) and with $\Delta t = T/N$ the differential dt,

$$a_0 = \frac{2}{N} \sum_{i=1}^{N} g(t_i) \tag{1.22}$$

$$a_n = \frac{2}{N} \sum_{i=1}^{N} g(t_i) \cos \frac{2n\pi t_i}{T}, \qquad b_n = \frac{2}{N} \sum_{i=1}^{N} g(t_i) \sin \frac{2n\pi t_i}{T}$$

In practical applications, it is very difficult to determine the beginning and end of the period and thus the period itself. In this case we usually take samples over a number of periods, and as the period we use the sum of these periods, that is, the true length of the sample. If the function is true periodic and the sampling is over an exact multiple of the period, the first few terms will turn out to be zero, because a periodic function of period T cannot have harmonics of period greater than T. If we apply the formulas (1.22) over a period $3T$, for example, the first nonzero terms a and b will be at least a_3 and b_3. In the case of experimental determination of the values $x_i = x(t_i)$, small harmonics of period greater than T, *called subharmonics*, will always be present, due to numerical inaccuracies and the fact that no measured function of this kind is exactly periodic.

The several harmonic functions $a_n \cos n\omega t$ or $b_n \sin n\omega t$ are called harmonics of the periodic function $g(t)$ of order n in respect to the period T. The harmonic of order n therefore has a period T/n. These harmonics can be plotted as vertical lines on an amplitude versus frequency diagram called a *spectral diagram* or *frequency spectrum*, shown in Figure 1.3b. For example, the frequency spectrum of a square wave, Figure 1.9a, is shown in Figure 1.9b (see also Example 1.2).

One can observe an interesting feature for the square function in Figure 1.9: Only the sine *Fourier coefficients* $b_1, b_2, \ldots, b_n$, are nonzero. In general, when $f(t)$ is an even function [i.e., when $f(t) = f(-t)$], the coefficients b_r ($r = 1, 2, \ldots$) vanish and the series is known as a *Fourier cosine series*. On the other hand, when

---

[10]Runge, C. 1905. Über die Zerlegung empirish gegebener periodischer Functionen in Sinuswellen. *Z. Math. Phys.*, *48*.

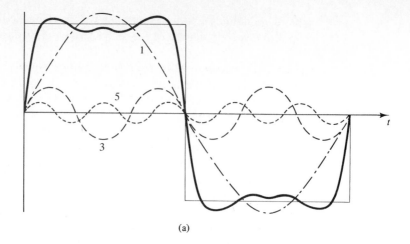

(a)

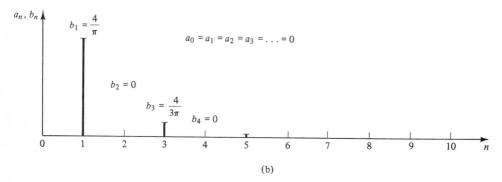

(b)

Figure 1.9 Frequency spectrum for a square function.

$f(t)$ is an odd function [i.e., when $f(t) = -f(-t)$], the coefficients a_r ($r = 0, 1, 2, \ldots$) vanish and the series is called a *Fourier sine series*. This can more conveniently be demonstrated by considering the interval $-\pi < t < \pi$ instead of $0 < t < 2\pi$.

If the function $f(t)$ is only piecewise continuous in a given interval, then a Fourier series representation using a finite number of terms approaches $f(t)$ in every interval that does not contain discontinuities. In the immediate neighborhood of a jump discontinuity, convergence is not uniform, and as the number of terms increases, the finite series approximation contains increasingly high-frequency oscillations which move closer to the discontinuity point. However, the total oscillation of the approximating curve does not approach the jump of $f(t)$, a fact known as the *Gibbs phenomenon* [Brigham, 1974].

☐ **Example 1.2 Fourier Analysis of a Square Wave**

Analyze the periodic square wave (Figure 1.9a) in a Fourier series.
Solution We will express the square wave as a function of time and will then apply equations (1.21) to find the Fourier coefficients. The function $g(t)$ of the square wave can be expressed as

$$g(t) = \begin{cases} 1 & \text{when } mT < t < (m + \frac{1}{2})T \\ -1 & \text{when } (m + \frac{1}{2})T < t < (m + 1)T \end{cases}, \quad m = 0, 1, 2, \ldots$$

To compute the Fourier coefficients a_n and b_n we apply equations (1.21):

$$a_n = \frac{2}{T}\int_0^T g(t) \cos n\omega t \, dt = \frac{2}{T}\int_0^{T/2} \cos n\omega t \, dt - \frac{2}{T}\int_{T/2}^T \cos n\omega t \, dt$$

$$= \frac{2}{n\omega t}\sin n\omega t \Big|_0^{T/2} - \frac{2}{n\omega T}\sin n\omega t \Big|_{T/2}^T = 0$$

Therefore, $a_0 = a_1 = a_2 = \cdots = 0$. If n is odd, the last of equations 11.21 yields similarly $b_n = 4/\pi n$. If n is even, $b_n = 0$. The frequency spectrum is shown in Figure 1.9b. All cosine terms are zero since the function is odd.

(This example demonstrates the Fourier analysis of functions that are available in algebraic form and that the integrals that give the Fourier coefficients can be integrated analytically.)

□

1.5 COMPLEX FORM OF THE FOURIER SERIES[11]

The Fourier series can also be expressed in terms of exponential functions, since

$$\cos \omega t = \frac{e^{i\omega t} + e^{-i\omega t}}{2} \quad \text{and} \quad \sin \omega t = \frac{e^{i\omega t} - e^{-i\omega t}}{2i} \tag{1.23}$$

Inserting equation (1.23) into equation (1.20), we obtain for a function $g(t)$ of time

$$g(t) = \tfrac{1}{2}a_0 + \tfrac{1}{2}\sum_{n=1}^{\infty} [a_n(e^{in\omega t} + e^{-in\omega t}) - ib_n(e^{in\omega t} - e^{-in\omega t})] \tag{1.24}$$

$$= \tfrac{1}{2}a_0 + \tfrac{1}{2}\sum_{n=1}^{\infty} [(a_n - ib_n)e^{in\omega t} + (a_n + ib_n)e^{-in\omega t}]$$

Introducing the notation

$$G_0 = \tfrac{1}{2}a_0, \quad G_n = \tfrac{1}{2}(a_n - ib_n), \quad G_{-n}$$

$$= \tfrac{1}{2}(a_n + ib_n), \quad r = 1, 2, \ldots \tag{1.25}$$

where G_{-n} is the complex conjugate of G_n, equation (1.20) reduces to

$$g(t) = \sum_{n=-\infty}^{\infty} G_n e^{in\omega t} \tag{1.26}$$

in which, using equations (1.21), the coefficients G_n have the form

[11]The study of Sections 1.5 to 1.7 can be postponed until the discussion of random vibration and machinery condition diagnosis (Chapters 10 and 14).

$$G_n = \frac{1}{2}(a_n - ib_n) = \frac{1}{T}\left[\int_0^T g(t)\cos n\omega t\, dt - i\int_0^T f(t)\sin n\omega t\, dt\right]$$

$$= \frac{1}{T}\int_0^T g(t)(\cos n\omega t - i\sin n\omega t)\, dt \qquad (1.27)$$

$$= \frac{1}{T}\int_0^T g(t)e^{-in\omega t}\, dt \qquad n = 0, 1, 2, \ldots$$

Equations (1.26) and (1.27) constitute the complex form, or exponential form, of the Fourier series in the form

$$g(t) = \sum_{n=-\infty}^{\infty} G_n e^{in\omega t} \qquad (a)$$

$$\qquad\qquad\qquad\qquad\qquad (1.28)$$

$$G_n = \frac{1}{T}\int_0^T g(t)e^{-in\omega t}\, dt \qquad (b)$$

All the results above apply to periodic signals, but it is possible to extend equations (1.28) to a more general case by letting $T \to \infty$, in which case the spacing $1/T$ between the harmonics tends to zero and G becomes a continuous function of $f = \omega/2\pi$. It can be shown that equation (1.28b) tends to

$$G^*(f) = \int_{-\infty}^{\infty} g^*(t)e^{-i2\pi ft}\, dt \qquad (1.29)$$

and equation (1.28a) becomes

$$g^*(t) = \int_{-\infty}^{\infty} G^*(f)e^{i2\pi ft}\, df \qquad (1.30)$$

where $G^*(f) = \lim_{T\to 0} TG_n$ and $g^*(t) = \lim_{T\to 0}(1/T)g(t)$. In the sequel, the asterisk in the definitions of equations (1.29) and (1.30) will be dropped.

Equation (1.29) is known as the *forward transform* and equation (1.30) as the *inverse transform*; together they form the *Fourier transform* pair. It can be seen that they are almost symmetrical. The only difference is the sign of the exponent of e. The most important thing about the symmetry is that results which apply to transformation in one direction generally also apply to transformation in the other direction. Figure 1.10a is a graphical representation of the Fourier integral transform.

☐ **Example 1.3   Partial Admission of Steam in Turbines**

One way of regulating the power output of turbines is by partial admission. Thus it is possible that steam is administered only for a part of the 360° arc in an array of nozzles. Quarter admission is when the steam projecting nozzles are one-fourth of the periphery. In this case, on every blade, there is force F_0 acting from the steam only during one-fourth of a revolution. Thus the force as a function of time is shown in Figure E1.3b. Determine the Fourier expansion of the force on the blades if the angular velocity of rotation is ω.

Solution   We will express the exciting force as a function of time and then apply equations (1.21) for the determination of the Fourier coefficients. The force is $F(t) = F_0$ for $0 < t < T/4$ and 0 for $T/4 < t < T$, where T is the duration of one revolution.

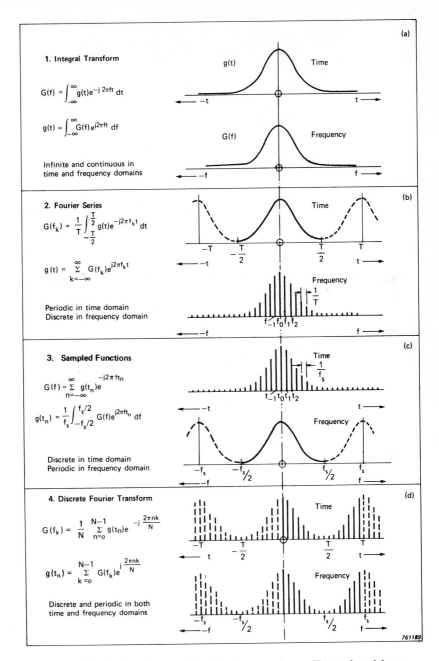

Figure 1.10 Various forms of the Fourier transform. (Reproduced from Randall, R. B. 1977. *Frequency Analysis*. Naerum, Denmark: Bruel & Kjaer. By permission.)

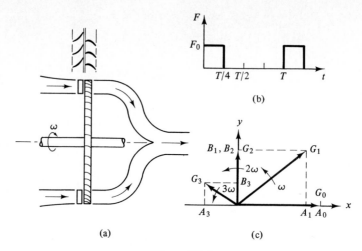

(a)   (c)

Figure E1.3

$$F(t) = \frac{a_0}{2} + \sum_{n=1}^{\infty} (a_n \cos n\omega t + b_n \sin n\omega t), \qquad \omega = \frac{2\pi}{T}$$

is the Fourier series expansion of the function $F(t)$, where

$$a_n = \frac{2}{T} \int_0^T F(t) \cos n\omega t \, dt$$

$$= \frac{2}{T} \int_0^{T/4} F_0 \cos n\omega t \, dt + \frac{2}{T} \int_{T/4}^T (0) \cos n\omega t \, dt$$

$$= \frac{2F_0}{n\omega T} \int_0^{T/4} \cos n\omega t \, d(n\omega t) = \frac{F_0}{\pi n} [\sin n\omega t]_0^{T/4}$$

$$= \frac{F_0}{\pi} \sin \frac{2\pi n}{4} = \frac{F_0}{\pi n} \sin \frac{\pi n}{2}$$

$$a_0 = \frac{2}{T} \int_0^T F(t) \, dt = \frac{2F_0}{T} \int_0^{T/4} dt = \frac{F_0}{2}$$

$$b_n = \frac{2}{T} \int_0^T F(t) \sin n\omega t \, dt = -\frac{F_0}{\pi n} [\cos n\omega t]_0^{T/4} = \frac{F_0}{\pi n} \left(1 - \cos \frac{\pi n}{2}\right)$$

Therefore,

$$a_0 = \frac{F_0}{2}, \quad a_1 = \frac{F_0}{\pi}, \quad a_2 = 0, \quad a_3 = -\frac{F_0}{3\pi}, \quad \ldots ,$$

$$a_{2m} = 0, \quad a_{2m+1} = (-1)^m \frac{F_0}{\pi(2m + 1)}$$

$$b_1 = \frac{F_0}{\pi}, \quad b_2 = \frac{F_0}{\pi}, \quad b_3 = \frac{F_0}{3\pi}, \quad \ldots ,$$

$$b_{2n} = \frac{F_0}{2\pi n[1 - (-1)^n]}, \quad b_{2n+1} = \frac{F_0}{\pi(2n + 1)}$$

The first three complex Fourier components are

$$G_0 = \frac{a_0}{2}$$

$$G_1 = \frac{a_1 + ib_1}{2}$$

$$G_2 = \frac{a_2 + ib_2}{2}$$

$$G_3 = \frac{a_3 + ib_3}{2}$$

The rotating vector diagram of the force components is shown in Figure E1.3c. The Fourier spectrum is shown in Figure E1.3d.

Finally,

$$F(t) = \frac{F_0}{4} + \frac{F_0}{\pi}\cos\omega t + \frac{F_0}{\pi}\sin\omega t + \frac{F_0}{\pi}\sin 2\omega t - \frac{F_0}{3\pi}\cos 3\omega t + \frac{F_0}{3\pi}\sin 3\omega t + \cdots$$

It can be seen from Figures E1.3c and d that the function is not symmetric; therefore, both real and complex (or cosine and sine) terms in the series have nonzero values.

(In this example, a periodic function of time was expressed in Fourier series. Since the function is not even or odd, both sine and cosine terms are computed.) □

1.6 SAMPLED TIME FUNCTIONS

Fourier transform pairs that apply to sampled time functions such as the ones in equations (1.22), that is, functions that are represented by a *time series*, a sequence of values at discrete equispaced points, are becoming very important with the increase in digital processing of experimental data. The discretization of the signal introduces not only a numerical error but also additional frequencies in the spectrum that must be identified; otherwise, they might lead to incorrect conclusions. To illustrate this, let us sample two periods of an exact sinusoidal function $A \sin \omega t$ (Figure 1.11a), which has a spectrum with one frequency, $\omega = \omega_0$. The sampling is at steps Δt approximately equal to $T/4$ (in fact, for the purpose of illustration, a little less). Assuming straight lines for the time function between sampling points, the sampled function that we measure (I) will be the sum of three components (Figure 1.11b):

1. The harmonic function itself $A \sin \omega t$ (II)
2. The error due to substitution of straight segments for the harmonic ones (III)
3. The error due to inaccurate estimation of the period T, which is usually not known exactly in advance (IV)

Consequently, in the spectrum shown in Figure 1.11c there are three corresponding components:

1. The main component $(2\pi/4\Delta t)$, with frequency a little above ω_0

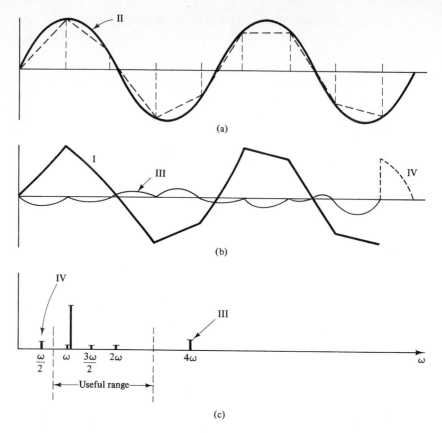

Figure 1.11 Harmonic analysis of a time-sampled function.

2. The component $2\pi/\Delta t$, due to linearization, with frequency a little above $4\omega_0$

3. The component $2\pi/8T$, due to the inaccuracy of Δt, with frequency a little above $\omega_0/2$

It is apparent that only the components of the spectrum between frequencies $2\pi/\Delta t > \omega > 2\pi/T_0$ have meaning, where T_0 is the duration of the signal sampled. The components in the spectrum due to substitution of the original function by the sampled one can be studied further in Figure 1.10.

The integral transform pair, as shown in Figure 1.10a, extends from $-\infty$ to ∞. Therefore, numerical handling of this range is impossible. Instead, an interval $(-T/2, T/2)$ is used and this is assumed to coincide with a period in the sense discussed above. Beyond this interval, in both directions, the function *is assumed to repeat itself* (Figure 1.10b). A fundamental frequency $f = 1/T$ corresponds to this period. A Fourier analysis of this function leads to a discrete frequency spectrum with an infinite number of harmonics, multiples of f (Figure 1.10b).

Sampled functions, with sampling taken at discrete intervals of time in steps of Δt, are sampled with a frequency $f_s = 1/\Delta t$. One expects that since a disturbance is happening every Δt seconds (very small), there is some inherent frequency in the spectrum $f_s = 1/\Delta t$ (very large). Indeed, this introduces in the frequency spectrum a periodicity of period f_s (Figure 1.10c).

In reality, there is a discrete sampling of the time function and a discrete frequency spectrum, both periodic. The only meaningful parts of both diagrams, however, are within one period, $1/f_s$ in the time domain and f_s in the frequency domain.

The Fourier transform, equations (1.29) and (1.30), for sampled time functions is computed by numerical integration, as follows:

$$G(f) = \sum_{n=-\infty}^{\infty} g(t_n)e^{-i2\pi f t_n} \tag{1.31}$$

$$g(t_n) = \frac{1}{f_s} \int_{-f_s/2}^{f_s/2} G(f)e^{i2\pi f t_n}\, df \tag{1.32}$$

where $t_n = N\,\Delta t$ (i.e., the time corresponding to the nth time sample).

If truncation is performed in the time series and discretization and truncation is similarly performed in the frequency function, the discrete transform becomes

$$G_k = \frac{1}{N}\sum_{n=0}^{N-1} g_n e^{-i(2\pi kn/N)} \tag{1.33}$$

$$g_n = \sum_{k=0}^{N-1} G_k e^{i(2\pi kn/N)} \tag{1.34}$$

Because the infinite continuous integrals of equations (1.29) and (1.30) have been replaced by finite summations, the transform pair above, known as the *discrete Fourier transform* (DFT), is much better adapted to digital computations. Even so, it can be seen that in order to obtain N frequency components from N time samples (or vice versa) requires N^2 complex multiplications. A calculation procedure known as the *fast Fourier transform* or *FFT algorithm* [Cooley and Tukey, 1965] obtains the same result with a much smaller number of complex multiplications. One way of expressing equation (1.33) is with the following matrix equation:

$$\{\mathbf{G}\} = \frac{1}{N}[\mathbf{A}]\{\mathbf{g}\} \tag{1.35}$$

where $\{\mathbf{G}\}$ is a column vector representing the N complex frequency components
$G_k,\ k = 0, 1, 2, \ldots, N-1$
N is the number of the equal time intervals between samples
$[\mathbf{A}]$ is an $N \times N$ square matrix of unit vectors which depend only on the number of samples, $a_{kn} = e^{-i(2\pi kn/N)}$
$\{\mathbf{g}\}$ is a column vector representing the N time samples $g_n,\ n = 0, 1, 2, \ldots, N-1$
For the particular case of $N = 8$, equation (1.35) may be written as follows:

$$\{\mathbf{G}\} = \frac{1}{8}\begin{bmatrix} 1.0 & 1.0 & 1.0 & 1.0 & 1.0 & 1.0 & 1.0 & 1.0 \\ 1.0 & 0.7 & 0.0 & -0.7 & -1.0 & -0.7 & -0.0 & 0.7 \\ 1.0 & 0.0 & -1.0 & -0.0 & 1.0 & 0.0 & -1.0 & -0.0 \\ 1.0 & -0.7 & -0.0 & 0.7 & -1.0 & 0.7 & 0.0 & -0.7 \\ 1.0 & -1.0 & 1.0 & -1.0 & 1.0 & -1.0 & 1.0 & -1.0 \\ 1.0 & -0.7 & 0.0 & 0.7 & -1.0 & 0.7 & -0.0 & -0.7 \\ 1.0 & -0.0 & -1.0 & 0.0 & 1.0 & -0.0 & -1.0 & 0.0 \\ 1.0 & 0.7 & -0.0 & -0.7 & -1.0 & -0.7 & 0.0 & 0.7 \end{bmatrix}$$

$$+ \; i\begin{bmatrix} 0.0 & 0.0 & 0.0 & 0.0 & 0.0 & 0.0 & 0.0 & 0.0 \\ 0.0 & 0.7 & 1.0 & 0.7 & 0.0 & -0.7 & -1.0 & -0.7 \\ 0.0 & 1.0 & 0.0 & -1.0 & -0.0 & 1.0 & 0.0 & -1.0 \\ 0.0 & 0.7 & -1.0 & 0.7 & 0.0 & -0.7 & 1.0 & -0.7 \\ 0.0 & 0.0 & -0.0 & 0.0 & -0.0 & 0.0 & -0.0 & 0.0 \\ 0.0 & -0.7 & 1.0 & -0.7 & 0.0 & 0.7 & -1.0 & 0.7 \\ 0.0 & -1.0 & 0.0 & 1.0 & -0.0 & -1.0 & 0.0 & 1.0 \\ 0.0 & -0.7 & -1.0 & -0.7 & 0.0 & 0.7 & 1.0 & 0.7 \end{bmatrix}\{\mathbf{g}\} \qquad (1.36)$$

Each element in the matrices represents the real or imaginary part of a unit vector $e^{-i2\pi kn/N}$ with a certain angular orientation, and multiplication by this vector results in a rotation through the angle $2\pi kn/N$. Each row in the matrix represents a different value of frequency ($k = 0, 1, 2, \ldots, 7$), while each column represents a different point in time ($n = 0, 1, 2, \ldots, 7$).

For either k or n equal to zero, the angle is always zero and thus multiplication is by unity. The first row of the matrix ($k = 0$) represents zero frequency, and since all elements equal 1, calculation of G_0 involves a simple addition of all the time samples (g_n) followed by division by 8 ($= N$). As would be expected, this result is the dc component. In fact, there are only eight different multiplications with components of the sampling function; the rest are additions and subtractions. The reduction factor in computation time is on the order of $N/\log_2 N$, which for the typical case of $N = 1024$ (2^{10}) is more than 100.

The second row ($k = 1$) represents the lowest nonzero frequency and it can be seen that for increasing values of n the angle changes by $2\pi/N$ [i.e., $(1/N)$th of a revolution]. Note that for the forward transform the negative sign of the exponent actually gives a rotation in the negative direction. For $k = 2$ the rotational frequency is $(2/N)$th of a revolution per time sample, and so on up to the last row, which represents $(N - 1)/N$ (in this case, $\frac{7}{8}$) revolution per time sample. Note that this can be more easily interpreted as a rotation, in the opposite direction, of $1/N$ revolution per time sample and thus equally well represents the frequency $-2\pi/N$ per time sample. In fact, all the frequencies above $k = N/2$ (in this case, 4) are more easily interpreted as negative frequencies, and this is perhaps the easiest way of understanding Shannon's sampling theorem [Brigham, 1974]. The negative-frequency components of a real-valued time function are determined with respect to the positive-frequency components, and thus if there were any frequency components above the Nyquist frequency, defined as half the sampling frequency (i.e., half a revolution per time sample), these would become inextricably mixed with the required negative frequency components, thus introducing an error. The periodicity of the spectrum for sampled time functions also becomes clear from examination of equation (1.36). The first row of **A** could equally well represent the sampling frequency f_s (one revolution per time sample) or $2f_s$, and so on, and thus the **A** matrix could equally represent the frequencies $k = 8$ to 15, 16 to 23, and so on. Since the rotating vectors are sampled only at discrete points in time, all information is lost about how many complete revolutions may occur between samples. However, restriction of frequency content to less than half the sampling frequency removes the ambiguity. It is apparent that the maximum frequency is $f_N = f_s/2$ and the resolution (distance of two successive frequency values) is $f_s/N = 2f_N/N$.

The misinterpretation of high frequencies (above half the sampling frequency) as lower frequencies is termed *aliasing*, and this is obviously one of the pitfalls to be avoided when digitizing continuous signals. To understand aliasing, it may help to consider two practical cases with which most people are familiar:

1. The cartwheels in Western films often appear to run backward (i.e., negative frequency) or too slowly forward because of the sampling involved in filming.
2. The stroboscope is in fact an aliasing device that is designed to represent high frequencies as low ones (even zero frequency when the picture is frozen).

SIMULAB Experiment 1.2

In file FFT.1 of your SIMULAB diskette, there is a time sample of the vibration of an electric generator. Use the SIMULAB spectrum analyzer to determine the predominant vibration frequency.

Solution   With the program diskette in drive A, we type

```
A> ANALYZER
```

The simulated spectrum analyzer and the tape recorder appear on the screen (Figure SE1.2a). The control buttons can be depressed by hitting the keyboard keys with the same starting letter on the button. Depressing X lights the button and transfers control to the tape recorder. Pressing the + button (+ of the keypad) advances the channel, and pressing the − button (− of the keypad) retracts the channel number. The tape recorder reads the data file FFT.N, where N is the channel number. We set the channel number to 1 and hit ENTER. It takes a few seconds to read the file with the time signal. The file window on the screen shows that there are 256 samples.

When the reels stop, control returns to the control console. We press the T key to view the time function. Figure SE1.2b appears on the screen. It is a clean signal and we can measure 10 periods to be .31 s; therefore, the period is 0.031 s and the frequency $1/0.031 = 32.2$ Hz.

Pressing the F key invokes the FFT analysis. After a few seconds, the light on the F button is off. We can now plot the results by pressing the A key for amplitude. The spectrum appears on the screen (Figure SE1.2c). The 32.2-Hz frequency is apparent. To zoom on this frequency, the Z key is pressed. The program asks for the frequency range. We specify $0 < f < 100$ Hz. The screen of Figure SE1.2d appears.

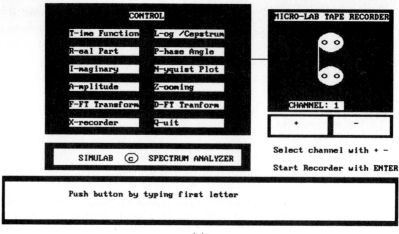

(a)

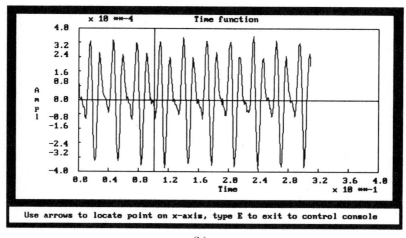

(b)

Figure SE1.2

SIMULAB Experiment 1.3

Figure SE1.3a shows the force diagram on a piston of a two-stroke internal combustion engine, rotating at 3000 rpm, found by pressure measurements. Perform a Fourier analysis to find the force harmonics. Use the SIMULAB spectrum analyzer to determine the fundamental vibration frequency.

Solution   The period of one revolution is $60/3000 = 0.020$ s. It is divided in 32 time intervals of width $dt = 0.020/32 = 6.25 \times 10^{-4}$ s. A file FFT.5 is made in a text editor having first line

$$32, 6.25E-4 \qquad \text{(number of points, time step)}$$

Then the values of the force at the sampling points are entered consecutively separated by commas:

$$2.35, \quad 2.46, \quad 2.80, \quad 3.05, \quad \ldots$$

and the file is saved.

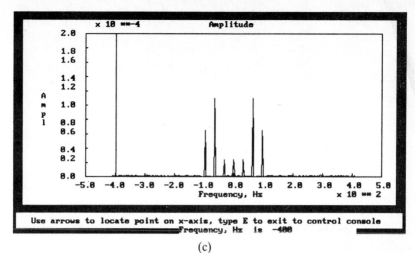

(c)

(d)

Figure SE1.2 *(cont.)*

With the program diskette in drive A, we type

A> ANALYZER

The simulated spectrum analyzer and the tape recorder appear on the screen (Figure SE1.2a). The control buttons can be depressed by hitting the keyboard keys with the same starting letter on the button. Depressing X lights the button and transfers control to the tape recorder. Pressing the + button (+ of the keypad) advances the channel, and pressing the − button (− of the keypad) retracts the channel number. We set channel number to 5 and hit ENTER. It takes a few seconds to read the file with the time signal. The file window on the screen shows that there are 32 samples. When the reels stop, control returns to the control console. We press the T key to view the time function. Figure SE1.3a appears on the screen.

Pressing the D key invokes the DFT analysis. On the lower window, the program asks for the desired vibration range. We specify $0 < f < 750$ Hz with a 50-Hz step.

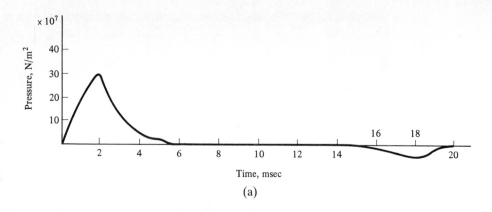

(a)

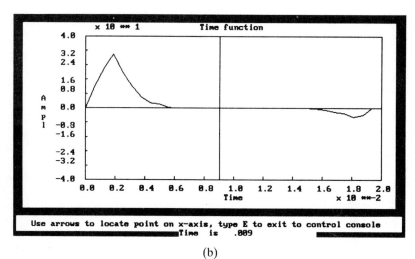

(b)

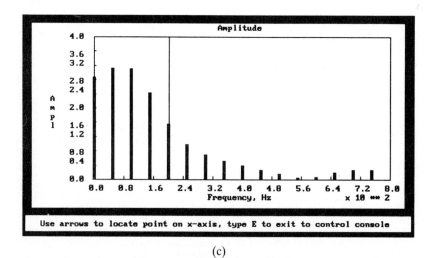

(c)

Figure E1.3

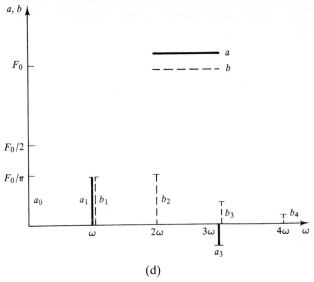

(d)

Figure E1.3 *(cont.)*

This includes the 50 Hz fundamental and a few subsequent harmonics. After a few seconds, the light on the D button is off. We can now plot the results by pressing the A key for amplitude. The discrete spectrum appears on the screen (Figure SE1.3b).

1.7 ALGORITHMS FOR FOURIER ANALYSIS

Given a sampled time function, the Fourier transform pair (1.33) and (1.34) can easily be implemented for machine computation. An FFT algorithm is more complicated and will not be discussed here. It is included, however, in the ANALYZER program of SIMULAB. An algorithm for the Fourier transform pair is described here so that students can program it themselves. Quick BASIC and FORTRAN code are given in the floppy diskette accompanying this book under the names FOURIER.BAS and FOURIER.FOR, respectively.

These computations require a substantial amount of computer time; therefore, computation economy and accuracy are major concerns. Equations (1.31) and (1.32) will be written in the form:

$$Re\{G(f)\} = \frac{1}{N} \sum_{n=-\infty}^{\infty} g(t_n) \cos 2\pi f t_n \tag{1.37}$$

$$Im\{G(f)\} = -\frac{1}{N} \sum_{n=-\infty}^{\infty} g(t_n) \sin 2\pi f t_n \tag{1.38}$$

$$Re\{g(t)\} = \sum_{f=-f_s/2}^{f_s/2} G(f) \cos 2\pi f t_n \tag{1.39}$$

$$Im\{g(t)\} = \sum_{f=-f_s/2}^{f_s/2} G(f) \sin 2\pi f t_n \tag{1.40}$$

Since a large number of harmonic functions are to be computed, the argument of each has been slightly different from the preceding one, and it is useful to note that

$$\cos(x + \Delta x) = \cos x \cos \Delta x - \sin x \sin \Delta x$$
$$\sin(x + \Delta x) = \sin x \sin \Delta x + \cos x \sin \Delta x$$

(1.41)

Since for sampling at constant time intervals Δt is constant and usually we want the frequency spectrum at constant frequency intervals, Δx in both cases is constant and a recursive computation of harmonic functions in equations (1.41) involves only two multiplications and one addition instead of the much slower computation of harmonic functions. The negative side is the accumulation of error since computation at each point uses the results of the preceding one; therefore, the truncation errors accumulate. However, very seldom are more than 1024 values computed, and this error removes two or three significant digits, which for most practical purposes is acceptable.

In writing algorithms in this book, standard algorithmic notation (pseudocode) will be used [Kruse, 1987], simplified for the purposes of the presentation. They resemble C-code structures. Coding them should be easy in any high-level language. A summary of algorithmic notation rules is given in Appendix I.

The algorithms for the forward [equations (1.37) and (1.38)] and inverse [equations (1.39) and (1.40)] transforms follow.

PROCEDURE FORWARDFT

```
GLOBAL    VARIABLES:
      gtre(1...Nsamples),gfre(1...Nsamples/2).    REAL ARRAYS
      gtim(1...Nsamples),gfim(1...Nsamples/2).    REAL ARRAYS
      Nsamples                INTEGER SCALAR
      fmin,fmax,fstep,tstep    REAL SCALAR
LOCAL    pi := 3.14159        REAL CONSTANT
/*gtre:      real part of the time function        (input)    */
/*gtim:      imaginary part of the time function   (input)    */
/*gfre:      real part of the frequency function   (output)   */
/*gfim:      imaginary part of the frequency function  (output) */
/*Nsamples:  number of samples of the time function    (input)    */
/*fmin,fmax: min, max frequencies of the spectrum   (input)    */
/*fstep,tstep:frequency and time steps            (input)    */
BEGIN PROCEDURE
   Nfreq := INT[fmax-fmin)/fstep];
   FOR ifreq := 1 to Nfreq DO BEGIN
     omega := 2*pi*(ifreq*fstep+fmin);
     cosomt := COS(0) ; sinomt := SIN(0);
     cosdomt := COS(2*pi*omega*tstep) ; sindomt :=
             SIN(2*pi*omega*tstep);
     gfre(ifreq) := 0 ; gfim(ifreq) := 0;
```

```
        FOR isample := 1 to Nsamples DO BEGIN
           cs := cosomt*cosdomt-sinomt*sindomt;
           ss := sinomt*cosdomt+cosomt*sindomt;
           cosomt := cs ; sinomt := ss;
           gfre(ifreq) := gfre(ifreq)+gtre(isample)*cosomt/Nsamples;
           gfim(ifreq) := gfim(ifreq)-gtre(isample)*sinomt/Nsamples;
        END DO
     END DO
END PROCEDURE
```

PROCEDURE INVERSEFT

```
GLOBAL VARIABLES: gtre(1...Nsamples*2),          REAL ARRAY
        gfre(1...Nsamples).                      REAL ARRAY
        gtim(1...Nsamples*2),gfim(1...Nsamples). REAL ARRAY
        Nsamples                                 INTEGER SCALAR
        fmin,fmax,fstep,tstep                    REAL SCALARS
LOCAL Constant pi := 3.14159                      REAL
/*gtre: real part of the time function              (output)    */
/*gtim: imaginary part of the time function         (output)    */
/*gfre: real part of the frequency function         (input)     */
/*gfim:  imaginary part of the frequency function   (output)    */
/*Nsamples: number of samples of the frequency      (input)     */
/*fmin,fmax:  min, max frequencies of the spectrum  (input)     */
/*fstep       :frequency step                       (input)     */
BEGIN PROCEDURE
   Ntime := 2*Nsamples ; tstep := 1/fmax;
   FOR itime := 1 to Ntime DO BEGIN
     omega    := 2*pi*fmin;
     cosomt   := COS(omega*itime*tstep) ; sinomt :=
                 SIN(omega*itime*tstep);
     cosdomt := COS(2*pi*itime*tstep*fstep);
     sindomt := SIN(2*pi*itime*tstep*fstep;
     gtre(itime) := 0 ; gtim(itime) := 0;
     ifrmin := INT(fmin/fstep) ; ifmax := INT(fmax/fstep);
     FOR ifreq := ifrmin to ifrmax DO BEGIN
        cs := cosomt*cosdomt-sinomt*sindomt;
        ss := sinomt*cosdomt+cosomt*sindomt;
        cosomt := cs ; sinomt := ss;
        gtre(itime) := gtre(itime)+[gfre(ifreq)*cosomt;
                    -gfim(ifreq)*sinomt]*fstep;
        gtim(itime) := gtim(itime)+[gfre(ifreq)*sinomt;
                     +gfim(ifreq)*cosomt]*fstep;

     END DO
   END DO
END PROCEDURE
```

TABLE 1.1. SUMMARY OF EQUATIONS OF CHAPTER 1

Properties of complex numbers

$$z = x + iy$$

$$z = re^{i\theta} = re^{i\omega t}$$

Beating oscillations

$$x = X \cos \omega t + X \cos(\omega + \delta\omega)t$$
$$= 2X \cos(\delta\omega/2)t \cos(\omega + \delta\omega/2)t$$

Fourier series

$$g(t) = a_0/2 + \sum_{n=1}^{\infty} (a_n \cos 2\pi nt/T + b_n \sin 2\pi nt/T)$$

$$a_0 = 2/T \int_0^T g(t) \, dt$$

$$a_n = (2/T) \int_0^T g(t) \cos n\omega t \, dt,$$

$$b_n = (2/T) \int_0^T g(t) \sin n\omega t \, dt$$

Fourier series of sampled functions

$$a_0 = (2/N) \sum_{i=1}^{N} g(t_i)$$

$$a_n = (2/N) \sum_{i=1}^{N} g(t_i) \cos(2n\pi t_i/T)$$

$$b_n = (2/N) \sum_{i=1}^{N} g(t_i) \sin(2n\pi t_i/T)$$

$$t_i = iT/N, \qquad T = 2\pi/\omega$$

Complex Fourier series

$$g(t) = \sum_{n=-\infty}^{\infty} G_n e^{in\omega t}$$

$$G_n = 1/T \int_0^T g(t) e^{-in\omega t} \, dt$$

Complex Fourier series of sampled functions

$$G_k = (1/N) \sum_{n=0}^{N-1} g_n e^{-i(2\pi kn/N)}$$

$$g_n = \sum_{k=0}^{N-1} G_k e^{i(2\pi kn/N)}$$

REFERENCES AND FURTHER READING

BRIGHAM, E. O. 1974. *The Fast Fourier Transform*. Englewood Cliffs, N.J.: Prentice Hall.

COOLEY, J. W., and TUKEY, J. W. 1965. An algorithm for the machine calculation of complex Fourier series. *J. Math. Comp.*, *19*(90): 297–301.

DEN HARTOG, J. P. 1934. *Mechanical Vibrations*. New York: McGraw-Hill.

HORT, W. 1910. *Technische Schwingungslehre*. Berlin: Springer-Verlag.

KLOTTER, K. 1951. *Technische Schwingungslehre*. Berlin: Springer-Verlag.

KRUSE, R. L. 1987. *Data Structures and Program Design*. Englewood Cliffs, N.J.: Prentice Hall.

RAMIREZ, R. W. 1985. *The FFT Fundamentals and Concepts*. Englewood Cliffs, N.J.: Prentice Hall.

RANDALL, R. B. 1977. *Frequency Analysis*. Naerum, Denmark: Bruel & Kjaer.

PROBLEMS

Sections 1.1 to 1.3

1.1. The motion of a certain point is governed by the equation $x = 0.5 + 3 \sin \omega t + 5 \cos \omega t$, where $\omega = 150$ rad/s. Find the amplitude and the phase angle of the equivalent harmonic motion, $A \cos(\omega t - \theta)$.

1.2. Plot the motion $g(t) = 2.5 \sin 32t - 1.2 \cos 64t$ on a vector diagram at $t = 0$, $T/4$, where T is the largest period of the two vibrations. Find the resulting displacements at these times.

1.3. A point A is known to be in harmonic motion along a straight line and about a point O, with frequency $\omega = 100$ rad/s. If at time $t = 0$, the point is at distance $OA = x_0 = 0.1$ m and has velocity $v_0 = 15$ m/s, determine the equation of the motion, vibration amplitude, and acceleration.

1.4. Two harmonic motions $x_1(t) = 1.05 \cos 377t$ and $x_2 = 2.1 \sin 375t$ are added. Find the maximum and minimum amplitude of the resulting beating vibration.

1.5. Find the phase angles between the following motions: **(a)** $Ae^{i\omega t}$ and $iAe^{i\omega t}$, and **(b)** $(2 + 3i)e^{i\omega t}$ and $(-1 + 4i)e^{i\omega t}$.

1.6. In a vibration test, an accelerometer measured maximum acceleration 14.5 m/s² and the period of vibration was measured on an oscilloscope to be 16.6 ms. Find the maximum vibration amplitude, assuming harmonic motion.

1.7. During a vibration test, the maximum velocity and the maximum acceleration of a vibrating machine were measured. They were $v_0 = 0.01$ m/s and $a_0 = 0.2g$, respectively. Assuming harmonic motion at a single frequency, determine the vibration frequency.

1.8. To locate the source of vibration aboard a ship, the maximum vibration amplitude and velocity were measured, $x_0 = 0.002$ m and $v_0 = 0.75$ m/s, respectively. There are two machines operating on the same floor: one at 3600 rpm, the other at 1720 rpm. Is the measured vibration compatible with any one of the two machines alone?

1.9. In addition to one at the running frequency, a turbocompressor rotating at 4000 rpm is suspected of having vibration at another frequency, 52 Hz. The available instrument can measure the sum of the squares of the rms values of displacement, velocity, and acceleration. On this basis, the measured rms amplitude is 0.02 mm and the rms velocity is 75 mm/s. Determine the vibration amplitudes at the two frequencies.

1.10. A velocity pickup (type of vibration measuring instrument) indicates a maximum velocity of 125 mm/s and a frequency of 120 Hz.

    (a) Determine the maximum values of displacement and acceleration.

    (b) Draw a vector diagram of displacement, velocity, and acceleration at $t = 0$.

Sections 1.3 to 1.7

1.11–1.18. Find analytically the Fourier series representation of the periodic functions shown in Figures P1.11 to P1.18, respectively.

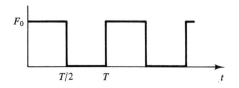

Figure P1.11

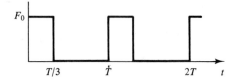

Figure P1.12

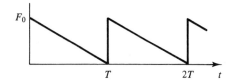

Figure P1.13

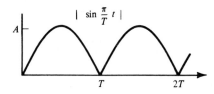

Figure P1.15

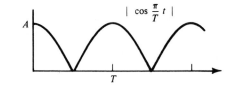

Figure P1.16

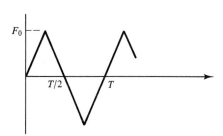

Figure P1.17

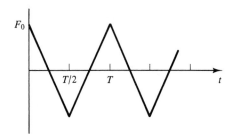

Figure P1.18

1.19. Prove that for even functions $[f(-t) = f(t)]$, the sine Fourier components are zero (i.e., $b_n = 0$, $n = 0, 1, 2, \ldots$). Prove also that for odd functions $[f(-t) = -f(t)]$, the cosine Fourier components are zero (i.e., $a_n = 0$, $n = 0, 1, 2, \ldots$).

1.20. Prove that equations (1.20) and (1.21) can be written as

$$F(t) = \frac{1}{T} \sum_{n=-\infty}^{\infty} c_n e^{in\omega t} \quad \text{where} \quad c_n = \int_0^T F(t) e^{-in\omega t}\, dt$$

1.21–1.28. Find analytically the complex Fourier series representation of the functions of Problems 1.11 to 1.18, respectively. Plot the first five terms on a spectral diagram.

SIMULAB PROBLEMS

S1.1–S1.8. Use the SIMULAB vibration analyzer to perform a FFT of the functions shown in Figures S1.1 to S1.8, respectively. To do this:
  (a) Divide one period into 16 equal time intervals and take 16 samples.
  (b) On your editor, make a file FFT.1. The first line contains the number of points and the time interval between sampling points. The subsequent lines contain successively the first 16 samples of the function. Using the Cut and Paste feature of your editor, repeat the 16 points eight times, for a total of 128 sampling points. Put the correct number of points (128) in the first line.
  (c) Run the ANALYZER program, simulating the vibration analyzer. Plot the time function and the spectrum amplitude. Perform a zoom about the fundamental frequency.

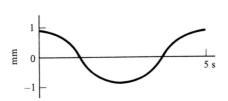

| Figure S1.1 | Figure S1.2 |

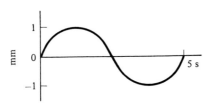

| Figure S1.3 | Figure S1.4 |

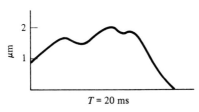

| Figure S1.5 | Figure S1.6 |

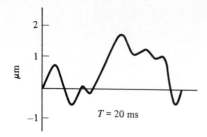

Figure S1.7

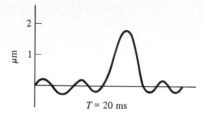

Figure S1.8

S1.9. Use the SIMULAB vibration analyzer to perform a FFT of the signal obtained from vibration measurements on a steam turbine running at 3600 rpm and contained in file FFT.9 of the SIMULAB program diskette.
(a) Run the ANALYZER program, simulating the vibration analyzer.
(b) Plot the time function and the spectrum amplitude. Perform a zoom about the fundamental frequency.
(c) Identify the fundamental frequency of vibration and its harmonics.

S1.10. Use the SIMULAB vibration analyzer to perform a FFT of the signal obtained from vibration measurements on an air compressor running at 1720 rpm and contained in file FFT.10 of the SIMULAB program diskette.
(a) Run the ANALYZER program, simulating the vibration analyzer.
(b) Plot the time function and the spectrum amplitude. Perform a zoom about the fundamental frequency.
(c) Identify the fundamental frequency of vibration and its harmonics.

S1.11. Use the SIMULAB vibration analyzer to perform a FFT of the signal obtained from vibration measurements on an aircraft engine, running at 7200 rpm and contained in FFT.11 of the SIMULAB program diskette.
(a) Run the ANALYZER program, simulating the vibration analyzer.
(b) Plot the time function and the spectrum amplitude. Perform a zoom about the fundamental frequency.
(c) Identify the fundamental frequency of vibration and its harmonics.

S1.12. A nozzle supplies steam to a small single-stage turbine of 500 hp. The stage has 18 blades at a pitch diameter of 0.3 m, of which one is missing. If the turbine is running at 4000 rpm, determine the harmonics of the torque applied to the shaft, assuming efficiency 75%. Make an appropriate file and use ANALYZER.

S1.13. The piston of a four-stroke engine operates with the compression closely following the perfect gas law. At the end of the compression, combustion increases the pressure by 120%. Determine the harmonics of the pressure force acting on the piston, assuming that the speed is 800 rpm, the piston diameter is 100 mm, and the stroke is 75 mm. The compression ratio is 9 and the atmospheric pressure is 0.1 MPa. Make an appropriate file and use ANALYZER.

DESIGN PROBLEMS

D1.1. Design a Scotch yoke (Figure 1.1) to produce a vibration of maximum velocity 3 m/s and frequency 120 rad/s for a sieve.

D1.2. Design a Scotch yoke (Figure 1.1) to produce a vibration of maximum amplitude 0.10 m and maximum velocity not to exceed 3 m/s for an aggregate separation machine.

D1.3. An almond transporter in a processing plant consists of a horizontal vibrating table forced by electrodynamic vibrators to move in an angle ϕ (Figure D1.3). Assuming that the static friction is high and slipping of the almond on the transporter surface is not possible, if the vertical component of the transporter deceleration as it moves toward the higher position is more than g, the acceleration of gravity, the almond will lose contact and move along a free path until it hits the transporter on its return from the lowest position. Assuming plastic impact, the almond keeps moving with the transporter until the cycle is repeated. Clearly, after every jump the almond moves to the right. Design a transporter for transporting speed 0.5 m/s.

D1.4. A vibrating feeder transporting bolts to an automatic assembly process consists of a helicoidal surface around a cylinder (Figure D1.4), which can have periodic torsional motion fast in one direction, slow in the other. Assuming that the static coefficient of friction is 0.2 and the kinetic coefficient of friction is 0.05, design the feeder for feeding speed 0.1 m/s.

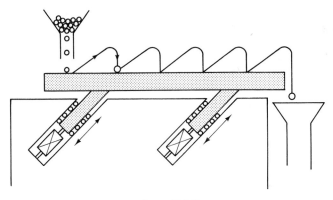

Figure D1.3

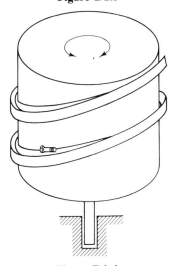

Figure D1.4

2

Natural Vibration about Equilibrium

2.1 VIBRATING SYSTEMS

Machines and structures are not rigid bodies, but rather, are systems of elastic components that respond to external or internal forces with finite deformations. In addition, there are relative motions between the components, giving rise to internal forces. Due to these and the external forces, the machine or structure moves. This motion, as a result of internal and external forces, is the subject of dynamics and vibration.

A complex consisting of several members with similar or dissimilar properties is called a *system*. A system is said to be in *equilibrium* when there is a complete balance of forces which could make the system move so that it is invariable with time. Figure 2.1 shows the equilibrium status of some systems. In Chapter 4, equilibrium is defined more precisely in energy terms. The systems considered here are capable of changing their configuration. Such systems possess *mobility*.

To describe a system that changes, the values of certain parameters that describe the geometry or the state of the system at any value of time must be

*Demolition by oscillation of a suspended bob. (Courtesy of Bruel & Kjaer. By permission.)

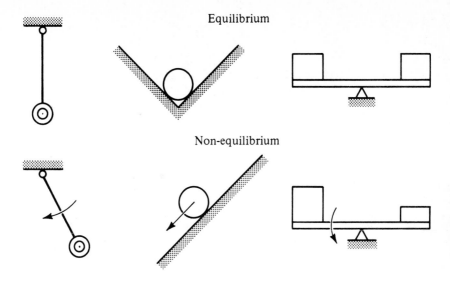

Equilibrium

Non-equilibrium

Figure 2.1   Mechanical system equilibrium.

specified. These parameters are called *coordinates*. Figure 2.2a shows possible coordinates for a cube that can slide along a flat bar. A number of coordinates can describe the configuration of the system, which is here the position of the cube on the bar at a certain time.

Another example is the piston shown in Figure 2.2b, which compresses a perfect gas adiabatically in a cylinder. The possible coordinates here are the distance x of the piston from some reference point, the volume of the compressed gas V, the gas temperature T, the gas pressure p, and so on. These coordinates are not all necessary to describe the system. In both systems, in fact, only one coordinate is sufficient to describe the system. The several coordinates are then said to be *dependent*. The relation between the dependent coordinates are called *equations of constraint*. For example, the equations

$$x_1 + x_2 = L \tag{2.1}$$

$$pV = nRT \tag{2.2}$$

are equations of constraint.

The minimum number of independent coordinates needed to describe a system completely is the number of *degrees of freedom* of the system. The systems in Figure 2.2 have one degree of freedom. A robot arm (Figure 2.3) has more than one degree of freedom.

A large class of systems can be described by a finite number of degrees of freedom, such as the simple systems in Figure 2.2. Other systems, such as those that include deformable members, have an infinite number of degrees of freedom. For example, to describe an elastic beam one needs an infinite number of coordinates (i.e., deflections of the infinite number of points on its elastic line). The elastic beam therefore has an infinite number of degrees of freedom. Almost all machines and structures have elastic members, and thus an infinite number of

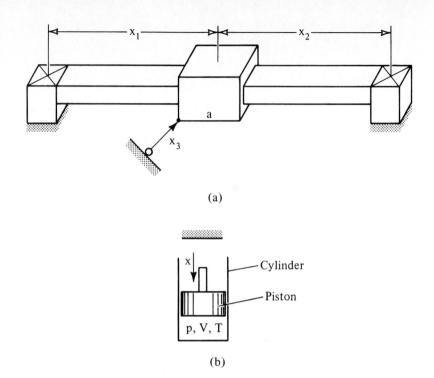

(a)

(b)

Figure 2.2 System coordinates.

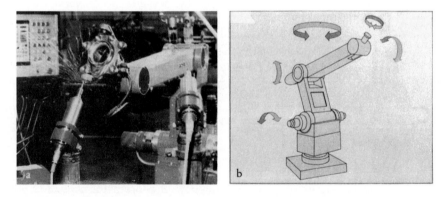

Figure 2.3 Degrees of freedom of an industrial robot arm.

degrees of freedom. Systems with a finite number of degrees of freedom are called *discrete systems*. Those with an infinite number of degrees of freedom are called *continuous systems*.

Although there are methods of dealing with continuous systems, engineering accuracy is reached, many times, by describing a continuous system with a finite number of coordinates. For example, an elastic beam can be described by a number of points along its elastic line. It appears that the larger number of coordinates yields the better approximation. The form of the elastic line between approximation points is assumed parabolic or any convenient curve.

At this point, without going into detail about the number of degrees of freedom required to describe a continuous system, it is assumed that every system, discrete or continuous, will be described by n independent coordinates: in other words, by n degrees of freedom. A *lumped system* is a particular case of a discrete system that consists of discrete elements, that is, elements which relate forces to displacements, velocities, and accelerations.

2.2 DISCRETE ELEMENTS OF VIBRATING SYSTEMS: STIFFNESS ELEMENTS

All properties of mechanical systems are distributed in space. In most cases, enough engineering accuracy can be achieved if these properties are approximately considered as lumped into single elements which, properly combined, can represent the dynamical properties of the system to sufficient accuracy. Such elements of vibrating systems are springs, dampers, and masses, which respond with reaction forces to displacement, velocity, and acceleration, respectively.

The linear spring is a device that reacts to its deformation with a force proportional to the deformation according to the relation

$$F = kx \qquad (2.3)$$

where F is the force, x the deformation, and k the coefficient of proportionality, called the *spring constant* or *stiffness coefficient* (units: FL^{-1}, N/m).

Similarly, a *rotational spring* is a device that reacts to rotational motion with a restoring torque proportional to the angle of rotation, $T = k_T\phi$. The coefficient of proportionality k_T is called the *rotational spring constant* (units: FL, $N \cdot m/rad$).

An elastic member can be deformed along many directions. Every relation between a force in one direction and a deformation in the same or another direction generates a spring constant. Equation (2.3) can therefore have the more general form

$$F_i = k_{ij}x_j \qquad (2.4)$$

where i and j can indicate, for example, translations or rotations along or about the three axis of a Cartesian coordinate system. Therefore, i and j can take six different values. In general, we shall have 6×6 independent coefficients k_{ij} for a physical component.

Consider, for example, the cantilever of Figure 2.4 with the coordinate system x, y, z, as indicated. If the cantilever has a circular section of diameter d and section area A, length L, modulus of elasticity E, shear modulus G, and section moments of inertia I_x, I_y, I_z, and if u, v, w are the deflections and θ, ϕ, ψ the rotations of its free end with respect to the coordinate system x, y, z, we shall have, from strength of materials,

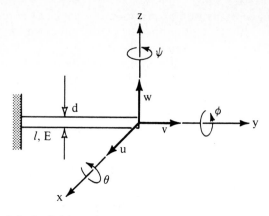

Figure 2.4 Definition of spring constants for a cantilever beam.

$$F_v = \frac{EAv}{L}, \qquad k_{vv} = \frac{EA}{L}$$

$$F_u = \frac{3EI_z u}{L^3}, \qquad k_{uu} = \frac{3EI_z}{L^3}$$

$$F_w = \frac{3EI_x w}{L^3}, \qquad k_{ww} = \frac{3EI_x}{L^3}$$

$$M_\phi = \frac{GI_y \phi}{L}, \qquad k_{\phi\phi} = \frac{GI_y}{L} \qquad (2.5)$$

$$M_\psi = \frac{EI_z \psi}{L}, \qquad k_{\psi\psi} = \frac{EI_z}{L}$$

$$M_\theta = \frac{EI_x \theta}{L}, \qquad k_{\theta\theta} = \frac{EI_x}{L}$$

where $I_x = I_z = \pi d^4/64$ and $I_y = \pi d^4/32$ for a circular cross section.

Systems with one degree of freedom have $i = j = 1$ and the subscript of the spring constant k is omitted. A tabulation of spring constants of typical elements is given in Appendix II.

There is a combination, in some systems, of several linear springs. These springs can be combined into an equivalent single spring element by using the following rules, which can easily be verified:

1. *Springs in parallel* (Figure 2.5). The spring constant of the equivalent spring is the sum of the spring constants of the springs (added forces with equal displacements):

$$k_{eq} = k_1 + k_2 + \cdots + k_n \qquad (2.6)$$

This relation is valid only when the springs deflect the same way; that is, the plates A and B remain parallel to one another or are very small.

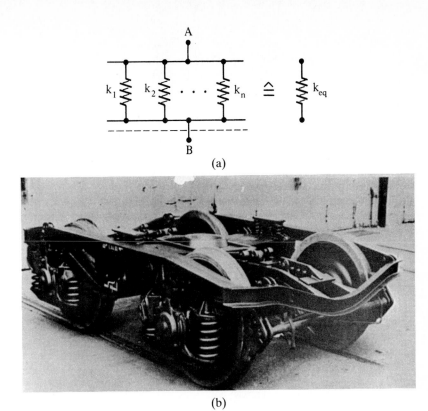

(a)

(b)

Figure 2.5 Linear springs connected in parallel.

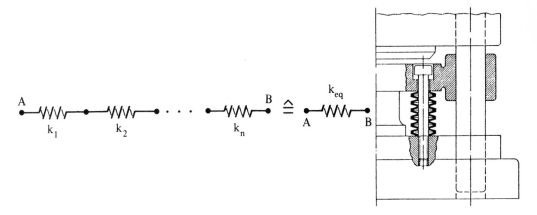

Figure 2.6 Linear springs connected in series.

2. *Springs in series* (Figure 2.6). The spring constant of the equivalent spring is (added displacements with equal forces)

$$\frac{1}{k_{eq}} = \frac{1}{k_1} + \frac{1}{k_2} + \cdots + \frac{1}{k_n} \tag{2.7}$$

Actual springs are not exactly linear, as required by equation (2.3), but their deformation is a certain function of the applied force. A spring is often linear within a certain range of deformation. Beyond this point, the stress exceeds the proportionality limit of Hooke's law (Figure 2.7) and after point 4, the deflection is no longer proportional to the force.

There are many other reasons that cause springs to be nonlinear. In the case of beams, for example, they behave as linear springs only if the deflections are small. It is known that for large deflections, simple beam theory is inadequate, and that between force and deflection there is a complicated nonlinear relationship.

Often, it is adequate for practical purposes to confine our analyses to small deformations. Then the force–deflection relationship of a nonlinear spring can be simplified by a process called *linearization*. Suppose that a spring has a nonlinear force–deflection relationship $F = f(x)$, as shown in Figure 2.8. A static force F_0 (called *preload*) is first applied on the spring, which forces it to deflect by x_0. At this position, we have an equilibrium if the spring reaction is exactly $-F_0$. Furthermore, an additional force ΔF will result in an incremental deflection Δx and we shall have $F_0 + \Delta F = f(x_0 + \Delta x)$. Expanding in Taylor series about the equilibrium position

$$F_0 + \Delta F = f(x_0 + \Delta x) = f(x_0) + f'\Delta x + \frac{f''}{2!}(\Delta x)^2 + \frac{f'''}{3!}(\Delta x)^3 + \cdots \qquad (2.8)$$

where all derivatives are evaluated at x_0.

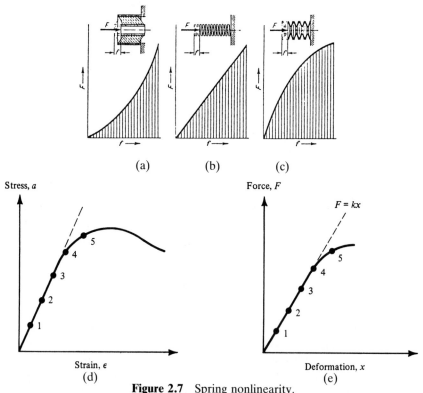

Figure 2.7 Spring nonlinearity.

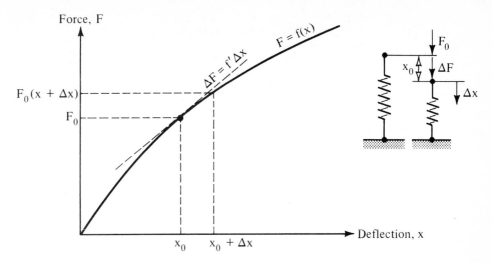

Figure 2.8   Spring linearization.

For small deflections Δx, if the function $f(x)$ is smooth enough, neglecting higher-order terms and taking $F_0 = f(x_0)$,

$$\Delta F \simeq f'(x_0)\,\Delta x \qquad (2.9)$$

This is a linear relationship and indicates that the spring constant k is the derivative of the function $F = f(x)$, evaluated at the equilibrium point. This has an obvious geometric interpretation. The tangent at the equilibrium point is substituted for the actual function.

There is no simple way of determining if the error of such simplification is acceptable. This depends on the function $f(x)$ itself, our standards of accuracy, and the purposes of the analysis. For example, later we show that the stability of a system might depend on the nonlinearity itself, and for the purpose of stability analysis, the approximation (2.9) many times is not acceptable, no matter how small we take Δx. An extensive list of spring constants for discrete elements is included in Appendix II.

☐ **Example 2.1   Compound Springs**

An overhead crane with a beam of length L and flexural rigidity EI is lifting a weight by using two cables of length l, diameter d, and modulus of elasticity E (Figure E2.1). Determine the spring constant between the hook and the ground in a vertical direction.
Solution   The beam is equivalent to a spring connected in series with the cable, which can also be modeled as a spring. The two springs will be combined to yield an equivalent single spring.

From Appendix II, for a simply supported beam loaded at the middle, $k_b = 48EI/L^3$. For the cable, which is a circular bar of length l, area $A = \pi d^2/4$, and Young's modulus E, we obtain $k_c = EA/l = \pi d^2 E/4l$. The model of the crane (Figure E2.1b) is simplified by combining the spring constants of the two cables according to equation (2.6). The model of Figure E2.1c, furthermore, can be simplified further by combining the springs k_b and $2k_c$ according to equation (2.7). Thus

$$\frac{1}{k_{eq}} = \frac{1}{2k_c} + \frac{1}{k_b}, \qquad k_{eq} = \frac{2k_b k_c}{k_b + 2k_c} = \frac{48\pi EId^2}{96Il + \pi d^2 L^3}$$

From this point on, we can deal with the equivalent spring constants for the dynamic response of the system.

(This example demonstrates the analysis of a complex structure by computing the stiffness of individual elements and combining them into a single equivalent spring.)

□

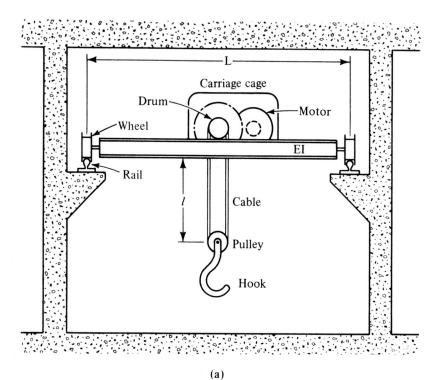

(a)

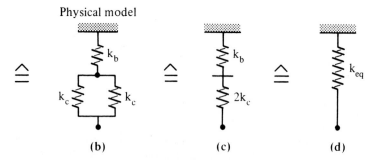

Figure E2.1

□ Example 2.2 Nonlinear Springs

An electronic instrument of mass $m = 8$ kg is placed on four elastic support pads of special rubber. The force–deflection relationship of each pad is given in Figure E2.2a. Determine the spring constant between the instrument and ground in the vertical direction.

Solution Since the spring is nonlinear, we will first find the equilibrium position, and at this point we will compute the equivalent linear spring as the slope at this point of the force–deflection curve.

Each pad will support a static preload of $8 \times 9.81/4 = 19.62$ N. The static deflection will be found from the force–deflection equation

$$19.62 \times 10^{-3} = 5x_0 + 1000x_0^2, \qquad x_0 = 2.6 \times 10^{-3} \text{ m}$$

The spring constant, evaluated at $x = x_0$, is then [equation (2.9)]

$$k = \left. \frac{dF}{dx} \right|_{x=x_0} = (5 + 2000x_0) \times 10^3 = 10.2 \times 10^3 \text{ N/m}$$

Since there are four springs in parallel,

$$k_{eq} = 4k = 40.8 \times 10^3 \text{ N/m}$$

(This example demonstrates the procedure of replacing a nonlinear spring with an equivalent linear spring.)

□

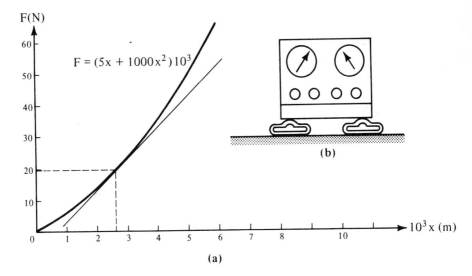

Figure E2.2

2.3 MODELING OF VIBRATING SYSTEMS

Engineering systems are always highly complex. As pointed out in Chapter 1, the vibration record obtained from measurements on a vibrating machine or structure is often very complicated, an indication of a complex system. To try to analyze the system in its finest detail—even more to try to predict its exact performance—is an impossible task. However, we are often interested only in the main features of the vibration record. Complete elimination of vibration is impossible and not important. Usually, elimination of the major vibration components is sufficient from an engineering standpoint. The vibration study of the system should therefore include only those features of the system that are related to the unwanted features of the response.

Vibration study of an engineering system is a systematic process and can be done either analytically or experimentally. In the former case the process is called *mathematical modeling* and consists of the following steps:

1. *Physical abstraction*. An engineering system is usually extremely complex. To find exactly how all the features of the system interact with one another is an impossible task. Moreover, all these interactions are not equally important. For example, there are several thousand components in an automobile. The vibration of the tailpipe certainly influences the vibration of the wheels. However, there is no evidence that this influence is substantial and that it affects the life of the tires. Therefore, one has to identify those elements of the system that are relevant to the problem under investigation and to the method of analysis that will be used. For example, let us suppose that we want to study the vertical vibration of an automobile as it goes over a road bump. It was stated above that an automobile has a great number of parts, all of which influence its dynamic performance. However, one can usually assume that, when going over a road bump, the deflection of the body of the car, together with its parts, engine, passengers, and so on, is much smaller than the motion of the car with respect to the road. Furthermore, the flexibility of the springs and the tires is much higher than the flexibility of the body of the car. We can therefore replace the car with a *model*—in other words, with an imaginary physical system which, for the purpose of determining the response of the car in going over a bump, will itself respond, to engineering accuracy, similar to the real system. It must be pointed out that a model which is adequate for one purpose might not be adequate for another purpose. Moreover, if higher accuracy is necessary, a model of higher fidelity might be needed (Figures 2.9 and 2.10). No general rule can be established, unfortunately, as to how to devise a model that adequately describes a real system. This is usually left to the experience and ingenuity of the engineer.

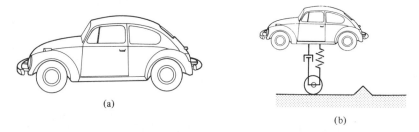

(a)

(b)

Figure 2.9 Modeling an automobile for vibration analysis with a single degree of freedom.

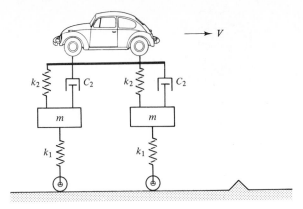

Figure 2.10 Two-degree-of-freedom model for an automobile.

2. *Mathematical formulation.* Using the model, one has to apply the appropriate laws of physics to derive one or more equations that describe the behavior of the system. The mathematical model is usually described by way of differential equations.

3. *Solution of the equations.* The equations of the mathematical model have to be solved to yield the desired results.

4. *Interpretation of the results.* Finally, the results obtained in step 3 are used to obtain solutions to the problem in the form of specific design decisions.

Steps 1 and 4 depend primarily on the engineer's experience and talent and are not properly the subject of a textbook. In this book we focus primarily on steps 2 and 3.

2.4 NEWTON'S SECOND LAW OF MOTION

Dynamics is the study of the motion of bodies and their mutual influences, in Newton's words *"from the phenomena of motion to investigate the forces of nature, and from these forces to investigate the other phenomena.*[1]

Newton's second law states that the change of motion is proportional to the motive force impressed and is made in the direction of the right line in which this force is impressed[2]:

$$\mathbf{F} = \frac{d(m\mathbf{v})}{dt} \tag{2.10}$$

and if the mass is invariable with time, for a material particle,

$$\mathbf{F} = \frac{md(\mathbf{v})}{dt} = m\ddot{\mathbf{r}} \tag{2.11}$$

[1] Newton, I. 1616. *Philosophiae naturalis principia mathematica.*

[2] Aristoteles' form of the second law was "if a double force moves a weight, it will move it to double distance at the same time."

Portrait of Newton. (Godfrey Kneller painting. Courtesy of the British Museum.)

SIR ISAAC NEWTON (1642–1727)

Professor of mathematics at Cambridge and president of the Royal Society of London, Newton's three laws of motion are included in his *Philosophia naturalis principia mathematica*, published in 1687, and have been the fundamental axioms of classical mechanics[3]:

Law I: Every body continues in its state of rest, or in uniform motion in a right line, unless it is compelled to change that state by forces imposed on it.

Law II: The change of motion is proportional to the motive force impressed and is made in the direction of the right line, in which this force is impressed.

Law III: To every action, there is always opposed an equal reaction, or the mutual actions of two bodies on each other are always directed to contrary parts.

(Translation: Motte, A. 1729. Berkeley, Calif.: University of California Press, 1960).

---

[3]Newton's laws were known in some forms before. In Newton's hands, however, they became a familiar tool of analysis.

where m is the mass of the particle, and $\mathbf{v}$ and $\mathbf{r}$ are the velocity and position vectors, respectively. It must be pointed out that the latter vectors have to be measured in an inertial coordinate system which, for common engineering purposes, it suffices to be fixed on the earth surface or on a large mass (a heavy vehicle, for example) which moves with constant velocity.

For rigid bodies of finite dimensions, Newton's second law has the form

$$\mathbf{F}_G = \frac{md(\mathbf{v}_G)}{dt} = m\ddot{\mathbf{r}}_G \tag{2.12}$$

$$\mathbf{M} = \dot{\mathbf{H}} \tag{2.13}$$

where the subscript G designates the center of mass of the body, $\mathbf{M}$ is the moment vector, and $\mathbf{H}$ is the angular momentum vector. This form of Newton's law of motion will be exemplified in Chapter 11. Here a particular form will be stated for rotation of the rigid body about an axis passing through a pivot point A, not necessarily the center of mass. If the axis of rotation coincides with the z-axis, equations (2.12) and (2.13) become

$$F_x = \frac{md(v_{xG})}{dt} = m\ddot{x}_G \tag{2.14}$$

$$F_y = \frac{md(v_{yG})}{dt} = m\ddot{y}_G \tag{2.15}$$

$$M_z = J_A\dot{\omega}_z \tag{2.16}$$

where F_x and F_y are the resultants of all forces (including the pivot forces) in the x and y directions, respectively, and M_z is the moment about the z-axis acting on the body in respect with point A.

2.5 D'ALEMBERT'S PRINCIPLE

It was pointed out by D'Alembert[4] that Newton's second law of motion could be considered from a different viewpoint in the form

$$\mathbf{F} + (-m\ddot{\mathbf{r}}) = 0 \tag{2.17}$$

treating the term $-m\ddot{\mathbf{r}}$ as if it were a force, called an *inertia force*. This facilitates application of the methods of statics, as equation (2.17) can be considered as an equation of static equilibrium. The idea can be extended to systems of particles and rigid bodies. For a rigid body, for example, performing plane motion, the equations of motion can be written in the form

$$\sum \mathbf{F} + (-m\ddot{\mathbf{r}}) = 0 \tag{2.18}$$

$$\sum \mathbf{M}_G + (-J_G\ddot{\boldsymbol{\theta}}) = 0 \tag{2.19}$$

[4]d'Alembert, J. M. 1743. *Traite de dynamique*. Paris. The concept of inertia force, the centrifugal force, was introduced by Huygens, C. 1673. 13 Theoremata de vi centrifuga. In *Horologium oscilatorium*, Chapter 5.

The equations can now be used with any coordinate system, provided that inertia forces have been expressed in respect to the centers of mass.

Portrait of D'Alembert. (Courtesy of the Bibliothèque Nationale, Paris.)

JEAN LE ROND D'ALEMBERT (1717–1783)

Abandoned as an infant at the doorsteps of a church, Le Rond managed to get a very broad education and collaborated with Diderot in the 28 volumes of the *Encyclopédie*. He played an important role in the ideological preparation of the French Revolution. At the age of 24 he was elected to the Académie des Sciences and in 1754 became its permanent secretary. His principle appeared in his treatise *Traite de dynamique* in 1743. He made important contributions to many branches of mathematics and developed the differential equation for the vibrating string.

2.6 UNDAMPED VIBRATION

It could be said that, in general, a system can have many degrees of freedom. The study of vibration should start with the simplest, the system with one degree of freedom. Some typical one-degree-of-freedom systems are shown in Figure 2.11. Figure 2.11a shows a mass m attached to a spring of constant k. The mass can slide in a guide without friction. It can, therefore, move along the vertical coordinate x only. From now on we shall omit the guide in the figure. The simple one-degree-of-freedom system shown in Figure 2.11c might represent a machine of mass M

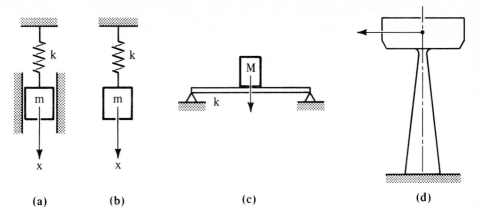

Figure 2.11 Typical single-degree-of-freedom systems.

supported by a flexible horizontal beam of stiffness k in the vertical direction. The beam, of course, has flexibility in a horizontal direction parallel to the plane of the paper and perpendicular to it. One could expect, however, that in most practical situations, the vertical motions are much larger than the horizontal ones. Thus the system can be considered, for all practical purposes, to have one degree of freedom. Another ground of simplification to one degree of freedom is the independence of motion. The water tower in Figure 2.11d can move horizontally in every direction. Its horizontal motion in the plane of the paper, however, does not affect its motion perpendicular to it. Its vertical motion affects it very little. If the water tower swings from left to right, for example, there is a slight vertical displacement. In most practical situations, the vertical motion is very small compared with the horizontal and can be considered negligible. These points will be exemplified further in the discussion of multidegree-of-freedom systems.

Let us begin with the system shown in Figure 2.11b. Because of the weight mg of the mass m, at the equilibrium position the spring has been stretched by an amount δ_0, so the restoring force $k\delta_0$ balances the weight mg. This is shown in the free-body diagram (Figure 2.12b).

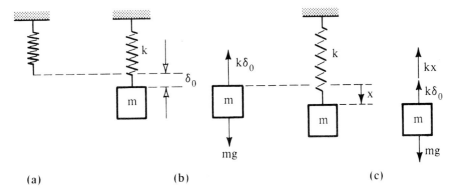

Figure 2.12 Dynamics of a single-degree-of-freedom system.

The position of the mass is expressed by the vertical coordinate x, measuring it from the static equilibrium position. Vertical displacement x results in an additional restoring force $-kx$. The free-body diagram is shown in Figure 2.12c. The forces are clearly not balanced in the vertical direction. Using $\ddot{x}$ to denote the acceleration d^2x/dt^2, the application of Newton's second law of motion in the vertical direction yields

$$m\ddot{x} = -kx - k\delta_0 + mg \tag{2.20}$$

Because $k\delta_0 = mg$ we obtain

$$m\ddot{x} + kx = 0 \tag{2.21}$$

This is the differential equation of motion for the system.

Equation (2.21) is an ordinary, second-order differential equation with constant coefficients. The general solution of this equation was obtained in Chapter 1 in the form $x(t) = a \cos \omega t + b \sin \omega t$ in a general way, using a series expansion of the solution. Using the complex expressions for the harmonic functions, $e^x = \cos x + i \sin x$, $i = (-1)^{1/2}$,[5] a harmonic solution can be written in the form[6]

$$x(t) = Xe^{pt} \tag{2.22}$$

where X and p are yet undetermined complex constants. Substitution into equation (2.21) yields

$$X(mp^2 + k) = 0 \tag{2.23}$$

Equation (2.23) implies that (2.22) is a solution of equation (2.21) for any value of X provided that $p^2 = -k/m$. Thus $p = \pm i\omega_n$, where $\omega_n = (k/m)^{1/2}$. Since both values of p yield solutions, the general solution of equation (2.21) is

$$x(t) = X_1 e^{i\omega_n t} + X_2 e^{-i\omega_n t} \tag{2.24}$$

where X_1 and X_2 are arbitrary constants.

Since $e^{i\phi} = \cos \phi + i \sin \phi$, equation (2.24) can be written as

$$x(t) = B_1 \cos \omega_n t + B_2 \sin \omega_n t \tag{2.25}$$

where B_1 and B_2 are two other arbitrary complex constants. This simply means that the system can vibrate harmonically at a single frequency that is a function of the system properties only and is independent of the mechanism of the vibration excitation. This is true for linear systems, in general, and was first observed by Pythagoras (sixth century B.C.) and later verified by Galileo.

In addition, the solution to equation (2.21) has to satisfy initial conditions

$$x(0) = x_0 \tag{2.26}$$

$$\dot{x}(0) = v_0$$

[5]This relation was essentially known to de Moivre, but it was popularized by Euler (*Introductio*, 1748).

[6]The solution for equation (2.21) was found by Daniel Bernoulli in 1739 and Euler developed the general theory of ordinary linear differential equations, usually encountered in vibration analysis (*Institutiones calculi integralis*, St. Petersburg, 1768–1770).

Portrait of Pythagoras. (From a statue of Pythagoras at the Archeological Museum of Napoli.)

PYTHAGORAS (ca. 580–ca. 500 B.C.)

From the Aegean island Samos, Pythagoras founded the Pythagorean school in southern Italy. He developed the mathematical relations of the music scale and discovered the laws of music harmony. He proved that every system has a natural frequency which depends on the system only and not on the vibration excitation mechanism. He experimented with the sound of hammers and also with vibrating strings. For the latter he proved that the natural frequency of vibration is a system property, dependent on the string length, diameter, and tension.

Equations (2.26) yield the values of the arbitrary constants B_1 and B_2, and then

$$x(t) = x_0 \cos \omega_n t + \frac{v_0}{\omega_n} \sin \omega_n t \qquad (2.27)$$

where $\omega_n = (k/m)^{1/2}$. The function $x(t)$ satisfies the differential equation in a certain interval of time (any time, in fact) and the initial conditions at $t = 0$. Such a solution is unique.[7]

Equation (2.27) represents a harmonic motion of angular velocity ω_n, frequency $f_n = \omega_n/2\pi$, and period $2\pi/\omega_n$. The quantity f_n is the *natural frequency of the vibration*. The *initial conditions* $x(0) = x_0$ and $\dot{x}(0) = v_0$ have a simple physical interpretation: they are the initial displacement and velocity of the mass m. Either one or both of these initial conditions actually start the motion. If $x_0 = 0$, $v_0 = 0$, equation (2.27) gives $x(t) = 0$, which verifies what should be expected from physical intuition. That is, the mass is never going to move unless someone disturbs it from

---

[7]Neumann, F. E. 1885. *Vorlesungen über die Theorie der Elastizität der fester Körper.* Leipzig.

the equilibrium position. Physically, x_0 can be initiated by holding the mass displaced by x_0 and releasing it. The initial velocity v_0 can be initiated, for example, with the blow of a hammer in the vertical direction.

The maximum amplitude of the vibration of equation (2.27) is

$$X = \left[x_0^2 + \left(\frac{v_0}{\omega_n} \right)^2 \right]^{1/2} \tag{2.28}$$

The time plot of the motion is shown in Figure 2.13. The motion is harmonic with amplitude X and continues indefinitely. This is an idealization because, in real systems, there are many factors that slow down this motion, as one can recall from his or her physical experiences. An important observation is that the natural frequency depends only on the physical properties k and m, and it is independent of the initial conditions.

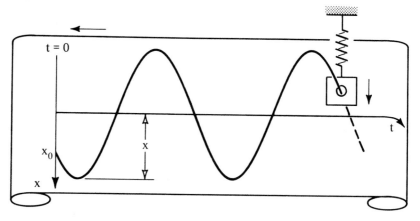

Figure 2.13 Response of a harmonic oscillator.

Vibration analysis of the one-degree-of-freedom system above is a systematic process and involves the following steps:

1. The system has one degree of freedom. Select the coordinate x.
2. Assume the mass displaced by x along the selected coordinate. Draw a free-body diagram showing forces on the mass.
3. Apply Newton's law:
   (a) *For linear motion:* Σ (forces) $= m\ddot{x}$, where $\ddot{x}$ is the acceleration *of the center of mass*.
   (b) *For rotary motion about a pivot point A:* Σ (moments)$_A = J_A \ddot{\theta}$, where $\ddot{\theta}$ is the angular acceleration and J_A the moment of inertia *about an axis through the pivot point A*, perpendicular to the plane of rotation.
4. Use static force equilibrium to remove unknown forces and geometric relationships to remove dependent coordinates, leaving only x and its derivatives.
5. Form a differential equation.
6. Find the natural frequency.

7. Apply the initial conditions to find the response of the system.

☐ **Example 2.3   Compound Frames**

Determine the natural frequency of vertical vibration of a mass attached to a flexible frame as shown in Figure E2.3.

Solution   The complex frame will be considered as two springs in series. These springs will be combined into a single one for further analysis. The model is shown in Figure E2.3a. For a simply supported beam (Appendix II), the spring constant for the lateral deflection at midspan $k = 48EI/L^3$.

Step 1: The system has one degree of freedom. We select the coordinate x.

Step 2: Assume the mass to be displaced by x. Applied forces are shown in Figure E2.3d. F is still unknown. The compatibility of displacements demands that

$$\delta_{12} = \frac{\delta_1 + \delta_2}{2}, \qquad \delta_1 = \frac{F/2}{k_1}, \qquad \delta_2 = \frac{F/2}{k_2}, \qquad x = \delta_{12} + \delta_3 = \frac{\delta_1 + \delta_2}{2} + \delta_3,$$

$$x = \frac{F}{4k_1} + \frac{F}{4k_2} + \frac{F}{k_3} = F\left(\frac{1}{4k_1} + \frac{1}{4k_2} + \frac{1}{k_3}\right)$$

Thus

$$F = \frac{x}{1/4k_1 + 1/4k_2 + 1/k_3}$$

Step 3: Newton's law demands that

$$m\ddot{x} = -F = \frac{-x}{1/4k_1 + 1/4k_2 + 1/k_3}$$

$$m\ddot{x} + \frac{1}{1/4k_1 + 1/4k_2 + 1/k_3} x = 0$$

Step 4: The natural frequency is

$$\omega_n = \left(\frac{1}{1/4k_1 + 1/4k_2 + 1/k_3}/m\right)^{1/2}$$

Alternative route: Using equivalent springs (Figure E2.3e), the equation of motion is $m\ddot{x} + k_{123}x = 0$. But

$$k_{12} = \frac{4k_1k_2}{k_1 + k_2}, \qquad k_{123} = \frac{k_{12}k_3}{k_{12} + k_3} = \frac{1}{1/4k_1 + 1/4k_2 + 1/k_3}$$

$$\omega_n = \left(\frac{k_{123}}{m}\right)^{1/2} = \left(\frac{1}{(1/4k_1 + 1/4k_2 + 1/k_3)m}\right)^{1/2}$$

Note that k_{12} is not $k_1 + k_2$ per equation (2.6) because the member AB does not remain horizontal and the displacements are not the same.

(This example demonstrates the modeling of a frame as a single spring by combining the equivalent springs of individual elements of the frame.)

☐

☐ **Example 2.4   Rotational Vibration of Solids**

A circular disk of radius R has a missing hole of radius r at a distance $\bar{r}$ from the center (Figure E2.4). The disk is free to rotate on the vertical plane about an axis

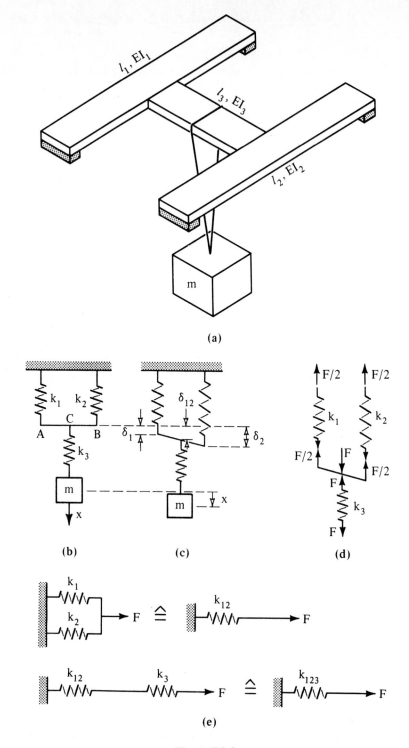

Figure E2.3

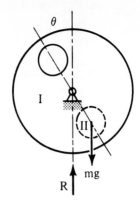

Figure E2.4

through its center and perpendicular to the plane of the disk. Determine the natural frequency of oscillations of the disk.

Solution   The system is dynamically equivalent to a disk I with two holes and a smaller disk II of mass m filling the lower hole. D'Alembert's principle is applied to determine the equation of motion. It states that inertia forces $-\Sigma\, m\ddot{x}$ (called *effective forces*) are in static equilibrium with the static forces on the system (Section 2.5). Consider the disk to be displaced from equilibrium by an angle θ. Consider effective forces on the disk, that is, the products $-m\ddot{x}$ (effective forces) and $-J_0\ddot{\theta}$ (effective moments). In addition, assume small displacements. D'Alembert's principle demands that

$$\left(\sum M\right)_0 + \left(\sum M_{\text{eff}}\right)_0 = 0$$

or

$$-mg(\bar{r}\sin\theta) \simeq -mg(\bar{r}\theta) = J_0\ddot{\theta}$$

If M is the mass of the disk without holes, we have

$$(J)_0 = \tfrac{1}{2}MR^2 - (\tfrac{1}{2}mr^2 + m\bar{r}^2)$$

and

$$(\tfrac{1}{2}MR^2 - \tfrac{1}{2}mr^2 - m\bar{r}^2)\ddot{\theta} + mg\bar{r}\theta = 0$$

or

$$\left(\frac{1}{2}\frac{M}{m}R^2 - \frac{r^2}{2} - \bar{r}^2\right)\ddot{\theta} + g\bar{r}\theta = 0$$

But

$$\frac{M}{m} = \frac{R^2}{r^2}$$

Therefore,

$$r^2\left[\frac{1}{2}\left(\frac{R}{r}\right)^4 - \left(\frac{\bar{r}}{r}\right)^2 - \frac{1}{2}\right]\ddot{\theta} + g\bar{r}\theta = 0$$

Thus

$$\omega_n = \left\{ \frac{g\bar{r}}{r^2[0.5(R/r)^4 - (\bar{r}/r)^2 - 0.5]} \right\}^{1/2}$$

(This example demonstrates the rotational vibration analysis of an unsymmetrical solid using D'Alembert's principle.)

☐

☐ **Example 2.5 Cantilever Beam**

A steel cantilever beam of length 1 m has a rectangular cross section of 0.01 × 0.12 m². A mass of 100 kg is attached to the free end of the beam as shown in Figure E2.5. Determine the natural frequency of the system for vertical vibration.

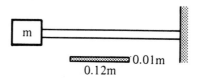

Figure E2.5

Solution   Assume that the mass of the beam is small. From Appendix II the deflection at the free end of the cantilever beam due to the end force P is $\delta = PL^3/3EI$. Therefore, for small oscillations, the spring constant is $k = P/\delta = 3EI/L^3$.

The moment of inertia I of the beam is $I = bh^3/12 = 0.12 \times 0.01^3/12 = 10^{-8}$ m⁴, and the modulus of elasticity of the steel is $E = 2.06 \times 10^{11}$ N/m². Therefore, $k = 3EI/L^3 = 3 \times 2.1 \times 10^{11} \times 10^{-8}/1^3 = 6300$ N/m.

The equation of motion for free undamped vibration is

$$m\ddot{x} + kx = 0$$

$$\omega_n = \left(\frac{k}{m}\right)^{1/2} = \left(\frac{6300}{100}\right)^{1/2} = 7.94 \text{ rad/s}$$

(This example demonstrates the vibration analysis of a mass at the end of a massless cantilever beam for lateral vibration.)

☐

☐ **Example 2.6 Mass on Taut String**

The string shown in Figure E2.6 is under tension T, which can be assumed to remain constant for small displacements. For small oscillations, find the natural frequency of the vertical vibration of the mass m. The effect of gravity is negligible.

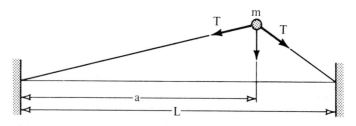

Figure E2.6

Solution Assume that the mass is displaced by x in the vertical direction. The tension in the string is the restoring force. Since the tension is essentially constant, the vertical components of the tension T on the mass is $T[x/a + x/(L - a)]$. Applying Newton's second law of motion, the equation of motion

$$m\ddot{x} + T\left(\frac{x}{a} + \frac{x}{L - a}\right) = 0 \quad \text{or} \quad m\ddot{x} + \left[\frac{TL}{a(L - a)}\right]x = 0$$

and

$$\omega_n = \left[\frac{TL}{ma(L - a)}\right]^{1/2}$$

(This example demonstrates the vibration analysis of a mass on a taut spring for lateral vibration.)

☐

☐ **Example 2.7 Physical Pendulum**

The physical pendulum is a rigid body of mass m and is pivoted at a point at distance d from its center of mass G (Figure E2.7). It is free to rotate under its own gravitational force. Find the frequency of oscillation of such a pendulum.

Solution Application of Newton's law for the moments about point O, the axis of rotation, yields

$$J_0\ddot{\theta} = -mgd \sin \theta$$

where J_0 is the moment of inertia of the body about the axis of rotation.
Considering only small oscillations, $\sin \theta \approx \theta$ and

$$\ddot{\theta} + \frac{mgd}{J_0} \theta = 0, \qquad \omega_n = \left(\frac{mgd}{J_0}\right)^{1/2}$$

But $\omega_n = (g/d)^{1/2}$ for a simple pendulum of length d. Hence J_0/md can be defined as the length of the equivalent simple pendulum.

Note: Measurements of the natural frequency of oscillation of compound bodies, supported from one pivot above their center of mass so that they become compound penduli, are used to yield the moment of inertia, which for complex geometry is

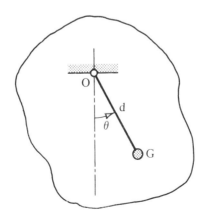

Figure E2.7

difficult to be calculated. To this end we measure the mass by weighing and the natural frequency of oscillation by causing the body to oscillate about a pivot as a compound pendulum. The center of mass is found at the intersection of two vertical lines drawn at static equilibrium from two different pivots. Therefore, the distance d is known. Then

$$J_0 = \frac{mgd}{\omega_n^2}, \qquad J_g = J_0 - md^2$$

are, respectively, the moments of inertia about the pivot O and about the center of mass. Instead of the mass moment of inertia, the *radius of gyration* is often used, $r_g = (J_g/m)^{1/2}$.

(This example demonstrates the vibration analysis of a solid pendulum, called a *physical pendulum*.)

□

□ **Example 2.8 Compound Physical Pendulum**

A system of pipe cutting in a continuous production process consists of a large disk of radius r and mass M that can rotate around its center O. At the end of a light rod of length L, a motor of mass m is attached with a cutting wheel (Figure E2.8). The system can oscillate on the plane of the disk about point O.

(a) Determine the period of the natural oscillation of the system for small angles.

(b) Determine the maximum linear velocity of the motor if the arm is displaced by an angle θ_0 and is released.

Solution (a) Newton's law about point O yields

$$J\ddot{\theta} = -mg(L + r) \sin \theta$$

For small θ,

$$J\ddot{\theta} + mg(L + r)\theta = 0$$

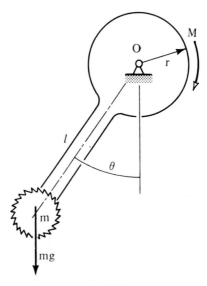

Figure E2.8

Natural Vibration about Equilibrium   Chap. 2

The moment of inertia

$$J = J_m + J_M = m(L + r)^2 + \tfrac{1}{2}Mr^2$$

The period of natural oscillation

$$T_n = 2\pi \left(\frac{m_{eq}}{k_{eq}}\right)^{1/2} \quad \text{or} \quad T_n = 2\pi \left[\frac{m(L + r)^2 + \tfrac{1}{2}Mr^2}{mg(L + r)}\right]^{1/2}$$

(b) For $\omega_n = 2\pi/T_n$, we shall have $\theta = \theta_0 \cos \omega_n t$. The angular velocity is $\dot{\theta} = -\theta_0 \omega_n \sin \omega_n t$. Its maximum value is

$$\dot{\theta}_{max} = \theta_0 \omega_n = \theta_0 \left[\frac{mg(L + r)}{m(L + r)^2 + \tfrac{1}{2}Mr^2}\right]^{1/2}$$

The maximum linear velocity will be $v_{max} = L(\dot{\theta}_{max})$.

(This example demonstrates the vibration analysis of a compound pendulum, which consists of several components rigidly connected.)

□

□ **Example 2.9   Vibration of Floating Bodies**

A solid cylinder of radius r is partially immersed in a bath of distilled water as shown in Figure E2.9. Find the natural frequency of oscillation of the cylinder in the vertical direction, assuming that it remains in an upright position. The densities of the cylinder and the water are ρ_c and ρ_w.

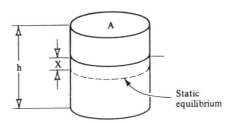

Figure E2.9

Solution   The vertical displacement of the cylinder is x. The weight of water displaced is $Ag\rho_w x$. This is the restoring force according to the Archimedes principle. The mass of the cylinder is $Ah\rho_c$. From Newton's law, the equation of motion is

$$\rho_c Ah\ddot{x} + Ag\rho_w x = 0$$

or

$$\ddot{x} + \frac{\rho_w g}{\rho_c h} x = 0$$

Therefore,

$$\omega_n = \left(\frac{\rho_w g}{\rho_c h}\right)^{1/2}$$

It should be pointed out that there is an error in this analysis. Part of the water is moving with the cylinder. Therefore, the natural frequency will be somewhat smaller.

(This example demonstrates the vibration analysis of a floating body for vertical vibration.)

□ Example 2.10   Vibration Due to Impact

A slab of mass m_1 is supported by a spring of stiffness k (Figure E2.10). A mass m falls from a height h on the slab with a perfectly plastic impact. Determine the resulting vibration.

Solution   We will first find the velocity of the falling mass at impact. Then, using impact theory and conservation of momentum, we will calculate the initial velocity of the two masses moving as a rigid body. This, in turn, will be the initial condition for the vibrating mass.

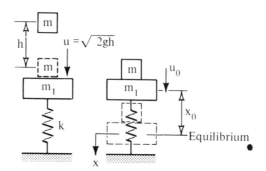

Figure E2.10

When mass m strikes the plate, it has a velocity of $u = (2gh)^{1/2}$. Conservation of momentum demands that $mu = (m_1 + m)u_0$, where u_0 is the velocity of the two masses after the impact. Therefore,

$$u_0 = (2gh)^{1/2}m/(m + m_1)$$

At this moment the system is not at static equilibrium. If we load the mass m_1 with an additional load mg, the static equilibrium would be $\delta_0 = mg/k$ lower. If displacement x is measured from this point, the initial conditions are

$$x_0 = \frac{-mg}{k}, \qquad u_0 = (2gh)^{1/2}m/(m + m_1)$$

Therefore,

$$x(t) = \frac{-mg}{k}\cos\omega_n t + \frac{(2gh)^{1/2}[m/(m + m_1)]}{\omega_n}\sin\omega_n t, \quad \omega_n = \left(\frac{k}{m + m_1}\right)^{1/2}$$

(This example demonstrates the vibratory response of a single-degree-of-freedom system to impact on the mass.)

2.7 DAMPING ELEMENTS

The viscous damper is a device that responds to a motion of one end relative to the other, with a force proportional to the associated relative velocity, according to the relation

$$F_x = c \frac{dx}{dt} \tag{2.29}$$

where c is a coefficient of proportionality called the *damping constant*. A simple viscous damper consists of two plates at a distance d and a fluid of viscosity η between the plates, which move relative to each other with a velocity v (Figure 2.14). Then the plates resist this motion with a force[8]

$$F = \frac{\eta A}{d} v \tag{2.30}$$

Therefore, the damping constant is

$$c = \frac{\eta A}{d} \tag{2.31}$$

where A is the wet area of the smaller plate. Linear dampers in series or parallel can be combined in the same way as the springs (Section 2.3).

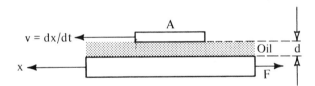

Figure 2.14   Simple shear flow model for damping.

This viscous damper model is linear within a wide range of velocities.[9] In other words, the damping constant is relatively insensitive to the velocity. Not all viscous dampers, however, have this property. With the damper shown in Figure 2.14, one needs a large plate area A or a very narrow clearance d to produce substantial damping. Because of this, other devices have been used, many times with a nonlinear force–velocity relationship (Figure 2.15).

The same procedure of linearization can be followed as that used in the case of the nonlinear spring. In general, if there is no constant velocity (similar to static load F_0), the derivative at zero is taken as the damping constant c[10]:

$$c = \left. \frac{df(\dot{x})}{d\dot{x}} \right|_{\dot{x}=0} \tag{2.32}$$

[8]Suggested by Newton for a class of common fluids called Newtonian. Explained by Maxwell as an interaction between molecules (1886, On the viscosity or internal friction of air and other gases, *Philos. Trans.*, *156:* 249).

[9]Stodola (see Chapter 11) published this in 1893. *Schweiz. Bauzg.*, *22:* 113.

[10]Lehr, E. 1930. *Schwingungstechnik*. Berlin: Verlag von Julius Springer.

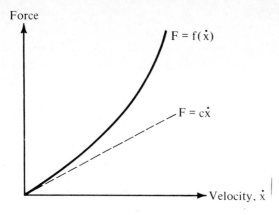

Force

$F = f(\dot{x})$

$F = c\dot{x}$

Velocity, $\dot{x}$

Figure 2.15 Nonlinear damper response.

Certain dampers have a resistance force that is proportional to the square of the velocity, $\dot{x}^2$. The derivative at $\dot{x}(0)$ is zero; therefore, linearization is not applicable and nonlinear methods have to be employed (Chapter 12). Extensive studies and design tables were reported by W. E. Milne.[11]

☐ **Example 2.11   Flat Plate Damper**

To dampen the motion of an instrument pointer, the latter is connected with a moving 30×30 mm steel strip placed between two plates, leaving a 0.1-mm gap on each side and confined to move in one direction and remain parallel to the plates. Determine the damping constant if the gap is filled with oil with viscosity $\eta = 20$ mPa·s.

Solution   The damping is due to Newtonian shear resistance and equation (2.31) should be used:

$$c = \frac{\eta A}{d}$$

where $A = 2 \times 0.03 \times 0.03 = 1.8 \times 10^{-3}$ m^2, $d = 0.0001$ m. Therefore,

$$c = 20 \times 10^{-3} \times 1.8 \times \frac{10^{-3}}{10^{-4}} = 0.36 \text{ N·s/m}$$

(This example demonstrates the computation of the linear damping constant for a flat damper.)

☐

☐ **Example 2.12   Torsional Viscous Damper**

A torsional damper consists of two concentric cylinders forming a clearance annulus and filled with a viscous fluid. The torsional damper shown in Figure E2.12 has radius $r = 0.50$ m, length $l = 0.75$ m, and clearance $e = 0.125$ mm. It is filled with oil with a viscosity of $\eta = 6 \times 10^{-3}$ N·s/m^2. Determine the torsional damping constant defined by the relation $M = -c_t\dot{\theta}$, where M is the damping torque and θ is the angle of rotation.

Solution   For angular velocity $\dot{\theta}$, the peripheral velocity is $v = r\dot{\theta}$. For an arc $d\theta$ of length $r\,d\theta$, there is a shear force due to the oil viscosity, as in equation (2.30):

---

[11]1929. *Univ. Oreg. Publ. Math. Ser.*, *1*(1). See also R. von Mises. 1914. *Elemente der technischen Hydromechanik*. Berlin: Julius Springer; and [Timoshenko, 1955].

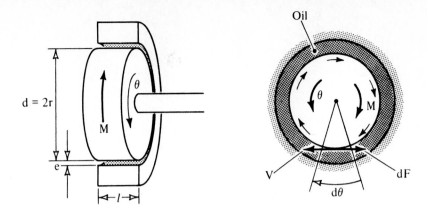

Figure E2.12

$$dF = -\frac{\eta A}{e} v = -\frac{\eta l r \, d\theta}{e} r\dot{\theta}$$

The moment about the axis of the cylinder is

$$dM = r \, dF = -\frac{\eta l r^3}{e} \dot{\theta} \, d\theta$$

Integrating yields

$$M = \int_0^{2\pi} dM = -\frac{\eta l r^3}{e} \dot{\theta} \int_0^{2\pi} d\theta = -\frac{2\pi \eta l r^3}{e} \dot{\theta}$$

Therefore, the damping constant will be

$$c_t = \frac{2\pi \eta l r^3}{e} = \frac{2\pi \times 6 \times 10^{-3} \times 0.75 \times 0.50^3}{0.125 \times 10^{-3}} = 28.3 \ \text{N} \cdot \text{m} \cdot \text{s}$$

□ **Example 2.13 Piston Damper**

A fluid damper, such as the type used for shock absorbers, consists of a piston of length L which has two holes of diameter D (Figure E2.13). Determine the damping constant assuming that the diameter of the piston is d and the oil has viscosity η and density ρ.[12]

Solution   For laminar flow in pipes, the pressure drop is

$$\Delta p = \eta \frac{L}{D} \frac{U^2}{2} f$$

where the friction factor

$$f = \frac{64\eta}{UD\rho}$$

Therefore,

$$\Delta p = \frac{32L\eta}{D^2} U$$

[12]Stodola, A. 1893. *Schweiz. Bauzg.*, *23*: 113.

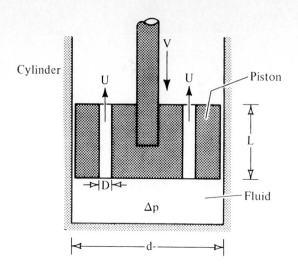

Figure E2.13

U is the average velocity of the oil flow through the holes. Due to continuity of the flow,

$$U = \frac{1}{2}\left(\frac{d}{D}\right)^2 V$$

if V is the piston velocity. Therefore,

$$\Delta p = \frac{32L\eta}{2D^2}\left(\frac{d}{D}\right)^2 V$$

If we assume that the holes are very small, the force on the piston, because of Δp, is

$$F = \frac{\pi d^2}{4}\Delta p = 4\pi L\eta\left(\frac{d}{D}\right)^4 V$$

The damping constant as defined by the relation $F = cV$, therefore, is

$$c = 4\pi L\eta\left(\frac{d}{D}\right)^4$$

For n holes,

$$c = \frac{8\pi L\eta}{n}\left(\frac{d}{D}\right)^4$$

□

2.8 DAMPED NATURAL VIBRATION

As pointed out already, a motion that has been started by an initial disturbance does not continue forever. Many factors contribute to this slowdown. The viscous damper, discussed in Section 2.7, is the most common one. A single-degree-of-freedom system with a viscous damper is shown in Figure 2.16. If the mass is in motion $x = x(t)$ at time t, the velocity will be $v = \dot{x}$. There will therefore be a

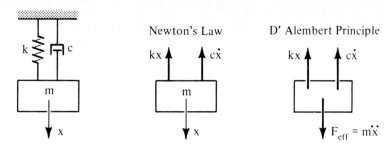

Figure 2.16 Damped harmonic oscillator (Courtesy of Tecumseh Products Co., Tecumseh, Michigan.)

force in the opposite direction $F = -c\dot{x}$, where c is the damping constant. This parameter depends on the physical properties of the damper, but it is independent of the displacement and time.

If x is measured from the equilibrium position, application of Newton's law in the vertical direction yields [Rayleigh, 1894]

$$m\ddot{x} = -c\dot{x} - kx \tag{2.33}$$

Using the natural frequency $\omega_n = (k/m)^{1/2}$ and the *fraction of critical damping* or *damping ratio* $\zeta = c/2m\omega_n$, the meaning of the latter to be explained later,

$$\ddot{x} + 2\zeta\omega_n\dot{x} + \omega_n^2 x = 0 \tag{2.34}$$

This is the differential equation of motion of the system. It is supplemented with the initial conditions

$$x = x_0, \quad \dot{x} = v_0, \quad \text{at } t = 0 \tag{2.35}$$

To find the solution of equation (2.34), the solution found for the harmonic oscillator $x(t) = e^{pt}$ will be used, where p is a still undetermined complex constant. Inserting this function into equation (2.34) yields

$$(p^2 + 2\zeta\omega_n p + \omega_n^2)e^{pt} = 0 \tag{2.36}$$

This yields the characteristic equation

$$p^2 + 2\zeta\omega_n p + \omega_n^2 = 0 \tag{2.37}$$

We therefore see that $x(t) = e^{pt}$ is a solution of equation (2.34) provided that p is a root of equation (2.37):

$$p_{1,2} = [-\zeta \pm (\zeta^2 - 1)^{1/2}]\omega_n \tag{2.38}$$

These roots are called *characteristic values* or *eigenvalues* of equation (2.34). They yield two solutions of this equation:

$$x_1(t) = e^{p_1 t}, \quad x_2(t) = e^{p_2 t} \tag{2.39}$$

where

$$p_1 = [-\zeta + (\zeta^2 - 1)^{1/2}]\omega_n, \quad p_2 = [-\zeta - (\zeta^2 - 1)^{1/2}]\omega_n \tag{2.40}$$

From the theory of the differential equations, it is known that the general solution of equation (2.34) will then be (see also Chapter 1)

$$x(t) = c_1 e^{p_1 t} + c_2 e^{p_2 t} \tag{2.41}$$

with c_1 and c_2 as arbitrary constants. Application of the initial conditions (2.35) to equation (2.41) yields

$$x(0) = c_1 + c_2 = x_0$$
$$\dot{x}\big|_{t=0} = p_1 c_1 + p_2 c_2 = v_0 \tag{2.42}$$

Equations (2.42) yield the values for the arbitrary constants c_1 and c_2:

$$c_1 = \frac{x_0 p_2 - v_0}{p_2 - p_1}$$
$$c_2 = \frac{v_0 - x_0 p_1}{p_2 - p_1} \tag{2.43}$$

With these values of the constants c_1 and c_2, the solution (2.41) is the unique solution of equation (2.34) for any time interval [Pontryagin, 1962].

To investigate the behavior of the solutions (2.39), the following four cases are considered.

I. $c_2 = 0$, $\zeta = 0$. The problem can be stated as

$$\ddot{x} + \omega_n^2 x = 0$$
$$x(0) = x_0, \qquad \dot{x}(0) = v_0 \tag{2.44}$$

The solution is equation (2.41), with

$$p_1 = i\omega_n, \qquad p_2 = -i\omega_n \tag{2.45}$$

If $m > 0$ and $k > 0$, $i = (-1)^{1/2}$, we obtain

$$x(t) = c_1 e^{i\omega_n t} + c_2 e^{-i\omega_n t} \tag{2.46}$$

Using the well-known relation $e^{i\theta} = \cos\theta + i\sin\theta$, we have

$$x(t) = (c_1 + c_2)\cos\omega_n t + (c_1 - c_2) i \sin\omega_n t \tag{2.47}$$

Because $p_1 = -p_2$, with equations (2.42) we obtain

$$x(t) = x_0 \cos\omega_n t + \frac{v_0}{\omega_n}\sin\omega_n t \tag{2.48}$$

This represents a harmonic oscillation with angular velocity $\omega_n = (k/m)^{1/2}$.

II. $c^2 < 4mk$ or $\zeta < 1$. Then $\zeta^2 - 1 < 0$ and

$$p_1 = [-\zeta + i(1 - \zeta^2)^{1/2}]\omega_n, \qquad p_2 = [-\zeta - i(1 - \zeta^2)^{1/2}]\omega_n \tag{2.49}$$

so that the two solutions will be

$$x_{1,2}(t) = e^{-\zeta\omega_n t}(\cos\omega_d t \mp i \sin\omega_d t) \tag{2.50}$$

where $\omega_d = \omega_n(1 - \zeta^2)^{1/2}$. The general solution is

$$x(t) = e^{-\zeta\omega_n t}[(c_1 + c_2)\cos\omega_d t + i(c_1 - c_2)\sin\omega_d t] \tag{2.51}$$

Using (2.41) and (2.43), we obtain

$$x(t) = e^{-\zeta\omega_n t}\left(x_0\cos\omega_d t + \frac{x_0\zeta\omega_n + v_0}{\omega_d}\sin\omega_d t\right) \tag{2.52}$$

Because of the term $e^{-\zeta\omega_n t}$, if $m > 0$ and $c > 0$, the solution is a harmonic oscillation (due to the expression in parentheses) but with an amplitude decreasing with time. The angular velocity of the oscillation is $\omega_d = \omega_n(1 - \zeta^2)^{1/2}$.

III. $c^2 > 4mk$ or $\zeta > 1$. Then the solution (2.41) is

$$x(t) = c_1 e^{p_1 t} + c_2 e^{p_2 t} \tag{2.53}$$

With

$$p_{1,2} = [-\zeta \mp (\zeta^2 - 1)^{1/2}]\omega_n,$$

$$c_1 = x_0\left[\frac{\zeta}{2(\zeta^2 - 1)^{1/2}} + \frac{1}{2}\right] + \frac{v_0}{2(\zeta^2 - 1)^{1/2}\omega_n} \tag{2.54}$$

$$c_2 = -x_0\left[\frac{\zeta}{2(\zeta^2 - 1)^{1/2}} - \frac{1}{2}\right] - \frac{v_0}{2(\zeta^2 - 1)^{1/2}\omega_n}$$

The solution given by equation (2.53) is a function decreasing monotonously with time.

IV. $c^2 = 4mk$, $\zeta = 1$. Then equation (2.37) has two equal roots,

$$p_1 = p_2 = -\zeta\omega_n$$

It is observed that $te^{\zeta\omega_n t}$ is also a solution and the general solution becomes

$$x(t) = e^{-\zeta\omega_n t}(c_1 + c_2 t) \tag{2.55}$$

$$c_1 = x_0, \qquad c_2 = v_0 + \zeta\omega_n x_0$$

This motion is also nonperiodic, decreasing with time.

These four cases are compared in Figure 2.17.

For most engineering systems not deliberately damped, the damping constant c is very small and $\zeta \ll 1$. For those systems, therefore, the frequency of the vibration does not differ much from the frequency of the undamped system. For systems with high damping, however, the frequency can decrease considerably. In the limit for value of damping $c_c^2 = 4mk$ or $\zeta = 1$, a condition called *critical damping*, the frequency becomes zero. This means that the period becomes infinite. In other words, the amplitude will never cross the t-axis but will only approach it asymptotically. For higher values of the damping constant, the result is merely that the amplitude approaches zero more slowly. This situation is shown in Figure 2.17. The meaning of ζ is apparent now: $\zeta = c/2\sqrt{mk}$ is the ratio of the damping constant c to the critical one $c_c = 2\sqrt{mk}$.

The roots of the characteristic equations, which in general are complex numbers, are shown in Figure 2.18 in the complex plane as functions of the damping ratio ζ. They trace a curve on the complex plane which we call the *root locus*.

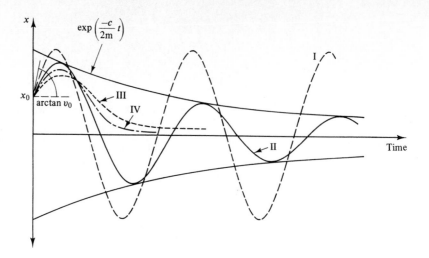

Figure 2.17 Dynamic response of a single-degree-of-freedom system.

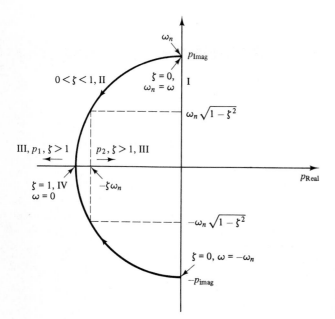

Figure 2.18 Root locus for the damped oscillator.

Another way of viewing the response of the damped oscillator is the *phase portrait* or *phase plot*. It is the plot of the response on the $(x, \dot{x} = v)$ plane. For the harmonic oscillator, $x = x_0 \cos \omega_n t$, $v = -x_0 \omega_n \sin \omega_n t$ describes an ellipse, or a circle with proper selection of scale (Figure 2.19). The phase portrait of a damped oscillator is a spiral curve asymptotically approaching the origin. The overdamped oscillator has motion toward the origin, which it also approaches asymptotically. From equation (2.52) we observe that the amplitude will be

Natural Vibration about Equilibrium   Chap. 2

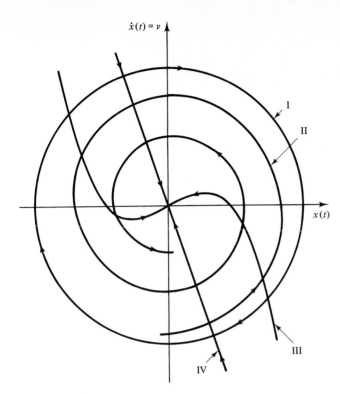

Figure 2.19 Response of an oscillator in the phase plane.

$$X = e^{-\zeta\omega_n t}\left[x_0^2 + \frac{(x_0\zeta\omega_n + v_0)^2}{\omega_d^2}\right]^{1/2} \tag{2.56}$$

The vibrations studied here are initiated by a disturbance of the system from its equilibrium position. This disturbance, of course, was made possible with the application of a force. After the system was released at time $t = 0$, however, no external force was acting on it, except its weight and the reaction of the spring support. The vibration is due to the initial disturbance only and is called *natural vibration*, undamped or damped. It has been seen that the system will vibrate with a certain frequency which is a function of the system properties m, c, and k. More precisely, it depends on two parameters ζ and ω_n and is independent of the magnitude of the disturbance.

The vibration amplitude of a damped oscillator due to an initial displacement x_0 is, from equation (2.56),

$$X = \frac{e^{-\zeta\omega_n t}x_0}{(1 - \zeta^2)^{1/2}} \tag{2.57}$$

The logarithm of the ratio of the amplitudes of two successive oscillations, differing in time by one period $2\pi/\omega_d$, is

$$\delta = log\,(e^{\zeta\omega_n T}) = \zeta\omega_n T = \frac{2\pi\zeta}{(1 - \zeta^2)^{1/2}} \tag{2.58}$$

δ is constant and related to the damping ratio, as in equation (2.58). It is called the *logarithmic decrement*[13] and it is an easily measured quantity. The ratio of the maximum vibration during one period is measured, and then, for more accuracy, another ratio after n cycles (Figure 2.20). Then

$$\delta = \frac{1}{n} \ln \frac{X_i}{X_{i+n}} = \zeta \omega_n T = \frac{2\pi\zeta}{(1 - \zeta^2)^{1/2}} \qquad (2.59)$$

The damping constant c is a *property of a specific device*, the damper. The damping factor and the logarithmic decrement are *system properties*.

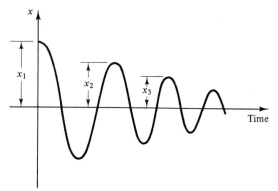

Figure 2.20 Logarithmic decrement.

For design purposes, we do not always want a system to be critically damped, because the higher the damping, the more time the system needs to return to equilibrium. The concept of *overshoot*, equal to $X_{1/2}/X_0$, is often used. The system is designed with damping less than the critical one, so that the maximum amplitude during oscillation will be a certain percentage of the initial displacement. Typical design values are 5% to 10% overshoot.

☐ **Example 2.14   Water Level Meter**

A water level meter consists of a light arm B and a floating cylinder of diameter $d = 50$ mm (Figure E2.14). Determine the value of the damping constant c of a dashpot for critical damping, assuming that the mass of the cylinder is 0.2 kg, $l = 75$ mm, $L = 250$ mm, and the water density is $\rho_w = 1000$ kg/m³.

Solution   The floating cylinder in water is equivalent to a spring of stiffness $k = Ag\rho_w$ (see Example 2.10). For rotation θ, the forces on the arm are shown in Figure E2.14c for small θ.

$$mL^2\ddot{\theta} = -cl^2\dot{\theta} - kL^2\theta \qquad \text{or} \qquad mL^2\ddot{\theta} + cl^2\dot{\theta} + kL^2\theta = 0$$

Therefore,

$$[cl^2]_c^2 = 4(mL^2)(kL^2)$$

$$c_c^2 = 4mk\left(\frac{L}{l}\right)^4 \qquad \text{or} \qquad c_c = \left(\frac{L}{l}\right)^2 (4mk)^{1/2}$$

[13]The idea is due to Helmholz. 1862. *Die Lehre von Tonempfindungen*.

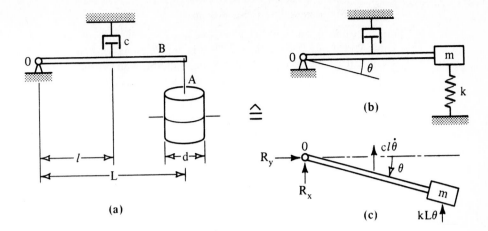

Figure E2.14

$$k = g\rho_w A = g\rho_w \frac{\pi d^2}{4}, \qquad g\rho_w = 10{,}000 \text{ N/m}^3,$$

$$k = 10{,}000 \times \pi \times \frac{0.05^2}{4} = 19.6 \text{ N/m}$$

$$c_c = \left(\frac{0.250}{0.075}\right)^2 (4 \times 0.2 \times 19.6)^{1/2} = 44 \text{ N} \cdot \text{s/m}$$

$\square$

☐ **Example 2.15 Automotive Suspension**

An automobile weighing 15,000 N is supported by four springs and four dampers (Figure E2.15). Assume that the static deflection of the car on the springs is 0.20 m, due to its own weight. Determine the required damping constant of each of the four shock absorbers in order to have critical damping. Assume that the car has only one degree of freedom and motion in the vertical direction.

Solution   Two dampers in parallel are equivalent to a damper with a damping constant $c_1 + c_2$. This is because the corresponding reaction is $c_1\dot{x} + c_2\dot{x} = (c_1 + c_2)\dot{x}$ to a velocity $\dot{x}$. The automobile can thus be modeled for vertical vibration as a harmonic oscillator $(m, 4c, 4k)$. Because $m = 15{,}000/g = 1529$ kg, we have $4k = w/\delta_{st} = 15{,}000/0.20 = 75{,}000$ N/m. The critical damping will be $4c_c = [4m(4k)]^{1/2}$, $c_c = (mk)^{1/2} = (1529 \times 75{,}000/4)^{1/2} = 5354$ N·s/m.

$\square$

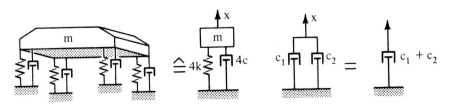

Figure E2.15

□ **Example 2.16 Damped System Response**

An electronic instrument has a mass $m = 1$ kg and is supported by springs that have equivalent spring constant $k = 2400$ N/m and damping constant $c = 2$ N·s/m (Figure E2.16). It is displaced by 20 mm from equilibrium and released. Determine the vibration amplitude after 5 oscillations, and then after 20 oscillations.

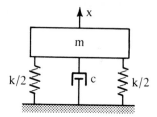

Figure E2.16

Solution

$$\zeta = \frac{c}{(4km)^{1/2}} = \frac{2}{(4 \times 2400 \times 1)^{1/2}} = 0.0204$$

and there will be a damped oscillation with

$$\omega_n = \left(\frac{k}{m}\right)^{1/2} = \left(\frac{2400}{1}\right)^{1/2} = 49 \text{ rad/s}, \qquad \omega_d = \omega_n(1 - \zeta^2)^{1/2}$$

$$= 49(1 - 0.02^2)^{1/2} = 49 \text{ rad/s}$$

Therefore,

$$x(t) = x_0 e^{-\zeta \omega_n t} \cos \omega_d t = 0.02 \times e^{-0.02 \times 49 t} \cos 49t$$

The amplitude is $X = 0.02 \times e^{-0.02 \times 49t}$; the period $T = 2\pi/\omega_d = 2\pi/49 = 0.13$ s. After 5 oscillations,

$$t = 5 \times T = 0.65 \text{ s}$$

$$X = 0.0106 \text{ m}$$

After 20 oscillations,

$$t = 20 \times T = 2.6 \text{ s}$$

$$X = 0.00156 \text{ m}$$

□

2.9 TORSIONAL VIBRATION

Many engineering systems, such as internal combustion engines, have rotational vibration about an axis, called *torsional vibration*. Consider a wheel of mass moment of inertia J_0, about its axis of symmetry, which is attached to a vertical rod of diameter d, length L, and shear modulus G. The upper end of the rod is fixed. This system can have torsional vibration about the axis of symmetry (Figure 2.21a).

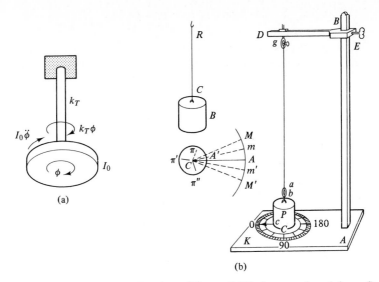

Figure 2.21 Torsional vibration. (Figure 2.21b is reproduced from S. Timoshenko, 1953. *History of Strength of Materials*. New York: McGraw-Hill Book Co. By permission.)

This problem was studied by Coulomb [Timoshenko, 1953]. The apparatus that he used is shown in Figure 2.21b.[14]

The torsional spring constant of the rod is determined from the relation of the torque to angular displacement:

$$\phi = \frac{TL}{I_p G}, \qquad T = k_T \phi, \qquad k_T = \frac{I_p G}{L} \qquad (2.60)$$

where I_p is the polar moment if inertia of the rod cross section. The equation of motion for rotation ϕ will be, by Newton's law for rotation about the center of mass,

$$J_0 \ddot{\phi} = -T = -k_T \phi \qquad \text{or} \qquad J_0 \ddot{\phi} + k_T \phi = 0 \qquad (2.61)$$

The natural frequency is

$$\omega_n = \left(\frac{k_T}{J_0}\right)^{1/2} = \left(\frac{I_p G}{J_0 L}\right)^{1/2} = \left(\frac{\pi d^4 G}{32 J_0 L}\right)^{1/2}$$

□ **Example 2.17   Torsional Vibration of Simple Rotors**

A high-speed turbine disk has mass $m = 60$ kg and polar mass moment of inertia $J_0 = 7$ kg·m². It is connected to the generator rotor, assumed to have constant angular velocity, with a two-section stub shaft of diameters 30 and 50 mm and lengths 500 and 400 mm, respectively, with shear modulus $G = 1.1 \times 10^{11}$ N/m² (Figure E2.17). Find the torsional natural frequency.

Solution   The rotational spring constants of the two cylindrical stub shafts are (Ap-

[14]Coulomb, A. 1784. *Recherches théoriques et expérimentales sur la force de torsion et sur l'élasticité des fils de metal.* Paris. In modern terms, torsional vibrations were advanced to today's level by Frahm, H. 1902. *Z. VDI*, pp. 779, 886.

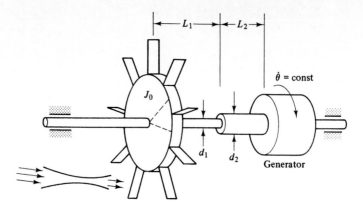

Figure E2.17

pendix II) $k_T = I_p G/L$, where the polar area moment of inertia of the section is $I_p = \pi d^4/32$. Therefore, for the two shaft sections,

$$I_{p1} = \pi \times \frac{0.030^4}{32} = 7.95 \times 10^{-8} \text{ m}^4, \qquad I_{p2} = \pi \times \frac{0.050^4}{32} = 61.3 \times 10^{-8} \text{ m}^4$$

$$k_{T1} = 7.95 \times 10^{-8} \times 1.1 \times \frac{10^{11}}{0.500} = 1.75 \times 10^4 \text{ N} \cdot \text{m}$$

$$k_{T2} = 61.3 \times 10^{-8} \times 1.1 \times \frac{10^{11}}{0.400} = 16.8 \times 10^4 \text{ N} \cdot \text{m}$$

The torsion stiffness of the compound shaft is

$$k_{T12} = \frac{k_{T1} k_{T2}}{k_{T1} + k_{T2}} = \frac{1.75 \times 16.8 \times 10^8}{(1.75 + 16.8) \times 10^4}$$

$$= 1.58 \times 10^4 \text{ N} \cdot \text{m}$$

and the torsional natural frequency is

$$\omega_n = \left(\frac{k_{T12}}{J}\right)^{1/2} = \left(\frac{15,850}{7}\right)^{1/2} = 47.6 \text{ rad/s}$$

$\square$

2.10 MASS ELEMENTS

Vibrating systems often consist of discrete mass elements comprising several solids either rigidly connected or in fixed kinematic relationship with one another. The equivalent mass of several masses rigidly connected, in rectilinear motion, is the sum of the masses:

$$m_{eq} = m_1 + m_2 + \cdots + m_n \tag{2.62}$$

The equivalent moment of inertia about point O of several solids rigidly connected for rotation about a point can be found by repeated application of *Steiner's rule* (see Appendix II):

$$J_0 = (J_{G1} + m_1 r_1^2) + (J_{G2} + m_2 r_2^2) + \cdots + (J_{Gn} + m_n r_n^2) \qquad (2.63)$$

where J_{G1}, J_{G2}, and J_{Gn} are the mass moments of inertia of the individual masses about their mass centers, $r_1, r_2, \ldots, r_n$ the distances of these centers from O, and $m_1, m_2, \ldots, m_n$ the respective masses.

The equivalent moment of inertia of several solids having a fixed ratio of angular velocities, geared systems for example, in respect to one axis of rotation $R_2, R_3, \ldots, R_n$ can be found by observing that for two gears (Figure 2.22) moving at a fixed ratio of angular rotations $\theta_1/\theta_2 = \omega_1/\omega_2 = R_2$, application of Newton's law for rotation of the two gears about their axes yields

$$J_1 \ddot{\theta}_1 + K_{T1}\theta_1 = T_1 - Fr_1 \qquad (2.64)$$

$$J_2 \ddot{\theta}_2 + K_{T2}\theta_2 = T_2 + Fr_2$$

Eliminating the unknown interaction force, $R_2 = r_2/r_1$, gives

$$\left(J_1 + \frac{J_2}{R_2^2}\right)\ddot{\theta}_1 + \left(K_{T1} + \frac{K_{T2}}{R_2^2}\right)\theta_1 = T_1 + \frac{T_2}{R_2} \qquad (2.65)$$

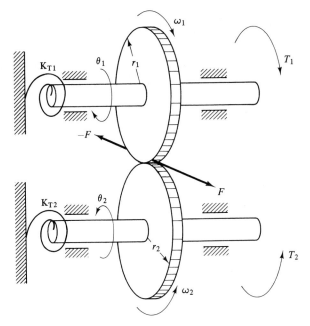

Figure 2.22 Torsional vibration of a gearing pair.

It is apparent that the equivalent moments of inertia, torsional stiffnesses, and applied torques are $J_1 + J_2/R_2^2$, $K_{T1} + K_{T2}/R_2^2$, and $T_1 + T_2/R_2$, respectively. This can be generalized for any number of rotating solids. Very complex combinations of solids in related rotational, translational, and other motions can be simplified with energy methods, which are discussed in Chapter 4.

SIMULAB Experiment 2.1

Use the simulated experiment LINESYS to compute the logarithmic decrement for a single-degree-of-freedom system with the properties $m = 10$ kg, $k = 1000$ N/m, and $c = 20$ N·s/m.

Solution   With the program diskette in drive A, we type

$$A> \text{LINESYS}$$

The page of Figure SE2.1a appears on the screen. We select a one-mass system. Then the menu page appears on the screen with the simulated one-degree-of-freedom system (Figure SE2.1b). Using the keypad arrow keys, we select menus. We set $m = 10$ kg, $k = 1000$ N/m, and $c = 20$ N·s/m using the + and − keys. We also specify a unit initial displacement. We select START from the menu and hit ENTER, which initiates the system motion.

The ratio of the amplitudes of the first over the tenth cycles is 2.7. Therefore,

$$\delta = \frac{ln(x_1/x_{10})}{n} = \frac{\ln(2.7)}{10} = 0.099 \simeq 0.1$$

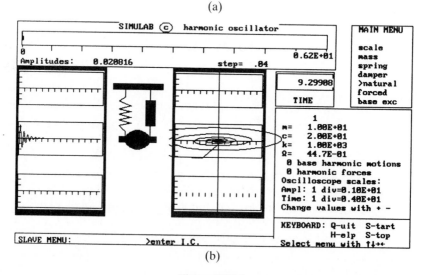

(a)

(b)

Figure SE2.1

TABLE 2.1. SUMMARY OF EQUATIONS IN CHAPTER 2

Definition of spring constant

$$F = kx$$

$$k_{ij} = \partial F_i/\partial x_j$$

Equivalent springs

Parallel: $k_{eq} = k_1 + k_2 + \cdots + k_n$

Serial: $1/k_{eq} = 1/k_1 + 1/k_2 + \cdots$

Newton's law

Particle: $\mathbf{F} = d(m\mathbf{v})/dt$

Rigid Body: $\mathbf{F}_G = md(\mathbf{v}_G)/dt = m\ddot{\mathbf{r}}_G$

$$\mathbf{M} = \dot{\mathbf{H}}$$

$$F_x = md(\mathrm{v}_{xG})/dt = m\ddot{x}_G$$

$$F_y = md(\mathrm{v}_{yG})/dt = m\ddot{y}_G$$

$$M_z = J\dot{\omega}_z$$

D'Alembert principle

$$\sum \mathbf{F} + (-m\ddot{\mathbf{r}}) = 0$$

$$\sum \mathbf{M}_G + (-J_G\ddot{\theta}) = 0$$

Undamped oscillator

$$m\ddot{x} + kx = 0$$

$$x(t) = x_0 \cos \omega_n t + v_0/\omega_n \sin \omega_n t$$

$$\omega_n = (k/m)^{1/2}$$

Damping constant

$$F_x = c(dx/dt)$$

$$c = df(\dot{x})/d\dot{x}\big|_{\dot{x}} = 0$$

Damping Factor $\zeta = c/2m\omega_n$.

Critical Damping

$$c_c = 2m\omega_n = 2\sqrt{mk}$$

Damped oscillator

$$\ddot{x} + 2\zeta\omega_n\dot{x} + \omega_n^2 x = 0$$

$$x_1(t) = e^{p_1 t}, \ x_2(t) = e^{p_2 t}$$

(continues)

TABLE 2.1. (cont.)

$$p_1 = [-\zeta + (\zeta^2 - 1)^{1/2}]\omega_n$$

$$p_2 = [-\zeta - (\zeta^2 - 1)^{1/2}]\omega_n$$

$$c_1 = (x_0 p_2 - v_0)/(p_2 - p_1)$$

$$c_2 = (v_0 - x_0 p_1)/(p_2 - p_1)$$

I : $c_2 = 0, \zeta = 0$.

$$\ddot{x} + \omega_n^2 x = 0$$

$$x(t) = x_0 \cos \omega_n t + v_0/\omega_n \sin \omega_n t$$

II : $c^2 < 4mk, \zeta < 1$

$$x(t) = e^{-\zeta \omega_n t}[x_0 \cos \omega_d t + (x_0 \zeta \omega_n + v_0)/\omega_d \sin \omega_d t]$$

III : $c^2 > 4mk, \zeta > 1$

$$x(t) = c_1 e^{p_1 t} + c_2 e^{p_2 t}$$

$$p_{1,2} = [-\zeta \mp (\zeta^2 - 1)^{1/2}]\omega_n,$$

$$c_1 = x_0(1/2\ \zeta/(\zeta^2 - 1)^{1/2} + 1/2) + v_0/[2(\zeta^2 - 1)^{1/2}\omega_n]$$

$$c_2 = -x_0(1/2\ \zeta/(\zeta^2 - 1)^{1/2} - 1/2) - v_0/2(\zeta^2 - 1)^{1/2}\omega_n]$$

IV: $c^2 = 4mk, \zeta = 1$

$$p_1 = p_2 = -\zeta \omega_n$$

$$x(t) = e^{-\zeta \omega_n t}(c_1 + c_2 t)$$

$$c_1 = x_0, c_2 = v_0 + \zeta \omega_n x_0$$

Logarithmic decrement

$$\delta = 1/n \ln(x_i/x_{i+n}) = \zeta \omega_n T = 2\pi \zeta/(1 - \zeta^2)^{1/2}$$

Torsional oscillator

$$J_0 \ddot{\phi} + k_T \phi = 0$$

Equivalent masses

Assembly of particles:
$$m_{eq} = m_1 + m_2 + \cdots + m_n$$
Assembly of rigid bodies:
$$J_0 = (J_{G1} + m_1 r_1^2) + (J_{G2} + m_2 r_2^2) + \cdots + (J_{Gn} + m_n r_n^2)$$
Serial Geared Systems:
$$J_{eq} = J_1 + J_2/R_2^2 + J_3/R_3^2 + \cdots$$
$$K_{Teq} = K_{T1} + K_{T2}/R_2^2 + K_{T3}/R_3^2 + \cdots$$
$$R_i = \omega_i/\omega_{i+1}$$

REFERENCES AND FURTHER READING

BEER, F. P., and JOHNSTON, E. R. 1984. *Vector Mechanics for Engineers: Dynamics*, 4th ed. New York: McGraw-Hill.

DEN HARTOG, J. P. 1952. *Mechanical Vibration*, 4th ed. New York: McGraw-Hill.

DIMAROGONAS, A. D. 1976. *Vibration Engineering*. St. Paul, Minn.: West Publishing.

DIMAROGONAS, A. D., and PAIPETIS, S. A. 1983. *Analytical Methods in Rotor Dynamics*. London: Elsevier–Applied Science.

HORT, W. 1910. *Technische Schwingungslehre*. Berlin: Julius Springer.

KLOTTER, K. 1960. *Technische Schwingungslehre*. Berlin: Springer-Verlag.

LALANNE, M., BERTHIER, P., and DER HAGOPIAN, J. 1983. *Mécanique des vibrations linéaires*. New York: Wiley.

PONTRYAGIN, L. S. 1962. *Ordinary Differential Equations*. Reading, Mass.: Addison-Wesley.

RAYLEIGH, J. W. S. 1894. *Theory of Sound*. New York: Dover Publications, 1946.

TIMOSHENKO, S. P. 1953. *History of the Strength of Materials*. New York: McGraw-Hill.

TIMOSHENKO, S. 1955. *Vibration Problems in Engineering*. Princeton, N.J.: D. Van Nostrand.

PROBLEMS

Sections 2.1 and 2.2

2.1. A multiple-strip spring consists of three steel strips of length $L = 0.30$ m, width $b = 0.10$ m, and thickness $t = 0.005$ m (Figure P2.1). Determine the spring constant for vertical deflection if the modulus of elasticity is $E = 2.1 \times 10^{11}$ N/m^2 and the connecting block is rigid.

2.2. A torsional spring connecting two shafts consists of eight bars of $d = 8$ mm, connected in a circle of radius $R = 100$ mm (Figure P2.2). If their length is $l = 250$ mm and the modulus of elasticity of the spring material is $E = 2.1 \times 10^{11}$ N/m^2, compute the torsional spring constant.

2.3. A restoring spring of a crank is a coil of six turns and is made of spring steel with $E = 2.1 \times 10^{11}$ N/m^2, $d = 3$ mm, and $D = 30$ mm (Figure P2.3). Determine the torsional spring constant.

2.4. A torsion bar consists of three segments with diameters 30, 40, and 50 mm and lengths 400, 600, and 500 mm, respectively. If $E = 2.1 \times 10^{11}$ N/m^2, determine the torsional spring constant.

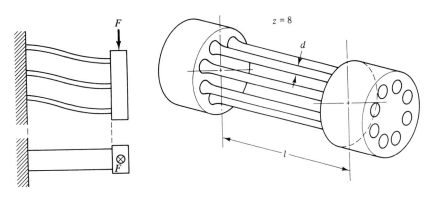

Figure P2.1   **Figure P2.2**

2.5. The torsional spring shown in Figure P2.5 consists of two tubes with dimensions d_1, d_2, $L = 60,65,400$ mm for the outer tube, and $40,45,350$ mm for the inner tube. The inner shaft has $d = 30$ mm and $L = 600$ mm. Material for the system is steel with $E = 2.1 \times 10^{11}$ N/m². Determine the torsional spring constant.

2.6. An electric motor is mounted on a simply supported steel beam of length $L = 0.8$ m, area moment of inertia $I = 8.33$ cm⁴, Young's modulus $E = 2.06 \times 10^{11}$ N/m², and negligible mass (Figure P2.6). To stiffen the support in the vertical direction, determine the stiffness of a spring needed below the beam in the midspan so that the spring constant is doubled.

2.7. In the system shown in Figure P2.6, the system is to be stiffened by inserting a piece of special rubber with dimensions 0.05×0.05 m² under the beam in the clearance space c, near the midspan. If the stress–strain curve of this rubber is given by the relationship $\sigma = 1.2 \times 10^8 \epsilon + 3.5 \times 10^6 \epsilon^2$ (σ in N/m²), determine the thickness of rubber needed to double the stiffness in the vertical direction, assuming a clearance of 0.02 m.

2.8. A flexible pedestal consists of two rigid plates P_1 and P_2, a rigid foot F, and two rubber pads R_1 and R_2 (Figure P2.8). The stress–strain relationship for this rubber is $\sigma = 1.2 \times 10^8 \epsilon$, σ in N/m². Assuming that the thickness of the rubber plates is 0.03 m and their area is 0.3×0.3 m², and that the plates P_1 and P_2 have been prestressed so that the rubber plates have been compressed to half their thickness, determine the spring constant of the pedestal. The hole in the upper rubber plate is negligible.

2.9. In Problem 2.8, determine the additional prestress needed to double the spring constant.

2.10. A refrigeration compressor will be mounted on four rubber pads of hollow cylinder shape with inner diameter 12 mm and length 30 mm. If the stress–strain relationship for the rubber is $\sigma = 1.2 \times 10^8 \epsilon$, σ in N/m², determine the outer diameter of the pads if the total spring constant is 10,000 N/m and the static load is negligible.

Sections 2.3 to 2.6

2.11. The cylinder of the servomechanism shown in Figure P2.11 has a piston with $m = 0.3$ kg and is supported by a helical spring of $d = 1$ mm, $D = 10$ mm, five turns, and $G = 1.05 \times 10^{11}$ N/m². Find the piston vibration natural frequency if there is no oil in the cylinder.

2.12. The tension pulley shown in Figure P2.12 is used to keep the belt prestressed. The

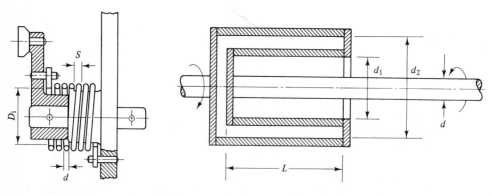

Figure P2.3   **Figure P2.5**

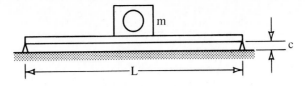

Figure P2.6

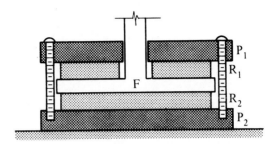

Figure P2.8

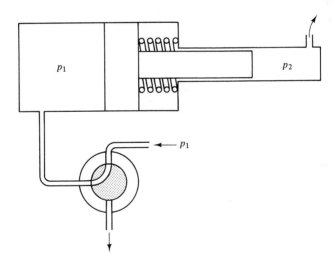

Figure P2.11

belt has section A = 500 mm^2, total length 2 m, E = 30,000 N/m^2, and L = 0.8 m. The loading mechanism has l_1 = 0.20 m, l_2 = 0.20 m, m = 3 kg, and M = 5 kg. Find the natural frequency of vibration of the loading arm about its pivot.

2.13. The cylinder of a fluid check valve shown in Figure P2.13 has a piston with m = 0.2 kg and is supported by a six-turn helical spring of d = 2 mm, D = 30 mm, and G = 1.05 × 10^{11} N/m^2. Find the piston vibration natural frequency if there is no fluid in the valve.

2.14. A cable system for lime transport consists of a steel cable, E = 2.06 × 10^{11} N/m^2, of length 1200 m and a trolley of mass m = 300 kg with the payload (Figure P2.14). Determine at the position shown the natural frequency of vertical vibration.

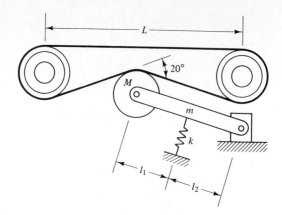

Figure P2.12

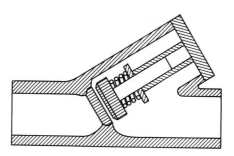

Figure P2.13

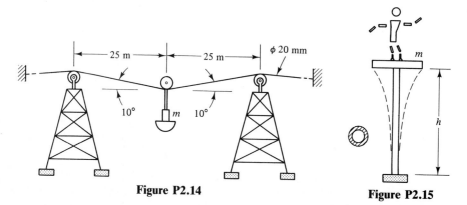

Figure P2.14

Figure P2.15

2.15. An acrobat performs an oscillation act on the top of a 20-m-high pole fixed on the ground (Figure P2.15). The pole is made out of steel pipe of 4 mm thickness, and the acrobat and foot plate weigh 78 kg. Determine the inner diameter of the pipe so that the period of natural vibration will be 4 s. $E = 2.06 \times 10^{11}$ N/m^2.

2.16. An air-conditioning unit is supported from the ceiling by four springs. The mass of the unit is 300 kg, and it is desired that the natural frequency for vertical vibration be between 32 and 40 Hz. Find the allowable range of the spring constant of each spring.

Natural Vibration about Equilibrium   Chap. 2

Figure P2.16 (Courtesy of Tecumseh Products Co., Tecumseh, Michigan.)

2.17. An air humidifier is supported from the ceiling by four 0.50-m-long bars (Figure P2.17). The mass of the unit is 200 kg and it is desired that the natural frequency for vertical vibration is less than 30 Hz and for horizontal between 10 and 15 Hz. Find the allowable range of the bar diameters. $E = 2.1 \times 10^{11} \, N/m^2$.

2.18. A glass-cloth dust collector is supported from the floor by four square tubular sections of thickness 5 mm and length 2 m (Figure P2.18). The mass of the unit is 500 kg and it is desired that the natural frequency of horizontal vibration be between 32 and 40 Hz. Find the allowable range of the tubular section width. $E = 2.1 \times 10^{11} \, N/m^2$.

2.19. An air cleaner (dust collector) unit is supported from the floor by six solid iron polls of rectangular shape, 100 mm wide in the direction of the wide side and 50 mm thick in the transverse direction, 2 m long, fixed both on the ground and the unit (Figure P2.19). The mass of the unit is 800 kg. Find the horizontal natural frequencies in the two directions. $E = 2.1 \times 10^{11} \, N/m^2$.

Figure P2.17 (Courtesy of American Air Filter, Louisville, Kentucky.)

Figure P2.18 (Courtesy of American Air Filter, Louisville, Kentucky.)

Figure P2.19 (Courtesy of American Air Filter, Louisville, Kentucky.)

2.20. A small refrigeration compressor is supported from the floor by four rubber springs which have spring constants 3000 N/m each in the vertical direction and 4000 N/m in the horizontal direction (Figure P2.20). The mass of the unit is 30 kg. Find the natural frequency for horizontal and vertical vibration, assuming that the unit does not tilt.

2.21. A drum with hazardous wastes hangs from a crane with a cable of length $L = 2$ m (Figure P2.21). The drum dimensions are diameter 0.60 m and length 1.20 m, and its mass is 70 kg. The density of the material in the drum is $\rho = 1080$ kg/m^3. The period of oscillation was measured, $T = 1.8$ s. Find the amount of liquid in the drum.

2.22. The reactor vessel shown in Figure P2.22 hangs from a crane with a cable of length $L = 20$ m. The vessel dimensions are diameter 5.60 m and length 12 m, and its mass is 70,000 kg. Find the period of oscillation if the distribution of the mass in the vessel is uniform.

2.23. The cover of an earth orbit simulator hangs from a crane with three cables of length $L = 5$, and the three lugs are on an 8-m diameter (Figure P2.23). The mass is 2000 kg and the moment of inertia is 30,000 Nm^2. Find the period of oscillation.

2.24. The heavy drum of a power plant is lifted from its permanent hangers by a crane at a vertical distance 15 m (Figure P2.24). The drum dimensions are diameter 1.60 m and length 3.20 m, and its mass is 5000 kg. The density of the material in the drum is $\rho = 1000$ kg/m^3. Determine the period of oscillation.

2.25. The head of a nuclear pressure vessel hangs from a crane with a cable of length $L = 6$ m (Figure P2.25). The period of oscillation was measured, $T = 5$ s. Then it was attached with a shorter cable 3 m long and the period of oscillation became 3.9 s. Find the mass of the head and the moment of inertia about the center of mass.

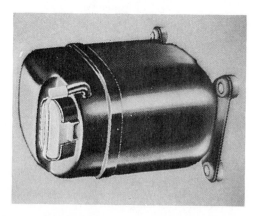

Figure P2.20 (Courtesy of Tecumseh Products Co., Tecumseh, Michigan.)

Figure P2.21

Figure P2.22

Figure P2.23

2.26. The moving core of the electromagnetic relay has mass $m = 12$ g and is supported by a spring with $k = 3000$ N/m (Figure P2.26). When energized, it closes the contacts, which are mounted on flexible strips 0.8 mm thick and 6 mm wide. The moving strip is 20 mm in length, and the stationary strips are 15 mm in length. Find the natural frequency with both an open and a closed relay. $E = 2.1 \times 10^{11}$ N/m^2.

2.27. A snap switch (Figure P2.27) consists of a pivot (1); a lever (2) of length 20 mm, width 10 mm, and thickness 3 mm; and a lever (4) of similar dimensions connected with (2) by a pin through a torsional spring (3) of stiffness $k_T = 10$ N·mm/rad. In the second lever, there is a slider supported by a spring (11) of stiffness 1 N/mm. The slider, of negligible mass, slides on a switching element (9) which in both positions is inclined by a 20° angle with the horizontal. If in either switch position the two levers are collinear, determine the natural frequency. $\rho = 7800$ kg/m^3.

2.28–2.30. Find the natural frequencies of the systems shown in Figures P2.28 to P2.30.

2.31–2.34. Determine the natural frequencies of the systems shown in Figures P2.31 to P2.34.

2.35–2.38. For the systems shown in Figures P2.35 to P2.38, a mass m falls on a mass m_1 and the collision is plastic; that is, the two masses move together. Determine the system response.

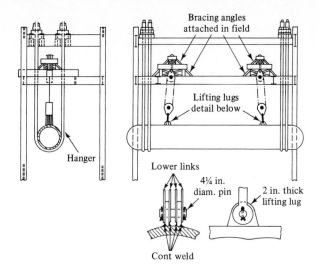

Figure P2.24

Figure P2.25

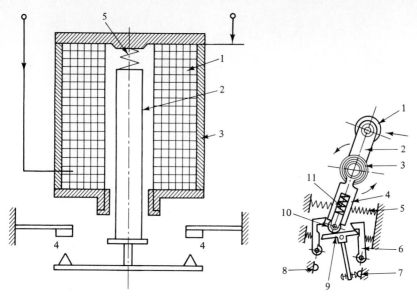

Figure P2.26

Figure P2.27

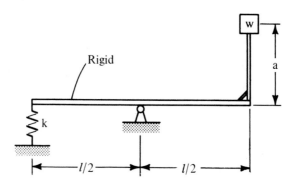

Figure P2.28

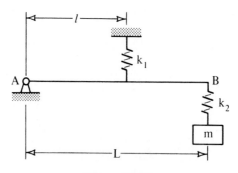

Figure P2.29

Natural Vibration about Equilibrium   Chap. 2

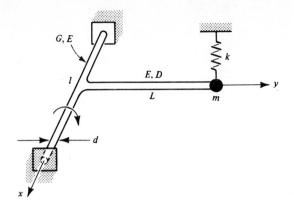

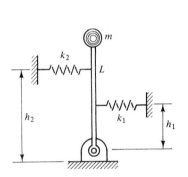

Figure P2.30 The pallograph of O. Schlick. 1893. *Trans. Inst. Nav. Arch.*, *34:* 167.

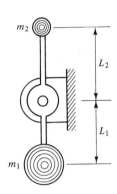

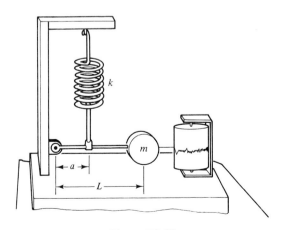

Figure P2.31

Figure P2.32

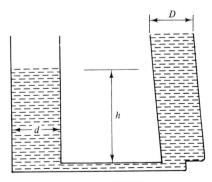

Figure P2.33

Figure P2.34

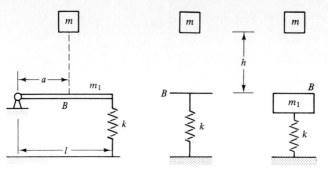

Figure P2.35   Figure P2.36   Figure P2.37

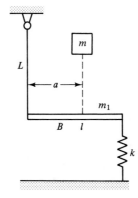

Figure P2.38

2.39–2.45. For the compound pendulums shown in Figures P2.39 to P2.45, determine the natural frequencies.

Sections 2.7 and 2.8

2.46. The cylinder of the servomechanism shown in Figure P2.11 has a piston with $m = 0.3$ kg and is supported by a helical spring of $d = 1$ mm, $D = 10$ mm, $G = 1.05 \times 10^{11}$ N/mm^2, and 10 turns. Find the piston vibration damped natural frequency and the logarithmic decrement if there is lubrication oil in the cylinder, filling the 51.5-mm-diameter 20-mm-long piston and the cylinder. The oil viscosity is 10×10^{-3} N·s/m^2 and the piston clearance is 0.1 mm.

2.47. A voltmeter (Figure P2.47) has a steel dial of length $l = 50$ mm, width 3 mm, and thickness 1.5 mm. The restoring spring has a rotational spring constant $k = 100$ N·mm/rad. At a radius $r = 8$ mm, there is a dashpot for critical damping. During a measurement, the instrument shows 80 V. Determine the time required for the dial to return to an indication of 1 V after the voltage is disconnected.

2.48. A water level meter (Figure P2.48) has a floating cylinder of 100 mm diameter and 0.5 kg mass and $l = 70$ mm, $L = 420$ mm. Determine the required damping constant of the damper for critical damping.

2.49. In the relay of Problem 2.26, light oil is introduced between the core and the cylinder

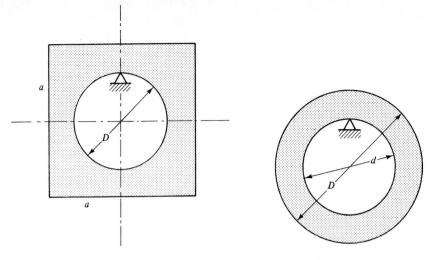

Figure P2.39   **Figure P2.40**

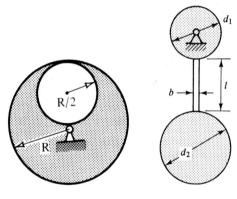

Figure P2.41   **Figure P2.42**

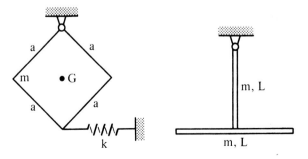

Figure P2.43   **Figure P2.44**

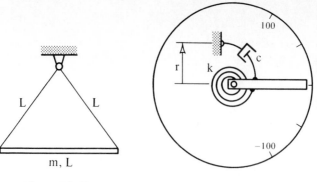

Figure P2.45   **Figure P2.47**

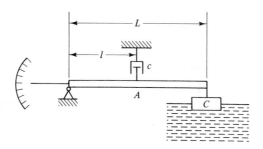

Figure P2.48

with viscosity 4×10^{-3} N·s/m². Determine the required clearance for critical damping if the cylinder has $L = 40$ mm and $d = 10$ mm.

2.50. In a test conducted with the refrigeration compressor of Problem 2.20, it was found that the logarithmic decrement was 0.90. Determine the damping constant of each of the four mounts.

2.51–2.54. If the systems of Problems 2.35 to 2.38 had 20% of the critical damping, determine the maximum amplitude during the first and second cycles of vibration.

Sections 2.9 and 2.10

2.55. The crankshaft of a single-cylinder internal combustion engine is connected to a heavy flywheel by a shaft of torsional stiffness 1000 N·m/rad (Figure P2.55). The crankshaft consists of two plates with mass moment of inertia about the axis of rotation $J_0 = 3 \times 10^{-4}$ kg·m⁴ each. The reciprocating system mass, which can be assumed concentrated at the crank 2, is 1.5 kg, the crank radius is 80 mm, and the flywheel can be assumed that rotates with constant angular velocity. Determine the torsional natural frequency of the crankshaft.

2.56. A centrifugal pump P rotor is connected to a motor running at constant angular velocity ω through a flexible coupling of torsional spring constant K_T and a pair of spur gears with radii r_1 and r_2 and polar mass moments of inertia J_1 and J_2, respectively (Figure P2.56). The pump rotor has polar mass moments of inertia J_p. Determine the natural frequency of torsional oscillation, assuming rigid connecting shafts.

2.57. Solve Problem 2.56 for the torsional flexibility of the connecting shafts if they both

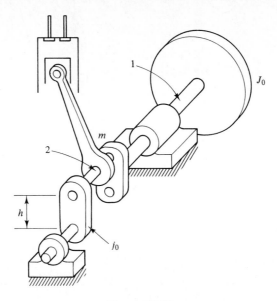

Figure P2.55

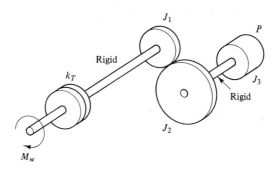

Figure P2.56

have length 1 m, diameter 20 mm, shear modulus of their material 0.9×10^{11} N/m^2 and $K_T = 2000$ N·m/rad, $J_1 = 0.0012$ kg·m^2, $J_2 = 0.0024$ kg·m^2, and $J_3 = 0.0050$ kg·m^2. The gear diameters are $d_1 = 100$ and $d_2 = 200$ mm.

2.58. The torsional vibration transducer shown in Figure P2.58 has a small flywheel of polar mass moment of inertia J and is connected to a very light pulley with a helical spring of torsional spring constant K_T. The pulley in turn is rotated by a flexible belt of cross section A and modulus of elasticity E. The distance between the shafts is 0.80 m. Assuming that the main shaft moves with constant angular velocity, determine the torsional natural frequency of the transducer.

2.59. A motor running at constant angular velocity, is connected to a worm-gear set through a flexible coupling of torsional stiffness K_T. The worm has negligible inertia and the gear has polar mass moment of inertia J_0 and speed ratio $R = 10$. The gear is rigidly connected to two spur gear pairs of speed ratios and polar mass moments of inertia R_{A1}, J_{A1}, J_{A2} and R_{B1}, J_{B1}, J_{B2}, respectively. Determine the torsional natural frequency of the system.

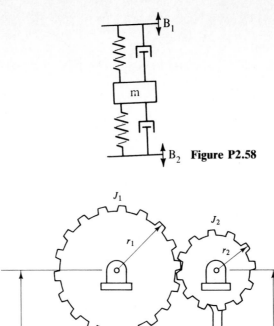

Figure P2.58

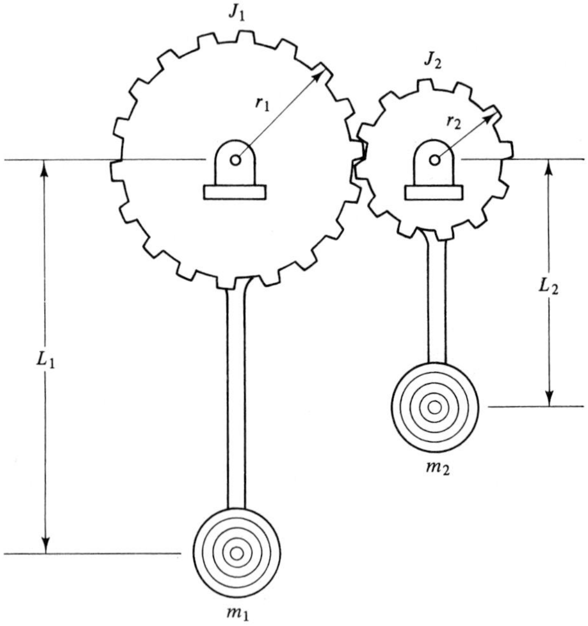

Figure P2.60

2.60. Find the natural frequency for small oscillations of the compound double pendulum shown in Figure P2.60 for small oscillations about the equilibrium position shown.

SIMULAB PROBLEMS

S2.1. Using the LINSYS program of your SIMULAB diskette, measure the natural frequency and logarithmic decrement of a single-degree-of-freedom system, observing the response to an initial displacement, with $m = 1$, $k = 1$, $c = 2$.

S2.2. Using the LINSYS program of your SIMULAB diskette, measure the natural frequency and logarithmic decrement of a single-degree-of-freedom system, observing the response to an initial displacement, with $m = 1$, $k = 1$, $c = 3.4$.

S2.3. Using the LINSYS program of your SIMULAB diskette, measure the vibration amplitude after 5 and 10 s of observing the response to an initial displacement of a system with $m = 1$, $k = 1$, $c = 4$.

S2.4. Using the LINSYS program of your SIMULAB diskette, measure the vibration amplitude after 5 and 10 s of observing the response to an initial displacement of a system with $m = 1$, $k = 1$, $c = 8$.

S2.5. Using the LINSYS program of your SIMULAB diskette, measure the natural frequency and logarithmic decrement of a single-degree-of-freedom system, observing the response to an initial displacement, with $m = 1$, $k = 1$, $c = 0.1$.

DESIGN PROBLEMS

D2.1. The mass of a mounting plate and a compressor on it is $m = 250$ kg and the frequency of rotation is 1800 rpm. Design four mounting rubber springs of prismatic shape such that the vertical and horizontal natural frequencies will be at least 25% above the running speed if the material of the springs can be considered linear with $E = 10,000$ Pa and $G = 5000$ Pa.

D2.2. The mass of a mounting plate and a compressor on it is $m = 250$ kg and the frequency of rotation is 1800 rpm. It is supported by four mounting rubber springs of prismatic shape, so that the vertical and horizontal natural frequencies, which will be above the running speed, are both 225 rpm. Design appropriate springs and dampers for the following design requirement (called *overshoot*): If the system is displaced vertically and released, the maximum vibration at the first cycle should not be greater then 10% of the initial displacement.

D2.3. Design the mount of Figure P2.1 if it will be used to mount on the plate a motor of mass 80 kg running at 3600 rpm of approximate dimensions $0.30 \times 0.25 \times 0.25$ m if we want it to have critical damping and a vertical natural frequency greater than 1.25 times the running frequency.

D2.4. The flexible coupling of Figure P2.2 will connect a motor of constant angular velocity 3000 rpm to a blower with polar mass moment of inertia $J = 0.12$ kg·m². Design the coupling for a torsional natural frequency of at least 3500 rpm.

D2.5. Design the mounting of Figure P2.6 to achieve natural frequency greater than 2200 rpm and overshoot at the most 10%. For the latter, you might have to use a separate damping device.

D2.6. Design the pedestal of Figure P2.8 if the stress–strain relationship for the rubber-like material used is $\sigma = 1.2 \times 10^8 \epsilon$ (N/m²), and the stiffness in the vertical direction should be $k = 50$ kN/m.

D2.7. In the check valve of Figure P2.13, it was found that the springs failed in fatigue. It was found that the valve piston of mass 0.2 kg was tuned to the frequency of the reciprocating pump of 1200 rpm. Redesign the system to have a natural frequency at least 2500 rpm if the piston diameter is 50 mm, the spring housing has inner diameter 80 mm and length 120 mm, and the opening pressure is 30 MPa.

D2.8. Design the seismograph of Figure P2.32 to have natural frequency below 2 rad/s and critical damping. As a fluid damper, use the mounting pin, assuming that you can use heavy oil with viscosity 300 mPa·s.

D2.9. Design the level recording devise of Figure P2.33 for 5% overshoot and natural frequency less than 5 rad/s if the mass of the floating bob is 0.650 kg.

D2.10. Redesign the damper of the voltmeter of Problem 2.47 for 8% overshoot.

3

Forced Harmonic Vibration

3.1 INTRODUCTION

Motion in engineering systems is associated with forces that change with time in magnitude or direction or both. Examples are the fluid forces on the blades of fluid machinery, the forces due to the combustion process in internal combustion engines, interaction forces in gears, inertia forces in rotors with unbalanced masses, and so on. The response of a mechanical system to such forces is called *forced response*. For the operation and the structural integrity of the machine or system it is very important to limit such response to acceptable levels.

Since it is very common in mechanical engineering practice to have constant angular velocity of certain machine members, the forces acting on a system may be harmonic. Response to harmonic excitation can be found for linear systems with very simple mathematical analysis, and it is of fundamental importance for the understanding of mechanical vibration.

Many systems, such as internal combustion engines operating at constant

*The above photo portraying a Jackhammer is a courtesy of Bruel & Kjaer Instruments, Inc. Reprinted by permission.

angular velocity, are acted upon by forces that are periodic in time but not harmonic, owing to the transformation of motion by the slider-crank mechanism. The response to general periodic excitation for linear systems can be obtained from the response to harmonic excitation by harmonic analysis and superposition, discussed in Section 3.3. A similar procedure will be followed for the response to exciting forces, which are general functions of time.

Due to the fundamental importance of understanding the forced response of the single-degree-of-freedom linear system to harmonic excitation, the remainder of the chapter is devoted to that subject.

3.2 THE EQUATION OF MOTION

A simple vibrating system with an external force F acting on the mass m is shown in Figure 3.1. This force F might be constant or varying with time, $F = f(t)$. From the free-body diagram of Figure 3.1, application of Newton's second law in the x direction yields

$$m\ddot{x} = -c\dot{x} - kx + f(t) \tag{3.1}$$

or

$$m\ddot{x} + c\dot{x} + kx = f(t) \tag{3.2}$$

This equation, in general, has to be supplemented with the initial conditions

$$x(0) = x_0, \qquad \dot{x}(0) = v_0 \tag{3.3}$$

The solution to this problem is expected to be a function $x(t)$ that satisfies the differential equation and the initial conditions.

Suppose that we can find a function $x_s(t)$ that satisfies the differential equation (3.2), but does not, in general, satisfy the initial conditions. This is a particular solution of the differential equation (3.2). Therefore,

$$m\ddot{x}_s + c\dot{x}_s + kx_s = f(t) \tag{3.4}$$

We shall try to find the solution $x(t)$ in the form

$$x(t) = x_s(t) + x_h(t) \tag{3.5}$$

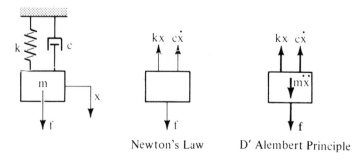

Newton's Law   D' Alembert Principle

Figure 3.1 Forced vibration of a single-degree-of-freedom system.

where $x_h(t)$ is a function to be determined. Substituting into equation (3.2), we obtain

$$m\ddot{x}_h + c\dot{x}_h + kx_h = 0 \tag{3.6}$$

Therefore, $x_h(t)$ is the solution of the homogeneous differential equation, that is, of the natural vibration problem. The initial conditions (3.3) become

$$x_h(0) = x_0 - x_s(0), \qquad \dot{x}_h\big|_{t=0} = v_0 - \dot{x}_s\big|_{t=0} \tag{3.7}$$

Equations (3.6) and (3.7) define a natural vibration problem, such as the one solved in Chapter 2. To solve the nonhomogeneous problem, therefore, one should proceed as follows:

1. Find a particular solution $x_s(t)$ of equation (3.2).
2. Using the value of the function x_s and its derivative at time $t = 0$, formulate the initial conditions (3.7).
3. Solve the homogeneous problem (3.6)–(3.7).
4. Add the two solutions $x_s(t) + x_h(t)$.

The problem now is to find particular solutions of equations (3.2). Some special forms of the function $f(t)$ will now be studied.

1. $f(t) = Fe^{i\omega t}$. A complex force has no physical meaning; nevertheless, it is a very useful concept in extracting responses to harmonic excitation. Here F and ω are two given constants, the harmonic force amplitude and angular velocity, respectively. Again the solution is found by trying for the function $x_s(t)$:

$$x_s = Xe^{i\omega t} \tag{3.8}$$

We obtain

$$X = \frac{F}{(-\omega^2 m + k) + i\omega c} \tag{3.9}$$

Multiplying the right-hand side of (3.9) by $(-\omega^2 m + k) - i\omega c$ and separating real from imaginary parts, we obtain

$$X = F\left[\frac{-\omega^2 m + k}{(-\omega^2 m + k)^2 + \omega^2 c^2} - \frac{i\omega c}{(-\omega^2 m + k)^2 + \omega^2 c^2}\right] \tag{3.10}$$

From the algebra of complex numbers, we know that any complex number $x + iy$ can be written as

$$x + iy = Re^{i\phi} \tag{3.11}$$

where $R = (x^2 + y^2)^{1/2}$ and $\tan \phi = y/x$. Therefore, (3.10) can be written as

$$X = \frac{F}{[(-\omega^2 m + k)^2 + \omega c^2]^{1/2}} e^{-i\phi}, \qquad \tan \phi = \frac{\omega c}{-\omega^2 m + k} \tag{3.12}$$

The solution $x_s(t)$, therefore, is

$$x_s(t) = \frac{F}{[(-\omega^2 m + k)^2 + (\omega c)^2]^{1/2}} e^{i\omega t - i\phi} \tag{3.13}$$

with $\tan \phi$ given in (3.12).[1]

2. $f(t) = F \cos \omega t$. Observing that $\cos \omega t = \text{Re}\{e^{i\omega t}\}$ and that $\cos(\omega t - \phi) = \text{Re}\{e^{i(\omega t - \phi)}\}$, the solution is readily obtained as

$$x_s(t) = \frac{F}{[(-\omega^2 m + k)^2 + (\omega c)^2]^{1/2}} \cos(\omega t - \phi) \tag{3.14}$$

3. $f(t) = F \sin \omega t$. Similarly, because $\sin \omega t = \text{Im}\{e^{i\omega t}\}$, we obtain

$$x_s(t) = \frac{F}{[(-\omega^2 m + k)^2 + (\omega c)^2]^{1/2}} \sin(\omega t - \phi) \tag{3.15}$$

In addition to a term $x_s(t)$ due to the external force $f(t)$, the solution of the forced vibration problem, equation (3.5), has another term, $x_h(t)$, which reflects the initial conditions. No matter what the initial conditions are, however, this function approaches zero after enough time has elapsed, if c and k are positive. This solution depends on the initial conditions and appears at short periods of time before damping causes it to slow down. It is a form of *transient vibration*. The solution $x_s(t)$ depends on the forcing function $f(t)$ only and not on the initial conditions. In time, the solution $x_h(t)$ diminishes. The remaining vibration after a long time, therefore, will be $x_s(t)$:

$$x(t) \underset{t \to \infty}{\rightarrow} x_s(t) \tag{3.16}$$

In many practical problems, $x_s(t)$ is a harmonic function with constant amplitude because the exciting force is such. The solution $x_s(t)$ and the vibration it represents, in this case, are called *steady-state solution* and *steady-state vibration*. In most vibration problems the solution is important only under forced excitation. For harmonic excitation, $f(t) = F \sin \omega t$, we found [equation (3.15)] that

$$x_s(t) = \frac{F}{[(-m\omega^2 + k)^2 + (\omega c)^2]^{1/2}} \sin(\omega t - \phi)$$

$$\tan \phi = \frac{\omega c}{-m\omega^2 + k} \tag{3.17}$$

$$x_h(t) = e^{-(c/2m)t} \left[x_0^* \cos \omega_d t + \left(\frac{x_0^* c}{2\omega_d m} + \frac{2v_0^*}{2\omega_d} \right) \sin \omega_d t \right]$$

where

$$\omega_d = \left[\frac{k}{m} - \left(\frac{c}{2m} \right)^2 \right]^{1/2} \tag{3.18}$$

$$x_0^* = x_0 - x_s(0), \quad v_0^* = v_0 - \dot{x}_s(0) \tag{3.19}$$

[1]In this expression, the numerator ωc is put in instead of $-\omega c$ since we have taken $e^{-i\phi}$ and $\tan(-\phi) = -\tan \phi$.

and the solution is also, using $\omega_n^2 = k/m$, $\zeta^2 = c^2/4km$, and equation (3.5),

$$x(t) = x_s(t) + x_h(t)$$

$$= e^{-\zeta\omega_n t}[A \cos(1 - \zeta^2)^{1/2}\omega_n t + B \sin(1 - \zeta^2)^{1/2}\omega_n t] \qquad (3.20)$$

$$+ \frac{F_0/k}{[(1 - (\omega/\omega_n)^2)^2 + (2\zeta\omega/\omega_n)^2]^{1/2}} \sin(\omega t - \phi)$$

$$\tan \phi = \frac{2\zeta(\omega/\omega_n)}{1 - (\omega/\omega_n)^2}, \quad A = x_0^*, \quad B = (x_0^*\zeta\omega_n + v_0^*)/\omega_d$$

Equation (3.20) gives the complete solution of the problem for a harmonic force $f(t) = F \sin \omega t$. It is observed that the initial conditions in the transient solution (3.18) are not exactly x_0 and v_0, but include the initial values of the steady-state solution $x_s(t)$. This is physically realizable if we consider that a sudden application of the force is a disturbance itself and gives rise to a transient phenomenon. The vibration represented by equation (3.20) is plotted in Figure 3.2. The following observations can be made at this point:

1. The two motions are *superposed*; that is, they are added algebraically.
2. After enough time has elapsed, practically only the steady-state solution is remaining. The transient solution, therefore, is important only for a limited period of time.
3. The effect of the initial conditions is diminished with time.

In the case of $c = 0$, it must be pointed out that both transient and steady-state vibrations coexist forever, since the transient solution has constant amplitude. In engineering systems, however, the damping is small enough to be neglected in some cases, while it is sufficient to diminish the transient vibration for a sufficiently long time. In this case, for $\zeta = 0$, equations (3.20) become

$$x(t) = x_s(t)$$

$$= \frac{F_0/k}{1 - (\omega/\omega_n)^2} \sin(\omega t - \phi) \qquad (3.21)$$

$$\tan \phi = 0, \qquad \phi = 0, \pi$$

For $\omega = (k/m)^{1/2}$, the steady-state vibration has infinite amplitude, as one can see from equation (3.21), no matter how small the exciting force F_0 is. Therefore, a machine operating at a rotating speed ω equal to its natural frequency will undergo high vibration, because there is always some, perhaps very small, excitation. This situation is called *resonance*.[2] Avoidance of resonance is one of the goals of design. In practical situations, there is always some damping c, so that the denominator in equation (3.21) cannot be zero. In this case one can prove that

---

[2]Forced vibration and resonance were studied experimentally by Galileo Galilei, 1638. *Discorsi e demonstrationi matematiche intorno a due nuove scienze attenenti alla mechanica e ai movimenti locali.* Translated by Crew, H., and de Salvio, A. 1914. New York: Macmillan.

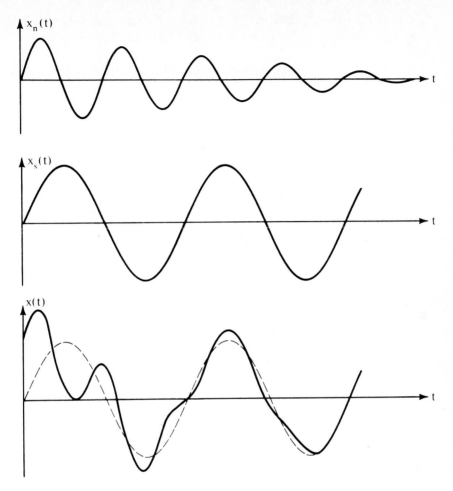

Figure 3.2 Superposition of transient and forced steady response.

the minimum of the denominator in equation (3.20), thus the maximum vibration amplitude, occurs for $\omega = \omega_n(1 - 2\zeta^2)^{1/2}$. For machines and structures, $\zeta < 0.1$ most of the time; therefore, resonance occurs at $1 > \omega/\omega_n > 0.98$. Usually, $\omega/\omega_n = 1$ is taken as resonance because even in highly damped systems the resonance curve is flat and the error of computing the maximum amplitude at $\omega/\omega_n = 1$ is negligible. This error is below 1% for $\zeta < 0.1$, up to 10% for $0.1 < \zeta < 0.5$, and greater for $\zeta > 0.5$.

At the resonant speed, the amplitude is not infinite, but for a low value of damping c it can have very high values. This will be exemplified further later in the chapter. A further observation is that the term $(F_0/k)/[(1 - (\omega/\omega_n)^2]$ in equation (3.21) is the vibration amplitude for steady-state vibration. It consists of two factors:

1. F_0/k is the static deflection if a static force F_0 would act on the system.
2. $1/[1 - (\omega/\omega_n)^2]$ is a factor that does not depend on the excitation but only on

the system properties *(Pythagoras's principle)* and will amplify the static deflection. For this reason it is called the *amplification factor*.

In the presence of damping, equation (3.20), the amplification factor is $H(\omega)$ = $1/[(1 - (\omega/\omega_n)^2)^2 + (2\zeta\omega/\omega_n)^2]^{1/2}$. At resonance, $\omega/\omega_n = 1$, the amplification factor becomes $Q = 1/2\zeta$. This quantity is called the *quality factor*. This factor can be quite large in machinery, where ζ is often as small as 0.01. For this reason, if resonant operation is expected, the machine should be designed with sufficient damping to limit the amplitude at resonance. Furthermore, one should observe that for low values of damping, the latter is important only near the resonance. For dynamic response at frequencies not near resonance, undamped system response analysis might be sufficient.

☐ **Example 3.1   Linear Oscillator Excited Harmonically**

A 55-N weight is suspended by a spring with a stiffness of 1100 N/m. It is forced to vibrate by a harmonic force of 5 N. Assuming a viscous damping of $c = 77$ N·s/m, find (a) the resonant frequency, (b) the amplitude at resonance, (c) the phase angle at resonance, and (d) the damped natural frequency.

Solution   The system properties are $k = 1100$ N/m, $c = 77$ N·s/m, $m = 55/g = 55/9.81 = 5.60$ kg. Therefore,

(a) $\omega_n = (k/m)^{1/2} = (1100/5.6)^{1/2} = 14$ rad/s, $\zeta^2 = 77^2/4 \times 1100 \times 5.6 = 0.24$, $\zeta = 0.49$

(b) At resonance,

$$\omega = \omega_n, \qquad X = \frac{F_0}{[(k - m\omega_n^2)^2 + (c\omega_n)^2]^{1/2}} \approx \frac{F_0}{c\omega_n}$$

$$= \frac{5}{77 \times 14} = 0.0046 \text{ m}$$

The static deflection is $F_0/k = 5/1100 = 0.0045$ m. It is observed that due to relatively high damping, the amplification factor is nearly 1.

(c) The phase angle is tan $\phi = 2\zeta(\omega/\omega_n)/[1 - (\omega/\omega_n)^2] \rightarrow \infty$ for $\omega \rightarrow \omega_n$ and $\phi = \pi/2$.

(d) This frequency is

$$\omega_d = \omega_n(1 - \zeta^2)^{1/2} = 13.16 \text{ rad/s}$$

(This example demonstrates the computation of the basic dynamic response parameters of the single-degree-of-freedom system under harmonic exciting force.)

☐

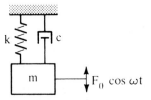

Figure E3.1

An electric motor is driving two wheels on both sides (Figure E3.2). They have very high inertia compared with the motor's rotor. For the purpose of torsional vibration analysis, the ends of the shafts can be considered clamped. There is a torque $T = T_0 \cos \omega t$ on the rotor, due to a high electrical harmonic, with $T_0 = 200$ N·m and $\omega = 500$ rad/s. Assuming that the inertia of the motor rotor is $J = 0.025$ kg·m² and the torsional spring constants of the shafts are $k_1 = k_2 = 3500$ N·m/rad, determine the amplitude of the resulting torsional vibration.

Solution   If the disk rotates by angle θ, the shafts will apply on the disk torques $-k_1\theta$ and $-k_2\theta$. Newton's law for the disk gives

$$J\ddot{\theta} = -k_1\theta - k_2\theta + T_0 \cos \omega t$$

or

$$J\ddot{\theta} + (k_1 + k_2)\theta = T_0 \cos \omega t$$

The response will be

$$\theta = \frac{T_0}{(k_1 + k_2 - J\omega^2)} \cos \omega t$$

The amplitude is

$$A = \left| \frac{T_0}{(k_1 + k_2 - J\omega^2)} \right| = \left| \frac{200}{(7000 - 0.025 \times 500^2)} \right| = 0.266 \text{ rad} = 15.28°$$

(This example demonstrates the computation of the dynamic response basic parameters of the single-degree-of-freedom torsional system under harmonic exciting torque.)

□

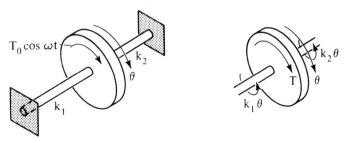

Figure E3.2

3.3 LINEARITY: PRINCIPLE OF SUPERPOSITION

The fact that the differential equations for the harmonic oscillator encountered up to this point are linear is important. Take, for example, the homogeneous equation

$$a\ddot{x} + b\dot{x} + cx = 0 \qquad (3.22)$$

This equation has the following properties:

1. The sum of the two solutions is also a solution. Suppose that two functions

$x_1(t)$ and $x_2(t)$ satisfy equation (3.22). This means that

$$a\ddot{x}_1 + b\dot{x}_1 + cx_1 = 0 \tag{3.23}$$

$$a\ddot{x}_2 + b\dot{x}_2 + cx_2 = 0 \tag{3.24}$$

When inserted into equation (3.22), the sum $x = x_1 + x_2$ yields

$$a(\ddot{x}_1 + \ddot{x}_2) + b(\dot{x}_1 + \dot{x}_2) + c(x_1 + x_2) = 0 \tag{3.25}$$

The sum $x(t) = x_1(t) + x_2(t)$, therefore, also satisfies the differential equation.

2. Since $x(t)$ is a solution of equation (3.22), it is clear that any function $hf(t)$, where h is a constant, is a solution. Similar properties hold for equation (3.2).

These two properties of the linear system constitute the *principle of superposition*.[3]

For a linear system, if the forcing function has the form

$$f(t) = f_1(t) + f_2(t) + \cdots + f_n(t) \tag{3.26}$$

and $x_1(t), x_2(t), \ldots, x_n(t)$ are steady-state solutions corresponding to forcing functions $f_1(t), f_2(t), \ldots, f_n(t)$, respectively, the solution will be

$$x_s(t) = x_1(t) + x_2(t) + \cdots + x_n(t) \tag{3.27}$$

If, for example,

$$f(t) = A \sin \omega t + B \cos \omega t \tag{3.28}$$

the solution will be the sum of the two solutions

$$x_s(t) = \frac{1}{[(-m\omega^2 + k)^2 + (c\omega)^2]^{1/2}} [A \sin(\omega t - \phi) + B \cos(\omega t - \phi)] \tag{3.29}$$

The phase angle ϕ is the same in both solutions since it depends only on the system and ω.

SIMULAB Experiment 3.1

Using the simulated experiment LINSYS, compute the motion of a harmonic oscillator with properties $m = 2$ kg, $k = 10$ N/m, $c = 5$ N·s/m acted upon by a force of amplitude 30 N and frequency 0.3 Hz applied suddenly at time 0 when the initial displacement was 1 mm.

Solution   With the program diskette in drive A, we type

A> LINSYS

The page of Figure SE3.1a appears on the screen. We select a one-mass system. Then the menu page appears on the screen with the simulated one-degree-of-freedom system (Figure SE3.1b). With the keypad arrow keys we select menus.

---

[3]It was introduced by Daniel Bernoulli (1700–1782) to explain observations of Joseph Sauveur (1653–1716) that a vibrating string could produce at the same time sounds that would correspond to several of its harmonics.

Portrait of Daniel Bernoulli. (By anonymous painter, kept at the Bernoulli–Thièbaud family in Basel.)

DANIEL BERNOULLI (1700–1782)

Daniel, the son of Johann Bernoulli, studied under him together with Euler. A professor of mathematics at St. Petersbourg, he is best known for his work in hydrodynamics (Bernoulli equation) and was the first to derive the differential equation for lateral vibration of beams. Many of his ideas were carried out by Euler, and he did experimental work that furnished Euler with new mathematical problems.

1. We set $m = 2$ kg, $k = 10$ N/m, and $c = 5$ N·s/m using the $+$ and $-$ keys.

2. We specify 1 mm initial displacement.

3. We select forced response, 1 force, amplitude, phase angle, angular velocity, 30, 0, and 1.89, because $\omega = 2\pi f = 2\pi \times 0.3 = 1.89$ rad/sec.

4. We hit S to start system operation. The resulting motion is shown in Figure SE3.1b.

(This example demonstrates the use of the linear simulator LINSYS to simulate the transient response of a single degree of freedom to an initial disturbance.)

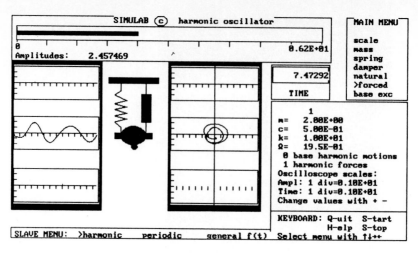

Figure SE3.1

3.4 VECTOR REPRESENTATION

It was shown in Chapter 1 that a harmonic motion can be represented with the aid of a rotating vector, a phasor. Let a harmonic exciting force be $f(t) = F_0 \cos \omega t$. The steady-state response of the system is given in equation (3.21). In an (x, t) diagram, the forcing function and the response appear as harmonic functions (Figure 3.3).

If the period is $T = 2\pi/\omega$, the forcing function has maxima at $t = nT$, $n = 0, 1, 2, 3, \ldots$. The response function $x_s(t)$ reaches a maximum when $\cos(\omega t - \phi)$ is 1 or $\omega t - \phi = 2\pi n$, $n = 0, 1, 2, \ldots$, or

$$t = \frac{2\pi n}{\omega} + \frac{\phi}{\omega} = nT + \frac{\phi}{\omega} \tag{3.30}$$

It therefore reaches a maximum at time ϕ/ω after the maxima of the forcing function. This is called *time lag* and has an obvious physical significance. This is explained by the fact that the system needs time to respond to an excitation. The quantity ϕ has the dimension of angle and it is called the *phase angle*. In general, however, this quantity is helpful merely for our computations. It is not a physical quantity, as one might think. This angle can be visualized from a vector diagram, as in Chapter 1.

Let the applied force be $F_0 \cos \omega t$. A vector of length F_0 rotating about its start point with an angular velocity ω has a tip on a circle of radius F_0, Figure 3.4. Considering that one rotation corresponds to one period $T = 2\pi/\omega$, the time can be measured on the periphery of this circle. The force $F_0 \cos \omega t$ is given by the projection of the vector on the horizontal axis. At time $t = 0$ (or any multiple of T), the vector representing the forcing function has the position OD. At the same time, the response will be

$$x_s(t) = X \cos(-\phi) \tag{3.31}$$

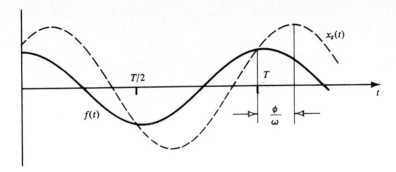

Figure 3.3 Phase angle between force and response.

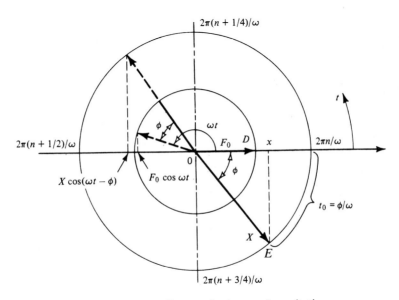

Figure 3.4 Polar diagram for harmonic excitation.

$$X = \frac{F_0}{[(-m\omega^2 + k)^2 + (c\omega)^2]^{1/2}}$$

A vector **OE** at an angle $-\phi$ has projection $X \cos(-\phi)$ on the horizontal axis. If we stipulate that the values of the projections on the horizontal axis are the real harmonic quantities $x(t)$ and $f(t)$, we can represent these quantities with the vectors **OD** and **OE**, rotating with an angular velocity ω. The force vector **OD** of length F_0 is ahead of the response vector **OE** of length X by an angle ϕ, the phase angle. This polar diagram is a useful tool for the study of harmonic motion. The velocity and acceleration can also be represented on the polar diagram (Figure 3.5).

Differentiating the relation $x = X \cos(\omega t - \phi)$, we obtain

$$\dot{x} = -\omega X \sin(\omega t - \phi) = -\omega X \cos\left(\omega t - \phi - \frac{\pi}{2}\right) \tag{3.32}$$

$$\ddot{x} = -\omega^2 X \cos(\omega t - \phi) \tag{3.33}$$

Thus the velocity and acceleration can also be represented by rotating vectors of length $-X$ and $-\omega^2 X$, respectively, and directions as shown in Figure 3.5. Equation (3.1), furthermore, can be written as

$$-m\omega^2 X \cos(\omega t - \phi) - c\omega X \sin(\omega t - \phi)$$

$$+ kX \cos(\omega t - \phi) = F_0 \cos \omega t \tag{3.34}$$

This equation is represented graphically as shown in the polygon of Figure 3.5.

The polar diagram is also useful in other ways: for example, in computing the response to a sum of harmonic exciting forces by adding them to a single harmonic force in the polar diagram and finding the response vector, or finding the responses to the various harmonic forces and adding them vectorially. The situation represented by equations (3.28) and (3.29) is shown in Figure 3.6. To perform amplitude vector additions, the exciting forces must have the same angular velocity ω.

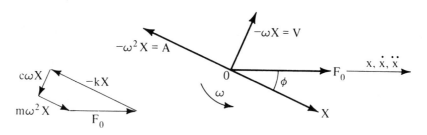

Figure 3.5 Vector representation of vibration.

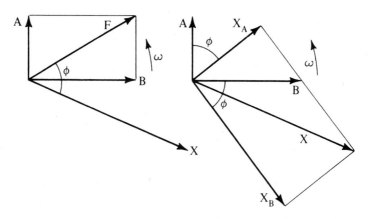

Figure 3.6 Superposition of harmonic excitation and response.

☐ **Example 3.3 Superposition and Vector Addition of Vibration**

The four equally spaced blades of a centrifugal blower rotating at 1200 rpm were weighted before assembly and were found to have masses 22.5, 22.3, 22.6, and 22.8 kg. Their mass center is at radius 0.40 m and the mass of the hub is 50 kg. Find the resulting vibration of the shaft of stiffness 500,000 N/m.

Solution The natural frequency is

$$\omega_n^2 = k/m = \frac{500,000}{50 + 22.5 + 22.3 + 22.6 + 22.8}, \qquad \omega_n = 59.7 \text{ rad/s or 549 rpm}$$

Since it is far from the rotating frequency, damping can be neglected. If M, the total mass, is 139.3 kg and $\omega = 1200 \times 2\pi/60 = 125.6$ rad/s, the response of the system to each one of the blades will be

$$X_1 = \frac{me\omega^2}{k - M\omega^2} = \frac{22.5 \times 0.40 \times 125.6^2}{500,000 - 139.3 \times 125.6^2} = 0.0836 \text{ m}$$

$$X_2 = \frac{me\omega^2}{k - M\omega^2} = \frac{22.3 \times 0.40 \times 125.6^2}{500,000 - 139.3 \times 125.6^2} = 0.0829 \text{ m}$$

$$X_3 = \frac{me\omega^2}{k - M\omega^2} = \frac{22.6 \times 0.40 \times 125.6^2}{500,000 - 139.3 \times 125.6^2} = 0.0840 \text{ m}$$

$$X_4 = \frac{me\omega^2}{k - M\omega^2} = \frac{22.8 \times 0.40 \times 125.6^2}{500,000 - 139.3 \times 125.6^2} = 0.0847 \text{ m}$$

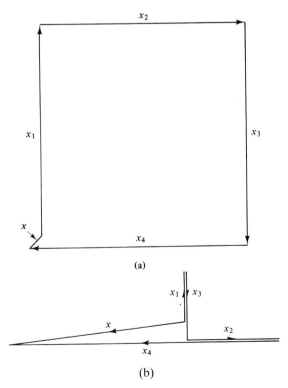

(a)

(b)

Figure E3.3

The response is shown in Figure E3.3. The vector sum of the response to the four blades separately is

$$X = [(0.0840 - 0.0836)^2 + (0.0847 - 0.0829)^2]^{1/2} = 0.00184 \text{ m}$$

(This example demonstrates the superposition principle and the vector addition of vibration of linear systems.)

□

3.5 OTHER HARMONIC EXCITATIONS

From equation (3.17), the following is concluded concerning harmonic excitation:

1. The phase angle depends on the system properties m, c, k and the angular velocity ω, but does not depend on the magnitude of the exciting force.
2. If $c = 0$, the denominator becomes zero for $\omega = (k/m)^{1/2} = \omega_n$. In an undamped system, therefore, the amplitude increases beyond bound if the frequency of the exciting harmonic force coincides with the natural frequency of the system. In engineering systems, this is an unwanted situation.
3. The denominator never vanishes in the presence of damping $c > 0$; therefore, the amplitude remains bounded. Differentiating with respect to ω and equating to zero, we can show that the denominator has a minimum. The response thus has a maximum at

$$\omega = \left(\frac{k}{m} - \frac{c^2}{2m^2}\right)^{1/2} = \omega_n(1 - 2\zeta^2) \tag{3.35}$$

It is apparent that one-degree-of-freedom linear systems always yield equations such as (3.2). Since equation (3.2) has a known solution, calculation of the response of the system is straightforward, at least for the harmonic exciting force $f(t)$. The problem in one-degree-of-freedom linear systems is therefore to determine the differential equation of motion. Some interesting cases of harmonic excitation follow.

Whirling of Rotating Shafts[4]

The most usual exciting force in rotating machines is the unbalance. An axisymmetric rotating member with mass m, with a geometric axis passing through its center of mass, has an additional unbalanced mass m_u attached to it at a distance e from the center of rotation. The center of mass of the system (rotor + unbalance) will be at a radius em_u/m and its vertical position will be $x + (em_u/m) \sin \omega t$ if we start measuring time when the unbalanced mass is in the horizontal position on

---

[4]Dunkerley, S., and Reynolds, O. 1883. On the whirling and vibration of shafts. *Philos. Trans.*, Ser. A, pp. 279–359. Föppl, A. 1895. Das Problem der DeLaval'schen Turbinenwelle. *Civilingenieur*, *61*: 333–342. This analysis has been wrongly credited to Jeffcott (*Philos. Mag.*, Mar. 1919, pp. 304–314).

the right (Figure 3.7). Application of Newton's second law for the motion of the mass center in the vertical direction yields

$$\frac{md^2[x + (em_u/m)\sin\omega t]}{dt^2} + c\frac{dx}{dt} + kx = 0 \qquad (3.36)$$

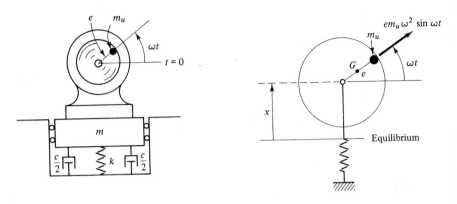

Figure 3.7 Unbalance excitation.

The first term on the left yields a term $-em_u\omega^2 \sin \omega t$, which appears as a force in the vertical direction. We shall call this the *unbalance force*. There is a horizontal component of the unbalance force, of course, but it is transferred on the sliders, which we assumed to confine the motion of the mass in the vertical direction only. The solution is equation (3.17) with

$$F = m_u e\omega^2, \quad H_u(\omega) = \frac{X}{e} = \frac{m_u}{m}\left(\frac{\omega}{\omega_n}\right)^2 H(\omega) \qquad (3.37)$$

☐ **Example 3.4   Unbalance Response of a Rotor**

A compressor rotor (Figure E3.4) consists of a slender shaft of stiffness $k = 1.8 \times 10^5$ N/m, a disk of diameter $d = 0.5$ m, and mass $m = 81.8$ kg. While rotating at 500 rpm, a steel bucket of length $l = 0.2$ m and constant section area $A = 1.125 \times 10^{-4}$ m^2 is broken at half its length. Determine the resulting steady-state vibration, assuming that the transients will eventually be damped out, due to the light damping that certainly must exist in the system.

Solution   Half of the opposite bucket will be the unbalanced mass. Its mass $m_e = \rho Al/2$. For steel, $\rho = 7.6 \times 10^4$ N/m. Thus $m_e = 0.855$ kg and $\omega = 500 \times 2\pi/60 = 52$ rad/s. The amplitude of the steady-state vibration will be

$$X = \frac{m_e e\omega^2}{|-m\omega^2 + k|} = \frac{0.855 \times 0.4 \times 52^2}{|-81.8 \times 52^2 + 1.8 \times 10^5|} = 2.03 \times 10^{-2} \text{ m}$$

(This example demonstrates the unbalance response of a single-degree-of-freedom whirling rotor.)

☐

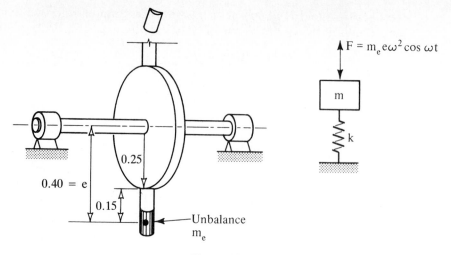

$F = m_e e\omega^2 \cos \omega t$

0.25

0.40 = e

0.15

—Unbalance
m_e

Figure E3.4

Base Excitation

A system, as shown in Figure 3.8, is excited by a harmonic motion of its support, $y = Y \sin \omega t$. If at time t the mass m is displaced from equilibrium by x, the spring is stretched by $x - y$ and the relative motion of the two members of the damper is $\dot{x} - \dot{y}$. For the forces shown in the free-body diagram, therefore, application of Newton's second law yields

$$m\ddot{x} = -k(x - y) - c(\dot{x} - \dot{y}) \tag{3.38}$$

Because $y = Y \sin \omega t$, $\dot{y} = \omega Y \cos \omega t$, equation (3.38) yields

$$m\ddot{x} + c\dot{x} + kx = kY \sin \omega t + c\omega Y \cos \omega t \tag{3.39}$$

The solution is as for a forced vibration problem with force

$$F = f(t) = A \sin \omega t + B \cos \omega t$$
$$A = kY, \qquad B = c\omega Y \tag{3.40}$$

Applying equations (3.17) and (3.20), we obtain

$$x(t) = \frac{kY \sin(\omega t - \phi_1)}{[(-m\omega^2 + k)^2 + (c\omega)^2]^{1/2}} + \frac{c\omega Y \cos(\omega t - \phi_1)}{[(-m\omega^2 + k)^2 + (c\omega)^2]^{1/2}} \tag{3.41}$$

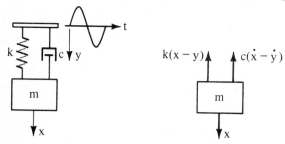

$k(x - y)$ $c(\dot{x} - \dot{y})$

Figure 3.8 Base excitation.

Forced Harmonic Vibration   Chap. 3

The phase angle ϕ_1 is the same for both terms because it depends on the system (m, c, k) and ω, but not on the amplitude of the excitation. Thus we obtain

$$x(t) = Y\left[\frac{k^2 + (c\omega)^2}{(-m\omega^2 + k)^2 + (c\omega)^2}\right]^{1/2} \cos(\omega t - \phi_1 + \phi_2) \qquad (3.42)$$

$$\tan \phi_1 = \frac{c}{k - m\omega^2}, \qquad \tan \phi_2 = \frac{k}{c\omega}$$

The amplitude X of the oscillation $x(t)$ will be

$$|X| = |Y|H_b(\omega), \qquad H_b(\omega) = \left[\frac{k^2 + (c\omega)^2}{(-m\omega^2 + k)^2 + c\omega)^2}\right]^{1/2} \qquad (3.43)$$

$$= [1 + (2\zeta\omega/\omega_n)^2]^{1/2}H(\omega)$$

The function $H_b(\omega)$ multiplies the amplitude of the base vibration and yields the system vibration amplitude. It is called the *amplification factor* and it is a property of the system.

□ **Example 3.5   Base Excitation**

A turbine foundation floor vibration in the vertical direction has an amplitude of $x = 2.5$ mm and a frequency of 60 Hz. A special telescope weighing $w_1 = 13.6$ N, mounted on a light tripod of stiffness $k_1 = 90,000$ N/m, is used to observe the foundation distortions on a scale that weighs $w_2 = 22.7$ N and is mounted on a light base of stiffness $k_2 = 180,000$ N/m (Figure E3.5). Determine the amplitude of the vibration of the mark on the scale that would be seen through the telescope by an observer.

Solution   The vibration of the scale as seen through the telescope is $x_2 - x_1$, where x_1 and x_2 are the motions of the masses m_1 and m_2, respectively. Assuming that the base has a motion $y = e \cos \omega t$, the differential equation of motion for the masses m_1 and m_2 are

$$m_1\ddot{x}_1 + k_1x_1 = k_1y = k_1e \cos \omega t$$

$$m_2\ddot{x}_2 + k_2x_2 = k_2y = k_2e \cos \omega t$$

The solutions, assumed in phase,

$$x_1 = \frac{k_1e}{k_1 - m_1\omega^2} \cos \omega t$$

$$x_2 = \frac{k_2e}{k_2 - m_2\omega^2} \cos \omega t$$

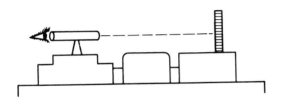
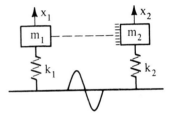

Figure E3.5

The observed vibration

$$x = x_2 - x_1 = e\left(\frac{k_2}{k_2 - m_2\omega^2} - \frac{k_1}{k_1 - m_1\omega^2}\right) \cos \omega t$$

The amplitude is

$$X = e\left(\frac{k_2}{k_2 - m_2\omega^2} - \frac{k_1}{k_1 - m_1\omega^2}\right)$$

It is given that $e = 2.5$ mm, $\omega = 60 \times 2\pi = 376.8$ rad/s, $m_1 = 13.6/g = 1.38$ kg, $k_1 = 90,000$ N/m, $m_2 = 22.7/g = 2.31$ kg, and $k = 180,000$ N/m. Therefore,

$$X = 0.0025\left(\frac{180,000}{180,000 - 2.31^3 \times 376.8^2} - \frac{90,000}{90,000 - 1.38 \times 376.8^2}\right)$$

$$= 0.914 \times 10^{-3} \text{ m}$$

(This example demonstrates the response of a single-degree-of-freedom system to harmonic base motion and the relative motion of two such systems.)

$\square$

SIMULAB Experiment 3.2

Using the simulated experiment LINSYS, compute the motion of a harmonic oscillator with properties $m = 2$ kg, $k = 10$ N/m, and $c = 5$ N·s/m acted on by a base motion of amplitude 0.01 m and frequency 0.3 Hz applied suddenly at time 0.
Solution With the program diskette in drive A, we type

$$\text{A> LINSYS}$$

The page of Figure SE3.1a appears on the screen. We select a one-mass system. Then the menu page appears on the screen with the simulated one-degree-of-freedom system (Figure SE3.1b). With the keypad arrow keys we select menus.

1. We set $m = 2$ kg, $k = 10$ N/m, and $c = 5$ N·s/m using the $+$ and $-$ keys.

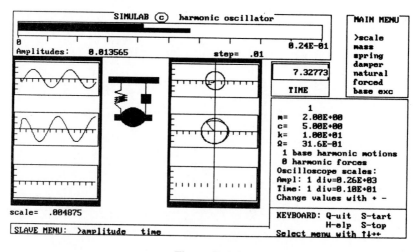

Figure SE3.2

Forced Harmonic Vibration   Chap. 3

2. We select base excitation, 1 motion, amplitude, phase angle, angular velocity, 0.01, 0, and 1.89.

3. We hit S, to start system operation. The resulting motion is shown in Figure SE3.2.

☐ **Example 3.6 Base Excitation**

A heat exchanger of mass m is supported midway between two floors of a building by vertical supports of spring constant k (Figure E3.6). If the upper floor is vibrating with an amplitude y_0 and frequency ω, determine the vertical vibration of the heat exchanger.

Solution Newton's law for the mass m yields

$$m\ddot{x} = -kx + k(y - x)$$

$$m\ddot{x} + 2kx = ky$$

Let $y = y_0 \cos \omega t$. Then

$$m\ddot{x} + 2kx = y_0 \cos \omega t$$

The solution for $F_0 = ky_0$ is

$$x_s(t) = \frac{2ky_0}{(2k) - m\omega^2} \cos \omega t$$

(This example demonstrates the response of a single-degree-of-freedom system to harmonic base motion through a linear spring.)

☐

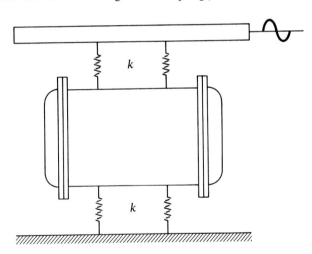

Figure E3.6

Higher Harmonics of Torque on Electric Machines

Electric machines sometimes impose higher harmonics of their synchronous frequency ω on their stator, which vibrates as a result. As an example,[5] consider a

[5]Soderberg, C. R. 1924. *Electr. J.*, *21:* 160.

single-phase machine, motor or generator. The analogy with the vibrating system is voltage to force and displacement to current. As the phase angle between force and displacement, there is a phase angle ϕ between voltage v and current i due to the electrical damping. Therefore, if the voltage is $v = V \sin \omega t$, the current will be $i = I \sin(\omega t - \phi)$. Therefore, the power is

$$W = VI \sin \omega t \sin(\omega t - \phi) = 0.5VI \cos \phi - 0.5VI \cos(2\omega t - \phi) \quad (3.43a)$$

The torque $T = W/\omega$ is transmitted from the rotor to the stator according to the principle of action and reaction. Therefore, the torque to the stator will be

$$T = \frac{0.5VI}{\omega} \cos \phi - \frac{0.5VI}{\omega} \cos(2\omega t - \phi) \quad (3.43b)$$

The first term is constant in time and results in a static twist of the stator in its elastic mounting. The second term causes torsional vibration of the stator at twice the speed of rotation.

If J is the polar moment of inertia of the stator (Figure 3.9) and k the spring constant of each of four springs, the differential equation for torsional vibration of the system and the torsional response will be

$$J\ddot{\theta} + 4k\left(\frac{l}{2}\right)^2 \theta = \frac{0.5VI}{\omega} \cos \phi - \frac{0.5VI}{\omega} \cos(2\omega t - \phi)$$

$$\theta(t) = \frac{0.5VI}{\omega} \frac{\cos \phi}{kl^2} - \frac{0.5VI}{\omega} \frac{\cos(2\omega t - \phi)}{kl^2 - J(2\omega)^2} \quad (3.43c)$$

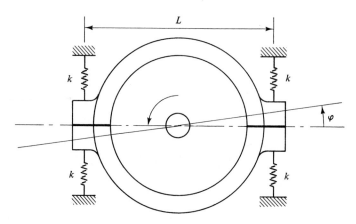

Figure 3.9 Elastic support of an electric stator. (From Soderberg, C. R. 1924. *Electrotech. J.*, *21*: 160.)

3.6 VIBRATION SPECTRA

The engineer is not always interested in the detailed response of a system in the form of all the vibration parameters, such as displacement, velocity, and acceleration. He or she might be interested, in particular, in a specific quantity such as maximum amplitude as a function of one of the design parameters of the system, such as speed of rotation. The engineer then plots the specific quantity versus the design parameter, maximum amplitude versus speed of rotation for example. The resulting plot is called *spectrum*.

On a one-degree-of-freedom system (m, c, k), for example, we apply a harmonic force $F_0 \cos \omega t$. Equation (3.13) implies that there is a relationship between the amplitude F_0 of the exciting harmonic force, called *input*, and the resulting amplitude X of the resulting steady-state vibration, called *output*:

$$X = H(\omega)(F_0/k) \tag{3.44}$$

where

$$H(\omega) = \frac{k}{[(-m\omega^2 + k)^2 + (c\omega)^2]^{1/2}}$$

is a function of the system parameters (m, c, k) and the circular frequency ω.

The function $H(\omega)$ is called the *amplification factor* or *receptance* of the system. The graph of $H(\omega)$ is called an *amplification factor spectrum*. Before plotting some interesting vibration spectra, the normalization scheme will be discussed.

A simple system itself is characterized by three parameters: m, c, k. In our studies the smallest number of independent parameters will be considered. To this end, the equation of motion is rewritten in the form

$$\ddot{x} + \frac{c}{m\dot{x}} + \frac{k}{mx} = \frac{1}{m} F(t) \tag{3.45}$$

That the value $c_c = 2(km)^{1/2}$ of the damping constant corresponds to the transition from the damped vibration to the nonoscillatory motion is already known as critical damping. The fraction of critical damping or damping factor has been already introduced:

$$\zeta = \frac{c}{c_c} = \frac{c}{2(km)^{1/2}} \tag{3.46}$$

Using, in addition to the undamped natural frequency $\omega_n = (k/m)^{1/2}$, the damping factor ζ, equation (3.46) can be rewritten for harmonic excitation as

$$\ddot{x} + 2\zeta\omega_n\dot{x} + \omega_n^2 x = \omega_n^2 \frac{F_0}{k} \sin \omega t \tag{3.47}$$

The number of parameters of the system was reduced from three (m, c, k) to two (ζ, ω_n). This normalization used to be very popular and helps in tabulating the solution of simple systems. For systems with more than a single degree of freedom, we shall work exclusively with the physical parameters. The solution of the problem (3.47) can be written as

$$x_h(t) = e^{-\zeta\omega_n t}[A \cos(1 - \zeta^2)^{1/2}\omega_n t + B \sin(1 - \zeta^2)^{1/2}\omega_n t] \tag{3.48}$$

$$x_s(t) = \frac{F_0/k}{[(1 - (\omega/\omega_n)^2)^2 + (2\zeta\omega/\omega_n)^2]^{1/2}} \sin(\omega t - \phi) \tag{3.49}$$

$$\tan \phi = \frac{2\zeta(\omega/\omega_n)}{1 - (\omega/\omega_n)^2} \tag{3.50}$$

The term $(1 - \zeta^2)^{1/2}$ is the factor that modifies the system's natural frequency with damping. In the steady-state response, F_0/k represents the static deflection of the system X_0 at zero speed, due to a static exciting force F_0. At speed ω, the static deflection is multiplied by a factor

$$H(\omega) = \frac{X}{X_0} = \frac{1}{[1 - (\omega/\omega_n)^2)^2 + (2\zeta\omega/\omega_n)^2]^{1/2}} \tag{3.51}$$

which is the amplification factor. To solve the problem for both transient and steady-state conditions, only two parameters of the system need to be specified. For the steady-state response, in fact, the amplitude and phase angle can be plotted for the harmonic excitation of constant amplitude in only two plots (Figures 3.10 and 3.11).

In the case of unbalance excitation, the force amplitude is $F_0 = m_u e\omega^2$, where m_u is the unbalance and e its eccentricity. The steady-vibration amplitude is not the one given in Figure 3.10, but rather, the one in Figure 3.12. We observe that in the case of constant harmonic excitation F_0, the vibration amplitude has a finite value F_0/k as $\omega \to 0$ and diminishes for $\omega \to \infty$. In the case of unbalance excitation, the amplitude diminishes for $\omega \to 0$ and tends to a constant value for $\omega \to \infty$.

Turning our attention now to the natural vibrations, we first observe that A and B in equation (3.48) should be determined by the initial conditions. The damping is sometimes given as the damping ratio ζ. A convenient way to express and measure damping is by the logarithmic decrement, since it is an easily measured quantity.

Let us examine a damped vibrating system with initial displacement x_0. The response, in normalized form, will be

$$x(t) = x_0 e^{-\zeta\omega_n t} \cos(1 - \zeta^2)^{1/2}\omega_n t \tag{3.52}$$

The maxima of this function roughly are, at the value of time t, when $\cos(1 - \zeta^2)^{1/2}\omega_n t = \max$ or $t = n\pi/[\omega_n(1 - \zeta^2)^{1/2}]$, $n = 1, 3, 5, \ldots$. The logarithmic decrement is defined in Chapter 2 as the logarithm of two successive maxima,

$$\delta = \ln\frac{x_1}{x_2} = \ln(e^{-\zeta\omega_n t_1}/e^{-\zeta\omega n t_2})$$
$$= \ln e^{\zeta\omega n(t_2 - t_1)} = \zeta\omega_n(t_2 - t_1) \tag{3.53}$$

$t_2 - t_1$ is the period of the vibration:

$$t_2 - t_1 = \frac{2\pi}{\omega_n(1 - \zeta^2)^{1/2}} \tag{3.54}$$

Therefore,

$$\delta = \frac{2\pi\zeta}{\sqrt{1 - \zeta^2}} \tag{3.55}$$

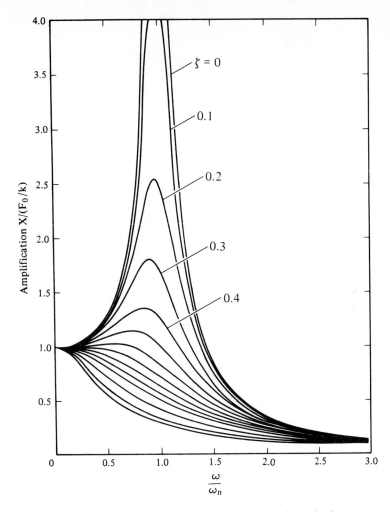

Figure 3.10 Amplification factor for harmonic excitation.

This is the same for any two successive oscillations. In many engineering systems, ζ is small and we can assume that

$$\sqrt{1 - \zeta^2} \approx 1$$

Therefore,

$$\delta \approx 2\pi\zeta \qquad (3.56)$$

Tabulations of typical logarithmic decrements for common structures appear in many handbooks. Some of these are given in Table 3.1.

The logarithmic decrement can be measured very easily by measuring the amplitude of two successive oscillations. For small damping, the difference is very

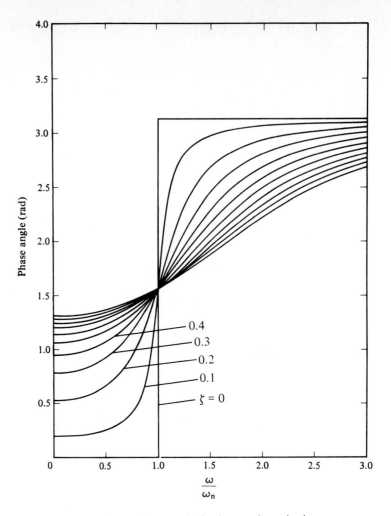

Figure 3.11 Phase angle for harmonic excitation.

TABLE 3.1 LOGARITHMIC DECREMENT OF STRUCTURES

Type of structure	Log decrement, δ
Multistory buildings (concrete)	0.12–0.18
Simple frames (concrete)	0.17–0.22
Bridges (concrete)	0.10–0.30
Bridges (steel)	0.02–0.15
Turbine foundations (concrete)	0.40

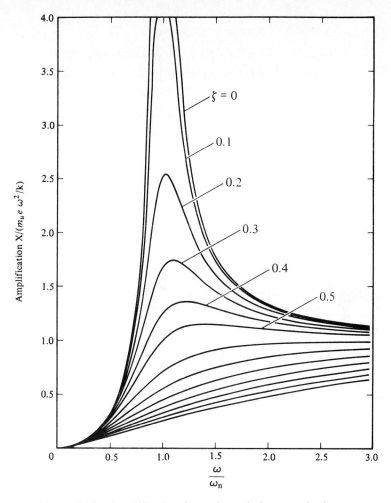

Figure 3.12 Amplification factor for unbalance excitation.

small. To improve accuracy, the maxima of two oscillations n periods apart are measured. Due to equation (3.32),

$$\frac{x_1}{x_2} = \frac{x_2}{x_3} = \cdots = \frac{x_{n-1}}{x_n} = e^{\delta}, \qquad \frac{x_1}{x_n} = e^{(n-1)\delta} \qquad (3.57)$$

Then

$$\delta = \frac{1}{n-1} \ln \frac{x_1}{x_n} \qquad (3.58)$$

It must be remembered that both the damping ratio and the logarithmic decrement *are not* properties of the damping device alone, but properties of the system. The same damping device used in conjunction with another mass and spring yields different values for ζ and δ. In some materials textbooks, logarithmic decrement for different materials is tabulated. In general, such tabulation has meaning only for specific system configurations.

☐ **Example 3.7 Unbalance Response of a Damped System**

A flexibly mounted base plate (Figure E3.7) with $k = 87{,}600$ N/m supports a small motor running at 1800 rpm and having an unbalance of 28.5 g at a radius of 0.15 m. With the motor still, the plate was initially displaced and released. The measured frequency of oscillation was 15 Hz. The ratio of amplitude of the first to the twenty-first oscillation was 1.1. What vibration amplitude will the system have with the motor running?

Solution The log decrement [equation (3.58)] is

$$\delta = \frac{1}{n-1} \ln \frac{x_1}{x_i} = \frac{1}{21-1} \ln 1.1 = \frac{1}{20} \times 0.095 = 4.7 \times 10^{-3}$$

The damping ratio is $\zeta = \delta/2\pi = 7.5 \times 10^{-4}$ [equation (3.56)]. Also, $f_n = 15$ Hz; thus $\omega_n = 2\pi f_n = 2\pi \times 15 = 94.2$ rad/s. The running frequency is $\omega = 1800 \times 2\pi/60 = 188.4$ rad/s. The amplitude for $F_0 = m_e e\omega^2$ and $\omega/\omega_n = 188.4/94.2 = 2$ is

$$X = \frac{m_e e\omega^2}{k[(1 - (\omega/\omega_n)^2)^2 + (2\zeta\omega/\omega_n)^2]^{1/2}}$$

$$= \frac{28.5 \times 10^{-3} \times 0.15 \times 188.4^2}{87600\,[(1 - 2^2)^2 + (2 \times 7.5 \times 10^{-4} \times 2)^2]^{1/2}} = 5.77 \times 10^{-4} \text{ m}$$

(This example demonstrates the response of a single-degree-of-freedom viscously damped system to harmonic exciting force due to unbalance.)

☐

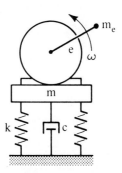

Figure E3.7

☐ **Example 3.8 Half-Power Bandwidth**

Half-power bandwidth is one of the methods used to measure very small damping. In a plot of power or amplitude versus frequency ratio (Figure E3.8), the bandwidth is measured at 0.707 of the maximum amplitude. Show that this is 2ζ.

Solution At $\omega/\omega_n = 1$, $X/X_0 = 1/2\zeta$ [equation (3.5)]. We seek a frequency such that $X/X_0 = 0.707\,(1/2\zeta)$:

$$\frac{1}{[(1 - (\omega/\omega_n)^2)^2 + (2\zeta\omega/\omega_n)^2]^{1/2}} = \frac{0.707}{2\zeta} = \frac{\sqrt{2}}{4\zeta}$$

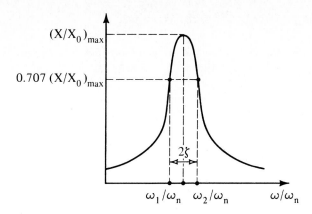

Figure E3.8

or

$$\left(\frac{\omega}{\omega_n}\right)^4 - 2(1 - 2\zeta^2)\left(\frac{\omega}{\omega_n}\right)^2 + (1 - 8\zeta^2) = 0$$

The solution of this quadratic equation is

$$\left(\frac{\omega}{\omega_n}\right)^2 = (1 - 2\zeta^2) \pm 2\zeta\sqrt{1 + \zeta^2}$$

Neglecting ζ^2 for small ζ, we obtain

$$\left(\frac{\omega}{\omega_n}\right)^2 = 1 \pm 2\zeta$$

There are two roots, ω_1 and ω_2.

$$\left(\frac{\omega_1}{\omega_n}\right)^2 = 1 - 2\zeta, \qquad \left(\frac{\omega_2}{\omega_n}\right)^2 = 1 + 2\zeta, \qquad \left(\frac{\omega_2}{\omega_n}\right)^2 - \left(\frac{\omega_1}{\omega_n}\right)^2 = 4\zeta$$

$$4\zeta = \frac{\omega_2^2 - \omega_1^2}{\omega_n^2} = \frac{(\omega_2 + \omega_1)(\omega_2 - \omega_1)}{\omega_n^2} \approx \frac{2\omega_n(\omega_2 - \omega_1)}{\omega_n^2} = \frac{2(\omega_2 - \omega_1)}{\omega_n}$$

Therefore,

$$\frac{\omega_2}{\omega_n} - \frac{\omega_1}{\omega_n} = 2\zeta$$

Some thought should be given to the question: Since we know that the maximum amplitude is $1/2\zeta$, why bother with the bandwidth?

(This example demonstrates a simple method for estimation of the damping factor of a single-degree-of-freedom viscously damped system, the half-power bandwidth.)

$\square$

3.7 MATERIAL AND INTERFACE DAMPING

From strength of materials it is known that for most materials the stress σ and strain ϵ are related within a certain range with a linear law $\sigma = E\epsilon$ (Hooke's law), where E is Young's modulus. This is only true, however, for stresses and strains which are invariable with time. If they vary with time, their relation for *viscoelastic materials* should include rates of change. Hooke's law $\sigma = E\epsilon$ is similar to the expression for the spring force $F = kx$, for example, because stress is normalized force and strain is normalized deformation. Similar to viscous damping, a rate of change of strain $\dot{\epsilon}$ would give rise to a damping force $c\dot{\epsilon}$, where c is an appropriate constant. For viscoelastic materials, higher derivatives of the strain also give rise to force, as do derivatives of the stress. In general, therefore, we can write[6]

$$a_0\sigma + a_1\frac{\partial\sigma}{\partial t} + \cdots + a_n\frac{\partial^n\sigma}{\partial t^n} + \cdots = b_0\epsilon + b_1\frac{\partial\epsilon}{\partial t} + \cdots + b_n\frac{\partial^n\epsilon}{\partial t^n} + \cdots \quad (3.59)$$

where $a_0, a_1, a_2, \ldots, b_0, b_1, b_2, \ldots$ are material constants.

Keeping in mind that stress represents force and strain represents displacement, relation (3.59) can be visualized as being generated by an equivalent combination of springs and dampers.

1. *Spring of stiffness k* (Figure 3.13a):

$$\sigma = \kappa k\epsilon \quad (3.60)$$

where κ is a constant. For example, for a spring consisting of an elastic rod in tension, $k = WA/l$. Therefore, $\kappa = l/A$.

2. *Damper of constant c* (Figure 3.13b):

$$\sigma = \kappa c\frac{\partial\epsilon}{\partial t} \quad (3.61)$$

3. *Spring–damper system* (Figure 3.13c):

$$\sigma = \kappa k\epsilon + \kappa\sigma\frac{\partial\epsilon}{\partial t} \quad (3.62)$$

4. *Two springs and a single damper* (Figure 3.13d):

$$\kappa_1\sigma + c_1\frac{\partial\sigma}{\partial t} = \kappa\left[\kappa k_1\epsilon + (k + k_1)c_1\frac{\partial\sigma}{\partial t}\right] \quad (3.63)$$

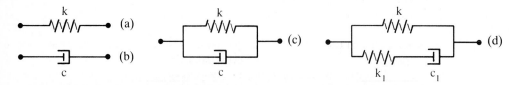

Figure 3.13 Viscoelastic material models.

[6]Skurdzyk, E. J. 1949. *Oester. Ing. Arch.*, 3: 356.

By employing a larger and larger number of springs and dampers, the constitutive equation (3.59) of the material can be approximated closer and closer.

Returning to equation (3.59), if the stress and strain are harmonic functions $\sigma_0 e^{i\omega t}$ and $\epsilon_0 e^{i(\omega t - \phi)}$, representing harmonic excitation and harmonic response $F e^{i\omega t}$ and $X e^{i\omega t}$, respectively, their time derivatives will be

$$\frac{\partial^n \sigma}{\partial t^n} = (i\omega)^n \sigma_0 e^{i\omega t}, \qquad \frac{\partial^n \epsilon}{\partial t^n} = (i\omega)^n \epsilon_0 e^{i\omega t} \tag{3.64}$$

Equation (3.59) therefore becomes

$$[a_0 + a_1(i\omega) + a_2(i\omega)^2 + \cdots]\sigma_0 = [b_0 + b_1(i\omega) + b_2(i\omega)^2 + \cdots]\epsilon_0 \tag{3.65}$$

which can be written as

$$\frac{\sigma_0}{\epsilon_0} = \frac{b'(\omega) + ib''(\omega)}{a'(\omega) + ia''(\omega)} \tag{3.66}$$

where a', a'', b', and b'' are appropriate functions of ω and of a_0, a_1, . . . , b_0, b_1, ...

The ratio of two complex numbers is itself a complex number. Equation (3.66) can therefore be written as

$$\sigma_0 = E^* \epsilon_0 \tag{3.67}$$

The complex number E^* is known as the *complex modulus of elasticity* and can be written as[7]

$$E^* = E(1 + i\gamma) \tag{3.68}$$

where E is Young's modulus and γ is the *material loss factor* (some authors use the symbol δ or η instead).

Similar expressions can be written for the shear modulus and the spring constant under certain assumptions:

$$G^* = G(1 + i\gamma) \tag{3.69}$$

$$k^* = k(1 + i\gamma) \tag{3.70}$$

Young's and the shear moduli E and G and the loss factor γ, in general, are functions of the frequency and the temperature for any specific material. For a limited range of frequency and temperature they can be considered as constants to an engineering accuracy.

Because the model of equation (3.68) has been derived on the assumption of harmonic excitation and response, it is applicable only to this case. In terms of the complex modulus of elasticity, the equation for the harmonic oscillator can be written as

$$m\ddot{x} + k(1 + i\gamma)x = F_0 e^{i\omega t} \tag{3.71}$$

For steady-state conditions, one can prove that the solution of equation (3.71)

---

[7]Myklestad, N. O. 1944. *Vibration Analysis*. New York: McGraw-Hill. Also, Snowdon, J. C. 1963. *J. Acoust. Soc. Am.*, 35(6).

is the same with the solution of

$$m\ddot{x} + \frac{\gamma k}{\omega}\dot{x} + kx = F_0 e^{i\omega t} \tag{3.72}$$

One can therefore assume an equivalent viscous damping constant $c_{eq} = \gamma k/\omega$.

The concept of complex modulus is very convenient. Since the loss factor is determined experimentally, we usually incorporate in it several mechanisms of energy dissipation in materials. Substituting the value of c_{eq} into (3.13), we obtain

$$X = \frac{F_0/k}{[1 - (\omega/\omega_n)^2]^2 + \gamma^2} \tag{3.73}$$

By comparison with equation (3.49), we conclude that the loss factor of the harmonic oscillator is related to other measures of damping as

$$\frac{\delta}{\pi} = \frac{\omega_n}{\omega}\gamma = \frac{2\zeta}{(1 - \zeta^2)^{1/2}} \approx 2\zeta \tag{3.74}$$

the last result for small damping, $(1 - \zeta^2)^{1/2} \approx 1$. One must use equation (3.74) with care, because γ is a *material property* and ζ and δ are *system properties*. Equation (3.74) is applicable only to *harmonically excited linear vibrating systems*.

The model of equation (3.71) can be used for several mechanisms of energy dissipation in structures.

1. *Viscoelastic materials.* Many polymers and other materials are classified under this category, and they closely follow the linear law (3.59). For these materials, the superposition principle holds because of the linearity. The parameters E, G, and γ are generally functions of frequency and temperature.

2. *Structural metals.* In metallic materials, in addition to linear viscoelastic effects, there are other mechanisms of energy dissipation, such as nonlinearities, plastic deformation, and internal Coulomb damping. The method of equivalent viscous damping has been used to obtain an average of these effects. In the presence of any type of damping, under harmonic excitation and response there is a time lag between a force and displacement. A hysteresis loop[8] thus appears in the stress–strain diagram (Figure 3.14). The area in the loop has the dimension of energy and, in fact, gives the energy dissipation per cycle. The hysteresis loop has also been related to fatigue strength.[9]

For viscous damping $c_v\dot{x}$, if $x = Xe^{i\omega t}$, then $\dot{x} = i\omega x$ and the energy dissipation per cycle (the loop area in Figure 3.14) is[10]

$$U_v = \int_{\text{cycle}} c_v\dot{x}\,dx = \int_0^{2\pi/\omega} (c_v i\omega x)(i\omega x)\,dt = \pi c_v X^2\omega \tag{3.75}$$

For structural materials experiments show that the energy dissipated per cycle is proportional to the stiffness and to the square of amplitude X^2 but is relatively

---

[8]Bairstow, L. 1912. *Trans. R. Soc. (London).* Values for different materials can be found in Lehr, E. 1925. Dissertation. Stuttgart. See also [Lazan, 1968].

[9]Extensive tests were reported by Föppl, A. 1931. *J. Iron Steel Inst. (London).*

[10]Crandal, S. H. 1969. *J. Sound & Vibrat.,* **11**(1).

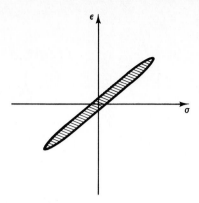

Figure 3.14 Material hysteresis.

unchanged with frequency ω. Therefore, we set up $U_h = \pi\gamma kX^2$, where γ is a material constant. Since the energy stored in the spring is $V = \frac{1}{2}kX^2$, the material constant γ has a simple physical meaning: It is the ratio

$$\gamma = \frac{U_h}{2\pi V} \tag{3.76}$$

which is the ratio of the energy dissipated over the energy stored in the spring at full deflection, per radian.

Furthermore, equating the two dissipated energies $U_v = U_h$, we find that the equivalent viscous damping constant is $c_{eq} = \gamma k/\omega$, as already found above. Thus the constant γ can be considered as the loss factor of the material.

Young's modulus and the loss factor remain in most cases relatively constant over narrow frequency ranges for structural materials but depend strongly on frequency and temperature for plastic materials.

3. *Interface damping.* In complex structures, there is energy dissipation during vibration due to Coulomb friction in contact surfaces, riveted joints,[11] and so on (Figure 3.15). To study this type of damping, a spring–mass system with Coulomb damping (Figure 3.16) is considered. The friction force is $F = mg\mu \, \text{sgn}(x)$. For harmonic vibration $Xe^{i\omega t}$, the energy dissipation per cycle is

$$U_F = \int F \, dx = F \int_{-X}^{X} dx + (-F) \int_{X}^{-X} dx = 4F|X| \tag{3.77}$$

The equivalent viscous damping and the equivalent damping ratio will be, from (3.75) and (3.77),

$$c_{eq} = \frac{4F}{\pi\omega|X|}, \quad \zeta_{eq} = \frac{2F}{\pi\omega_n m\omega|X|} \tag{3.78}$$

Since it depends on the amplitude, it does not fit the models of viscous or viscoelastic damping.[12]

[11]Pian, T. H. H., and Hallowell, F. C., Jr. 1952. *Proc. First U.S. Natl. Congr. Appl. Mech.* New York: ASME.

[12]Vibration with constant Coulomb friction is discussed by Escolt, W. 1926. *Z. Tech. Phys.*, 7: 226.

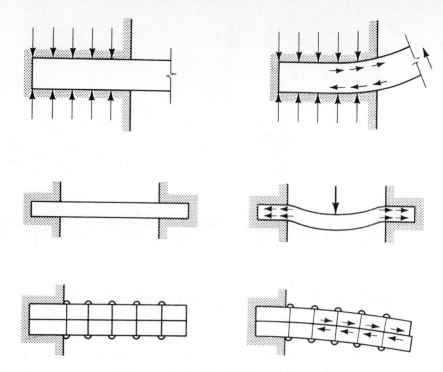

Figure 3.15 Different damping mechanisms.

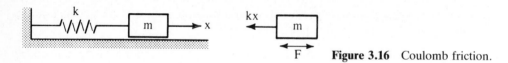

Figure 3.16 Coulomb friction.

The concept of energy dissipation per cycle, however, can be used for approximate computation of vibration amplitude. For example, substituting the equivalent viscous damping (3.78) into equation (3.31) and solving for X, we obtain

$$X = \left[\frac{F_0^2 - (4F/\pi)^2}{k - m\omega^2} \right]^{1/2} \tag{3.79}$$

Satisfactory results can be achieved by an equivalent loss factor γ, provided that light interface damping is accompanied by substantial material damping, and also assuming that it has been measured on similar structures.

3.8 MEASURING INSTRUMENTS IN VIBRATION ANALYSIS[13]

As in every branch of engineering, in vibration engineering analytical methods are not always adequate to help the engineer solve all vibration problems in the design

[13]Geiger, J. 1927. *Mechanische Schwingungen und ihre Messe.* Berlin: Springer-Verlag.

or application stage. Vibration measurements always accompany the application of analytical and numerical methods.

Transducers are elements of measurement systems that sense the quantity to be measured and transform it into a signal, usually electrical, which can easily be measured. Transducers are classified in several different ways:

1. According to the measured quantity: force, displacement, velocity, and acceleration transducers.
2. According to the principle used: electromagnetic, capacitance, inductance, piezoelectric, and optical transducers.
3. According to their position in respect to the measured element: contact and noncontact probes.

All transducers need to have some common characteristics:

1. The relation of the measured quantity, acceleration for example, to the output quantity, voltage for example, is generally nonlinear. For a range of the measured quantity, this relationship is nearly linear to engineering requirements. It is desirable that this range be as wide as possible.
2. The relationship above is, in general, influenced by the frequency of the quantity measured. By "flat response" is meant a wide range of frequencies within which the input–output relationship is not affected by frequency, to engineering accuracy.
3. The transducers themselves are vibrating systems and as such they exhibit the known properties of resonance and phase shift. Property 2 above is related to the requirement that the operating range should be far from resonance. Moreover, the phase shift should be zero, if possible, and certainly constant over the range of measured frequencies.

Proximity probes, in general, measure the distance of the vibrating mass from another component. The three most common ways of measuring this distance are shown in Figure 3.17. In Figure 3.17a, a linear voltage differential transducer

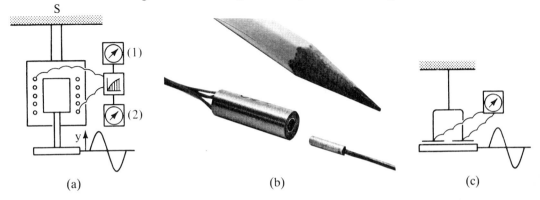

(a)   (b)   (c)

Figure 3.17   Vibration transducers (d, f, j, l, courtesy of Bruel and Kjaer Instruments, Inc. By permission.)

Sec. 3.8   Measuring Instruments in Vibration Analysis

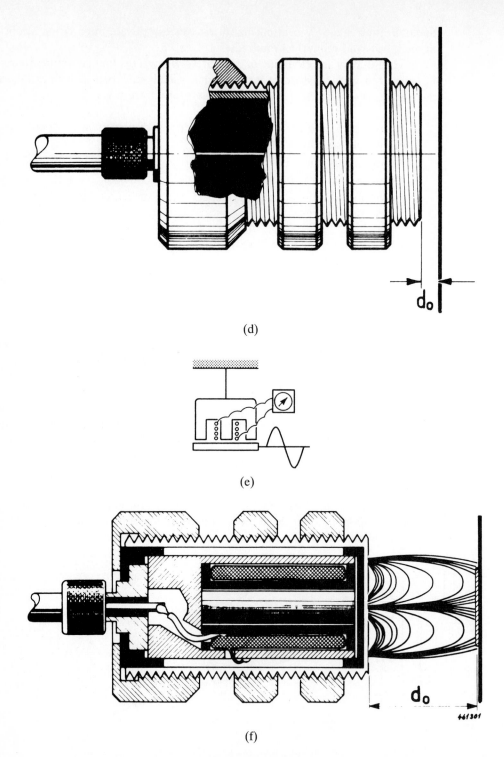

(d)

(e)

(f)

Figure 3.17   (cont.)

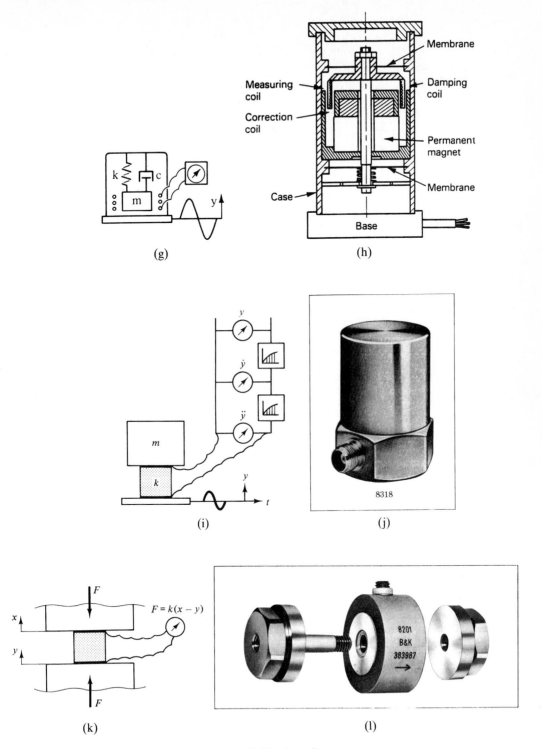

Figure 3.17 (cont.)

Sec. 3.8 Measuring Instruments in Vibration Analysis 151

(LVDT) is shown. One end is attached to the stationary member and the other to the vibrating one. The relative position of the two ends is made proportional to the voltage of a magnet and coil system. Change in the voltage is proportional to the motion y of the vibrating member. An integrator is provided to measure the vibration amplitude y_{max} rather than the motion $y(t)$. Due to its inertia, a voltmeter connected before the integrator will not be able to follow the motion $y(t)$. It will thus indicate the average or rms voltage, which will be proportional to the static relative displacement of the two members.

The capacitance probe is shown in Figure 3.17c. Two small plates are placed at some distance from the vibrating surface. The capacitance will be inversely proportional to the distance. The average and maximum capacitance fluctuation will again be proportional to the static displacement and vibration amplitude, respectively.

The inductance probe is shown in Figure 3.17d. An electromagnet excited by a high-frequency voltage is placed close to the vibrating member. As the gap changes, the inductance of the coil changes due to the change in the air gap and the change in the eddy currents in the vibrating member. Measurement of the coil inductance will yield the displacement, which can have both constant and varying components.

Seismic probes are harmonic oscillators excited by the vibrating body. These transducers are mounted on the vibrating member and measure absolute vibration with respect to the inertial coordinate system. A mass m is supported on the instrument base by a spring k and a damper c. The instrument is placed on the vibrating member with vibration $y = y(t)$ (Figure 3.17g). If the motion of the mass is $x = x(t)$, application of Newton's law for the mass m yields

$$m\ddot{x} = -k(x - y) - c(\dot{x} - \dot{y}) \tag{3.80}$$

Let $z = x - y$. Then equation (3.80) becomes

$$m\ddot{z} + c\dot{z} + kz = -M\ddot{y} \tag{3.81}$$

We note that the quantity $z = x - y$ is the relative motion between the seismic mass m and the cage. The mass m is moving in a coil to produce a voltage that is proportional to the velocity $\dot{z}$. This type of instrument is called a *velocity probe*.

Piezoelectric transducers are instruments with a piezoelectric crystal that gives an electric charge proportional to its deformation z (Figure 3.17i). If the transducer is placed between two solid elements (Figure 3.17k), it measures the force between them and is called a *force transducer*. If one end is fixed on the vibrating element and the other end is a free mass attached to the crystal, the crystal acts as a spring and the transducer is an oscillator with very high natural frequency, called an *accelerometer*.

If the measured motion $y = Y \sin \omega t$, we obtain

$$m\ddot{z} + c\dot{z} + kz = m\omega^2 Y \sin \omega t \tag{3.82}$$

The normalized steady-state solution will be $z = Z \sin(\omega t - \phi)$, where

$$Z = \frac{(\omega/\omega_n)^2}{[(1 - (\omega/\omega_n)^2)^2 + [2\zeta(\omega/\omega_n)]^2]^{1/2}} Y \tag{3.83}$$

$$\phi = \arctan \frac{2\zeta(\omega/\omega_n)}{1 - (\omega/\omega_n)^2}$$

where ω_n is the natural frequency $(k/m)^{1/2}$. From equations (3.83) it is observed that when $\omega_n \ll \omega$, we have $Z \approx Y$. This type of instrument is the seismic probe discussed above. To have low natural frequency, the transducer must have a heavy mass on a soft spring. The measured quantity Z therefore equals the absolute vibration to be measured. This situation is common in velocity transducers. In such transducers, therefore, the voltage is proportional to the velocity $\dot{Z}$ and is thus proportional to $\dot{y}$. An integrator is used to yield the displacement y.

If $\omega_n \gg \omega$, which is the case for piezoelectric accelerometers, equations (3.83) yield

$$Z = \frac{\omega^2 Y}{\omega_n^2} \tag{3.84}$$

But ω_n is a constant and $\omega^2 Y$ is the acceleration of the vibrating member. The quantity Z is therefore proportional to the acceleration $\ddot{y}$. Since high natural frequency is required, a small mass on a stiff piezoelectric crystal is used, and then the output is proportional to the acceleration. The voltage is integrated twice to yield velocity and displacement. The natural frequency of the accelerometers is usually in the range of kilohertz. The range of the velocity transducers is 1 to 50 Hz.

The calibration of the instruments is made on a very high or a very low value of the ratio ω_n/ω, respectively. If the two frequencies approach each other, the instrument indication should be corrected with division by the following, which are the instrument amplification factors. For a velocity transducer,

$$\frac{(\omega/\omega_n)^2}{[(1 - (\omega/\omega_n)^2)^2 + (2\zeta(\omega/\omega_n))^2]^{1/2}} \tag{3.85}$$

For an accelerometer,

$$\frac{1}{[(1 - (\omega/\omega_n)^2)^2 + (2\zeta(\omega/\omega_n))^2]^{1/2}} \tag{3.86}$$

These corrections are usually provided by the instrument manufacturer.

Phase distortion. Equation (3.83) yields the correction due to the phase shift, that is the instrument phase shift that should be subtracted from the measured one to yield the actual phase shift. For a velocity transducer,

$$\phi \xrightarrow[\omega/\omega_n \to \infty]{} \pi \tag{3.87}$$

For an accelerometer,

$$\phi \xrightarrow[\omega/\omega_n \to 0]{} 0 \tag{3.88}$$

These equations are used to correct the phase angle of the vibration measured. To this end, the phase angle between the voltage output of the instrument and a reference signal is measured. This is the vibration phase angle, adjusted according to equations (3.87) and (3.88). In the case of a rotating shaft, the reference signal is provided with a signal generator mounted on the shaft.

When the frequencies ω and ω_n approach each other, equations (3.87) and (3.88) are no longer adequate. Instead, the phase angles indicated by the instrument have to be decreased as follows. For a velocity transducer,

$$\arctan \frac{2\zeta(\omega/\omega_n)}{1 - (\omega/\omega_n)^2} \qquad (3.89)$$

For an accelerometer,

$$\arctan \frac{2\zeta(\omega/\omega_n)}{1 - (\omega/\omega_n)^2} \qquad (3.90)$$

These corrections are also given by instrument manufacturers as functions of ω/ω_n. It is always preferable to refer to the manufacturer's specifications rather than to attempt to compute the corrections. The reason for this is that the electronics (amplifiers, integrators, etc.) have their own resonances; therefore, they introduce the need for additional corrections.

□ **Example 3.9 Velocity Transducer**

Determine the phase shift of a velocity transducer with a natural frequency of 25 Hz and damping ratio 0.4, assuming that it has a vibration of 60 Hz and the amplitude error.

Solution   For a velocity transducer, the phase shift is given by equation (3.89):

$$\phi = \arctan \frac{2\zeta(\omega/\omega_n)}{1 - (\omega/\omega_n)^2} = \arctan(-0.40)$$

Therefore, $\phi = -0.38$ rad $= -21.8°$. From the indication of the instrument, this angle is subtracted, thereby obtaining the correct phase angle of the system. The error factor is

$$\frac{(\omega/\omega_n)^2}{[(1 - (\omega/\omega_n)^2)^2 + (2\zeta(\omega/\omega_n))^2]^{1/2}} = 1.12$$

Therefore, the error is $(1.12 - 1)/1 = 12\%$.

(This example demonstrates the basic corrections to accelerometer measurements.)

□

3.9 FOUNDATIONS: VIBRATION ISOLATION

Machines that are expected to transmit substantial static or dynamic forces through their pedestal are installed on foundations.[14] A usual arrangement of this kind is shown in Figure 3.18. The machine, of mass m, mounted on a massive foundation of mass M, rests directly on soil or some other elastic material, such as cork, rubber, springs, and so on. It can therefore be represented as a single-degree-of-freedom system. The spring constant k can be determined from the dimension and properties

---

[14]The problem was studied by Sommerfeld (Lorenz, 1902. *Technische Mechanik*. Munich.)

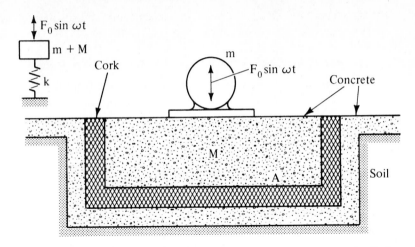

Figure 3.18 Machine vibration isolation. (From Hort, W. 1910. *Technische Schwingunslehre.* Berlin: Julius Springer.)

of the elastic material. If the mass M rests on soil, the spring constant will be

$$k = Ak_s \qquad (3.91)$$

where A is the footing surface and k_s is a constant called the *coefficient of compression of the soil*[15] (N/m^3 or similar). Typical properties of relevant materials and soils are given in Table 3.2.

The purpose of this construction is to keep at a miminum the force transmitted through the foundation to the surroundings. We are therefore interested in the

TABLE 3.2

Soil type	Allowable load (MPa)	k_s (MPa/m)	
		Vertical	Horizontal
Gray plastic silty clay and organic silt	1	0.2	0.30
Brown, saturated silty clay with sand	1.5	0.3	0.46
Dense silty clay with sand	5.2	0.5	0.86
Medium moist sand	2.1	0.3	0.54
Dry sand with gravel	2.1	0.3	0.54
Fine saturated sand	2.4	0.41	0.92
Medium sand	2.4	0.41	0.92
Loesial, natural moisture	3.1	0.46	1.06
Gravel	6.2	1.02	2.72
Sandstone	10.4	0.95	2.72
Limestone	11.4	1.08	3.26
Granite, partly decomposed	41.5	2.72	9.52
Granite, sound	58.7	2.85	9.75

[15]Barkan, D. D. 1962. *Dynamics of Bases and Foundations.* New York: McGraw-Hill.

force transmitted. This is the force carried through the spring and damper and has the value

$$f = kx + c\dot{x} \qquad (3.92)$$

Here x is the displacement, which can be computed from the imposed force $F_0 \sin \omega t$ and the properties of the system $(m + M)$ and k. The amplitude of this force will be

$$F_T = kX + ic\omega X \qquad (3.93)$$
$$= \frac{F_0[1 + (2\zeta\omega/\omega_n)^2]^{1/2}}{[(1 - (\omega/\omega_n)^2)^2 + (2\zeta\omega/\omega_n)^2]^{1/2}}$$

Therefore, the ratio of the transmitted to the imposed force is

$$\text{TR} = \frac{F_T}{F_0} = \frac{[1 + (2\zeta\omega/\omega_n)^2]^{1/2}}{[(1 - (\omega/\omega_n)^2)^2 + (2\zeta\omega/\omega_n)^2]^{1/2}} = H_b(\omega) \qquad (3.94)$$

This ratio, called *transmissibility*, is given in Figure 3.19 and the phase angle in Figure 3.20.

For quick calculations, we sometimes use the static deflection δ_{st} of the system, due to its own weight. This can readily be measured by loading the foundation with a static weight and multiplying the resulting static deflection by $(M + m)g/w$. The natural frequency is

$$\omega_n = \sqrt{\frac{kg}{(m + M)g}} = \sqrt{\frac{g}{\delta_{st}}} \qquad (3.95)$$

The following formula is popular for the natural frequency:

$$f_n(\text{cpm}) = \frac{300}{\sqrt{\delta_{st}}} \quad \text{cm} \qquad (3.96)$$

In terms of the frequency $f = \omega \times 60/2\pi$ (cpm), for $\zeta = 0$, we obtain a transmissibility

$$\text{TR} = \frac{1}{(2\pi f)^2 \delta_{st}/g - 1} \qquad (3.97)$$

From equation (3.94) we observe that for $\zeta = 0$ we have effective isolation (TR < 1) only if $(\omega/\omega_n)^2 - 1 > 1$ or $\omega > \omega_n\sqrt{2}$. From Figure 3.19, however, we observe that for $\omega/\omega_n = 3.0$ the transmissibility is about 0.1 and decreases very slowly after that. The value $\omega = 3\omega_n$ is very often used for foundation design.

In Figure 3.19 we observe that damping has no effect on the point where effective isolation (TR < 1) starts. This point is $\omega/\omega_n = \sqrt{2} = 1.41$. For frequencies $\omega < 1.41\omega_n$, damping reduces the transmissibility ratio and thus the transmitted force. For frequencies $\omega > 1.41\omega_n$, however, damping increases the transmissibility ratio and the transmitted force. Therefore, in the case that $\omega \gg \omega_n$ steel springs are used for foundation isolation with very low damping. This observation shows that damping is not always a desirable feature in engineering systems.

The spectra of Figure 3.19 can also be used to study the transmissibility of motion in the case of base excitation. Thus if the base has a motion $y = Y \cos \omega t$,

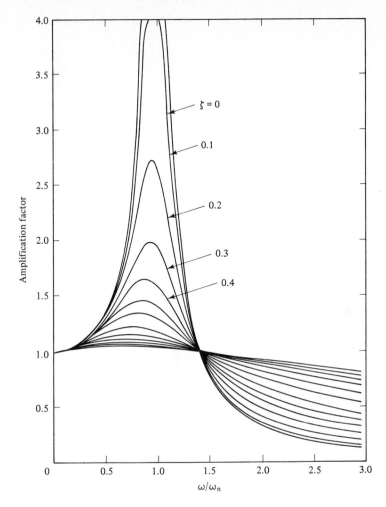

Figure 3.19 Amplification factor for transmissibility.

the mass will have a motion $x = X \cos(\omega t - \phi)$. The ratio of resulting to imposed motion (output/input) is $|X|/|Y| = H_b(\omega)$, where $H_b(\omega)$ is given by equation (3.43). This is exactly the function plotted in Figure 3.19.

Physically, the situation might represent the isolation of, for example, an instrument mounted on a vibrating floor. The engineer often has to design a flexible mounting for an instrument, to minimize its vibration when the floor or base vibrates. Here the motion transmissibility ratio equals the force transmissibility ratio in the previous case of a harmonic force imposed on the mass. They are both plotted in Figure 3.19. From this figure we again observe that effective motion isolation is incurred for $\omega > 1.41\omega_n$ only, regardless of damping. For higher exciting frequencies, damping has an adverse affect. As a general rule, therefore, for a given ω one must try to provide a flexible mounting (low ω_n, thus high ω/ω_n) with light damping.

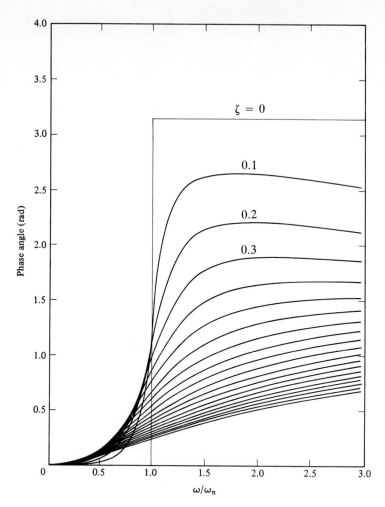

Figure 3.20 Phase angle for transmissibility.

3.10 MICROCOMPUTED-ASSISTED VIBRATION MEASUREMENTS

An increasing utilization of microcomputers for handling of vibration measurements is the result of the rapid decrease in price and increase in speed of microcomputers. A microcomputer-assisted vibration measuring system is shown in Figure 3.21. An *interface* device, usually a plug-in card, translates a measured analog signal to a digital one, representing the instantaneous voltage of the signal at the time of the measurement. A range of voltages is usually specified with the interface device. The measured signal has to be of the same order of magnitude as the specified range, because this range is digitized, that is, divided in a number of discrete points, depending on the interface device. This is expressed by the number of bits (0's and 1's) used to translate the signal. An 8-bit interface device uses $2^8 = 256$ divisions, a 12-bit device uses $2^{12} = 4096$, and so on. Therefore, to make full use of the

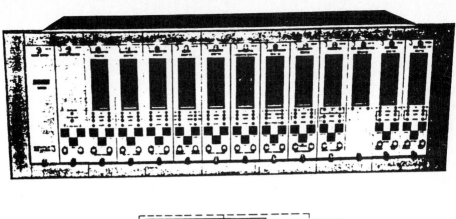

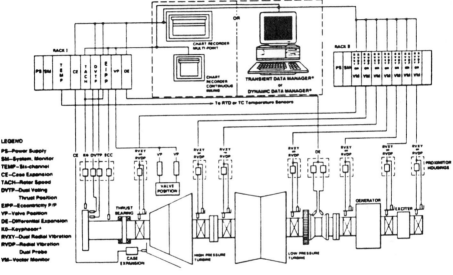

☐ **Rotor speed:**   ☐ **Valve position:**   ☐ **Case expansion:**

☐ **Absolute vibration:**   ☐ **Thrust position:**   ☐ **Differential expansion:**

☐ **Radial vibration:**   ☐ **Eccentricity:**

Figure 3.21 Turbine supervisory instrumentation.

resolution of the device, the signal has to be as close to the limit as possible. For example, an 8-bit interface with maximum voltage 10 V can measure a signal of ∓ 1 V but with resolution $1/(256 \times 1/10) = 0.04$ V, an error of 4%.

The vibration transducers deliver, in general, voltages that are different from the ones for which the interface devices are rated. The matching of the two requirements is called *signal conditioning*. The signal from the vibration transducer is passed to the signal conditioning instrumentation, which includes one or both of the following signal conditioning features:

1. *Amplification*, to match the voltages, as described above
2. *Filtering*, because measurements are associated with *noise*, voltages due to other vibration sources, inducted electric currents, or electrical effects of the measuring devices themselves

The programming of the interface devices to read signals in an orderly manner and store them in the computer storage device (main memory, virtual disk, hard disk, floppy disk, etc.) will depend on the device, and it will be assumed that the reader can use the device to read data and put them in a stored record, which, to be compatible with the software of this book, has to be an ASCII text file in the following form:

> *First line:* problem identification
> *Second line:* title of the x-axis
> *Third line:* title of the y-axis
> *Fourth line:* number of measurement points equally spaced in time. N; time step *DeltaT*; scale factor to multiply data to yield the measured quantity *ScaleFac*
> *Fifth line:* N measured data separated by commas
> More lines as needed

For Example, a file TEST.DAT will be

> System Test Measurements
>
> Time (s)
>
> Displacement (mm)
>
> 16, 0.01, 1000
>
> 0, 11, 30, 42, 50, 58, 60, 59, 51, 43
>
> 31, 1, -5, -20, -12, 0

Such files with data will be produced by several software items developed in this book. Readers can use either routines available to them to plot the data in the devices available, or the ANALYZER program included in the programs diskette for a screen plot.

3.11 ALGORITHM FOR HARMONIC VIBRATION ANALYSIS OF SINGLE-DEGREE-OF-FREEDOM SYSTEM

The algorithm OSCILLATOR inputs system data and excitation in the form of harmonic force, harmonic support motion, and initial conditions for a single-degree-of-freedom oscillator and produces a file with the vibration amplitude at a specified number of points and time step.

OSCILLATOR

```
begin
     /*                      INPUT DATA                        */
     input name of ResultsFile
     OPEN ResultsFile for WRITING data
        Input  Problem Identification
        Input Title of the x-axis
        Input Title of the y-axis
        input NumberOfPoints
        Input mass m
        Input spring constant k
        Input damping constant c
        Input Harmonic Force Amplitude  Force, Phase Angle PhaseF,
              Angular Velocity omegaF
        Input Harmonic Base Motion Amplitude Base, Phase Angle PhaseB,
              Angular Velocity omegaB.
        Input Initial Displacement x0, Velocity v0
        Input NumberOfPoints, DeltaT
        WRITE on ResultsFile Problem Identification
        WRITE on ResultsFile Title of the x-axis
        WRITE on ResultsFile Title of the y-axis
        WRITE on ResultsFile NumberOfPoints,DeltaT,1
        Loop until Ipoint >=NumberOfPoints
        Ipoint=Ipoint+1
        Time=Time+DeltaT
        /*                                 Response to Force */
        PhiF=arctan[omegaF*c/(k-m*omegaF²)]
        cosfun=cos(omegaF*Time-PhiF-PhaseF)
        AmplitudeF= Force*cosfun/[(k-m*omega²)²+(omega*c)²]^(1/2)
        /*                            Response to Base Motion */
        PhiB1=arctan[omegaB*c/(k-m*omegaB²)]
        PhiB2=arctan[k/c*omegaB]
        cosfun=cos(omegaB*Time-PhiB1-PhiB2-PhaseB)
        AmplitudeB= Base*cosfun/{[k²+(omegaB*c)²]/[(k-m*omegaB²)²
                    +(omegaB*c)²]}^(1/2)
        /*                            Transient Response */
        omegaD=[k/m-(c/2m)²]^(1/2)
        AmplitudeT=e^(-c*Time/2m)[x0*cos(omegaD*Time)
                   +(x0*c/m+2*v0)*sin(omegaD*Time)/2*omegaD]
        Amplitude=AmplitudeF+AmplitudeB+AmplitudeT
        WRITE Amplitude on file ResultsFile
        if Ipoint<NumberOfPoints then
        WRITE , (a comma) on file ResultsFile
     End Loop
   CLOSE file ResultsFile
 end
```

TABLE 3.3 SUMMARY OF EQUATIONS OF CHAPTER 3

Differential equation

$$m\ddot{x} + c\dot{x} + kx = f(t)$$

Harmonic excitation

$$f(t) = Fe^{i\omega t}$$

Damped system response

$$x(t) = F/[(-\omega^2 m + k)^2 + \omega c)]^{1/2} e^{i\omega t - \phi}$$

$$\tan \phi = \omega c/(-m\omega^2 + k)$$

$$x(t) = (F_0/k)/[(1 - (\omega/\omega_n)^2)^2$$

$$+ (2\zeta\omega/\omega_n)^2]^{1/2} e^{i(\omega t - \phi)}$$

$$\tan \phi = 2\zeta(\omega/\omega_n)/[1 - (\omega/\omega_n)^2]$$

Undamped system response

$$x(t) = (F_0/k)/[(1 - (\omega/\omega_n)^2] \sin(\omega t - \phi)$$

$$\tan \phi = 1, \quad \phi = \pm\pi/2$$

Amplification factor

Force of constant amplitude:

$$H(\omega) = 1/[[1 - (\omega/\omega_n)^2]^2 + (2\zeta\omega/\omega_n)^2]^{1/2}$$

Unbalance:

$$H_u(\omega) = \frac{X}{e} = \frac{m_u}{m}\left(\frac{\omega}{\omega_n}\right)^2 H(\omega)$$

Base excitation:

$$H_b(\omega) = \frac{X}{Y} = \left[1 + \left(2\zeta\frac{\omega}{\omega_n}\right)^2\right]^{1/2} H(\omega)$$

Quality factor, at $\omega/\omega_n = 1$

$$Q = 1/2\zeta$$

Whirling of shafts

$$F_0/k = m_u/m\, e(\omega/\omega_n)^2$$

Principle of superposition

$$f(t) = f_1(t) + f_2(t) + \cdots + f_n(t)$$

$$x(t) = x_1(t) + x_2(t) + \cdots + x_n(t)$$

Time derivatives,

$$x = X\cos(\omega - \phi)$$

(continues)

TABLE 3.3 *(continued)*

$$\dot{x} = -\omega X \sin(\omega t - \phi)$$
$$\ddot{x} = -\omega^2 X \cos(\omega t - \phi)$$

Base excitation $Y \sin \omega t$

$$m\ddot{x} + c\dot{x} + kx = kY \sin \omega t + c\omega Y \cos \omega t$$
$$x(t) = Y\{[k^2 + (c\omega)^2]/[(-m\omega^2 + k)^2$$
$$+ (c\omega)^2]\}^{1/2} \cos(\omega t - \phi_1 + \phi_2)$$
$$\tan \phi_1 = c/(k - m\omega^2), \qquad \tan \phi_2 = k/c\omega$$

Damping parameters

$$\delta = \ln x_1/x_2$$
$$\delta = 2\pi\zeta/\sqrt{1 - \zeta^2} \approx 2\pi\zeta$$

Complex moduli

$$E^* = E(1 + i\gamma), \qquad G^* = G(1 + i\gamma),$$
$$k^* = k(1 + i\gamma)$$
$$\delta/\pi = (\omega_n/\omega)\gamma = 2\zeta/(1 - \zeta^2)^{1/2} \approx 2\zeta$$

Coulomb friction force F

$$c_{eq} = 4F/\pi\omega|X|$$

Structural damping

$$c_{eq} = \gamma k/\omega$$
$$\gamma = U_h/\pi k X^2$$

Measurement corrections

Velocity transducer:
$$(\omega/\omega_n)^2/[[1 - (\omega/\omega_n)^2]^2$$
$$+ [2\zeta(\omega/\omega_n)]^2]^{1/2}$$

Accelerometer:
$$1/[[1 - (\omega/\omega_n)^2]^2 + [2\zeta(\omega/\omega_n)]^2]^{1/2}$$

Phase distortion

Velocity transducer:
$$\arctan 2\zeta(\omega/\omega_n)/[1 - (\omega/\omega_n)^2]$$

Accelerometer:
$$\arctan 2\zeta(\omega/\omega_n)/[1 - (\omega/\omega_n)^2]$$

REFERENCES AND FURTHER READING

DEN HARTOG, J. P. 1952. *Mechanical Vibration*, 4th ed. New York: McGraw-Hill.

DIMAROGONAS, A. D. 1976. *Vibration Engineering*. St. Paul, Minn.: West Publishing.

HORT, W. 1910. *Technische Schwingunslehre*. Berlin: Julius Springer.

HOUSNER, G. W., and HUDSON, D. E. 1950. *Applied Mechanics: Dynamics*. Princeton, N.J.: D. Van Nostrand.

KLOTTER, K. 1960. *Technische Schwingungslehre*. Berlin: Springer-Verlag.

LAZAN, B. J. 1968. *Damping of Materials and Members in Structural Mechanics*. Oxford: Pergamon Press.

PONTRYAGIN, L. S. 1962. *Ordinary Differential Equations*. Reading, Mass.: Addison-Wesley.

RAYLEIGH, J. W. S. 1894. *Theory of Sound*. New York: Dover Publications, 1946.

Spectral Dynamics Corp. 1990. *Vibration Handbook*.

TIMOSHENKO, S. P. 1953. *History of the Strength of Materials*. New York: McGraw-Hill.

PROBLEMS

Sections 3.1 to 3.5

3.1. A refrigeration compressor is mounted on four springs of stiffness $k = 20,000$ N/m and has mass $m = 55$ kg (Figure P3.1). The mounts are made of spring steel with very low damping. Due to the design of the compressor, there is a vertical harmonic force of 12 N oscillating at the operating frequency of 1750 rpm. Determine the vertical vibration of the compressor.

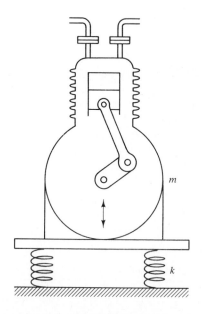

Figure P3.1

3.2. To measure the unbalanced horizontal harmonic force in a piston-type air compressor of mass $m = 80$ kg, an engineer mounted it on a platform of mass $M = 50$ kg that could vibrate horizontally without friction and had an elastic support with stiffness in the horizontal direction $k = 3500$ N/m (Figure P3.2). At a running frequency of 1150 rpm the engineer measured horizontal vibration as 0.0005 m. Compute the amplitude of the unbalanced horizontal force, neglecting damping.

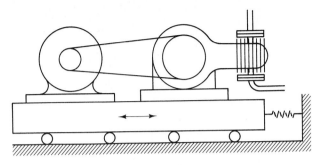

Figure P3.2

3.3. An electric motor of mass $m = 22$ kg is mounted at the midspan of a simply supported steel beam of rectangular cross section, length $L = 1$ m, width $b = 0.2$ m, and thickness $t = 10$ mm (Figure P3.3). The amplitude of the unbalanced harmonic vertical force of the motor in the vertical direction is known to be 55 N at 58 Hz. Determine the resulting vibration, neglecting damping.

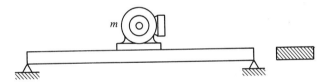

Figure P3.3

3.4. The moving core of an electromagnetic relay (Figure P3.4) has mass $m = 12$ g and is supported at one end with a coil spring with $k = 3000$ N/m and at the other end, at the closed contact position, by the elastic strips of the electric contacts, which have stiffness 1200 N/m in the direction of the coil motion (Figure P2.6). There is an unbalanced oscillating force due to the electric field along the core axis of 1.3 N at the synchronous electric frequency of 60 Hz. Determine the core vibration, neglecting damping.

3.5. The dc voltmeter of an ac-to-dc converter has a steel dial of length $l = 50$ mm, width 3 mm, and thickness 1 mm. The restoring spring has a constant $k = 0.006$ N·m/rad. The maximum indication is 100 V at a 50° angle. On the dc voltage there is a ripple due to noncomplete rectification of the ac. Determine its magnitude in volts if the observed vibration of the dial was 2° at 100 V dc and the frequency is 60 Hz, neglecting damping (see Figure P2.47).

3.6. A single-blade wind turbine rotor is shown in Figure P3.6. The blade has mass 35 kg and the center of mass is at radius $R = 6$ m. As it rotates, the static weight imposes a torque on the shaft varying harmonically with the rotation. If the polar moment of inertia of the rotor about the axis of rotation is 1400 kgm², the shaft connecting the

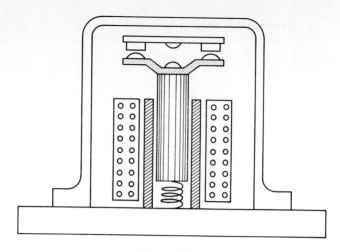

Figure P3.4

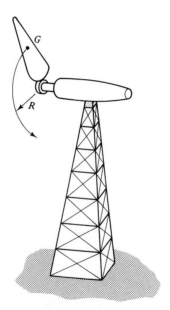

Figure P3.6

rotor has a torsional stiffness $k_T = 4 \times 10^5$ Nm/rad, determine the resulting torsional vibration of the rotor, assuming no damping and that the generator has infinite inertia due to the connection with the electric grid.

3.7. The tail rotor of a helicopter has four blades (Figure P3.7). The tail section was tested while the helicopter was on the ground by placing a static load of 500 N at the tail rotor location. A deflection of 4 mm was measured while keeping the cabin fixed. Determine the vertical vibration of the tail rotor if one blade of mass 2 kg and center of mass at radius 200 mm brakes while running at 1000 rpm, assuming no damping and that the mass of the tail rotor assembly is 50 kg, neglecting the tail structure mass and assuming that the cabin has too high mass and inertia to be affected.

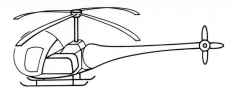

Figure P3.7

3.8. A compressor rotor consists of a slender shaft of stiffness $k = 9000$ N/m and a disk of diameter $d = 0.5$ m and mass $m = 100$ kg (Figure P3.8). While rotating at 500 rpm, two steel blades 450 apart are broken. They have length $l = 0.2$ m and constant section $A = 900$ mm². Determine the resulting steady-state vibration. Assume that the light damping which certainly exists in the system eliminates the transient vibration.

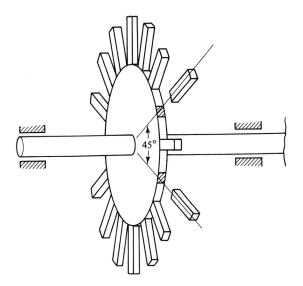

Figure P3.8

3.9. One type of hydraulic shaker consists of a cylinder C, piston of mass m, and a spring k (Figure P3.9). A fluid fills the lower part, which is under a fluctuating pressure $p = p_0 \cos \omega t$. Determine the force acting on the base, disregarding damping.

3.10. A centrifugal shaker[16] consists of two counter-rotating rotors with unbalanced masses m_e at distance e from the axis of rotation (Figure P3.10). The horizontal forces cancel each other and purely vertical harmonic forces $2m_e e\omega^2 \cos \omega t$ are transmitted to the base if the mounting is rigid. Otherwise, if the mounting has a flexibility k and the shaker a total mass m, the transmitted force is different. Determine the amplitude and phase angle of the transmitted force as a function of the parameters m, e, m_e, ω, and k.

3.11. Solve Problem 3.1 assuming that there is damping in the system measured as a logarithmic decrement of 5% during damped natural vibration tests.

[16]Späth, W. 1929. *Z. VDI*, 73: 963.

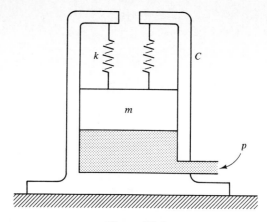

Figure P3.9

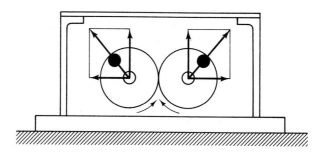

Figure P3.10

3.12. Solve Problem 3.2 assuming that there is damping in the system measured as a logarithmic decrement of 3% during damped natural vibration tests.

3.13. Solve Problem 3.3 assuming that there is damping in the system measured as a logarithmic decrement of 6% during damped natural vibration tests.

3.14. Solve Problem 3.4 assuming that the system is critically damped.

3.15. Solve Problem 3.5 assuming that the system is critically damped, as shown in Figure P2.52.

3.16. To reduce torsional vibration in Problem 3.6, an elastic coupling was used between the shaft and the motor of torsional stiffness $K_T = 1200$ N·m/rad and damping $C_T = 100$ N·m·s/rad. Determine the resulting torsional vibration under the same other conditions.

3.17. Solve Problem 3.7 assuming that there is damping in the system measured as a logarithmic decrement of 3% during damped natural vibration tests.

3.18. Solve Problem 3.8 assuming that there is damping in the system measured as a logarithmic decrement of 6% during damped natural vibration tests.

3.19. Solve Problem 3.9 assuming that there is clearance c between the piston and cylinder, piston length L, and there is lubricant in the cylinder of viscosity η.

3.20. Solve Problem 3.10 assuming that the system is critically damped.

3.21–3.25. In the system shown in Figures P3.21 to P3.25, respectively, the massless base B is forced to a harmonic motion of amplitude e and frequency ω in the direction

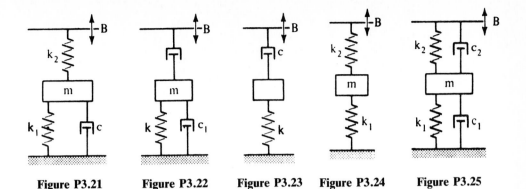

| Figure P3.21 | Figure P3.22 | Figure P3.23 | Figure P3.24 | Figure P3.25 |

shown. Find the resulting motion of the mass and the phase angle with the motion of B.

3.26. A turbine foundation floor, vibrating in the vertical direction, has an amplitude of $x = 0.25$ mm and frequency 30 Hz. A special telescope weighing $w_1 = 30$ N is mounted on a light tripod of stiffness $k_1 = 90,000$ N/m (Figure P3.26). It is used to observe the foundation distortions on a scale weighing $w_2 = 10$ N and is mounted on a base of stiffness $k_2 = 180,000$ N/m. Determine the amplitude of the vibration of a mark on the scale that would be seen through the telescope by an observer.

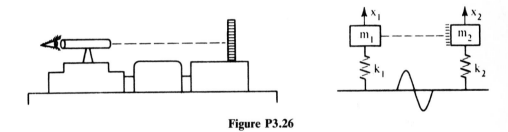

Figure P3.26

3.27. A cam and follower mechanism (Figure P3.27) consists of an eccentric circular disk rotating around P with an angular velocity $\omega = 300$ rad/s. Assuming that the disk radius is $r = 50$ mm and the follower mass 1.3 kg, determine the spring constant k required to maintain contact between the cam and follower at all times if damping can be neglected. Disregard gravity. Suppose that this device is used as a vibrator. Compute the force transmitted to the base under the foregoing condition of continuous contact. Will the force of the vibrator be harmonic?

3.28. An elevator cable (Figure P3.28) is $l = 30$ m long. It is moving upward at a speed of $v = 1.5$ m/s by a pulley A of radius $r = 50$ mm, which, due to nonuniform wear, has an eccentricity of $e = 2$ mm. The mass of the cabin is $m = 300$ kg. Determine **(a)** the vibration amplitude of the cabin at a cable length $L = 100$ m, and **(b)** the length l at which there will be resonance between the rotating speed of the pulley and the system.

3.29. An automobile of mass $m = 1000$ kg is driven at 40 mph on a rough road, assumed to have a sinusoidal surface with maximum bumps (amplitude) 60 mm and distance between them of 0.3 m. If the natural frequency of the car is 0.8 Hz and it is critically damped, determine its vertical vibration assuming that the wheels move simultaneously (base length multiple of the bump distance).

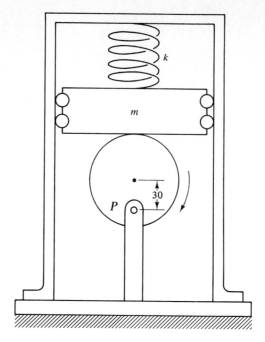

Figure P3.27

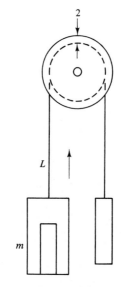

Figure P3.28

3.30. A torsiograph is an instrument that measures torsional vibration. It consists of an inextensible belt B that is wrapped around the shaft S, the torsional vibration that is to be measured, and a light pulley P (Figure P3.30). A torsional spring k connects a heavy disk of inertial J to the pulley. The relative torsional motion of the disk with respect to the pulley shaft is measured by electric means and gives an account of the

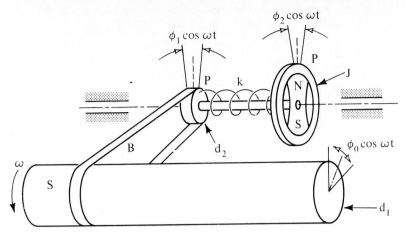

Figure P3.30

torsional vibration of the shaft. Derive a relation between this relative motion and the torsional vibration of the shaft.

3.31. An electric motor of mass $m = 80$ kg is mounted on a steel cantilever beam (Figure P3.31), welded to the vertical bulkhead of the ship, of length $L = 1.2$ m, height $h = 0.01$ m, and width $b = 0.12$ m. The motor is vibrating due to an unbalance. It was measured that the amplitude at resonance was 3.5×10^{-6} m. The plate was covered with a damping material and the vibration amplitude was reduced to 0.8×10^{-6} m. If Young's modulus for steel is $E = 2.06 \times 10^{-11}$ N/m^2 and the loss factor for steel is 0.001, determine the loss factor for the damping material, the equivalent viscous damping in both cases, and the logarithmic decrement.

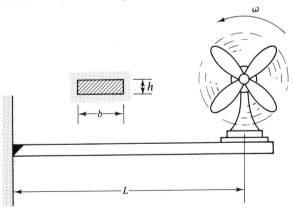

Figure P3.31

3.32. The system of Problem 3.31 developed a fatigue crack at the weld after being operated for a length of time and the vibration amplitude at resonance was increased to 2.1×10^{-6} m. Determine the depth of the crack, knowing that the crack introduces a torsional stiffness at the weld of magnitude $k_T = E(h - \delta)^2/16.67$, where δ is the depth of the crack. Torsional stiffness k_T is per unit width.

3.33. A machine of mass $m = 500$ kg, mounted on a concrete block of mass $M = 1500$ kg, rests on gravel. If the concrete block has a base area of 1.2×0.60 m, determine the resonant frequency and the transmissibility ratio at resonance, assuming that the logarithmic decrement of the system is 0.22.

Section 3.7

3.34. Solve Problem 3.33, assuming that the damping is hysteretic and the loss factor is 0.01.

3.35. A turbine rotor was tested with a centrifugal shaker. The force–deflection curve is shown in Figure P3.35. From the area of the hysteresis loop, determine the material loss factor γ.

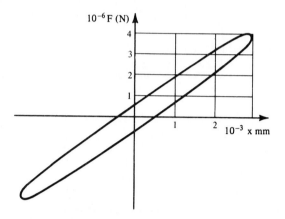

Figure P3.35

3.36. A small aluminum beam was covered with a special viscoelastic coating and tested. The force-deflection curve is shown in Figure P3.36. From the area of the hysteresis loop, determine the material loss factor of the composite beam.

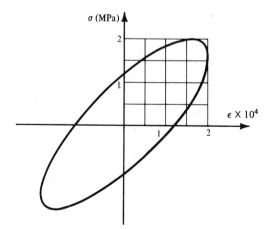

Figure P3.36

3.37. An engineer tested a structure to determine the loss factor. She measured the energy dissipated per cycle from the force–deflection hysteresis loop and plotted it for several vibration amplitudes. She then fitted a second-degree polynomial to the results and found that with good accuracy, the energy dissipated per cycle was related to amplitude as $U = 3X + 500X^2$. If the stiffness of the structure was $k = 380,000$ N/m and the units in the energy expression are N·m and m, interpret the results.

Section 3.8

3.38. A seismic transducer has a natural frequency of 1.2 Hz and a 0.8 damping ratio. Determine the percent error in amplitude measurements and the phase-angle error to measure vibration at **(a)** 12 Hz, and **(b)** 1.2 Hz.

3.39. An accelerometer has a natural frequency of 10 kHz and a damping ratio 0.05. Determine the percent error in amplitude measurements and the phase-angle error to measure vibration at **(a)** 1200 Hz, and **(b)** 11 kHz.

3.40. A seismograph has a natural frequency of 0.2 Hz and it is practically undamped. Determine the maximum percent error in amplitude measurements of an earthquake with a range of the frequency spectrum between 0.3 and 3 Hz.

3.41. Vibration measured with a seismic transducer was found to be 25 μm at a phase angle of 125° and 280 rpm. The seismic transducer has a natural frequency of 1.2 Hz and a 0.8 damping ratio. Apply the appropriate correction to the vibration measurement.

3.42. Vibration measured with an accelerometer was found to be 40 μm at a phase angle of 75° and 8500 rpm. The accelerometer has a natural frequency of 1.2 kHz and 0.1 damping ratio. Apply the appropriate correction to the vibration measurement.

3.43. Vibration measured with a vibration transducer was found to be 0.1 mm at a phase angle of 140° and 3600 rpm. The seismic transducer has a natural frequency of 8 Hz and is critically damped. Apply the appropriate correction to the vibration measurement.

3.44. A refrigeration compressor is mounted on four springs and has mass $m = 55$ kg. Due to the design of the compressor, there is a vertical harmonic force of 12 N oscillating at the operating frequency of 1750 rpm. Design an appropriate mount for the compressor so that the harmonic force transmitted in the vertical direction to the structure will be less than 1 N.

Section 3.9

3.45. A 2000-kg foundation of a machine operating at 3600 rpm is mounted with a footing of 1.2×1.2 m directly on clay. Determine the transmissibility ratio.

3.46. A refrigeration compressor is mounted on four springs 40 mm in diameter and 50 mm long, made of special rubber with $E = 3 \times 10^6$ N/m². Find the transmissibility ratio for vertical vibration if $m = 10$ kg and $\omega = 160$ rad/s.

3.47. A refrigeration compressor is mounted on four springs 40 mm in diameter and 50 mm long, made of special rubber with $E = 3 \times 10^6$ N/m² and damping factor $\gamma = 0.05$. Find the rms value of the transmitted force in the vertical direction if the excitation forces are 180 N at 1800 rpm and 130 N at 3600 rpm and $m = 320$ kg.

3.48. The mounting of a 12-kg electronic instrument was measured to give vertical natural frequency 15 Hz and 0.03 damping factor. Find the maximum acceleration of the instrument if the base vibrates at 25 Hz.

3.49. A turbine-generator is based on a concrete foundation of dimensions 30×100 m founded on clay soil. The total weight is 3000 tons. Determine the maximum acceleration on the turbine-generator to a ground motion of frequency 2.5 Hz and amplitude 10 mm in the longitudinal and 5 mm in the vertical direction.

SIMULAB PROBLEMS

S3.1. Using the LINSYS program of your SIMULAB diskette, find the response of a single-degree-of-freedom system for a harmonic force of amplitude 2 and angular velocity 5 rad/s, with $m = 1$, $k = 1$, $c = 0$.

S3.2. Using the LINSYS program of your SIMULAB diskette, find the response of a single-degree-of-freedom system for a harmonic force of amplitude 2 and angular velocity 5 rad/s, with $m = 1$, $k = 1$, $c = 2$.

S3.3. Using the LINSYS program of your SIMULAB diskette, find the response of a single-degree-of-freedom system for a harmonic force of amplitude 2 and angular velocity 5 rad/s, with $m = 1$, $k = 1$, $c = 2$, and $x_0 = 1$.

S3.4. Using the LINSYS program of your SIMULAB diskette, find the response of a single-degree-of-freedom system for base motion of amplitude 1 and angular velocity 5 rad/s, with $m = 1$, $k = 1$, $c = 1$.

S3.5. Using the LINSYS program of your SIMULAB diskette, find the response of a single-degree-of-freedom system for base motion of amplitude 2 and angular velocity 5 rad/s, with $m = 1$, $k = 1$, $c = 2$, and $x_0 = 1$.

DESIGN PROBLEMS

D3.1. For the mounting of the motor in Problem 3.3, design the beam using carbon steel material with damping factor $\gamma = 0.005$ to limit the response to a 15-μm vibration amplitude.

D3.2. Design the elastic strips of the relay in Problem 3.4 to limit the vibration of the core to 10 μm.

D3.3. Design the hydraulic shaker of Problem 3.9 to deliver a harmonic force of amplitude 100 to 5000 N at frequencies of 5 to 25 Hz.

D3.4. Design the centrifugal shaker of Problem 3.10 to deliver a harmonic force of amplitude 1000 to 100,000 N at frequencies of 25 to 80 Hz.

D3.5. Design the cam–follower shaker of Problem 3.27 to deliver a harmonic force of amplitude 100 to 1000 N at frequencies of 25 to 5 Hz.

D3.6. Design the torsiograph of Problem 3.30 to measure torsional vibration of internal combustion engines in the range 500 to 5000 rpm speed of rotation and up to the third harmonic of the speed of rotation.

D3.7. A machine weighs 25,000 N, rotates at 1500 rpm, and has maximum allowable unbalance 0.01 kg·m. Determine the size of the concrete foundation block, resting directly on gravel, to achieve a transmissibility ratio less than 0.02.

D3.8. Design a velocity transducer to measure machine vibration in the range 1 to 100 Hz with error less than 1% on magnitude and 3% on phase distortion for measured magnitudes 0.1 to 1 mm.

D3.9. Design an accelerometer to measure 10 to 10,000 Hz vibration of magnitude 0.1 to 2 g, with error less than 1% on magnitude and 3% on phase distortion.

D3.10. The dc voltmeter of an ac-to-dc converter has a steel dial of length $l = 50$ mm and a restoring spring and a damper, both rotational. The maximum indication is 100 V at 50° angle. On the dc voltage, there is a ripple due to the noncomplete rectification of the ac. Design the dial system so that the ripple of magnitude 5 V will not result in vibration of the dial of more than 1° at 100 V dc and the frequency is 60 Hz (see Figure P2.52).

4

Energy
Methods:
Stability

4.1 CONSERVATION OF ENERGY

As any natural science, mechanics is founded on a set of principles and axioms the universal validity of which has been established by experience. Every additional piece of knowledge is based on these axioms through mathematical analysis.[1] The derivation of the differential equations of motion in Chapters 2 and 3 was based on Newton's laws, as basic axioms, and balance of forces and rates of change of momenta. An alternative used extensively in dynamics is the energy balance. Using the axiom that energy is conserved, the equations of motion of a system are determined by observing its energy dealings with its surroundings and within itself.

Newton's second law for a particle acted on by force $\mathbf{F}$ can be written in the form $\mathbf{F} = m\ddot{\mathbf{r}}$. Forming the dot product of each side with the displacement $d\mathbf{r}$ and integrating [Housner and Hudson, 1950], we obtain the first law of thermodynamics:

*The above photo portrays unstable oscillations of the Tacoma Narrows Bridge.

[1]"Newton admits nothing but what he gains from experiments and accurate observations. From this foundation, whatever is further advanced, is deducted by strict mathematical reasoning" (Emerson, W., 1754, *The Principles of Mechanics*).

$$\int_{r_1}^{r_2} \mathbf{F} \cdot d\mathbf{r} = \int_{r_1}^{r_2} m\ddot{\mathbf{r}} \cdot d\mathbf{r} = \int_{t_1}^{t_2} m\ddot{\mathbf{r}} \cdot \frac{d\mathbf{r}}{dt} dt = \frac{1}{2}m \int_{t_1}^{t_2} \frac{d(\dot{\mathbf{r}} \cdot \dot{\mathbf{r}})}{dt} dt$$
$$= \frac{1}{2}m \int_{t_1}^{t_2} d(v^2) = \tfrac{1}{2}mv_2^2 - \tfrac{1}{2}mv_1^2 \tag{4.1}$$

The integral on the left is called *work*, done by the force $\mathbf{F}$, and the term $\frac{1}{2}mv^2$ is called *kinetic energy*. The right-hand side of equation (4.1) depends only on the velocities of the particle at the two endpoints 1 and 2. The value of the left side, in general, will depend on the path of integration unless $\mathbf{F} \cdot d\mathbf{r}$ is an exact differential. If this is so, there exists some function V such that $V = \int \mathbf{F} \cdot d\mathbf{r}$. In terms of this function, called the *potential energy function*, equation (4.1) can be written in the form

$$V_1 + \tfrac{1}{2}mv_1^2 = V_2 + \tfrac{1}{2}mv_2^2 \tag{4.2}$$

expressing the fact that the sum of the potential and kinetic energy remains constant in systems for which the function V exists.[2] The system energy is said to be conserved and the system is *conservative*.

The work expression $\mathbf{F} \cdot d\mathbf{r}$ for a conservative system is compatible with a system of forces $F_x = \partial V/\partial x$, $F_y = \partial V/\partial y$, $F_z = \partial V/\partial z$, which can thus be obtained from the potential energy function. Indeed, many vibrating systems do not exchange energy with their surroundings. An example of this is an undamped harmonic oscillator. At any time the energy of such an oscillator has two parts: the kinetic energy $T = \tfrac{1}{2}m\dot{x}^2$ and the potential energy $V = \tfrac{1}{2}kx^2$, which is stored in the spring as strain energy. The principle of conservation of energy, equation (4.2), demands that

$$\frac{d}{dt}\left(\frac{1}{2}m\dot{x}^2 + \frac{1}{2}kx^2\right) = m\dot{x}\ddot{x} + kx\dot{x} = 0 \tag{4.3}$$

Canceling $\dot{x}$ we obtain, as expected, the equation for the harmonic oscillator:

$$m\ddot{x} + kx = 0 \tag{4.4}$$

In arriving this way at equation (4.4), we did not have to deal with vectors and their directions and signs. Often, these are sources of error. Moreover, for the application of Newton's law, the motion of the mass center must be used as the coordinate. Application of the conservation-of-energy method can be applied with any coordinate system, provided that kinetic and potential energies can be expressed as functions of these coordinates. Moreover, forces of constraint do not, in general, produce work; therefore, they are not involved in the equations of energy directly, as is the case with direct application of Newton's law, where forces of constraint have to be taken into account and then eliminated. These forces are treated as unknown quantities, and more equations than the number of degrees of freedom have to be developed to eliminate the unknown forces. This point is exemplified in the example that follows.

[2]Alexander of Afrodisias (early third century B.C.). *Commentary on Aristoteles' "Metaphysics."* Translated by Sir Thomas Heath. London: J. M. Dent. See Historical Introduction.

□ **Example 4.1 Soil Compacting Roller**

A rolling soil compacter consists of a circular cylinder of mass m and radius r that is connected to the prime mover by a spring of constant k as shown in Figure E4.1. Find its frequency of oscillation, assuming that it is free to roll on the rough horizontal surface without slipping.

Solution *Energy method:* We will use the fact that the total energy of the system consists of both kinetic energy (KE) (rotational and translational) and potential energy (PE) and should constantly remain the same.

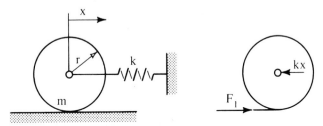

Figure E4.1

The translational KE $= \frac{1}{2}m\dot{x}^2$ and the rotational KE $= \frac{1}{2}J_0\dot{\theta}^2$, where the moment of inertia of the cylinder is $J_0 = \frac{1}{2}mr^2$. Also, $r\theta = x$, or $r\dot{\theta} = \dot{x}$; thus for the system at any time,

$$KE = \frac{1}{2}m\dot{x}^2 + \frac{1}{2}\left[\left(\frac{1}{2}mr^2\right)\left(\frac{\dot{x}}{r}\right)^2\right] = \frac{3}{4}m\dot{x}^2$$

$$PE = \frac{1}{2}kx^2$$

and

$$\frac{d}{dt}(KE + PE) = 0 \text{ or } \left(\frac{3}{2}m\ddot{x} + kx\right)\dot{x} = 0$$

Since $\dot{x}$ is not always zero, the equation of motion becomes

$$\frac{3}{2}m\ddot{x} + kx = 0$$

so

$$\omega_n = \left(\frac{2k}{3m}\right)^{1/2}$$

Use of Newton's second law of motion: Applying Newton's law of motion to the cylinder in respect to the center of mass,

$$F = ma$$

or

$$m\ddot{x} = -kx + F_f$$

where F_f is the friction force, yet unknown.
Using the torque equation, $\Sigma M = J_0\dot{\theta}$,

$$J_0\ddot{\theta} = -F_f r$$

or

$$\left(\frac{1}{2}mr^2\right)\left(\frac{\ddot{x}}{r}\right) = -F_f r$$

and hence $F_f = -\frac{1}{2}m\ddot{x}$. We substitute this expression for F_f into the force equation to obtain

$$m\ddot{x} = -kx - \tfrac{1}{2}m\ddot{x}$$

or

$$\tfrac{3}{2}m\ddot{x} + kx = 0$$

so

$$\omega_n = \left(\frac{2k}{3m}\right)^{1/2}$$

(This example demonstrates the alternative routes to obtaining the equations of motion of a conservative system through Newton's second law and the first law of thermodynamics. Moreover, the elimination of forces of constraint in constrained systems is demonstrated.)

☐

4.2 RAYLEIGH'S METHOD[3]

An undamped harmonic oscillator having a vibration amplitude x_0 is shown in Figure 4.1. The system is shown at times $T/4$ apart, where T is the period of the

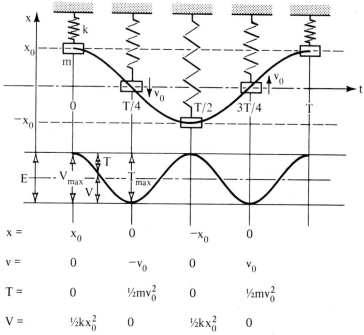

Figure 4.1 Energy exchange in an undamped harmonic oscillator.

[3]Lord Rayleigh. June 1873. *Proc. Math. Soc. London.*

Portrait of Rayleigh. (Courtesy of
the Library of Congress.)

LORD RAYLEIGH (1842–1919)

Rayleigh, born John William Strutt, the eldest son of the second Baron
Rayleigh of Terling Place, Essex, is a rare example of a nobleman who
devoted his life to science. He entered Trinity College at 20, where he
studied mathematics under E. J. Routh and profited greatly from the Cambridge lectures of G. G. Stokes. In 1868 he was elected a fellow of Trinity
College. His first experimental investigations were in electricity, but soon
he turned to acoustics and vibration. He started writing *Theory of Sound*
on a boat trip up the Nile in 1872 and the book was published in 1877. He
succeeded Clerk Maxwell in the Cavendish professorship in 1879. He became President of the Royal Society and Chancellor of Cambridge in 1908
and served until his death. He won the Nobel Prize in Physics in 1904. He
was instrumental in the promotion of experimental research in England.

oscillation, starting when the mass is at its highest position. It is apparent that the
equation of motion is $x = x_0 \cos \omega t$. At the position $t = 0$ the system has maximum
potential energy $V_{max} = \frac{1}{2} k x_0^2$ and zero kinetic energy. At $t = T/4$ the potential
energy is zero and the kinetic energy is maximum (because the velocity is maxi-
mum), $T_{max} = \frac{1}{2} m v_0^2$. But $v_0 = \omega x_0$; therefore, $T_{max} = \frac{1}{2} m \omega^2 x_0^2$. Since no energy

is transmitted to the environment,

$$T_{\max} = V_{\max} \quad \text{or} \quad \tfrac{1}{2} m \omega^2 x_0^2 = \tfrac{1}{2} k x_0^2 \tag{4.5}$$

Therefore, $\omega^2 = k/m$, as we expected. This result shows two things. First, it shows that the natural frequency is independent of the amplitude x_0. Second, the natural frequency can be determined without first finding and solving the differential equations of motion.

From Figure 4.1 one can see that the potential and kinetic energy oscillate from 0 to $V_{\max} = T_{\max}$ with frequency 2ω and with an $180°$ phase angle. Their sum is always constant. Thus an oscillation can be described as a continuous conversion of energy process.

4.3 LAGRANGE'S EQUATION

The conservation of energy and the Rayleigh method are limited to conservative systems. In many engineering systems, there are forces that result in energy dissipation. To handle such systems, our energy equation (4.2) has to be replaced. To this end we start from Newton's law, $m\ddot{x} = F$, where F is the sum of all the external forces in the direction of x. Since we do not want to confine our analysis to the motion x of the center of mass, we assume any coordinate q that can describe the position of the particle. Of course, we cannot write $m\ddot{q} = F$. To find the differential equation in terms of the dependent variable q, we proceed as follows,[4] confining our analysis to a system for which a relationship $x = f(q)$ exists: more specifically, to systems for which the number of degrees of freedom is equal to the number of independent coordinates necessary to describe the system configuration, called *holonomic*. By virtue of Newton's second law we can write $m\ddot{x} = F$, or

$$m\ddot{x}\,\frac{\partial f}{\partial q} = F\,\frac{\partial f}{\partial q} \tag{4.6}$$

Because $\dot{x}$ is a function of q and t, we obtain

$$\frac{\partial \dot{x}}{\partial \dot{q}} = \frac{\partial}{\partial \dot{q}} \left(\frac{\partial f}{\partial q} \dot{q} + \frac{\partial f}{\partial t} \right) \tag{4.7}$$

Because $f(q)$ does not depend explicitly on t, $\partial f/\partial t = 0$. Therefore,

$$\frac{\partial \dot{x}}{\partial \dot{q}} = \frac{\partial}{\partial \dot{q}} \left(\frac{\partial f}{\partial q} \dot{q} \right) = \frac{\partial f}{\partial q} \tag{4.8}$$

Then we can write

$$
\begin{aligned}
\ddot{x}\,\frac{\partial f}{\partial q} &= \ddot{x}\,\frac{\partial \dot{x}}{\partial \dot{q}} = \frac{d}{dt}\left(\dot{x}\,\frac{\partial \dot{x}}{\partial \dot{q}} \right) - \dot{x}\,\frac{d}{dt}\left(\frac{\partial x}{\partial q} \right) \\
&= \frac{d}{dt}\left(\dot{x}\,\frac{\partial \dot{x}}{\partial \dot{q}} \right) - \dot{x}\,\frac{\partial \dot{x}}{\partial q} \\
&= \frac{d}{dt}\left[\frac{\partial}{\partial \dot{q}}\left(\frac{1}{2}\dot{x}^2 \right) \right] - \frac{\partial}{\partial q}\left(\frac{1}{2}\dot{x}^2 \right)
\end{aligned}
\tag{4.9}
$$

[4]Lagrange, J. L. 1765. *Mécanique analytique*. Paris. See also [Whittaker, 1904].

For $T = \frac{1}{2}m\dot{x}^2$, equations (4.8) and (4.9) give

$$\frac{d}{dt}\left(\frac{\partial T}{\partial \dot{q}}\right) - \frac{\partial T}{\partial q} = F\frac{\partial f}{\partial q} \qquad (4.10)$$

The right-hand side of equation (4.10) can be written in the form

$$F_q = F\frac{\partial f}{\partial q} \qquad (4.11)$$

This is equivalent to $\partial W/\partial q$, the gradient of the work of the external forces with respect to the coordinate q.

Some of the forces can be expressed by a potential, such as spring reactions. Such forces are gradients of the potential

$$F = \frac{-\partial V}{\partial q} \qquad (4.12)$$

The remaining forces are designated as F_{qn} and Lagrange's equation is written as

$$\frac{d}{dt}\left(\frac{\partial T}{\partial \dot{q}}\right) - \frac{\partial T}{\partial q} + \frac{\partial V}{\partial q} = F_{qn} \qquad (4.13)$$

Because the potential energy does not depend explicitly on time, we can write

$$\frac{d}{dt}\left(\frac{\partial L}{\partial \dot{q}}\right) - \frac{\partial L}{\partial q} = F_{qn} \qquad (4.14)$$

here $L = T - V$ is the *Lagrangian function* or *kinetic potential*.

Rayleigh introduced the dissipation function $D = \frac{1}{2}c\dot{q}^2$ to include linear damping forces proportional to the velocity $\dot{q}$. In terms of this function, Lagrange's equation can be written in the form

$$\frac{d}{dt}\left(\frac{\partial L}{\partial \dot{q}}\right) - \frac{\partial L}{\partial q} + \frac{\partial D}{\partial \dot{q}} = F_{qn} \qquad (4.15)$$

For the determination of F_{qn}, the work of the remaining external forces is expressed as function of q and $\dot{q}$ and differentiated. In the expression of work, only these forces that do work need be considered. Thus the particle at position q has velocity $\dot{q}$. A small displacement δq will result in work, as given by the expression $\{\cdot\}\delta q$. The expression in braces is the generalized force.

There are several classes of forces that occur frequently in vibration systems that do no work on the system during the motion [Whittaker, 1904]:

1. The reactions of fixed smooth surfaces, *smooth* meaning that the reaction is normal to the surface. This is the case when the friction force can be neglected during sliding (coefficient of friction $\mu \rightarrow 0$).
2. The reactions of perfectly rough surfaces, *rough* meaning that slipping is not possible and that only rolling is encountered ($\mu \rightarrow \infty$).
3. The mutual reaction of two particles or rigid bodies connected together rigidly.

Portrait of Lagrange. (Courtesy of
the Bibliothèque Nationale, Paris.)

JOSEPH-LOUIS LAGRANGE (1736–1813)

Born in Turin, where at a young age he became professor of mathematics
at the military academy, Lagrange was founder of the Turin Academy,
where he published important contributions in the calculus of variations.
Euler nominated him as a foreign member of the Berlin Academy, where
he later succeeded Euler as a full member, and he moved to Berlin in
1766. In 1788, he published his *Mécanique analytique* after years of searching for publisher, due to its advanced level. In 1787, he moved to Paris,
where he taught at the Ecole Polytechnique until his death.

4. The internal stresses in a rigid body, *rigid* meaning that the relative distance
   of its particles remains invariant during motion.

5. The reaction at a fixed pivot about which one rigid body can rotate with
   respect to another without friction.

6. An interesting case is the contact of two rigid bodies with a finite friction
   coefficient μ at the interface. If the driving force is less than the friction force,
   there is no sliding and the friction force does no work. If the driving force
   exceeds the friction force, sliding will occur and the friction force does work
   on the system.

The forces described above, called *forces of constraint*, do not appear explicitly in Lagrange's equations and their determination is the subject of *kinetostatics*.

Following are some illustrations of the application of Lagrange's equation.

1. *Viscous damper of constant c* (Figure 4.2):

$$q = x$$

$$T = \tfrac{1}{2}m\dot{x}^2, \qquad V = \tfrac{1}{2}kx^2, \qquad U_{x \to x + \delta x} = \{-c\dot{x}\}\delta x, \qquad F_{qn} = -c\dot{x}$$

Alternatively, $D = \tfrac{1}{2}c\dot{q}^2$ and $\partial D/\partial \dot{q} = c\dot{q}$.

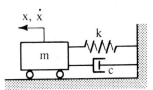

Figure 4.2 System with viscous damper of constant c.

2. *Sliding under dry friction of coefficient μ* (Figure 4.3):

$$q = x, \qquad W = mg$$

$$T = \tfrac{1}{2}m\dot{x}^2, \qquad V = \tfrac{1}{2}kx^2, \qquad U_{x \to x + \delta x} = -\mu W|\delta x| = \{-\mu W \operatorname{sgn}(x)\}\,\delta x$$

$$F_{qn} = -\mu W \operatorname{sgn}(x)$$

(*Note*: The displacement δq or $-\delta x$ must be *virtual*, in other words, a possible or feasible displacement of the system without violation of the constraints. If the velocity is positive in this example, x must be positive. If $x < 0$, the work must be computed for $\delta x < 0$.)

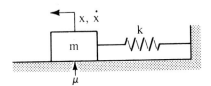

Figure 4.3 System with dry friction of coefficient μ.

3. *Constrained motion* (Figure 4.4):

(a) $q = x, \ \omega = \dfrac{\dot{x}}{r}$,

$$T = \tfrac{1}{2}m\dot{x}^2 + \tfrac{1}{2}I\omega^2 = \tfrac{1}{2}(m + I/r^2)\dot{x}^2,$$

$$V = \tfrac{1}{2}kx^2, \ U_{x \to x + \delta x} = \{-c\dot{x}\}\delta x, \ F_x = -c\dot{x}$$

(b) $q = \theta, \ T = \tfrac{1}{2}I\dot{\theta}^2 + \tfrac{1}{2}mr^2\dot{\theta}^2, \ V = \tfrac{1}{2}kr^2\theta^2,$

$$U_{\theta \to \theta + \delta\theta} = -M\delta\theta = \{-cr^2\dot{\theta}\}\,\delta\theta, \ F_\theta = -cr^2\dot{\theta}$$

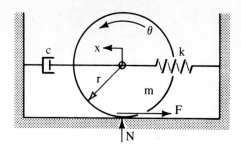

Figure 4.4 Constrained motion.

□ **Example 4.2   Rectilinear and Rotational Motion**

Determine the differential equation of motion of the system in Example 2.4.
Solution

Generalized coordinate: the angle θ
Geometric relations: $y = r(1 - \cos \theta)$
Kinetic energy: $T = \frac{1}{2}J_0\dot{\theta}^2$
Potential energy: $V = mgy$
Lagrangian:

$$L = T - V = \tfrac{1}{2}J_0\dot{\theta}^2 - mgr(1 - \cos \theta)$$

$$\frac{\partial L}{\partial \dot{\theta}} = J_0\dot{\theta}$$

$$\frac{\partial L}{\partial \theta} = -mgr \sin \theta$$

$$\frac{d}{dt}\left(\frac{\partial L}{\partial \dot{\theta}}\right) - \frac{\partial L}{\partial \theta} = J_0\ddot{\theta} + mgr \sin \theta = 0$$

Small-angle approximation: $\sin \theta \approx \theta$
Equation of motion: $J_0\ddot{\theta} + mgr\theta = 0$

But

$$J_0 = \frac{1}{2}MR^2 - \frac{1}{2}mr^2 - m\bar{r}^2 = mr^2\left[\frac{1}{2}\frac{M}{m}\left(\frac{R}{r}\right)^2 - \left(\frac{\bar{r}}{r}\right)^2 - \frac{1}{2}\right]$$

Therefore,

$$r^2\left[\frac{1}{2}\left(\frac{R}{r}\right)^4 - \frac{1}{2} - \left(\frac{\bar{r}}{r}\right)^2\right]\ddot{\theta} + g\bar{r}\theta = 0$$

(This example demonstrates the mathematical modeling of a single-degree-of-freedom system that has rotational and translational motion through Lagrange's equation.)

□

□ Example 4.3 Constrained Rectilinear and Circular Motion

Determine the differential equation of motion for the system shown in Figure E4.3. This system consists of a cylinder of mass M and radius r that rolls without slipping on a rail R. It carries a mass m by way of a light arm of length l, which is rigidly attached to the center of the cylinder as shown.

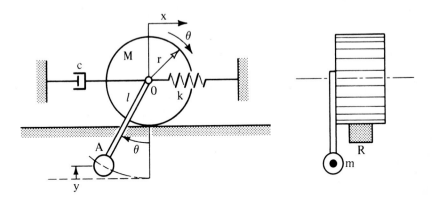

Figure E4.3

Solution

Geometric relations: $x = r\theta$, $y = l(1 - \cos\theta)$

Kinematic relations:

$$\dot{x} = r\dot{\theta}, \qquad u_A = u_0 - r_{OA}\dot{\theta}$$

or

$$|u_A|^2 = \dot{x}^2 + (\dot{\theta}l)^2 - 2\dot{x}(\dot{\theta}l)$$

Kinetic energy:

$$T = \tfrac{1}{2}M\dot{x}^2 + \tfrac{1}{2}J_0\dot{\theta}^2 + \tfrac{1}{2}mu_A^2$$

$$= \frac{1}{2}M\dot{x}^2 + \frac{1}{2}\left(\frac{1}{2}Mr^2\right)\frac{\dot{x}^2}{r^2} + \frac{1}{2}m\left[\dot{x}^2 + \left(\frac{l}{r}\right)^2\dot{x}^2 - \frac{2l}{r}\dot{x}^2\right]$$

$$= \frac{1}{2}\left[\frac{3}{2}M + m\left(1 - \frac{l}{r}\right)^2\right]\dot{x}^2$$

Potential energy:

$$V = mg\,y + \frac{1}{2}kx^2 = mgl\left(1 - \cos\frac{x}{r}\right) + \frac{1}{2}kx^2$$

Generalized force: $F_x = -c\dot{x}$

Lagrangian function:

$$L = T - V$$

$$\frac{\partial L}{\partial \dot{x}} = \left[\frac{3}{2} M + m \left(1 - \frac{l}{r} \right)^2 \right] \dot{x}$$

$$\frac{\partial L}{\partial x} = -mg \frac{l}{r} \sin \frac{x}{r} - kx$$

$$\frac{d}{dt} \left(\frac{\partial L}{\partial \dot{x}} \right) - \frac{\partial L}{\partial x} = \left[\frac{3}{2} M + m \left(1 - \frac{l}{r} \right)^2 \right] \ddot{x} + kx + \frac{l}{r} mg \sin \frac{x}{r} = -c\dot{x}$$

Small-angle approximation:

$$\sin \theta \approx \theta = \frac{x}{r}$$

Differential equation of motion:

$$\left[\frac{3}{2} M + m \left(1 - \frac{l}{r} \right)^2 \right] \ddot{x} + c\dot{x} + \left(k + \frac{lmg}{r^2} \right) x = 0$$

(This example demonstrates the mathematical modeling of a single-degree-of-freedom system with constraints that has rotational and translational motion through Lagrange's equation.)

□

4.4 STABILITY

From statics we know that the equilibrium of a system can be *structurally stable* or *unstable*, and the potential energy has a minimum or maximum, respectively. For one-degree-of-freedom systems, depending on the coordinate x, if the expression for the potential energy is $V(x)$, we can state that

$$\text{minimum} \rightarrow \frac{dV}{dx} = 0 \quad \text{and} \quad \frac{d^2V}{dx^2} > 0 \rightarrow \text{stability} \tag{4.16}$$

$$\text{maximum} \rightarrow \frac{dV}{dx} = 0 \quad \text{and} \quad \frac{d^2V}{dx^2} < 0 \rightarrow \text{instability} \tag{4.17}$$

The *threshold of instability* is at $d^2V/dx^2 = 0$ and separates the regions of stability and instability.

In an undamped harmonic oscillator, expressed by the equation

$$m\ddot{x} + kx = 0 \tag{4.16}$$

the potential energy is

$$V(x) = \tfrac{1}{2} kx^2 \tag{4.17}$$

The equilibrium is at $x = 0$. The threshold of instability is therefore given by

$$\frac{dV}{dx} = kx = 0 \tag{4.18}$$

$$\frac{d^2V}{dx^2} = k = 0 \tag{4.19}$$

Equation (4.18) is always satisfied at $x = 0$. Equation (4.19) is satisfied for $k = 0$ only. In a harmonic oscillator, therefore, instability is implied by $k \leq 0$.

Physically, $k = 0$ means that there is no spring to control the motion. Once the mass is given a disturbance x_0 or v_0 from equilibrium, it will never return to equilibrium. Another definition of stability is that once a system is disturbed from equilibrium, it tends to return to the equilibrium position.

Since $k = 0$ implies structural instability in this case, we observe that $\omega_n = \sqrt{k/m} = 0$. Thus when natural frequency equals zero, it is implied that the system is at the threshold of instability. The value $k = 0$ separates the stable region ($k > 0$) and the unstable region ($k < 0$).

☐ **Example 4.4 Inverted Pendulum Seismograph**

Investigate the stability of a seismograph that consists of the inverted pendulum of Figure E4.4. Applied forces are the weight, the reaction of the spring, and the reaction at O.

Solution Taking moments about O and applying Newton's law, we obtain

$$mL^2\ddot{\theta} = -ka \sin \theta \, a \cos \theta + mgL \sin \theta$$

$$\ddot{\theta} + \left(\frac{ka^2}{mL^2} - \frac{g}{L}\right)\theta = 0$$

$$\omega_n = \sqrt{\frac{ka^2}{mL^2} - \frac{g}{L}}$$

$g/L = ka^2/mL^2$ implies that $\omega_n = 0$, which indicates unstable motion. Physically, it indicates that the pendulum will slowly start swinging downward without returning.

(This example demonstrates the properties of the inverted pendulum used as a seismograph, since by adjusting the spring stiffness one can make the natural frequency very small, as needed for measuring earthquakes, usually below 1 rad/s. To use a regular pendulum for this purpose, we need a length $l = g\omega^2 = 9.81$ m, which is impractical.)

☐

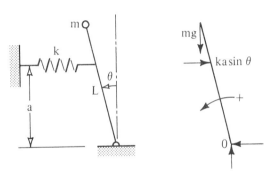

Figure E4.4

4.5 DYNAMIC INSTABILITY: SELF-EXCITED VIBRATION

Structural instability is usually not of particular concern in vibrating systems since this type of instability is ruled out at the static design stage. A far more dangerous situation develops when there is a mechanism of supplying energy into the system, as it vibrates. The vibrations are sustained, then amplified, until they reach dangerous and sometimes destructive levels. The flutter of airplane wings, instability of rotating shafts, and flow-induced vibration in nuclear power plants are typical examples of this type of vibration. Vibrations of this type are known as *self-excited vibrations*.

In this chapter our study of self-excited vibration will be confined to linear one-degree-of-freedom systems (m, c, k) expressible by an ordinary linear differential equation of the second order. We start our study with the behavior of the solution of such an equation.

The nonhomogeneous problem expressed by equation (3.2) and the initial conditions (3.3) is solved completely if the functions $x_s(t)$ and $x_h(t)$ are determined. Of these functions, $x_s(t)$ depends on the function $f(t)$. For the examples studied here, the function $x_s(t)$ is periodic and remains bounded; that is, the value of the function remains below a certain level at all times. The same is not true for the function $x_h(t)$. Due to exponential terms, under certain circumstances, $x_h(t)$ could increase continuously with time beyond any bound.

To study the circumstances that give rise to instability, we examine the behavior of the solution $x_h(t)$ of the homogeneous problem. Without loss in generality, we can assume that $m > 0$. [If it is not, we multiply equation (3.6) by -1 so that the mass multiplying the second derivative is positive.] The following observations are made:

1. If the initial conditions are $x_0 = 0$, $v_0 = 0$, the solution to $x_h(t)$ is zero for all t.

2. If $x_0 \neq 0$ or $v_0 \neq 0$, the behavior of the solution depends on the value of m, c, and k.

3. In the case $c = 0$, the solution is given by equation (2.16) and is bounded ($\sin \omega_n t$ and $\cos \omega_n t$ cannot become greater than 1).

4. If $c^2 < 4km$, the solution is equation (2.50). We observe that $e^{-c/2m \, t}$ multiplies a bounded function of time. The behavior therefore depends on the value of $e^{-c/2m \, t}$. We can see immediately that $m > 0$, $c > 0$ implies stability because the function $e^{-c/2m \, t}$ decreases with time. In the case $c < 0$, however, the exponential function increases continuously with time and can become greater than any preset value for a sufficient period of time.[5] It might be argued that with zero for initial conditions, the solution remains zero no matter what c is. In physical systems, however, there are always small disturbances, which give rise to small initial conditions. If $c < 0$, once x_0 or v_0 are given even a small value, the function $x_h(t)$ increases continuously with time beyond bound. This is related to the physical definition of stability. A system is stable if once it has been displaced from its equilibrium, it tends to return to equilibrium.

[5]Lord Rayleigh. 1883. On maintained vibration. *Philos. Mag.*

5. $c^2 > 4mk$. We observe that the exponents are

$$-\frac{c}{2m} \pm \left[\left(\frac{c}{2m}\right)^2 - \frac{k}{m}\right]^{1/2}$$

which are always negative if $c > 0$. The system is therefore stable. When $c < 0$, the system is unstable, because

$$-\frac{c}{2m} + \left[\left(\frac{c}{2m}\right)^2 - \frac{k}{m}\right]^{1/2} > 0$$

6. We observe that $c = 0$ is the value that separates the regions of stability and instability. For $c = 0$, the function becomes neither zero nor unbounded, but rather, oscillates between two values. This value of c is the threshold of instability.

7. From what has been said up until now, it appears that stability depends on the value of the coefficients m, c, and k of the homogeneous system.

In conclusion, we can state that for an (m, c, k) system with $m > 0$, instability is implied either by $c < 0$ or $k < 0$.

☐ **Example 4.5   Instability of Fluid-Carrying Pipes**

A steel pipe of length l and flexural rigidity EI is simply supported at its two ends (Figure E4.5). It has negligible mass compared with a mass m on its midspan. There is a fluid flowing with velocity V through the pipe. Determine the velocity V at which the system will be unstable.

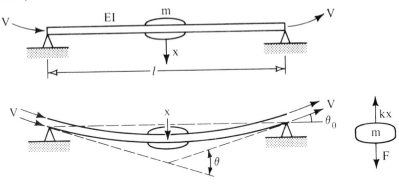

Figure E4.5

Solution   If the mass moves downward by x, the relative angle between the two ends of the pipe will be

$$\theta = 2\theta_0 = \frac{6x}{l}$$

On the mass m there will be a spring force kx, where $k = 48EI/l^3$, and a force due to change in direction of the flow,

$$F = 2A\rho V^2 \sin \theta_0$$

where A is the cross-sectional area of the pipe and ρ the fluid density. For small θ,

$$F = A\rho V^2\theta = \frac{6A\rho V^2}{l}x$$

The equation of motion, therefore, will be

$$m\ddot{x} + kx = F$$

or

$$m\ddot{x} + \left(\frac{48EI}{l^3} - \frac{6A\rho V^2}{l}\right)x = 0$$

When the coefficient of x becomes negative, we shall have instability for

$$V > \left(\frac{8EI}{A\rho l^2}\right)^{1/2}$$

(This example demonstrates the structural instability in fluid-carrying pipes at high flow velocities.)

☐

☐ **Example 4.6   Stability of Flow Valves**

Investigate the stability of a stop valve as shown in Figure E4.6a and b in the open position. Assume that there is a constant flow Q of viscous fluid of viscosity μ and density ρ. The pressure drop in the clearance is given by the formula

$$\Delta p = \frac{12L\mu}{h^3}\left(\frac{Uh}{2} + \frac{Q}{W}\right)$$

where W is the perimeter and h the clearance. This is valid for a plate moving over a plane surface, as shown in Figure E4.6c.

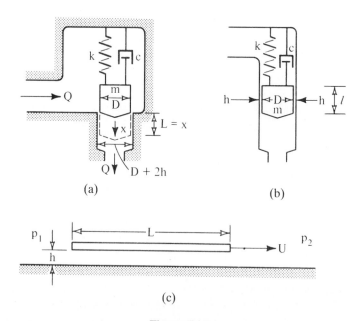

(a)

(b)

(c)

Figure E4.6

Solution   (a) If the stem moves downward by x, we shall have $L = x$, $W = \pi D$. Disregarding the effect of the velocity U, we have

$$\Delta p = \frac{12x\mu}{h^3} \frac{Q}{\pi D}$$

Due to the pressure difference, there will be a downward force

$$F = A\,\Delta p = \frac{\pi D^2}{4} x \frac{12\mu Q}{\pi D} = 3\mu DQx$$

The equation of motion will be

$$m\ddot{x} + c\dot{x} + kx = 3\mu DQx$$

or

$$m\ddot{x} + c\dot{x} + (k - 3\mu DQ)x = 0$$

The coefficient of x will become negative for $k - 3\mu DQ < 0$. Instability will therefore occur at a flow

$$Q_{\text{crit}} = \frac{k}{3\mu D}$$

(b) In this configuration, the force does not depend on x, but rather, on $U = \dot{x}$. We have

$$\Delta p = \frac{12L\mu}{h^3}\left(\frac{h}{2}\dot{x} + \frac{Q}{\pi D}\right)$$

The equation of motion is

$$m\ddot{x} + c\dot{x} + kx = \frac{3\pi D^2 L}{2h^2}\dot{x} + \frac{3DL\mu Q}{h^3}$$

or

$$m\ddot{x} + \left(c - \frac{3\pi D^2 L\mu}{2h^2}\right)\dot{x} + kx = \frac{3DL\mu Q}{h^3}$$

For stable operation, the coefficient of $\dot{x}$ must be positive. For stable operation, therefore, we need an external damping device of damping constant

$$c \geq \frac{3\pi D^2 L\mu}{2h^2}$$

(This example demonstrates the instability of different designs of stop valves at high flow velocities and the design value of external damping.)   □

4.6 PARAMETRICALLY EXCITED VIBRATION

Up to this point we have considered vibration that can be described mathematically by a second-order linear differential equation with constant coefficients. This has a sound physical basis, because in most physical situations these constant coefficients correspond to properties such as mass, stiffness, and damping. These properties usually remain unchanged with time. There are situations where this feature does not exist, but these properties are functions of time. In this case, the problem is still linear, but the solutions we have presented are not valid.

Consider, for example, a simple pendulum of mass m and length l. This is anchored at a point A that is not stationary but moves vertically (Figure 4.5). Let $\mathbf{a}_A$ be acceleration of point A. The acceleration of point B will be[6]

$$\mathbf{a}_B = \mathbf{a}_A + \mathbf{a}_{B/A} = \mathbf{a}_A + \mathbf{a}_t + \mathbf{a}_r \tag{4.20}$$

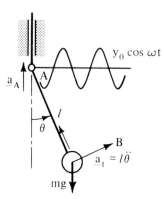

Figure 4.5 Pendulum with moving pivot.

If we apply Newton's law about the point A' of the stationary coordinate system, coinciding with A at time t, we obtain

$$J_A\ddot{\theta} + ma_Al \sin\theta = -mgl \sin\theta \tag{4.21}$$

For small oscillations, we assume that $\sin\theta \approx \theta$. Assume, in addition, that point A has a vertical harmonic motion $y = y_0 \cos t$. Thus $a_A = \ddot{y} = -\omega^2 y_0 \cos\omega t$,

$$\ddot{\theta} + \frac{g}{l}\left(1 - \frac{\omega^2}{g} y_0 \cos\omega t\right)\theta = 0 \tag{4.22}$$

The coefficient of θ is not constant, but rather is a function of time. Equation (4.22) has a particular solution $\theta = 0$ (equilibrium). Furthermore, this equation takes the form

$$y'' + (a - 2q \cos 2\pi t)y = 0 \tag{4.23}$$

where $a = 4g/l\omega^2$, $q = 2y_0/g$, which is the Mathieu equation. This equation can be solved by using the tabulated Mathieu functions.[7]

One important property of a system expressed by the Mathieu equation is that instabilities are exhibited for certain ranges of the parameters a and q.[8] In

---

[6]Beer, F., and Johnston, E. 1984. *Vector Mechanics for Engineers*. New York: McGraw-Hill.

[7]Treffetz, E. 1930. *Aachener Voträge 1929*. Berlin: Julius Springer. For an extensive tabulation, see Abramowitz, M., and Stegun, I. A., 1964. *Handbook of Mathematical Functions*. Washington, D.C.: U.S. Department of Commerce.

[8]Lord Rayleigh. 1831. *Philos. Trans. London*. Also, 1887. On the maintainance of vibrations by forces of double frequency. *Philos. Mag.*, Aug.

other words, if the system is displaced from the equilibrium $y = 0$, a motion of increasing amplitude is followed. This can be verified by the reader by moving a simple pendulum up and down with several amplitudes or frequencies until this instability is experienced. The instability regions are shown in the shaded areas of the chart shown in Figure 4.6.

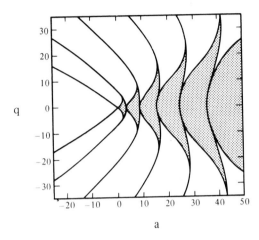

Figure 4.6 Mathieu equation stability chart.

In some systems, the spring stiffness does not change harmonically, as in the case of the Mathieu function, but in another periodic way. Such is the case with a massless cracked cantilever beam supporting a mass when there is a large harmonic axial force of frequency Ω (Figure 4.7). The crack opens and closes regularly; therefore, if the open crack has a torsional flexibility $c_T = 1/k_T$ and the cantilever beam flexibility $c = 1/k$, the flexibility of the spring for lateral vibration is $c + (lc_T \mp lc_T)/2$, where the $+$ sign is for one-half of the longitudinal vibration cycle (open crack) and the minus sign is for the other half of the longitudinal vibration cycle (closed crack). Several unilateral contact problems behave in this way. The differential equation expressing them is of the form

$$m\ddot{x} + (\lambda \mp \gamma)x = 0 \qquad (4.24)$$

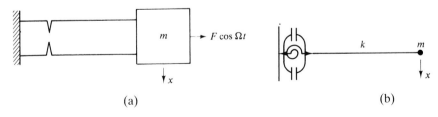

Figure 4.7 Unilateral contact cracked beam.

This is known as the Meissner equation [Klotter, 1951] and has the regions of stability and instability shown in Figure 4.8.

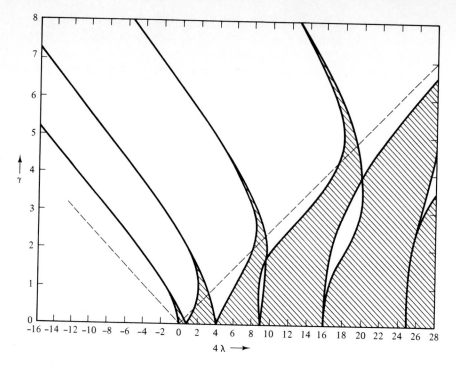

Figure 4.8 Stability chart for Meisner's equation. (From Klotter, K. 1951. *Technische Schwingungslehre*. Berlin: Springer-Verlag. By permission.)

☐ Example 4.7 Parametrically Excited Physical Pendulum

A heat exchanger in a nuclear power plant has a mass $m = 500$ tons. It is attached to the upper slab of the turbogenerator foundation by tie rods of length $l = 0.2$ m (Figure E4.7). Assuming the rotating speed of the machine is 1800 rpm, determine the vibration amplitude of the upper slab that would initiate unstable vibration of the heat exchanger.

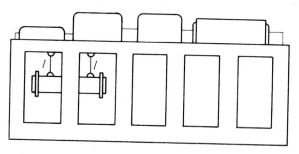

Figure E4.7

Solution The system can be considered as a simple pendulum of length l. If y_0 is the vibration amplitude of the upper slab, the equation of motion will be

Energy Methods: Stability   Chap. 4

$$\ddot{\theta} + \frac{g}{l}\left(1 - \frac{\omega^2}{g} y_0 \cos \omega t\right)\theta = 0$$

To bring this equation to the standard form (4.24), we substitute $\omega t = 2\tau$. Then

$$\frac{d^2\theta}{dt^2} = \frac{\omega^2 d^2\theta}{4d\tau^2} \text{ and } \frac{d^2\theta}{d\tau^2} + \left(\frac{4g}{\omega^2 l} - \frac{4y_0}{l}\cos 2\tau\right)\theta = 0$$

Thus, $a = 4g/\omega^2 l$, and $q = 2y_0/l$. Substituting $g = 9.81$ m/s², $l = 0.2$ m, and $\omega = 1800 \times 2\pi/60 = 188.4$ rad/s, we obtain $a = 0.5 \times 10^{-4}$ m/s². Utilizing Figure 4.6, the threshold of instability for $a = 0.5 \times 10^{-4}$ m/s² is $q = 0.2$. Therefore,

$$y_0 = \frac{ql}{2} = \frac{0.1 \times 0.2}{2} = 0.01 \text{ m} = 10 \text{ mm}$$

The system will become unstable at a vibration of the upper slab of 10 mm. The heat exchange will then start swinging.

(This example demonstrates the instability of a physical pendulum, parametrically excited by harmonic motion of its pivot.) □

4.7 FLUID-FLOW-INDUCED VIBRATION

A common source of self-excited vibration is fluid flow about elastic structures. The flow is a source of energy and it can maintain substantial vibration in the case of instability.

Galloping of Electric Lines

A typical example of flow-induced vibration is the galloping of transmission lines [Den Hartog, 1952]. An arbitrary cylindrical shape is located in a cross-flow of velocity V. The cylinder is elastically supported and has mass m. If there is a vertical motion of velocity $v = \dot{x}$, the apparent velocity of the flow changes direction and it is inclined by an angle α, $\tan \alpha = v/V$. The drag and lift forces have directions along and perpendicular to the apparent flow velocity, respectively (Figure 4.9).

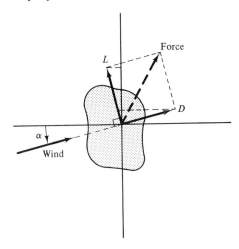

Figure 4.9 Lift and drag forces on a section.

The vertical force is

$$F = 0.5\rho V^2 bD(C_L \cos \alpha + C_D \sin \alpha) \tag{4.25}$$

where b is the width and D the hydraulic diameter, $4 \times$ area of cross section/ perimeter. For small angles of attack, $\alpha = -\dot{x}/V$ and the linearization of the vertical force will yield

$$F = \frac{-0.5\rho V^2 bD(\partial C_L/\partial \alpha + C_D)\dot{x}}{V} \tag{4.26}$$

The system will become unstable for

$$c_v - 0.5\rho VbD\left(\frac{\partial C_L}{\partial \alpha} + C_D\right) < 0 \tag{4.27}$$

where c_v is the linear viscous damping coefficient. The parameter $\partial C_L/\partial \alpha + C_D$ can be found using wind tunnel tests for particular sections [Blevins, 1977].

Perfectly circular sections are stable. Ice deposits in electric transmission lines can make the section unstable. From equation (4.27) one can observe that stability in this case is wind-velocity dependent. In fact, instability can occur at very low wind speeds. The shape of the ice deposits is very uncertain. Instead of studying the different shapes further, dampers are added to the line to balance negative values of $\partial C_L/\partial \alpha + C_D$ (Figure 4.10).

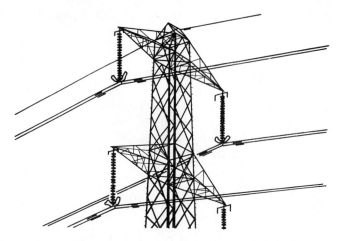

Figure 4.10 Stabilization of electric power lines. (Adapted from Davison, A. E. 1932. Canadian Electric Association.)

Von Kármán Vortex Trails

When a fluid flows by a cylindrical obstacle, the wake behind the obstacle is no longer regular but will have vortices as shown in Figure 4.11. The vortices alternate clockwise and anticlockwise in a regular pattern and create an alternating lateral force on the section. Experiments showed that for every section there is a similitude

Figure 4.11 Von Kármán vortices for flow past a cylinder. (From Hort, W. 1922. *Technische Schwingungslehre*, 2ⁿᵈ ed., Berlin: Julius Springer.)

parameter, the *Strouhal number*, relating the frequency of the alternating force f, the flow velocity V, and the section characteristic dimension (diameter for a circular section) D:

$$\text{Sr} = \frac{fD}{V} \tag{4.28}$$

The lateral force amplitude is

$$F_K = \tfrac{1}{2}\rho V^2 A C_K \tag{4.29}$$

where ρ is the fluid density, A the cross-sectional area, and C_K the von Kármán coefficient, approximately equal to 1 for a great range of Reynolds numbers. The Strouhal number is 0.20 for the cylinder, 0.12 to 0.15 for open-channel sections, 0.18 to 0.22 for flat thin sections, and 0.12 for rectangular sections [Blevins, 1977].

The lateral alternating force can lead to fatigue failure, but more dangerous is a coincidence of the Strouhal frequency with the natural frequency of the structure. Very substantial vibration can be initiated even at small values of the lateral force. Tall chimneys are subject to vortex-induced vibration. Helicoidal spoilers have been used to brake the vortices (Figure 4.12). This phenomenon was studied extensively in relation to the destruction of the Tacoma Narrows Bridge in the

Figure 4.12 Helicoidal spoilers to stabilize tall chimneys.

state of Washington. The bridge had been in operation for about a year when, in 1940, it was destroyed completely by a moderate wind. The section was I-shaped, as shown in Figure 4.13. Experiments showed that at the 42-mph wind speed there was a coincidence of the Strouhal frequency with the torsional natural frequency of the bridge. A new box-shaped section was designed to stiffen the bridge in torsion, and a trussed structure allowed for flow through the structure and reduced the von Kármán vortices.

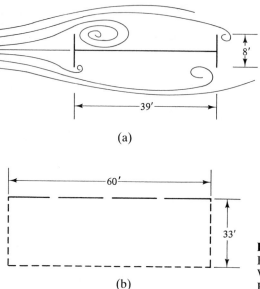

(a)

(b)

Figure 4.13 Tacoma Narrow Bridge structural section. (Univ. of Washington, Eng. Exp. Station, Bulletin No. 116.)

☐ **Example 4.8 Flow-Induced Structural Vibration**

A thermocouple probe (Figure E4.8) consists of a cylindrical head of outer diameter $d = 25.4$ mm supported by a wire. The natural frequency in the direction of flow is 213 Hz. Find the flow velocity of the helium coolant at which vortex-induced vibration will occur.

Solution The Strouhal number for the cylinder is 0.20. Therefore, the maximum velocity is

$$U = \frac{fD}{\text{Sr}} = \frac{213 \times 0.0254}{0.2} = 27.51 \text{ m/s}$$

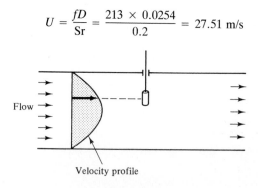

Velocity profile

Figure E4.8

(This example demonstrates the self-excited vibration conditions of a structure in cross-flow.)

□

4.8 FRICTION-INDUCED VIBRATION AND STICK-SLIP

It is known from mechanics that the kinetic friction coefficient is very different from the static one and depends strongly on the sliding velocity. Elastic systems that are acted upon by static friction forces often develop instabilities due to the sliding friction. To understand this instability mechanism, consider a weight W on a plane moved by a much larger element A which moves with constant linear velocity V to the right (Figure 4.14a). Between them is a spring of stiffness k, initially unstretched. As long as $kx < \mu_0 W$, where μ_0 is the static coefficient of friction, the spring force cannot overcome static friction and the weight does not move. As x increases, at some point kx becomes $> \mu_0 W$ and the weight starts moving to the right under the force $F_0 = kx - \mu_0 W$. When motion begins, the friction coefficient becomes kinetic and $\mu \ll \mu_0$. Then the accelerating force becomes $F = kx - \mu W \gg F_0$ and the weight starts accelerating rapidly to the right, faster than the constant velocity of A. If this happens, at some point the spring gets stretched and the direction of the spring force is reversed. Then the weight starts decelerating, until it stops at point C. The friction force jumps to the static friction value and the cycle is repeated. This phenomenon is called *stick-slip*. Many phenomena associated with friction and a squeaking sound are due to stick-slip.

The behavior of the $\mu(V)$ function is related with instability. For example, in lubricated bearings this relationship is as shown in Figure 4.14b. The part of the function on the left, corresponding to boundary or mixed lubrication, has a negative slope. Any increase in velocity will lead to decreasing friction, thus accelerating the motion. This clearly implies instability. On the contrary, on the right of the diagram the friction increases with increasing velocity (hydrodynamic lubrication). Any increase in velocity will lead to increasing friction, thus decelerating the motion. This clearly leads to stability.

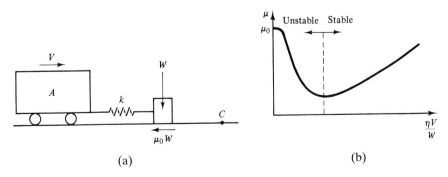

(a)   (b)

Figure 4.14   Stick-slip mechanism.

☐ **Example 4.9 Metal-Cutting Instability**

A spring-supported mass m, representing the cutting head of a plane mill, cuts on a surface moving with a constant speed V (Figure E4.9). The coefficient of friction within a practical velocity range is $\mu = a - bU$, where U is the cutting velocity. Determine the critical value of the system viscous damping constant c_v at which self-excited vibration will start.

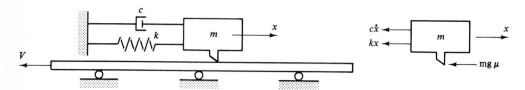

Figure E4.9

Solution   The friction force will be

$$F = -mg[a - b(V + \dot{x})]$$

The equation of motion, therefore, is

$$m\ddot{x} + c_v\dot{x} + kx = -mga + mgb(V + \dot{x})$$

or

$$m\ddot{x} + (c_v - mgb)\dot{x} + kx = -mga + mgbV$$

We note the following:

1. If $b < 0$, the coefficient of friction increases with increasing velocity. The system is stable for all values of c_v because the coefficient of x is always positive.
2. If $b > 0$ (as usually happens), instability (stick-slip chatter) will occur when

$$c_v < mgb$$

(This example demonstrates the instability during metal cutting due to stick-slip.)

☐

4.9 STABILITY OF MECHANICAL SYSTEMS

Mechanical and electromechanical networks and control systems exhibit dynamic behavior very similar to that of vibrating mechanical systems. Consider, for example, the centrifugal governor of a gasoline engine, steam or gas turbine, diesel engine, and so on. In its simpler form, the governor consists of a rhomboidal mechanism with a mass m attached to point A. There is a constant gear ratio η of the angle of rotation ϕ of the governor to that of the shaft θ (Figure 4.15): $\phi = \eta\theta$. The slider B moves the valve that controls the working fluid (steam, gas, fuel) so that there is a relationship $T = \xi T_0 s/s_0$, where T is the applied torque, T_0 the rated torque, and ξ the throttle ratio $0 < \xi < 1$, $s = 2r(1 - \cos\psi)$.

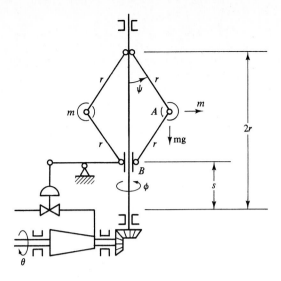

Figure 4.15 Centrifugal governor.

For angular velocity of the governor $\dot\phi$, there is a centrifugal force $m\dot\phi^2 r \sin\psi$ and a weight mg on the governor's mass. Equilibrium about the pivot yields

$$\cos\psi_0 = \frac{g}{\dot\phi^2 r} = \frac{g}{\eta r\dot\theta} \tag{4.30}$$

The torque on the shaft is

$$T = \frac{\xi T_0(1 - \cos\psi)}{1 - \cos\psi_0} \tag{4.31}$$

If the rotating system has moment of inertia J and the inertia of the governor is negligible, the equation for rotary motion of the shaft, considering the external torque $-\xi T_0$ and the generated torque T, is

$$J\ddot\theta = \frac{-\xi T_0 + \xi T_0(1 - \cos\psi)}{1 - \cos\psi_0} \tag{4.32}$$

For $\psi = \psi_0$ the acceleration is zero and the shaft runs with constant angular velocity at rated load and speed. This is the equilibrium position and it is interesting to study its stability. Using equation (4.32), we have

$$J\ddot\theta - f(\theta) = -\xi T_0 \tag{4.33}$$

where $f(\theta) = \xi T_0(1 - g/r\eta^2\dot\theta^2)/(1 - g/r\eta^2\dot\theta_0^2)$. This function is linearized about the equilibrium point $\theta = \theta_0$ by expanding in Taylor series and keeping only the first-order term:

$$f(\theta) \simeq \xi T_0 + \frac{df(\dot\theta)}{d\dot\theta}\dot\theta \tag{4.34}$$

At $\theta = \theta_0$,

$$f(\dot{\theta}) \simeq \xi T_0 + \frac{2\xi\eta T_0}{r\eta^2\dot{\theta}_0^2/g - 1}\dot{\theta} \qquad (4.35)$$

Equation (4.33), also assuming viscous damping $c\dot{\theta}$, becomes

$$J\ddot{\theta} + \left(c - \frac{2\xi\eta T_0}{r\eta^2\dot{\theta}_0^2/g - 1}\right)\dot{\theta} = 0 \qquad (4.36)$$

From equation (4.31) it is apparent that in the absence of viscous damping the system is unstable. In fact, this instability has long been known as *governor hunting*. The viscous damping constant needed to sustain stable operation is

$$c = \frac{2\xi\eta T_0}{r\eta^2\dot{\theta}_0^2/g - 1} \qquad (4.37)$$

☐ **Example 4.10   Centrifugal Governor of a Steam Turbine Drive**

A small steam turbine drive is rated at 1500 kW at 4000 rpm. The governor has an arm length $r = 0.20$ m and speed ratio $\eta = 0.25$. Determine the viscous camping constant at the shaft necessary for stable operation up to 10% overload.
Solution   The rated torque is $T_0 = 1.5 \times 10^6/(4000 \times 2\pi/60) = 3581$ N·m and $\xi = 1.1$. The rated angular velocity is $\dot{\theta}_0 = 4000 \times 2\pi/60 = 419$ rad/s. Therefore, the viscous damping constant should be, from equation (4.37),

$$c = \frac{2 \times 1.1 \times 0.25 \times 3581}{2 \times 0.2 \times 0.25^2 \times 419^2/9.81 - 1} = 4.4 \text{ N·m·s/rad}$$

(The design equation for the viscous damping needed for stable governor operation is developed in this example.)

☐

SIMULAB Experiment 4.1

Using the simulated experiment LINSYS, compute the motion of a harmonic oscillator with properties $m = 2$ kg, $k = 10$ N/m, and $c = 5$ N·s/m, the mass of which is acted upon by an aerodynamic force $c_a\dot{x}$, where $c_a = -6.5$ N·s/m.
Solution   With the program diskette in drive A, we type

A> LINSYS

The page of Figure SE4.1a appears on the screen. We select a one-mass system. Then the menu page appears on the screen with the simulated one-degree-of-freedom system (Figure SE4.1b). With the keypad arrow keys we select menus.

1. We set $m = 2$ kg, $k = 10$ N/m, and $c = 5 - 6.5 = -1.5$ N·s/m using the $+$ and $-$ keys.
2. We specify a unit initial displacement.
3. We hit S to start system operation. The resulting motion is shown in Figure SE4.1b.

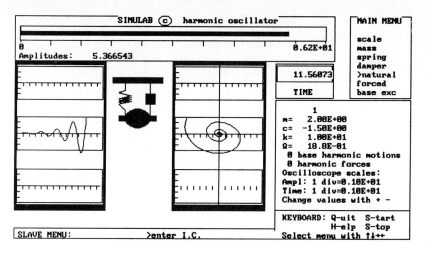

Figure SE4.1

TABLE 4.1. SUMMARY OF EQUATIONS OF CHAPTER 4

Kinetic energy of particle

$$KE = T = \tfrac{1}{2}mv^2$$

Potential energy

Spring: $PE = V = \tfrac{1}{2}kx^2$

Gravity field: $PE = V = mgh$

Conservation of energy

$$\frac{d}{dt}(T + V) = 0, \text{ or}$$

$$V_1 + \frac{1}{2}mv_1^2 = V_2 + \frac{1}{2}mv_2^2$$

Rayleigh's method

$$T_{\max} = V_{\max}$$

Lagrange's equation

$$\frac{d}{dt}\left(\frac{\partial T}{\partial \dot{q}}\right) - \frac{\partial T}{\partial q} + \frac{\partial V}{\partial q} = F_{qn}$$

$$\frac{d}{dt}\left(\frac{\partial L}{\partial \dot{q}}\right) - \frac{\partial L}{\partial q} = F_{qn}$$

Lagrangian function

$$L = T - V$$

(continues)

TABLE 4.1. *(cont.)*

Stability (structural)

$$\frac{dV}{dx} = 0 \text{ and } \frac{d^2V}{dx^2} > 0 \rightarrow \text{stability}$$

$$\frac{dV}{dx} = 0 \text{ and } \frac{d^2V}{dx^2} < 0 \rightarrow \text{instability}$$

instability: $k < 0$

Dynamic instability

$c < 0$

Mathieu equation

$$y'' + (a - 2q \cos 2\omega t)\, y = 0$$

Meissner equation

$$m\ddot{x} + (\lambda \mp \gamma)\, x = 0$$

Galloping of electric lines

if $c_v - 0.5\rho VbD(\partial C_L/\partial \alpha + C_D) < 0$

Von Kármán vortex trails

Strouhal Number $Sr = fD/V$

$$F_K = \tfrac{1}{2}\rho V^2 A C_K$$

C_K = von Kármán coefficient ≈ 1

REFERENCES AND FURTHER READING

BISHOP, R. E. D. 1979. *Vibration*, 2nd ed. Cambridge: Cambridge University Press.

BLEVINS, R. D. 1977. *Flow-Induced Vibration*. New York: Van Nostrand Reinhold.

DEN HARTOG, J. P. 1952. *Mechanical Vibration*, 4th ed. New York: McGraw-Hill.

DIMAROGONAS, A. D. 1976. *Vibration Engineering*. St. Paul, Minn.: West Publishing.

DIMAROGONAS, A. D., and PAIPETIS, S. A. 1983. *Analytical Methods in Rotor Dynamics*. London: Elsevier–Applied Science.

HORT, W. 1910. *Technische Schwingungslehre*. Berlin: Springer-Verlag.

HOUSNER, G. W., and HUDSON, D. E. 1950. *Applied Mechanics: Dynamics*. Princeton, N.J.: D. Van Nostrand.

KLOTTER, K. 1960. *Technische Schwingungslehre*. Berlin: Springer-Verlag.

PONTRYAGIN, L. S. 1962. *Ordinary Differential Equations*. Reading, Mass.: Addison-Wesley.

RAYLEIGH, J. W. S. 1894. *Theory of Sound*. New York: Dover Publications, 1946.

TIMOSHENKO, S. P. 1953. *History of the Strength of Materials*. New York: McGraw-Hill.

TIMOSHENKO, S., and YOUNG, D. H. 1948. *Advanced Dynamics*. New York: McGraw-Hill.

WHITTAKER, E. T. 1904. *Analytical Dynamics of Particles and Rigid Bodies*. London: Cambridge University Press.

PROBLEMS

Sections 4.1 and 4.2

4.1–4.5. Solve Problems 2.39 to 2.45 using the energy method.

4.6. A simple pendulum rotates about its vertical axis with an angular velocity ω (Figure P4.6). Assuming small angle θ:
   (a) Determine the differential equation of motion.
   (b) Investigate the stability of the system.

Figure P4.6

4.7. Solve Problem 4.6, assuming that the mass has been replaced with a circular disk of mass m and moment of inertia J and is perpendicular to the pendulum line.

4.8–4.11. Using the energy method, determine the differential equation of motion and the natural frequency of the systems shown in Figures P4.8 to P4.11. Assume small amplitude wherever applicable.

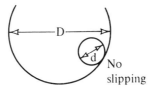

Figure P4.8

4.12. Determine the natural frequency of a door with an axis of rotation inclined with respect to the vertical by a small angle θ (Figure P4.12).

4.13. A relative rigid beam B of a crane is operated by a cable of stiffness k (Figure P4.13). It has a length l and moment of inertia I_0. At a 45° angle and when the upper branch of the cable is horizontal, the vertical branch of the cable, supporting weight W, breaks. Determine the resulting motion of the beam, assuming that the crane motion is negligible.

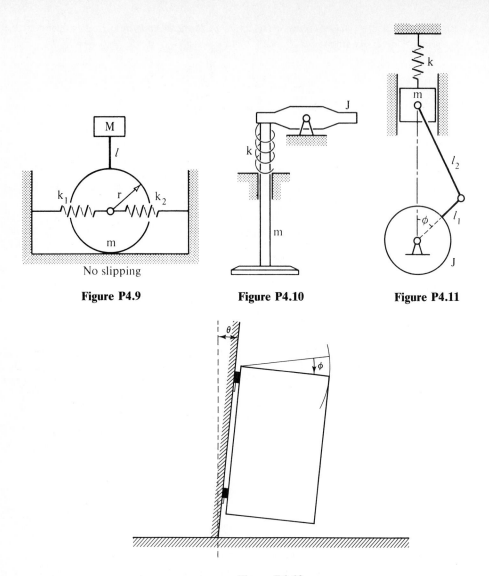

Figure P4.9

Figure P4.10

Figure P4.11

Figure P4.12

4.14–4.17. Determine the natural frequency of the system shown in Figures P4.14 to P4.17 to torsional vibration.

4.18. Find the natural frequency of the oscillation of a rigid beam of length L and height h, rolling on a stationary cylinder of diameter D without slipping (Figure P4.18). Investigate the stability of the system.

4.19. An empty rectangular container of length 6 m and width 1 m rests on a cylinder with a radius of 3 m, as shown in Figure P4.19. Determine the height up to which we fill the container with plywood ($\gamma = 0.8$) without tilting. Consider the mass of the container to be negligible.

4.20. Solve Problem 4.19 assuming that the container is filled with distilled water of free surface.

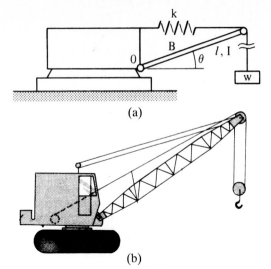

(a)

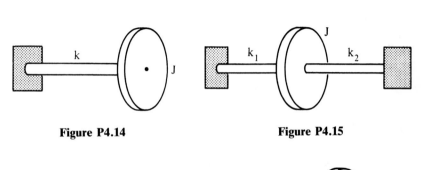

(b)

Figure P4.13

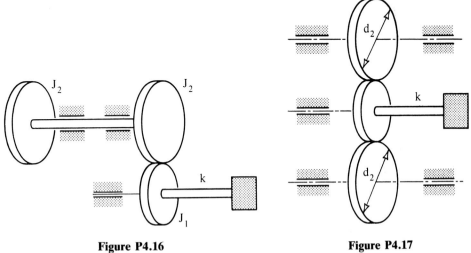

Figure P4.14

Figure P4.15

Figure P4.16

Figure P4.17

4.22–4.34. Using Rayleigh's method, determine the frequency of oscillation of the systems of Problems 2.52 to 2.55, 4.8 to 4.11, and 4.14 to 4.18.

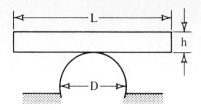

Figure P4.18

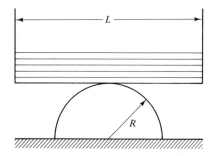

Figure P4.19

Section 4.3

4.35. Determine the differential equation of motion for the system of Example 4.4 using Lagrange's equation.

4.36. Determine the differential equation of motion for the system shown in Figure P4.8, based on the observation that there is, because of the damping, a logarithmic decrement of oscillations $\delta = 0.1$, using Lagrange's equation.

4.37. Determine the differential equation of motion for the system shown in Figure P4.9, using Lagrange's equation. Determine an appropriate dissipation function to account for a material loss factor $\gamma = 0.001$ in the two springs k_1 and k_2.

4.38. The system shown in Figure P4.10 has a material loss factor $\gamma = 0.001$ in the spring k and viscous damping in the bearing and guide, which alone results in a logarithmic decrement $\delta = 0.2$. Determine an appropriate dissipation function and the differential equation of motion. Use Lagrange's equation.

4.39. Determine the differential equation of motion for the system of Figure P4.11, at the position shown, using Lagrange's equation. Assume that the spring k has been replaced with a cylinder of diameter d and compression ratio r and that the compression is adiabatic. Determine the differential equation for $\phi = 0$ and comment on its stability.

Sections 4.4 to 4.9

4.40. An electric motor has a rotor of mass $m = 120$ kg supported by a bearing of stiffness $k = 3.5 \times 10^5$ N/m. Because of the electric field, there is a force in the direction of displacement of the rotor x of magnitude $k_e = 0.9 \times 10^5 W$ N/m, where W is the power in kW. Determine the load at which the rotor will become unstable.

4.41. In Problem 4.40, assume that the motor is running at 3600 rpm at a load of 0.2 kW. Also assume that a piece of insulation of mass $m_e = 0.010$ kg at a distance from the axis of rotation $e = 0.20$ m is missing. Determine the vibration amplitude.

4.42. In a power plant it is expected that the several slabs of the plant will have a maximum vibration of 5 mils (0.005 in.) at the running frequency of 3600 rpm. Determine the maximum length of the lighting fixture brackets attached from the ceilings (Figure P4.42) in order to avoid swinging.

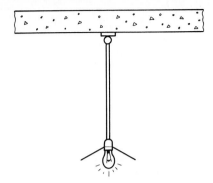

Figure P4.42

4.43. Prove that the system shown in Figure P4.43 can exhibit unstable vibration in the direction x assuming that there is a harmonic force $F_0 \cos \omega t$ on it [Klotter, 1960].

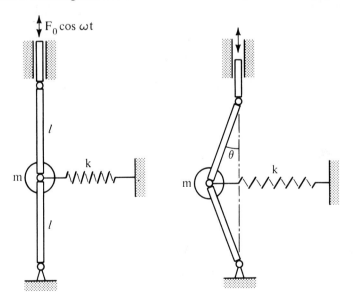

Figure P4.43

4.44. A light horizontal beam of length l supports a mass m at the center (Figure P4.44). If the support at the right is subject to a harmonic force in the horizontal direction, determine the differential equation of motion of the mass m for large lateral vibrations.

4.45. A cubical water reservoir of size $a = 4$ m and mass 3×10^5 kg is supported by a framed steel tower (Figure P4.45) that presents little resistance to the wind. The system has a natural frequency 1.5 Hz and a fraction of critical damping $\zeta = 0.03$ for horizontal

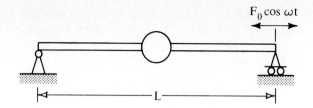

Figure P4.44

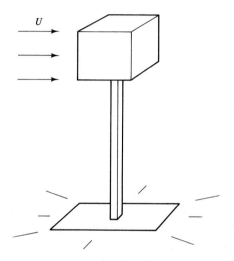

Figure P4.45

motion. For a square shape, $\partial C_L/\partial \alpha + C_D = -2.7$. If the air density is 1.23 kg/m³, find the equivalent viscous damping. Then find the wind velocity U for which galloping of the reservoir will occur.

4.46. A pedestrian bridge of height $D = 1$ m, width 2 m, and mass 12,000 kg is supported by a concrete structure (Figure P4.46) that presents little resistance to the wind. The system has a natural frequency 3.2 Hz and a fraction of critical damping $\zeta = 0.05$ for horizontal motion. For a rectangular shape 2×1, $\partial C_L/\partial \alpha + C_D = -3$. If the air density is 1.23 kg/m³, find the equivalent viscous damping. Then find the wind velocity U for which galloping of the bridge will occur.

4.47. For Problem 4.45, determine the wind velocity that will result in vortex-induced vibration if the Strouhal number for a square shape is 0.12.

4.48. For Problem 4.46, determine the wind velocity that will result in vortex-induced vibration if the Strouhal number for a rectangular shape is 0.22.

4.49. A cylindrical container of diameter $D = 3$ m and length $L = 20$ m is transported by a helicopter to which it is attached by a 35-m-long cable. Find the helicopter speed in a direction perpendicular to the cylinder axis for which the cylinder will oscillate due to the vortex flow.

4.50. The steel tubes of a gas-cooled reactor have diameter 0.15 m and natural frequency

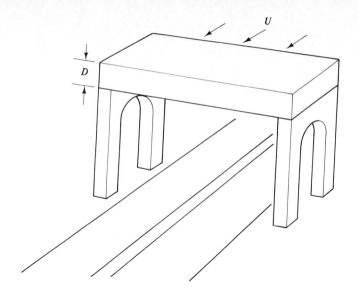

Figure P4.46

of lateral vibration 320 rad/s. Determine the velocity of cross-flow that will result in vortex-induced vibration on the tubes if the Strouhal number is 0.34.

4.51. A tanker terminal in a river estuary is supported by eight cylindrical concrete piles of diameter 1 m and length 20 m embedded in the riverbed and clamped on the terminal platform. The piles were designed for a compressive stress of 10 MPa due to the terminal weight. Determine the natural frequency for horizontal motion of the terminal platform and the velocity of the water flow that will result in vortex-induced vibration if the Strouhal number for circular sections is 0.20 and the modulus of elasticity of the concrete is 14×10^9 Pa.

4.52. A rigid cylindrical rod is placed on two cylinders of diameter D that rotate with a constant angular velocity ω in opposite directions (Figure P4.52). The coefficient of friction between the rod and the cylinder is $f = a - bU$, where U is the sliding velocity. It has been noted that for some speeds, the rod starts oscillating in the axial direction. Investigate this problem.

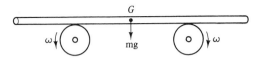

Figure P4.52

4.53. A pendulum oscillates about a pivot A of diameter d (Figure P4.53), and the coefficient of friction is $f = a - bU$, where U is the peripheral sliding speed and there is a viscous rotary damping due to the pivot c_T. Investigate the conditions for development of stick-slip motion of the pendulum.

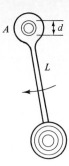

Figure P4.53

DESIGN PROBLEMS

D4.1. A pedestrian bridge of width 2 m and mass 12,000 kg is supported by a concrete structure that presents little resistance to the wind. The system has a natural frequency of 3.2 Hz and a damping factor $\zeta = 0.05$ for horizontal motion. If the air density is 1.23 kg/m^3 and the wind velocity U has maximum value 60 mph, design the shape of the deck so that galloping of the bridge will not occur.

D4.2. For Problem D4.1, design, in addition, for avoidance of vortex-induced vibration if the Strouhal number for a rectangular shape is 0.22.

D4.3. A cylindrical container of diameter $D = 3$ m and length $L = 20$ m is transported at 130 mph by a helicopter to which it is attached by a cable. Design the cable so that the cylinder will not oscillate due to the vortex flow.

D4.4. The steel tubes of a gas-cooled reactor have diameter 0.15 m and minimum thickness 10 mm. Design their supports to avoid vortex-induced vibration on the tubes if the Strouhal number is 0.34.

D4.5. A tanker terminal in a river estuary is supported by eight cylindrical concrete piles of diameter 1 m and length 20 m embedded in the riverbed and clamped on the terminal platform. The piles were designed for a compressive stress of 10 MPa due to the terminal weight. Design appropriate connections along their length so that the 1-m/s velocity of the water flow will not result in vortex-induced vibration if the Strouhal number for circular sections is 0.20 and the modulus of elasticity of the concrete is 14×10^9 Pa.

5

Nonharmonic Excitation

5.1 INTRODUCTION

Harmonic exciting forces on a vibrating system is the most common cause of vibration, especially in mechanical systems. It was found in Chapter 3 that in the case of such an excitation, there is always a closed-form solution. This solution is also harmonic. In many engineering problems, there are exciting forces that vary with time in a nonharmonic way. In general in such cases, we do not have a closed-form solution, but approximations can always be found to engineering requirements. We shall distinguish the general excitation in two categories: *periodic* and *nonperiodic excitation*.

5.2 PERIODIC EXCITATION

Let $F(t)$, a periodic function of time, represent the exciting force on a single-degree-of-freedom system. In Chapter 1 we found that any periodic function $F(t)$ of period

*Building damaged during the 1990 San Francisco earthquake, a nonharmonic ground motion.

T can be represented by an infinite sum (called a *Fourier series*)

$$F(t) = \frac{a_0}{2} + \sum_{n=1}^{\infty} (a_n \cos n\omega t + b_n \sin \omega t) \qquad (5.1)$$

where $\omega = 2\pi/T$ and

$$a_0 = \frac{2}{T} \int_0^T F(t)\, dt \qquad (5.2)$$

$$a_n = \frac{2}{T} \int_0^T F(t) \cos n\omega\, dt \qquad (5.3)$$

$$b_n = \frac{2}{T} \int_0^T F(t) \sin n\omega t\, dt \qquad (5.4)$$

The equation of motion of the system becomes

$$m\ddot{x} + c\dot{x} + kx = \frac{a_0}{2} + \sum_{n=1}^{\infty} (a_n \cos n\omega t + b_n \sin n\omega t) \qquad (5.5)$$

The right-hand side of the equation is a sum of harmonic functions. Because of the linearity, the superposition principle can be applied and the steady-state solution will be the sum of the solutions of the equations

$$m\ddot{x}_n + c\dot{x}_n + kx_n = a_n \cos n\omega t + b_n \sin \omega t \qquad (5.6)$$

for $n = 1, 2, \ldots$. Therefore, we obtain

$$x_s = \frac{a_0}{2k} + \sum_{n=1}^{\infty} \frac{a_n \cos(n\omega t - \phi_n) + b_n \sin(n\omega t - \phi_n)}{[(k - n^2\omega^2 m)^2 + (n\omega c)^2]^{1/2}} \qquad (5.7)$$

$$\phi_n = \arctan \frac{cn\omega}{k - mn^2\omega^2} \qquad (5.8)$$

Given an exciting function $F(t)$, one should obtain the coefficients a_n and b_n, computing the integrals in equations (5.2), (5.3), and (5.4), and then form the solution of equations (5.7) and (5.8). For practical evaluation of the response, one has to take a finite number of terms in the series: in other words, perform *truncation*. To assume that the truncation error is small, the series must converge uniformly. This, however, should not be always taken for granted. For large enough n, the series indeed converge, usually uniformly. The sin and cos terms remain bounded; therefore, the performance of the series depends on the function

$$f_n = \frac{a_n \quad \text{or} \quad b_n}{[(k - n^2\omega^2 m)^2 + (nc\omega)^2)]^{1/2}} \qquad (5.9)$$

We observe that around some value

$$n_c \approx \left(\frac{k}{\omega^4 m} \right)^{1/2}$$

this function has a maximum. For smooth functions $F(t)$, it is usually sufficient to take three to five times this value of n_c as the maximum n in the truncated series.

Truncation should always be performed with caution and careful consideration should always be given to convergence.

☐ **Example 5.1   Reciprocating Engine Mount**

On the reciprocating engine mount shown in Figure E5.1a the exciting force is given in Figure E5.1b. Determine the response of the system, in particular the first three harmonics, and plot them on a rotating vector diagram.

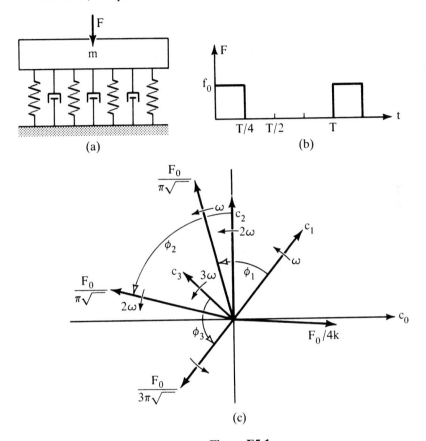

Figure E5.1

Solution   The forcing function will be expressed in a Fourier series and the response will be the sum of the responses to each harmonic term of the series. The response of a system (m,c,k) to a harmonic force $a_n \cos n\omega t$ will be (similar for $\sin n\omega t$)

$$x_n(t) = \frac{a_n}{[(k - m(n\omega)^2)^2 + (c(n\omega))^2]^{1/2}} \cos(n\omega t - \phi_n)$$

$$\tan \phi_n = \frac{c(n\omega)}{k - m(n\omega)^2}$$

The Fourier expansion of the square wave was obtained in Example 1.3. The complex Fourier components of the square-wave force $c_n = a_n + ib_n$ are shown in

Figure E5.1c. Due to linearity the response will be

$$x(t) = \frac{F_0}{4k} = \sum_{n=1}^{\infty} \frac{a_n \cos(n\omega t - \phi_n) + b_n \sin(n\omega t - \phi_n)}{[(k - m\omega^2 n^2)^2 + (cn\omega)^2]^{1/2}}$$

where a_n and b_n are the coefficients of the Fourier expansion of the forcing function. Using the results for a_n and b_n from Example 1.3,

$$x(t) = \frac{F_0}{4k} + \frac{\pi[\cos(\omega t - \phi_1) + \sin(\omega t - \phi_1)]F_0}{[(k - m\omega^2)^2 + (c\omega)^2]^{1/2}}$$

$$+ \frac{\pi[\cos(\omega t - \phi_1) + \sin(2\omega t - \phi_1)]F_0}{[(k - 4m\omega^2)^2 + (2c\omega)^2]^{1/2}}$$

$$+ \frac{3\pi[\cos(3\omega t - \phi_3) + \sin(3\omega t - \phi_3)]F_0}{[(k - 9m\omega^2)^2 + (3c\omega)^2]^{1/2}} + \cdots$$

The harmonic components of the response are shown in Figure E5.1c.

(In this example, a typical reciprocating excitation of a machine mount is presented.)

$\square$

$\square$ **Example 5.2   Periodic Excitation of Turbomachinery Blades**

A one-stage steam turbine runs a synchronous electric generator (Figure E5.2). The rotor of the generator is rotating at a constant speed of 3600 rpm. The turbine stage has 17 blades and the eighteenth is missing, due to the way they mount the blades on the rotor. If the moment of inertia of the turbine stage is 1 kg·m², and the power generated at the stage is 611 kW, determine the torsional vibration of the system due to the discontinuity of the power generation because of the missing blade. The shaft is 50 mm in diameter and 640 mm in length; the shear modulus $G = 1.1 \times 10^{11}$ N/m².

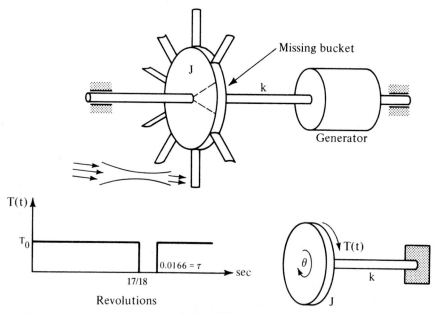

Figure E5.2

Solution Power on the wheel is produced only during 17/18 of the revolution. Therefore, the steam applies a torque that has the interrupted form shown. The average torque is

$$T_{av}(\text{Nm}) = \frac{P}{\omega} = \frac{60 \times 611 \times 10^3}{2\pi \times 3600}$$

$$= 1.62 \times 10^3 \text{ Nm}$$

But $T_0 = \frac{18}{17} T_{av} = 1.72 \times 10^3$ Nm. The function $T(t)$ must be expressed in the Fourier series

$$T(t) = \frac{A_0}{2} + \sum_{n=1}^{\infty} (A_n \cos n\omega t + B_n \sin n\omega t)$$

where $\omega = 2\pi \times \text{rpm}/60 = 376.8$ rad/s and $\tau = 2\pi/\omega = 0.0166$ s. The coefficients A_n and B_n are similar to the ones in Example 5.1, but the upper limits in all the integrations will be $17\tau/18$ instead of $\tau/4$. Thus

$$A_0 = \frac{2T_0}{\tau} \int_0^{17\tau/18} dt = \frac{2T_0}{\tau} \left(\frac{17\tau}{18}\right) = 3240 \text{ Nm}$$

$$A_n = \frac{T_0}{\pi n} (\sin n\omega t)\big|_0^{17\tau/18}$$

$$= \frac{T_0}{\pi n} \sin\left[n\left(\frac{2\pi}{\tau}\right)\frac{17\tau}{18}\right] = \frac{546}{n} \sin(5.93n)$$

$$B_n = -\frac{T_0}{\pi n} (\cos n\omega t)\big|_0^{17\tau/18}$$

$$= \frac{T_0}{\pi n}\left[1 - \cos\left(\frac{n2\pi}{\tau}\frac{17\tau}{18}\right)\right] = \frac{546}{n}[1 - \cos(5.93n)]$$

We are now ready for the dynamic problem. The torque $T(t)$ applies on the disk. The differential equation of motion is

$$J\ddot{\theta} + k\theta = T(t) = \frac{A_0}{2} + (A_n \cos n\omega t + B_n \sin n\omega t)$$

The torsional spring constant k of the shaft is

$$k = \frac{\pi d^4 G}{32l} = \frac{\pi \times 0.05^4 \times 1.1 \times 10^{11}}{32 \times 0.64} = 1.07 \times 10^5 \text{ N} \cdot \text{m/rad}$$

The response to the nth harmonic will be

$$\theta_n(t) = \frac{A_n \cos n\omega t + B_n \sin n\omega t}{k - J(n\omega)^2}$$

$$= \frac{A_n \cos n\omega t + B_n \sin n\omega t}{1.07 \times 10^5 - 1 \times (n\omega)^2}$$

Therefore,

$$\theta(t) = \frac{A_0}{2k} + \sum_{n=1}^{\infty} \frac{546/n \sin(5.39n) \cos n\omega t + [1 - \cos(5.39n)] \sin n\omega t}{1.07 \times 10^5 - 1.42 \times 10^5 n^2}$$

Harmonics:

Zero order (static): $\dfrac{A_0}{2k} = \dfrac{3240}{2 \times 1.07 \times 10^5} = 1.49 \times 10^{-2}$ rad

First order: $\dfrac{[546 \times \sin^2 5.39 + (1 - \cos 5.39)^2]^{1/2}}{1.07 \times 10^5 - 1.42 \times 10^5} = -13 \times 10^{-3}$ rad

(This example demonstrates periodic impulsive excitation on turbomachinery blades.)

□

□ **Example 5.3 Dynamics of a Punch Press**

The periodic excitation shown in Figure E5.3 is applied to the base of a punch press. Determine the resulting motion.

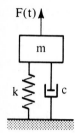
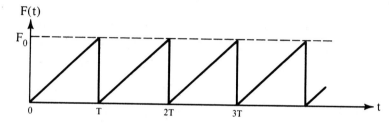

Figure E5.3

Solution The function $F(t)$ can be represented by the Fourier series

$$F(t) = \sum_{n=0}^{\infty} (a_n \cos n\omega t + b_n \sin n\omega t)$$

For this excitation, $F(t) = (F_0/T)t$. Then

$$a_0 = \frac{1}{T} \int_0^T F(t)\, dt = \frac{F_0}{T^2} \int_0^T t\, dt = \frac{F_0}{2}$$

$$a_n = \frac{2}{T} \int_0^T F(t) \cos n\omega t\, dt = \frac{2F_0}{T^2} \int_0^T t \cos n\omega t\, dt = 0$$

$$b_n = \frac{2}{T} \int_0^T F(t) \sin n\omega t\, dt = \frac{2F_0}{T^2} \int_0^T t \sin n\omega t\, dt = -\frac{F_0}{n\pi}$$

Thus the series expansion of $F(t)$ is

$$F(t) = \left(\frac{1}{2} - \frac{1}{\pi} \sum_{n=1}^{\infty} \frac{1}{n} \sin n\omega t \right) F_0$$

Now the differential equation of motion becomes

$$m\ddot{x} + c\dot{x} + kx = \left(\frac{1}{2} - \frac{1}{\pi} \sum_{n=1}^{\infty} \frac{1}{n} \sin n\omega t \right) F_0$$

The solution is

$$x(t) = \frac{F_0}{2k} - \frac{F_0}{\pi} \sum_{n=1}^{\infty} \frac{1}{n} \frac{\sin (n\omega t - \phi_n)}{[(k - n^2\omega^2 m)^2 + (nc\omega)^2]^{1/2}}$$

(This example presents the dynamic analysis of impact-loaded machine mounts, such as a punch press.)

□

5.3 NONPERIODIC EXCITATION

A nonperiodic exciting force usually has the form of a force that acts for a specific period of time only and then stops. In other cases, the force can have considerable duration and magnitude with known but nonperiodic time history. The entire time history of the response is important here and steady-state is either zero or not important. The response during a period of time related with the duration of the exciting force, called *transient*, reaches maximum value. The simplest form of such a force is the impulsive force. For a very short period of time Δt, a rather large constant force F is applied. As we know from dynamics,[1] the impulse equals the difference in the momenta before and after the impulse; in other words,

$$F \, \Delta t = \text{Imp}_{1 \to 2} = mv_2 - mv_1 \tag{5.10}$$

If we designate with $\hat{F}$ the impulse $F \, \Delta t$, we can write in general

$$\hat{F} = \int_t^{t + \Delta t} F \, dt \tag{5.11}$$

A unit impulse will be defined as

$$\hat{f} = \lim_{\Delta t \to 0} \int_t^{t + \Delta t} F \, dt = F \, dt = 1 \tag{5.12}$$

Such a function has no physical meaning but is a convenient tool in our computations. It is called a *Dirac[2] delta function*, designated as $\delta(t - \tau)$, with the properties

$$\delta(t - \tau) = 0 \qquad \text{for } t \neq \tau \tag{5.13}$$

$$\int_{t=0}^{\infty} \delta(t - \tau) \, dt = 1 \tag{5.14}$$

$$\int_0^{\infty} f(t)\delta(t - \tau) \, dt = f(\tau) \tag{5.15}$$

where $f(t)$ is any function of time.

From the impulse–momentum theorem, the value $\hat{F}$ of the impulse at time $\tau = 0$ is

$$\hat{F} = m\dot{x}\big|_{+0} - m\dot{x}\big|_{-0} = mv_0 \tag{5.16}$$

which means that because of the impulse the mass m will have an initial velocity $v_0 = \hat{F}/m$.

---

[1]Beer, F., and Johnston, E. 1984. *Vector Mechanics for Engineers*. New York: McGraw-Hill.

[2]Used long before by Kirchhoff, has been misnamed the "Dirac function." Lewis, T. 1937. Some applications of the Kirchhoff–Dirac delta function. *Philos. Mag.*, 7(24): 329–360.

From previous discussions we know that for a one-degree-of-freedom system the response due to initial velocity will be [see equation (2.52)]

$$x = \frac{v_0 e^{(-c/2m)t}}{\omega_d} \sin \omega_d t \tag{5.17}$$

or

$$x = \hat{F} \frac{e^{(-c/2m)t}}{m\omega_d} \sin \omega_d t$$

$$\omega_d = \left[\frac{k}{m} - \left(\frac{c}{2m} \right)^2 \right]^{1/2} \tag{5.18}$$

Furthermore, we can write, for $t > 0$,

$$x(t) = \hat{F}g(t), \qquad g(t) = \frac{1}{m\omega_d} e^{(-c/2m)t} \sin \omega_d t \tag{5.19}$$

We note that the response is the product of the impulse $\hat{F}$ and a function $g(t)$ that is characteristic of the system (m,c,k), called the *unit impulse response function*, and does not depend on the magnitude of the impulse.

Let now examine the response of the system to a general excitation $F(t)$ (Figure 5.1). At a time τ the force is $F(\tau)$ and during a period of time $\Delta\tau$ gives to the system an impulse $F(\tau) \Delta\tau$. At time t, the elapsed time since the impulse is $t - \tau$. Therefore, the response at time t will be $\Delta x(t) \simeq F(\tau) \Delta\tau g(t - \tau)$. The total response at time t will be found by summing all the responses due these elementary impulses at all times τ. This is equivalent with integration, equation (5.20), with respect to τ:

$$x(t) = \int_0^t F(\tau)g(t - \tau) \, d\tau \tag{5.20}$$

If the force was acting from time zero to t_0, we can write, for $t > t_0$,

$$x(t) = \int_0^{t_0} F(\tau)g(t - \tau) \, d\tau \tag{5.21}$$

For many cases, the function $F(\tau)$ has a form that allows an explicit integration in equation (5.20). In case such integration is not possible, the integral is proper

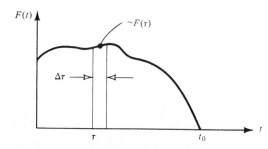

Figure 5.1 General excitation $f(t)$. (From Lord Rayleigh. 1894. *Theory of Sound.*)

and can be evaluated numerically with modest effort. The integral in equation (5.20) is called a *convolution integral* or *Duhamel's integral*.[3]

It must be pointed out that the solution given by equations (5.20) and (5.21) is a particular solution of the differential equation of motion and has to be supplemented by the general solution of the homogeneous differential equation $x_h(t)$ and appropriate initial conditions. Therefore, the most general form of the response of the harmonic oscillator with a force $F(t)$ and initial conditions x_0 and v_0 is

$$x(t) = \int_0^{t_0} \frac{F(\tau)}{m\omega_d} e^{(-c/2m)(t-\tau)} \sin \omega_d(t - \tau) \, d\tau$$

$$+ \frac{e^{-(c/2m)t}[x_0 \cos \omega_d t + (x_0 c/m + 2v_0)}{2\omega_d} \sin \omega_d t \qquad (5.22)$$

In most practical problems, shock is applied while the system is at rest, and use of the general solution is not necessary.

☐ **Example 5.4   Soil Compaction Machine**

On a soil compaction roller of mass m and radius r that can roll *without slipping* on a horizontal plane as shown, a harmonic force $F(t) = F_0 \sin \omega t$ is applied only for one period, $T = 2\pi/\omega$ (Figure E5.4). Determine the response of the system after the time T.

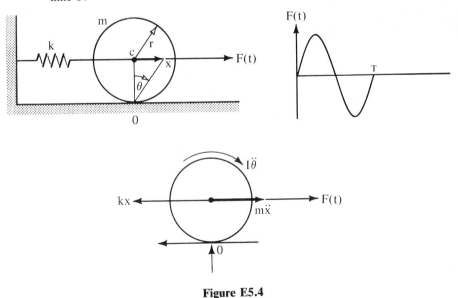

Figure E5.4

[3]The idea was introduced by Euler in the solution of differential equations and used by Cauchy in the multiplication of the power series. Poisson used superposition and step functions in 1826. Duhamel (1797–1872), used convolution in vibration analysis in 1834, in Mémoire sur les vibrations d'un système quelconque de points matériels. *J. Ec. Polytech.*, 25: 1–36. It was actually applied by St.-Venant [Timoshenko, 1953].

Solution We will first find the differential equation of motion and then apply the convolution integral to find the forced response. We take as coordinate the displacement x. The center of rotation is O. Therefore,

$$\theta = \frac{x}{r} \qquad \dot{\theta} = \frac{\dot{x}}{r} \tag{a}$$

Take moments about O:

$$M_O = (M_O)_{\text{eff}}, \qquad -(kx)r + Fr = I\ddot{\theta} + (m\ddot{x})r$$

or because of (a) and $J = \frac{1}{2}mr^2$,

$$\tfrac{3}{2}mr\ddot{x} + krx = Fr$$

$$\tfrac{3}{2}m\ddot{x} + kx = F(t)$$

The response is given by the convolution integral,

$$x(t) = \int_0^t F(\tau)g(t - \tau)\, d\tau$$

with

$$g(t) = \frac{\sin \omega_n t}{m\omega_n}, \qquad \omega_n = \sqrt{\frac{k}{1.5m}}$$

Thus

$$x(t) = \int_0^T + \int_T^t = \int_0^T \frac{F_0 \sin \omega\tau \sin \omega_n(t - \tau)}{m\omega_n}\, d\tau$$

$$= \frac{F_0}{m\omega_n} \int_0^T \sin \omega\tau \sin \omega_n(t - \tau)\, d\tau$$

$$= \frac{F_0}{m\omega_n} \int_0^T \frac{1}{2}[\cos(\omega\tau - \omega_n t + \omega_n\tau) - \cos(\omega\tau + \omega_n t - \omega_n\tau)]\, d\tau$$

$$= \frac{F_0}{m\omega_n} \left\{ \frac{1}{\omega + \omega_n} \sin[(\omega + \omega_n)\tau - \omega_n t] - \frac{1}{\omega - \omega_n}[\sin(\omega - \omega_n)\tau + \omega_n t] \right\} \Bigg|_0^T$$

$$x(t) = \frac{F}{m\omega_n} \left\{ \frac{1}{\omega + \omega_n} \{\sin[(\omega - \omega_n)T - \omega_n t] + \sin \omega_n t\} \right.$$

$$\left. - \frac{1}{\omega - \omega_n} \{\sin[(\omega - \omega_n)T + \omega_n t] - \sin \omega_n t\} \right\}$$

(This example demonstrates the response of a soil compaction machine, modeled as a single-degree-of-freedom system, to an impulse in the form of a half-period of a sinusoidal function.)

□

☐ **Example 5.5 Response of a Machine Mount to Intermittent Force**

An undamped harmonic oscillator representing a machine mount is acted upon by a constant force F_0 for a period of t_0 seconds, twice at time 0 and t_1 (Figure E5.5).

(a) Find the response after the time $t_1 + t_0$. Comment on the timing needed to minimize the response after time $t_1 + t_0$.

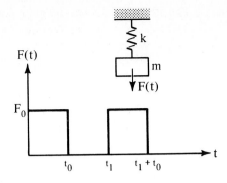

Figure E5.5

(b) Find the response if only the first pulse is applied.

Solution The impulse response is

$$g(t) = \frac{1}{m\omega_n} \sin \omega_n t, \qquad \omega_n = \left(\frac{k}{m}\right)^{1/2}$$

(a) The response after time $t_1 + t_0$ will be (convolution integral)

$$x(t) = \int_0^t F(\tau)g(t - \tau)\, d\tau$$

$$= -\int_0^t F(\tau) \frac{1}{m\omega_n} \sin \omega_n(t - \tau)\, d(t - \tau)$$

$$= \int_0^{t_0} + \int_{t_0}^{t_1} + \int_{t_1}^{t_1+t_0} + \int_{t_1+t_0}^t$$

$$= \frac{F_0}{m\omega_n} [\cos \omega_n(t - t_0) - \cos \omega_n t + \cos \omega_n(t - t_1 + t_0) - \cos \omega_n(t - t_1)]$$

It is apparent that if $\omega_n t_1 = \pi$, that is, the second pulse is applied at a time π/ω_n after the first, the final response will be zero!

(b) If only the first pulse is applied, the solution takes the form

$$x(t) = \frac{F_0}{k} [\cos \omega(t - t_0) - \cos \omega t] \qquad \text{for } t > t_0$$

For $t < t_0$,

$$x(t) = -\int_0^t F(\tau) \frac{\sin \omega_n(t - \tau)d(t - \tau)}{m\omega_n}$$

$$= \frac{F_0}{k} (\cos \omega_n (t - t_0) - \cos \omega t)$$

(This example demonstrates the response of a machine mount to a periodic intermittent force.) □

□ **Example 5.6 Step Force Excitation**

Find the time T corresponding to the peak response of the damped spring–mass system excited by a step force F_0.

Sec. 5.3   Nonperiodic Excitation   **223**

Solution The response of an (m,c,k) system is computed from the convolution integral,

$$x(t) = \int_0^t \frac{F_0}{m\omega_d} e^{-(c/2m)(t-\tau)} \sin \omega_d(t - \tau) \, d\tau$$

But

$$\int e^{\alpha t} \sin \beta t \, dt = \frac{\alpha e^{\alpha t}}{\alpha^2 + \beta^2} \left(\cos \beta t - \frac{\beta}{\alpha} \sin \beta t \right)$$

Therefore,

$$x(t) = \frac{-F_0}{m\omega_d} \left\{ \frac{\alpha e^{-c/2m(t-\tau)}}{\alpha^2 + \beta^2} \left[\cos \omega_d(t - \tau) - \frac{\beta}{\alpha} \sin \omega_d(t - \tau) \right] \right\} \Big|_0^t$$

$$= \frac{-\alpha F_0}{m\omega_d(\alpha^2 + \beta^2)} \left[1 - e^{-c/2mt} \left(\cos \omega_d t - \frac{\beta}{\alpha} \sin \omega_d t \right) \right]$$

$$\alpha = \frac{-c}{2m}, \qquad \beta = \omega_d$$

This function has maximum when the derivative is zero:

$$\frac{-c}{2m} e^{-c/2mt} \left(\cos \omega_d t - \frac{\beta}{\alpha} \sin \omega_d t \right) + e^{-c/2mt} \left(-\omega_d \sin \omega_d t - \frac{\beta}{\alpha} \cos \omega_d t \right) = 0$$

$$\left(\frac{-c}{2m} - \frac{\beta \omega_d}{\alpha} \right) \cos \omega_d t + \left(\frac{c\beta}{2m\alpha} - \omega_d \right) \sin \omega_d t = 0$$

$$\tan \omega_d t = \frac{c/2m + \beta \omega_d/\alpha}{c\beta/2m\alpha - \omega_d}$$

$$t = \frac{1}{\omega_d} \arctan \left(\frac{2m\omega_d^2/c - c/2m}{2\omega_d} \right)$$

(This example demonstrates the response of a single-degree-of-freedom system to a step force.)

□

□ **Example 5.7 Impulsive Excitation**

Determine the peak displacement for the impulsively excited one-degree-of-freedom system shown in Figure E5.7.

Solution For an impulse $F_0 \, \Delta t$ the response is $(\hat{F} = F_0 \, \Delta t)$

$$x(t) = \frac{\hat{F} e^{-(c/2m)t}}{m\omega_d} \sin \omega_d t$$

The maximum value occurs, approximately, at $\sin \omega_d t = 1$, or $\omega_d t = \pi/2$ or $t = \pi/2\omega_d$. At this time, the response is

$$X_{max} = \frac{\hat{F} e^{-\pi c/4m\omega_d}}{m\omega_d}$$

(This example further investigates the response of a single-degree-of-freedom system to an impulsive force.)

□

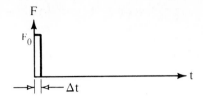 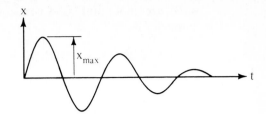

Figure E5.7

☐ **Example 5.8 Step Force Excitation**

Find the response of the damped spring–mass system excited by a step force F_0 of duration t_0.

Solution   The response of an (m,c,k) system is given by the convolution integral (see Example 5.6) for $t < t_0$,

$$x(t) = \frac{\alpha F_0}{m\omega_d(\alpha^2 + \beta^2)}\left[1 - e^{-(c/2m)t}\left(\cos \omega_d t - \frac{\beta}{\alpha}\sin \omega_d t\right)\right]$$

For $t > t_0$, we have

$$x(t) = \int_0^t = \int_0^{t_0} + \int_{t_0}^t$$

$$x(t) = \int_0^{t_0} \frac{F_0}{m\omega_d} e^{-(c/2m)(t-\tau)} \sin \omega_d(t - \tau)\, d\tau$$

$$= \frac{-F_0}{m\omega_d} \frac{\alpha e^{-(c/2m)(t-\tau)}}{\alpha^2 + \beta^2}\left[\cos \omega_d(t - \tau) - \frac{\beta}{\alpha}\sin \omega_d(t - \tau)\right]\Big|_0^{t_0}$$

$$x(t) = \frac{-F_0}{m\omega_d}\left\{\frac{e^{-(c/2m)(t-t_0)}}{\alpha^2 + \beta^2}\left[\cos \omega_d(t - \tau) - \frac{\beta}{\alpha}\sin \omega_d(t - t_0)\right]\right\}$$

$$+ \frac{F_0}{m\omega_d}\frac{\alpha e^{-(c/2m)t}\left(\cos \omega_d t - \frac{\beta}{\alpha}\sin \omega_d t\right)}{(\alpha^2 + \beta^2)}$$

(This example further investigates the response of a single-degree-of-freedom system to a step force.)   ☐

5.4 THE METHOD OF LAPLACE TRANSFORM

The equation for a single-degree-of-freedom linear system with general excitation $f(t)$ is

$$m\ddot{x} + c\dot{x} + kx = f(t) \tag{5.23}$$

We assume that $\lim e^{-st}f(t) = 0$ for $t \to \infty$, where s is any positive constant. This is a restriction on $f(t)$ that is justifiable for almost any physically realizable excitation. We expect that the response obeys a similar restriction, as $t \to \infty$, $\lim e^{-st} x(t)$

$= 0$. We multiply both sides of the equation with $e^{-st}\, dt$ and integrate from zero to infinity. We obtain

$$m \int_0^\infty \ddot{x}(t)e^{-st}\, dt + c \int_0^\infty \dot{x}(t)e^{-st}\, dt + k \int_0^\infty x(t)e^{-st}\, dt$$

$$= \int_0^\infty f(t)e^{-st}\, dt = F(s) \qquad (5.24)$$

Integrating by parts, we can prove that

$$\int_0^\infty \dot{x}(t)e^{-st}\, dt = sx(s) - x(0) \qquad (5.25)$$

$$\int_0^\infty \ddot{x}(t)e^{-st}\, dt = s^2x(s) - sx(0) - \dot{x}(0) \qquad (5.26)$$

where

$$L\{f(t)\} = F(s) = \int_0^\infty f(t)e^{-st}\, dt \qquad (a)$$

$$\qquad\qquad\qquad\qquad\qquad\qquad\qquad\qquad (5.27)$$

$$L\{x(t)\} = X(s) = \int_0^\infty x(t)e^{-st}\, dt \qquad (b)$$

Substituting in equation (5.24), we obtain

$$X(s) = \frac{F(s) + (ms + c)x(0) + m\dot{x}(0)}{ms^2 + cs + k} \qquad (5.28)$$

For zero initial conditions, equation (5.28) becomes

$$X(s) = G(s)F(s) \qquad (5.29)$$

where $G(s) = 1/(ms^2 + cs + k)$ is the system *transfer function*. The roots of the denominator, called *poles*, are the complex natural frequencies and $G(i\omega)/k = H(i\omega)$ is the complex amplification factor (Chapter 3) if $f(t)$ is harmonic.

Integrations such as in equations (5.27) have been carried out for various usual forms of the functions and tabulated. Such a tabulation is given in Appendix VI. We observe that on the right-hand side of the table are functions of time $f(t)$ and on the left are the corresponding integrals [equation (5.27)] $F(s)$. It is also true that for every function $F(s)$ corresponds a function $f(t)$ which integrated according to equation (5.27) gives $F(s)$. This property gives a method for finding $x(t)$. From equation (5.28) we have $X(s)$ if we can compute or find $F(s)$. If we can find this function on the left of the table, the function on the right is the solution $x(t)$. It is obvious that in equation (5.28), $x(0)$ and $\dot{x}(0)$ are the initial displacement and velocity of the system at time $t = 0$, respectively.

The integral in (5.27a) is called the *Laplace transform*[4] of the function $f(t)$.

[4]Laplace, P. S. 1779. Sur les suites. *Mem. Acad. Sci.* Laplace, P. S. (1812). *Théorie analytique des probabilités*, in *Oeuvres complètes, 1878–1912*. Paris: Gauthier-Villars. There is a wide impression, probably due to a misinterpreted remark in Den Hartog's *Mechanics*, that Heaviside preceded Laplace in the development of the Laplace transform. This is impossible, since Laplace preceded Heaviside by one century. Heaviside, however, used ideas of both Laplace and Cauchy in the development of his operational calculus.

Portrait of Laplace. (Courtesy of Bibliothèque Nationale, Paris.)

PIERRE-SIMON LAPLACE (1749–1827)

Laplace was born of poor parents, but his mathematical abilities early won him good teaching posts. Unlike other great contemporary mathematicians, such as Carnot, he was a political opportunist during the stormy days of the French Revolution, which won him power and the title of marquis. His monumental work *Traite de mécanique céleste* won him the title the "Newton of France." To Napoleon's remark that God was not mentioned in his treatise, Laplace answered: "I did not need this hypothesis." Lagrange remarked later: "Ah, but it is a beautiful hypothesis."

In summary, we have seen that we can take the Laplace transform of a function or a sum of functions or an ordinary differential equation with constant coefficients such as equation (5.23). This leads to the determination of a function $X(s)$ which is the Laplace transform of the solution $x(t)$. If $X(s)$ appears on the left-hand side of the tables, the solution $x(t)$ follows immediately. Several rules given in part I of Appendix VI can assist in transforming the function to forms that are tabulated in part II.

We observe that in transforming equation (5.23), we have taken into account both the initial conditions x_0 and v_0 and the forcing functions $f(t)$. Therefore, the resulting function $x(t)$ is the total response of the system, transient and steady state.

This method is very general and can be used for a large class of problems which can be described with linear (partial or ordinary) differential equations provided that the forcing functions $f(t)$ are restricted to being piecewise continuous, and they do not go to infinity with time very rapidly. Also, they are nonzero only for positive values of time, although extension of the Laplace transform can be defined from $-\infty$ and $+\infty$.

□ **Example 5.9   Simple Demonstration of the Laplace Transform Method**

Determine the response of mass m sliding without friction on a horizontal plane acted upon by a constant horizontal force F_0 from time t_1 to t_2.

Solution   *Using the convolution integral:* Unit impulse at time $t = 0$ will result in initial velocity $F/m = 1/m$. Therefore, the motion will be

$$g(t) = \frac{t}{m}$$

The response to the force $f(t)$ will be

$$x(t) = \int_0^t F(\tau)g(t - \tau)\, d\tau, \qquad t < t_1, \quad F(\tau) = 0, \quad x(t) = 0, \quad t_1 < t < t_2$$

$$x(t) = \int_0^{t_1} + \int_{t_1}^t = \int_{t_1}^t \frac{F_0(t - \tau)}{m}\, d\tau = \frac{F_0}{m}\frac{(t - t_1)^2}{2}, \qquad t_2 < t,$$

$$x(t) = \int_0^t \frac{F_0(t - \tau)}{m}\, d\tau = \frac{F_0}{m}\left[\frac{(t - t_1)^2}{2} - \frac{(t - t_2)^2}{2}\right]$$

Using the method of Laplace transform: The equation of motion is $m\ddot{x} = f(t)$. Taking the transforms, we obtain

$$s^2 m x = F(s) = \frac{F_0(e^{-t_1 s} - e^{-t_2 s})}{s}$$

$$x = \frac{F_0}{m}\frac{e^{-t_1 s} - e^{-t_2 s}}{s^3}$$

The shifting property (see Appendix III) is

$$L^{-1}[e^{-as}F(s)] = \begin{cases} F(t - a) & \text{for } t > a \\ 0 & \text{for } t < a \end{cases}$$

Therefore,

$$L^{-1}\left[\frac{e^{-as}}{s^3}\right] = \begin{cases} 0, & 0 < t < a \\ \dfrac{F_0}{m}\dfrac{(t - t_1)^2}{2}, & a < t \end{cases}$$

and

$$x(t) = \begin{cases} 0, & 0 < t < t_1 \\ \dfrac{F_0}{m} \dfrac{(t - t_1)^2 - (t - t_2)^2}{2}, & t_2 < t \end{cases}$$

same as with the convolution integral.

(This is a simple demonstration of the method of Laplace transform and the convolution integral, for comparison.)

☐

☐ **Example 5.10   Pulse-Excited Single-Degree-of-Freedom System with the Laplace Transform Method**

Determine the response of the harmonic oscillator (m,c,k) to the excitation shown in Figure E5.10, a constant force F_0 acting for time t_0.

Figure E5.10

Solution   The differential equation of motion is $m\ddot{x} + c\dot{x} + kx = f(t)$. We take the Laplace transform of the equation, observing that $L\{f(t)\} = (1 - e^{-t_0 s})/s$. Therefore,

$$m[-sx(0) - \dot{x}(0) + s^2 X(s)] + c[-x(0) + sX(s)] + kX(s) = \frac{F_0(1 - e^{-t_0 s})}{s}$$

or

$$(ms^2 + cs + k)X(s) = \frac{F_0(1 - e^{-t_0 s})}{s} + (ms + c)x(0) + m\dot{x}(0)$$

But $x(0) = 0$ and $\dot{x}(0) = 0$; therefore,

$$X(s) = \frac{(1 - e^{-t_0 s})F_0}{s(ms^2 + cs + k)} = X_1(s) + x_2(s)$$

$$X_1(s) = \frac{F_0}{s(ms^2 + cs + k)}, \qquad X_2(s) = \frac{-F_0 e^{-t_0 s}}{s(ms^2 + cs + k)}$$

We shall find the inverse transforms of $X_1(s)$ and $X_2(s)$. We note that $X_1(s)$ is of the form

$$\frac{P(s)}{Q(s)}, \qquad P(s) = F, \ Q(s) = ms^3 + cs^2 + ks$$

Using property 15 of Appendix III, we obtain

$$Q' = 3ms^2 + 2cs + k$$

and the roots of $Q(s)$ are

$$a_1 = 0, \qquad a_{2,3} = \frac{-c}{2m} \mp \frac{(c^2 - 4km)^{1/2}}{2m}$$

$$x_1(t) = F_0\left(\frac{1}{k} + \frac{e^{a_2 t}}{3ma_2^2 + 2ca_2 + k} + \frac{e^{a_3 t}}{3ma_3^2 + 2ca_3 + k}\right)$$

For the $X_2(s)$ we note that it is of the form $X_2(s) = -e^{at}X_1(s)$. Using property (4) of Appendix IV, we obtain

$$x_2(t) = -F_0\left(\frac{1}{k} + \frac{e^{a_2(t-t_0)}}{3ma_2^2 + 2ca_2 + k} + \frac{e^{a_3(t-t_0)}}{3ma_3^2 + 2ca_3 + k}\right),$$

$$t > t_0, \quad x_2(t) = 0 \quad \text{for } t < t_0$$

(This example demonstrates the Laplace transform method applied to a pulse excitation on a single-degree-of-freedom system.)

□

5.5 DIRECT NUMERICAL INTEGRATION

For many forcing functions, analytical evaluation of the convolution integral is possible. In other cases this might be difficult or impossible, especially when the force is known as a sampled function measured experimentally. Let there be $n + 1$ values of this function, $f_0, f_1, f_2, \ldots, f_n$, at equidistant times with a time step Δt. The integral of this function over the time 0 to $n\,\Delta t$ can be approximated using the trapezoidal rule, summing up the areas of the trapezoids formed by successive values of the force and the t-axis,

$$I = \sum_{i=0}^{n} \epsilon_i f(t_i)\,\Delta t \tag{5.30}$$

where $\epsilon_i = 0.5$ for $i = 0$ or n and 1 otherwise.

The response of the system (m,c,k) at time $i\,\Delta t$ can be evaluated with the convolution integral using the trapezoidal rule as

$$x(i\,\Delta t) = \sum_{j=0}^{i} \epsilon_j f(j\,\Delta t) g(i\,\Delta t - j\,\Delta t)\,\Delta t \tag{5.31}$$

An alternative to the numerical evaluation of the convolution integral is direct integration of the differential equation with a numerical method that we discuss later when dealing with nonlinear systems.

There is no a priori preference for one or the other method; it depends on the problem. In general, for smooth forcing functions and if the force sampling time step Δt is of the same or one less order of magnitude with the period of natural vibration of the system T_n, and the length of the total time record desired is small, numerical evaluation of the convolution integral is faster and more accurate.

PROCEDURE 5.1: Trapezoidal Rule (TR)

/* Numerical evaluation of the integral of a function f(x)

```
given at n+1 points at equal distances Δx                            */
TrapezoidalRule[n,Δt,f(n+1),Value]
Declarations:
  Integer n
  Floating Value
  Floating or Integer Array f(0 TO n)
begin
  Value=0
  For i=0 TO n Step 1 Do Loop
  If i=0 OR i=n Then ε_i=0.5 Else ε_i=1
  Value=Value+ε_i*f(i)
  End Loop
  Value=Value*Δt
end
```

PROCEDURE 5.2: Response to General Excitation (GE)

```
\* Response of a harmonic oscillator to General Excitation
The force is defined at n+1 equidistant points, It
returns the maximum response xmax                            *\
GeneralExc[m,c,k,n,Δt,f(n+1),Nmax,xmax]
Declarations:
  Integer n
  Floating Tmax,n,c,k,xmax
  Floating or Integer Arrays f(0 TO n), F(0 TO Nmax)
begin
  ω_d = sqrt[k/m-(c/2m)²]
  xmax=0
  Loop i=0 TO Nmax Step 1
  t=i*Δt
  Loop j=1 to i step 1
  τ=jΔt
  If i<=n then
  g=exp[(-c/(2m))(t-τ)]sin(ω_d(t-τ))/(mω_d)
  F(j)=f(i)g
  else
  F(j)=0
  end if
  End Loop j
  Call TrapezoidalRule[n,Δt,F(i+1),Value]
  print scrn: t, value
  print file: t, value
  if |value|> xmax then xmax=|value|
  End Loop i
end
```

5.6 RESPONSE SPECTRA

Generally, the engineer is not interested in the response at all values of time but
in the important features of the response that will affect the operation or integrity
of the machine or structure. Such an important feature is the maximum displace-

ment; notable others are maximum velocity and acceleration. Since the response of the system will depend on the system parameters T_n and ζ, response features can be plotted against T_n with ζ as a parameter. Such plots are called *response spectra*.

An important response spectrum is the *shock spectrum*, defined as the maximum displacement amplification factor $x_{max}/(F_0/k)$ due to an impulsive force of a certain maximum value F_0 and duration T as function of T/T_n and ζ.

A typical assumed form of a shock is a half-sinusoidal function $f(t) = \sqrt{2}F_0 \sin \pi t/T$ for $0 < t < T$ and 0 otherwise. The response spectrum is shown in Figure 5.2, normalized in respect to the impulse $F_m T$, where F_m is the average value of the force during the shock. If the shock is a square pulse, $F_0 = F_m$ and the response spectrum has the constant value 2 in Figure 5.2 for $T > T_n/2$ and $2 \sin \omega T_n/2$ otherwise (see Example 5.5). It is apparent that for short durations of the shock, smaller than half the period of natural vibration of the system, the amount of impulse is important and the exact history of the shock has a small effect on the resulting vibration.

Support motion $y(t)$ can introduce shock, due mainly to elastic support of equipment. The differential equation in terms of the relative motion $z = x - y$ was developed in Chapter 3:

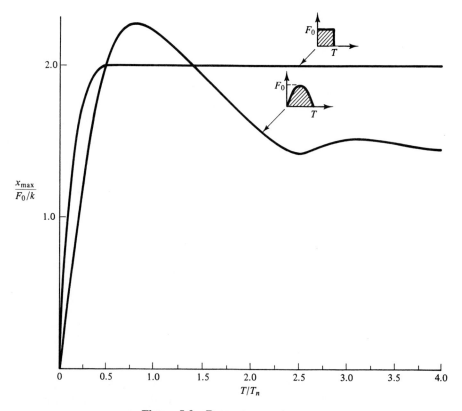

Figure 5.2 Response spectrum.

$$m\ddot{z} + c\dot{z} + kz = -m\ddot{y} \qquad (5.32)$$

$$\ddot{z} + 2\omega_n\zeta\dot{z} + \omega_n^2 z = -\ddot{y}$$

The shock spectrum is the same as with the force excitation with the scale $z_{max}/(\ddot{y}_{max}/\omega_n^2)$ in Figure 5.2 instead of $x_{max}/(F_0/k)$.

5.7 EARTHQUAKE RESPONSE

Earthquakes are violent and rather irregular ground motions that can severely damage machines and structures. When a structure or a component piece of equipment is subjected to earthquake motions, its base or support tends to move with the ground. If the component is rigid, it moves with the motion of base. If it is flexible, the component may be subject to deflections and loads considerably different from those of the base motion.

Unfortunately, the exact form of the earthquakes is subject to a high degree of uncertainty because of the very few available observations of strong earthquakes for long periods. One earthquake record used extensively as a basis of design in the United States is the May 18, 1940 El Centro earthquake. A typical ground motion measured during this earthquake is shown in Figure 5.3.

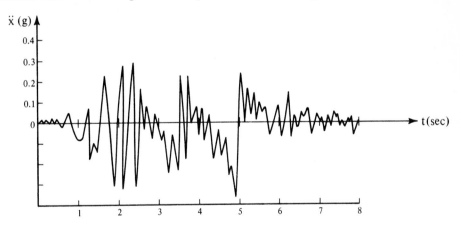

Figure 5.3 Motion measured during a strong earthquake.

An earthquake can have any direction, vertical or horizontal. The simplest way to consider the system is on the basis of one degree of freedom. Typical one-degree-of-freedom modeling of systems is shown in Figure 5.4. The system should be considered separately for horizontal and vertical vibration. Assume, in general, vibration in the x direction with an assumed ground motion y in the same direction. The equation of motion will be

$$m\ddot{x} + c\dot{x} + kx = c\dot{y} + ky \qquad (5.33)$$

or

$$\ddot{x} + 2\zeta\omega_n\dot{x} + \omega_n^2 x = 2\zeta\omega_n\dot{y} + \omega_n^2 y \qquad (5.34)$$

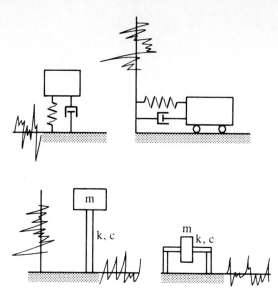

Figure 5.4 Typical one-degree-of-freedom models of systems for earthquake analysis.

where ζ and ω_n are the damping ratio and the natural frequency, respectively. If the ground displacement y and velocity $\dot{y}$ are known, from Figure 5.3, for example, one can compute the maximum value of the response $x(t)$ using the convolution integral or any other suitable method.[5] For the El Centro earthquake, this computation has been performed by numerical integration, and maximum incurred velocities have been plotted in Figure 5.5. versus the natural period $T_n = 2\pi/\omega_n$ of the structure. To obtain one point on this graph, one should do the following:

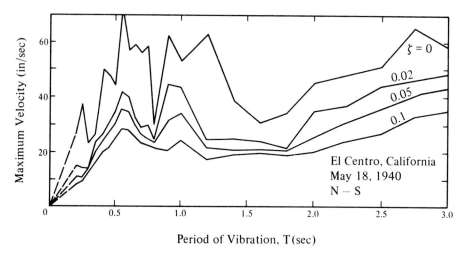

Figure 5.5 El Centro earthquake response spectrum.

[5]Housner, G. S. 1961. Vibration of structures induced by seismic waves. Chapter 50 in Harris, C. M., and Crede, C. E. *Shock and Vibration Handbook*. New York: McGraw-Hill.

1. Assume a value of ω_n and ζ.
2. Solve equation (5.33) by way of the convolution integral with input y and $\dot{y}$ of a certain record of the El Centro earthquake, for the duration of the earthquake.
3. Find the maximum velocity v_{max} for the duration of the motion.
4. Plot v_{max} versus T_n with parameter ζ.

As we can see from equation (5.33), the parameters governing the response are the damping ratio ζ and the undamped natural frequency ω_n. One has to estimate them and from Figure 5.5 compute the structure response (displacement, velocity, or acceleration). Values of the damping ratio (100ζ) are given in Tables 3.1. and 5.1. The natural frequency of the structure can be determined with the usual methods.

TABLE 5.1 DESIGN VALUES FOR STRUCTURAL DAMPING

Stress level	Type and conditions of Structure	Percentage of critical, ζ	Loss factor, γ	Logarithmic decrement, δ
1. Low, well below proportional limit, stresses below $\frac{1}{4}$ yield point	a. Vital piping b. Steel, reinforced, or prestressed concrete, wood; no cracking, no joint slip	0.5 0.5–1.0	0.01 0.01–0.02	0.03 0.03–0.06
2. More than about $\frac{1}{2}$ yield point	a. Vital piping b. Welded steel, prestressed concrete, well-reinforced concrete (only slight cracking) c. Reinforced concrete with considerable cracking d. Bolted and/or riveted steel, wood structures with nailed or bolted joints	0.5–1.0 3–5 5–7 5–7	0.01–0.02 0.04 0.06–0.10 0.1–0.14	0.03–0.06 0.125 0.20–0.30 0.30–0.45
3. At or just below yield point	a. Vital piping b. Welded steel, prestressed concrete (without complete loss in prestress) c. Prestressed concrete with no prestress left d. Reinforced concrete e. Bolted and/or riveted steel, wood structures with bolted joints	2 5 7 7–10 10–15	0.04 0.10 0.14 0.14–20	0.125 0.30 0.45 0.45–0.63

(continues)

TABLE 5.1 *(cont.)*

Stress level	Type and conditions of Structure	Percentage of critical, ζ	Loss factor, γ	Logarithmic decrement, δ
	f. Wood structures with nailed joints	15–20	0.30–40	1–1.25
4. Beyond yield point, with permanent strain greater than yield point limit strain	a. Piping	5	0.10	0.3
	b. Welded steel	7–10	0.14–0.20	0.45–0.63
	c. Prestressed concrete, reinforced concrete	10–15	0.20–0.30	0.63–1
	d. Bolted and/or riveted steel or wood structures	20	0.40	1.25
5. All ranges	Rocking of entire structure			
	a. On rock	2–5	0.04–10	0.125–0.30
	b. On firm soil	5–7	0.10–14	0.30–0.45
	c. On soft soil	7–10	0.14–20	0.45–0.63

In Figure 5.6, a convenient nomogram is shown which can provide displacement, velocity, and acceleration on the same graph. This nomogram, called the *Housner design spectrum*, is derived from an earthquake analysis such as that in

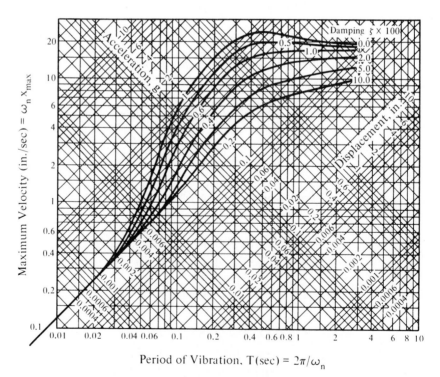

Figure 5.6 Housner's design spectrum.

Figure 5.5 by taking a smooth envelope of the response of 0.2g at low periods of natural vibration of the structure, a typical design value.

In certain cases, light equipment are attached to structures with considerably higher mass: for example, piping in a large reactor building. In this case we calculate the response of the *primary component* (building, reactor, etc.) and use this as base excitation for the *secondary component*.

☐ **Example 5.11   Earthquake Response of a Water Reservoir**

A 4545-kg water reservoir is supported by a 150-mm-diameter steel pipe 6 mm thick and 3.6 m long (Figure E5.11). From similar constructions we know that the damping ratio is 15%.

(a) Find the maximum horizontal displacement of the reservoir due to a strong horizontal earthquake according to the Housner design spectrum.

(b) Find the maximum stress at the root A of the support for this displacement.

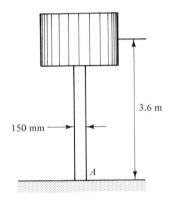

3.6 m

150 mm

A

Figure E5.11

Solution   (a) The area moment of inertia of the pipe section is $I = 2\pi r^3 t = 17.6 \times 10^6$ mm^4. Therefore, the spring constant of the cantilever beam pipe is $k = 3EI/l^3 = 2.3 \times 10^5$ N/m. The natural frequency is

$$\omega_n = \left(\frac{k}{m}\right)^{1/2} = 7.02 \text{ rad/s}, \qquad T = 0.8945 \text{ s}$$

From Figure 5.6, $(x_{max}) = 38.1$ mm.

(b) The horizontal force $F = kx_{max} = 8539$ N, and the moment at A, $M = Fh = 49.2 \times 10^6$ N·m.

$$\sigma_{max} = \frac{M}{I} r = \frac{49.2 \times 10^3}{17.6 \times 10^6} \times 75 = 13.460 \text{ N/mm}^2$$

(In this example, the earthquake spectrum is used to obtain the maximum earthquake loading of a single-degree-of-freedom system.)

☐

☐ **Example 5.12   Earthquake Response of a Turbine-Generator Foundation**

A large turbine-generator foundation (Figure E5.12) has a horizontal natural frequency of 15 Hz. Determine the clearance from the neighboring structure, of natural frequency 40 Hz, necessary to avoid collision during a strong horizontal earthquake

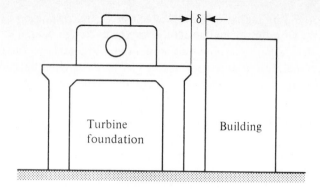

Figure E5.12

according to the Housner design spectrum. Assume that the logarithmic decrement $\delta = 0.4$.

Solution For concrete foundations, $\delta = 0.4$; therefore, $\delta/2\pi = 0.06$. For the foundation, $T_1 = 1/f_1 = 1/15 = 0.67$ s; for the neighboring structure, $T_2 = 1/f_2 = 1/40 = 0.025$ s. From Figure 5.6, for $\zeta = 6\%$ we obtain maximum displacements

$$(x_1)_{max} = 0.28 \text{ mm}, \qquad (x_2)_{max} = 0.025 \text{ mm}$$

The worse case is when they vibrate at a 180° angle. Therefore, the clearance should be

$$(x_1)_{max} + (x_2)_{max} = 0.305 \text{ mm}$$

(In this example, the earthquake spectrum is used to obtain the maximum earthquake loading of a simple structure.)

□

□ **Example 5.13   Earthquake Response of a Train on a Bridge**

A train is over a concrete bridge of $\omega_n = 10$ rad/s when an earthquake shakes the bridge in the vertical direction (Figure E5.13). Assuming the train to be a rigid mass, determine whether the train will derail or not, according to the Housner design spectrum.

Solution From Table 3.1, for steel bridges, $\delta = 0.02$ to 0.15. We select the most conservative value, $\delta = 0.02$, and get $\zeta = \delta/2\pi = 0.003 = 0.3\%$. The natural period of vibration $T_n = 2\pi/\omega = 0.628$ s. From Figure 5.6, we obtain maximum acceleration $0.6g$ or $(\ddot{x})_{max} = 5.9$ m/s². Clearly, this is less than g; therefore, the train will not derail.

(In this example, the earthquake spectrum is used to obtain the maximum earthquake loading of a structure, considered as a single-degree-of-freedom system.)

□

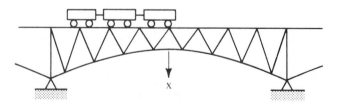

Figure E5.13

5.8 PIECEWISE LINEARIZATION

To obtain the response of a system to a given earthquake or other experimentally determined forcing function, one has to resort to numerical methods. The trapezoidal rule (Section 5.5) can be used, but the computation effort is excessive. For some problems the method of piecewise linearization reduces the computation effort considerably.[6]

The forcing function is assumed to be a sampled function of time, known by its values F_1, F_2, . . . at discrete values of time t_1, t_2, ... The forcing function between the sample points is assumed to be linear (Figure 5.7). Therefore, the forcing function between times t_j and t_{j+1} is

$$F(s) = F_j + \frac{\Delta F_j}{\Delta t} s \qquad (5.35)$$

where $\Delta t_j = t_{j+1} - t_j$, $s = t - t_j$, and $\Delta F_j = F_{j+1} - F_j$.

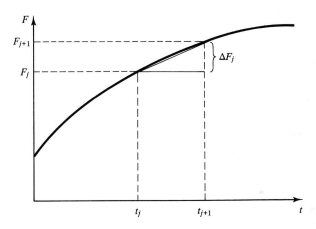

Figure 5.7 Piecewise linearization procedure.

The response of a single-degree-of-freedom system (m, c, k) to such a function is given by equation (5.22), with $F(s)$ given by equation (5.35).

$$x(t) = \int_0^s \frac{F(\tau)}{m\omega_d} e^{-(c/2m)(t-\tau)} \sin \omega_d(t - \tau) \, d\tau$$

$$+ e^{-(c/2m)s} \frac{x_j \cos \omega_d s + (x_j c/m + 2v_j)}{2\omega_d} \sin \omega_d s \qquad (5.36)$$

where x_j and v_j are the displacement and velocity at time t_j. Using equation (5.35) and integrating yields

$$x(s) = x_1(s) + x_2(s) + x_3(s) \qquad (5.37)$$

[6]Newmark, N. M. 1959. A method of computation for structural dynamics. *ASCE J. Eng. Mech. Div.*, 85(EM3): 67–94.

where

$$x_1(s) = \frac{F_j}{k}\left[1 - E\left(C + \frac{c}{2m\omega_d}S\right)\right]$$

$$x_2(s) = \frac{\Delta f_j}{\Delta t_j k\omega_d^2}\left\{\omega_d^2 s - \frac{c}{m} + E\left[\frac{c}{m}C - \frac{(\omega_d^2 - c^2/4m^2)S}{\omega_d}\right]\right\}$$

$$x_3(s) = E\left(x_j C + \frac{(x_j c/m + 2v_j)S}{2\omega_d}\right)$$

$$E = e^{-(c/2m)s}$$

$$S = \sin \omega_d s$$

$$C = \cos \omega_d s$$

The velocity at time s is obtained by differentiation of equations (5.37). At point $j + 1$, $s = \Delta t_j$ and then

$$x_{j+1} = a_{11}x_j + a_{12}v_j + a_{13}F_j + a_{14}F_{j+1}$$
$$v_{j+1} = a_{21}x_j + a_{22}v_j + a_{23}F_j + a_{24}F_{j+1}$$
(5.38)

where

$$a_{11} = E\left(C + \frac{Sc}{2m\omega_d}\right), \qquad a_{12} = ES$$

$$a_{13} = 1 - E(C + Sc/2m\omega_d) - \frac{1 - E(C + Sc/2m\omega_d)}{k\,\Delta t_j}$$

$$a_{14} = \frac{1 - E(C + Sc/2m\omega_d)}{k\,\Delta t_j}$$

$$a_{21} = -ES\left(\omega_d + \frac{c^2}{4m^2\omega_d}\right), \qquad a_{22} = EC$$

$$a_{23} = ES\omega_d/k - \frac{1 - E(C + Sc/2m\omega_d)}{k\,\Delta t_j}$$

$$a_{24} = \frac{1 - E(C + Sc/2m\omega_d)}{k\,\Delta t_j}$$

$$E = e^{-(c/2m)\Delta t_j}$$

$$S = \sin \omega_d\,\Delta t_j$$

$$C = \cos \omega_d\,\Delta t_j$$

Equations (5.38) can be used for a successive forward integration if the function $f(t)$ is known in advance. In some problems $f(t)$ depends on $x(t)$, and then the value F_{j+1} is not known in advance. In this case one starts with a linear extrapolation F_{j+1}^* from the previous values and then repeats the application of equations (5.38), computing F_{j+1} with the value of x_{j+1} obtained with the extrapolated value F_{j+1}^*.

Usually, the sampling time step Δt is constant. This renders the linear coefficients a_{ij} constant with time. This facilitates computations considerably because a_{ij} are computed only once and then the linear formulas (5.38) are applied successively. The procedure PWL implements the piecewise linearization method.

PROCEDURE 5.3: Response to General Excitation with Piecewise Linearization (PWL)

```
\* Response of a harmonic oscillator to General Excitation
with piecewise linearization of the forcing function.
The force is defined at n+1 equidistant points. It
returns the maximum response xmax             *\
PWL[m,c,k,n,Δt,f(n+1),Nmax,xmax]
Declarations:
Integer n,Nmax
Floating Tmax,n,c,k,xmax,xj,xj+1,vj,vj+1,value
a12,a12,a13,a14,a21,a22,a23,a24,E,S,C
Floating or Integer Array f(0 TO n)
begin
open an ASCII file named "PWL.RES" as #1
ωd = sqrt[k/m-(c/2m)²]
E=e-(c/2m)Δt
S=sinωdΔt
C=cosωdΔt
a11=E(C+Sc/2mωd), a12=ES:
a13=1-E(C+Sc/2mωd)-[-E(C+Sc/2mωd)]/kΔt
a14=[1-E(C+Sc/2mωd)]/kΔt
a21=-ES(ωd+c²/4m²ωd), a22=EC,
a23=ESωd/k-[1-E(C+Sc/2mωd)]/kΔt
a24=[1-E(C+Sc/2mωd)]/kΔt
xmax=0
Loop j=0 TO Nmax Step 1
t=j*Δt
xj+1=a11xj + a12vj + a13Fj + a14Fj+1
vj+1=a21xj + a22vj + a23Fj + a24Fj+1
value=xj+1
print scrn: t, value; prompt and clear screen every 20 lines
print in file #1: t, value
if |value|> xmax then xmax=|value|
End Loop j
close file #1
```

SIMULAB Experiment 5.1

Using the simulated experiment LINSYS, compute the motion of an electronic device with mass $m = 2$ kg elastically supported with $k = 10$ N/m and $c = 5$ N·s/m in a vehicle that crashes when deceleration of the vehicle is known from experiments to be

Time (s)	Acceleration (g)
0.1	0
0.2	0.5
0.3	1
0.4	1.5
0.5	2.0
0.6	1.5
0.7	1
0.8	0.5
0.9	0
1	0

Solution   The differential equation of motion of the system is equation (3.81). We prepare a file with the following data:

Number of records, Δt: 10, .1
10 values of $-m\ddot{y}$: 0, 1, 2, 3, 4, 3, 2, 1, 0, 0

We save it with the name S5_1.DAT. With the program diskette in drive A, we type

A> LINESYS

The page of Figure SE5.1a appears on the screen. We select a one-mass system. Then the menu page appears on the screen with the simulated one-degree-of-freedom system (Figure SE5.1b). With the keypad arrow keys we select menus.

1. We set $m = 2$ kg, $k = 10$ N/m, and $c = 5$ N·s/m using the + and − keys.
2. We specify base excitation, nonharmonic. At the prompt we enter the file name S5_1.DAT.
3. We hit S to start system operation. The resulting motion is shown in Figure SE5.1b.

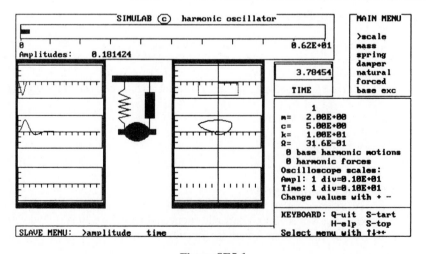

Figure SE5.1

TABLE 5.2 SUMMARY OF EQUATIONS OF CHAPTER 5

Periodic f(t)

$$m\ddot{x} + c\dot{x} + kx = a_0/2 + \sum_{n=1}^{\infty} (a_n \cos n\omega t + b_n \sin n\omega t)$$

Response

$$x = a_0/2k + \sum_{n=1}^{\infty} \frac{[a_n \cos(n\omega t - \phi_n) + b_n \sin(n\omega t - \phi_n)]}{[(k - n^2\omega^2 m)^2 + (n\omega c)^2]^{1/2}}$$

$$\phi_n = \arctan \frac{c n\omega}{(k - mn^2\omega^2)}$$

Duhamel's Integral

$$x(t) = \int_0^t F(\tau) g(t - \tau)\, d\tau$$

General excitation

$$x(t) = \int_0^t \frac{F(\tau)}{m\omega_d} e^{(-c/2m)(t-\tau)} \sin \omega_d(t - \tau)\, d\tau$$

$$+ e^{-(c/2m)t}\left[x_0 \cos \omega_d t + \frac{(x_0 c/m + 2v_0)}{2\omega_d} \right] \sin \omega_d t$$

Laplace Transform

$$L\{x(t)\} = X(s) = \int_0^\infty x(t) e^{-st}\, dt$$

REFERENCES AND FURTHER READING

Bykhovsky, I. I. 1969. *Fundamentals of Vibration Engineering*. Moscow: Machinostroienye.

Den Hartog, J. P. 1952. *Mechanical Vibration*, 4th ed. New York: McGraw-Hill.

Dimarogonas, A. D. 1976. *Vibration Engineering*. St. Paul, Minn.: West Publishing.

Housner, G. W., and Hudson, D. E. 1950. *Applied Mechanics: Dynamics*. Princeton, N.J.: D. Van Nostrand.

Klotter, K. 1960. *Technische Schwingungslehre*. Berlin: Springer-Verlag.

LaLanne, M., Berthier, P., and Der Hagopian, J. 1983. *Mécanique des vibrations linéaires*. New York: Wiley.

Pontryagin, L. S. 1962. *Ordinary Differential Equations*. Reading, Mass.: Addison-Wesley.

Rayleigh, J. W. S. 1894. *Theory of Sound*. New York: Dover Publications, 1946.

Timoshenko, S. P. 1953. *History of the Strength of Materials*. New York: McGraw-Hill.

Weaver, W., Jr., Timoshenko, S. P., and Young, D. H. 1990. *Vibration Problems in Engineering*, 5th ed. New York: Wiley.

PROBLEMS

Sections 5.1 and 5.2

5.1–5.8. On the harmonic oscillator without damping (spring–mass), determine the response to an exciting force of the form shown in Figures P1.11 to P1.18, respectively. Plot the first three harmonics on a frequency spectrum diagram.

5.9. On a harmonic oscillator of the type shown in Figure E5.3a, the exciting force is given in Figure E1.3b. Determine the response of the system, in particular the first three harmonics, and plot them on a rotating vector diagram, $\omega/\omega_n = 2$, $\zeta = 0.1$.

5.10–5.17. Solve Problem 5.9 for the excitations of Figures P1.11 to P1.18.

5.18. On a large concrete block of mass $m = 350$ kg resting on four springs of constant $k = 8900$ N/m each, there is a one-cylinder internal combustion engine (Figure P5.18). If it rotates at 1000 rpm, the piston weighs 0.8 kg, and the other masses can be neglected, determine the vertical vibration of the block if $l = 100$ mm and $r = 36$ mm.

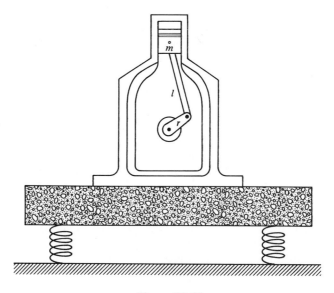

Figure P5.18

5.19. A single-stage steam turbine is connected to a synchronous electric generator. The rotor of the generator can be considered as rotating at a constant speed 3600 rpm. The turbine stage has seven blades and the eighth is missing, due to the way they mount the blades on the rotor. If the moment of inertia of the turbine stage is 2 kgm², the shaft has a diameter of 150 mm and length 635 mm, the shear modulus is $G = 85 \times 10^{11}$ N/m², and the power generated at the stage is 600 kW, determine the torsional vibration of the system due to the discontinuity of the power generation because of the missing blade.

Section 5.3

5.20. An automobile of mass 1400 kg is running over a road bump of shape giving the wheels a vertical motion as shown in Figure S1.2. If the shock absorbers of the car are worn ($c = 0$) and the period of natural vibration is 1.2 s, determine the response of the car assuming it to be a model as shown in Figure 2.2.

5.21–5.29. Solve Problem 5.20 for the excitations of Figures P5.34 to P5.42.

5.30. Find the time T corresponding to the peak response of the damped spring–mass system excited by a step force F_0 (Figure E5.10) for time t_0.

5.31. Show that the response $h(t)$ to a unit step function $F_0 = 1$ is related to the impulsive response $g(t)$ [equation (5.19)] as

$$g(t) = \frac{dh(t)}{dt}$$

5.32. Find the time T corresponding to the peak response for an impulsively excited one-degree-of-freedom system (m,c,k).

5.33. Determine the peak displacement for an undamped one-degree-of-freedom system excited by a force $F = F_0 e^{-\lambda t}$.

5.34–5.42. Determine the response of the harmonic oscillator (m,c,k) to the excitations shown in Figures P5.34 to P5.42.

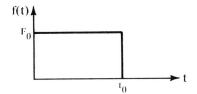

Figure P5.34

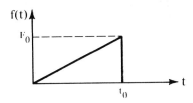

Figure P5.35

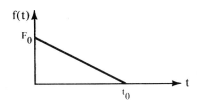

Figure P5.36

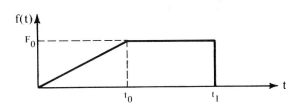

Figure P5.37

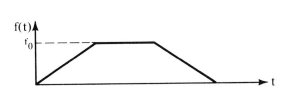

Figure P5.38

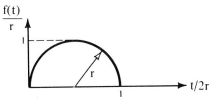

Figure P5.39

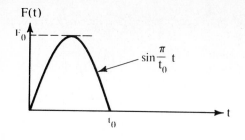

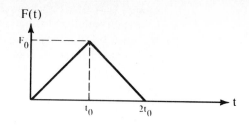

Figure P5.40

Figure P5.41

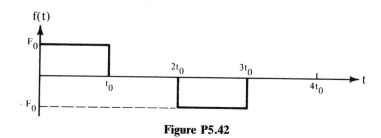

Figure P5.42

Section 5.4

5.43–5.21. Solve Problems 5.34 to 5.42 using the method of the Laplace transform.

Sections 5.5 to 5.8

5.52. On a strip-mining operation, the excavator consists of a beam of length $L = 50$ m, moment of inertia $I = 38$ cm^4, Young's modulus $E = 2.1 \times 10^{11}$ Pa, and an excavator head of weight $W = 1.5$ kN, diameter $D = 2$ m, and having six blades (Figure P5.52). If the wheel runs with a constant frequency of rotation 75 rpm and the static deflection during coal mining is 0.25 m, determine the lateral vibration of the wheel if one of the blades is damaged and presents 50% higher cutting force.

5.53. A plane mill is cutting a horizontal surface with a speed of 75 mm/s (Figure P5.53). When passing over a groove of width $b = 6$ mm, the cutting force drops from 1500 N to zero. If the weight of the cutting head is 200 N and the spring constant 400 kN/m, determine the maximum cutting error after the cutting tool passes over the groove, assuming the cutting force independent of the cutting depth and that the system has a damping constant 0.1 of critical damping c_c.

5.54. What will be the maximum acceleration from a strong earthquake of a steel bridge with a natural frequency of 3 Hz, based on the standard $0.20g$ Housner design spectrum?

5.55. Calculate the maximum stress on a circular on a pipe of inner diameter $d = 150$ mm, outer diameter $D = 162$ mm, and length $l = 1.60$ m, simply supported at the ends, filled with water, which is in a building during an earthquake according to the Housner design spectrum if the building's natural frequency is 1.5 Hz and logarithmic decrement is 0.05. The direction of the earthquake is perpendicular to the pipe.

5.56. For the building of Problem 5.55 there is an electric motor of mass 45 kg supported by springs of total constant $k = 356,000$ N/m. Assuming that the earthquake is harmonic and at the natural frequency of the building, calculate the maximum force

(a)

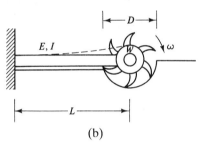

(b)

Figure P5.52

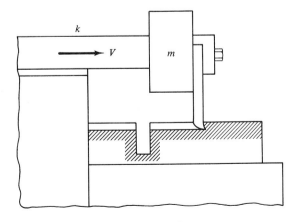

Figure P5.53

that the motor transmits to its base during the earthquake according to the Housner design spectrum.

5.57. A train is over a concrete bridge of $\omega_n = 2.5$ rad/s when an earthquake shakes the bridge in the vertical direction. Assuming the train to be rigid, determine whether the train will derail or not according to the Housner design spectrum.

5.58. A large turbine-generator foundation has a horizontal natural frequency of 2.5 Hz. Determine what should be the clearance from the neighboring foundation of natural frequency 40 Hz to avoid collision during a strong horizontal earthquake according to the Housner design spectrum.

5.59. Determine the maximum stress on a relatively rigid containment vessel of a nuclear reactor subject to a vertical strong earthquake according to the Housner design spectrum. Assume the vessel to be a hollow vertical cylinder of concrete, inner diameter 6 m, thickness 0.6 m, and height 18 m, bearing an additional mass at the top of 5 tons, a natural frequency of the vessel on its support of 25 Hz, and a logarithmic decrement of 0.1.

DESIGN PROBLEMS

D5.1. The maximum stress on a circular pipe of inner diameter $d = 150$ mm, outer diameter $D = 162$ mm, and length $l = 1.60$ m filled with water, which is in a building during an earthquake should not exceed 120 MPa, if the building natural frequency is 4 Hz and the logarithmic decrement is 0.05. The direction of the earthquake is perpendicular to the pipe. Design the elastic support of the pipe for an earthquake according to the Housner design spectrum.

D5.2. In the building of Problem D5.1, there is an electric motor of mass 45 kg elastically supported. Assuming that the earthquake is harmonic and at the natural frequency of the building, design the elastic support to reduce the motor acceleration during the earthquake to $0.1g$ according to the Housner design spectrum.

D5.3. An air compressor has unbalanced periodic horizontal forces 240 N at 25 Hz, 80 N at 50 Hz, and 40 N at 75 Hz. If the compressor mass is 200 kg, design a proper mounting to reduce the rms value of the transmitted force to 50 N.

D5.4. An electronic instrument has a 4-kg mass and must be packaged with elastic support so that when it falls from 3 m, the acceleration will not exceed $0.5g$. Design the packaging.

D5.5. A utility vehicle has a 300-kg mass. Design an elastically supported bumper so that a collision at a 10-mph speed will result in deceleration on the vehicle of less than $2g$.

D5.6. An electronic instrument has a 2-kg mass, and the acceptance test of the instrument on its mount is to be subject to a vertical sweep with a square waveform of 2 mm amplitude at frequencies of 10 to 100 Hz without the acceleration on the instrument exceeding $0.5g$. Design the mount.

6

Coupled
Vibrating
Systems

6.1 INTRODUCTION

Machines and structures consist, in general, of elements with distributed mass and elasticity. There are systems, however, which have massive and stiff elements connected with elements that have relatively small mass but high flexibility. Consider, for example, the beam of Figure 6.1, modeling an automobile suspension.

A beam of 100-kg mass and a stiffness, if simply supported and loaded at midspan, of 10^6 N/m is supported by two springs of 1-kg mass and stiffness 1000 N/m each. If a vertical force of 1000 N is imposed in the midspan of the beam, the two springs will deform by 1 m each, but the beam itself will deform only by $1000 \times 10^{-6} = 10^{-3}$ m. Similarly, assume that the beam and the springs have a vertical harmonic rigid-body motion of amplitude 1 mm at a 100-rad/s frequency. The inertia forces due to the beam and spring masses are $m\omega^2 x$, that is, $100 \times 100^2 \times 0.001 = 1000$ N and $2 \times 100^2 \times 0.001 = 20$ N, respectively. Therefore, for most practical purposes and for conditions near the assumed ones in the order-of-magnitude

*The above photo portrays a Sikorsky Helicopter lifting a truck through a flexible cable.

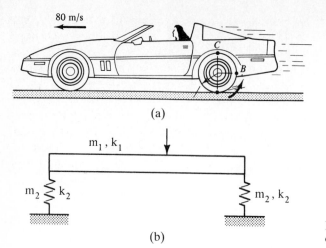

(a)

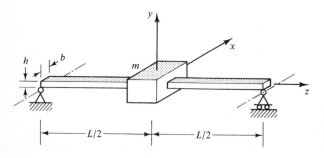

(b)

Figure 6.1 Simple elastic beam on elastic supports.

calculation above, we can consider this particular beam as a rigid body supported by two massless springs.

Another approximation discussed in Chapter 2 is the number of degrees of freedom of the system. The minimum system configuration would be a single mass of small dimensions, which in mechanics is called a *material particle*. This has three degrees of freedom. It seems that a one-degree-of-freedom system cannot exist unless the two degrees of freedom are constrained. Consider, however, the beam of Figure 6.2 of length L that has a rectangular cross section of height h and width b with a mass m much larger than the beam mass, which can thus be neglected, at midspan. If x and y are the lateral, horizontal and vertical, motions and z the axial motion of the mass, as indicated, Newton's law yields

$$m\ddot{x} + k_x x = 0$$

$$m\ddot{y} + k_y y = 0 \tag{6.1}$$

$$m\ddot{z} + k_z z = 0$$

where $k_x = 48E(hb^3/12)/L^3$, $k_y = 48E(bh^3/12)/L^3$, $k_z = E(bh)/(L/2)$, and E is the modulus of elasticity of the beam material.

Equations (6.1) show that the three motions x, y, and z are not related and the three differential equations can be dealt with separately. The system and the

Figure 6.2 Heavy mass at the mids pan of an elastically supported light beam.

three motions are said to be *uncoupled*. This observation leads to one-degree-of-freedom system modeling in cases for which we have an a priori knowledge that the motions along different coordinates are uncoupled.

This is not always the case. Consider, for example, the case of an overhead crane (Figure 6.3). The carriage has a mass m_2 and the payload has a mass m_1. If the beam and the cable are assumed massless elastic elements, the two motions x_1 and x_2 are not independent—they are coupled. Since the coupling is due to the elastic cable, a structural element, this type of coupling is called *structural coupling*.

Consider now the motion of the system in Figure 6.4. It can be described by the vertical motion of the two ends, for example, x_1 and x_2. These two motions

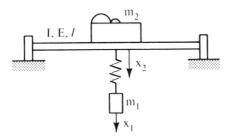

Figure 6.3 Model of an overhead crane for vibration analysis.

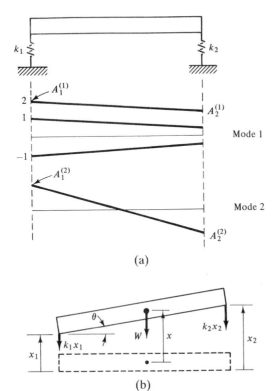

(a)

(b)

Figure 6.4 Vibration modes of an elastically supported rigid beam.

are not independent due to the system geometry; the system is coupled. This type of coupling is called *geometric coupling*.[1]

6.2 EQUATIONS OF MOTION FOR TWO-DEGREE-OF-FREEDOM COUPLED SYSTEMS

We shall first derive the differential equations of motion. Our strategy for the derivation of the equations of motion will follow closely the methods used for the one-degree-of-freedom system.

Application of Newton's Laws of Motion

1. The system is modeled first as an assembly of rigid bodies connected with massless elastic members.
2. The static equilibrium of the system is determined.
3. A number of coordinates that are necessary to describe the geometry of the system are identified. (In general, for a rigid body six coordinates in the space or three coordinates on the plane are needed, minus the number of coordinate constraints. It is generally convenient to select as coordinates the translations of the center of mass and the rotations about it.) Positive directions are designated.
4. The system is displaced from equilibrium to finite values of the selected coordinates, without violation of the imposed constraints.
5. At this position, the several forces and moments acting on the masses due to deflection of elastic elements, gravity, and so on, as functions of the selected coordinates are computed. Free-body diagrams for every mass are drawn.
6. The accelerations of the centers of mass of the rigid bodies of the system and their angular accelerations are expressed as functions of the coordinates selected. (One can express these accelerations in any convenient coordinates first and then transform into the coordinates selected.)
7. Newton's second law for the linear motion of the center of mass and the rotations about it is now applied. This must be done as many times as the number of degrees of freedom and the number of the system coordinates, plus the number of yet unknown forces of constraint.
8. The equations of motion can also be derived based on D'Alembert's principle. To this end, *effective forces* are applied on the mass centers equal to the products (mass $\times$ acceleration) and *effective moments* about the mass centers equal to the products (mass moment of inertia $\times$ angular acceleration). Then the conditions of static equilibrium $\Sigma F = \Sigma F_{\text{Eff}}$ and $\Sigma M = \Sigma M_{\text{Eff}}$ are expressed.

---

[1] Coupled vibration, though known to the ancients, was discussed by Galileo in his *Discorsi*, where he termed them "sympathetic vibrations." Galileo Galilei. 1638. *Discorsi e demonstrationi matematiche intorno a due nuove scienze attenenti alla mechanica e ai movimenti locali.* Translated by Crew, H., and de Salvio, A. 1939. Chicago: Rand McNally. However, the term was known to Polybios, 2nd Century B.C. and was used in relation to vibrating strings and vessels by Theon of Smyrna, 2nd Century A.D.

In the latter, also, the moments of the effective forces in the right-hand side should be included.

9. Unknown forces of constraint are then eliminated.

10. The terms containing the coordinates and their derivatives are placed on the left-hand side of the equations, and the constant terms and functions of time, on the right-hand side of the equations.

Consider the system in Figure 6.4, which models [Hort, 1910] the suspension system of an automobile.

Step 1: Assume the beam to be rigid and the springs k_1 and k_2 massless.

Step 2: Assume that the equilibrium is in the horizontal position and that the springs have been compressed initially to balance the weight of the beam.

Step 3: As coordinates, select the vertical displacements of the ends of the beam from equilibrium. Assume no motion in the horizontal direction.

Step 4: Assume the system to be initially in the dashed position. The ends are displaced by x_1 and x_2, Figure 6.4b.

Step 5: Draw a free-body diagram, with the forces k_1x_1, k_2x_2, and W and the reactions of the springs to the static load W.

Step 6: The vertical displacement x of the center of mass and the rotation are

$$x = \frac{x_1 + x_2}{2}, \qquad \theta = \frac{x_2 - x_1}{L} \tag{6.2}$$

where L is the length of the beam.

Differentiating twice with respect to time yields

$$a = \ddot{x} = \frac{\ddot{x}_1 + \ddot{x}_2}{2}, \qquad \alpha = \ddot{\theta} = \frac{\ddot{x}_2 - \ddot{x}_1}{L} \tag{6.3}$$

Step 7: Newton's law is applied:

$$ma = F \qquad \text{and} \qquad J_G\alpha = \Sigma M \tag{6.4}$$

or

$$\frac{m(\ddot{x}_1 + \ddot{x}_2)}{2} = -k_1x_1 - k_2x_2 - W + \frac{W}{2} + \frac{W}{2}$$

$$\frac{J_G(\ddot{x}_2 - \ddot{x}_1)}{L} = k_1Lx_1 - k_2Lx_2 - \frac{W}{2}L + \frac{W}{2}L \tag{6.5}$$

Step 8:

$$m\ddot{x}_1 + m\ddot{x}_2 + 2k_1x_1 + 2k_2x_2 = 0 \tag{6.6}$$

$$-J_G\ddot{x}_1 + J_G\ddot{x}_2 - k_1L^2x_1 + k_2L^2x_2 = 0$$

(In steps 7 and 8, observe that one could omit the static equilibrium forces W and $W/2$ in the first place.)

A different formulation.   Going back to step 3, select as coordinates x and θ. Then:

Step 5:

$$x_1 = x - L\theta, \qquad x_2 = x + L\theta \tag{6.7}$$

Therefore,

$$k_1 x_1 = k_1(x - L\theta), \qquad k_2 x_2 = k_2(x + L\theta) \tag{6.8}$$

Step 6:

$$a = \ddot{x}, \qquad \alpha = \ddot{\theta} \tag{6.9}$$

Step 7:

$$m\ddot{x} = -k_1(x - L\theta) - k_2(x + L\theta)$$
$$J_G\ddot{\theta} = k_1(x - L\theta)L - k_2(x + L\theta)L \tag{6.10}$$

Step 8:

$$m\ddot{x} + (k_1 + k_2)x + (k_2 - k_1)L\theta = 0$$
$$J_G\ddot{\theta} + (k_2 - k_1)Lx + (k_1 + k_2)L^2\theta = 0 \tag{6.11}$$

Use of Lagrange's Equations

In Chapter 4 an alternative formulation of the equations of motion was developed by way of Lagrange's equations. For the system at hand, the kinetic and potential energies have the form, neglecting gravity,

$$T = \tfrac{1}{2}(m\dot{x}^2 + J_G\dot{\theta}^2)$$
$$V = \tfrac{1}{2}(k_1 x_1^2 + k_2 x_2^2) \tag{6.12}$$

Substituting $x_1 = x - L\theta$ and $x_2 = x + L\theta$ gives us

$$V = \tfrac{1}{2}[k_1(x - L\theta)^2 + k_2(x + L\theta)^2]$$

$$L = T - V$$

$$\frac{\partial L}{\partial \dot{x}} = m\dot{x}, \qquad \frac{d(\partial L/\partial \dot{x})}{dt} = m\ddot{x}$$

$$\frac{\partial L}{\partial \dot{\theta}} = J_G\dot{\theta}, \qquad \frac{d(\partial L/\partial \dot{\theta})}{dt} = J_G\ddot{\theta}$$

$$\frac{\partial L}{\partial x} = -(k_1 + k_2)x + (k_2 - k_1)L\theta$$

$$\frac{\partial L}{\partial \theta} = (k_2 - k_1)Lx + (k_1 + k_2)L^2\theta$$

Application of Lagrange's equations gives

$$\frac{d(\partial L/\partial \dot{x})}{dt} - \frac{\partial L}{\partial x} = m\ddot{x} + (k_1 + k_2)x + (k_2 - k_1)L\theta = 0 \tag{6.13}$$

$$\frac{d(\partial L/\partial \dot{\theta})}{dt} - \frac{\partial L}{\partial \theta} = J_G\ddot{\theta} + (k_2 - k_1)Lx + (k_1 + k_2)L^2\theta = 0$$

These are identical to equations (6.11).

6.3 NATURAL VIBRATION OF COUPLED SYSTEMS

In the absence of external forces and damping, the systems of differential equations (6.6) or (6.11) are of the type

$$m_{11}\ddot{x}_1 + m_{12}\ddot{x}_2 + k_{11}x_1 + k_{12}x_2 = 0 \tag{6.14}$$

$$m_{21}\ddot{x}_1 + m_{22}\ddot{x}_2 + k_{21}x_1 + k_{22}x_2 = 0$$

The two equations are not independent, due to the coupling terms m_{12}, m_{21}, k_{12}, and k_{21}.

Another way to characterize coupling is if the coupling is due to the mass terms m_{12} or m_{21}—then we have *dynamic coupling*—or if the coupling is due to the stiffness terms k_{12} or k_{21}—when we have *static coupling*. As is apparent from equations (6.6) and (6.11), which describe the same system, the type of the coupling is not a system property but depends mainly on the selection of the coordinates. Later we shall prove that in undamped systems there always exist appropriate coordinates that yield uncoupled systems of differential equations [Rayleigh, 1894]. In general, we shall assume that we have a coupled system of differential equations, of the equation (6.14) type.

As we did for the harmonic oscillator, we shall study the feasibility of harmonic natural vibration of the system at a frequency yet unknown ω. Therefore, we try the solution

$$x_1 = A_1 \cos \omega t \tag{6.15}$$

$$x_2 = A_2 \cos \omega t$$

where the amplitudes A_1 and A_2 are yet undetermined. Substitution in (6.14) yields

$$(-m_{11}\omega^2 + k_{11})A_1 + (-m_{12}\omega^2 + k_{12})A_2 = 0 \tag{6.16}$$

$$(-m_{21}\omega^2 + k_{21})A_1 + (-m_{22}\omega^2 + k_{22})A_2 = 0$$

If this system of linear algebraic equations has a solution, (6.15) is indeed a solution of (6.14). The system (6.16) is homogeneous; therefore, it has a solution only if

$$\begin{vmatrix} -m_{11}\omega^2 + k_{11} & -m_{12}\omega^2 + k_{12} \\ -m_{21}\omega^2 + k_{21} & -m_{22}\omega^2 + k_{22} \end{vmatrix} = 0 \tag{6.17}$$

The determinant in equation (6.17) is called a *frequency determinant* and equation (6.17) is called a *frequency equation*. This equation contains only one unknown, ω. Because, in general, this equation can have two distinct roots ω_1 and ω_2, we can conclude that the system can undergo harmonic vibration at two frequencies, ω_1 and ω_2. Furthermore, we observe from equation (6.17) that the frequencies ω_1 and ω_2 depend on the system properties m and k *only*.

The amplitudes A_1 and A_2 of the harmonic vibration are not altogether arbitrary. Indeed, for every ω satisfying equation (6.17), there is a solution of the system (6.16). In fact, equation (6.17) is merely a condition that equations (6.16) are dependent. Therefore, for $\omega = \omega_1$ or $\omega = \omega_2$, there is only one independent equation relating A_1 and A_2. Assigning any arbitrary value to, say, A_1, we can compute A_2 from any one of (6.16). Using the first one, we have

$$A_2^{(1)} = \frac{m_{11}\omega_1^2 - k_{11}}{-m_{12}\omega_1^2 + k_{12}}, \qquad A_1^{(1)} = 1$$

$$A_2^{(2)} = \frac{m_{11}\omega_2^2 - k_{11}}{-m_{12}\omega_2^2 + k_{12}}, \qquad A_1^{(2)} = 1$$

(6.18)

We use superscripts 1 or 2 to indicate the respective circular frequencies ω_1 and ω_2 with which the amplitudes A are associated.

We conclude that the two-degree-of-freedom system can have harmonic *natural vibration* at two different frequencies ω_1 and ω_2, and at either of the two frequencies the amplitudes will have a certain ratio $A_2^{(1)}/A_1^{(1)}$ and $A_2^{(2)}/A_1^{(2)}$, respectively, which depends on the system properties.

The circular frequencies ω_1 and ω_2 are called *natural frequencies* and the couples $(A_1^{(1)}, A_2^{(1)})$ and $(A_1^{(2)}, A_2^{(2)})$ are *called natural modes* of the vibration of the system.

The natural mode of vibration has the meaning that if the system has a natural vibration at frequency ω_j, if at time t_1 the displacement along the coordinate x_1 is 1, the system coordinates will be

$$x_1 = 1, \quad x_2 = A_2^{(j)}, \quad \ldots$$

Moreover, the amplitudes related with the system coordinates will be in fixed proportions to each other during the motion. Furthermore, if the system is displaced from equilibrium initially by $x_1(0) = A_1^{(j)}$, $x_2(0) = A_2^{(j)}$, $j = 1$ or 2, it is apparent that the functions of equation (6.18) will also satisfy the initial conditions; therefore, the system will vibrate at frequency ω_j and amplitudes equal to the natural mode corresponding to ω_j.

The natural modes are usually plotted in diagrams such as Figure 6.4. Every mode has not fixed amplitudes, but the ratio of the amplitudes at any two coordinates remain constant for the same mode. Thus the shape of the mode always remains the same, regardless of the absolute value of the amplitudes. This is shown for the first mode of Figure 6.4, plotted for values of $A_1^{(1)} = 1, -1, 2$.

The following functions satisfy the differential equations: $\cos \omega_1 t$, $\cos \omega_2 t$, $\sin \omega_1 t$, $\sin \omega_2 t$, provided that the amplitudes at any frequency are related as the vibration modes. Therefore, the general solution of the natural vibration of the problem will be

$$x_1(t) = A_1^{(1)} \cos \omega_1 t + A_1^{(2)} \cos \omega_2 t + B_1^{(1)} \sin \omega_1 t + B_1^{(2)} \sin \omega_2 t$$

$$x_2(t) = A_2^{(1)} \cos \omega_1 t + A_2^{(2)} \cos \omega_2 t + B_2^{(1)} \sin \omega_1 t + B_2^{(2)} \sin \omega_2 t$$

(6.19)

There are four arbitrary parameters in equations (6.19): $A_1^{(1)}$, $A_1^{(2)}$, $B_1^{(1)}$, $B_1^{(2)}$, the arbitrary values of the first component of the vibration mode, as explained above. $A_2^{(1)}$, $A_2^{(2)}$, $B_2^{(1)}$, and $B_2^{(2)}$ are the components of the modes, which can be computed from $A_1^{(1)}$, $A_1^{(2)}$, $B_1^{(1)}$, and $B_1^{(2)}$. The modes $B_1^{(1)}$, $B_1^{(2)}$, $B_2^{(1)}$, and $B_2^{(2)}$ are essentially the same modes as the As, but different symbols are used because of the arbitrary selection of one of their components. The arbitrary parameters can be computed from the four initial conditions: $x_1(0)$, $\dot{x}_1(0)$, $x_2(0)$, and $\dot{x}_2(0)$, and the problem of natural vibration is solved. The solution can be written in the form

$$\begin{bmatrix} x_1(t) \\ x_2(t) \end{bmatrix} = c_1 \begin{bmatrix} A_1^{(1)} \\ A_2^{(1)} \end{bmatrix} \cos \omega_1 t + c_2 \begin{bmatrix} A_1^{(2)} \\ A_2^{(2)} \end{bmatrix} \cos \omega_2 t$$

$$+ c_3 \begin{bmatrix} A_1^{(1)} \\ A_2^{(1)} \end{bmatrix} \sin \omega_1 t + c_4 \begin{bmatrix} A_1^{(2)} \\ A_2^{(2)} \end{bmatrix} \sin \omega_2 t \qquad (6.19a)$$

where c_1, c_2, c_3, and c_4 are arbitrary constants. It can be seen that the vibration at any time consists of a sum of the two vibrating modes multiplied by factors that depend on the initial conditions.

The natural modes of vibration contain a great deal of information about the way the system vibrates and they have obvious engineering significance. Moreover, they are useful tools for further analysis, as we shall see in a subsequent chapter. There are many practical uses for the natural modes of vibration. For example, if the system has forced vibration, the resonance occurs when the exciting force is at a frequency equal to one of the natural frequencies. In this case, the vibration amplitudes are, in general, very similar to the corresponding natural mode. This does not reveal the actual vibration amplitudes; however, it shows where in the system these amplitudes are high.

☐ **Example 6.1   Vehicle–Trailer Dynamics**

An automobile of mass 1000 kg pulls a trailer of mass 2000 kg (Figure E6.1). If the flexibility of the hitch is 20,000 N/m, determine the natural frequencies and natural modes of the system.

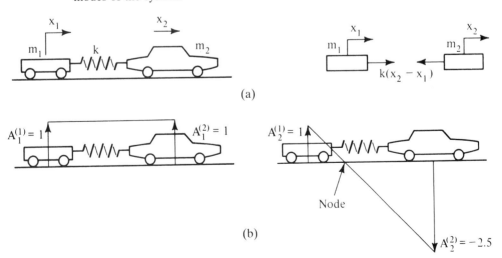

Figure E6.1

Solution From the free-body diagram on the right in Figure E6.1a, application of Newton's law in the horizontal direction yields

$$m_1\ddot{x}_1 = k(x_2 - x_1), \qquad m_1\ddot{x}_1 + kx_1 - kx_2 = 0$$

$$m_2\ddot{x}_2 = -k(x_2 - x_1), \qquad m_2\ddot{x}_2 - kx_1 + kx_2 = 0$$

We look for solutions $x_1 = A_2 \cos \omega t$, $x_2 = A_2 \cos \omega t$. We obtain

$$(-m_1\omega^2 + k)A_1 - kA_2 = 0$$

$$-kA_1 + (-m_2\omega^2 + k)A_2 = 0$$

The condition for solution of this homogeneous system is

$$\begin{vmatrix} -m_1\omega^2 + k & -k \\ -k & -m_2\omega^2 + k \end{vmatrix} = 0$$

or

$$(-m_1\omega^2 + k)(-m_2\omega^2 + k) - k^2 = 0$$

or

$$m_1 m_2 \omega^4 - k(m_1 + m_2)\omega^2 = 0$$

or

$$[m_1 m_2 \omega^2 - k(m_1 + m_2)]\omega^2 = 0$$

The roots of this equation are

$$\omega_1 = 0, \qquad \omega_2 = \left[\frac{k(m_1 + m_2)}{m_1 m_2} \right]^{1/2}$$

$\omega_1 = 0$ expresses the fact that the system can have rigid-body motion, and ω_2 is the natural frequency. For the data given

$$\omega_2 = \left[\frac{20,000(1000 + 2000)}{1000 \times 2000} \right]^{1/2} = 5.47 \text{ rad/s}$$

For $\omega_1 = 0$, the first of the homogeneous equations yields $A_1 = A_2$. Therefore, for $A_1^{(1)} = 1$ it is $A_2^{(1)} = 1$. Thus in the first natural mode, the two masses move by equal amplitudes, which is a rigid-body motion.

For $\omega = \omega_2$, we obtain

$$A_2 = \left(-\frac{m_1}{k}\omega_2^2 + 1 \right)A_1$$

For $A_1^{(2)} = 1$ it is

$$A_2^{(2)} = -\left(\frac{2000}{20,000} \right)5.47^2 + 1 = -2$$

(This example demonstrates the salient features of the static coupling of two masses with a connecting linear spring.)

$\square$

Example 6.2 Seismograph Dynamics [Hort 1910]

One type of seismograph, a device that records earthquakes, can be modeled as shown in Figure E6.2. For this model, determine (a) the differential equations of motion, (b) the frequency equation and the natural frequencies, and (c) the natural modes of vibration. (Assume small angles θ.)

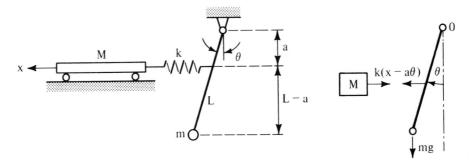

Figure E6.2

Solution (a) In a displaced position, the forces acting on the system are shown in the free-body diagram (Figure E6.2). Newton's law for the mass M yields, for small θ,

$$\Sigma F_x = M\ddot{x} \quad \text{or} \quad M\ddot{x} + kx - ka\theta = 0$$

For the pendulum

$$\Sigma M_o = J_o\ddot{\theta} \quad \text{or} \quad mL^2\ddot{\theta} + mgL\theta - ak(x - a\theta) = 0$$

(b) The frequency equation is

$$\begin{vmatrix} -M\omega^2 + k & -ka \\ -ka & -mL^2\omega^2 + mgL + a^2k \end{vmatrix} = 0$$

or

$$mML^2\omega^4 + [M(mgL + a^2k) - mL^2k]\omega^2 + mgLk = 0$$

$$\omega_{1,2} = \frac{-[M(mgL + a^2k) - mL^2k] \mp [[M(mgL + a^2k) - mL^2k]^2 - 4m^2MgL^3k]^{1/2}}{2mML^2}$$

(c) The natural modes are given by the equations

$$(-M\omega^2 + k)X - ka\Theta = 0$$

$$-kaX + (-m_L^2\omega^2 + mgL + a^2k)\Theta = 0$$

for $x = X \cos \omega t$ and $\theta = \Theta \cos \omega t$.

Therefore, the natural modes are

$$\left(\frac{X}{\Theta}\right)^{(1),(2)} = \frac{ka}{-M\omega_{1,2}^2 + k}$$

(This example demonstrates the determination of the natural frequencies and natural modes of a coupled two-degree-of-freedom system.) □

6.4 HARMONIC EXCITATION

If external forces are acting on a system, the equations of motion will be, for a two-degree-of-freedom coupled system, for example, with static coupling,

$$m_1\ddot{x}_1 + k_{11}x_1 + k_{12}x_2 = f_1(t)$$
$$m_2\ddot{x}_2 + k_{21}x_1 + k_{22}x_2 = f_2(t) \tag{6.20}$$

Assume, in addition, that the exciting forces $f_1(t)$ and $f_2(t)$ are harmonic, $f_1(t) = F_1 \cos \omega t$ and $f_2(t) = F_2 \cos \omega t$.

We shall seek a harmonic steady-state solution of the form

$$x_1 = A_1 \cos \omega t$$
$$x_2 = A_2 \cos \omega t \tag{6.21}$$

Substituting in (6.19) and canceling $\cos \omega t$, we obtain

$$(-m_1\omega^2 + k_{11})A_1 + k_{12}A_2 = F_1$$
$$k_{21}A_1 + (-m_2\omega^2 + k_{22})A_2 = F_2 \tag{6.22}$$

The system (6.22) is linear and nonhomogeneous and can be solved with Cramer's rule[2]:

$$A_1 = \frac{\begin{vmatrix} F_1 & k_{12} \\ F_2 & -m_2\omega^2 + k_{22} \end{vmatrix}}{\begin{vmatrix} -m_1\omega^2 + k_{11} & k_{12} \\ k_{21} & -m_2\omega^2 + k_{22} \end{vmatrix}}$$

$$A_2 = \frac{\begin{vmatrix} -m_1\omega^2 + k_{11} & F_1 \\ k_{21} & F_2 \end{vmatrix}}{\begin{vmatrix} -m_1\omega^2 + k_{11} & k_{12} \\ k_{21} & -m_2\omega^2 + k_{22} \end{vmatrix}} \tag{6.23}$$

We observe that the determinants in the denominators are the same with the frequency determinant of the system. Therefore, the vibration amplitudes A_1 and A_2 will be infinite if ω coincides with either one of the natural frequencies ω_1 and ω_2. Therefore, the natural frequencies are equal with the resonant frequencies.

In the case that in (6.20) we have viscous damping, the treatment is similar. Assuming for simplicity that $c_{12} = c_{21} = 0$,

$$m_1\ddot{x}_1 + c_1\dot{x}_1 + k_{11}x_1 + k_{12}x_2 = F_1e^{i\omega t}$$
$$m_2\ddot{x}_2 + c_2\dot{x}_2 + k_{21}x_1 + k_{22}x_2 = F_2e^{i\omega t} \tag{6.24}$$

[2]Gabriel Cramer (1704–1752). 1750. *Introduction à l'analyse des lignes courbes algebriques*. This rule was known to Maclaurin as the determinant method for solving linear equations (Maclaurin, C., 1748, *Treatise of Algebra*).

Substituting harmonic responses $x_1 = X_1 e^{i\omega t}$ and $x_2 = X_2 e^{i\omega t}$, we obtain

$$(-m_1\omega^2 + \omega i c_1 + k_{11})X_1 + \qquad\qquad k_{12}X_2 = F_1 \qquad (6.25)$$

$$k_{21}X_1 + (-m_2\omega^2 + i\omega c_2 + k_{22})X_2 = F_2$$

Cramer's rule can also be used here for the complex amplitudes X_1 and X_2:

$$X_1 = \frac{\begin{vmatrix} F_1 & k_{12} \\ F_2 & -m_2\omega^2 + i\omega c_2 + k_{22} \end{vmatrix}}{\begin{vmatrix} -m_1\omega^2 + i\omega c_1 + k_{11} & k_{21} \\ k_{21} & -m_2\omega^2 + i\omega c_2 + k_{22} \end{vmatrix}}$$

$$X_2 = \frac{\begin{vmatrix} -m_1\omega^2 + i\omega c_1 + k_{11} & F_1 \\ k_{21} & F_2 \end{vmatrix}}{\begin{vmatrix} -m_1\omega^2 + i\omega c_1 + k_{11} & k_{12} \\ k_{21} & -m_2\omega^2 + i\omega c_2 + k_{22} \end{vmatrix}}$$

To separate real and imaginary parts and compute the vibration amplitude and the phase angle, one can proceed using the methods of Chapter 3 (see Example 6.3).

☐ **Example 6.3 Vibration of a Milling Machine**

Determine the steady-state vertical vibration of the elastically supported milling machine modeled as shown in Figure E6.3.
Solution The equations of motion for forced damped vibration are [Rayleigh 1984]

$$m_1\ddot{x}_1 + (c_1 + c_2)\dot{x}_1 + (k_1 + k_2)x_1 - c_2\dot{x}_2 - k_2 x_2 = F_0 \sin \omega t$$

$$m_2\ddot{x}_2 + c_2\dot{x}_2 + k_2 x_2 - c_2\dot{x}_1 - k_2 x_1 = 0$$

We substitute $F_0 e^{i\omega t}$ for $F_0 \sin \omega t$, $X_1 e^{i\omega t}$ for x_1, and $X_2 e^{i\omega t}$ for x_2. Rearranging and dividing through by $e^{i\omega t}$, the equations of motion become

$$[(k_1 + k_2) - m_1\omega^2 + i(c_1 + c_2)\omega]X_1 - (k_2 + ic_2\omega)X_2 = F_0$$

$$-(k_2 + ic_2\omega)X_1 + (k_2 - m_2\omega^2 + ic_2\omega)X_2 = 0$$

Solving by Cramer's rule gives

$$X_1 = \frac{\begin{vmatrix} F_0 & -(k_2 + ic_2\omega) \\ 0 & (k_2 - m_2\omega^2 + ic_2\omega) \end{vmatrix}}{[(k_1 + k_2) - m_1\omega^2 + i(c_1 + c_2)\omega][k_2 - m_2\omega^2 + ic_2\omega] - (k_2 + ic_2\omega)^2}$$

which is of the form $(A + iB)/(C + iD)$ or $(G + iH)$, and

$$X_1 = \frac{(A + iB)(C - iD)}{(C + iD)(C - iD)}$$

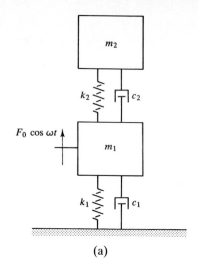

(a)

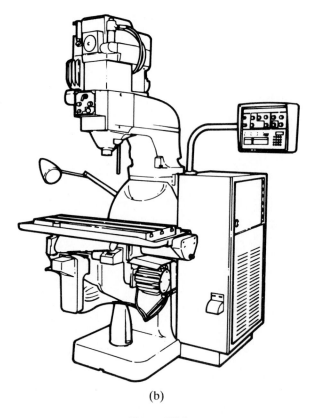

(b)

Figure E6.3

where

$$A = k_2 - m_2\omega^2$$

$$B = c_2\omega$$

$$C = [(k_1 + k_2) - m_1\omega^2](k_2 - m_2\omega^2) - (c_1 + c_2)c_2\omega^2 - k_2^2 + c_2^2$$

$$D = [(k_1 + k_2) - m_1\omega^2]c_2\omega + (k_2 - m_2\omega^2)(c_1 + c_2)\omega - 2k_2c_2\omega$$

or

$$X_1 = \frac{F_0[(AC - BD) + i(AD + BC)]}{C^2 + D^2}$$

Similarly,

$$X_2 = \frac{F_0[(aC - bD) + i(aD + bC)]}{C^2 + D^2}$$

where $a = -k_2$ and $b = -c_2\omega$.

From the theory of complex numbers,

$$X_1 = F_0(G + iH) = R_1e_1^{i\phi} = R_1(\cos\phi_1 + i\sin\phi_1)$$

$$X_2 = F_0(K + iL) = R_2e_2^{i\phi} = R_2(\cos\phi_2 + i\sin\phi_2)$$

where

$$R_1 = \frac{F_0[(AC - BD)^2 + (AD + BC)^2)]^{1/2}}{C^2 + D^2}$$

$$R_2 = \frac{F_0[(aC - bD)^2 + (aD + bC)^2)]^{1/2}}{C^2 + D^2}$$

$$\phi_1 = \tan^{-1}\left(\frac{AD + BC}{AC - BD}\right)$$

$$\phi_2 = \tan^{-1}\left(\frac{aD + bC}{aC - bD}\right)$$

But the forcing function is

$$F_0 \sin\omega t = \text{Im}(F_0e^{i\omega t})$$

so

$$x_1 = \text{Im}(R_1e^{i\omega t}) = \text{Im}(R_1e^{i(\omega t + \phi_1)}) = R_1\sin(\omega t + \phi_1)$$

$$x_2 = \text{Im}(R_2e^{i\omega t}) = \text{Im}(R_2e^{i(\omega t + \phi_2)}) = R_2\sin(\omega t + \phi_2)$$

(This example presents an analytical solution for the response of the two-degree-of-freedom system under harmonic excitation.)

□

□ **Example 6.4   Response of an Automobile to Ground Motion**

An automobile was modeled for vertical vibration as in Figure E6.4. The mass of the automobile m_2 is 1909 kg, the wheels and axles $m_1 = 227$ kg, the constant of the spring $k_2 = 5270$ N/m, and the spring constant of the tires $k_1 = 17,700$ N/m. If the car is put on a test stand vibrating at 300 cpm with an amplitude of 25 mm, determine

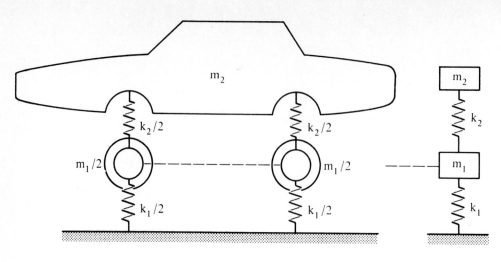

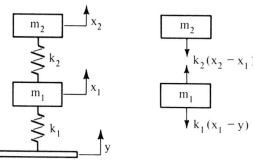

Figure E6.4

the vertical motion of the body assuming that the car does not have rocking motion and the wheels have the same motion.

Solution   Referring to the free-body diagram, application of Newton's law yields

$$m_1\ddot{x}_1 = k_2(x_2 - x_1) - k_1(x_1 - y), \qquad m_1\ddot{x}_1 + (k_1 + k_2)x_1 - k_2x_2 = k_1y$$

$$m_2\ddot{x}_2 = -k_2(x_2 - x_1), \qquad\qquad m_2\ddot{x}_2 - k_2x_1 + k_2x_2 = 0$$

Since $y = y_0 \cos \omega t$, this is a forced vibration problem. We seek solutions in the form $x_1 = A_1 \cos \omega t$ and $x_2 = A_2 \cos \omega t$. Substitution and division by $\cos \omega t$ yields

$$(-m_1\omega^2 + k_1 + k_2)A_1 \qquad\qquad - k_2A_2 = k_1y_0$$

$$-k_2A_1 + (-m_2\omega^2 + k_2)A_2 = 0$$

$\omega = 300/60 \times 6.28 = 31.4$ rad/s and

$$A_2 = \frac{\begin{vmatrix} -m_1\omega^2 + k_1 + k_2 & y_0k_1 \\ -k_2 & 0 \end{vmatrix}}{\begin{vmatrix} -m_1\omega^2 + k_1 + k_2 & -k_2 \\ -k_2 & -m_2\omega^2 + k_2 \end{vmatrix}} = 0.006 \text{ mm}$$

The motion of the body will be

$$x_2 = 0.006 \cos (31.4t)$$

☐

6.5 MATRIX FORMULATION

In matrix form the differential equations of motion for a coupled system can be written in the form

$$\mathbf{M}\ddot{x} + \mathbf{C}\dot{x} + \mathbf{K}x = \mathbf{F}(t) \tag{6.26}$$

where

$$\mathbf{M} = \begin{bmatrix} m_{11} & m_{12} \\ m_{21} & m_{22} \end{bmatrix}, \quad \mathbf{C} = \begin{bmatrix} c_{11} & c_{12} \\ c_{21} & c_{22} \end{bmatrix}, \quad \mathbf{K} = \begin{bmatrix} k_{11} & k_{12} \\ k_{21} & k_{22} \end{bmatrix}$$

are the mass, damping, and stiffness matrices, respectively, and $x = \{x_1 \quad x_2\}^3$ and $\mathbf{F} = \{F_1 \quad F_2\}$ the displacement and force vectors.[4] A harmonic excitation $\mathbf{F}_0 e^{i\omega t}$ will lead to the equation

$$\mathbf{M}\ddot{x} + \mathbf{C}\dot{x} + \mathbf{K}x = \mathbf{F}_0 e^{i\omega t} \tag{6.27}$$

The steady-state solution will be found with the substitution

$$\mathbf{x} = \mathbf{X}e^{i\omega t} \tag{6.28}$$

Then equation (6.27) after division by $e^{i\omega t}$ and matrix operation yields

$$\mathbf{X} = [-\omega^2 \mathbf{M} + i\omega \mathbf{C} + \mathbf{K}]^{-1}\mathbf{F}_0 \tag{6.29}$$

provided that the matrix $\mathbf{B} = [-\omega^2 \mathbf{M} + i\omega \mathbf{C} + \mathbf{K}]$ is not singular.

The solution (6.29) will yield an amplitude vector $\mathbf{X} = \{X_1 \quad X_2\}$ with complex components, in general. Therefore, the solution on a vector diagram at time $t = 0$ will be as shown in Figure 6.5. The components of the forcing vector $\mathbf{F}_0 =$

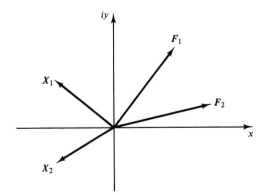

Figure 6.5 Superposition of harmonic forces.

[3]Braces used with vectors designate column vectors. (See Appendix IV for a review of matrices and vectors.)

[4]Zurmühl, B. R. 1958. *Matrizen*. Berlin: Springer-Verlag. The first extensive treatment of matrices for structural and vibration analysis.

Portrait of Cayley. (Courtesy of the
Library of Congress.)

Arthur Cayley (1821–1895)

After receiving his degree from Trinity College, which had almost a mo-
nopoly in algebra, Cayley took to the law for 14 years. During that time
he published hundreds of papers, mostly on algebraic invariants. In 1863
he became Sadlerian Professor at Cambridge and developed matrix algebra. He was friend and rival with Sylvester, who was given professorships in the United States. Cayley's *Collected Mathematical Papers*, published between 1889 and 1898, consist of 14 volumes.

$\{F_1 \quad F_2\}$ might be complex, which means that they might have phase angles from
each other. The same will then be true for the solution vector $\mathbf{X}$.

For periodic but not harmonic forcing functions, one can analyze them in
Fourier series, as in Chapter 5.

☐ **Example 6.5   Identification of a Two-Degree-of-Freedom System**

To predict the dynamic response of a two-degree-of-freedom system having 2 masses
$m_1 = 0.5$ kg and $m_2 = 1$ kg, an engineer performed a static test by applying known
displacements and measuring the static forces. Thus he constructed the stiffness matrix
for the system (Figure E6.5). What will be the vibration amplitude if he applies at
the mass m_1 a harmonic force of amplitude $F_0 = 3$ N and frequency 5 rad/s in the
direction x_1?

Solution   The differential equation of motion will be

$$\begin{bmatrix} 0.5 & 0 \\ 0 & 1 \end{bmatrix} \begin{bmatrix} \ddot{x}_1 \\ \ddot{x}_2 \end{bmatrix} + \begin{bmatrix} 6 & 2 \\ 2 & 3 \end{bmatrix} \begin{bmatrix} x_1 \\ x_2 \end{bmatrix} = \begin{bmatrix} 3 \\ 0 \end{bmatrix} \cos 5t$$

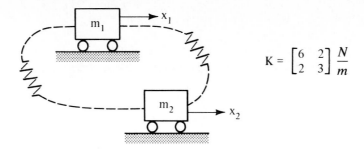

Figure E6.5

$$K = \begin{bmatrix} 6 & 2 \\ 2 & 3 \end{bmatrix} \frac{N}{m}$$

Let $x_1 = A_1 \cos 5t$ and $x_2 = A_2 \cos 5t$. Then

$$(-0.5 \times 5^2 + 6)A_1 + 2A_2 = 3, \qquad -6.5A_1 + 2A_2 = 3$$
$$2A_1 + (-1 \times 5^2 + 3)A_2 = 0, \qquad 2A_1 - 22A_2 = 0$$

By Cramer's rule, we obtain

$$A_1 = \frac{\begin{vmatrix} 3 & 2 \\ 0 & -22 \end{vmatrix}}{\begin{vmatrix} -6.5 & 2 \\ 2 & -22 \end{vmatrix}} = \frac{3 \times (-22) - 2 \times 0}{(-6.5) \times (-22) - 2 \times 2} = -0.474 \text{ mm}$$

$$A_2 = \frac{\begin{vmatrix} -6.5 & 3 \\ 2 & 0 \end{vmatrix}}{\begin{vmatrix} -6.5 & 2 \\ 2 & -22 \end{vmatrix}} = -0.043 \text{ mm}$$

(This example demonstrates the dynamic response of an undamped two-degree-of-freedom system to harmonic excitation.) ☐

6.6 DYNAMIC VIBRATION ABSORBERS

Undamped absorber[5]

The dynamic response of the two-degree-of-freedom system shown in Figure E6.3 for a harmonic force $F_1 \cos \omega t$ and $F_2 = 0$ is obtained from Example 6.3, setting $F_2 = 0$ and damping $= 0$:

$$A_1 = \frac{(F_1/m_1)(k_2/m_2 - \omega^2)}{[(k_1 + k_2)/m_1 - \omega^2](k_2/m_2 - \omega^2) - k_2^2/m_1 m_2} \tag{6.30}$$

$$A_2 = \frac{(k_2/m_2)A_1}{[(k_1 + k_2)/m_1 - \omega^2](k_2/m_2 - \omega^2) - k_2^2/m_1 m_2} \tag{6.31}$$

[5] H. Frahm (1911, *Jahrb. Schiffsbautech. Ges.*) introduced the double-pendulum vibration absorber for stabilization of rocking oscillation of ships and developed the vibration absorber theory.

From equation (6.30) it is apparent that for $k_2/m_2 - \omega^2 = 0$, $A_1 = 0$ and the mass 1 does not vibrate.

This observation resulted in the development of devices based on this principle, called *dynamic vibration absorbers*. If a system has a mass, called a *primary component*, forced by an exciting force or base motion to vibrate, it is possible to design an attached spring–mass system (m_2, k_2) so that the vibration of the primary component is eliminated. The design equation is

$$\frac{k_2}{m_2} - \omega^2 = 0 \tag{6.32}$$

which means that the natural frequency of the secondary component should coincide with the exciting frequency.

There are, however, more design considerations. For example, for the case that this device is used to reduce vibration at resonance of the primary component, it is designed to have $k_2/m_2 = k_1/m_1$; that is, the absorber natural frequency coincides with the one of the primary component. Such design is very sensitive to frequency variations. To investigate this, we observe that the denominator in equation (6.30) is a polynomial in ω and vanishes at two values of ω. If equation (6.30) is written in dimensionless form, for $k_2/m_2 = k_1/m_1$,

$$A_1 = \frac{(F_1/k_1)(1 - r^2)}{(1 - r^2)(1 - r^2 + \mu) - \mu} \tag{6.33}$$

where $r = \omega/\omega_n$ and $\mu = m_2/m_1$, $\omega_n^2 = k_1/m_1$.

Equation (6.33) is plotted in Figure 6.6. It is apparent that the resonance ω_n of the primary component at $r = 1$ has degenerated to two resonances, due to the absorber, ω_{n1} and ω_{n2} and $\omega_{n1} > \omega_n > \omega_{n2}$. To compute them, the denominator in equation (6.33) is equated to zero and the resulting biquadratic equation has roots

$$\left(\frac{\omega}{\omega_n}\right)^2 = 1 + \frac{\mu}{2} \pm \left(\mu + \frac{\mu^2}{4}\right)^{1/2} \tag{6.34}$$

This equation cannot be used for design because near the limiting values of the frequency the amplitude has high values. Instead, a reduction factor R is defined, the ratio of the amplitude of the primary component with the absorber, given by equation (6.33), to the one without the absorber, $A_0 = (F_1/k_1)/(1 - r^2)$:

$$R = \frac{A_1}{A_0} = \frac{(1 - r^2)(1 - r^2 + \mu) - \mu}{(1 - r^2)^2} \tag{6.35}$$

Improvement exists clearly when $|R| < 1$. The limit values are obtained by setting $R = \pm 1$. The value $R = 1$ yields the trivial solution $r = 0$. $R = -1$ yields

$$r^4 - (2 + \mu)r^2 + 1 = 0 \tag{6.36}$$

$$r_{1,2}^2 = \left(\frac{\omega}{\omega_n}\right)_{1,2}^2 = 1 + \frac{\mu}{2} \pm \left(\mu + \frac{\mu^2}{4}\right)^{1/2}$$

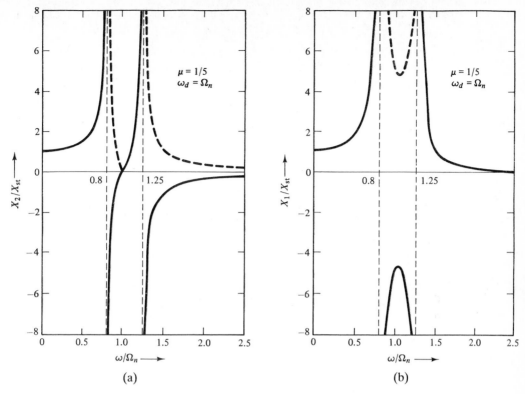

Figure 6.6 Motion of an undamped dynamic vibration absorber. (From Ormonroyd, J., and den Hartog, J. P. 1928. The theory of the dynamic vibration absorbers. *Trans. ASME, 50:* APM-241.)

It is apparent that increase in the mass ratio μ yields a wider range of safe operating frequencies. In selecting the mass ratio the designer has to satisfy the requirements for a range of variation of the operating frequency, μ large, and for a compact design, μ small. A reduction factor $|R| < 1$ is always required, usually about 0.2, and the corresponding range of frequencies is computed by equation (6.35). The foregoing type of absorber was invented by Frahm in 1909 [Den Hartog, 1952], but the theoretical basis dates back to Lord Rayleigh (1894).[6]

Damped Absorber[7]

The undamped dynamic vibration absorber absorbs vibration by transferring the energy from the primary to the secondary component. This leads to very high vibration levels of the absorber itself and fatigue failures. Therefore, it is often

---

[6]Zurmühl, op. cit.

[7]Hort, W. 1922. *Technische Schwingungslehre*, 2nd ed. Berlin: Springer-Verlag. Hahnkamm, E., 1932. Untersuchungen über das reibungs-und trägheitsgekoppelte Schwingungs-system Schiff und Schlingertank. Ingenieur Archiv, III, pp. 251–261.

necessary to add damping to the absorber. The response is obtained from Example 6.3, setting $c_1 = 0$ and $F_2 = 0$, in the form of the amplification factor:

$$\frac{A_1}{F_1 k_1} =$$

$$\left\{ \frac{(k_2 - m_2\omega^2)^2 + c_2^2\omega^2}{[(-m_1\omega^2 + k_1)(-m_2\omega^2 + k_2) - m_2\omega^2 k_2]^2 + \omega^2 c_2^2(-m_1\omega^2 + k_1 - m_2\omega^2)^2} \right\}^{1/2} k_1 \qquad (6.37)$$

In dimensionless form, the amplification factor has the form

$$H(\omega) = \left\{ \frac{(2\zeta)^2 r^2 + (r - f)^2}{(2\zeta)^2 r(r - 1 + \mu r)^2 + [\mu f r - (r - 1)(r - f)]^2} \right\}^{1/2} \qquad (6.38)$$

where ζ is the damping factor, defined as $\zeta = c_2/2m_1\omega_n$, $f = (k_2/m_2)/(k_1/m_1)$, and r and μ are as before.

Two characteristic cases are of interest. If $c = 0$ and $f = 1$, the absorber is undamped, tuned to the natural frequency of the primary component. This case was discussed above. If $c = \infty$, which means that the absorber mass is rigidly attached to the mass of the primary component, the response is the one for a one-degree-of-freedom system, with mass $= m_1 + m_2$. The response of both systems is shown in Figure 6.7. The two curves meet at P and Q. Their location depends on the choice of μ. If damping has a finite value, it can be proved that the corresponding response curve passes through P and Q if μ is the same. It can be seen that for the limiting values $c_2 = 0$ and ∞, the amplitude becomes infinite at different speeds. There is a finite value of c_2 which yields the smallest of the possible maximum values of the amplitude.

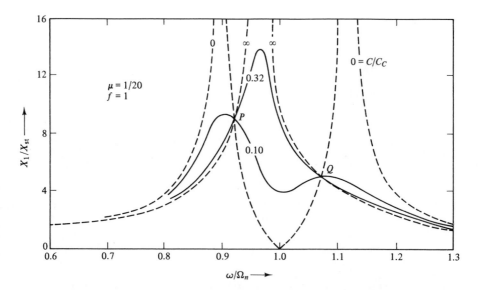

Figure 6.7 Motion of a damped dynamic vibration absorber. (From Hahnkamm, E., 1932, Ingenieur Archiv, page 261. By permission of the Springer Verlag.)

Optimal Design of Dynamic Vibration Absorbers

There are three independent parameters involved in the design of a damped vibration absorber: μ, f, ζ. In general, an optimization procedure can be utilized to yield their values, which minimize the function $H(\omega)$ of equation (6.38) within a range of frequencies $\omega_1 < \omega < \omega_2$ and suitable limits on the design parameters μ, f, ζ with any design optimization algorithm [Dimarogonas, 1988]. In general, the range of frequencies might or might not include the natural frequency of the primary component. A wider operating range can be achieved with a damped absorber.

For absorbers that operate in the vicinity of the natural frequency of the primary component, a nearly optimal design can be achieved with analytical design computations. This design is achieved by two selections: First, the absorber is tuned not at the natural frequency of the primary component, $f = 1$, but at

$$f = \frac{1}{(1 + \mu)^2} \tag{6.39}$$

This results in the amplitudes at points P and Q being equal [Den Hartog, 1952] and the maximum amplitude is smaller. Second, the damping factor, an average of the values that yield zero slope at P and Q is [Den Hartog, 1952] curve (c) in Figure 6.8:

$$\zeta^2 = \frac{3\mu}{8(1 + \mu)^3} \tag{6.40}$$

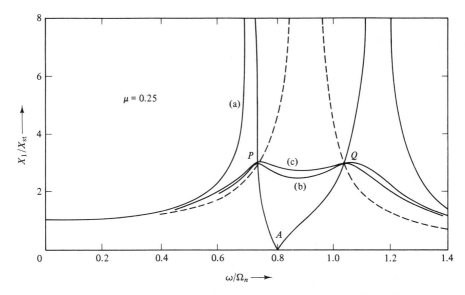

Figure 6.8 Optimized damped dynamic vibration absorber. (From Hahn-kamm, E., 1932, Ingenieur Archiv, page 261. By permission of the Springer Verlag.)

Torsional Absorber[8]

Internal combustion and other reciprocating machines have inherent torsional vibration due to unbalanced harmonic components of the torque, and vibration absorbers are used for this purpose (Figure 6.9). The design procedure is the same, with proper definition of the mass, stiffness, and damping terms. However, some of these machines, such as ship propulsion engines, usually operate in a wide range of frequencies. In such cases tuning at any operating frequency is achieved with the pendulum absorber (Figure 6.9b).

The natural frequency of the pendulum is $(g/L)^{1/2}$, where g is the acceleration of the gravity field and L the length. If the pendulum rotates, the acceleration has value $\omega^2(a + L)$ at the pendulum bob, usually $>>g$, where a is the distance of the pivot from the center of the shaft. Therefore, the natural frequency is $\omega(a/L + 1)^{1/2}$. If a is small, the natural frequency of the absorber nearly coincides with the frequency of rotation, and the undamped absorber is correctly tuned at any engine speed. Higher harmonics can be absorbed by proper selection of the distance a. For example, to absorb the second harmonic of the rotational speed, $\omega(a/L + 1)^{1/2} = 2\omega$ and $a = 3L$.

An important design is the Lanchester absorber (Figure 6.9), in which there is no stiffness in the secondary system; thus $f = 0$. The optimal viscous damping is then

$$\zeta^2 = \tfrac{1}{2}(2 + \mu)(1 + \mu) \tag{6.41}$$

The Lanchester damper of Figure 6.9c is the Coulomb friction type, and as explained in Chapter 3, the equivalent ζ is valid only at a specific vibration amplitude, at which the absorber is exactly tuned.

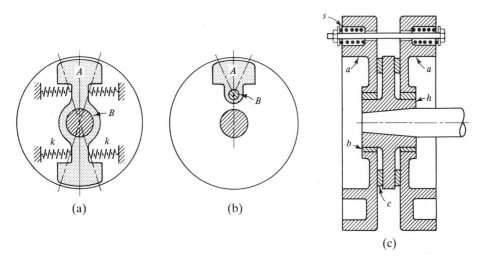

Figure 6.9 Torsional vibration absorbers.

[8]Taylor, E. S. 1936. *J. SAE*, *38*: 81.

An electromechanical relay consists of an electromagnet M powered by 60-Hz alternating voltage (Figure E6.6). To prevent transmission of the 60-Hz vibration to the base, an undamped dynamic vibration absorber should be designed. The mass of the relay is 1 kg and the mass of the absorber should not be greater than 10% of it. The spring will be made of a steel strip 0.5 mm thick and 5 mm wide with a Young's modulus of 2.1×10^5 MPa. Find the length of the strip.

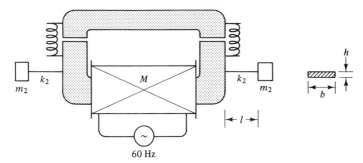

60 Hz

Figure E6.6

Solution   The absorber should be tuned to the operating frequency, $\omega = 2\pi N = 2\pi \times 60 = 377$ rad/s. The mass of the absorber $m_2 = 0.1m_1 = 0.1 \times 1 = 0.1$ kg. The stiffness $k_2 = \omega^2 m_2 = 377^2 \times 0.1 = 14{,}198$ N/m. For a cantilever beam

$$k = \frac{3EI}{l^3} = \frac{3Ebh^3}{12l^3}$$

$$l = \left(\frac{Ebh^3}{4k}\right)^{1/3} = \left(\frac{2.1 \times 10^5 \times 5 \times 0.5^3}{4 \times 14.198}\right)^{1/3}$$

$$l = 13.2 \text{ mm}$$

(This example demonstrates the mechanical design of a dynamic vibration absorber for an electromagnetic relay modeled as a two-degree-of-freedom system.)

☐

SIMULAB Experiment 6.1

Using the simulated experiment LINSYS, simulate the motion of a harmonic oscillator with properties $m_1 = 2$ kg, $k_1 = 10$ N/m, and $c_1 = 5$ Ns/m acted upon by a harmonic force of amplitude 30 N and frequency 0.3 Hz, with an undamped dynamic vibration absorber attached to it, designed for the particular exciting frequency with a mass ratio 0.2.

Solution   The absorber will have mass $m_2 = 2 \times 0.2 = 0.4$ kg and spring constant $k_2 = \omega^2 m_2 = (0.3 \times 2\pi)^2 \times 0.4 = 1.42$ N/m. With the program diskette in drive A, we type

A› LINESYS

The page of Figure SE6.1a appears on the screen. We select a two-mass system. Then the menu page appears on the screen with the simulated two-degree-of-freedom system (Figure SE6.1b). With the keypad arrow keys we select menus.

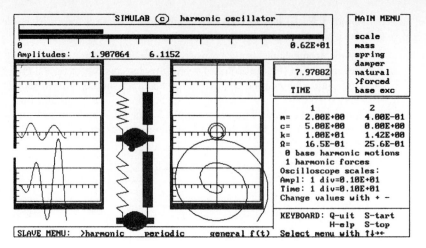

Figure SE6.1

1. We set $m_1 = 2$ kg, $k_1 = 10$ N/m, $c_1 = 5$ N·s/m, $m_2 = 0.4$ kg, and $k_2 = 1.42$ N/m using the + and − keys.
2. We select forced response, 1 force, with amplitude, phase angle, angular velocity, 30, 0, and 01.89, respectively, entered at the prompt.
3. We hit S to start system operation. The resulting motion is shown in Figure SE6.1b.

6.7 COUPLED ELECTROMECHANICAL SYSTEMS

Electrical systems are known to have dynamic behavior similar in many ways to the one of mechanical systems.[9] Electric generators and motors are mechanical systems themselves, coupled to the electric supply system. This coupling causes some problems worth discussing.

The amount of energy stored in the inductance coils of an electrical circuit due to the motion of the electric charge is

$$T = \tfrac{1}{2} L \dot{q}^2 \tag{6.42}$$

in direct analogy to the kinetic energy of a mechanical system, where $\dot{q} = i$ is the electric flow, also defined as electric current i. The electric circuit stores energy in its capacitors

$$V = \frac{1}{2} \frac{q^2}{C} \tag{6.43}$$

in direct analogy to the potential energy of mechanical systems, where C is the capacitance and q the electric charge.

[9]Firestone, F. A. 1933., *J. Acoust. Soc. Am.*, 4: 249.

The rate at which electromagnetic energy is converted into heat is

$$D = R\dot{q}^2 = Ri^2 \tag{6.44}$$

in direct analogy to viscous damping in a mechanical system, where R is the electric resistance in the circuit. The voltage e is the electromotive force that forces the electric charge to flow, in direct analogy to the force in a mechanical system.

The rate of change of the kinetic and potential energy in the system should equal the rate of energy dissipation:

$$\frac{d}{dt}\left(\frac{1}{2}L\dot{q}^2 + \frac{1}{2}\frac{q^2}{C}\right) = -R\dot{q}^2 \tag{6.45}$$

$$L\ddot{q}q + \frac{q\dot{q}}{C} = R\dot{q}^2 \tag{6.46}$$

$$L\ddot{q} + R\dot{q} + \frac{q}{C} = 0 \tag{6.47}$$

If a voltage is applied,

$$L\ddot{q} + R\dot{q} + \frac{q}{C} = e(t) \tag{6.48}$$

Equation (6.48) shows the similarity to mechanical systems. All results apply with the analogies listed in Table 6.1. Equation (6.48) could be obtained with Lagrange's equation, using the Lagrangian function $T - V = \frac{1}{2}L\dot{q}^2 - \frac{1}{2}q^2/C$ and the dissipation function D.

The coupling mechanism is an electrical machine that relates the mechanical torque T to the electric current i by a function

$$T = f(i, \omega) \tag{6.49}$$

where the function $f(i, \omega)$ depends on the current, the machine, its angular velocity ω, and the speed control mechanism. A typical electromechanical system is depicted in Figure 6.10 and consists of a single electric circuit, an electric machine E, a shaft with a torsional stiffness k_T, and a disk with moment of inertia J, much larger than the moment of inertia of the electric machine, which is neglected. It is a two-

TABLE 6.1

Mechanical system		Electrical system	Units (SI)
Linear	Torsional		
Displacement x	Angle ϕ	Electric charge q	coulomb
Velocity $v = \dot{x}$	Angular velocity $\dot{\phi}$	Electric current $i = \dot{q}$	ampere
Mass m	Moment of inertia J	Inductance L	henry
Damping c	Torsional damping c_T	Electric resistance R	ohm
Compliance $1/k$	Torsional compliance $1/k_T$	Capacitance C	farad
Force F	Torque T	Voltage e	volt

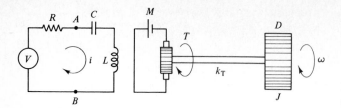

Figure 6.10 Model of a coupled electromechanical system.

degree-of-freedom system; electric charge q and angular displacement ϕ of the disk are the system coordinates. The equations of the system are

$$L\ddot{q} + R\dot{q} + \frac{q}{C} = e \tag{6.50}$$

$$J\ddot{\phi} + k_T\phi = f(\dot{q}, \phi) \tag{6.51}$$

Near the operating point, the function $f(i, \omega)$ is nearly linear and further analysis will depend on its form.

☐ **Example 6.7 Dynamics of an Electromechanical Drive**

The electric machine of Figure 6.10 is a field-controlled dc motor for which the function $T = K_f i$, where K_f is a constant. The rotor of the driven machine is a disk D with moment of inertia $J = 3$ kg·m², and the connecting shaft has torsional stiffness $k_T = 360{,}000$ N·m/rad. The electric circuit has an electric resistance $R = 10\ \Omega$, capacitance $C = 2 \times 10^{-6}$ F, inductance 4 H, and operates at a voltage of 220 V. The motor is operating at 15 kW and 3000 rpm, with an efficiency $\eta = 75\%$ and a 60-Hz harmonic voltage of $V_1 = 15$ V due to incomplete rectification, superposed to $V_0 = 220$ V dc. Determine the torsional vibration amplitude of the disk and the maximum dynamic torque on the shaft.

Solution At normal operation, the current is $i = W/\eta V = 15{,}000/0.75 \times 220 = 91$ A. The electric frequency $\Omega = 60 \times 2\pi = 377$ rad/s. The operating angular velocity is $\omega = 3000 \times 2\pi/60 = 314$ rad/s. The operating torque is $T = W/\omega = 15{,}000/314 = 47.8$ N·m. Therefore, the motor constant is $K_f = T/i = 47.8/91 = 0.525$ N·m/A. Equations (6.50) and (6.51) will yield

$$L\ddot{q} + \frac{q}{C} = V_0 + V_1 \cos \Omega t \tag{a}$$

$$J\ddot{\phi} + k_T\phi - K_f\dot{q} = 0 \tag{b}$$

The static voltage does not excite vibration. Assuming the harmonic excitation to be in the form $V_1 e^{i\Omega t}$, the solution will be harmonic of the form

$$q = Q e^{i\Omega t}$$

$$\phi = \Phi e^{i\Omega t}$$

Equations (a) and (b) give

$$\left(-\Omega^2 L + \frac{1}{C}\right)Q = V_1 \tag{c}$$

$$-\Omega^2 J\Phi + k_T\Phi - i\Omega K_f Q = 0 \tag{d}$$

Solving (c) for Q and using (d) yields

$$\Phi = \frac{i\Omega K_f Q}{-\Omega^2 J + k_T} = \frac{i\Omega K_f V_1}{(-\Omega^2 J + k_T)(-\Omega^2 L + 1/C)} \qquad (e)$$

The vibration amplitude is

$$|\Phi| = \frac{377 \times 0.525 \times 15}{(-377^2 \times 4 + \frac{1}{2} \times 10^6)(-377^2 \times 3 + 360{,}000)}$$

$$= 0.666 \times 10^{-6} \text{ rad}$$

The maximum dynamic torque at the shaft is

$$T_{d\max} = k_T \Phi_{\max} = 360{,}000 \times 0.666 \times 10^{-6} = 0.240 \text{ N·m}$$

(This example demonstrates the dynamics of an electromechanical system consisting of a dc motor and a rotating disk.)

□

TABLE 6.2. SUMMARY OF EQUATIONS FOR CHAPTER 6

Two-degree-of-freedom, no damping

$$m_{11}\ddot{x}_1 + m_{12}\ddot{x}_2 + k_{11}x_1 + k_{12}x_2 = 0$$

$$m_{21}\ddot{x}_1 + m_{22}\ddot{x}_2 + k_{21}x_1 + k_{22}x_2 = 0$$

Frequency equation

$$\begin{vmatrix} -m_{11}\omega^2 + k_{11} & -m_{12}\omega^2 + k_{12} \\ -m_{21}\omega^2 + k_{21} & -m_{22}\omega^2 + k_{22} \end{vmatrix} = 0$$

Natural vibration

$$\begin{bmatrix} x_1(t) \\ x_2(t) \end{bmatrix} = c_1 \begin{bmatrix} A_1^{(1)} \\ A_2^{(1)} \end{bmatrix} \cos \omega_1 t + c_2 \begin{bmatrix} A_1^{(2)} \\ A_2^{(2)} \end{bmatrix} \cos \omega_2 t$$

$$+ c_3 \begin{bmatrix} A_1^{(1)} \\ A_2^{(1)} \end{bmatrix} \sin \omega_1 t + c_4 \begin{bmatrix} A_1^{(2)} \\ A_2^{(2)} \end{bmatrix} \sin \omega_2 t$$

Forced response, harmonic force

$$\{F(t)\} = \{F_1 \quad F_2\} \cos \omega t$$

$$A_1 = \frac{\begin{vmatrix} F_1 & k_{12} \\ F_2 & -m_2\omega^2 + k_{22} \end{vmatrix}}{\begin{vmatrix} -m_1\omega^2 + k_{11} & k_{12} \\ k_{21} & -m_2\omega^2 + k_{22} \end{vmatrix}}$$

$$A_2 = \frac{\begin{vmatrix} -m_1\omega^2 + k_{11} & F_1 \\ k_{21} & F_2 \end{vmatrix}}{\begin{vmatrix} -m_1\omega^2 + k_{11} & k_{12} \\ k_{21} & -m_2\omega^2 + k_{22} \end{vmatrix}}$$

(continues)

TABLE 6.2 *(cont.)*

Damped system

$$m_1\ddot{x}_1 + c_1\dot{x}_1 + k_{11}x_1 + k_{12}x_2 = F_1e^{i\omega t}$$

$$m_2\ddot{x}_2 + c_2\dot{x}_2 + k_{21}x_1 + k_{22}x_2 = F_2e^{i\omega t}$$

$$X_1 = \frac{\begin{vmatrix} F_1 & k_{12} \\ F_2 & -m_2\omega^2 + i\omega c_2 + k_{22} \end{vmatrix}}{\begin{vmatrix} -m_1\omega^2 + i\omega c_1 + k_{11} & k_{21} \\ k_{21} & -m_2\omega^2 + i\omega c_2 + k_{22} \end{vmatrix}}$$

$$X_2 = \frac{\begin{vmatrix} -m_1\omega^2 + i\omega c_1 + k_{11} & F_1 \\ k_{21} & F_2 \end{vmatrix}}{\begin{vmatrix} -m_1\omega^2 + i\omega c_1 + k_{11} & k_{12} \\ k_{21} & -m_2\omega^2 + i\omega c_2 + k_{22} \end{vmatrix}}$$

Matrix notation

$$\mathbf{M}\ddot{x} + \mathbf{C}\dot{x} + \mathbf{K}x = \mathbf{F}(t)$$

$$\mathbf{M} = \begin{bmatrix} m_{11} & m_{12} \\ m_{21} & m_{22} \end{bmatrix}, \quad \mathbf{C} = \begin{bmatrix} c_{11} & c_{12} \\ c_{21} & c_{22} \end{bmatrix}, \quad \mathbf{K} = \begin{bmatrix} k_{11} & k_{12} \\ k_{21} & k_{22} \end{bmatrix}$$

$$x = \mathbf{X}e^{i\omega t}, \quad \mathbf{F} = \mathbf{F}_0e^{i\omega t},$$

$$\mathbf{X} = [-\omega^2\mathbf{M} + i\omega\mathbf{C} + \mathbf{K}]^{-1}\mathbf{F}_0$$

Dynamic vibration absorber
Undamped

$$A_1 = \frac{(F_1/m_1)(k_2/m_2 - \omega^2)}{[(k_1 + k_2)/m_1 \times \omega^2](k_2/m_2 - \omega^2) - k_2^2/m_1m_2}$$

$$A_2 = \frac{(k_2/m_2)A_1}{[(k_1 + k_2)/m_1 \times \omega^2](k_2/m_2 - \omega^2) - k_2^2/m_1m_2}$$

Design equations

$$k_2/m_2 - \omega^2 = 0$$

$$k_2/m_2 = k_1/m_1,$$

$$A_1 = \frac{(F_1/k_1)(1 - r^2)}{(1 - r^2)(1 - r^2 + \mu) - \mu}$$

$$r = \omega/\omega_n, \quad \mu = m_2/m_1, \quad \omega_n^2 = k_1/m_1$$

$$(\omega/\omega_n)^2 = 1 + \frac{\mu}{2} \pm \left(\mu + \frac{\mu^2}{4}\right)^{1/2}$$

$$A_0 = \frac{(F_1/k_1)}{(1 - r^2)},$$

(continues)

TABLE 6.2 *(cont.)*

$$R = \frac{A_1}{A_0} = \frac{(1 - r^2)(1 - r^2 + \mu) - \mu}{(1 - r^2)^2}$$

$$r^4 - (2 + \mu/2)r^2 + 1 = 0$$

$$r_{1,2}^2 = \left(\frac{\omega}{\omega_n}\right)_{1,2}^2 = 1 + \frac{\mu}{2} \pm \left(\mu + \frac{\mu^2}{4}\right)^{1/2}$$

Damped absorber

$$H^2(\omega) = \frac{(2\zeta)^2 r^2 + (r - f)^2}{(2\zeta)^2 r(r - 1 + \mu r)^2 + [\mu f r - (r - 1)(r - f)]^2}$$

$$\zeta = \frac{c_2}{2m_1\omega_n,} \qquad f = \frac{(k_2/m_2)}{(k_1/m_1),}$$

Design equations

$$f = 1/(1 + \mu)^2$$

$$\zeta^2 = 3\mu/8(1 + \mu)^3$$

REFERENCES AND FURTHER READING

BYKHOVSKY, I. I. 1969. *Fundamentals of Vibration Engineering*. Moscow: Machinostroienye.

DEN HARTOG, J. P. 1952. *Mechanical Vibration*, 4th ed. New York: McGraw-Hill.

DIMAROGONAS, A. D. 1976. *Vibration Engineering*, St. Paul, Minn. West Publishing.

DIMAROGONAS, A. D., 1988. Computer Aided Machine Design. Englewood Cliffs, N.J.: Prentice Hall.

HORT, W. 1910. *Technische Schwingungslehre*. Berlin: Julius Springer. 2nd Edition, 1922.

LALANNE, M., BERTHIER, P., and DER HAGOPIAN, J. 1983. *Mécanique des vibrations Linéaires*. New York: Wiley.

RAYLEIGH, J. W. S. 1894. *Theory of Sound*. New York: Dover Publications, 1946.

REED, F. E. 1961. Dynamic vibration absorbers and auxiliary mass dampers. Chapter 6 in Harris, C. M., and Crede, C. E., eds. *Shock and Vibration Handbook*. New York: McGraw-Hill.

VAN DE VEGTE, J. 1990. *Feedback Control Systems*. Englewood Cliffs, N.J.: Prentice Hall.

PROBLEMS

Sections 6.1 to 6.3

6.1. A turbine-generator foundation can be modeled as the system of Figure P6.1. If the spring constant of the soil $k_1 = 80 \times 10^6$ N/m, the mass of the base slab $m_1 = 2.27 \times 10^5$ kg, the mass of the machine and upper slab $m_2 = 1.59 \times 10^5$ kg, and the flexibility of the foundation is $k_2 = 140.8 \times 10^6$ N/m, determine the two natural frequencies and vibration modes of the foundation.

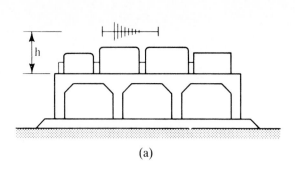

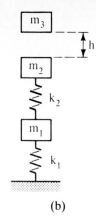

(a)

(b)

Figure P6.1

6.2. Express the arbitrary parameters c in equation (6.19) in terms of the initial conditions.

6.3. An automobile weighting 1818 kg pulls a trailer that weighs 3567 N. If the hitch consists of two springs with constants $k_1 = 17,000$ N/m and $k_2 = 22,000$ N/m connected in series, determine the natural frequencies of the system.

6.4. The system of Figure P6.4 models the coupling J_1 and the exciter J_2 of a large electric generator, operating at constant angular velocity, for torsional vibration. The first shaft has diameter $d_1 = 250$ mm and length $l_1 = 750$ mm, the second shaft has $d_2 = 250$ mm and $l_2 = 1$ m, the intermediate coupling has a diameter $D_1 = 0.630$ m and length $L_1 = 0.500$ m, and the exciter rotor has dimensions $D_2 = 0.75$ m and $L_2 = 1$ m. Assuming that the mass of the shafts is negligible and the flexibility of the coupling and exciter rotor are also negligible, determine for shear modulus $G = 0.82 \times 10^{11}$ N/m² **(a)** the torsional natural frequencies, **(b)** the torsional vibration modes, and **(c)** the nodal points on the shaft for the two modes.

6.5. The system of Figure P6.5 models a motor driving a centrifugal pump. If their moments of inertia are $J_1 = 2$ kg·m² and $J_2 = 0.4$ kg·m² and the connecting shaft has diameter $d = 25$ mm and length $l = 0.6$ m, determine **(a)** the torsional natural frequencies, and **(b)** the point of the shaft where we should put the coupling in order to be insensitive to torsional vibration $(G = 0.82 \times 10^{11}$ Pa).[10]

6.6. In a gearbox (Figure P6.6), the primary shaft has diameter $d_1 = 25$ mm and length $l_1 = 0.5$. The secondary shaft has $d_2 = 17$ mm and $l_2 = 0.625$ m. The gears have masses $m_1 = 2$ kg and $m_2 = 4$ kg, radii of gyration $\bar{r}_1 = 25$ mm and $\bar{r}_2 = 100$ mm, and radii $r_1 = 75$ mm and $r_2 = 150$ mm. If $G = 0.82 \times 10^{11}$ Pa, determine the torsional natural frequency.

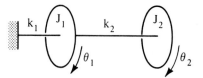

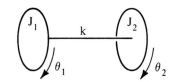

Figure P6.4

Figure P6.5

[10]The equivalent problem of the torsional vibration of the engine–propeller system in a ship was studied first by Frahm, H. 1902. Neue Untersuchungen über dynamischen Forgänge in den Wellenleitungen von Schiffsmaschinen. *J. VDI*, p. 797.

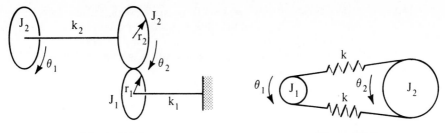

Figure P6.6	Figure P6.7

6.7. A system of two pulleys and a rubber transmission belt is shown in Figure P6.7. If the pulley diameters are $d_1 = 250$ mm, $d_2 = 500$ mm, the moments of inertia are $J_1 = 0.08$ kg·m^2 and $J_2 = 1.6$ kg·m^2, the stiffness of the belt is $k = 500$ N/m, and the transmission ratio is 2:1, determine the system natural frequencies and the vibration mode.

6.8. The rigid compressor shaft of Figure P6.8 has mass $m = 200$ kg, diameter $d = 50$ mm, and length $l = 1.12$ m. At one-third and two-thirds of its length it bears two masses $m_1 = 300$ kg and $m_2 = 500$ kg with radii of gyration $r_1 = 250$ mm and $r_2 = 490$ mm, respectively. If the end bearings have constants $k_1 = 3 \times 10^6$ N/m and $k_2 = 5 \times 10^6$ N/m, determine **(a)** the system natural frequencies, and **(b)** the vibration modes.

6.9. Solve Problem 6.8 if we place the second bearing in the midspan (Figure P6.9).

6.10. A moderately tall building can be modeled as in Figure P6.10 for horizontal earthquakes. If the mass of the building is $m = 500,000$ kg, height is 30 m, moment of inertia $J_G = 50 \times 10^6$ kg·m^2, soil horizontal stiffness $k = 500,000$ N/m, and rotary stiffness $k_T = 25 \times 10^6$ N·m/rad, determine the natural frequencies and the vibration modes of the building.

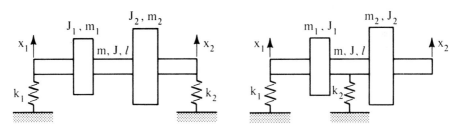

Figure P6.8	Figure P6.9

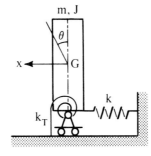

Figure P6.10

6.11. A bridge (Figure P6.11) consists of two rigid sections of moments of inertia $J_A = J_B = J = 3.6 \times 10^6$ kg·m² about their respective pivots A and B, and masses $m_1 = m_2 = 30,000$ kg are interconnected by way of springs of stiffness $k = 2.5 \times 10^6$ N/m. If the center of mass is at $h_G = 20$ m and the height of the bridge is 25 m, determine the natural frequencies and vibration modes.

6.12. The bridge of Problem 6.11 is redesigned with a link at point c and the system shown at A and B with $k = 2.5 \times 10^7$ N/m (Figure P6.12). Determine the natural frequencies and the vibration modes.

6.13. A steam generator (Figure P6.13) is mounted on a flexible column and can have horizontal and rotational vibration in the plane shown. If $m = 10,000$ kg, $J = 23,000$ kg·m², $h = 15$ m, $l = 5$ m, $I = 0.2$ m⁴, and $E = 2.1 \times 10^{11}$ Pa, determine the natural frequencies and the vibration modes.

6.14. An elevator cabin has a mass $m_1 = 600$ kg and a counterweight $m_2 = 300$ kg. They are connected by a cable with stiffness 50,000 N/cm if it has a length of 100 m. If the length of the cable $h_1 + h_2 = 60$ m, determine the position of the cabin h_1 where vertical vibration of the cabin will not result in torsional vibration of the driving pulley (Figure P6.14).

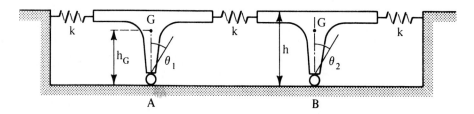

Figure P6.11

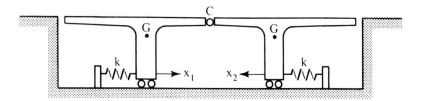

Figure P6.12

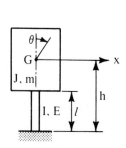

Figure P6.13

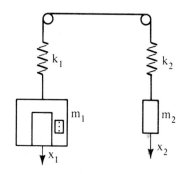

Figure P6.14

6.15. A fire truck has a mass $m = 3000$ kg and moment of inertia $J = 6000$ kg·m² and center of mass at the midspan between wheels and 2 m above ground. A man of mass $m_2 = 80$ kg is at the end of the ladder at a height $h = 25$ m, $L = 6$ m. The wheels of the truck have a distance $l = 5$ m and stiffness $k = 50,000$ N/m (Figure P6.15). Determine the natural frequencies and the vibration modes if the man carrying the carriage is **(a)** at $h = 25$ m, and **(b)** at $h = 12$ m and $L = -6$ m.

6.16. A construction crane (Figure P6.16) consists of a vertical beam with $h = 35$ m, $I = 0.0025$ m⁴, $E = 2.1 \times 10^{11}$ Pa, a counterweight $m = 500$ kg, and $L = 10$ m. It is lifting a mass $m_2 = 100$ kg, and the stiffness of the cable is 50,000 N/m. Determine the natural frequencies and the vibration modes.

6.17. An overhead crane has a beam with $I = 1.2 \times 10^{-3}$ m⁴, $E = 2.1 \times 10^{11}$ Pa, $L = 50$ m. The truck has a mass $m_2 = 1.2$ tons and it is lifting a mass $m_1 = 5$ tons through a cable of stiffness $k = 200,000$ N/m. Determine the natural frequencies of the system and the vibration modes (Figure P6.17).

6.18. The system of Problem 6.3 is moving with a constant speed 30 km/h when the automobile brakes momentarily and reduces its speed to 25 km/h. Determine the resulting motion of the trailer.

6.19. During a repair job, the turbine foundation of Problem 6.1 was hit by a turbine rotor that was dropped from a height $h = 4.5$ m and the impact was plastic. If the weight of this rotor is 111,000 N, determine the resulting vibration, assuming plastic impact.

6.20. The shaft in Problem 6.4 is suddenly fixed on the left as shown, due to a generator failure, while rotating at 3600 rpm. Determine the maximum angle of twist and torque in the shaft.

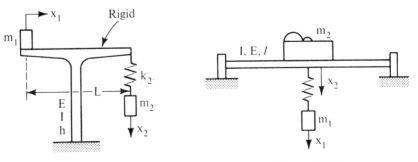

Figure P6.15

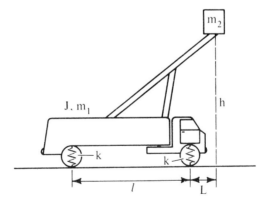

Figure P6.16   **Figure P6.17**

6.21. Shaft 1 in Problem 6.6 is suddenly fixed at the right, as shown, when it was running at 850 rpm. Determine the amplitudes of the resulting vibration and the maximum torques on the shafts.

6.22. The pulleys of Problem 6.7 are at a distance $l = 2$ m. To test the tension of the belt, one pushes the upper branch of the belt perpendicular to the belt by 100 mm in the middle and measures the force. Determine the resulting vibration of the pulleys when the belt is released if the pulleys have diameters $d_1 = 250$ mm and $d_2 = 500$ mm.

6.23. A small aircraft weighing 800 kg collides with the top of the building of Problem 6.10 while flying with a horizontal speed of 150 mph. Determine the motion of the building.

6.24. An automobile weighing 1500 kg applies its brakes and stops very fast when running at 50 mph on the midspan of the first segment of the bridge of Problem 6.11. Determine the motion of the bridge.

6.25. An automobile weighing 1500 kg applies its brakes and stops very fast when running at 50 mph on the midspan of the first segment of the bridge of Problem 6.12. Determine the motion of the bridge.

6.26. The steam generator of Problem 6.13 was connected with a horizontal steel pipe with the stiff end of a turbine at its center of mass. When the pipe of length $L = 33$ m and area of cross-section $A = 0.002$ m² was heated to 550°C, it failed due to thermal expansion. Determine **(a)** the initial displacement of the steam generator, and **(b)** the system vibration after the failure.

6.27. On the ladder of Problem 6.15 the man jumps from a height 1 m above the end. Determine the resulting motion of the system.

6.28. In Problem 6.16, while the crane is lifting a weight with a speed 0.15 m/s, the motor suddenly stops. Determine the resulting motion.

6.29. In Problem 6.17, while the crane is lifting the weight with a speed 0.15 m/s, the motor suddenly stops. Determine the resulting motion.

Sections 6.4 and 6.5

6.30. A turbine-generator foundation can be modeled as the system of Figure P6.1. The spring constant of the soil $k_1 = 80 \times 10^6$ N/m, the mass of the base slab $m_1 = 2.27 \times 10^5$ kg, the mass of the machine and upper slab $m_2 = 1.59 \times 10^5$ kg, and the flexibility of the foundation is $k_2 = 140.8 \times 10^6$ N/m. To test the foundation, an engineer installs a centrifugal shaker on the top slab. The shaker has an eccentric mass of 3 kg at a radius of 0.1 m. Determine the resulting vibration of the top slab at the maximum speed of the shaker, which is 850 rpm, assuming that the mass of the shaker itself can be neglected.

6.31. The system of Figure P6.4 models the coupling J_1 and the exciter J_2 of a large electric generator, operating at constant angular velocity, for torsional vibration. The first shaft has diameter $d_1 = 250$ mm and length $l_1 = 750$ mm, the second shaft has $d_2 = 250$ mm and $l_2 = 1.0$ m, the intermediate coupling has a diameter $D_1 = 0.630$ m and length $L_1 = 0.500$ m, and the exciter rotor has dimensions $D_2 = 0.75$ m and $L_2 = 1$ m. The mass of the shafts is negligible, the flexibility of the coupling and exciter rotor are also negligible, and the shear modulus of the shaft material is $G = 2.1 \times 10^{11}$ N/m². Due to unbalanced electrical harmonics, the generator has a torsional vibration with amplitude 0.00012 rad at 120-Hz frequency. Determine the torsional vibration of the exciter.

6.32. The system of Figure P6.5 models a motor driving a centrifugal pump. Their moments of inertia are $J_1 = 2$ kg·m² and $J_2 = 0.4$ kg·m², the connecting shaft has diameter d

$= 25$ mm and length $l = 0.6$ m, and the shear modulus of the shaft material is $G = 0.82 \times 10^{11}$ Pa. Due to the wear of the pump rotor, there is harmonic torque applied to it of amplitude 1.3 N·m at the frequency of rotation, which is 1800 rpm. Determine the resulting torsional vibration.

6.33. In a gearbox (Figure P6.6), the primary shaft has diameter $d_1 = 25$ mm and length $l_1 = .5$ m. The secondary shaft has $d_2 = 17$ mm and $l_2 = 0.625$ m. The gears have masses $m_1 = 2$ kg and $m_2 = 4$ kg, radii of gyration $\bar{r}_1 = 25$ mm and $\bar{r}_2 = 100$ mm, and radii $r_1 = 75$ mm and $r_2 = 150$ mm. The shear modulus of the shaft material is $G = 0.82 \times 10^{11}$ Pa. If the base element at the right has a torsional vibration of amplitude 0.002 rad at frequency 45 Hz, determine the resulting torsional vibration of the gear at the left end.

6.34. A system of two pulleys and a rubber transmission belt is shown in Figure P6.7. The moments of inertia are $J_1 = 0.08$ kg·m^2 and $J_2 = 1.6$ kg·m^2, the stiffness of the belt is $k = 500$ N/m, and the transmission ratio is 2:1 and $d_1 = 250$ mm. Pulley 1 on the left is subject to a torque of amplitude 1.2 N·m at frequency 600 rpm. Determine the torsional vibration of the pulleys.

6.35. The rigid compressor shaft of Figure P6.8 has mass $m = 200$ kg, diameter $d = 50$ mm, and length $l = 1.12$ m. At one-third and two-thirds of its length it bears two masses $m_1 = 300$ kg and $m_2 = 500$ kg with radii of gyration $r_1 = 250$ mm and $r_2 = 490$ mm, respectively. The end bearings have spring constants $k_1 = 3 \times 10^6$ N/m and $k_2 = 5 \times 10^6$ N/m and damping constants $c = 4000$ N·s/m and 5000 N·s/m, respectively. The compressor rotates at 3600 rpm and has an unbalanced mass of 0.050 kg at a radius of 0.25 m on mass m_1. Determine the resulting vertical vibration at the bearings.

6.36. A refrigeration compressor is mounted on four springs of stiffness $k = 20,000$ N/m and has mass $m = 55$ kg. The mounts are made of spring steel with very low damping. Due to the design of the compressor, there is a vertical harmonic force of 12 N oscillating at the operating frequency of 1750 rpm. An undamped dynamic vibration absorber of mass $m/10$ and $\omega_n = 183$ rad/s is attached to the compressor. Determine the resulting vibration of the compressor.

6.37. A piston-type air compressor of mass $m = 80$ kg is rigidly mounted on a platform of mass $M = 50$ kg which could vibrate horizontally without friction and has an elastic support with stiffness in the horizontal direction $k = 3500$ N/m. At the running frequency of 1150 rpm the measured horizontal vibration is 0.005 m. Then an undamped dynamic vibration absorber of mass $(m + M)/10$ and $\omega_n = 120$ rad/s is attached to it. Determine the resulting vibration of the compressor.

6.38. An electric motor of mass $m = 22$ kg is mounted at the midspan of a simply supported steel beam of rectangular cross section, length $L = 1$ m, width $b = 0.2$ m, and thickness $t = 10$ mm. The amplitude of the unbalanced harmonic vertical force of the motor in the vertical direction is known to be 55 N at 58 Hz. Then an undamped dynamic vibration absorber of mass $m/10$ and $\omega_n = 183$ rad/s is attached to it. Determine the resulting vibration of the compressor.

6.39. The moving core of an electromagnetic relay (Figure P6.39) has mass $m = 12$ g and is supported at one end with a coil spring with $k = 3000$ N/m and at the other end, at the closed contact position, by the elastic strips of the electric contacts, which have stiffness 1200 N/m in the direction of the coil motion (Figure P2.6). There is an unbalanced oscillating force due to the electric field along the core axis of 1.3 N at the synchronous electric frequency of 60 Hz. Then an undamped dynamic vibration absorber of mass $m/10$ and $\omega_n = 377$ rad/s is attached to it. Determine the resulting vibration of the relay core.

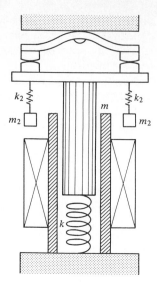

Figure P6.39

Section 6.6

6.40. The dc voltmeter of an ac-to-dc converter has a steel dial of length $l = 50$ mm, width 3 mm, and thickness 1 mm. The restoring spring has a constant $k = 0.006$ N·m/rad. The maximum indication is 100 V at a 50° angle. On the dc voltage there is a 2-V ripple of frequency 60 Hz due to the noncomplete rectification of the ac. Then a torsional damped dynamic vibration absorber of mass $J/10$, $\zeta = 0.1$, and $\omega_n = 377$ rad/s is attached to it. Determine the resulting vibration of the dial.

6.41. A refrigeration compressor is mounted on four springs of stiffness $k = 20,000$ N/m and has mass $m = 55$ kg. The mounts are made of spring steel with very low damping. Due to the design of the compressor, there is a vertical harmonic force of 12 N oscillating at the frequency of rotation 1750 rpm. Then a damped dynamic vibration absorber of mass $m/10$, $\zeta = 0.1$, and $\omega_n = 183$ rad/s is attached to it. Determine the resulting vibration of the compressor.

6.42. A piston-type 3600-rpm air compressor has mass $m = 80$ kg and is rigidly mounted on a platform of mass $M = 50$ kg which can vibrate horizontally without friction and has an elastic support with stiffness in the horizontal direction $k = 3500$ N/m. It vibrates with amplitude 35 μm. Then a damped dynamic vibration absorber of mass $m/10$, $\zeta = 0.1$, and $\omega_n = 377$ rad/s is attached to it. Determine the resulting vibration of the compressor.

6.43. A 1800-rpm electric motor of mass $m = 22$ kg is mounted at the midspan of a simply supported steel beam of rectangular cross section, length $L = 1$ m, width $b = 0.2$ m, and thickness $t = 10$ mm. The motor vibration is 22 μm. Then a damped dynamic vibration absorber of mass $m/10$, $\zeta = 0.1$, and $\omega_n = 180$ rad/s is attached to it. Determine the resulting vibration.

6.44. The moving core of a 60-Hz electromagnetic relay (Figure 6.39) has mass $m = 12$ g and is supported at one end with a coil spring with $k = 3000$ N/m and at the other end, at the closed contact position, by the elastic strips of the electric contacts, which have stiffness 1200 N/m in the direction of the coil motion. It vibrates with amplitude

5 μm. Then a damped dynamic vibration absorber of mass $m/10$, $\zeta = 0.1$, and $\omega_n = 377$ rad/s is attached to it. Determine the resulting vibration.

6.45. The dc voltmeter of an ac-to-dc converter has a steel dial of length $l = 50$ mm, width 3 mm, and thickness 1 mm. The restoring spring has a constant $k = 0.006$ N·m/rad. The maximum indication is 100 V at a 50° angle. On the dc voltage there is a 2-V ripple due to noncomplete rectification of the 50-Hz ac. Then a damped dynamic vibration absorber of mass $m/10$, $\zeta = 0.1$, and $\omega_n = 377$ rad/s is attached to it. Determine the resulting vibration of the dial.

Section 6.7

6.46. The electric machine of Figure 6.10 is a field-controlled dc motor for which the function $T = K_f i$, where K_f is a constant. The rotor of the driven machine is a disk D with moment of inertia $J = 3$ kg·m², and the connecting shaft has torsional stiffness $k_T = 360,000$ N·m/rad. The electric circuit has an electric resistance $R = 10\ \Omega$, capacitance $C = 2 \times 10^{-6}$ F, inductance 4 H, and operates at a voltage of 220 V. The motor is operating at 15 kW and 3000 rpm with an efficiency $\eta = 75\%$ when there is a short circuit between points A and B. Determine the vibration amplitude of the disk, the maximum current, and the maximum dynamic torque on the shaft.

6.47. The electric machine of Figure P6.47 is an armature-controlled dc motor for which the function $T = K_f i$, where K_f is a constant. In addition, a counter-electromotive force voltage $V_c = K_e \dot{\phi}$ is developed, which is added to the left-hand side of equation (6.50). At the operating speed $V_c = 100$ V, the rotor of the driven machine is a disk D with moment of inertia $J = 3$ kg·m² and the connecting shaft has torsional stiffness $k_T = 360,000$ N·m/rad. The electric circuit has an electric resistance $R = 10\ \Omega$, capacitance $C = 2 \times 10^{-6}$ F, inductance 4 H, and operates at a voltage of 220 V. The motor is operating at 15 kW and 3000 rpm with an efficiency $\eta = 75\%$ with a 60-Hz harmonic voltage of $V_1 = 15$ V due to incomplete rectification, superposed to $V_0 = 220$ V dc. Determine the torsional vibration amplitude of the disk and the dynamic torque on the shaft.

6.48. In the system of Problem 6.47 there is a short circuit between points A and B (Figure P6.47) while the machine is operating at rated conditions. Determine the resulting maximum current and shaft torque.

6.49. The electric machine of Figure P6.49 is an ac servomotor for which the function $T = K_f i - K_\omega \dot{\phi}$, where K_f and K_ω are constants and the two terms at operating conditions are $(K_f i)_0 = (2K_\omega \dot{\phi})_0$. The rotor of the driven machine is a disk D with moment of inertia $J = 3$ kg·m², and the connecting shaft has torsional stiffness 360,000 N·m/rad. The electric circuit has electric resistance $R = 10\ \Omega$, capacitance $C = 2 \times 10^{-6}$ F, inductance 4 H, and operates at a voltage of 220 V. The motor is operating at 15 kW and 3000 rpm with an efficiency $\eta = 75\%$ with a 60-Hz harmonic voltage

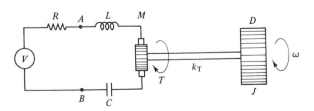

Figure P6.47

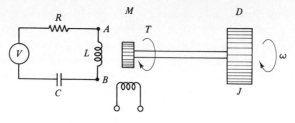

Figure P6.49

of $V_1 = 15$ V due to incomplete rectification, superposed to $V_0 = 220$ V dc. Determine the vibration amplitude of the disk, the maximum current, and the maximum dynamic torque on the shaft.

6.50. In the system of Problem 6.49, there is a short circuit between the two ends of the motor armature, A and B (Figure P6.49), while the machine is operating at rated conditions. Determine the resulting maximum current and shaft torque.

DESIGN PROBLEMS

Sections 6.1 to 6.5

D6.1. A refrigeration compressor is mounted on four springs of stiffness $k = 20,000$ N/m and has mass $m = 55$ kg. The mounts are made of spring steel with very low damping. Due to the design of the compressor, there is a vertical harmonic force of 12 N oscillating at the operating frequency of 1750 rpm. Design an undamped dynamic vibration absorber of mass up to one-tenth of the compressor mass to absorb vibration at the operating speed.

D6.2. A piston-type air compressor with mass $m = 80$ kg is rigidly mounted on a platform of mass $M = 50$ kg which could vibrate horizontally without friction and has an elastic support with stiffness in the horizontal direction $k = 3500$ N/m. At the running frequency of 1150 rpm the measured horizontal vibration is 0.0005 m. Design an undamped vibration absorber with mass ratio ≤ 0.2 to absorb vibration at the operating speed.

D6.3. An electric motor of mass $m = 22$ kg is mounted at the midspan of a simply supported steel beam of rectangular cross section, length $L = 1$ m, width $b = 0.2$ m, and thickness $t = 10$ mm. The amplitude of the unbalanced harmonic vertical force of the motor in the vertical direction is known to be 55 N at 58 Hz. Design an undamped vibration absorber with mass ratio ≤ 0.2 to absorb vibration at the operating speed.

D6.4. The moving core of an electromagnetic relay (Figure P6.39) has mass $m = 12$ g and is supported at one end by a coil spring with $k = 3000$ N/m and at the other end, at the closed contact position, by the elastic strips of the electric contacts, which have stiffness 1200 N/m in the direction of the coil motion. There is an unbalanced oscillating force due to the electric field along the core axis of 1.3 N at a synchronous electric frequency of 60 Hz. Design an undamped vibration absorber with mass ratio ≤ 0.3 to absorb vibration at the operating speed.

D6.5. The dc voltmeter of an ac-to-dc converter has a steel dial of length l = 50 mm, width 3 mm, and thickness 1 mm. The restoring spring has a constant k = 0.006 N·m/rad. The maximum indication is 100 V at a 50° angle. On the dc voltage, there is a ripple of frequency 60 Hz due to the noncomplete rectification of the ac. Design an undamped vibration absorber with mass ratio ≤ 0.1 to eliminate torsional vibration of the dial at the frequency of 60 Hz.

D6.6. A refrigeration compressor is mounted on four springs of stiffness k = 20,000 N/m and has mass m = 55 kg. The mounts are made of spring steel with very low damping. Due to the design of the compressor, there is a vertical harmonic force of 12 N oscillating at the frequency of rotation 1200 rpm. Design a damped dynamic vibration absorber of mass up to one-tenth of the compressor mass to absorb the vibration of the compressor when it goes through the natural frequency.

D6.7. A piston-type 1500-rpm air compressor has mass m = 80 kg and it is rigidly mounted on a platform of mass M = 50 kg that could vibrate horizontally without friction and has an elastic support with stiffness in the horizontal direction k = 3500 N/m. Design a damped vibration absorber with mass ratio ≤ 0.2 to absorb vibration of the compressor when it goes through the natural frequency.

D6.8. A 3500-rpm electric motor of mass m = 22 kg is mounted at the midspan of a simply supported steel beam of rectangular cross section, length L = 1 m, width b = 0.2 m, and thickness t = 10 mm. Design a damped vibration absorber with mass ratio ≤ 0.2 to absorb vibration of the motor when it goes through the natural frequency.

D6.9. The moving core of an electromagnetic relay (Figure P6.39) has mass m = 12 g and is supported at one end with a coil spring with k = 3000 N/m and at the other end, at the closed contact position, by the elastic strips of the electric contacts, which have stiffness 1200 N/m in the direction of the coil motion. Design a damped vibration absorber with mass ratio ≤ 0.3 to absorb vibration of the relay for a wide range of excitation frequencies.

D6.10. The dc voltmeter of an ac-to-dc converter has a steel dial of length l = 50 mm, width 3 mm, and thickness 1 mm. The restoring spring has a constant k = 0.006 N·m/rad. The maximum indication is 100 V at a 50° angle. On the dc voltage, there is a ripple due to noncomplete rectification of the ac. Design a damped vibration absorber with mass ratio ≤ 0.1 to absorb the vibration of the dial for wide excitation frequencies.

7

Lumped Mass Systems: Natural Vibration

7.1 INTRODUCTION

In Chapter 2, the minimum number of independent coordinates needed to describe a system completely was called the *number of degrees of freedom* of the system. The systems in Figure 2.2 have one degree of freedom. A robot arm (Figure 2.3) has more than one degrees of freedom. A large class of systems can be described by a finite number of degrees of freedom, such as the simple systems in Figure 6.1. Other systems, such as those that include deformable members, have an infinite number of degrees of freedom. For example, to describe an elastic beam, one needs an infinite number of coordinates (i.e., the deflections of the infinite number of points of its elastic line). The elastic beam therefore has an infinite number of degrees of freedom. Almost all machines and structures have elastic members and thus an infinite number of degrees of freedom. Systems with a finite number of degrees of freedom are called *discrete systems*. Those with an infinite number of degrees of freedom are called *continuous systems*.

Although there are methods of dealing with continuous systems, engineering

*The above photo portrays a set of chime bells from the 5th Century B.C. China. Hupei Provincial Museum, Wuhan.

accuracy is often reached by describing a continuous system with a finite number of coordinates. For example, the deflection of an elastic beam can be described by the displacement of a number of points along its elastic line. It appears that a larger number of coordinates yields a better approximation. The form of the elastic line between approximation points is assumed parabolic or any convenient curve satisfying certain requirements of smoothness.

At this point, without going into detail about the number of degrees of freedom required to describe a continuous system, it is assumed that a system, discrete or continuous, will be described by *n* independent coordinates—in other words, by *n* degrees of freedom. A *lumped mass system* is a particular case of a discrete system that consists of discrete elements, inertias, dampers, and springs, that is, elements that relate forces to displacements, velocities, and accelerations.

7.2 DIRECT METHODS OF ANALYSIS

For the lumped mass system we first derive the differential equations of motion based on Newton's second law of motion, discussed in Chapter 2.

Equations of Motion

Our strategy for the derivation of the equations of motion will follow closely the methods used for the one- and two-degree-of-freedom systems:

1. We model the system as a complex of rigid bodies connected with massless elastic members.
2. We find the static equilibrium of the system.
3. We select a number of coordinates that are necessary to describe the geometry of the system. (In general, for a rigid body, we need six coordinates if it has spatial motion or three coordinates if it is confined to move on a plane. It is generally convenient to select as coordinates the translations of the center of mass and the rotations about it.) We designate positive directions.
4. We assume the system to be displaced from equilibrium to small but finite values of the selected coordinates.
5. At this position, we compute the several forces and moments acting on the masses due to deflection of elastic elements, rate of deformation of dampers, gravity, and so on, as functions of the coordinates selected. We draw free-body diagrams for every mass.
6. We express the accelerations of the centers of mass of the masses of the system and their angular accelerations as functions of the coordinates selected. (We can express these accelerations in any convenient coordinates first and then transform them into the coordinates selected.)
7. We apply Newton's law for the linear motion of the centers of mass and the rotations about them, for every rigid body of the system. This must be done as many times as the number of rigid bodies in the system, plus the number of unknown forces of constraint.

8. We can also derive the equations of motion based on D'Alembert's principle. To this end, we apply *effective forces* on the mass centers equal to the products (*mass × acceleration*) and *effective moments* about the mass centers equal to the products (*mass moment of inertia × angular acceleration*). Then we write the conditions of static equilibrium $\Sigma F = \Sigma F_{\text{Eff}}$ and $\Sigma M = \Sigma M_{\text{Eff}}$ for every rigid body in the system. In the latter equation we include also the moments of the effective forces on the right-hand side. We eliminate forces of constraint (e.g., forces of reaction between connected rigid bodies) using an equal number of equations.

9. We keep the terms containing the coordinates and their derivatives on the left-hand side of the equations and the constant terms and functions of time on the right-hand side of the equations.

Consider, for example, the system in Figure 7.1, which models a tall building, the twin towers of the New York World Trade Center. It consists of n floors and the slabs can be considered rigid masses, while the vertical columns can be considered massless elastic members with flexibility in bending much higher than their

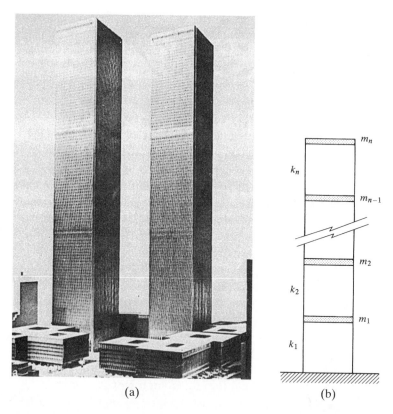

(a)   (b)

Figure 7.1 (a) Twin buildings of New York World Trade Center. (b) Line model of a multistory building.

flexibility in tension–compression. Therefore, the slabs have mainly parallel horizontal motion (Figure 7.2a). The motion of the slabs x_i, $i = 1, 2, \ldots, n$, are the system coordinates.

The stiffness of each column, for the deformation shown in Figure 7.2a, is $k = 12\,EI/L^3$, where L is the length and EI the flexural rigidity of the column. Usually, there is more than one column. The columns can be considered as springs in parallel; therefore, the total stiffness for each floor is the sum of the stiffnesses of the several columns of that floor, $k_i = \Sigma_{j=1}^{n}\, 12\,EI_j/L_j^3$, where n is the number of columns.

We are now ready to model the system:

Step 1: We assume the system to consist of n masses with parallel motion, connected by n springs (Figure 7.1).

Step 2: We assume that the equilibrium is in the position shown and that no static forces act in the horizontal direction.

Step 3: As coordinates we select the horizontal displacements of the slabs from equilibrium. We assume significant motion only in the horizontal direction.

Step 4: We assume the system to be in the displaced position. The slabs are displaced by $x_1, x_2, \ldots, x_n$, which are the system coordinates (Figure 7.2a).

Step 5: We draw a free-body diagram for each mass (Figure 7.2b).

Step 6: The coordinates selected are assumed to be displacements of the mass centers of the slabs.

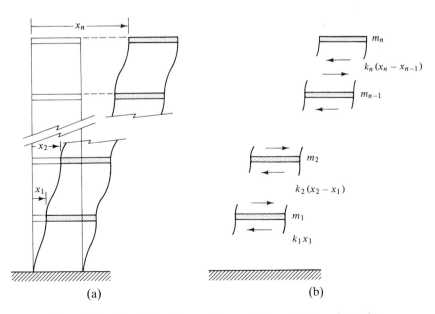

Figure 7.2 Free-body diagram for multistory building dynamics.

Step 7: Newton's second law for the horizontal motion yields:

$$\text{mass 1: } m_1\ddot{x}_1 = -k_1x_1 + k_2(x_2 - x_1)$$

$$\text{mass 2: } m_2\ddot{x}_2 = -k_2(x_2 - x_1) + k_3(x_3 - x_2)$$

$$\text{mass 3: } m_3\ddot{x}_3 = -k_3(x_3 - x_2) + k_4(x_4 - x_3) \qquad (7.1)$$

$$\vdots$$

$$\text{mass } n: m_n\ddot{x}_n = -k_n(x_n - x_{n-1})$$

Step 8: Rearranging, we obtain the differential equations of motion:

$$m_1\ddot{x}_1 + (k_1 + k_2)x_1 \quad -k_2x_2 \qquad\qquad\qquad\qquad = 0$$

$$m_2\ddot{x}_2 \quad -k_2x_1 + (k_2 + k_3)x_2 \quad -k_3x_3 \qquad\qquad = 0$$

$$m_3\ddot{x}_3 \qquad\qquad -k_3x_2 + (k_3 + k_4)x_3 - k_4x_4 \qquad = 0$$

$$\vdots$$

$$m_n\ddot{x}_n \qquad\qquad\qquad\qquad\qquad -k_nx_{n-1} + k_nx_n = 0$$

$$\qquad\qquad\qquad\qquad\qquad\qquad\qquad\qquad\qquad (7.2)$$

Natural Vibration: Method of the Frequency Determinant

Application of the procedure outlined above will yield for a lumped mass system, in the absence of external forces and damping, a system of linear differential equations with constant coefficients. We observe that these equations are not independent, due to the coupling terms k_{ij}, $i \neq j$. Therefore, we have to solve the differential equations simultaneously. The resulting vibration is coupled. As stated in Chapter 6, if the coupling is due to the mass terms m_{ij}, we have *dynamic coupling*; if the coupling is due to the stiffness terms k_{ij}, we have *static coupling*. It was shown in Chapter 6 that the type of the coupling is not a system property but depends mainly on the selection of the coordinates. We shall prove later that in undamped systems there exist appropriate coordinates which yield uncoupled systems of differential equations. In general, we shall assume that the system can be modeled by a coupled system of differential equations, of the equation (7.2) type.

As we did for the harmonic oscillator, we shall study the feasibility of harmonic natural vibration of the system at a single frequency (yet unknown) ω. Therefore, we try the solution [Rayleigh, 1894]

$$x_1 = A_1 \cos \omega t$$

$$x_2 = A_2 \cos \omega t$$

$$\vdots \qquad\qquad\qquad (7.3)$$

$$x_n = A_n \cos \omega t$$

where $A_1, A_2, \ldots, A_n$ are amplitudes and ω the circular frequency, undetermined yet. Substitution in (7.2) yields

$$
\begin{aligned}
(-m_1\omega^2 + k_1 + k_2)A_1 & & -k_2 A_2 & & & = 0 \\
-k_2 A_1 & + (-m_2\omega^2 + k_2 + k_3)A_2 & & -k_3 A_3 & & = 0 \\
& -k_3 A_2 & + (-m_3\omega^2 + k_3 + k_4)A_3 & -k_4 A_4 & = 0 \\
& & & \vdots \\
& & -k_n A_{n-1} + (-m_n\omega^2 + k_n)A_n & = 0
\end{aligned}
$$

$$(7.4)$$

If this system of linear algebraic equations has a solution, (7.3) is indeed a solution of (7.2). The system of linear algebraic equations (7.4) is homogeneous; therefore, it has a solution only if[1]

$$\det(\mathbf{A}) = 0 \qquad\qquad (7.5)$$

where

$$
\mathbf{A} = \begin{bmatrix}
-m_1\omega^2 + k_1 + k_2 & -k_2 & 0 & 0 & \cdots & 0 \\
-k_2 & -m_2\omega^2 + k_2 + k_3 & -k_3 & 0 & \cdots & 0 \\
0 & -k_3 & -m_3\omega^2 + k_3 + k_4 & -k_4 & \cdots & 0 \\
\vdots & & & & & \\
0 & 0 & & \cdots & -k_n & -m_n\omega^2 + k_n
\end{bmatrix}
$$

The determinant in equation (7.5) is called a *frequency determinant*[2] and equation (7.5) is called a *frequency equation*. This equation contains only one unknown, ω. Because, in general[3] (*but not always*), this equation has n distinct real roots, $\omega_1, \omega_2, \ldots, \omega_n$, we can conclude that the system can undergo harmonic vibration at n frequencies $\omega_1, \omega_2, \ldots, \omega_n$. Furthermore, we observe from equation (7.5) that the frequencies ω_i depend only on the system property m and k values.[4] Some of the frequencies might be multiple.[5] We will exemplify on this point later.

---

[1]Maclaurin, C., 1748. *Treatise of Algebra*.

[2]Routh, E. J. 1892. *Dynamics of a System of Rigid Bodies*, 5th ed., New York: Dover Publications.

[3]Full discussion in Routh, E. J. 1877. *Adams Prize Essay*. See also *Dynamics of a System of Rigid Bodies*, 5th ed., Dover Publications, New York, 1955.

[4]Ibid.

[5]See [Rayleigh, 1894] and Routh, E. J. 1892. *Dynamics of a System of Rigid Bodies*, 5th ed., Dover Publications, New York, 1955.

The amplitudes $A_1, A_2, \ldots, A_n$ of the harmonic vibration are not altogether arbitrary. Indeed, for every ω satisfying equation (7.5) there is a solution of the system (7.4). In fact, equation (7.5) is merely a condition that equations (7.4) are linearly dependent; that is, at least one of them can be found to be a linear combination of the others. Therefore, for $\omega = \omega_i$, we have $n - 1$ equations for

---

EDWARD JOHN ROUTH (1831–1907)

Routh, born in Canada son of a high-ranking army officer, attended University College in London. He entered the Peterhouse, a prestigious Cambridge division, together with J. C. Maxwell. The latter moved to Trinity College, it is said to avoid the competition with Routh. In the mathematical contest of 1854, Routh was senior wrangler and Maxwell was second. They shared the Smith Prize in the same year. Routh became a famous "coach" preparing students for the Cambridge Examinations and for mathematical competitions. Forty of his students won the Smith Prize. He wrote very popular treatises and wrote few research works, but he made a distinctive impression on classical mechanics. He won the Adams Prize and was elected fellow of the Royal Society of London. He was the first to receive the Sc.D. degree from Cambridge.

---

the amplitudes $A_1, A_2, \ldots, A_n$. Assigning any arbitrary value to, say, A_1, we can compute $A_2, \ldots, A_n$ from $n - 1$ equations from the system (7.4):

$$
\begin{aligned}
-k_2 A_2 \qquad\qquad\qquad\qquad\qquad\qquad &= -(-m_1\omega^2 + k_1 + k_2)A_1 \\
(-m_2\omega^2 + k_2 + k_3)A_2 \qquad\quad -k_3 A_3 \qquad\quad &= k_2 A_1 \\
-k_3 A_2 \quad + (-m_3\omega^2 + k_3 + k_4)A_3 - k_4 A_4 \quad &= 0 \\
\cdots\cdots\cdots\cdots\cdots\cdots\cdots\cdots\cdots\cdots\cdots\cdots\cdots\cdots&\cdots\cdots\cdots\cdots \\
-k_n A_{n-1} + (-m_n\omega^2 + k_n)A_n &= 0
\end{aligned}
$$

$$(7.6)$$

We shall use the superscript j to indicate the circular frequency ω_j with which the amplitudes $A_i^{(j)}$, $i = 1, 2, \ldots, n$, are associated.

We conclude that the n-degree-of-freedom system can have harmonic natural vibration at n different frequencies $\omega_1, \omega_2, \ldots, \omega_n$, and at every frequency ω_j the amplitudes will have certain ratios $A_2^{(j)}/A_1^{(j)}$, $A_3^{(j)}/A_1^{(j)}$, $\ldots$, $A_n^{(j)}/A_1^{(j)}$, which depend on the system properties.

The circular frequencies ω_j are called *natural frequencies* and the vectors $\{A_1^{(j)}, A_2^{(j)}, \ldots, A_n^{(j)}\}$ are called *natural modes*. As a general procedure, once the system of differential equations has been obtained, we determine the natural frequencies and natural modes as follows:

Step 1: We set up the system of differential equations in the form

$$m_{11}\ddot{x}_1 + m_{12}\ddot{x}_2 + \cdots + m_{1n}\ddot{x}_n + k_{11}x_1 + k_{12}x_2 + \cdots + k_{1n}x_n = 0$$

$$m_{21}\ddot{x}_1 + m_{22}\ddot{x}_2 + \cdots + m_{2n}\ddot{x}_n + k_{21}x_1 + k_{22}x_2 + \cdots k_{2n}x_n = 0$$

$$\vdots \tag{7.7}$$

$$m_{n1}\ddot{x}_1 + m_{n2}\ddot{x}_2 + \cdots + m_{nn}\ddot{x}_n + k_{n1}x_1 + k_{n2}x_2 + \cdots k_{nn}x_n = 0$$

Step 2: We substitute in (7.7):

$$x_1 = A_1 \cos \omega t, \qquad \ddot{x}_1 = -A_1\omega^2 \cos \omega t$$

$$x_2 = A_2 \cos \omega t, \qquad \ddot{x}_2 = -A_2\omega^2 \cos \omega t$$

$$\vdots \qquad\qquad \vdots \tag{7.8}$$

$$x_n = A_n \cos \omega t, \qquad \ddot{x}_n = -A_n\omega^2 \cos \omega t$$

Step 3: We obtain the system of algebraic equations:

$$(-m_{11}\omega^2 + k_{11})A_1 + (-m_{12}\omega^2 + k_{12})A_2 + \cdots + (-m_{1n}\omega^2 + k_{1n})A_n = 0$$

$$(-m_{21}\omega^2 + k_{21})A_1 + (-m_{22}\omega^2 + k_{22})A_2 + \cdots + (-m_{2n}\omega^2 + k_{2n})A_n = 0$$

$$\vdots \tag{7.9}$$

$$(-m_{n1}\omega^2 + k_{n1})A_1 + (-m_{n2}\omega^2 + k_{n2})A_2 + \cdots + (-m_{nn}\omega^2 + k_{2nn})A_n = 0$$

Step 4: We set up the condition for existence of a solution of (7.9):

$$\begin{vmatrix} -m_{11}\omega^2 + k_{11} & -m_{12}\omega^2 + k_{12} & \cdots & -m_{1n}\omega^2 + k_{1n} \\ -m_{21}\omega^2 + k_{21} & -m_{22}\omega^2 + k_{22} & \cdots & -m_{2n}\omega^2 + k_{2n} \\ \cdots\cdots\cdots\cdots\cdots\cdots\cdots\cdots\cdots\cdots\cdots\cdots \\ -m_{n1}\omega^2 + k_{n1} & -m_{n2}\omega^2 + k_{n2} & \cdots & -m_{nn}\omega^2 + k_{nn} \end{vmatrix} = 0 \tag{7.10}$$

Step 5: We expand the determinant (7.10) into a polynomial in ω and find its roots[6]:

$$\omega_1, \omega_2, \ldots, \omega_n$$

[6]This is used in systems with relatively small number of degrees of freedom, usually up to 3. For larger systems, numerical methods are used, such as the ones discussed in Chapter 10.

Step 6: For every value ω_j, $j = 1, 2, \ldots, n$, we set $A_1^{(j)} = 1$. Then we take $n - 1$ of the equations (7.9), transfer the $A_1^{(j)}$ terms to the right-hand side, and solve for $A_2^{(j)}, A_3^{(j)}, \ldots, A_n^{(j)}$.

The vector $\{A_1^{(j)}, A_2^{(j)}, \ldots, A_n^{(j)}\}$ is the natural mode of vibration corresponding to the natural frequency ω_j. *Natural mode of vibration* means that if the system has

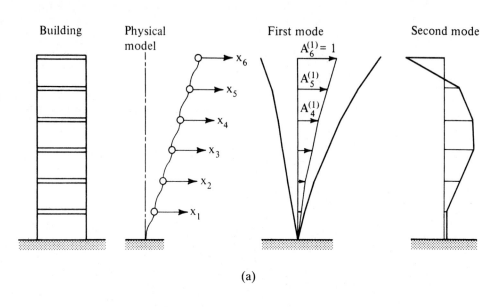

(a)

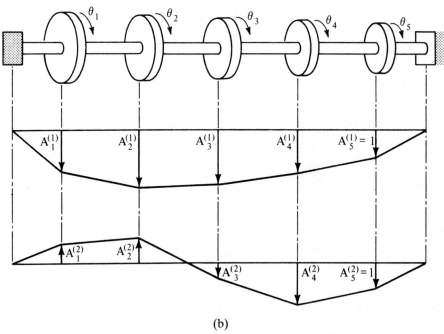

(b)

Figure 7.3 Torsional vibration of a multidisk shaft.

a natural vibration at frequency ω_j, if at time t the displacement along coordinate j is $\cos \omega_j t$, the other displacements will be

$$x_2 = A_2^{(j)} \cos \omega_j t, \quad x_3 = A_3^{(j)} \cos \omega_j t, \quad \cdots$$

Furthermore, if the system is displaced from equilibrium initially by $x_1(0) = cA_1^{(j)}$, $x_2(0) = cA_2^{(j)}, \ldots, x_n(0), = cA_n^{(j)}$, where c is an arbitrary constant, it is apparent that the functions of equation (7.8) will also satisfy the initial conditions; therefore, the system will vibrate at frequency ω_j and amplitudes equal to the natural mode corresponding to ω_j, multiplied by c.

The natural modes are usually plotted in diagrams such as Figure 7.3. Every mode has no fixed amplitudes, but the ratio of the amplitudes at any two coordinates remains constant for the same mode. Thus the shape of the mode always remains the same, regardless of the absolute value of the amplitudes. This is shown for the first mode of the building in Figure 7.2, plotted for values of $A_n^{(1)} = 1, -1, 2$ (Figure 7.3a).

The natural modes of vibration are useful tools for further analysis, as we shall see in a later chapter. However, there are many immediate practical uses of the natural modes of vibration. For example, if the system has forced vibration, the danger occurs when the exciting force is at a frequency near one of the natural frequencies. In this case the vibration amplitudes are very near the natural mode, which does not reveal the actual vibration amplitudes; however, it shows where in the system these amplitudes are high. If, for example, the shaft of Figure 7.3b has torsional vibration at the second natural frequency, the place to measure it is at the fourth disk. Also, the place to expect failure is between the second and third disks because the twist of the shaft is maximum there.

☐ **Example 7.1 Torsional Vibration of a Gas Turbine Rotor**

A gas turbine rotor is arranged as in Figure E7.1. The moments of inertia of the component rotors are J_1, J_2, and J_3, and the torsional spring constants are as indicated, (a) Write the equations of motion in matrix form.

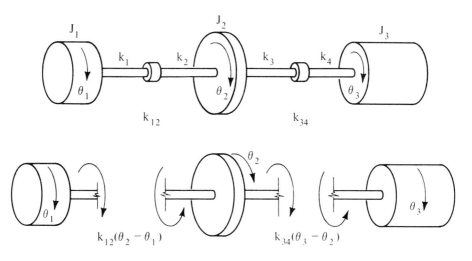

Figure E7.1

(b) Determine the torsional natural frequencies and the vibration modes.

(c) If $J_1 = 10$ kg·m², $J_2 = 5$ kg·m², $J_3 = 15$ kg·m², and $k_1 = k_2 = k_3 = k_4 = 20 \times 10^6$ N·m/rad, determine the rotating speeds that might cause torsional resonances if the turbine stages have 12 blades, the compressor has 16 blades, and the generator has four poles.

Solution (a) To find the rotor torsional natural frequencies, we model it as a system of rigid disks connected by way of massless elastic shafts. First we will find the single equivalent spring constants between rotors. Then we will write the equations of motion, set up the frequency equation, and find its roots.

We substitute equivalent springs

$$k_{12} = \frac{k_1 k_2}{k_1 + k_2} = 10^7, \qquad k_{34} = \frac{k_3 k_4}{k_3 + k_4} = 10^7 \text{ N·m/rad}$$

We assume the system to be in the displaced position θ_1, θ_2, θ_3. Application of Newton's law about the axis of rotation yields

$$J_1 \ddot{\theta}_1 = k_{12}(\theta_2 - \theta_1)$$

$$J_2 \ddot{\theta}_2 = -k_{12}(\theta_2 - \theta_1) + k_{34}(\theta_3 - \theta_2)$$

$$J_3 \ddot{\theta}_3 = -k_3(\theta_3 - \theta_2)$$

In matrix form

$$\begin{bmatrix} J_1 & 0 & 0 \\ 0 & J_2 & 0 \\ 0 & 0 & J_3 \end{bmatrix} \begin{bmatrix} \ddot{\theta}_1 \\ \ddot{\theta}_2 \\ \ddot{\theta}_3 \end{bmatrix} + \begin{bmatrix} k_{12} & -k_{12} & 0 \\ -k_{12} & k_{12} + k_{34} & -k_{34} \\ 0 & -k_{34} & k_{34} \end{bmatrix} \begin{bmatrix} \theta_1 \\ \theta_2 \\ \theta_3 \end{bmatrix} = 0$$

(b) The frequency determinant is then

$$\begin{vmatrix} k_{12} - J_1\omega^2 & -k_{12} & 0 \\ -k_{12} & k_{12} + k_{34} - J_2\omega^2 & -k_{34} \\ 0 & -k_{34} & k_{34} - J_3\omega^2 \end{vmatrix} = 0$$

The vibration modes are a solution of

$$(k_{12} - J_1\omega^2)A_1 - k_{12}A_2 = 0$$
$$-k_{12}A_1 + (k_{12} + k_{34} - J_2\omega^2)A_2 - k_{34}A_3 = 0$$
$$-k_{34}A_3 + (k_{34} - J_3\omega^2)A_3 = 0$$

with $A_1 = 1$.

(c) The frequency determinant becomes

$$\begin{vmatrix} 10^7 - 10\omega^2 & -10^7 & 0 \\ -10^7 & 2 \times 10^7 - 5\omega^2 & -10^7 \\ 0 & -10^7 & 10^7 - 15\omega^2 \end{vmatrix} = 0$$

$$= (10^7 - 10\omega^2)(2 \times 10^7 - 5\omega^2)(10^7 - 15\omega^2) - 10^{14}(10^7 - 10\omega^2) - 10^{14}(10^7 - 15\omega^2) = 0$$

Dividing by 10^{21}, we simplify it to

$$(1 - 10x^2)(2 - 5x^2)(1 - 15x^2) - 2 + 25x^2 = 0, \qquad \text{where } x^2 = 10^{-7}\omega^2$$

$$(2 - 25x^2 + 50x^4)(1 - 15x^2) - 2 + 25x^2 = 0$$

$$2 - 30x^2 - 25x^2 + 375x^4 + 50x^4 - 750x^6 - 2 + 25x^2 = 0$$

Excluding the root $x = 0$, which corresponds to zero angular velocity and rigid-body motion, we have

$$750x^4 - 425x^2 + 30 = 0, \qquad x^2 = \frac{425 \pm (425^2 - 4 \times 750 \times 30)^{1/2}}{2 \times 750}$$

Thus

$$\omega_1 = 909 \text{ rad/s}, \qquad \omega_2 = 2200 \text{ rad/s}$$

The dangerous rotating speeds will be equal to the natural frequencies and their fraction by 4, 12, 16 (values in rad/s):

	1	2
ω	909	2200
$\omega/4$	227	550
$\omega/12$	76	183
$\omega/16$	57	137

(This example demonstrates the torsional vibration analysis of a multimass compound rotor using the frequency determinant method.)

□

7.3 INFLUENCE COEFFICIENTS

In machines and structures we usually select coordinates that are the displacements at certain points. These points are called *nodes*. The information associated with each node, such as displacements, transmitted forces, and so on, is called *state* at the node. Variations of these parameters at the nodes due to variations of parameters at other nodes are called *influence coefficients*.[7] These coefficients are associated with linear systems only.

Stiffness Influence Coefficients

Consider two coordinates i and j of a system. The force generated along the coordinate i due to a unit displacement along the coordinate j while all other coordinates are kept fixed is called *stiffness influence coefficient k_{ij}*. Due to displacement x_j, the value of the force F_i for a linear system will then be $F_i = k_{ij}x_j$. If the system has displacements $x_1, x_2, \ldots, x_n$, the force at node i will be the sum of the contributions of all displacements x_j, according to the superposition principle[8]:

$$F_i = k_{i1}x_1 + k_{i2}x_2 + \cdots + k_{in}x_n, \qquad i = 1, 2, \ldots, n \tag{7.11}$$

Equations (7.11) can be written in matrix form as

$$\mathbf{F} = \mathbf{Kx} \tag{7.12}$$

[7]Introduced by Maxwell, J. C. 1850. *On the Equilibrium of Elastic Solids*. Royal Society, Edinburgh.

[8]Bresse, J. A. C. 1854. *Recherches analytiques sur la flexion et la résistance des pièces courbes*.

where $\mathbf{F} = \{F_1 \quad F_2 \quad \cdots \quad F_n\}$ is a vector of the forces at the n nodes, $\mathbf{x} = \{x_1 \quad x_2 \quad \cdots \quad x_n\}$ is the vector of the displacements of the nodes and $\mathbf{K}$ is the matrix of the influence coefficients k_{ij}, called a *stiffness matrix*. The coefficients with the same indices are called *direct coefficients* and represent forces along one coordinate for unit variation of the same coordinate. All the others are called *coupling coefficients*. In the corresponding matrix, direct coefficients represent elements along the main diagonal, while coupling terms represent the off-diagonal elements.

For the single-degree-of-freedom system, the stiffness matrix degenerates to a scalar, the spring constant. Calculation of the stiffness influence coefficients is a static problem. Variation of one coordinate by one unit generates n stiffness influence coefficients (forces, in general, along the n coordinates). Since we can vary sequentially n coordinates, we can generate n^2 coefficients. (It must be remembered that calculation of stiffness influence coefficient involves variation of one coordinate *keeping all the others fixed*.)

Flexibility Influence Coefficients

The displacement along the coordinate i due to a unit force along the coordinate j, without other force acting on the system, is called a *flexibility influence coefficient* a_{ij}. We can repeat here that for n degree of freedom, there are n^2 flexibility influence coefficients, which can be grouped in a matrix $\mathbf{A}$ called *flexibility matrix*. We can also distinguish direct and coupling coefficients.

Due to force F_i the value of the displacement x_j for a linear system will then be $x_i = a_{ij}F_j$. If the system is loaded by several forces $F_1, F_2, \ldots, F_n$, the displacement at node i will be the sum of the contributions of all forces F_j, according to the superposition principle:

$$x_i = a_{i1}F_1 + a_{i2}F_2 + \cdots + a_{in}F_n, \quad i = 1, 2, \ldots, n \quad (7.13)$$

Equations (7.13) can be written in matrix form as

$$\mathbf{x} = \mathbf{AF} \quad (7.14)$$

where $\mathbf{F} = \{F_1 \quad F_2 \quad \cdots \quad F_n\}$ is a vector of the forces at the n nodes, $\mathbf{x} = \{x_1 \quad x_2 \quad \cdots \quad x_n\}$ is the vector of the displacements of the nodes and $\mathbf{A}$ is the matrix of the influence coefficients a_{ij}, called a *flexibility matrix*. The coefficients with the same indices are called direct coefficients and represent displacements along one coordinate for unit force on the same coordinate. All the others are called coupling coefficients. In the matrix, direct coefficients represent elements along the main diagonal, while coupling terms represent the other elements.

For the single-degree-of-freedom system, the flexibility matrix degenerates to a scalar, the *compliance a* or the inverse of the spring constant $1/k$. Comparison of equations (7.12) and (7.14) yields the following relation between stiffness and flexibility influence coefficients:

$$\mathbf{K} = \mathbf{A}^{-1} \quad \text{or} \quad \mathbf{A} = \mathbf{K}^{-1} \quad (7.15)$$

This relation is very useful because it allows one to calculate the more convenient of these coefficients. And in most engineering systems, the computation of one type of coefficients might be easier than the other. The matrix inversion (7.15) is usually a modest job for a digital computer. A list of influence coefficients for the usual simple system is given in Appendix II.

Potential Energy

We now consider a system at static equilibrium. If the system is stable, in order to displace the system from the equilibrium to a different configuration, we have to apply forces. These forces will move their points of application during the deformation of the system; therefore, we shall put some energy into the system. Some of this energy can be recovered and some will be irreversibly lost. For example, energy to deform springs and other elastic members is stored in them as potential energy. Energy that was dissipated in damping devices or energy dissipated in the material because it is not perfectly elastic, and so on, cannot be recovered. Accordingly, we distinguish systems as *conservative* or *nonconservative*. In conservative mechanical systems, the energy is usually stored as the potential energy of raising a weight in a gravity field or as potential energy of elastic deformation of elastic elements. From your study of dynamics, you know the following:

1. The potential energy is related to the initial and final positions only and is independent of the path of the deformation.
2. The potential energy of a weight displaced in a gravity field is $V_g = mgh$, where m is the mass, g the acceleration of the gravity, and h the vertical distance between the final and initial positions. h is positive when the mass is raised to a higher elevation; in other words, it moves against the gravity force field.
3. The potential energy of elastic deformation of an elastic spring is $V_e = \frac{1}{2}Fx = \frac{1}{2}kx^2$, where k is its spring constant, F the spring force, and x the spring deformation (Chapter 4).

A very interesting member in most machines and structures is the slender beam. If such a beam has been deformed to a known configuration, it is important to know its energy of elastic deformation, because this energy can be related to the stiffness of the beam. If we think of the beam as spring deflected by x, its potential energy of elastic deformation will be $V_e = \frac{1}{2}kx^2$. If we possess an independent way to compute the potential energy, the spring constant will be $k = 2V_e/x^2$.

We consider a beam of flexural rigidity EI and we isolate a piece of length ds (Figure 7.4). This part of beam is deformed by an angle $d\phi$ due to a bending moment M. From strength of materials we know that $M = EI(d^2x/ds^2)$ and $d\phi = ds/\rho = (d^2x/ds^2)\,ds$, where ρ is the radius of curvature of the bent beam. The work of the moment M during a rotation $d\phi$ will be (Figure 7.4b) $dV_e = \frac{1}{2}M\,d\phi$. Therefore, $dV_e = \frac{1}{2}EI\,(d^2x/ds^2)^2\,ds$. Integrating over the length L of the beam, we obtain[9]

$$V_e = \frac{1}{2}\int_0^L EI\left(\frac{d^2x}{ds^2}\right)^2 ds \qquad (7.16)$$

[9]Euler, L. 1736. *Mechanica sine motus scientia analytice exposita*. St. Petersburg.

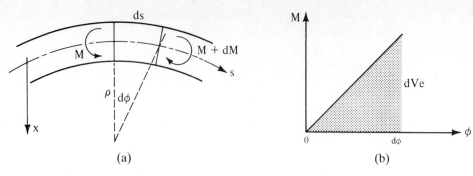

Figure 7.4 Energy of elastic deformation.

This is a fundamental formula that allows computation of the potential energy of elastic deformation once the equation of the elastic line is known. Usually, we do not have to solve the beam equation explicitly for the elastic line. For example, consider the lateral motion of a uniform beam simply supported at the two ends. The vertical deflection of the midspan will be used as the coordinate x_1. If $x(s)$ is the deflection along the axial coordinate s, we know that at the ends is zero, at midspan $x(L/2) = \delta$ and also the derivative dx/ds is zero. For constant shear force, the third derivative of the deflection is zero. Therefore, we can assume a third-degree polynomial between the left and the midspan for the shape of the elastic line,

$$x(s) = c_1 s^3 + c_2 s^2 + c_3 s + c_4 \tag{7.17}$$

where c_1, c_2, c_3, and c_4 constants to be determined from the conditions

$$x(0) = 0$$

$$x\left(\frac{L}{2}\right) = \delta$$

$$x'\left(\frac{L}{2}\right) = 0 \text{ (due to symmetry)}$$

$$x''(0) = 0 \text{ (moment zero)}$$

$$\tag{7.18}$$

We obtain for $L = 2l$

$$x(s) = \frac{\delta}{L^3}(s^3 - 3ls^2 + 3l^2 s) \tag{7.19}$$

Due to symmetry, we integrate between 0 and l and multiply by 2:

$$V_e = 2\frac{EI}{2}\int_0^l \frac{\delta^2}{l^6}(s^3 - 3ls^2 + 3l^2 s)^2 \, ds \tag{7.20}$$

to obtain

$$V_e = \frac{1}{2}\frac{48EI}{(2l)^3}\delta^2 \tag{7.21}$$

We immediately recognize the term

$$k = \frac{48EI}{(2l)^3} \tag{7.22}$$

as the spring constant of the beam at the midspan.

The computation effort here was not less then calculating the elastic line with other methods and determining the spring constant. However, later, the utility of this method will be apparent. We can say here that the error involved in the assumption of the shape of the beam will be largely corrected in the integration (7.16). This property is very useful because in the case of complicated beams (stepped, for example) we can assume a polynomial shape and still get a very good approximation.

Assume, for example, that our polynomial matches the end conditions but that in between we have a maximum deviation ϵ_{max} because the shear force is not constant or the section of the beam is not constant along the length. Assume, then, that along the beam the error is

$$\epsilon = \frac{s}{L} \left(1 - \frac{s}{L}\right) \epsilon_{max} \tag{7.23}$$

Integrating yields

$$V_\epsilon = \frac{16\epsilon_{max}^2}{L^4} \int_0^L (sL - s^2)^2 \, ds = \frac{16}{30} L\epsilon_{max}^2 \tag{7.24}$$

We observe that the error in the potential energy, thus in the spring constant, is proportional to the square of the error in the assumed form of the elastic line. For example, if the polynomial deviates 10% from the elastic line, the error in the potential energy will be on the order of 1% only. If the case studied here, there was no error because we know that a beam with constant EI without external load between two points has an elastic line between these points that is a polynomial of the third degree. We will return to this methodology later in more detail and in a more formal way.

Potential Energy in Terms of Influence Coefficients

In a system with one degree of freedom, the influence coefficients can be evaluated directly from the potential energy. In fact, we know that the form $V_e = \frac{1}{2}kx^2$ suggests immediately that k is the spring constant. Thus if we obtain an expression for the potential energy $V_e = \frac{1}{2}[\cdot]x^2$, we can immediately identify the expression in brackets as the spring constant. Moreover, one can see immediately that

$$k = \frac{d^2V_e}{dx^2} \tag{7.25}$$

To generalize these results, let us assume that on the nodes of an n-degree-of-freedom system we gradually apply n static forces $F_1, F_2, \ldots, F_n$ along the n

coordinates of the system $x_1, x_2, \ldots, x_n$. For a linear system, if only F_i is present, the potential energy will be

$$V_i = \tfrac{1}{2} F_i x_i \tag{7.26}$$

If we consider all the forces $F_1, F_2, \ldots, F_n$, the potential energy will be

$$V = \sum_{i=1}^{n} V_i = \tfrac{1}{2} \sum_{i=1}^{n} F_i x_i \tag{7.27}$$

But the displacement x_i consists of the contribution of all the forces multiplied by the flexibility influence coefficients

$$x_i = \sum_{j=1}^{n} a_{ij} F_j \tag{7.28}$$

Therefore,

$$V = \tfrac{1}{2} \sum_{i=1}^{n} \sum_{j=1}^{n} a_{ij} F_i F_j \tag{7.29}$$

We may also write the potential energy in the form

$$V = \tfrac{1}{2} \sum_{j=1}^{n} F_j x_j \tag{7.30}$$

from which, in a similar vein, we obtain

$$V = \tfrac{1}{2} \sum_{i=1}^{n} \sum_{j=1}^{n} a_{ji} F_j F_i \tag{7.31}$$

Comparison of (7.29) and (7.31) yields

$$a_{ij} = a_{ji} \tag{7.32}$$

This is the *Maxwell–Betti theorem of reciprocity*.[10] Similarly, using (7.15), we can prove that

$$k_{ij} = k_{ji} \tag{7.33}$$

In words, the force along coordinate x_j due to a unit variation of coordinate x_i equals the force along coordinate x_i due to a unit variation of the coordinate x_j. A similar statement can be made for the flexibility influence coefficients. This property is useful and can be used (1) to save computations by calculating only one-half of the indirect terms, and (2) to check the computation by computing independently and comparing symmetric coefficients.

Furthermore, considering that the forces F_i are due to all displacements, x_j is

$$F_i = \sum_{j=1}^{n} k_{ij} x_j \tag{7.34}$$

[10]It was proved for two forces by Maxwell and for the general case by Betti, E. 1872. *Nuovo Cimento, 7, 8*. It was generalized for dynamic response by Rayleigh, J. W. S. 1874, 1875. *Philos. Mag., 48, 49.*

and we obtain [Rayleigh 1894]

$$V = \tfrac{1}{2} \sum_{i=1}^{n} \sum_{j=1}^{n} k_{ij} x_i x_j \qquad (7.35)$$

Going backwards, if one can compute the potential energy for a state $x_1, x_2, \ldots, x_n$ of the system in the form

$$V = \tfrac{1}{2} \sum_{i=1}^{n} \sum_{j=1}^{n} [\cdot] x_i x_j \qquad (7.36)$$

the quantity in brackets will be the stiffness influence coefficient k_{ij}. Similarly, if a set of forces $F_1, F_2, \ldots, F_n$ result in potential energy

$$V = \tfrac{1}{2} \sum_{i=1}^{n} \sum_{j=1}^{n} [\cdot] F_i F_j \qquad (7.37)$$

the quantity in brackets will be the flexibility influence coefficient a_{ij}. The reader can also verify that for linear systems,

$$k_{ij} = \epsilon_{ij} \frac{\partial^2 V}{\partial x_i \, \partial x_j} \qquad (7.38)$$

and

$$a_{ij} = \epsilon_{ij} \frac{\partial^2 V}{\partial F_i \, \partial F_j} \qquad (7.39)$$

where $\epsilon_{ij} = 1$ if $i = j$ and $\epsilon_{ij} = \tfrac{1}{2}$ if $i \neq j$.

☐ **Example 7.2 Influence Coefficients**

Determine the stiffness influence coefficients of the system shown in Figure E7.2 (a) with the direct method, and (b) with the potential energy method.

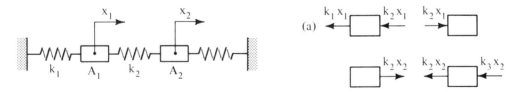

Figure E7.2

Solution (a) If we give a unit displacement along x_1 keeping the point A_2 fixed ($x_2 = 0$), the forces on the two nodes A_1 and A_2 will be $(k_1 + k_2)x_1$ and $-k_2 x_1$, respectively. Therefore,

$$k_{11} = k_1 + k_2, \; k_{12} = -k_2$$

Similarly, until displacement of node A_2 along x_2 with A_1 fixed ($x_1 = 0$) will yield on the nodes A_1 and A_2 forces $-k_2 x_2$ and $(k_2 + k_3)x_2$, respectively. Therefore,

$$k_{21} = -k_2, \; k_{22} = k_2 + k_3$$

(b) The potential energy consists only of energy of elastic deformation because variation of the coordinates x_1 and x_2 will not result in vertical displacement of the system masses.

The energy of elastic deformation of a spring of constant k for a deflection x is $\frac{1}{2}kx^2$. For displacements x_1 and x_2 of the nodes A_1 and A_2, the three springs will be deformed by x_1, $x_2 - x_1$, and x_3, respectively. Therefore, their potential energy of elastic deformation will be

$$V_e = \tfrac{1}{2}k_1x_1^2 + \tfrac{1}{2}k_2(x_2 - x_1)^2 + \tfrac{1}{2}k_3x_2^2$$

or

$$V_e = \tfrac{1}{2}[(k_1 + k_2)x_1^2 - 2k_2x_1x_2 + (k_2 + k_3)x_2^2]$$

Therefore, using equation (7.38), we obtain

$$k_{11} = k_1 + k_2, \qquad k_{12} = -k_2$$
$$k_{21} = -k_2, \qquad k_{22} = k_2 + k_3$$

(This example demonstrates the computation of the influence coefficients with static analysis techniques.)

□

□ **Example 7.3   Influence Coefficient of a Rotating Shaft on Elastic Bearings**

For the system shown in Figure E7.3 (a flexible shaft on two linear bearings), determine the stiffness influence coefficients.

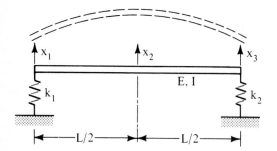

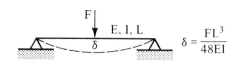

$$\delta = \frac{FL^3}{48EI}$$

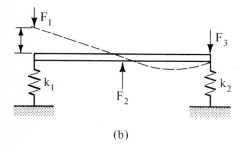

(b)

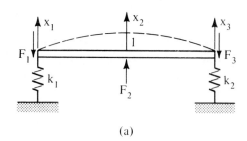

(a)

Figure E7.3

Solution   We will give unit displacement along each coordinate successively, keeping the others zero, and the reaction forces will be the stiffness influence coefficients.

(a) Let $x_1 = 0$, $x_2 = 1$, and $x_3 = 0$. If the applied forces are F_1, F_2, and F_3,

$$F_2 = \frac{48EI}{l^3}$$

$$F_1 = \frac{-F_2}{2}, \qquad F_3 = \frac{-F_2}{2}$$

By definition,

$$k_{22} = F_2 = \frac{48EI}{l^3}, \qquad k_{12} = F_1 = \frac{-24EI}{l^3}$$

$$k_{32} = F_3 = \frac{-24EI}{l^3} = k_{23}, \qquad k_{21} = k_{12}$$

(b) Let $x_1 = 1$ and $x_2 = x_3 = 0$:

$$F_1 = +1 \times k_1 + \frac{3EI}{2(l/2)^3} = +k_1 + \frac{12EI}{l^3} = F_3$$

Taking moments about the bearing 1 yields

$$F_2 = -2 \times \frac{12EI}{l^3}$$

$$k_{11} = +k_1 + \frac{12EI}{L^3}, \qquad k_{21} = F_2 = -\frac{24EI}{L^3}, \qquad k_{31} = \frac{12EI}{L^3} = k_{13}$$

(c) By symmetry,

$$k_{33} = +k_2 + \frac{12EI}{L^3}$$

Finally,

$$K = \begin{bmatrix} k_1 + k & -2k & k \\ -2k & 4k & -2k \\ k & -2k & k_2 + k \end{bmatrix}, \qquad k = \frac{12EI}{L^3}$$

(This example demonstrates the computation of the influence coefficients for bending of a rotating shaft with static analysis techniques.)

□

□ **Example 7.4   Flexibility Influence Coefficients of a Three-Mass Beam**

For the system shown in Figure E7.4 (a simply supported uniform flexible beam with three equidistant lumped masses), determine the flexibility influence coefficients for lateral deflection.

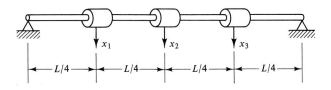

Figure E7.4

Solution We shall apply unit forces along each coordinate successively and statically compute the displacements along the designated coordinates, which will be the stiffness influence coefficients.

From Appendix II:

For $F_1 = 1$ and $F_2 = F_3 = 0$: $\alpha_{11} = 9L^3/12EI$ and $\alpha_{21} = 11L^3/12EI$.
For $F_2 = 1$ and $F_1 = F_3 = 0$: $\alpha_{22} = 16L^3/12EI$.
For $F_3 = 1$ and $F_1 = F_2 = 0$: $\alpha_{13} = 7L^3/12EI$.

Due to symmetry and reciprocity,

$$\alpha_{33} = \alpha_{11} = \frac{9L^3}{12EI}, \qquad \alpha_{32} = \alpha_{23} = \alpha_{12} = \alpha_{21} = \frac{11L^3}{12EI}, \qquad \alpha_{31} = \alpha_{13} = \frac{7L^3}{12EI}$$

The flexibility matrix, for $\mathbf{x} = \{x_1 \quad x_2 \quad x_3\}$, is

$$\mathbf{A} = \begin{bmatrix} 9 & 11 & 7 \\ 11 & 16 & 11 \\ 7 & 11 & 9 \end{bmatrix} \frac{L^3}{12EI}$$

(This example demonstrates computation of the flexibility influence coefficients as displacements due to applied unit static loads.)

□

7.4 MATRIX METHODS

Development of the differential equations of motion for the lumped mass system by Newton's second law can also be applied in conjunction with matrix methods.[11] Here we discuss two such methods: the stiffness and flexibility methods.

Stiffness Matrix Method

In an earlier section, the stiffness influence coefficient k_{ij} was defined as the force along the coordinate i due to a unit displacement along the coordinate j. The force along the coordinate i due to a displacement x_j will be $k_{ij}x_j$. If all coordinates are displaced $x_1, x_2, \ldots, x_n$, the total reaction force along i will be

$$-F_i = \sum_{j=1}^{n} k_{ij}x_j \tag{7.40}$$

If the coordinates selected are displacements of the centers of mass and rotations about it, Newton's law yields

$$m_1\ddot{x}_1 = -(k_{11}x_1 + k_{12}x_2 + \cdots + k_{1n}x_n) + f_1(t)$$
$$m_2\ddot{x}_2 = -(k_{21}x_1 + k_{22}x_2 + \cdots + k_{2n}x_n) + f_2(t)$$
$$\vdots \tag{7.41}$$
$$m_n\ddot{x}_n = -(k_{n1}x_1 + k_{n2}x_2 + \cdots + k_{nn}x_n) + f_n(t)$$

---

[11]The matrix algebra was used for structural analysis since the 1930s. Duncan, J. W., and Collar, A. R. 1934. A method for solution of vibration problems by matrices. *Philos. Mag.*, 7(17): 865–909.

In matrix notation

$$\mathbf{M\ddot{x} + Kx = f} \tag{7.42}$$

where[12]

$\mathbf{M} = \text{diag}(m_1, m_2, \ldots, m_n)$, the mass matrix

$\mathbf{K} = $ stiffness influence coefficient matrix $= [k_{ij}]$

$\mathbf{f} = \{f_1(t) \quad f_2(t) \quad \cdots \quad f_n(t)\}$, the external force vector, function of time in general

If we define the damping influence coefficient c_{ij} as the force along the co-ordinate x_i due to a unit velocity along the coordinate x_j, and define a damping matrix $C = [c_{ij}]$, similar reasoning yields[13]

$$\mathbf{M\ddot{x} + C\dot{x} + Kx = f} \tag{7.43}$$

☐ **Example 7.5 Rigid Beam on Elastic Supports**

A uniform rigid beam (Figure E7.5) is supported at the ends with linear springs of stiffness k_1 and k_2 and viscous dampers c_1 and c_2. Write the equations of motion for the beam using influence coefficients.

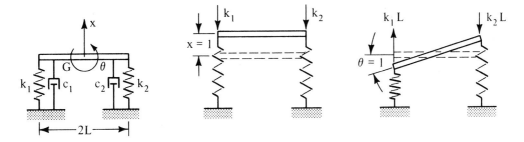

Figure E7.5

Solution Now we have no freedom in selecting coordinates. We have to select x and θ. We will now find the stiffness influence coefficients using unit displacements. A unit displacement along x with $\theta = 0$ results in forces

$$k_{11} = k_1 + k_2, \text{ in the } x \text{ direction}$$

$$k_{21} = -k_1 L + k_2 L, \text{ moment in the } \theta \text{ direction}$$

A unit rotation along θ with $x = 0$, yields forces

$$k_{12} = -k_1 L + k_2 L, \text{ in the } x \text{ direction}$$

$$k_{22} = k_1 L^2 - k_2 L_2, \text{ moment in the } \theta \text{ direction}$$

---

[12]Quade, W. 1933. *Klassification der Schwingungsvorgänge in gekoppelten Stromkreisen*. Leipzig.

[13]Duncan, W. J., and Collar, A. R. 1935. Matrices applied to the motion of damped systems. *Philos. Mag.*, 7(19): 197–219.

Similarly, unit velocity $\dot{x}$ in the x direction yields

$$c_{11} = -c_1 - c_2, \text{ force in the } x \text{ direction}$$

$$c_{21} = c_2 L - c_1 L, \text{ moment in the } \theta \text{ direction}$$

Unit angular velocity $\dot{\theta}$ yields

$$c_{12} = c_1 L - c_2 L, \text{ force in the } x \text{ direction}$$

$$c_{22} = -c_1 L^2 + c_2 L^2, \text{ moment in the } \theta \text{ direction}$$

Therefore,

$$\mathbf{K} = \begin{bmatrix} k_1 + k_2 & -k_1 L + k_2 L \\ -k_1 L + k_2 L & k_1 L^2 - k_2 L^2 \end{bmatrix}$$

$$\mathbf{C} = \begin{bmatrix} c_1 + c_2 & -c_1 L + c_2 L \\ c_2 L - c_1 L & c_1 L^2 - c_2 L^2 \end{bmatrix}$$

$$\mathbf{M} = \text{diag } [m, J_G]$$

where m is the mass and J_G the mass moment of inertia about the mass center of the beam, and the system equations of motion are

$$\mathbf{M\ddot{x}} + \mathbf{C\dot{x}} + \mathbf{Kx} = \mathbf{0}$$

(This example demonstrates the use of influence coefficients for the development of the equations of motion of a rigid beam on elastic bearings.)

□

Flexibility Matrix Method

In an earlier section the flexibility influence coefficient a_{ij} was defined as the displacement along the coordinate i due to a unit force along the coordinate j. If along x_j, instead of unit force we have a force F, the displacement will be $a_{ij}F$. Since the inertia force along the coordinate j is $-m_j\ddot{x}_j$, the displacement along x_i will be $-a_{ij}m_j\ddot{x}_j$. The displacement x_i will consist of the contributions to x_i of all the inertia forces. Therefore,

$$
\begin{aligned}
x_1 &= -a_{11}m_1\ddot{x}_1 - \cdots - a_{1n}m_n\ddot{x}_n \\
x_2 &= -a_{21}m_1\ddot{x}_1 - \cdots - a_{2n}m_n\ddot{x}_n \\
&\vdots \\
x_n &= -a_{n1}m_1\ddot{x}_1 - \cdots - a_{nn}m_n\ddot{x}_n
\end{aligned}
\tag{7.44}
$$

In matrix notation $\mathbf{x} = -\mathbf{AM\ddot{x}}$, or with the external force vector $\mathbf{f}$,

$$\mathbf{AM\ddot{x}} + \mathbf{x} = \mathbf{Af} \tag{7.45}$$

Multiplying by $\mathbf{A}^{-1}$ from the left, we obtain

$$\mathbf{M\ddot{x}} + \mathbf{A}^{-1}\mathbf{x} = \mathbf{f} \tag{7.46}$$

Equation (7.45) is the general form of the equations for the flexibility matrix method.

Both the stiffness matrix and the flexibility matrix methods, suffer from the limitation of the coordinate selection, which have to be the displacements of the mass centers. However, they are convenient for automatic computations and they are used extensively.

☐ **Example 7.6 Differential Equations of Motion Using Flexibility Influence Coefficients**

For the beam of Example 7.4 write the equations of motion using flexibility influence coefficients.

Solution The flexibility matrix for $\mathbf{x} = \{x_1 \quad x_2 \quad x_3\}$ is (Example 7.4)

$$
\mathbf{A} = \begin{bmatrix} 9 & 11 & 7 \\ 11 & 16 & 11 \\ 7 & 11 & 9 \end{bmatrix} \frac{L^3}{12EI}
$$

The equations of motion, equations (7.45), are

$$
\frac{L^3}{12EI} \begin{bmatrix} 9 & 11 & 7 \\ 11 & 16 & 11 \\ 7 & 11 & 9 \end{bmatrix} \begin{bmatrix} m_1 & 0 & 0 \\ 0 & m_2 & 0 \\ 0 & 0 & m_3 \end{bmatrix} \begin{bmatrix} \ddot{x}_1 \\ \ddot{x}_2 \\ \ddot{x}_3 \end{bmatrix} + \begin{bmatrix} x_1 \\ x_2 \\ x_3 \end{bmatrix} = \begin{bmatrix} 9 & 11 & 7 \\ 11 & 16 & 11 \\ 7 & 11 & 9 \end{bmatrix} \begin{bmatrix} f_1(t) \\ f_2(t) \\ f_3(t) \end{bmatrix}
$$

(This example demonstrates the use of the flexibility matrix for the development of the differential equations of motion of a system.)

☐

Lagrange's Equations

From equation (4.1), where the kinetic energy of a particle was defined, the kinetic energy for the assembly of a system of particles can be written in the form $T = \frac{1}{2} \sum_{i=1}^{n} m_i \dot{\mathbf{r}}_i \cdot \dot{\mathbf{r}}_i$, where the system of particles is described by n generalized coordinates x_i and $\mathbf{r}_i = \{x_1, x_2, \ldots, x_n\}$ is the position vector of the particle i. The time derivative is

$$
\dot{\mathbf{r}}_i = \left[\left(\frac{\partial \mathbf{r}_i}{\partial x_1}\right)\dot{x}_1 \quad \left(\frac{\partial \mathbf{r}_i}{\partial x_2}\right)\dot{x}_2 \quad \cdots \quad \left(\frac{\partial \mathbf{r}_i}{\partial x_n}\right)\dot{x}_n \right]
$$

Therefore, the kinetic energy of the system is the sum of the kinetic energy of the particles:

$$
T = \frac{1}{2} \sum_{i=1}^{n} m_i \left(\sum_{j=1}^{n} \frac{\partial \mathbf{r}_i}{\partial x_j}\dot{x}_j \right) \left(\sum_{k=1}^{n} \frac{\partial \mathbf{r}_i}{\partial x_k}\dot{x}_k \right)
$$

$$
= \frac{1}{2} \sum_{j=1}^{n} \sum_{k=1}^{n} \left(\sum_{i=1}^{n} m_i \left(\frac{\partial \mathbf{r}_i}{\partial x_j}\right)\left(\frac{\partial \mathbf{r}_i}{\partial x_k}\right) \dot{x}_j \dot{x}_k \right) \tag{7.47}
$$

$$
= \frac{1}{2} \sum_{j=1}^{n} \sum_{k=1}^{n} m_{jk} \dot{x}_j \dot{x}_k
$$

One can verify immediately that equation (7.47) (the form on the right-hand side is called a *quadratic form*) is equivalent to

$$
T = \frac{1}{2} \dot{\mathbf{x}}^T \mathbf{M} \dot{\mathbf{x}} \tag{7.48}
$$

where

$$\mathbf{x} = \{x_1 \quad x_2 \quad \cdots \quad x_n\}$$

$$\mathbf{M} = [m_{jk}] = \left[\sum_{i=1}^{n} m_i \left(\frac{\partial \mathbf{r}_i}{\partial x_j} \right) \cdot \left(\frac{\partial \mathbf{r}_i}{\partial x_k} \right) \right]$$

Since the kinetic energy does not depend on the sequence of selection of the indices j and k in (7.47), the result will not change if we interchange j and k. This will result in the same matrix $\mathbf{M}$, which then will not change if *we transpose* it, that is, interchange rows with columns, $\mathbf{M} = \mathbf{M}^T$. This type of matrix is called *symmetric*.

The kinetic energy is a function of the system coordinates and is a quadratic form. It is useful at this point to mention some properties of such functions, which in turn will be useful in deriving some qualitative results for the response of vibrating systems.

It can be seen from equation (4.1) that the kinetic energy function is always positive except when all variables are zero; in this case it is zero. Such a function is called *positive definite*. If it can be zero with nonzero values of some variables, the function is *positive semidefinite*. Matrices with elements that are the coefficients of positive definite or positive semidefinite quadratic forms are called *positive definite* or *positive semidefinite matrices*, respectively. Therefore, the mass matrix is a *positive definite matrix*.

Equations (7.31) and (7.35) can be written in a matrix form (their right-hand side is a quadratic form) as

$$V = \tfrac{1}{2}\mathbf{F}^T\mathbf{A}\mathbf{F} = \tfrac{1}{2}\mathbf{x}^T\mathbf{K}\mathbf{x} \tag{7.49}$$

where

$$\mathbf{F} = \{F_1 \quad F_2 \quad \cdots \quad F_n\}$$

$$\mathbf{x} = \{x_1 \quad x_2 \quad \cdots \quad x_n\}$$

and $\mathbf{A}$ and $\mathbf{K}$ the flexibility and stiffness matrices, respectively.

The flexibility and stiffness matrices $\mathbf{A}$ and $\mathbf{K}$ are derived from quadratic forms, but their character depends on the system. At this point we shall assume that vibrating systems have positive definite stiffness matrix, and rigid-body motion implies a positive semidefinite stiffness matrix.

A lumped system of solids with masses or moments of inertia $m_1, m_2, \ldots,$ m_n that are moving along the coordinates $x_1, x_2, \ldots, x_n$ with velocities (linear or angular) $\dot{x}_1, \dot{x}_2, \ldots, \dot{x}_n$ possesses the following kinetic energy:

$$T = \tfrac{1}{2}m_1\dot{x}_1^2 + \tfrac{1}{2}m_2\dot{x}_2^2 + \cdots + \tfrac{1}{2}m\dot{x}_n^2 \tag{7.50}$$

This matrix is diagonal and, of course, symmetric.

Lagrange's equation, which was developed in Chapter 4, can be applied to a system in respect to more than one coordinates, under the same conditions:

$$\frac{d}{dt}\left(\frac{\partial L}{\partial \dot{x}_i}\right) - \frac{\partial L}{\partial x_i} + \frac{\partial U}{\partial \dot{x}_i} = F_{xi}, \qquad i = 1, 2, \ldots, n \tag{7.51}$$

where $L = T - V$ is the Lagrangian function, U is the dissipation function [Rayleigh, 1894], and F_{xi} the nonconservative, nonviscous generalized external force along the coordinate x_i.

Recalling that the derivative of a matrix is defined as the matrix of the derivatives of the matrix elements, we can write, because $\partial V / \partial \dot{x}_i = 0$,

$$\frac{d}{dt}\left(\frac{\partial T}{\partial \dot{x}_i}\right) - \frac{\partial T}{x_i} + \frac{\partial V}{\partial x_i} + \frac{\partial U}{\partial \dot{x}_i} = F_{xi} \qquad (7.52)$$

The reader can easily verify that $[(\partial T / \partial \dot{x}_i)] = \mathbf{M}\dot{\mathbf{x}}$, $(\partial T / \partial x_i) = 0$, $[\partial U / \partial \dot{x}_i] = \mathbf{C}\dot{\mathbf{x}}$, and $[(\partial V / \partial x_i)] = \mathbf{K}\mathbf{x}$. Therefore,

$$\mathbf{M}\ddot{\mathbf{x}} + \mathbf{C}\dot{\mathbf{x}} + \mathbf{K}\mathbf{x} = \mathbf{F}_x \qquad (7.53)$$

which are the equations of motion of the system, which we have obtained already with the matrix force method.

The difference here is that we are not confined to the coordinates of the center of mass, but we use any set of coordinates, which we call *general coordinates*, and they unambiguously describe the geometry of the system. We can then write expressions for the kinetic and potential energy and apply equation (7.51) or (7.52).

It must be pointed out that the force vector $\mathbf{F}_x$ does not include forces of constraint, since they do not contribute to the virtual work, as described in Chapter 4.

☐ **Example 7.7 Equations of Motion Using Lagrange's Equations**

For the beam of Example 7.5, determine the differential equations of motion with Lagrange's equations.

Solution   We will develop expressions for the kinetic and potential energy and apply Lagrange's equations. The kinetic energy will be

$$T = \frac{1}{2}m\dot{x}^2 + \frac{1}{2}J_G\dot{\theta}^2 = \frac{1}{2}m\left(\dot{x}_1 + \frac{\dot{x}_2}{2}\right)^2 + \frac{1}{2}J_G\left(\dot{x}_2 - \frac{\dot{x}_1}{L}\right)^2$$

$$\frac{\partial T}{\partial \dot{x}_1} = m\left(\dot{x}_1 + \frac{\dot{x}_2}{2}\right) - J_G\left(\dot{x}_2 - \frac{\dot{x}_1}{L}\right), \qquad \frac{\partial T}{\partial x_1} = 0$$

$$\frac{\partial T}{\partial \dot{x}_2} = m\left(\dot{x}_1 + \frac{\dot{x}_2}{2}\right) + J_G\left(\dot{x}_2 - \frac{\dot{x}_1}{L}\right), \qquad \frac{\partial T}{\partial x_2} = 0$$

The potential energy

$$V = \tfrac{1}{2}k_1 x_1^2 + \tfrac{1}{2}k_2 x_2^2$$

$$\frac{\partial V}{\partial x_1} = k_1 x_1, \qquad \frac{\partial V}{\partial x_2} = k_2 x_2$$

Equations (7.51) yield

$$m\left(\ddot{x}_1 + \frac{\ddot{x}_2}{2}\right) - J_G\left(\ddot{x}_2 - \frac{\ddot{x}_1}{L}\right) + k_1 x_1 = 0$$

$$m\left(\ddot{x}_1 + \frac{\ddot{x}_2}{2}\right) + J_G\left(\ddot{x}_2 - \frac{\ddot{x}_1}{L}\right) + k_2 x_2 = 0$$

If we add and subtract these two equations, we obtain the equations of Example 7.4.

(This example demonstrates the application of Lagrange's equations to find the differential equations of motion of a system.)

□

7.5 NATURAL MODES OF VIBRATION: THE EIGENVALUE PROBLEM

In Section 7.2 we have investigated the feasibility of natural vibration of a system with n degrees of freedom. We have proved that an undamped linear system can in fact have harmonic natural vibration in n distinct natural frequencies, in general. While vibrating at any natural frequency ω_j, it vibrates at a specific natural mode that is associated with this natural frequency. The natural frequencies and natural modes are properties of the system and depend on its constant parameters only.

If the equations are available in matrix form, we determine the natural frequencies and natural modes with the method of Section 7.2. Thus we seek a solution

$$\mathbf{x} = \mathbf{z} \cos \omega t \tag{7.54}$$

where $\mathbf{x} = [x_1 \quad x_2 \quad \cdots \quad x_n]^T$ and $\mathbf{z} = [z_1 \quad z_2 \quad \cdots \quad z_n]^T$. The solution (7.54) has the form of a vector of spatial elements multiplied by a time function. It is equivalent with the separation of variables in partial differential equations (see Chapter 9).

The system of differential equation for natural undamped vibration is

$$\mathbf{M\ddot{x}} + \mathbf{Kx} = 0 \tag{7.55}$$

Substituting (7.54) into (7.55), we obtain

$$(-\omega^2\mathbf{M} + \mathbf{K})\mathbf{z} = 0 \tag{7.56}$$

Equations (7.56) are a system of linear algebraic equations. This system is homogeneous. In order that it will have a nontrivial solution, we must have

$$\left|-\omega^2\mathbf{M} + \mathbf{K}\right| = 0 \tag{7.57}$$

This is an algebraic equation in ω and yields, in general, n roots $\omega_1, \omega_2, \ldots, \omega_n$, the system's natural frequencies. For every natural frequency ω_j, the system (7.56) has a solution with one of the unknown components of z being specified arbitrarily. This procedure (as in Section 7.2) yields n natural frequencies and n natural modes:

$$\omega_1 \qquad\qquad \omega_2 \quad \cdots \qquad\qquad \omega_n$$

$$\mathbf{z}^{(1)} = \begin{bmatrix} z_1^{(1)} \\ z_2^{(1)} \\ \vdots \\ z_n^{(1)} \end{bmatrix} \quad \mathbf{z}^{(2)} = \begin{bmatrix} z_1^{(2)} \\ z_2^{(2)} \\ \vdots \\ z_n^{(2)} \end{bmatrix} \quad \cdots \quad \mathbf{z}^{(n)} = \begin{bmatrix} z_1^{(n)} \\ z_2^{(n)} \\ \vdots \\ z_n^{(n)} \end{bmatrix}$$

The problem of determining the n values of ω (alternatively, the values of $\lambda = \omega^2$ called *eigenvalues*)[14] and the n associated natural modes $\mathbf{z}$ (called *eigenvectors*) of equation (7.56) is called an *eigenvalue problem*. Solution of the eigenvalue problem involves, in principle, finding the n roots of an algebraic equation and n solutions of a system of $(n - 1)$ linear algebraic equations. For small systems, $n = 2$ or 3, for example, one can use the determinant form of the frequency equation (7.57) directly. For larger problems, this becomes progressively more cumbersome. For large systems, a variety of numerical methods have been developed which solve the eigenvalue problem with efficiency. Some of them will be discussed in Chapter 10.

☐ **Example 7.8   Torsional Vibration of a Motor–Pump**

A motor M is driving a pump P. The rotor of the motor has a moment of inertia J_1 and the rotor of the pump J_2. They are connected with two stub shafts, as shown. Determine, using matrix methods, (a) the equivalent spring constant of the shaft between the two rotors, and (b) the location of the nodal point (i.e., the zero-amplitude point) if both stub shafts have the same diameter. The shear modulus is G.

Solution   (a) For a cylinder the torsional spring constant is $k = \pi d^4/32l$. The equivalent torsional spring constant (springs in series)

$$k_{12} = \frac{k_1 k_2}{k_1 + k_2} = \frac{\pi G}{32} \frac{d_1^4 d_2^4}{l_1 l_2 (d_1^4/l_1 + d_2^4/l_2)}$$

$$= \frac{\pi G}{32} \frac{d_1^4 d_2^4}{l_2 d_1^4 + l_1 d_2^4}$$

(b) The equations of motion are

$$J_1 \ddot{\theta}_1 + k_{12}\theta_1 - k_{12}\theta_2 = 0$$

$$J_2 \ddot{\theta}_2 - k_{12}\theta_1 + k_{12}\theta_2 = 0$$

$$\begin{bmatrix} J_1 & 0 \\ 0 & J_2 \end{bmatrix} \begin{bmatrix} \ddot{\theta}_1 \\ \ddot{\theta}_2 \end{bmatrix} + k_{12} \begin{bmatrix} 1 & -1 \\ -1 & 1 \end{bmatrix} \begin{bmatrix} \theta_1 \\ \theta_2 \end{bmatrix} = 0$$

Let

$$\theta_1 = A_1 \cos \omega t, \qquad \theta_2 = A_2 \cos \omega t$$

Then

$$(-\omega^2 J_1 + k_{12})A_1 - k_{12}A_2 = 0$$

$$-k_{12}A_1 + (-\omega^2 J_2 + k_{12})A_2 = 0$$

The condition for the existence of a solution is

$$\begin{vmatrix} -\omega^2 J_1 + k_{12} & -k_{12} \\ -k_{12} & -\omega^2 J_2 + k_{12} \end{vmatrix} = 0$$

[14]The names *eigenvalue* and *eigenvector* were used first by German authors and they became standard terminology in many languages.

or

$$\omega^4 J_1 J_2 - \omega^2 k_{12}(J_1 + J_2) = 0$$

$$\omega_1 = 0 \text{ (rigid-body motion)}$$

and

$$\omega_2^2 = \frac{(J_1 + J_2)k_{12}}{J_1 J_2}$$

From the first of equations (a) for $A_1^{(2)} = 1$, we obtain

$$A_2^{(2)} = \frac{-\omega_2^2 J_1 + k_{12}}{k_{12}}$$

$$= \frac{-J_1 + J_2}{J_2} + 1 = \frac{-J_1}{J_2}$$

$$\mathbf{z}^{(1)} = \begin{bmatrix} 1 \\ 1 \end{bmatrix}, \qquad \mathbf{z}^{(2)} = \begin{bmatrix} 1 \\ \dfrac{-J_1}{J_2} \end{bmatrix}$$

The location of the nodal point is shown in Figure E7.8.

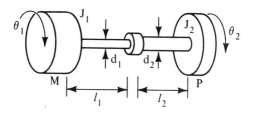

Figure E7.8

(This example demonstrates the solution of the torsional vibration problem of a motor–pump system.)

$\square$

7.6 FREE VIBRATION: THE INITIAL VALUE PROBLEM

In a manner similar to the treatment of the one- or two-degree-of-freedom systems, one can set up the *general solution* of equation (7.55) as the sum of all the *particular solutions* (7.54). As in single-degree-of-freedom systems (Chapter 2), the form of the solution depends on the nature of the eigenvalues $\omega_1, \omega_2, \ldots, \omega_n$.

Distinct Eigenvalues

It is obvious that if instead of equations (7.54) we use $\mathbf{x}(t) = \mathbf{z} \sin \omega t$ in equations (7.55), we shall obtain the same system of linear equations (7.56). Therefore, the general solution is the sum of all particular solutions

$$\mathbf{x}(t) = c_1 \mathbf{z}^{(1)} \cos \omega_1 t + c_2 \mathbf{z}^{(2)} \cos \omega_2 t + \cdots + c_n \mathbf{z}^{(n)} \cos \omega_n t \qquad (7.58)$$
$$+ c_{n+1} \mathbf{z}^{(1)} \sin \omega_1 t + c_{n+2} \mathbf{z}^{(2)} \sin \omega_2 t + \cdots + c_{2n} \mathbf{z}^{(n)} \sin \omega_n t$$

where $c_1, c_2, \ldots, c_{2n}$ are $2n$ arbitrary constants, to be determined from $2n$ initial conditions

$$\mathbf{x}(0) = \mathbf{x}^{(0)} \qquad (7.59)$$

$$\dot{\mathbf{x}}(0) = \mathbf{v}^{(0)}$$

where $\mathbf{x}(0)$ is the initial displacement vector and $\mathbf{v}(0)$ the initial velocity vector at the beginning of time. Any solution of (7.55) is given by (7.58). The solution of the system of equations (7.55) with the initial conditions (7.59) is called the *initial value problem*.

It is important to show that equation (7.58) is indeed the general solution of the problem and every solution can be expressed in the form of equations (7.58) by proper selection of the constants c. This will be done later.

Another way of setting up the general solution can be obtained by rearranging equations (7.58) and using elementary trigonometric identities:

$$\mathbf{x}(t) = \mathbf{z}^{(1)} (c_1 \cos \omega_1 t + c_{n+1} \sin \omega_1 t)$$

$$+ \mathbf{z}^{(2)} (c_2 \cos \omega_2 t + c_{n+2} \sin \omega_2 t) + \cdots \qquad (7.60)$$

$$+ \mathbf{z}^{(n)} (c_n \cos \omega_n t + c_{2n} \sin \omega_n t)$$

or

$$\mathbf{x}(t) = \mathbf{z}^{(1)} x_1(t) + \mathbf{z}^{(2)} x_2(t) + \cdots + \mathbf{z}^{(n)} x_n(t) \qquad (7.61)$$

where $x_1(t) = A_1 \cos (\omega_1 t + \theta_1)$, $x_2(t) = A_2 \cos (\omega_2 t + \theta_2)$, $\ldots$, $x_n(t) = A_n \cos (\omega_n t + \theta_n)$ and $A_1, A_2, \ldots, A_n, \theta_1, \theta_2, \ldots, \theta_n$ are arbitrary constants, to be determined from the initial conditions.

One can see from equation (7.58) that the solution to the problems consists of adding all the modes of natural vibration $\mathbf{z}^{(1)}, \mathbf{z}^{(2)}, \ldots, \mathbf{z}^{(n)}$, multiplied by some arbitrary constants, $c_1, c_2, \ldots, c_{2n}$ [or the equivalent formulation of equation (7.61)], which depend only on the initial conditions. Therefore, the initial conditions (7.59) specify just how much of each mode is added in order to establish the natural vibration. We say that the eigenvectors $\mathbf{z}^{(j)}$ form a *basis set* with respect to the matrices $\mathbf{M}$ and $\mathbf{K}$ in equations (7.55). Any solution of (7.55) can be expressed in terms of these eigenvectors (the *expansion theorem*).

One can recognize the similarity with the expansion of a periodic function in Fourier series. Indeed, the system response is expanded in a finite series of the natural modes of vibration. One can observe that for some systems the vibration modes resemble harmonic functions, and in fact some continuous systems can be modeled by lumped mass systems, and if the number of coordinates tends to infinity, the expansions (7.58) or (7.61) become Fourier expansions in the axial coordinate.

If the motion is initiated only by initial displacement

$$\mathbf{x}^{(0)} = \{ x_1^{(0)} \quad x_2^{(0)} \quad \cdots \quad x_n^{(0)} \}$$

from equilibrium (then $\mathbf{v}^{(0)} = \{0\}$), we shall have in equations (7.58)

$$c_{n+1} = c_{n+2} = \cdots = c_{2n} = 0$$

or in equations (7.61) we shall have

$$\theta_1 = \theta_2 = \cdots = \theta_n = 0$$

If the motion is initiated by an impact ($\mathbf{x}^{(0)} = \{0\}$),

$$c_1 = c_2 = \cdots = c_n = 0$$

and

$$\theta_1 = \theta_2 = \cdots = \frac{\pi}{2}$$

In the case that $\mathbf{v}^{(0)} = \{0\}$, we have, at $t = 0$,

$$\mathbf{x}^{(0)} = \mathbf{Zc} \qquad (7.62)$$

where

$$\mathbf{c} = \{c_1 \quad c_2 \quad \cdots \quad c_n\}$$

and

$$\mathbf{Z} = [\mathbf{z}^{(1)} \quad \mathbf{z}^{(2)} \quad \cdots \quad \mathbf{z}^{(n)}]$$

is an $n \times n$ square matrix called the *modal matrix*. In terms of the modal matrix, the vector of the arbitrary constants $\mathbf{c}$ is, from equation (7.62),

$$\mathbf{c} = \mathbf{Z}^{-1}\mathbf{x}^{(0)} \qquad (7.63)$$

Thus the arbitrary constant vector $\mathbf{c}$ can be obtained by solving a system of n linear algebraic equations in n unknowns with the right-hand-side constant vector being the initial displacements.

One can proceed the same way if initial velocities are specified. Finally, by superposition of the two solutions,

$$\mathbf{x}(t) = \mathbf{Z}(\mathbf{A}\mathbf{Z}^{-1}\mathbf{x}^{(0)} + \mathbf{B}\mathbf{Z}^{-1}\mathbf{v}^{(0)}) \qquad (7.64)$$

where

$$\mathbf{A} = \text{diag}[\cos \omega_i t], \qquad \mathbf{B} = \text{diag}[\sin \omega_i t]$$

Multiple Eigenvalues

It is possible that some of the roots of the frequency equation are *multiple* or *repeated*; therefore, the total number of distinct roots is less than n. This is inherent in some types of problems, especially in symmetric structures. For example, a circular shaft carrying a number of disks has the same natural frequencies in the vertical and horizontal directions, if the bearing supports have the same properties in both directions.

In this case the solution can be obtained by assigning arbitrary constants c_j which are functions of time. By multiplication of the second equation by $\mathbf{z}^{(j+1)T}$, the third by $\mathbf{z}^{(j+2)T}$, and so on, and using orthogonality, the reader can verify that if a root j has multiplicity k, the k eigenvectors with eigenvalue ω_j^2 satisfy the following equations, which can be used for their successive determination:

$$\mathbf{Kz}^{(j)} = \omega_j^2 \mathbf{Mz}^{(j)}$$

$$\mathbf{Kz}^{(j+1)} = \omega_j^2 \mathbf{Mz}^{(j+1)} + \mathbf{Mz}^{(j)}$$

$$\vdots \tag{7.65}$$

$$\mathbf{Kz}^{(j+k-1)} = \omega_j^2 \mathbf{Mz}^{(j+k-1)} + \mathbf{Mz}^{(j+k-2)}$$

Furthermore, the particular solutions corresponding to this multiple root are

$$\mathbf{x}_{j+r} = \mathbf{z}^{(j)} + \mathbf{z}^{(j+1)}t + \cdots + \mathbf{z}^{(j+r-1)}\frac{t^{r-1}}{(r-1)!}\cos(\omega_j t - \theta_j), \tag{7.66}$$

$$r = 1, 2, \ldots, k$$

Again, the general solution is the sum of all particular solutions:

$$\mathbf{x}(t) = \mathbf{z}^{(1)}x_1(t) + \mathbf{z}^{(2)}x_2(t) + \cdots + \mathbf{z}^{(n)}x_n(t) \tag{7.67}$$

Zero Eigenvalues and Rigid-Body Motion

Zero eigenvalues were encountered earlier in torsional vibration examples. In fact, this situation is inherent in any system that is not restrained against rigid-body motion. Such a system should have at least one zero eigenvalue. The corresponding eigenvector is the solution of the static problem $\mathbf{Kz}^{(1)} = \{0\}$, and the time function has constant $\mathbf{A}^{(1)}$ with $\theta_1 = 0$. Multiple zero eigenvalues are treated similarly.

☐ **Example 7.9 Impact Excitation of a Two-Degree-of-Freedom Machine Mount**

A flexible machine on a elastically supported foundation is modeled as a two-degree-of-freedom spring–mass system (Figure E7.9b). The machine (mass 2) is hit by a mass

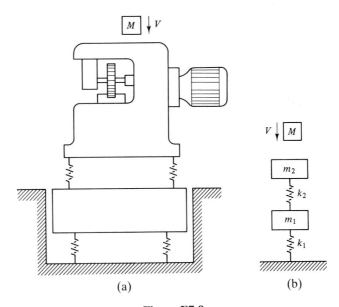

(a) (b)

Figure E7.9

$M \ll m_2$ moving at constant vertical speed V with a purely plastic impact. Determine the resulting motion.

Solution We shall first determine the initial velocity of the machine due to impact and then solve the initial value problem.

Newton's Law yields

$$m_1 \ddot{x}_1 = -k_1 x_1 - k_2(x_1 - x_2)$$

$$m_2 \ddot{x}_2 = -k_2(x_2 - x_1)$$

or

$$m_1 \ddot{x}_1 + (k_2 + k_1)x_1 - k_2 x_2 = 0$$

$$m_2 \ddot{x}_2 + k_2 x_2 - k_2 x_1 = 0$$

In matrix form

$$\mathbf{M}\ddot{\mathbf{x}} + \mathbf{K}\mathbf{x} = 0, \qquad \mathbf{M} = \begin{bmatrix} m_1 & 0 \\ 0 & m_2 \end{bmatrix}, \qquad \mathbf{K} = \begin{bmatrix} k_1 + k_2 & -k_2 \\ -k_2 & k_1 \end{bmatrix}$$

The characteristic determinant

$$|-\omega^2 \mathbf{M} + \mathbf{K}| = \begin{vmatrix} k_1 + k_2 - \omega^2 m_1 & -k_2 \\ -k_2 & k_2 - \omega^2 m_2 \end{vmatrix} = 0$$

or

$$\omega^4 - \left(\frac{k_1 + k_2}{m_1} + \frac{k_2}{m_2} \right) \omega^2 + \frac{k_1 k_2}{m_1 m_2} = 0$$

$$\omega^2 = \frac{k_1 + k_2}{2m_1} + \frac{k_2}{2m_2} \pm \left\{ \frac{[(k_1 + k_2)/m_1 + k_2/m_2]^2}{4} - \frac{k_1 k_2}{m_1 m_2} \right\}^{1/2}$$

The eigenvectors (natural modes) will be determined from

$$[-\omega^2 \mathbf{M} + \mathbf{K}]\mathbf{z} = 0, \qquad \mathbf{z} = \{z_1 \quad z_2\}$$

Assuming that $z_2 = 1$, we have

$$z_1 = \frac{k_2}{k_1 + k_2} - \omega^2 m_1$$

Therefore,

$$z_1^{(1)} = \frac{k_2}{k_1 + k_2} - \omega_1^2 m_1, \qquad z_1^{(2)} = \frac{k_2}{k_1 + k_2} - \omega_2^2 m_1$$

$$z_2^{(1)} = 1 \qquad\qquad\qquad z_2^{(1)} = 1$$

The general solution will be

$$\begin{bmatrix} x_1(t) \\ x_2(t) \end{bmatrix} = c_1 \begin{bmatrix} z_1^{(1)} \\ z_2^{(1)} \end{bmatrix} \sin(\omega_1 t + \psi_1) + c_2 \begin{bmatrix} z_1^{(2)} \\ z_2^{(2)} \end{bmatrix} \sin(\omega_2 t + \psi_2)$$

where c_1, c_2, ψ_1, and ψ_2 should be determined from the initial conditions $x_1(0) = 0$, $x_2(0) = 0$, $\dot{x}_1(0) = MV/m_2$, and $\dot{x}_2(0) = 0$.

The response will be, from equation (7.64),

$$\mathbf{x}(t) = \mathbf{Z}\mathbf{B}\mathbf{Z}^{-1}\mathbf{v}^{(0)}$$

where

$$\mathbf{Z} = \begin{bmatrix} \dfrac{k_2}{k_1} + k_2 - \omega_1^2 m_1 & \dfrac{k_2}{k_1} + k_2 - \omega_2^2 m_1 \\[2mm] 1 & 1 \end{bmatrix}, \quad \mathbf{Z}^{-1}\mathbf{v}(0) = \begin{bmatrix} \dfrac{a}{a+b} \\[2mm] 1 - \dfrac{a}{a+b} \end{bmatrix}$$

$$a = \frac{k_2}{k_1 + k_2} - \omega_1^2 m_1, \qquad b = \frac{k_2}{k_1 + k_2} - \omega_2^2 m_1$$

(This example demonstrates the matrix solution of the initial value problem for a two-degree-of-freedom system machine mount.)

$\square$

7.7 ORTHOGONALITY OF THE NATURAL MODES: NORMAL COORDINATES AND MODAL ANALYSIS

While due to the Maxwell–Betti theorem of reciprocity the stiffness matrix $\mathbf{K}$ is always symmetric, the same is true, in general, for the mass matrix $\mathbf{M}$. One of the consequences of symmetry is the *orthogonality of the natural modes*.[15] Thus if ω_i and ω_j are two eigenvalues (natural frequencies) and $\mathbf{z}^{(i)}$ and $\mathbf{z}^{(j)}$ are the corresponding eigenvectors (natural modes), they must satisfy (7.56):

$$\omega_i^2 \mathbf{M} \mathbf{z}^{(i)} = \mathbf{K} \mathbf{z}^{(i)}, \qquad \omega_j^2 \mathbf{M} \mathbf{z}^{(j)} = \mathbf{K} \mathbf{z}^{(j)} \tag{7.68}$$

We multiply from the left with $[\mathbf{z}^{(j)}]^T$ and $[\mathbf{z}^{(i)}]^T$, respectively, to obtain

$$\omega_i^2 [\mathbf{z}^{(j)}]^T \mathbf{M} \mathbf{z}^{(i)} = [\mathbf{z}^{(j)}]^T \mathbf{K} \mathbf{z}^{(i)} \tag{7.69a}$$

$$\omega_j^2 [\mathbf{z}^{(i)}]^T \mathbf{M} \mathbf{z}^{(j)} = [\mathbf{z}^{(i)}]^T \mathbf{K} \mathbf{z}^{(j)} \tag{7.69b}$$

Due to symmetry of $\mathbf{M}$ and $\mathbf{K}$, if we transpose (7.69b), we obtain

$$\omega_j^2 [\mathbf{z}^{(j)}]^T \mathbf{M} \mathbf{z}^{(i)} = [\mathbf{z}^{(j)}]^T \mathbf{K} \mathbf{z}^{(i)} \tag{7.70}$$

Subtracting (7.70) from (7.69a), we obtain

$$(\omega_i^2 - \omega_j^2)[\mathbf{z}^{(j)}]^T \mathbf{M} \mathbf{z}^{(i)} = 0 \tag{7.71}$$

For $\omega_i \neq \omega_j$ we must have

$$[\mathbf{z}^{(j)}]^T \mathbf{M} \mathbf{z}^{(i)} = 0, \qquad i \neq j \tag{7.72}$$

Equation (7.72) is the orthogonality relation for the natural modes, similar to one that will be found later for the continuous systems. This property is very useful. For example, one can check the accuracy of the computation of normal modes by application of equation (7.72).

If $\mathbf{M}$ is diagonal, the orthogonality condition becomes simply

$$\sum_{r=1}^{n} m_r \mathbf{z}_r^{(i)} \mathbf{z}_r^{(j)} = 0, \qquad i \neq j \tag{7.73}$$

[15]For an extensive discussion of the orthogonality principle, see von Kármán, T., and Biot, M. A. 1939. *Mathematical Methods in Engineering*. New York: McGraw-Hill.

The orthogonality of the natural modes can have a direct geometric interpretation for $n = 2$ or 3. It is known from analytic geometry that in a Cartesian coordinate system (x, y, z), a vector $\mathbf{a}$ is represented by its components a_1, a_2, a_3 along x, y, z. Let two vectors be $\mathbf{a}(a_1, a_2, a_3)$ and $\mathbf{b}(b_1, b_2, b_3)$. The two vectors are orthogonal (means at *normal angles*, that is, perpendicular to one another) if $\mathbf{a} \cdot \mathbf{b} = a_1 b_1 + a_2 b_2 + a_3 b_3 = 0$, which is the same as equation (7.72) or (7.73) with the matrix $\mathbf{M}$ equal to the identity matrix $\mathbf{I}$.

What happens to the orthogonality relations if $i = j$? Apparently, the left-hand sides in equation (7.72) or (7.73) are not zero; in fact, they are always positive, equal to a scalar:

$$\mathbf{z}^{(i)T} \mathbf{M} \mathbf{z}^{(i)} = m_i \tag{7.74}$$

or for diagonal mass matrix,

$$\sum_{r=1}^{n} m_r [z_r^{(i)}]^2 = m_i \tag{7.75}$$

It will become apparent later that normalizing each mode in a way that will yield $m_i = 1$ is very convenient in further computations. To this end we note that if we divide each mode with the constant $(m_i)^{1/2}$, we shall obtain the normalized mode

$$\boldsymbol{\phi}^{(i)} = \frac{\mathbf{z}^{(i)}}{(m_i)^{1/2}} \tag{7.76}$$

such that

$$\boldsymbol{\phi}^{(i)T} \mathbf{M} \boldsymbol{\phi}^{(i)} = \frac{1}{m_i} \mathbf{z}^{(i)T} \mathbf{M} \mathbf{z}^{(i)} = 1 \tag{7.77}$$

We shall call the modes $\boldsymbol{\phi}^{(i)}$, which are simply the natural modes $\mathbf{z}^{(i)}$ in a convenient scale, *normal modes*.

The matrices $\mathbf{M}$ and $\mathbf{K}$ can be diagonal or fully populated, depending on our choice of coordinates. Having diagonal matrices is useful from a practical standpoint. Inversion of a diagonal matrix is trivial:

$$\text{diag}[m_1, m_2, \ldots, m_n]^{-1} = \text{diag}[1/m_1, 1/m_2, \ldots, 1/m_n] \tag{7.78}$$

If both matrices $\mathbf{M}$ and $\mathbf{K}$ in equations (7.55) are diagonal, the system of equations has a very simple form. For example,

$$\begin{bmatrix} m_1 & 0 \\ 0 & m_2 \end{bmatrix} \begin{bmatrix} \ddot{x}_1 \\ \ddot{x}_2 \end{bmatrix} + \begin{bmatrix} k_1 & 0 \\ 0 & k_2 \end{bmatrix} \begin{bmatrix} x_1 \\ x_2 \end{bmatrix} = 0 \tag{7.79}$$

means simply that

$$\begin{aligned} m_1 \ddot{x}_1 + k_1 x_1 &= 0 \\ m_2 \ddot{x}_2 + k_2 x_2 &= 0 \end{aligned} \tag{7.80}$$

Equations (7.80) are independent; therefore, they can be solved separately. This results in the solution of two separate one-degree-of-freedom systems. As discussed in Chapter 6, if the matrix $\mathbf{M}$ has nondiagonal terms, the system has *dynamic coupling*. If the $\mathbf{K}$ matrix is nondiagonal, the system has *static coupling*. The same problem can be expressed with or without one or both kinds of coupling, depending on the selection of coordinates. If, by successful choice of the coordinates, the system is uncoupled, we call the coordinates chosen *normal*.[16]

In a way we have already chosen normal coordinates in the one-degree-of-freedom system. Saying that we consider only vertical vibration of a simply supported massless beam carrying a mass at the midspan, we imply that the vertical displacement is a normal coordinate.

We have already seen that the solution to the natural vibration problem can be obtained by adding all the natural coordinates with proper weighing factors, which depend on the initial conditions. We can show that the normal coordinates $\phi(i)$ can be used to uncouple the equations of motion. To this end, we assemble an $n \times n$ matrix of the normal modes, the *modal matrix*

$$\mathbf{\Phi} = [\phi^{(1)} \quad \phi^{(2)} \quad \cdots \quad \phi^{(n)}]$$

Because of equations (7.77) and (7.72),

$$\mathbf{\Phi}^T \mathbf{M} \mathbf{\Phi} = \mathbf{I} = \text{diag}[1 \quad 1 \quad \cdots \quad 1] \tag{7.81}$$

Equation (7.81) multiplied from the right side by $\mathbf{\Phi}^{-1}$ yields

$$\mathbf{\Phi}^T \mathbf{M} = \mathbf{\Phi}^{-1} \tag{7.82}$$

Furthermore, from equations (7.69) and (7.81) we obtain

$$\omega_i^2 = \phi^{(i)T} \mathbf{K} \phi^{(i)} \tag{7.83}$$

$$0 = \phi^{(j)T} \mathbf{K} \phi^{(i)}, \quad i \neq j \tag{7.84}$$

Therefore,

$$\mathbf{\Phi} \mathbf{K} \mathbf{\Phi} = \text{diag}[\omega_1^2, \omega_2^2, \ldots, \omega_n^2] = \mathbf{\Omega} \tag{7.85}$$

We now define a new set of variables q as

$$\mathbf{x} = \mathbf{\Phi} \mathbf{q} \tag{7.86}$$

where

$$\mathbf{q} = \{q_1 \quad q_2 \quad \cdots \quad q_n\}$$

Equation (7.86) defines a linear relationship between the components of vectors $\mathbf{x}$ and $\mathbf{q}$, called *linear transformation*. It is important that for the *inverse transformation*, from $\mathbf{q}$ to $\mathbf{x}$ as $\mathbf{q} = \mathbf{\Phi}^{-1}\mathbf{x}$, one does not have to invert the modal matrix. The inverse $\mathbf{\Phi}^{-1}$ can be computed from (7.82) with a matrix multiplication.

Substituting (7.86) into (7.55) and multiplying from the left with $\mathbf{\Phi}^T$, we obtain

$$\mathbf{\Phi}^T \mathbf{M} \mathbf{\Phi} \ddot{\mathbf{q}} + \mathbf{\Phi}^T \mathbf{K} \mathbf{\Phi} \mathbf{q} = 0 \tag{7.87}$$

---

[16]Thomson, W., Tait, P. G. 1867. Treatise on *Natural Philosophy*. Cambridge Univ. Press. The transformation to normal coordinates is always a real transformation. [Sylvester, 1852, *Philos. Mag.*, 4(iv): 138].

which, due to equations (7.82) and (7.85), becomes, with $\boldsymbol{\Omega} = \text{diag}[\omega_1^2 \quad \omega_2^2 \quad \cdots \quad \omega_n^2]$,

$$\ddot{\mathbf{q}} + \boldsymbol{\Omega}_i\mathbf{q} = 0 \tag{7.88}$$

or

$$\ddot{q}_1 + \omega_1^2 q_1 = 0$$
$$\ddot{q}_2 + \omega_2^2 q_2 = 0$$
$$\vdots \tag{7.89}$$
$$\ddot{q}_n + \omega_n^2 q_n = 0$$

which is a system of decoupled equations. Each equation can be solved separately. The linear transformation (7.86) applied on equations (7.55) does not change the eigenvalues associated with matrices $\mathbf{M}$ and $\mathbf{K}$ since matrix $\boldsymbol{\Phi}$ is constant, and due to the property $\det(\mathbf{AB}) = \det(\mathbf{A})\det(\mathbf{B})$, the frequency equation will be simply multiplied by a constant, rendering its roots invariant upon the linear transformation.

Of course, there is no sense in solving the uncoupled equations for the determination of the natural frequencies because it was already done in order to arrive at equations (7.89). However, this transformation is useful for a general method of forced vibration analysis, called *modal analysis*, which is discussed further in Chapter 8.

In terms of the normal modes, one can considerably simplify the general solution of the initial value problem as follows. The general solution (7.58) can be written as

$$\mathbf{x}(t) = \boldsymbol{\Phi}\mathbf{Aa} + \boldsymbol{\Phi}\mathbf{Bb} \tag{7.90}$$

where

$$\mathbf{A} = \text{diag}[\cos \omega_i t], \qquad \mathbf{B} = \text{diag}[\sin \omega_i t]$$
$$\mathbf{a} = \{c_1 \quad c_2 \quad \cdots \quad c_n\}, \qquad \mathbf{b} = \{c_{n+1} \quad c_{n+2} \quad \cdots \quad c_{2n}\}$$

with the initial conditions

$$\mathbf{x}^{(0)} = [x_1(0) \quad x_2(0) \quad \cdots \quad x_n(0)]^T$$
$$\mathbf{v}^{(0)} = [\dot{x}_1(0) \quad \dot{x}_2(0) \quad \cdots \quad \dot{x}_{n(0)}]^T \tag{7.91}$$

Application of the initial conditions yields

$$\boldsymbol{\Phi}\mathbf{a} = \mathbf{x}(0)$$
$$\boldsymbol{\Phi}\boldsymbol{\Omega}\mathbf{b} = \mathbf{v}^{(0)} \tag{7.92}$$

where $\boldsymbol{\Omega} = \text{diag}[\omega_i]$.

Multiplying (7.92) by $\boldsymbol{\Phi}^T$ from the right, we obtain

$$\boldsymbol{\Phi}^T\mathbf{M}\boldsymbol{\Phi}\mathbf{a} = \boldsymbol{\Phi}^T\mathbf{M}\mathbf{x}^{(0)}$$
$$\boldsymbol{\Phi}^T\mathbf{M}\boldsymbol{\Phi}\boldsymbol{\Omega}\mathbf{b} = \boldsymbol{\Phi}^T\mathbf{M}\mathbf{v}^{(0)} \tag{7.93}$$

In view of (7.82) we have

$$\mathbf{a} = \mathbf{\Phi T M x}(0) \qquad (7.94)$$

$$\mathbf{b} = \mathbf{\Omega}^{-1}\mathbf{\Phi}^T\mathbf{M}\mathbf{v}^{(0)}$$

where

$$\mathbf{\Omega}^{-1} = \text{diag}\left[\frac{1}{\omega_i}\right]$$

because $\mathbf{\Omega}$ is diagonal.

Equations (7.94) give **a** and **b** with only matrix multiplications if the solution of the eigenvalue problem is known. Furthermore,

$$\mathbf{x}(t) = \mathbf{\Phi}[\mathbf{A}\mathbf{\Phi}^T\mathbf{M}\mathbf{x}^{(0)} + \mathbf{B}\mathbf{\Omega}^{-1}\mathbf{\Phi}^T\mathbf{M}\mathbf{v}^{(0)}] \qquad (7.95)$$

□ **7.10 Decoupling the Equations of Motion**

Using the normal modes, write the differential equations of motion for the system in Figure E7.9 in decoupled form and determine the normal coordinates.

Solution   We shall first obtain the mass and stiffness matrices, then the natural frequencies and the vibration modes, and eventually, the equations in uncoupled form.
The coupled differential equations of motion are

$$J_1\ddot{\theta}_1 + (k_1 + k_2)\theta_1 - k_2\theta_2 = 0, \qquad \ddot{\theta}_1 + \frac{(k_1 + k_2)}{J_1}\theta_1 - \frac{k_2}{J_1}\theta_2 = 0$$

and

$$J_2\ddot{\theta}_2 - k_2\theta_1 + k_2\theta_2 = 0, \qquad \ddot{\theta}_2 - \frac{k_2}{J_2}\theta_1 + \frac{k_2}{J_2}\theta_2 = 0$$

The frequency equation for $\theta_1 = A_1 \cos \omega t$, $\theta_2 = A_2 \cos \omega t$ is

$$\begin{vmatrix} -J_1\omega^2 + k_1 + k_2 & -k_2 \\ -k_2 & -J_2\omega^2 + k_2 \end{vmatrix} = 0 \quad \text{or} \quad \begin{vmatrix} \dfrac{k_1 + k_2}{J_1} - \omega^2 & \dfrac{-k_2}{J_1} \\ \dfrac{-k_2}{J_2} & \dfrac{k_2}{J_2} - \omega^2 \end{vmatrix} = 0$$

$$\omega^4 + \left(\frac{k_1 + k_2}{J_1} + \frac{k_2}{J_2}\right)\omega^2 + \frac{k_1 k_2}{J_1 J_2} = 0$$

$$\omega^2 = -\frac{(k_1 + k_2)/J_1 + k_2/J_2}{2} \pm \frac{\{[(k_1 + k_2)/J_1 + k_2/J_2)]^2 - 4k_1 k_2/J_1 J_2\}^{1/2}}{2}$$

The equations for normal modes are:
First mode:

$$A_1^{(1)} = 1$$

$$A_2^{(1)} = \frac{(k_1 + k_2)/J_1 - \omega_1^2}{k_2/J_1}$$

Second mode:

$$A_1^{(2)} = 1$$

$$A_2^{(2)} = \frac{(k_1 + k_2) - \omega_2^2}{k_2/J_1}$$

Transformation equations:

$$\begin{bmatrix} \theta_1 \\ \theta_2 \end{bmatrix} = \begin{bmatrix} A_1^{(1)} & A_2^{(1)} \\ A_2^{(1)} & A_2^{(2)} \end{bmatrix} \begin{bmatrix} q_1 \\ q_2 \end{bmatrix}$$

Finally, the equations in decoupled form are

$$\ddot{q}_1 + \omega_1^2 q_1 = 0$$

$$\ddot{q}_2 + \omega_2^2 q_2 = 0$$

(This example demonstrates the procedure for the uncoupling of the equations of motion using the natural vibration modes.)

☐

7.8 DAMPED NATURAL VIBRATION

The undamped systems considered so far are merely a convenient idealization. All real systems are damped. However, many system features can be found from the analysis of undamped systems. It was shown, for example, in Chapter 2 that for light damping, usual in a system not deliberately damped, the natural frequencies of the undamped system do not differ appreciably from the ones of the damped system. Furthermore, it was shown in Chapter 3 that if we are not close to resonance, the vibration amplitudes are not affected materially by light damping. Whenever one requires such features, an undamped system analysis will suffice. There are cases, however, where damping has to be considered.[17] The differential equations of motion of the damped system are

$$\mathbf{M\ddot{x}} + \mathbf{C\dot{x}} + \mathbf{Kx} = 0 \tag{7.96}$$

An assumption that can be made many times (but not always) for engineering systems is that the modal matrix uncouples the damping matrix $\mathbf{C}$ too:

$$\mathbf{\Phi}^T \mathbf{C} \mathbf{\Phi} = \text{diag}[c_1 \quad c_2 \quad \cdots \quad c_n] \tag{7.97}$$

This is a rather severe restriction. It can be justified in certain systems: for example, in structural damping or in cases where the damping matrix $\mathbf{C}$ is diagonal, as often happens. In general, equation (7.97) is satisfied for damping of the form $\mathbf{C} = \alpha\mathbf{M} + \beta\mathbf{K}$, where α, β are scalars, called *Rayleigh damping*.[17] If the damping matrix $\mathbf{C}$ satisfies equation (7.97) or if the off-diagonal terms in the matrix $\mathbf{\Phi}^T\mathbf{C}\mathbf{\Phi}$ can be neglected, we set up

$$\mathbf{\Phi}^T \mathbf{C} \mathbf{\Phi} = \text{diag}[2\zeta_i\omega_i] \tag{7.98}$$

---

[17]Lord Rayleigh, 1873. Some general theorems relating to vibrations. *Proc. Math. Soc.* London.

where $\zeta_i = c_i/2\omega_i$ is the fraction of critical damping (damping factor) of the mode i. Then the equations (7.96) can be uncoupled with the linear transformation $\mathbf{x} = \mathbf{\Phi q}$:

$$\ddot{q}_i + 2\zeta_i\omega_i\dot{q}_i + \omega_i^2 q_i = 0 \tag{7.99}$$

and the solution is

$$q_i = e^{-\zeta_i\omega_i t}[a_i \cos \omega_{di}t + b_i \sin \omega_{di}t] \tag{7.100}$$

where $\omega_{di} = \omega_i(1 - \zeta_i^2)^{1/2}$ with the appropriate initial conditions.

The damping factor ζ_i of the mode i can be measured and tabulated for similar structures. Then application of this method is greatly facilitated.

In the general case that the damping matrix $\mathbf{C}$ deviates considerably from equation (7.97), one has to obtain the natural vibration the hard way. We start with equation (7.96) and seek a solution

$$\mathbf{x} = \mathbf{X}e^{pt} \tag{7.101}$$

Equation (7.96) becomes

$$[p^2\mathbf{M} + p\mathbf{C} + \mathbf{K}]\mathbf{X} = 0 \tag{7.102}$$

The problem of finding (complex) eigenvalues p and (complex) eigenvectors $\mathbf{X}$ that satisfy equation (7.102) is called a *complex eigenvalue problem*. For small systems one can solve the problem with the usual procedure. Equation (7.102) is a system of linear complex homogeneous algebraic equations in n unknowns X_1, $X_2, \ldots, X_n$. The condition for existence of nontrivial solution is the frequency equation

$$\left|p^2\mathbf{M} + p\mathbf{C} + \mathbf{K}\right| = 0 \tag{7.103}$$

Expanding the determinant, one can find a polynomial of degree $2n$. For usual systems, this will yield n pairs of complex conjugate roots $a_j \pm ib_j$. These roots, substituted in equation (7.102), will yield n pairs of complex eigenvectors which can be used to construct the general solution of the problem. For large systems this is a very tedious procedure. The reader is referred to special texts on applied mathematics or numerical analysis for further details on more efficient numerical methods.

The general solution will be

$$\mathbf{x}(t) + c_1\mathbf{z}^{(1)}e^{p_1 t} + c_2\mathbf{z}^{(2)}e^{p_2 t} + \cdots + c_{2n}\mathbf{z}^{(2n)}e^{p_{2n}t} \tag{7.104}$$

$2n$ initial conditions, the initial condition vectors $\mathbf{x}(0)$ and $\dot{\mathbf{x}}^\circ(0)$, must be supplied for the determination of the $2n$ constants c_j. These initial conditions will yield

$$\begin{aligned}\mathbf{x}(0) &= c_1\mathbf{z}^{(1)} + c_2\mathbf{z}^{(2)} + \cdots + c_{2n}\mathbf{z}^{(2n)} \\ \mathbf{x}(0) &= c_1 p_1 \mathbf{z}^{(1)} + c_2 p_2 z^{(2)} + \cdots + c_{2n}p_{2n}\mathbf{z}^{(2n)}\end{aligned} \tag{7.105}$$

We define

$$\mathbf{Z}^* = \begin{bmatrix} \mathbf{z}^{(1)} & \mathbf{z}^{(2)} & \cdots & \mathbf{z}^{(2n)} \\ \hline p_1\mathbf{z}^{(1)} & p_2\mathbf{z}^{(2)} & \cdots & p_n\mathbf{z}^{(2n)} \end{bmatrix} \quad \text{(dimension } 2n \times 2n)$$

$$\mathbf{E} = \mathrm{diag}[e^{p_i t}] \quad \text{(dimension } 2n \times 2n\text{)}$$

$$\mathbf{x}_0^* = \begin{bmatrix} \mathbf{x}(0) \\ \dot{\mathbf{x}}(0) \end{bmatrix} \quad \text{(dimension } 2n \times 1\text{)}$$

$$\mathbf{Z} = [\mathbf{z}^{(1)} \quad \mathbf{z}^{(2)} \quad \cdots \quad \mathbf{z}^{(2n)}] \quad \text{(dimension } n \times 2n\text{)}$$

$$\mathbf{c}^* = \{c_1 \quad c_2 \quad \cdots \quad c_{2n}\} \quad \text{(dimension } 2n \times 1\text{)}$$

Equations (7.105) can be written in matrix notation,

$$\mathbf{x}_0^* = \mathbf{Z}^* \mathbf{c}^* \tag{7.106}$$

or if $\mathbf{Z}^*$ is nonsingular,

$$\mathbf{c}^* = \mathbf{Z}^{*-1} \mathbf{x}_0^* \tag{7.107}$$

Equation (7.104) can be written in the form

$$\mathbf{x}(t) = \mathbf{Z}^* \mathbf{E} \mathbf{Z}^{*-1} \mathbf{x}_0^* \tag{7.108}$$

Equations (7.108) require the inversion of the matrix $\mathbf{Z}^*$, which for large systems will require substantial computation effort. One can obtain a more convenient form of the general solution using the orthogonality relations for a system with a special $\mathbf{C}$ matrix that can be diagonalized with the modal matrix.

We set the general solution in the form

$$\mathbf{q}(t) = \mathbf{E}(\mathbf{A}\mathbf{a} + \mathbf{B}\mathbf{b}) \tag{7.109}$$

where

$$\mathbf{E} = \mathrm{diag}[e^{-\zeta_i \omega_i t}], \quad \mathbf{A} = \mathrm{diag}[\cos \omega_i t], \quad \mathbf{B} = \mathrm{diag}[\sin \omega_i t],$$

$$\mathbf{a} = \{a_1 \quad a_2 \quad \cdots \quad a_n\}, \quad \mathbf{b} = \{b_1 \quad b_2 \quad \cdots \quad b_n\}$$

Therefore,

$$\mathbf{x}(t) = \mathbf{\Phi}\mathbf{q}(t) = \mathbf{\Phi}\mathbf{E}(\mathbf{A}\mathbf{a} + \mathbf{B}\mathbf{b})$$
$$\mathbf{x}(t) = \mathbf{\Phi}\mathbf{E}(\mathbf{A}\mathbf{a} + \mathbf{B}\mathbf{b}) + \mathbf{\Phi}\mathbf{E}[-\mathbf{\Omega}\mathbf{B}\mathbf{a} + \mathbf{A}\mathbf{b}] \tag{7.110}$$

where

$$\mathbf{\Omega} = \mathrm{diag}[\omega_i], \quad \mathbf{E} = -\mathbf{\Delta}\mathbf{\Omega}\mathbf{E}, \quad \mathbf{\Delta} = \mathrm{diag}[\zeta_i]$$

Application of the initial conditions yields

$$\mathbf{x}^{(0)} = \mathbf{\Phi}\mathbf{a}$$
$$\mathbf{v}^{(0)} = -\mathbf{\Phi}\mathbf{\Delta}\mathbf{\Omega}\mathbf{a} + \mathbf{\Phi}\mathbf{\Omega}\mathbf{b} = -\mathbf{\Delta}\mathbf{\Omega}\mathbf{x}^{(0)} + \mathbf{\Delta}\mathbf{\Omega}\mathbf{b} \tag{7.111}$$

or

$$\mathbf{\Phi}\mathbf{a} = \mathbf{x}^{(0)}$$
$$\mathbf{\Phi}\mathbf{\Omega}\mathbf{b} = \mathbf{v}^{(0)} + \mathbf{\Delta}\mathbf{\Omega}\mathbf{x}^{(0)} \tag{7.112}$$

Multiplying from the right with $\mathbf{\Phi}^T \mathbf{M}$, we obtain

$$\mathbf{a} = \mathbf{\Phi}^T \mathbf{M} \mathbf{x}^{(0)} \mathbf{b} = \mathbf{\Omega}^{-1} \mathbf{\Phi}^T \mathbf{M} [\mathbf{v}^{(0)} + \mathbf{\Delta}\mathbf{\Omega}\mathbf{x}^{(0)}] \tag{7.113}$$

where simply

$$\mathbf{\Omega}^{-1} = \text{diag}\left[\frac{1}{\omega_1}\right]$$

The solution becomes

$$\mathbf{x}(t) = \mathbf{\Phi E}[(\mathbf{A} + \mathbf{B\Delta})\mathbf{\Phi}^T\mathbf{Mx}^{(0)} + \mathbf{B\Omega}^{-1}\mathbf{\Phi}^T\mathbf{Mv}^{(0)}] \qquad (7.114)$$

This form of the solution requires only matrix multiplications and is very convenient.

□ **Example 7.11   Damped Two-Degree-of-Freedom Machine Mount**

Derive the equations of motion and their solution for the system shown in Figure E7.11.

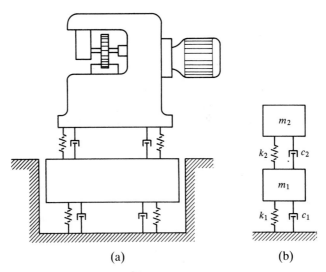

(a)   (b)

Figure E7.11

Solution   Application of Newton's law yields

$$m_1\ddot{x}_1 + (c_1 + c_2)\dot{x}_1 + (k_1 + k_2)x_1 - c_2\dot{x}_2 - k_2x_2 = 0$$

$$m_2\ddot{x}_2 + c_2\dot{x}_2 + k_2x_2 - c_2\dot{x}_1 - k_2x_1 = 0$$

Let

$$x_1 = Ae^{pt}, \qquad x_2 = Be^{pt}$$

Substituting these expressions into the differential equations and dividing by e^{pt}, we have

$$[m_1p^2 + (c_1 + c_2)p + (k_1 + k_2)]A - (c_2p + k_2)B = 0$$

$$- (c_2p + k_2)A + (m_2p^2 + c_2p + k_2)B = 0$$

The frequency equation

$$\begin{vmatrix} m_1p^2 + (c_1 + c_2)p + (k_1 + k_2) & - (c_2p + k_2) \\ - (c_2p + k_2) & m_2p^2 + c_2p + k_2 \end{vmatrix} = 0$$

Expanding the determinant, we obtain the frequency equation

$$[m_1 p^2 + (c_1 + c_2)p + k_1 + k_2](m_2 p^2 + c_2 p + k_2) - (c_2 p + k_2)^2 = 0$$

If p_1, p_2, p_2, p_3, and p_4 are the roots, the general solution is

$$x_1(t) = A_1 e^{p_1 t} + A_2 e^{p_2 t} + A_3 e^{p_3 t} + A_4 e^{p_4 t}$$

$$x_2(t) = B_1 e^{p_1 t} + B_2 e^{p_2 t} + B_3 e^{p_3 t} + B_4 e^{p_4 t}$$

The natural modes are found from either of the two algebraic equations in A and B:

$$\frac{A_i}{B_i} = \frac{c_2 p_i + k_2}{m_1 p_i^2} + (c_1 + c_2)p_i + k_1 + k_2)$$

$$= \frac{m_2 p_i^2 + c_2 p_i + k_2}{c_2 p_i + k_2}$$

where $i = 1, 2$.

(This example demonstrates the procedure for natural vibration analysis of the damped two-degree-of-freedom system.)

□

7.9 NUMERICAL METHODS

The methods presented in this chapter are purely analytical and can be used for up to a certain size of the system. For example, evaluation of the roots of the characteristic equation by purely algebraic methods is practicable for systems with up to four degrees of freedom. For such systems the characteristic determinant can be expanded to yield the characteristic polynomial, and from the latter the roots can be obtained by the formulas given in Appendix III. For larger systems one has to resort to numerical methods, either to evaluate the roots of the characteristic determinant or to apply numerical methods from the beginning. The former case will be dealt with in this section, the latter in a later chapter.

Undamped Systems

In an undamped system with n degrees of freedom, one expects n real roots. Some of them might be equal and some might be zero, for systems with possible rigid-body motion. The problem is given in the form of the characteristic equation

$$f(\lambda) = |A - \lambda I| = 0 \qquad (7.115)$$

where $\lambda = \omega^2$, I the identity matrix $\text{diag}[1 \quad 1 \quad 1 \quad \cdots \quad 1]$, and A the system matrix $= M^{-1}K$. The value of the determinant is computed with the methods of Appendix IV. Alternatively, the coefficients of the characteristic polynomial should be computed, and the function $f(\lambda)$ will be

$$f(\lambda) = a_0 \lambda^n + a_1 \lambda^{n-1} + \cdots + a_{n-1}\lambda + a_n = 0 \qquad (7.116)$$

Finding the roots of either equation (7.115) or (7.116) can be done in several ways. Since most of the methods used are available in standard software packages, only one will be described, a combination of Newton's iteration with the interpolation (*regula falsi*) method.

A sequence of values of $f(\lambda)$ is computed for $\lambda = \lambda_0 + \Delta\lambda$, $\lambda_0 + 2\Delta\lambda$, $\lambda_0 + 3\Delta\lambda$, . . ., with appropriate values of λ_0 and $\Delta\lambda$. These values are selected on the basis of an a priori knowledge of the order of magnitude of the natural frequencies ω_j. For example, λ_0 has to be less than the lowest expected values of λ and $\Delta\lambda$ a search step much smaller of the expected difference between the values of λ. In engineering problems, selecting these values is not a difficult task.

If at two successive points $\lambda_0 + k\Delta\lambda$ and $\lambda_0 + (k + 1) \Delta\lambda$, the function changes sign, a zero or a root is bracketed (Figure 7.5a). An approximate value of the root can be found assuming that the function $f(\lambda)$ is linear between the two points:

$$\lambda_{k0} = \lambda_0 + k \Delta\lambda + \frac{f_k \Delta\lambda}{f_k + f_{k+1}} \tag{7.117}$$

For most practical purposes this value has sufficient accuracy. However, more accuracy can usually be obtained with Newton's iteration. To this end, the function $f(\lambda)$ is expanded in a Taylor series about λ_{k0} keeping only the linear term:

$$f(\lambda) = f(\lambda_{k0}) + f'(\lambda_{k0})\delta\lambda \tag{7.118}$$

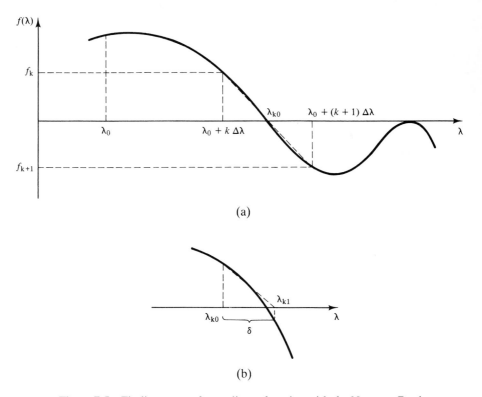

(a)

(b)

Figure 7.5 Finding roots of a nonlinear function with the Newton–Raphson method.

Geometrically, this means that the tangent is substituted for the function and cuts the λ-axis at $\lambda = \lambda_{k0} + \delta\lambda$, where $f(\lambda) = 0$, for a linear function. From (7.118),

$$\delta\lambda = -\frac{f(\lambda_{k0})}{f'(\lambda_{k0})} \tag{7.119}$$

and then the better approximation to the root will be

$$\lambda_{k1} = \lambda_{k0} + \delta\lambda = \lambda_{k0} - \frac{f(\lambda_{k0})}{f'(\lambda_{k0})} \tag{7.120}$$

Further improvement can be achieved by the sequence $\lambda_{k2}, \lambda_{k3}, \ldots$, which can be computed by repeated application of equation (7.120).

Application of such methods should be done with care. A graph of the function $f(\lambda)$ is always useful in obtaining a good understanding of its behavior. For example, the second root of the function shown in Figure 7.5a is apparently double, and the interpolation method would not locate it since the function does not change sign in its vicinity.

```
PROCEDURE ROOTS(A,λ,n,λ₀,Δλ,δmin)

*/ Computation of n roots of the determinant |A-λI| with the
interpolation and Newton's methods /*
Global Real variable Arrays A(1...n,1...n),
B(1...n,1,,,n),λ(1...n)
Global Real variables λ₀,Δλ,δmin,fk+1
Global Integer variable n
Local Real Variables λk₀, λk1, fk
Local Integer Variable i
For i=1 to n do loop:
   λ=λ₀;fk=0;fk+1=0
   While fk+1fk› or =0 do
      OPTION: Set Pointer for maximum number of iterations
      λ=λ+Δλ
      fk=fk+1
      for j=1 to n do loop:
         B(j,j)=A(j,j)-λ
      end loop j
      DETERMINANT(B,n,fk+1)
   End While fk+1fk
   λk₀=λ-Δλ+fkΔλ/(fk+fk+1)*/ interpolation /*
*/ Newton Iteration: /*
   While δ›δmin do:
      OPTION: Set Pointer for maximum number of iterations
      for j=1 to n do loop:
         B(j,j)=A(j,j)-λ
      end loop j
      DETERMINANT(B,n,fk)
      for j=1 to n do loop:
         B(j,j)=A(j,j)-λ-Δλ/10
      end loop j
```

```
      DETERMINANT(B,n,f_k+1)
      f'=(f_k+1-f_k)*10/Δλ; δλ=-f_k/f'; λ_k1=λ_k0+δ
   End While δλ<δλ_0
   λ(i)=λ_k1; λ_0=λ_k1+δλ
   End Loop i
End
```

Damped Systems

In a lightly damped stable system with n degrees of freedom, the system characteristic determinant will be a complex polynomial of degree $2n$ and one expects n complex conjugate roots with negative real parts, where the eigenvalue λ is now of a different nature, giving information about the vibration frequency and rate of decay. Some of the roots might be equal and some might have real or imaginary parts zero, the latter for systems with possible rigid-body motion.

The problem is stated in the form of the characteristic equation

$$f(p) = |p^2\mathbf{M} + p\mathbf{C} + \mathbf{K}| = 0 \tag{7.121}$$

The value of the determinant is computed using the methods of Appendix IV. Alternatively, the complex coefficients of the characteristic polynomial

$$f(p) = a_0 p^n + a_1 p^{n-1} + \cdots + a_{n-1}p + a_n = 0 \tag{7.122}$$

are computed.

Finding the roots of either equation (7.121) or (7.122) can be done in several ways. Since most of the methods used are available in standard software packages, only one will be described, for the case of real roots: a combination of Newton's iteration with the interpolation (*regula falsi*) method.

A sequence of the values of the function $[\text{Re}\{f(p)\}]^2 + [\text{Im}\{f(p)\}]^2$ is computed for $p = p_0 + \Delta p, p_0 + 2\Delta p, p_0 + 3\Delta p, \ldots,$ with appropriate values of p_0 and Δp. This function never changes sign, but between two successive points $p_0 + k\,\Delta p$ and $p_0 + (k+1)\,\Delta p$ the function has a minimum; a root is bracketed. An approximate value of the root can be found:

$$p_{k0} = p_0 + (k + 0.5)\,\Delta p \tag{7.123}$$

This can be used as a starting value for the Newton's iteration in two parameters: the real and imaginary parts of $p = x + jy$. To this end, the function $f(p)$ is expanded in a Taylor series about p_{k0}, keeping only the linear term:

$$f(p_k) = f(p_{k0}) + \frac{\partial f(p_{k0})}{\partial x}\,\delta x + \frac{\partial f(p_{k0})}{\partial y}\,\delta y \tag{7.124}$$

Geometrically, this means that the tangent plane is substituted for the function. For this approximation, both the real and imaginary parts of the function $f(x + jy)$ are zero. This gives two scalar equations in δx and δy which can be used to improve the root. Let

$$a_{11} = \text{Re}\left\{\frac{\partial f(p_{k0})}{\partial x}\right\}, \qquad a_{12} = \text{Re}\left\{\frac{\partial f(p_{k0})}{\partial y}\right\},$$

$$a_{21} = \text{Im}\left\{\frac{\partial f(p_{k0})}{\partial x}\right\}, \qquad a_{22} = \text{Im}\left\{\frac{\partial f(p_{k0})}{\partial y}\right\} \tag{7.125}$$

Then

$$a_{11}\delta x + a_{12}\delta y = -\text{Re}\{p_{k0}\}$$
$$a_{21}\delta x + a_{22}\delta y = -\text{Im}\{p_{k0}\} \tag{7.126}$$

By Cramer's rule,

$$\delta x = \frac{-\text{Re}\{p_{k0}\}a_{22} + \text{Im}\{p_{k0}\}a_{12}}{a_{11}a_{22} - a_{21}a_{12}}$$

$$\delta y = \frac{\text{Re}\{p_{k0}\}a_{21} - \text{Im}\{p_{k0}\}a_{11}}{a_{11}a_{22} - a_{21}a_{12}} \tag{7.127}$$

and the improved value of the complex root is

$$\text{Re}\{p_{k1}\} = \text{Re}\{p_{k0}\} + \delta x$$
$$\text{Im}\{p_{k1}\} = \text{Im}\{p_{k0}\} + \delta y \tag{7.128}$$

Further improvement can be achieved by the sequence $p_{k2}, p_{k3}, \ldots$, which can be computed by repeated application of equations (7.125) and (7.128).

PROCEDURE CROOTS(M,C,K,p,n,p$_0$,Δp,δ_{min})

```
*/ Computation of n complex roots of the determinant
|p²M+pC+K| with the Newton's methods /*
Global Real variable Arrays M(1...n,1...n),
C(1...n,1...n),K(1...n,1...n),p(1...n)
Global Complex Variable Array A(1...n,1...n)
Global Real variables Δp,δmin
Global Complex Variable fk+1
Global Integer variable n
Local Complex Variable fk,fk-1,p0
Local Integer Variable i
For i=1 to n do loop:
  p=p0;fk=(0,0);fk+1=(0,0);fk-1=(0,0)
  While f'k+1f'k› or =0 do
    OPTION: Set Pointer for maximum number of iterations
    p=p+Δp
    fk-1=fk; fk=fk+1
    for j=1 to n do loop:
      A(j,j)=-p²M+jpC+K
    end loop j
    CDETERMINANT(A,n,fk+1)
    f'k=(Re{fk}-Re{fk-1})/Δp; f'k+1=(Re{fk+1})Re{fk})/Δp
    if p‹p+3*Δp then f'k+1f'k=0
  End While f'k+1f'k
```

```
        p₀=[Re{p}-0.5*Δp,Im{p}]
*/ Newton Iteration: /*
   While |δ|>δ_min do:
       OPTION: Set Pointer for maximum number of iterations
       for j=1 to n do loop:
          A(j,j)=[Re{-p²*M(i,j)+K(i,j)}-Im{p*C(i,j),
                  Im{-p²*M(i,j)+K(i,j)}+Re{p*C(i,u)}
       end loop j
       DETERMINANT(A,n,f_k)
       p₀=p
       p=(Re{p₀}+Δp/10,Im{p₀})
       for j=1 to n do loop:
          A(j,j)=[Re{-p²*M(i,j)+K(i,j)}-Im{p*C(i,j),
                  Im{-p²*M(i,j)+K(i,j)}+Re{p*C(i,j)}
       end loop j
       DETERMINANT(A,n,f_{k+1})
       p=(Re{p₀},Im{p₀}+Δp/10
       for j=1 to n do loop:
          A(j,j)=[Re{-p²*M(i,j)+K(i,j)}-Im{p*C(i,j),
                  Im{-p²*M(i,j)+K(i,j)}+Re{p*C(i,j)]
       end loop j
       DETERMINANT)a,n,f_{k-1})
       Dx=10*(f_{k+1}-f_k)/Δp
       Dy=10*(f_{k+1}-f_k)/Δp
       a₁₁=Re{Dx}, a₁₂=Re{Dy}, a₂₁=Im{Dx}, a₂₂=Im{Dy}
       δx=[-Re{p₀}*a₂₂+Im{p₀}*a₁₂/[a₁₁*a₂₂-a₂₁*a₁₂]
       δy=[-Re{p₀}*a₂₁+Im{p₀}*a₁₁]/[a₁₁*a₂₂-a₂₁*a₁₂]
       p=[Re{p₀}+δx,Im(p₀)+δy]
       δ=(δx²+δy²)^{1/2}
    End While |δ|
    p(i)=p; p₀=[Re{p}+|δx|,Im{p}]
  End Loop i
End
```

TABLE 7.1. SUMMARY OF EQUATIONS OF CHAPTER 7

Stiffness and flexibility matrices

$$\mathbf{K} = \mathbf{A}^{-1} \text{ or } \mathbf{A} = \mathbf{K}^{-1}$$

Reciprocity

$$a_{ij} = a_{ji}, \; k_{ij} = k_{ji}$$

Energy of elastic deformation

$$V_e = \frac{1}{2} \int_0^L EI \left(\frac{d_2 x}{ds^2} \right)^2 ds$$

$$V_e = \frac{1}{2} \sum_{i=1}^n \sum_{j=1}^n k_{ij} x_i x_j$$

$$k_{ij} = \epsilon_{ij} \frac{\partial^2 V}{\partial x_i \partial x_j}$$

$$a_{ij} = \epsilon_{ij} \frac{\partial^2 V}{\partial F_i \partial F_j}$$

Equations in matrix form

$$\mathbf{M\ddot{x}} + \mathbf{Kx} = \mathbf{F}$$

$$\mathbf{AM\ddot{x}} + \mathbf{x} = \mathbf{Af}$$

Lagrange's equations

$$T = \frac{1}{2} \mathbf{\dot{x}}^T \mathbf{M\dot{x}}$$

$$V = \frac{1}{2} \mathbf{F}^T \mathbf{AF} = \frac{1}{2} \mathbf{x}^T \mathbf{Kx}$$

$$\frac{d}{dt} \left(\frac{\partial L}{\partial \dot{x}_i} \right) - \frac{\partial L}{\partial x_i} + \frac{\partial U}{\partial \dot{x}_i} = F_{xi}, \; i = 1, 2, \ldots, n$$

Damped system

$$\mathbf{M\ddot{x}} + \mathbf{C\dot{x}} + \mathbf{Kx} = \mathbf{F}_x$$

$$\mathbf{x}(0) = \mathbf{x}^{(0)}$$

$$\mathbf{\dot{x}}(0) = \mathbf{v}^{(0)}$$

$$\mathbf{Z} = [\mathbf{z}^{(1)} \quad \mathbf{z}^{(2)} \quad \ldots \quad \mathbf{z}^{(n)}]$$

$$\mathbf{x}(t) = \mathbf{Z}(\mathbf{AZ}^{-1} \mathbf{x}^{(0)} + \mathbf{BZ}^{-1} \mathbf{v}^{(0)})$$

$$\mathbf{A} = \text{diag}[\cos \omega_i t], \; \mathbf{B} = \text{diag}[\sin \omega_i t]$$

$$[\mathbf{z}^{(i)}]^T \mathbf{Mz}^{(i)} = 0, \; i \neq j$$

$$\sum_{r=1}^n m_r z_r^{(i)} z_r^{(j)} = 0, \; i \neq j$$

TABLE 7.1 *(cont.)*

$$\mathbf{z}^{(i)T}\mathbf{M}\mathbf{z}^{(i)} = m_i$$

$$\sum_{r=1}^{n} m_r \, [z_r^{(i)}]^2 = m_i$$

$$\boldsymbol{\phi}^{(i)} = \frac{\mathbf{z}^{(i)}}{(m_i)^{1/2}}$$

$$\boldsymbol{\Phi} = [\boldsymbol{\phi}^{(1)} \vdots \boldsymbol{\phi}^{(2)} \vdots \cdots \vdots \boldsymbol{\phi}^{(n)}]$$

$$\boldsymbol{\Phi}^T \mathbf{M} \boldsymbol{\Phi} = \mathbf{I} = \mathrm{diag}[1 \quad 1 \quad \cdots \quad 1]$$

$$\boldsymbol{\Phi}^T \mathbf{M} = \boldsymbol{\Phi}^{-1}$$

$$\omega_1^2 = \boldsymbol{\phi}^{(i)T} \mathbf{K} \boldsymbol{\phi}^{(i)}$$

$$0 = \boldsymbol{\phi}^{(j)T} \mathbf{K} \boldsymbol{\phi}^{(i)}, \; i \neq j$$

$$\boldsymbol{\Phi} \mathbf{K} \boldsymbol{\Phi} = \mathrm{diag}[\omega_1^2, \omega_2^2, \ldots, \omega_n^2] = \boldsymbol{\Omega}$$

$$\mathbf{x} = \boldsymbol{\Phi} \mathbf{q}$$

$$\mathbf{q} = \{q_1 \quad q_2 \quad \cdots \quad q_n\}.$$

$$\ddot{\mathbf{q}}_i + \boldsymbol{\Omega}_i \mathbf{q} = 0$$

$$\mathbf{A} = \mathrm{diag}[\cos \omega_i t], \; \mathbf{B} = \mathrm{diag}[\sin \omega_i t]$$

$$\mathbf{a} = \{c_1 \quad c_2 \quad \cdots \quad c_n\}, \; \mathbf{b} = \{c_{n+1} \quad c_{n+2} \quad \cdots \quad c_{2n}\},$$

$$\mathbf{x}^{(0)} = [x_1(0) \quad x_2(0) \quad \cdots \quad x_n(0)]^T$$

$$\mathbf{v}^{(0)} = [x_1(0) \quad x_2(0) \quad \cdots \quad x_n(0)]^T$$

$$\boldsymbol{\Omega}^{-1} = \mathrm{diag}\left[\frac{1}{\omega_i}\right]$$

$$\mathbf{x}(t) = \boldsymbol{\Phi}[\mathbf{A}\boldsymbol{\Phi}^T\mathbf{M}\mathbf{x}^{(0)} + \mathbf{B}\boldsymbol{\Omega}^{-1}\boldsymbol{\Phi}^T\mathbf{M}\mathbf{v}^{(0)}]$$

$$\mathbf{M}\ddot{\mathbf{x}} + \mathbf{C}\dot{\mathbf{x}} + \mathbf{K}\mathbf{x} = 0$$

$$\boldsymbol{\Phi}^T\mathbf{C}\boldsymbol{\Phi} = \mathrm{diag}[c_1 \quad c_2 \quad \cdots \quad c_n]$$

$$\boldsymbol{\Phi}^T\mathbf{C}\boldsymbol{\Phi} = \mathrm{diag}[2\zeta_i\omega_i]$$

$$\zeta_i = \frac{c_i}{2\omega_n}$$

$$\ddot{q}_i + 2\zeta_i\omega_i\dot{q}_i + \omega_i^2 q_i = 0$$

$$q_i = e^{-\zeta_i\omega_i t}[a_i \cos \omega_{di} t + b_i \sin \omega_{di} t]$$

$$\omega_{di} = \omega_i (1 - \zeta_i^2)^{1/2}$$

$$\mathbf{x} = \mathbf{X}e^{pt}$$

$$[p^2\mathbf{M} + p\mathbf{C} + \mathbf{K}]\mathbf{X} = 0$$

$$|p^2\mathbf{M} + p\mathbf{C} + \mathbf{K}| = 0$$

(continues)

TABLE 7.1 *(cont.)*

$$\mathbf{Z}^* = \begin{bmatrix} \mathbf{z}^{(1)} & \mathbf{z}^{(2)} & \cdots & \mathbf{z}^{(2n)} \\ \hline p_1\mathbf{z}^{(1)} & p_2\mathbf{z}^{(2)} & \cdots & p_n\mathbf{z}^{(2n)} \end{bmatrix}$$

$$\mathbf{E} = \mathrm{diag}[e^{p_i t}]$$

$$\mathbf{x}_0^* = \begin{bmatrix} \mathbf{x}(0) \\ \hline \dot{\mathbf{x}}(0) \end{bmatrix}$$

$$\mathbf{Z} = [\mathbf{z}^{(1)} \quad \mathbf{z}^{(2)} \quad \cdots \quad \mathbf{z}^{(2n)}]$$

$$\mathbf{E} = \mathrm{diag}[e^{-\zeta_i \omega_i t}]$$

$$\mathbf{A} = \mathrm{diag}[\cos \omega_i t]$$

$$\mathbf{B} = \mathrm{diag}[\sin \omega_i t]$$

$$\mathbf{\Omega} = \mathrm{diag}[\omega_i], \quad \mathbf{E} = -\mathbf{\Delta}\mathbf{\Omega}\mathbf{E}, \quad \mathbf{\Delta} = \mathrm{diag}[\zeta_i]$$

$$\mathbf{x}(t) = \mathbf{\Phi}\mathbf{E}[(\mathbf{A} + \mathbf{B}\mathbf{\Delta})\mathbf{\Phi}^T\mathbf{M}\mathbf{x}^{(0)} + \mathbf{B}\mathbf{\Omega}^{-1}\mathbf{\Phi}^T\mathbf{M}\mathbf{v}^{(0)}]$$

$$\lambda_{k0} = \lambda_0 + k\Delta\lambda + f_k\Delta\lambda/(f_k + f_{k+1})$$

$$\lambda_{k1} = \lambda_{k0} + \delta\lambda = \lambda_{k0} - f(\lambda_{k0})/f'(\lambda_{k0})$$

REFERENCES AND FURTHER READING

BISHOP, R. E. D., and JOHNSON, D. C. 1979. *Vibration*, 2nd ed. Cambridge: Cambridge University Press.

CLOUGH, R. W., and PENZIEN, J. 1975. *Dynamics of Structures*. New York: McGraw-Hill.

DEN HARTOG, J. P. 1952. *Mechanical Vibration*, 4th ed. New York: McGraw-Hill.

DIMAROGONAS, A. D. 1976. *Vibration Engineering*. St. Paul, Minn.: West Publishing.

DIMAROGONAS, A. D., and PAIPETIS, S. A. 1983. *Analytical Methods in Rotor Dynamics*. London: Elsevier–Applied Science.

KLOTTER, K. 1960. *Technische Schwingungslehre*. Berlin: Springer-Verlag.

PONTRYAGIN, L. S. 1962. *Ordinary Differential Equations*. Reading, Mass.: Addison-Wesley.

RAYLEIGH, J. W. S. 1894. *Theory of Sound*. New York: Dover Publications, 1946.

TIMOSHENKO, S. P. 1953. *History of the Strength of Materials*. New York: McGraw-Hill.

TIMOSHENKO, S., and YOUNG, D. H. 1955. *Vibration Problems in Engineering*. Princeton, N.J.: Van Nostrand.

WEAVER, W., TIMOSHENKO, S., and YOUNG, D. H. 1990. *Vibration Problems in Engineering*. New York: McGraw-Hill.

PROBLEMS

Sections 7.1 and 7.2

7.1. A gas turbine rotor is arranged as in Figure P7.1. If the moments of inertia of the component rotors are J_1, J_2, and J_3 and the torsional spring constants are as indicated.

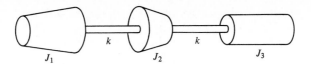

Figure P7.1

(a) write the equations of motion; and **(b)** determine the torsional natural frequencies and the vibration modes. **(c)** If $J_1 = 10$ kg·m², $J_2 = 5$ kg·m², $J_3 = 15$ kg·m², and $k = 20 \times 10^6$ N·m/rad, determine the dangerous rotating speeds if the turbine stages have 12 blades, the compressor has 16 blades, and the generator has two poles.

7.2. A marine diesel engine runs a propeller of inertia $J_1 = 100$ kg·m², and each of the six cranks of the crankshaft has inertia 10 kg·m². If we lump the crankshaft inertia at its two ends, the spring constant between them is 5×10^5 N·m/rad and between the end of the crankshaft and the propeller is 10^6 N·m/rad. Determine the torsional natural frequencies and the associated natural modes.

7.3. A compressor rotor consists of a shaft of uniform diameter $d = 0.15$ m and length $L = 1.4$ m. At the two ends there are disks with moment of inertia $J = 1$ kg·m² and at midspan a disk of inertia 30 kg·m². If the shear modulus of the shaft material is $G = 8.5 \times 10^{10}$ Pa, determine the torsional natural frequencies and vibration modes.

7.4. A motor is powering a pump through a gearbox (Figure P7.4). The rotors have moments of inertia $J_1 = 1.5$ kg·m², $J_2 = 0.5$ kg·m², $J_3 = 1.0$ kg·m², and $J_4 = 2.0$ kg·m², while the connecting shafts have torsional stiffnesses $k_1 = 1200$ N·m/rad and $k_2 = 2000$ N·m/rad. The gear ratio is 2:1. Determine the natural frequencies and the vibration modes.

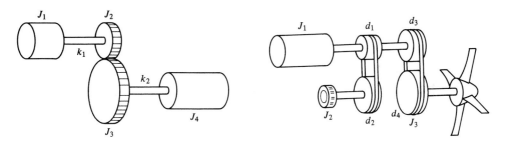

Figure P7.4   **Figure P7.5**

7.5. An electric motor is moving a pump and a blower, as shown in Figure P7.5, by way of belt drives. The motor rotor, shaft, and pulleys are considered rigid and have moment of inertia $J_1 = 1.2$ kg·m². The belts have equivalent free length 0.6 m and the pulley diameters are $d_1 = 0.15$ m and $d_2 = 0.20$ m. The pump and the blower rotors have $J_2 = 0.3$ kg·m² and $J_3 = 0.6$ kg·m² and have speed ratios to the motor speed 1:2, $d_3 = 0.2$ m, $d_4 = 0.3$ m. The belts have sections 5×40 mm each and the modulus of elasticity of the belt material is 50,000 N/m². Write the equations of motion and determine the natural frequencies of torsional vibration of the system and the vibration modes.

7.6. The fuselage of an aircraft weights $w = 10{,}000$ kg. The half wing has a length of 7 m and the flexural rigidity is 6×10^6 N·m². At each of the wing tip tanks there is 500 kg of fuel. Determine the natural frequency of the flexural vibration of the system and the vibration modes, assuming that the fuselage has no rotation about the longitudinal axis of the aircraft.

7.7. A train consists of three.cars of 40,000 kg each (Figure P7.7). Each car has spring coupling at both sides of constant $k = 10,000$ N/m. Determine the natural frequencies for oscillations along the train axis and the vibration modes.

Figure P7.7

7.8. A three-story building (Figure P7.8) consists of three slabs of 5000 kg each, and at each level there are six columns of section 0.4×0.4 m and 3.5 m high. Assuming that the shear modulus of the reinforced concrete is $G = 5 \times 10^9$ N/m², determine the natural frequencies of the building oscillations and the vibration modes.

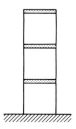

Figure P7.8

7.9. A machine is rigidly anchored on a rigid concrete block and the mass of both is 3000 kg. The block has height $h = 1$ m and square horizontal section with $a = 1.2$ m and is supported at the corners by 12 springs of stiffness $k = 2000$ N/m (Figure P7.9). Determine the natural frequencies and the vibration modes.

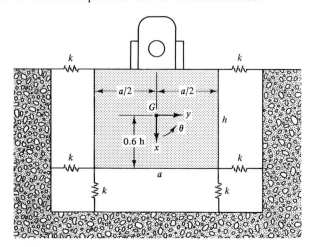

Figure P7.9

7.10. A mass $m = 2000$ kg is raised by a crane that is on a barge, as shown in Figure P7.10. The barge has square shape and the mass center is at G. The moment of inertia about axes G_y and G_z of barge and crane is 3×10^5 kg·m² and the mass 3500 kg. Horizontal

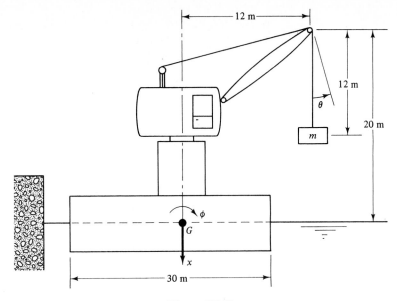

Figure P7.10

motion of the barge is restrained and motion is expected along x, θ, and ϕ. Determine the natural frequencies and vibration modes.

Sections 7.3 to 7.5

7.11–7.18. For the systems of Problems 7.1 to 7.8, determine the force influence coefficients. Then set up the differential equations in matrix form and determine the natural frequencies and vibration modes.

7.19–7.20. For the systems of Problems 7.9 and 7.10, determine the flexibility influence coefficients. Then set up the differential equations in matrix form and determine the natural frequencies and vibration modes.

7.21–7.22. Determine the force influence coefficients for the systems shown in Figures P7.21 and P7.22, respectively. Then set up the differential equations in matrix form and determine the natural frequencies and vibration modes.

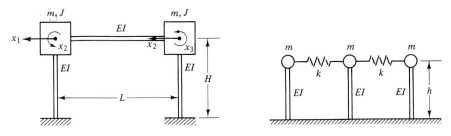

Figure P7.21   **Figure P7.22**

7.23–7.25. Determine the flexibility influence coefficients for the systems shown in Figures P7.23 to P7.25, respectively. Then set up the differential equations in matrix form and determine the natural frequencies and vibration modes.

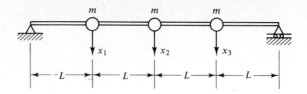

Figure P7.23

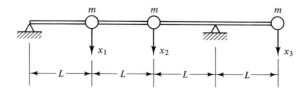

Figure P7.24

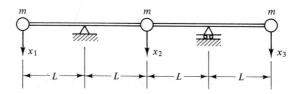

Figure P7.25

Section 7.6

7.26. While moving at 5 mph, the train of Problem 7.7 hits the rail end and latches. Determine the resulting vibration.

7.27. To test the natural frequencies of the building of Problem 7.8, the upper slab is hit with the bob of a pendulum of 30 kg mass moving at 3 m/s speed, and the impact is perfectly plastic. Determine the resulting motion.

7.28. A 200-kg mass falls on the middle of one edge of the block of Problem 7.9 and stays in contact with it. Neglecting the change in the mass, determine the resulting motion.

7.29. The mass m of Problem 7.10 is displaced from the vertical position by 15°. Determine the resulting motion.

7. 30. A small boat of 300 kg mass runs onto the middle of the side of the barge of Problem 7.10 with speed 10 mph, and the impact is purely elastic. Determine the resulting motion.

Sections 7.7 and 7.8

7.31–7.40. The systems of Problems 7.1 to 7.10, respectively, have modal damping 5% at all modes. Write the differential equations of motion in terms of the normal coordinates.

8

Lumped Mass Systems: Forced Vibration

8.1 EXACT METHODS

A large class of systems can be described by a finite number of degrees of freedom, under certain assumptions that were discussed in Chapter 7, by lumping their masses and moments of inertia. To study the natural vibration of the lumped mass system, we have derived the differential equations of motion, based on Newton's second law of motion or Lagrange's equations, in Chapter 7. Here we shall study the forced vibration of such systems.

Steady-state solutions due to harmonic excitations and general excitation will be developed. For moderately large systems, exact methods can be used, such as the ones used in Chapter 6, while for large systems approximate methods, such as modal analysis, are more appropriate using only a number of the lower modes.

Consider, for example, the system in Figure 7.1, which models a moderately tall building. It consists of n floors and the slabs can be considered rigid masses, while the vertical columns can be considered massless elastic members with flex-

*The above photo portrays an exhaust gas turbocharger for a ship's diesel engine. Courtesy of FAG Kugelfischer Georg Schäffer KGaA. Reprinted by permission.

ibility in bending much higher than their flexibility in tension–compression. Therefore, the slabs have mainly parallel horizontal motion (Figure 7.2a). The horizontal motions of the slabs x_i, $i = 1, 2, \ldots, n$, are the system coordinates.

The stiffness of each column for the deformation shown in Figure 7.2a is $k = 12EI/L^3$, where L is the length and EI the flexural rigidity of the column. Usually, there is more than one column. The columns can be considered as springs in parallel; therefore, the total stiffness is the sum of the stiffnesses of the several columns, $k_i = \sum_{j=1}^{n} 12EI_j/L_j^3$, where n is the number of columns.

Assuming that on each slab i there is an external horizontal force $f_i(t)$, going back to the step 7 of the equations development procedure of Section 7.2, application of Newton's second law at each slab in the horizontal direction yields:

$$\text{mass 1: } m_1\ddot{x}_1 = -k_1 x_1 + k_2(x_2 - x_1) + f_1(t)$$

$$\text{mass 2: } m_2\ddot{x}_2 = -k_2(x_2 - x_1) + k_3(x_3 - x_2) + f_2(t)$$

$$\text{mass 3: } m_3\ddot{x}_3 = -k_3(x_3 - x_2) + k_4(x_4 - x_3) + f_3(t) \tag{8.1}$$

$$\vdots$$

$$\text{mass } n: m_n\ddot{x}_n = -k_n(x_n - x_{n-1}) + f_n(t)$$

Step 8: Rearranging, we obtain the differential equations of motion[1]:

$$
\begin{aligned}
m_1\ddot{x}_1 + (k_1 + k_2)x_1 \quad\quad -k_2 x_2 \quad\quad\quad\quad\quad\quad\quad\quad &= f_1(t) \\
m_2\ddot{x}_2 \quad -k_2 x_1 \quad + (k_2 + k_3)x_2 \quad\quad -k_3 x_3 \quad\quad\quad &= f_2(t) \\
m_3\ddot{x}_3 \quad\quad\quad\quad -k_3 x_2 \quad + (k_3 + k_4)x_3 - k_4 x_4 \quad &= f_3(t) \\
\quad\quad\quad\quad\quad\quad\quad\quad\quad\quad\quad\quad\quad\quad\quad\quad\quad &\vdots \\
m_n\ddot{x}_n \quad\quad\quad\quad\quad\quad\quad\quad\quad\quad -k_n x_{n-1} + k_n x_n &= f_n(t)
\end{aligned}
\tag{8.2}
$$

Application of the procedure outlined above to a lumped mass linear system will yield a system of linear differential equations, usually with constant coefficients. We observe that these equations are not independent, due to the coupling terms k_{ij}, $i \neq j$. Therefore, we have to solve the differential equations simultaneously. The resulting vibration is coupled.

It is important to study the response of the system to harmonic excitation. As for the single degree of freedom system, this solution is fundamental for two reasons:

1. In most cases of forced vibration, especially in rotating machinery, the excitation is indeed harmonic.
2. The solution to harmonic excitation can be used to obtain the response to other time histories of the exciting forces.

---

[1][Rayleigh. 1894]; Gümbel, E. 1912. *Z. VDI*, 56: 1025; Holzer, H. 1921. *Die Berechnung der Drehschwingungen*. Berlin.

The forcing functions will be assumed in the form

$$f_1(t) = F_1 \cos \omega t$$

$$f_2(t) = F_2 \cos \omega t$$

$$\vdots \qquad\qquad (8.3)$$

$$f_n(t) = F_n \cos \omega t$$

As we did for the harmonic oscillator, we shall study the feasibility of harmonic natural vibration of the system at frequency ω. Therefore, we try the steady-state solution[2]

$$x_1(t) = A_1 \cos \omega t$$

$$x_2(t) = A_2 \cos \omega t$$

$$\vdots \qquad\qquad (8.4)$$

$$x_n(t) = A_n \cos \omega t$$

where $A_1, A_2, \ldots, A_n$ are amplitudes yet undetermined. Substitution in (8.2) yields

$$(-m_1\omega^2 + k_1 + k_2)A_1 \qquad -k_2A_2 \qquad\qquad\qquad\qquad = F_1$$

$$-k_2A_1 + (-m_2\omega^2 + k_2 + k_3)A_2 \qquad -k_3A_3 \qquad\qquad = F_2$$

$$-k_3A_2 + (-m_3\omega^2 + k_3 + k_4)A_3 \;-k_4A_4 \qquad = F_3$$

$$\cdots\cdots\cdots\cdots\cdots\cdots\cdots\cdots\cdots\cdots\cdots\cdots\cdots\cdots\cdots\cdots\cdots\cdots\cdots\cdots\cdots$$

$$\vdots$$

$$-k_nA_{n-1} + (-m_n\omega^2 + k_n)A_n = F_n$$

$$(8.5)$$

If this system of linear algebraic equations has a solution, (8.4) is indeed a solution of (8.2). The system of linear algebraic equations (8.5) is nonhomogeneous; therefore, by Cramer's rule it has a solution

$$A_1 = \frac{\det(\mathbf{D}_1)}{\det(\mathbf{D})}$$

$$A_2 = \frac{\det(\mathbf{D}_2)}{\det(\mathbf{D})}$$

$$\vdots \qquad\qquad (8.6)$$

$$A_n = \frac{\det(\mathbf{D}_n)}{\det(\mathbf{D})}$$

[2]It was suggested by Laplace and proved by J. Hershel [Rayleigh, 1894].

only if $\det(\mathbf{D}) \neq 0$, where

$$\mathbf{D} = \begin{bmatrix} -m_1\omega^2 + k_1 + k_2 & -k_2 & 0 & 0 & \cdots & 0 \\ -k_2 & -m_2\omega^2 k_2 + k_3 & -k_3 & 0 & \cdots & 0 \\ 0 & -k_3 & -m_3\omega^2 k_3 + k_4 & -k_4 & \cdots & 0 \\ \vdots & & & & & \\ 0 & 0 & & \cdots & -k_n & -m_n\omega^2 + k_n \end{bmatrix}$$

and the determinants $\det(\mathbf{D}_j)$, $j = 1, \ldots, n$, result from $\det(\mathbf{D})$ by substitution of the column j by the force vector of the right-hand side of equations (8.5).

We observe that the determinant $\det(\mathbf{D})$ in equation (8.6) is the frequency determinant of the system. This determinant has several roots, the natural frequencies of the system. It is apparent from equations (8.6) that if the frequency of the exciting force is equal to any one of the system natural frequencies, the amplitude will become infinitely large. Therefore, for a multidegree-of-freedom system there are several resonances that are identical with the system natural frequencies.

In addition to the steady-state solution given by equations (8.4) and (8.6), there is a transient due to the system initial conditions. In general, this response is short-lived due to damping, which exists in the system, even if it is small-enough to be neglected in the computation of the forced vibration.

☐ **Example 8.1   Forced Torsional Vibration of a Gas Turbine Rotor**

A gas turbine rotor is arranged as in Figure E7.1. If the moments of inertia of the component rotors are J_1, J_2, and J_3, the torsional spring constants are as indicated, and on the third rotor (the generator) there is an unbalanced electric torque $T_0 \cos \omega t$:

(a) Determine the torsional vibration amplitudes.

(b) If $J_1 = 10$ kg·m^2, $J_2 = 5$ kg·m^2, $J_3 = 15$ kg·m^2, $k_1 = k_2 = k_3 = k_4 = 20 \times 10^6$ N·m/rad, $T_0 = 100$ N·m, and $\omega = 2 \times 377$ rad/s, determine the vibration amplitudes.

Solution   (a) We will first find the equivalent three-degree-of-freedom system, then its differential equations, and finally the solution.

For the two-section shafts we substitute equivalent springs:

$$k_{12} = \frac{k_1 k_2}{k_1 + k_2} = 10^7, \qquad k_{34} = \frac{k_3 k_4}{k_3 + k_4} = 10^7 \text{ N·m/rad}$$

We assume the system in the displaced position θ_1, θ_2, θ_3. Application of Newton's law about the axis of rotation yields

$$J_1\ddot{\theta}_1 = k_{12}(\theta_2 - \theta_1)$$

$$J_2\ddot{\theta}_2 = -k_{12}(\theta_2 - \theta_1) + k_{34}(\theta_3 - \theta_2)$$

$$J_3\ddot{\theta}_3 = -k_3(\theta_3 - \theta_2) + T_0 \cos \omega t$$

or

$$J_1\ddot{\theta}_1 - k_{12}(\theta_2 - \theta_1) = 0$$

$$J_2\ddot{\theta}_2 + k_{12}(\theta_2 - \theta_1) - k_{34}(\theta_3 - \theta_2) = 0$$

$$J_3\ddot{\theta}_3 + k_3(\theta_3 - \theta_2) = T_0 \cos \omega t$$

(b) The vibration amplitudes $A_1, A_2, \ldots, A_n$ for $\theta_j = A_j \cos \omega t$ are the solution of

$$(k_{12} - J_1\omega^2)A_1 \qquad - k_{12}A_2 \qquad\qquad\qquad = 0$$

$$-k_{12}A_1 + (k_{12} + k_{34} - J_2\omega^2)A_2 \qquad -k_{34}A_3 = 0$$

$$-k_{34}A_2 + (k_{34} - J_3\omega^2)A_3 = T_0$$

which is

$$A_1 = \begin{vmatrix} 0 & -k_{12} & 0 \\ 0 & k_{12} + k_{34} - J_2\omega^2 & -k_{34} \\ T_0 & -k_{34} & k_{34} - J_3\omega^2 \end{vmatrix} / D = \frac{T_0 k_{12} k_{34}}{D}$$

$$A_2 = \begin{vmatrix} k_{12} - J_1\omega^2 & 0 & 0 \\ -k_{12} & 0 & -k_{34} \\ 0 & T_0 & k_{34} - J_3\omega^2 \end{vmatrix} / D = \frac{T_0(k_{12} - J_2\omega^2)k_{34}}{D}$$

$$A_3 = \begin{vmatrix} k_{12} - J_1\omega^2 & -k_{12} & 0 \\ -k_{12} & k_{12} + k_{34} - J_2\omega^2 & 0 \\ 0 & -k_{34} & T_0 \end{vmatrix} / D =$$

$$\frac{T_0[(k_{12} - J_1\omega^2)(k_{12} + k_{34} - J_2\omega^2) - k_{12}^2]}{D}$$

$$D = \begin{vmatrix} k_{12} - J_1\omega^2 & -k_{12} & 0 \\ -k_{12} & k_{12} + k_{34} - J_2\omega^2 & -k_{34} \\ 0 & -k_{34} & k_{34} - J_3\omega^2 \end{vmatrix}$$

$$= \begin{vmatrix} 10^7 - 10 \times 754^2 & -10^7 & 0 \\ -10^7 & 10^7 + 10^7 - 5 \times 754^2 & -10^7 \\ 0 & -10^7 & 10^7 - 15 \times 754^2 \end{vmatrix}$$

$$= 10^{21} \begin{vmatrix} 0.43 & -1 & 0 \\ -1 & 1.7 & -1 \\ 0 & -1 & 0.9 \end{vmatrix}$$

$$= -0.67 \times 10^{21}$$

$$A_1 = 100 \times 10^7 \times \frac{10^7}{-0.67 \times 10^{21}} = -21.3 \times 10^{-6} \text{ rad}$$

$$A_2 = 100 \times (10^7 - 5 \times 754^2) \times \frac{10^7}{-0.67 \times 10^{21}} = -9.19 \times 10^{-6} \text{ rad}$$

$$A_3 = 100 \times \frac{(10^7 - 10 \times 754^2)(10^7 - 5 \times 754^2)}{-0.67 \times 10^{21}} = -5.5 \times 10^{-6} \text{ rad}$$

(In this example we find the dynamic response of a three-disk gas turbine rotor to torsional harmonic excitation.)

□

8.2 MATRIX FORMULATION: PERIODIC EXCITATION

The differential equations of motion of a linear system such as the one described by equations (8.2), including linear damping, can be written in matrix form,[3] as explained in Chapter 7, as

$$\mathbf{M\ddot{x}} + \mathbf{C\dot{x}} + \mathbf{Kx} = \mathbf{f}(t) \tag{8.7}$$

where $\mathbf{M}$, $\mathbf{C}$, and $\mathbf{K}$ are the system mass, damping, and stiffness matrices, respectively, and $\mathbf{x}$ and $\mathbf{f}$ are the response and force vectors. A harmonic excitation $\mathbf{f}(t) = \mathbf{F}_0 e^{i\omega t}$ will lead to the equations

$$\mathbf{M\ddot{x}} + \mathbf{C\dot{x}} + \mathbf{Kx} = F_0 e^{i\omega t} \tag{8.8}$$

The steady-state solution will be found with the substitution

$$\mathbf{x} = \mathbf{X} e^{i\omega t} \tag{8.9}$$

Then equation (8.8) after division by $e^{i\omega t}$ and matrix operations yields

$$\mathbf{X} = [-\omega^2 \mathbf{M} + i\omega \mathbf{C} + \mathbf{K}]^{-1} \mathbf{F}_0 \tag{8.10}$$

provided that the matrix $\mathbf{B} = [-\omega^2 \mathbf{M} + i\omega \mathbf{C} + \mathbf{K}]$ is not singular.

The solution (8.10) will yield an amplitude vector $\mathbf{X} = [X_1 \quad X_2 \quad \cdots \quad X_n]$ with complex components, in general. Therefore, the solution, on a vector diagram at time $t = 0$, will be as shown in Figure 8.1. The components of the forcing vector $\mathbf{F}_0 = [F_1 \quad F_2 \quad \cdots \quad F_n]$ might be complex, which means that they might be at different phase angles in respect to one another. The same will then be true for the solution vector $\mathbf{X}$.

For periodic but not harmonic forcing functions, one can analyze them in Fourier series, as in Chapter 5, and superpose the responses. The same method is used if the different elements of the forcing vector are harmonic functions but at different angular velocities.

[3]Duncan, W. J., and Collar, A. R. 1935. Matrices applied to the motion of a damped system. *Philos. Mag.*, 7(19): 197–219.

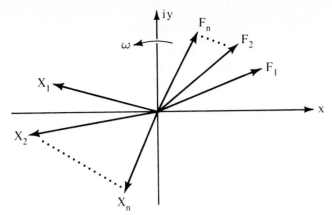

Figure 8.1 Superposition of harmonic functions.

☐ **Example 8.2   Forced Harmonic Vibration of a Three-Story Building**

A three-level building is shown in Figure E8.2. The masses of floor slabs are m_1, m_2, and m_3, and the spring constants are as indicated. On the first floor there is an internal combustion engine running at 1500 rpm which has unbalanced horizontal forces 150 $\cos \omega t$, 250 $\cos 2\omega t$, and 80 $\cos 4\omega t$ N, where ω is the running angular velocity. Determine the vibration amplitude at the top floor if $m_1 = 10{,}000$ kg, $m_2 = 8000$ kg, $m_3 = 7000$ kg, $k_1 = k_2 = k_3 = 2 \times 10^6$ N/m.

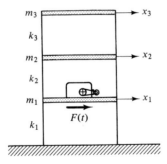

Figure E8.2

Solution   We consider a three-slab system, as in Figure 8.1, with three degrees of freedom. Application of Newton's law in the horizontal direction yields

$$m_1\ddot{x}_1 = -k_1 x_1 + k_2(x_2 - x_1) + F(t)$$

$$m_2\ddot{x}_2 = -k_2(x_2 - x_1) + k_3(x_3 - x_2)$$

$$m_3\ddot{x}_3 = -k_3(x_3 - x_2)$$

or

$$m_1\ddot{x}_1 + k_1 x_1 - k_2(x_2 - x_1) = F(t)$$

$$m_2\ddot{x}_2 + k_2(x_2 - x_1) - k_3(x_3 - x_2) = 0$$

$$m_3\ddot{x}_3 + k_3(x_3 - x_2) = 0$$

Let $F(t) = F_1 \cos \omega t + F_2 \cos 2\omega t + F_3 \cos 4\omega t$.

First harmonic of the response: The vibration amplitudes A_1, A_2, A_3 for $x_j = A_j \cos \omega t$ are the solution of

$$(k_1 + k_2 - m_1\omega^2)A_1 \qquad\qquad -k_2 A_2 \qquad\qquad\qquad\qquad = F_1$$

$$-k_2 A_1 \qquad +(k_2 + k_3 - m_2\omega^2)A_2 \qquad -k_3 A_3 \qquad = 0$$

$$-k_3 A_2 \qquad +(k_3 - m_3\omega^2)A_3 = 0$$

which is for $\omega = 1500 \times 2\pi/60 = 157$ rad/sec,

$$A_3^{(1)} = \begin{vmatrix} k_1 + k_2 - m_1\omega^2 & -k_2 & F_1 \\ -k_2 & k_2 + k_3 - m_2\omega^2 & 0 \\ 0 & -k_3 & 0 \end{vmatrix} = \frac{k_2 k_3 F_1}{D}$$

$$D = \begin{vmatrix} k_1 + k_2 - m_1\omega^2 & -k_2 & 0 \\ -k_2 & k_2 + k_3 - m_2\omega^2 & -k_2 \\ 0 & -k_3 & k_3 - m_3\omega^2 \end{vmatrix}$$

$$= \begin{vmatrix} -2.42 \times 10^8 & -2 \times 10^6 & 0 \\ -2 \times 10^6 & -1.93 \times 10^8 & -2 \times 10^6 \\ 0 & -2 \times 10^6 & -1.72 \times 10^8 \end{vmatrix} = 8.033 \times 10^{24}$$

Therefore,

$$A_3^{(1)} = \frac{k_2 k_3 F_1}{D} = \frac{2 \times 10^6 \times 2 \times 10^6 \times 150}{8.033 \times 10^{24}} = 7.5 \times 10^{-11} \text{ m}$$

Second harmonic of the response: The vibration amplitudes A_1, A_2, A_3 for $x_j = A_j \cos 2\omega t$ are the solution of

$$(k_1 + k_2 - m_1 4\omega^2)A_1 \qquad\qquad -k_2 A_2 \qquad\qquad\qquad\qquad = F_1$$

$$-k_2 A_1 \qquad +(k_2 + k_3 - m_2 4\omega^2)A_2 \qquad -k_3 A_3 \qquad = 0$$

$$-k_3 A_2 \qquad +(k_3 - m_3 4\omega^2)A_3 = 0$$

$$D = \begin{vmatrix} k_1 + k_2 - m_1 4\omega^2 & -k_2 & 0 \\ -k_2 & k_2 + k_3 - m_2 4\omega^2 & -k_3 \\ 0 & -k_3 & k_3 - m_3 4\omega^2 \end{vmatrix}$$

$$= \begin{vmatrix} -9.81 \times 10^8 & -2 \times 10^6 & 0 \\ -2 \times 10^6 & -7.72 \times 10^8 & 0 \\ 0 & -2 \times 10^6 & -3.66 \times 10^8 \end{vmatrix}$$

Therefore,

$$A_3^{(2)} = \frac{k_2 k_3 F_1}{D} = \frac{2 \times 10^6 \times 2 \times 10^6 \times 250}{3.74 \times 10^{26}} = 1.17 \times 10^{-12} \text{ m}$$

Third harmonic of the response: The vibration amplitudes A_1, A_2, A_3 for $x_j = A_j \cos 4\omega t$ are the solution of

$$(k_1 + k_2 - m_1 16\omega^2)A_1 \qquad\qquad -k_2 A_2 \qquad\qquad\qquad\qquad\qquad = F_3$$
$$-k_2 A_1 \qquad\qquad +(k_2 + k_3 - m_2 16\omega^2)A_2 \qquad -k_3 A_3 \qquad = 0$$
$$-k_3 A_2 \qquad\qquad +(k_3 - m_3 16\omega^2)A_3 = 0$$

$$D = \begin{vmatrix} k_1 + k_2 - m_1 16\omega^2 & -k_2 & 0 \\ -k_2 & k_2 + k_3 - m_2 16\omega^2 & -k_3 \\ 0 & -k_3 & k_3 - m_3 16\omega^2 \end{vmatrix}$$

$$= \begin{vmatrix} -3.93 \times 10^{11} & -2 \times 10^6 & 0 \\ -2 \times 10^6 & -3.09 \times 10^{11} & \\ 0 & -2 \times 10^6 & -2.75 \times 10^{11} \end{vmatrix}$$

Therefore,

$$A_3^{(3)} = \frac{k_2 k_3 F_1}{D} = \frac{2 \times 10^6 \times 2 \times 10^6 \times 80}{33.4 \times 10^{33}} = 1.8 \times 10^{-14} \text{ m}$$

Finally,

$$x_3(t) = A_3^{(1)} \cos \omega t + A_3^{(2)} \cos 2\omega t + A_3^{(3)} \cos 4\omega t$$

(In this example we find the dynamic response of a three-story building to harmonic excitation. Moreover, the diminishing of the vibration amplitudes at excitation frequencies far from the natural frequencies is demonstrated.)

☐

☐ **Example 8.3   Lateral Vibration of a Rotor on Linear Bearings**

A turbine rotor can be modeled by way of a three-mass system m_1, m_2, m_3 connected with a massless shaft of diameter d (Figure E8.3). The two end masses represent the journals, which are supported by linear bearings having stiffnesses k_1 and k_2 and linear viscous damping constants c_1 and c_2. Determine the response at midspan for an unbalance mass m on the mass m_2 at a radial distance r from the center of the shaft if ω is the speed of rotation. The horizontal and vertical vibration are not coupled. Numerical application: $m_1 = 300$ kg, $m_2 = 800$ kg, $m_3 = 400$ kg, $k_1 = 5 \times 10^6$ N/m, $k_2 = 4 \times 10^6$ N/m, $c_1 = 5000$ N·s/m, $c_2 = 4000$ N·s/m, $\omega = 300$ rad/s, $d = 0.15$ m, $L = 3$ m, $E = 2 \times 10^{11}$ Pa, $m = 0.1$ kg, and $r = 0.2$ m.

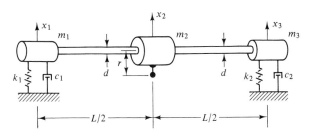

Figure E8.3

Solution The system is modeled with three degrees of freedom, as shown. The vertical spring constant of the shaft itself with respect to rigid bearings is, as a simply supported beam, $k = 48EI/L^3 = 48 \times 2 \times 10^{11}(\pi \times 0.15^4/64)/3^3 = 8.83 \times 10^6$ N/m.

If the masses are displaced by x_1, x_2, and x_3, as shown, the force on the mass m_2 due to the deflection of the shaft only will be $-k[x_2 - (x_1 + x_3)/2]$ and on the masses m_1 and m_3, equal forces $k[x_2 - (x_1 + x_3)/2]/2$, due to symmetry.

Application of Newton's law in the vertical direction, with the nomenclature of Figure E8.3, yields the differential equations of motion:

$$m_1\ddot{x}_1 + c_1\dot{x}_1 + \left(\frac{k}{4} + k_1\right)x_1 - \frac{kx_2}{2} + \frac{kx_3}{4} = 0$$

$$m_2\ddot{x}_2 \qquad -\frac{kx_1}{2} + kx_2 \qquad -\frac{kx_3}{2} = mr\omega^2 \cos \omega t$$

$$m_3\ddot{x}_3 + c_3\dot{x}_3 \qquad +\frac{kx_1}{4} - \frac{kx_2}{2} + \left(\frac{k}{4} + k_2\right)x_3 = 0$$

In matrix form,

$$\mathbf{M\ddot{x} + C\dot{x} + Kx = F_0}e^{i\omega t}$$

where

$$\mathbf{M} = \begin{bmatrix} m_1 & 0 & 0 \\ 0 & m_2 & 0 \\ 0 & 0 & m_3 \end{bmatrix}, \qquad \mathbf{C} = \begin{bmatrix} c_1 & 0 & 0 \\ 0 & 0 & 0 \\ 0 & 0 & c_2 \end{bmatrix},$$

$$\mathbf{K} = \begin{bmatrix} \dfrac{k}{4} + k_1 & -\dfrac{k}{2} & \dfrac{k}{4} \\ -\dfrac{k}{2} & k & -\dfrac{k}{2} \\ \dfrac{k}{4} & -\dfrac{k}{2}\dfrac{k}{4} + k_2 \end{bmatrix},$$

$$\mathbf{x} = \{x_1 \ x_2 \ x_3\}, \qquad \mathbf{F}_0 = \{0 \ mr\omega^2 \ 0\}$$

The solution is given by equations (8.10):

$$[-\omega^2\mathbf{M} + i\omega\mathbf{C} + \mathbf{K}]\mathbf{X} = \mathbf{F}_0$$

This is a system of complex linear algebraic equations:

$$\begin{bmatrix} -m_1\omega^2 + i\omega c_1 + \dfrac{k}{4} + k_1 & -\dfrac{k}{2} & \dfrac{k}{4} \\ -\dfrac{k}{2} & -m_2\omega^2 + k & -\dfrac{k}{2} \\ \dfrac{k}{4} & -\dfrac{k}{2} & -m_3\omega^2 + i\omega c_2 + \dfrac{k}{4} + k_2 \end{bmatrix} \begin{bmatrix} X_1 \\ X_2 \\ X_3 \end{bmatrix} = \begin{bmatrix} 0 \\ mr\omega^2 \\ 0 \end{bmatrix}$$

The solution is obtained with Cramer's rule:

$$X_2 = \begin{vmatrix} -m_1\omega^2 + i\omega c_1 + \dfrac{k}{4} + k_1 & 0 & \dfrac{k}{4} \\[2ex] -\dfrac{k}{2} & mr\omega^2 & -\dfrac{k}{2} \\[2ex] \dfrac{k}{4} & 0 & -m_3\omega^2 + i\omega c_3 + \dfrac{k}{4} + k_3 \end{vmatrix} \Big/ D$$

where

$$D = \begin{vmatrix} -m_1\omega^2 + i\omega c_1 + \dfrac{k}{4} + k_1 & -\dfrac{k}{2} & \dfrac{k}{4} \\[2ex] -\dfrac{k}{2} & -m_2\omega^2 + k & -\dfrac{k}{2} \\[2ex] \dfrac{k}{4} & -\dfrac{k}{2} & -m_3\omega^2 + i\omega c_3 + \dfrac{k}{4} + k_3 \end{vmatrix}$$

$$= \begin{vmatrix} -19.8 + 1.5i & -4.4 & 2.2 \\ -4.4 & -63 & -4.4 \\ 2.2 & -4.4 & -9.8 + 1.2i \end{vmatrix} \times 10^6 = (-11{,}147 + 380i) \times 10^6$$

The force is $mr\omega^2 = 0.1 \times 0.2 \times 300^2 = 1800$ N; therefore,

$$X_2 = \begin{vmatrix} -19.8 + 1.5i & 0 & 2.2 \\ -4.4 & 0.0018 & -4.4 \\ 2.2 & 0 & -9.8 + 1.2i \end{vmatrix} \times 10^6 \Big/ D = \dfrac{0.346 - 0.07i}{-11{,}147 + 380i}$$

$$= \dfrac{(0.346 - 0.07i)(-11{,}147 - 380i)}{(-11{,}147 + 380i)(-11{,}147 - 380i)}$$

$$= \dfrac{-3830 + 911i}{1.24 \times 10^8} = (-30.9 + 7.34i) \times 10^{-6} \text{ m}$$

The amplitude is

$$|X_2| = (30.9^2 + 7.34^2)^{1/2} \times 10^{-6} = 31.7 \times 10^{-6} \text{ m} = 31.7 \ \mu\text{m}$$

The phase angle is

$$\theta = \arctan \dfrac{7.34}{-30.9} = -9.92°$$

(In this example we find the dynamic response of a three-mass flexible rotor on elastic linear bearings under unbalance excitation.) ☐

8.3 NONPERIODIC EXCITATION: DIRECT METHOD

Since nonperiodic excitation and shock are encountered in multidegree-of-freedom systems, we shall extend the results of Chapter 5 for nonperiodic excitation of such systems.

Because of the linearity and the validity of superposition,[4] we shall assume that the vector $f(t) = [F_1(t) \; F_2(t) \cdots F_n(t)]^T$ has only one nonzero component, say $F_j(t)$.

We assume the excitation force to be an impulse, $F_j(t) = \hat{F}_j$. Due to this force, there will be an initial velocity of the mass m_j, assuming a diagonal mass matrix, equal to $v_j(0) = \hat{F}_j/m_j$.

In terms of equation (7.105), the initial conditions are written as

$$\mathbf{x}_0 = \{\underbrace{0 \quad 0 \quad \cdots \quad 0}_{n} \quad \underbrace{0 \quad 0 \quad \cdots \quad 0 \quad 1}_{j} \quad \underbrace{0 \quad \cdots \quad 0}_{n-j}\} \frac{\hat{F}_j}{m_j} \tag{8.11}$$

If we designate with $\mathbf{u}_j$ the vector in braces above,

$$\mathbf{x}_0 = \mathbf{u}_j \frac{\hat{F}_j}{m_j} \tag{8.12}$$

Equation (7.108) can be written as

$$\mathbf{x}(t) = \mathbf{ZEc}^* \tag{8.13}$$

where

$$\mathbf{E} = \text{diag}[e^{P_1 t}, e^{P_2 t}, \dots, e^{P_{2n} t}]$$

but

$$\mathbf{c}^* = \mathbf{Z}^{*-1}\mathbf{x}_0^{*-1} = \mathbf{Z}^{*-1}\mathbf{u}_j \frac{\hat{F}_j}{m_j} \tag{8.14}$$

Therefore,

$$\mathbf{x}(t) = \mathbf{Z}^*\mathbf{E}\mathbf{Z}^{*-1}\mathbf{u}_j \frac{\hat{F}_j}{m_j} \tag{8.15}$$

which is of the form

$$\mathbf{x}(t) = \mathbf{G}(t)\hat{\mathbf{f}}_j \tag{8.16}$$

where

$$\hat{\mathbf{f}}_j = \frac{\hat{F}_j}{m_j}\mathbf{u}_j$$

[4]It was introduced by Daniel Bernoulli (1700–1782) to explain observations of Joseph Sauveur (1653–1716) that a vibrating string could produce at the same time sounds which would correspond to several of its harmonics.

and

$$G(t) = \mathbf{Z}^*\mathbf{E}\mathbf{Z}^{*-1}$$

is a property of the system and does not depend on the particulars of the excitation. The matrix $\mathbf{G}(t)$ of dimension $2n \times 2n$ will be called the *impulse response matrix* of the system.

For a general vector $\mathbf{f}(t)$ of the excitation at mass m_j we can extend Duhamel's integral, defining $\int_0^t [w_{ij}] \, dt = [\int_0^t w_{ij} \, dt]$,

$$\mathbf{x}(t) = \int_0^t \mathbf{G}(t - t^*)\hat{\mathbf{f}}_j(t^*) \, dt^* \tag{8.17}$$

Thus for an exciting force $\mathbf{f}(t) = \{f_1(t) \quad f_2(t) \quad \cdots \quad f_n(t)\}$ we shall have, if the matrix $\mathbf{M}$ is diagonal,

$$\mathbf{x}(t) = \mathbf{M}^{-1} \int_0^t \mathbf{G}(t - t^*)\mathbf{f}(t^*) \, dt^* \tag{8.18}$$

where

$$\mathbf{f}(t) = \{\underbrace{0 \quad 0 \quad \cdots \quad 0}_{n} \quad f_1(t) \quad f_2(t) \quad \cdots \quad f_n(t)\}$$

☐ **Example 8.4 Transient Response of a Two-Degree-of-Freedom System**

On the mass m_2 of a damped two-degree-of-freedom system (Figure E8.4), a constant force F is applied for a short time T. Determine the system response.

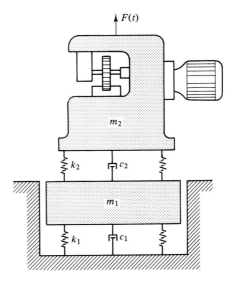

Figure E8.4

Solution We will determine first the differential equations of motion, find the natural frequencies and modes, and then use the convolution integral with the modal equations.

Application of Newton's law in the vertical direction yields

$$m_1\ddot{x}_1 + (c_1 + c_2)\dot{x}_1 - c_2\dot{x}_2 + (k_1 + k_2)x_1 - k_2x_2 = 0$$

$$m_2\ddot{x}_2 - c_2\dot{x}_1 + c_2\dot{x}_2 - k_2x_1 + k_2x_2 = F(t)$$

Denoting

$$\mathbf{M} = \begin{bmatrix} m_1 & 0 \\ 0 & m_2 \end{bmatrix}, \quad \mathbf{C} = \begin{bmatrix} c_1 + c_2 & -c_2 \\ -c_2 & c_2 \end{bmatrix}, \quad \mathbf{K} = \begin{bmatrix} k_1 + k_2 & -k_2 \\ -k_2 & k_2 \end{bmatrix},$$

$$\mathbf{x} = \{x_1 \quad x_2\}, \quad \mathbf{f} = \{0 \quad F(t)\}$$

$$\mathbf{M}\ddot{\mathbf{x}} + \mathbf{C}\dot{\mathbf{x}} + \mathbf{K}\mathbf{x} = \mathbf{f}$$

To solve the natural vibration problem, let $\mathbf{x} = \mathbf{X}e^{pt}$,

$$[p^2m_1 + (c_1 + c_2)p + (k_1 + k_2)]X_1 - (c_2p + k_2)X_2 = 0$$

$$-(c_2p + k_2)X_1 + (p^2m_2 + c_2p + k_2)X_2 = 0$$

The frequency equation is

$$\begin{vmatrix} p^2m_1 + (c_1 + c_2)\,p + (k_1 + k_2)] & -(c_2p + k_2) \\ -c_2p - k_2 & p^2m_2 + c_2p + k_2 \end{vmatrix} = 0$$

$$[p^2m_1 + (c_1 + c_2)p + (k_1 + k_2)](p^2m_2 + c_2p + k_2) - (c_2p + k_2)^2 = 0$$

This equation has four complex conjugate roots, p_1, p_2, p_3, and p_4, computed numerically, in general. The eigenvectors will be, for $X_1 = 1$,

$$Z_1^{(i)} = 1$$

$$Z_2^{(i)} = \frac{-c_2p_i - k_2}{p_i^2m_2 + c_2p_i + k_2}, \quad i = 1, 2, 3, 4$$

For the definition of the modal matrix $\mathbf{Z}$,

$$Z_3^{(i)} = p_i,$$

$$Z_4^{(i)} = \frac{p_i(-c_2p_i - k_2)}{p_i^2m_2 + c_2p_i + k_2}$$

$$\mathbf{Z}^* = [\mathbf{Z}^{(1)} \quad \mathbf{Z}^{(2)} \quad \mathbf{Z}^{(3)} \quad \mathbf{Z}^{(4)}]$$

To apply equation (8.18), we need $\mathbf{G} = \mathbf{Z}^*\mathbf{E}\mathbf{Z}^{*-1}$ where $\mathbf{E} = \text{diag}[e^{p_1t} \quad e^{p_2t} \quad e^{p_3t} \quad e^{p_4t}]$ and $\mathbf{M}^{-1} = \text{diag}[1/m_1 \quad 1/m_2]$. Then

$$\mathbf{x}(t) = \mathbf{M}^{-1}\int_0^t \mathbf{G}(t - t^*)\mathbf{f}(t^*)\,dt^*$$

$$= \mathbf{M}^{-1}\mathbf{f}\int_0^T \mathbf{G}(t - t^*)\,dt^*$$

$$= \mathbf{M}^{-1}\mathbf{f}\int_0^T \mathbf{Z}^*\mathbf{E}(t - t^*)\mathbf{Z}^{*-1}\,dt^*$$

$$= \mathbf{M}^{-1}\mathbf{f}\mathbf{Z}^*\left[\int_0^T \mathbf{E}(t - t^*)\,dt^*\right]\mathbf{Z}^{*-1}$$

Because $\int_0^T e^{p_i(t-t^*)}\,dt^* = e^{p_i t}(1 - e^{p_i T})/p_i,$

$$\mathbf{x}(t) = \mathbf{M}^{-1}\mathbf{f}\mathbf{Z}^* \operatorname{diag}\left[\frac{e^{p_i t}(1 - e^{p_i T})}{p_i}\right]\mathbf{Z}^{*-1}$$

(In this example, an analytical solution is developed for the response of the two-degree-of-freedom system to general excitation in terms of the natural modes.)

$\square$

8.4 GENERAL EXCITATION: MODAL ANALYSIS

The method of the previous section is applicable to small systems. For large ones, the computation effort is prohibitive and the numerical accuracy not always sufficient. Because in many large systems the excitation has only low frequency components, it seems natural to use the modal analysis method with only a small number of the lower modes. In Chapter 7 the coupled system of differential equations of an undamped vibrating system was decoupled using the modal matrix

$$\mathbf{\Phi} = [\mathbf{\phi}^{(1)} \quad \mathbf{\phi}^{(2)} \quad \cdots \quad \mathbf{\phi}^{(n)}] \tag{8.19}$$

and the orthogonality relations

$$\mathbf{\Phi}^T\mathbf{M}\mathbf{\Phi} = \mathbf{I} = \operatorname{diag}[1 \quad 1 \quad \cdots \quad 1] \tag{8.20}$$
$$\mathbf{\Phi}^T\mathbf{K}\mathbf{\Phi} = \operatorname{diag}[\omega_1^2, \omega_2^2, \ldots, \omega_n^2]$$

with the linear transformation

$$\mathbf{x} = \mathbf{\Phi}\mathbf{q} \tag{8.21}$$

and the inverse transformation

$$\mathbf{q} = \mathbf{\Phi}^{-1}\mathbf{x} = \mathbf{\Phi}^T\mathbf{M}\mathbf{x} \tag{8.22}$$

where $\mathbf{q} = \{q_1 \quad q_2 \quad \cdots \quad q_n\}$ is a new set of coordinates, called *normal coordinates*, in terms of which the system coupling can be removed.

A damped linear system can be expressed, in general, by the system of equations

$$\mathbf{M}\ddot{\mathbf{x}} + \mathbf{C}\dot{\mathbf{x}} + \mathbf{K}\mathbf{x} = \mathbf{f}(t) \tag{8.23}$$

Equation (8.23), multiplied from the left by $\mathbf{\Phi}^T$ and using (8.20) and (8.21), yields

$$\ddot{q}_i + 2\zeta_i\omega_i\dot{q}_i + \omega_i^2 q_i = \mathbf{\phi}_i^T\mathbf{f}(t), \qquad i = 1, 2, \ldots, n \tag{8.24}$$

or

$$\ddot{q}_1 + 2\zeta_1\omega_1\dot{q}_1 + \omega_1^2 q_1 = \mathbf{\phi}_1^T\mathbf{f}(t)$$
$$\ddot{q}_2 + 2\zeta_2\omega_2\dot{q}_2 + \omega_2^2 q_2 = \mathbf{\phi}_2^T\mathbf{f}(t)$$
$$\vdots \tag{8.25}$$
$$\ddot{q}_n + 2\zeta_n\omega_n\dot{q}_n + \omega_n^2 q_n = \mathbf{\phi}_n^T\mathbf{f}(t)$$

which is a system of decoupled equations, provided that the damping matrix is of such structure that

$$\Phi^T C \Phi = \text{diag}[\mu_1 \quad \mu_2 \quad \cdots \quad \mu_n] \tag{8.26}$$

In this case we define the *modal damping factor* ζ_i with the equation

$$\mu_i = \phi_i^T C \phi_i = 2\zeta_i \omega_i \tag{8.27}$$

where ϕ_i is the normal mode corresponding to ω_i.

Equation (8.27) can be used to compute ζ_i if the damping matrix is known. As we shall see later, this is a quantity that can be measured and, in some cases, tabulated for similar structures.

Equation (8.26) poses a severe restriction. Systems including discrete damping devices, such as bearings in rotating machinery, do not satisfy this equation and have to be dealt with by the direct method. In systems where the only source of damping is distributed structural damping in the stiffness elements, this condition is satisfied in general and the modal matrix can decouple the damping matrix. Rayleigh has shown that if the damping matrix can be set in the form *(Rayleigh damping)*

$$C = a M + b K \tag{8.28}$$

where a and b are any constants, the damping matrix can be decoupled. Indeed, equation (8.26) becomes

$$\Phi^T C \Phi = \Phi^T(a M + b K)\Phi = a\Phi^T M \Phi + b\Phi^T K \Phi = \text{diag}[a + \omega_i^2 b] \tag{8.29}$$

For example, viscoelastic structural damping described using the complex modulus of elasticity concept (see Chapter 3) and the loss factor γ will have an equivalent damping matrix, assuming harmonic excitation at frequency ω,

$$C = \frac{\gamma}{\omega} K, \qquad a = 0, \quad b = \frac{\gamma}{\omega} \tag{8.30}$$

$$\Phi^T C \Phi = \text{diag}\left[\frac{\gamma \omega_i^2}{\omega}\right] \tag{8.31}$$

It is apparent from equation (8.29) that if $a = 0$ (damping proportional to the stiffness matrix) the modal damping ratio $\zeta_i = (\Phi_i^T C \Phi_i / 2\omega_i)\omega_i^2$ increases with frequency; thus it is higher at high modes. If $b = 0$ (damping proportional to the mass matrix) the modal damping ratio $\zeta_i = \Phi_i^T C \Phi_i / 2\omega_i = \gamma \omega_i / \omega$ decreases with frequency; thus it is smaller at higher modes.

It must be pointed out that equation (8.26) is only a sufficient condition. There are other types of damping for which the damping matrix can be decoupled.[5]

Equations (8.25) describe the system in terms of the normal coordinates rather than the physical coordinates initially selected. It is apparent that there is always a set of coordinates which yields an uncoupled system of differential equations, provided that the damping matrix is of a special form that can be decoupled.

---

[5]Caugey, T. K. 1960. Classical normal modes in damped linear systems. *J. Appl. Mech., 27*: 269–271.

Each of the decoupled equations (8.25) can be solved separately. The linear transformation (7.86) applied on equations (7.55) does not change the eigenvalues associated with matrices $\mathbf{M}$ and $\mathbf{K}$ since matrix $\boldsymbol{\Phi}$ is constant, and due to the property $\det(\mathbf{AB}) = \det(\mathbf{A})\det(\mathbf{B})$, the frequency equation will be simply multiplied by a constant, rendering its roots invariant upon the linear transformation.

Of course, there is no sense in solving the uncoupled equations for the determination of the natural frequencies because it was already done in order to arrive at equations (7.89). However, this transformation is useful for a general method of forced vibration analysis, called *modal analysis*. Indeed, since each of the equations (8.25) is a mathematical description of the one-degree-of-freedom system, the general response in normal coordinates from each is,

$$q_i(t) = e^{-\zeta_i\omega_i t}\left\{ q_i(0)\cos\omega_{di}t + \left[2\zeta_i\omega_i q_i(0) + \frac{\dot{q}_i(0)}{\omega_{di}} \right]\sin\omega_{di}t \right\}$$

$$+ \int_0^t F_i(t^*)g_i(t - t^*)\,dt^* \qquad (8.32)$$

where

$$\omega_{di} = \omega_i^2(1 - \zeta_i^2)^{1/2}$$

$$g_i(t) = \frac{1}{\omega_{di}}e^{-\zeta_i\omega_i t}\sin\omega_{di}t$$

$$F_i(t) = \boldsymbol{\phi}_i^T f(t)$$

Further, $q_i(0)$ and $\dot{q}_i(0)$ are the initial values of the normal coordinates, which can be computed from the initial conditions using the transformation (8.22):

$$\mathbf{q}(0) = \boldsymbol{\Phi}^T\mathbf{M}\mathbf{x}(0) \qquad (8.33)$$

$$\dot{\mathbf{q}}(0) = \boldsymbol{\Phi}^T\mathbf{M}\dot{\mathbf{x}}(0) \qquad (8.34)$$

The observation is made that if the damping is proportional to the stiffness matrix, the modal damping ratio $\zeta_i = (\boldsymbol{\Phi}_i^T\mathbf{C}\boldsymbol{\Phi}_i/2)\omega_i$ increases with frequency; thus it is higher at high modes, which is very important. Indeed, since the proportional damping is usually due to different mechanisms of structural damping, thus proportional to the structural stiffness, the higher modes are highly damped (in proportion to ω) and thus there amplitude is diminished due to high damping. Therefore, one does not make a severe error by omitting the very high nodes and instead uses a small number of modes corresponding to the lower natural frequencies. This procedure is called *truncation,* in direct analogy to the truncation performed on the Fourier series.

There is no general rule for the number of modes necessary to capture the dynamic response of the system with sufficient fidelity. It takes experience and good knowledge of the forcing function and the system to determine this. For example, if the spectrum of the exciting force is limited to a band of frequencies, the modes corresponding to natural frequencies much higher (say, two to four times) than the higher frequency in the spectrum of the exciting force can usually be neglected.

The response of an undamped system is given by equation (8.32) with zero modal damping:

$$q_i(t) = q_i(0) \cos \omega_{ni} t + \frac{\dot{q}_i(0)}{\omega_{ni}} \sin \omega_{ni} t + \int_0^t F_i(t^*) g_i(t - t^*) \, dt^* \qquad (8.35)$$

where

$$g_i(t) = \frac{1}{\omega_{ni}} \sin \omega_{ni} t$$

$$F_i(t) = \phi_i^T f(t)$$

☐ **Example 8.5   Dynamic Response of a Motor Packaging**

The packaging of two electric motors in a container is modeled as shown in Figure E8.5. Determine the response to a sudden stop when the container moves on a conveyor belt with velocity $v_0 = 0.5$ m/s if $k = 16{,}000$ N/m , $m = 80$ kg, and there are maximum accelerations on the motor.

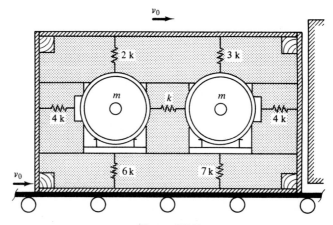

Figure E8.5

Solution   First we obtain the equations of motion, then we determine the natural frequencies and modes. Finally, the dynamic response is computed in terms of the natural coordinates.

Using Newton's law, the equations of motion are

$$m\ddot{x}_1 + 5kx_1 - kx_2 = 0$$

$$m\ddot{x}_2 + 5kx_2 - kx_1 = 0$$

$$m\ddot{x}_3 + 8kx_3 = 0$$

$$m\ddot{x}_4 + 10kx_4 = 0$$

In matrix form,

$$\mathbf{M\ddot{x}} + \mathbf{Kx} = [0]$$

where

$$\mathbf{M} = \text{diag}[1 \quad 1 \quad 1 \quad 1]m$$

$$\mathbf{K} = k \begin{bmatrix} 5 & -1 & 0 & 0 \\ -1 & 5 & 0 & 0 \\ 0 & 0 & 8 & 0 \\ 0 & 0 & 0 & 10 \end{bmatrix}$$

The frequency determinant is

$$|-\omega^2\mathbf{M} + \mathbf{K}| = \begin{vmatrix} 5 - \dfrac{m\omega^2}{k} & -1 & 0 & 0 \\ -1 & 5 - \dfrac{m\omega^2}{k} & 0 & 0 \\ 0 & 0 & 8 - \dfrac{m\omega^2}{k} & 0 \\ 0 & 0 & 0 & 10 - \dfrac{m\omega^2}{k} \end{vmatrix} \left(\dfrac{k}{m}\right)^4 = 0$$

$$\left[\left(5 - \dfrac{m\omega^2}{k}\right)\left(5 - \dfrac{m\omega^2}{k}\right) - 1\right]\left(8 - \dfrac{m\omega^2}{k}\right)\left(10 - \dfrac{m\omega^2}{k}\right)\left(\dfrac{k}{m}\right)^4 = 0$$

The natural frequencies are

$$\omega_1 = \left(\dfrac{4k}{m}\right)^{1/2}, \qquad \omega_2 = \left(\dfrac{6k}{m}\right)^{1/2}, \qquad \omega_3 = \left(\dfrac{8k}{m}\right)^{1/2}, \qquad \omega_4 = \left(\dfrac{10k}{m}\right)^{1/2}$$

The vibration modes are solutions of

$$\begin{bmatrix} 5 - \dfrac{m\omega^2}{k} & -1 & 0 & 0 \\ -1 & 5 - \dfrac{m\omega^2}{k} & 0 & 0 \\ 0 & 0 & 8 - \dfrac{m\omega^2}{k} & 0 \\ 0 & 0 & 0 & 10 - \dfrac{m\omega^2}{k} \end{bmatrix} \left(\dfrac{k}{m}\right)^4 \begin{bmatrix} A_1 \\ A_2 \\ A_3 \\ A_4 \end{bmatrix} = [0]$$

with one of the A_i's arbitrary. It is worth noting that the third and fourth equations are already uncoupled, due to the selected coordinates x_3 and x_4, which happen to be normal. Therefore, the natural modes are

$$\omega = \omega_1, \quad A_1 = 1, \quad A_2 = 5 - \dfrac{m\omega^2}{k} = 1, \quad A_3 = 0, \quad A_4 = 0$$

$$\omega = \omega_2, \quad A_1 = 1, \quad A_2 = 5 - \dfrac{m\omega^2}{k} = -1, \quad A_3 = 0, \quad A_4 = 0$$

$$\omega = \omega_3, \quad A_1 = 0, \quad A_2 = 0, \qquad\qquad A_3 = 1, \quad A_4 = 0$$

$$\omega = \omega_4, \quad A_1 = 0, \quad A_2 = 0, \qquad\qquad A_3 = 0, \quad A_4 = 1$$

and the modal matrix is

$$\mathbf{Z} = \begin{bmatrix} 1 & 1 & 0 & 0 \\ 1 & -1 & 0 & 0 \\ 0 & 0 & 1 & 0 \\ 0 & 0 & 0 & 1 \end{bmatrix}$$

To normalize it, we form the product

$$\mathbf{Z}^T\mathbf{MZ} = m \operatorname{diag}(1 \ \ 1 \ \ 1 \ \ 1)$$

Therefore, the normalized modal matrix is

$$\boldsymbol{\Phi} = \begin{bmatrix} 1 & 1 & 0 & 0 \\ 1 & -1 & 0 & 0 \\ 0 & 0 & 1 & 0 \\ 0 & 0 & 0 & 1 \end{bmatrix} m^{-1/2}$$

The principal coordinates are given by the transformation

$$\mathbf{x} = \begin{bmatrix} 1 & 1 & 0 & 0 \\ 1 & -1 & 0 & 0 \\ 0 & 0 & 1 & 0 \\ 0 & 0 & 0 & 1 \end{bmatrix} m^{-1/2}\mathbf{q}$$

where $\mathbf{q} = \{q_1 \ \ q_2 \ \ q_3 \ \ q_4\}$, the vector of the normal coordinates. The solution for $\mathbf{q}$ is [equation (8.22)]

$$\mathbf{q} = \boldsymbol{\Phi}^{-1}\mathbf{x} = \boldsymbol{\Phi}^T\mathbf{Mx} = \begin{bmatrix} 1 & 1 & 0 & 0 \\ 1 & -1 & 0 & 0 \\ 0 & 0 & 1 & 0 \\ 0 & 0 & 0 & 1 \end{bmatrix} m^{1/2}\mathbf{x}$$

or

$$q_1 = (x_1 + x_2)m^{1/2}, \qquad q_2 = (x_1 - x_2)m^{1/2}, \qquad q_3 = x_3 m^{1/2}, \qquad q_4 = x_4 m^{1/2}$$

The initial conditions are

$$\dot{q}_1(0) = [\dot{x}_1(0) + \dot{x}_2(0)]m^{1/2} = 2v_0 m^{1/2}$$

$$\dot{q}_2(0) = [\dot{x}_1(0) - \dot{x}_2(0)]m^{1/2} = 0$$

$$\dot{q}_3(0) = 0$$

$$\dot{q}_4(0) = 0$$

Therefore,

$$q_1 = \frac{2v_0 m^{1/2}}{\omega_1} \sin \omega_1 t, \qquad q_2 = 0, \qquad q_3 = 0, \qquad q_4 = 0$$

and the system dynamic response will be

$$x_1 = (q_1 + q_2)m^{-1/2} = v_0\left(\frac{m}{k}\right)^{1/2} \sin\left(\frac{4k}{m}\right)^{1/2} t$$

$$x_2 = (q_1 - q_2)m^{-1/2} = v_0\left(\frac{m}{k}\right)^{1/2} \sin\left(\frac{4k}{m}\right)^{1/2} t$$

$$x_3 = 0, \qquad x_4 = 0$$

The maximum amplitudes are

$$X_1 = X_2 = 0.5 \times \left(\frac{80}{16,000}\right)^{1/2} = 0.035 \text{ m}$$

The maximum accelerations a_i will be

$$a_1 = 4v_0\left(\frac{k}{m}\right)^{1/2} = 4 \times 0.5 \times \left(\frac{16,000}{80}\right)^{1/2} = 28.3 \text{ m/s}^2$$

$$a_2 = 4v_0\left(\frac{k}{m}\right)^{1/2} = 4 \times 0.5 \times \left(\frac{16,000}{80}\right)^{1/2} = 28.3 \text{ m/s}^2$$

$$a_3 = 0, \qquad a_4 = 0.$$

(In this example, a packaging design problem is presented, the response of a motors packaging to an impact.)

□

□ **Example 8.6   Dynamic Response of a Packaged Motor with Damping**

The packaging material of the system in Example 8.5 has viscoelastic damping with loss factor $\gamma = 0.01$. Determine the response under the same conditions as in Example 8.5.

Solution   Equation (8.32) will now be applied with damping factor $\zeta = \gamma\omega_i/2$ [from equation (8.35)]:

$$q_1(t) = e^{-\zeta_1\omega_1 t}\left[\frac{q_1(0)}{\omega_{d1}}\right]\sin \omega_{d1} t$$

$$= e^{-\gamma(4k/m)t/2}\frac{2v_0 m}{2(k)^{1/2}}\sin\left[\frac{4k}{m}(1 - \gamma^2)\right]^{1/2} t$$

$$q_2 = 0, \qquad q_3 = 0, \qquad q_4 = 0$$

and the system dynamic response will be, for $\omega_n = 4k/m$,

$$x_1 = (q_1 + q_2)m^{-1/2} = e^{-\gamma\omega_n t/2}\left[v_0\left(\frac{m}{k}\right)^{1/2}\right]\sin\left[\frac{4k}{m}(1 - \gamma^2)\right]^{1/2} t$$

$$x_2 = (q_1 - q_2)m^{-1/2} = e^{-\gamma\omega_n t/2}\left[v_0\left(\frac{m}{k}\right)^{1/2}\right]\sin\left[\frac{4k}{m}(1 - \gamma^2)\right]^{1/2} t$$

$$x_3 = 0, \qquad x_4 = 0$$

For the numerical values given,

$$\left[\frac{4k}{m}(1 - \gamma^2)\right]^{1/2} = \left[\left(4 \times \frac{16,000}{80}\right)(1 - 0.01^2)\right]^{1/2} = 28.3 \text{ rad/s}$$

$$v_0\left(\frac{m}{k}\right)^{1/2} = 0.035$$

$$\frac{\gamma\omega_n}{2} = \frac{\gamma(4k/m)^{1/2}}{2} = \frac{0.1(4 \times 16,000/80)^{1/2}}{2} = 0.141$$

Therefore,

$$x_1 = 0.035e^{-0.141t}\sin(28.3t)$$

$$x_2 = 0.035e^{-0.141t}\sin(28.3t)$$

$$x_3 = x_4 = 0$$

(In this example, a packaging design problem is presented, the response of the damped packaging of a motor to a impact. Damping did not reduce the vibration amplitude due to the impact.)

□

8.5 RANDOM EXCITATION

Random Processes

In recent times, studies of vibration and high cycle fatigue in machine elements and aircraft structures and earthquake analyses of structures involved excitations that have no repeated pattern, for a variety of reasons, and thus they could not be given an analytical expression of their magnitude as function of time. Such excitations are called *random* [Clough and Penzien, 1975]. Several records of some of these excitations showed, however, that certain statistical characteristics were common, lending themselves to analytical treatment of the problem of finding the dynamic response.

Some examples of mechanisms that generate random excitations on mechanical systems are shown in Figure 8.2. Figure 8.2a shows the turbulent flow past a cylinder at high velocity. The fluid vortices, first noticed by Leonardo da Vinci, are random in location and intensity, rendering the pressure on the cylinder in that region random. Figure 8.2b shows wind flow past a tall building. The direction and velocity of the wind vary in a random way. The result is a random force on the building. Figure 8.2c shows a structure affected by an earthquake. An earth fault *A* under a relative sliding motion creates random vibration, of the sort produced if one tries to write on the blackboard holding the chalk perpendicular to the blackboard. Figure 8.2d shows a car going over a dirt road. The surface of the road and the size of stones over which the car will pass are highly irregular, leading to a vertical random loading on the car due to base excitation.

The majority of random signals of interest in vibration analysis will be in the form of continuous time histories $x(t)$ which represent the forcing function or the

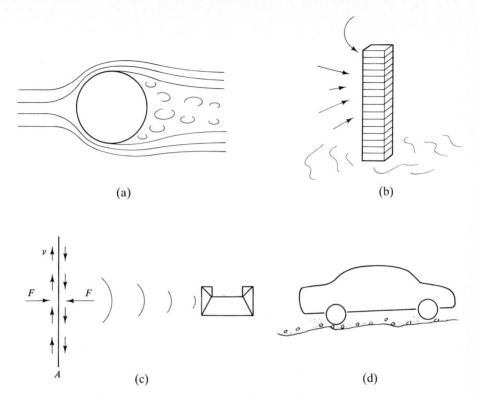

(a)

(b)

v

F F

A

(c)

(d)

Figure 8.2 Random excitation mechanisms for engineering systems.

response of some part of the mechanical system and will be related directly or indirectly to a mechanical process. In general, the signals will be corrupted to some extent by noise, and the component of interest may be either an impulse, a single or repetitive transient, or a continuous harmonic or random signal. The analysis method is therefore generally based on the procedures developed for continuous random time histories.

We suppose that a function $x(t)$, say the vibration amplitude at the body of a race car going over a rough road at time t, was measured a great number of times N and of these times, $n(x_k)$ times the measured values were between x_k and $x_k + \Delta x_k$ (Figure 8.3a). We define as frequency of x_k the ratio

$$h(x_k) = \frac{n(x_k)}{N} \tag{8.36}$$

Alternatively, if the time difference Δt_i is observed every time the function has value x_k and changes by Δx_k (Figure 8.3b), we can define

$$h(x_k) = \sum \frac{\Delta t_i}{T} \tag{8.37}$$

where T is the total length of the time record.

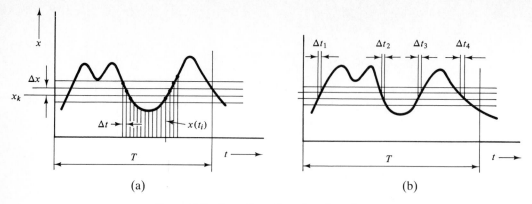

Figure 8.3 Sampling of random functions.

The limit of the function $h(x_k)$ for $\Delta x_k \to 0$ is defined as the *probability density function* $p(x_k)$. It is apparent that the *probability* (the ratio of the number of values having the property to the total number of measured values, the latter assumed infinitely large) of the function x being (Figure 8.4) below or above x, or between x_1 and x_2, is as follows:

$$\text{below } x: \qquad P^x_{-\infty} = \int_{-\infty}^{x} p(x)\, dx \qquad (8.38)$$

$$\text{above } x: \qquad P^{\infty}_{x} = \int_{x}^{\infty} p(x)\, dx \qquad (8.39)$$

$$\text{between } x_1 \text{ and } x_2: \qquad P^{x_2}_{x_1} = \int_{x_1}^{x_2} p(x)\, dx \qquad (8.40)$$

The identities

$$P^{\infty}_{-\infty} = 1$$
$$P^{x}_{-\infty} + P^{\infty}_{x} = 1 \qquad (8.41)$$

are obvious.

In Chapter 1 the linear and quadratic averages of periodic functions have been defined. For a nonperiodic function, the period $T \to \infty$ and we can define the *temporal averages*

$$\overline{x(t)} = \lim_{T \to \infty} \frac{1}{T} \int_{0}^{t} x(t)\, dt \qquad (8.42)$$

$$\overline{x^2(t)} = \lim_{T \to \infty} \frac{1}{T} \int_{0}^{t} x^2(t)\, dt \qquad (8.43)$$

$\overline{x(t)}$ is the mean value and $\overline{x^2(t)}$ is the *mean-square value* of the variable $x(t)$.

For a random variable x with probability density function $p(x)$, we define the mean and the mean-square values as

$$\overline{x} = E[x] = \int_{-\infty}^{\infty} x p(x)\, dx \qquad (8.44)$$

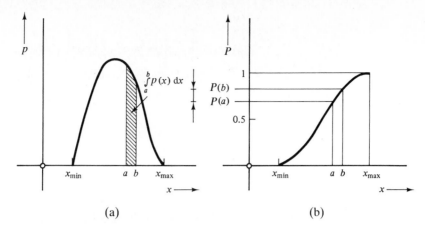

Figure 8.4 Probability density and probability distribution functions.

$$\overline{x^2} = E[x^2] = \int_{-\infty}^{\infty} x^2 p(x)\,dx \tag{8.45}$$

where, in general, the mathematical expectation of the function $f(x)$ is defined as $E[f(x)] = \overline{f(x)} = \int_{-\infty}^{\infty} f(x)p(x)\,dx$.

In the case of a vibration signal, the mean gives a measure of the static component of the signal. To measure the severity of the vibration, we must find a way to sum the deviation of the vibration signal from the mean, which is accomplished by summing the square of the deviation from the mean value

$$\overline{\sigma^2} = \int_{-\infty}^{\infty} (x - \overline{x})^2 p(x)\,dx = \overline{x^2} - \overline{x}^2 \tag{8.46}$$

called *variance*. The square root σ of the variance is called *standard deviation* of the process. This quantity, the rms value of $x(t)$, is generally used as an amplitude measure. In practice, in equations (8.42) and (8.43), the integration time must be long compared with the period of the lowest-frequency component in the signal.

Mean and standard deviation are only two parameters describing a process qualitatively. To have a full representation of the process, we need a great number of measured values. However, a great deal of information about certain class of random processes can be derived if only these two parameters are known.

Let us suppose that we observe the record of the vibration of the body of a race car going over a rough road. If certain statistical characteristics of several parts of the record of durations $t_1, t_2, \ldots, t_n$ (e.g., of a long run or several runs over the same road) have the same statistical value (Figure 8.5), we call the measured process *stationary*. If, in addition, several records of a family (or an *ensemble*) of similar measurements $x_1(t), x_2(t), \ldots, x_n(t)$ (e.g., on a group of racing cars) have the same statistical characteristics, we call the process *ergodic*. The vibration processes that we will study in the sequel will be assumed (although not always necessary) stationary and ergodic.

A sine wave is one example of a stationary signal. Many random phenomena, such as wind and wave loading on structures, are not strictly stationary but can

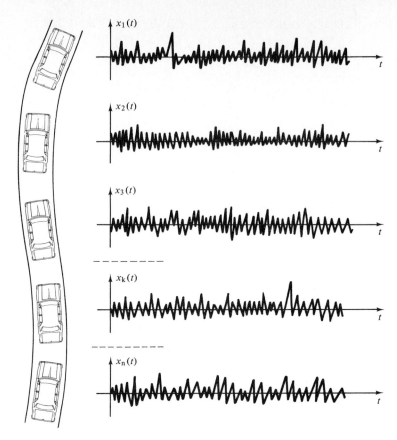

Figure 8.5 Stationary processes.

often be considered so for relatively short periods of time. Some of the results of the analysis of stationary signals can be applied to nonstationary or transient data.

We shall now consider the types of description that can be applied to a stationary signal. It is convenient to consider the general case of stationary random data and then show that harmonic signals can be represented as a special case.

It is also convenient to separate the steady or mean value from the signal and concentrate on the fluctuating part. The steady component can be treated by the standard static analyses. Thus the initial step in any vibration data-processing procedure is to determine the mean value and then separate out this "dc" component before subsequent processing.

The description of the random process with the mean value and standard deviation [equations (8.44) and (8.46)] contains amplitude information only and thus is known as *amplitude description*. It gives no information for the way the process changes in time. For example, two harmonic motions with the same maximum amplitude have the same probability density function, mean, and standard deviation, regardless of their frequency.

The rotor of a blower consists of three blades (Figure E8.7a), and it is commissioned in service while it is well balanced by the selection of equal-mass blades. At startup, one of the blades was damaged and was replaced by a new one. The blades are rejected during quality control if the have mass below m_1 or above m_2. In between, the quality control records show that the mass has uniform probability density: that is, $p(m) = 0$ for $m < m_1$ or $m > m_2$ and $p(m) = 1/(m_2 - m_1)$ for $m_1 < m < m_2$ (Figure E8.7b). If the radius R of the mass center and the angular velocity ω are constant, determine (a) the mean value and the standard deviation of the centrifugal force on the shaft due to the unbalanced mass of the blade, assuming that the broken blade had the mean mass; and (b) the probability that the unbalance force exceeds F_0.

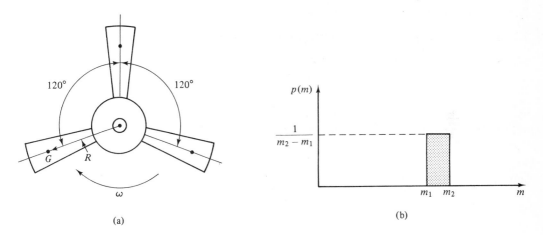

(a)

(b)

Figure E8.7

Solution   We will first find the mean value and standard deviation of the unbalance and then the probability.

(a) The unbalanced centrifugal force will be $F = \omega^2 Rm$. The mean value will be

$$\overline{F} = \int_{-\infty}^{\infty} Fp(F)\, dF = \omega^2 R \int_{m1}^{m2} m\, \frac{1}{m_2 - m_1}\, dm = \frac{\omega^2 R(m_1 + m_2)}{2}$$

The mean square will be

$$\overline{F^2} = \int_{-\infty}^{\infty} F^2 p(F)\, dF = (\omega^2 R)^3 \int_{m_1}^{m_2} \left(m - \frac{m_1}{2} - \frac{m_2}{2}\right)^2 \frac{1}{m_2 - m_2}\, dm$$

$$= (\omega^2 R)^3 \frac{(m_2/2 - m_1/2)^3/3 + (m_2/2 - m_1/2)^3/3}{m_2 - m_1}$$

$$= \frac{(\omega^2 R)^3 (m_2 - m_1)^2}{12}$$

The standard deviation will be $\sigma_F = [(\omega^2 R)^3 (m_2 - m_1)/12]^{1/2}$.

(b) The probability of the excess force F_0 is $P^{\infty}_{F_0} = \int_{F_0}^{\infty} p(F)\, dF$. The unbalanced force F_0 corresponds to unbalanced mass $m_0 - (m_1 + m_2)/2 = F_0/[\omega^2 R(m_1 + m_2)/2]$.

Portrait of Gauss. (Library of Congress.)

CARL FRIEDRICH GAUSS (1777–1855)

One of the greatest mathematicians of all times, Gauss was the son of a poor Brunswick laborer. He discovered an error in his father's books at the age of 3. Became famous since high school and with the help of the Duke of Brunswick, he entered Göttingen in 1795 and received his doctorate from Helmdtaedt in 1799, at 22, for developing the fundamental theorem of algebra. Two years later he published his famous *Disquisitiones arithmeticae*. In 1807 he was appointed Director of the Göttingen Observatory. He has developed many original ideas which he left unpublished, such as the elliptic integral. He enjoyed helping a few devotees but he disliked classroom teaching. His famous assertion is: *Mathematics is the queen of sciences and theory of numbers is the queen of mathematics*.

Assuming that $m_1 < m_0 < m_2$ (otherwise, the probability is 0 or 1), the probability that the unbalanced force exceeds F_0 will be

$$P_{F_0}^{\infty} = \int_{F_0}^{\infty} p(F) \, dF = \int_{m_0}^{m_1} p(m) \, dm = \frac{m_2 - m_0}{m_2 - m_1}$$

$$= 0.5 + \frac{F_0}{\omega^2 R(m_1 + m_2)/2}$$

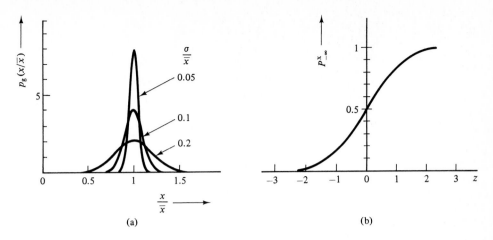

Figure 8.6 Normal distribution.

(This example demonstrates the unbalance response of a blower rotor if the blades are assembled in random and the unbalance is thus random.)

□

The Normal Distribution

Many physical phenomena, including vibration,[6] are known to have a probability density function,

$$p_g(x) = \frac{1}{\sigma(2\pi)^{1/2}} e^{(x-\bar{x})^2/2\sigma^2} \tag{8.47}$$

This is called *normal* or *Gaussian distribution*, and σ is the standard deviation [equation (8.46)]. It has the shape of an inverted bell the sharpness of which is expressed by σ. Figure 8.6a shows the Gaussian distribution for several values of σ. The probability function for this distribution will be

$$P^x_{-\infty}(x) = \frac{1}{\sigma(2\pi)^{1/2}} \int_{-\infty}^{x} e^{(x-\bar{x})^2/2\sigma^2}\, dx \tag{8.48}$$

and it is plotted in Figure 8.6b as a function of $z = (x - \bar{x})/\sigma$.

□ **Example 8.8   Random Unbalance of Motor Rotors**

Analysis of balance records for a class of small motors revealed that the distribution of the unbalanced mass is Gaussian with $\bar{m} = 20$ g and $\sigma_m = 10$ g and occurs at the slotted area, at fixed radius. The tolerable unbalance is 30 g. Determine the percentage of motors that will require balancing.

Solution   We will first determine the value of the normal distribution parameter z and then from tables we will find the probability.

$$z = \frac{m - \bar{m}}{\sigma} = \frac{30 - 20}{10} = 1$$

[6]Lord Rayleigh, 1880. *Philos. Mag.*, Aug.

From Figure 8.6b, $P_z^\infty = 0.1587$. Therefore, the percentage of motors requiring balancing is 15.87%.

(This example demonstrates the determination of the number of motor units in production requiring balancing if the distribution of the unbalance is normal and the statistical parameters in production are known.)

$\square$

The Rayleigh Distribution

The Gaussian distribution admits values of the random variable x in the range $-\infty < x < \infty$. Many physical processes, however, do not admit negative values of the random variable x. Many such processes can be described by the probability density function

$$p_r(x) = \frac{x}{\alpha^2} e^{-(x/\alpha)^2/2} \tag{8.49}$$

This is called *Rayleigh distribution*, and α is a distribution parameter. The mean is

$$\bar{x}_r = \int_0^\infty \left(\frac{x}{\alpha}\right)^2 e^{-(x/\alpha)^2/2} \, dx = 1.253\alpha \tag{8.50}$$

The variance is

$$\bar{x}_r^2 = \alpha \int_0^\infty \left(\frac{x}{\alpha}\right)^3 e^{-(x/\alpha)^2/2} \, dx = 2\alpha^2 \tag{8.51}$$

Figure 8.7a shows the Rayleigh distribution for several values of α. The probability function for this distribution will be

$$P_{r0}^x = \int_0^\infty \frac{x}{\alpha^2} e^{-(x/\alpha)^2/2} \, dx \tag{8.52}$$

and it is plotted in Figure 8.7b as a function of $u = x/\alpha$.

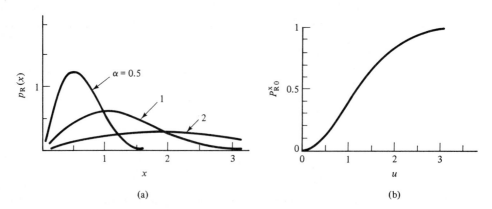

Figure 8.7 Rayleigh distribution.

☐ **Example 8.9 Statistical Unbalance of Pump Rotors**

Analysis of balance records for a class of centrifugal pumps revealed that the distribution of the unbalanced mass follows the Rayleigh distribution with $\alpha = 20$ g. The tolerable unbalance is 30 g. Determine the percentage of pumps that will require balancing, assuming that unbalance occurs at a fixed radius.

Solution We will first determine the value of the Rayleigh distribution parameter u and then from tables we will find the probability:

$$u = \frac{x}{\alpha} = \frac{30}{20} = 1.5$$

From Figure 8.7b, $P^*_{1.5} = 1 - 0.672 = 0.328$. Therefore, the percentage of pump rotors requiring balancing is 32.8%.

(This example demonstrates the determination of the number of pump rotor units in production requiring balancing if the distribution of the unbalance is Rayleigh and the statistical parameters in production are known.)

☐

Time-Domain Description

As stated above, amplitude description cannot give information about the waveform of a vibration signal itself, such as the vibration period. A mathematically rigorous analysis can be built up from a study of the time history of the waveform. By comparing the value $x(t)$ of the signal with its value $x(t + \tau)$ a short time τ later, we can derive a function that indicates the typical periods present in the waveform. We take the two values mentioned above, multiply them together, and then build up a sum of such products each taken at different reference points t sequentially along the record. We finally set up the *autocorrelation function,* which is defined as

$$R(\tau) = \lim_{T \to \infty} \frac{1}{T} \int_0^T x(t)x(t + \tau)\, dt \tag{8.53}$$

In practice, of course, we have a finite signal length T and so can only produce an estimate of $R(\tau)$. Intuitively, we can see that the closeness of this estimate to the true value of $R(\tau)$ will depend on the length of the sample T. Examples of autocorrelation functions (e) of different time functions (a) of interest, to be discussed later, are shown in Figure 8.8.

It is apparent from its definition [equation (8.53)] that the autocorrelation functions has the following properties:

$$R_x(0) = \sigma_x^2 \quad \text{(variance)} \tag{8.54}$$

$$R_x(\tau) = R_x(-\tau) \tag{8.55}$$

$$R_x(\tau) \le R_x(0) \tag{8.56}$$

Equation (8.54) results directly from equation (8.53) for $\tau = 0$. Equation (8.55) is proven for a stationary process by replacing t with $t - \tau$. Equation (8.56) results from the observation that the following mean-square value is positive:

$$\overline{[x(t) - x(t + \tau)]^2} = \lim_{T \to \infty} \frac{1}{T} \int_0^t [x(t) - x(t + \tau)]^2 \, dt$$

$$= \lim_{T \to \infty} \frac{1}{T} \int_0^t [x^2(t) - 2x(t + \tau)x(t) + x^2(t + \tau)]^2 \, dt \qquad (8.57)$$

$$= 2R_x(0) - 2R_x(\tau)$$

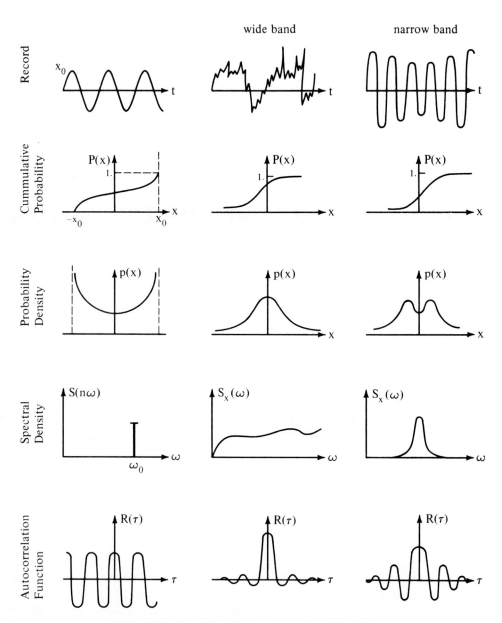

Figure 8.8 Random processes and related functions.

Lumped Mass Systems: Forced Vibration   Chap. 8

☐ **Example 8.10   Autocorrelation Function and Variance of Periodic Functions**

Find the autocorrelation function and the variance of (a) a harmonic, and (b) a periodic vibration record expressed in the form of a Fourier series of period $T = 2\pi/\omega_0$.
Solution   (a) Equation (8.53) gives

$$R_x(\tau) = \lim_{T \to \infty} \frac{1}{T} \int_0^T x(t) x(t + \tau) \, dt$$

$$= \lim_{T \to \infty} \frac{1}{T} \int_0^T A^2 \cos \omega_0 t \cos \omega_0(t + \tau) \, dt$$

$$= \lim_{T \to \infty} \frac{1}{T} \int_0^T 0.5 A^2 [\cos \omega_0(2t + \tau) + \cos \omega_0 \tau] \, dt$$

Because the function $x(t)$ is periodic,

$$R_x(t) = \frac{1}{T} \int_0^T 0.5 A^2 [\cos \omega_0(2t + \tau) + \cos \omega_0 \tau] \, dt = A^2 \cos \omega_0 \tau$$

a harmonic oscillation in τ. The variance $\overline{x^2(t)} = R_x(0) = A^2/2$ [equation (8.54)].
(b) Let $x(t) = A_0 + \sum_{n=1}^{\infty} (A_n \cos n\omega t + B_n \sin n\omega t)$. Then

$$R_x(\tau) = \lim_{T \to \infty} \frac{1}{T} \int_0^T x(t) x(t + \tau) \, dt$$

$$= \lim_{T \to \infty} \frac{1}{T} \int_0^T \left[A_0 + \sum_{n=1}^{\infty} (A_n \cos n\omega t + B_n \sin n\omega t) \right]$$

$$\left[A_0 + \sum_{n=1}^{\infty} (A_n \cos n\omega(t + \tau) + B_n \sin n\omega(t + \tau)) \right] dt$$

The integrals of the products $A_m \cos m\omega t B_n \sin n\omega(t + \tau)$ are zero and the same is true for $A_m \cos m\omega t A_n \cos n\omega(t + \tau)$ or $B_m \sin m\omega t B_n \sin n\omega(t + \tau)$, unless $m = n$. Therefore,

$$R_x(t) = \frac{A_0^2 \sum_{n=1}^{\infty} (A_n^2 \cos n\omega \tau + B_n^2 \sin n\omega \tau)}{2}$$

Note that the series in the autocorrelation function converges much faster (due to the A^2 and B^2 terms) than the Fourier expansion of the measured signal.

(This example demonstrates the analytical computation of the autocorrelation function and variance of harmonic and periodic functions.)

☐

Frequency-Domain Description

From an engineer's point of view many nonharmonic signals appear to have many different frequency components within them. Thus it would seem reasonable to try to express the signal as some sort of assembly of individual frequency components.

In Chapter 1 we have seen that any periodic function $x(t)$ can be expressed in the form of a complex Fourier series

$$x(t) = \sum_{n=-\infty}^{\infty} C_n e^{in\omega_0 t} \tag{8.58}$$

where the complex coefficients C_n are

$$C_n = \frac{1}{T} \int_{-T/2}^{T/2} x(t) e^{-in\omega_0 t} \, dt \tag{8.59}$$

Thus the function $x(t)$, of period T, has been expressed as a sum of an infinite number of harmonics corresponding to $n = 0, 1, 2, \ldots$. The frequency of each harmonic is a multiple of the fundamental $f = 1/T$ and the amplitude is given by (8.59). The distance between two harmonics is $\Delta\omega = 2\pi/T = \omega_0$, the circular frequency of the fundamental harmonic that corresponds to the period T. It is apparent that the larger the period T, the denser the frequency spectrum is. Considering any nonperiodic function $x(t)$ as having an infinite period, the frequency spectrum becomes continuous. Thus $\Delta\omega = 2\pi/T \to d\omega$ and equations (8.58) and (8.59) become

$$x(t) = \frac{1}{2\pi} \int_{-\infty}^{\infty} X(i\omega) e^{i\omega t} \, d\omega \tag{8.60}$$

and

$$X(i\omega) = \int_{-\infty}^{\infty} x(t) e^{-i\omega t} \, dt \tag{8.61}$$

The quantity X is the amplitude at the frequency ω. We see here that there is a component at any frequency and the sum of harmonics became the integral (8.61), a continuous function of the frequency.

The pair (8.58)–(8.59) is called *discrete Fourier transform,* and the pair (8.60)–(8.61) is called *integral Fourier transform.* They are associated with a *discrete frequency spectrum C_n* or a *continuous frequency spectrum $X(i\omega)$.*

It is of importance to know the concentration of harmonics in the frequency spectrum of the record of the vibration of a system. A measure of this concentration is the *spectral density function.* For a discrete spectrum C_n of a periodic function $x(t)$, the discrete spectral density function $G(n\omega_0)$ is defined as

$$G(n\omega_0) = \frac{|C_n|^2}{\Delta\omega} = \frac{C_n \overline{C}_n}{2\Delta\omega} = \frac{T}{4\pi} C_n \overline{C}_n \tag{8.62}$$

Here we observe that integrals of the products $\int e^{in\omega_0 t} e^{in\omega_0 t} \, dt$ vanish for $m \neq n$, as one can easily verify. Therefore, taking the square of the absolute value of both sides of equation (8.58) and integrating, we obtain

$$\overline{x}^2 = \sum_n \frac{C_n \overline{C}_n}{2} = \sum_n G(n\omega_0) \, \Delta\omega \tag{8.63}$$

For $T \to \infty$, we define the continuous spectral density function $G_x(\omega)$ by replacing the summation with integration:

$$\overline{x}^{-2} = \lim_{T \to \infty} \frac{1}{T} \int_{-T/2}^{T/2} |x(t)|^2 \, dt = \int_{-\infty}^{\infty} G_x(\omega) \, d\omega \tag{8.64}$$

and in a manner similar to equation (8.62),

$$G_x(\omega) = \lim_{\Delta\omega \to 0} \frac{|X(i\omega)|^2}{\Delta\omega} \tag{8.65}$$

where $X(i\omega)$ is taken as a mean value over the interval $\Delta\omega$.

Electrical engineering applications have prompted use of the term *power spectral density* function by many authors.

If the spectral density function G is known (for given records can be measured or computed), equation (8.63) or (8.64) gives the mean-square values and thus from equation (8.47) one can find the probability density function if the process is assumed to be Gaussian.

Often, instead of the spectral density, the autocorrelation function $R(\tau)$ is more convenient to measure. Related to it is a two-sided spectral density function defined by the equation

$$R_x(\tau) = \frac{1}{2\pi} \int_{-\infty}^{\infty} S_x(\omega) e^{i\omega t} \, dt \tag{8.66}$$

where $S_x(\omega) = 2\pi G(\omega)$, as one can verify by using the Fourier transform of the function $x(t)$ and $x(t + \tau)$ in the definition of the autocorrelation function. Thus the spectral density function $S_x(\omega)$ is the integral Fourier transform of the autocorrelation function

$$S_x(\omega) = \int_{-\infty}^{\infty} R_x(\tau) e^{-i\omega\tau} \, d\tau \tag{8.67}$$

The reason for using $S(\omega)$ instead of $G(\omega)$ is only numerical convenience, since it is the Fourier transform of the autocorrelation function. The readers must familiarize themselves with the different definitions of the spectral density function $G(\omega)$ and $S(\omega)$ to avoid confusion.

A very common process in engineering applications is *white noise,* having

$$S(\omega) = S_0 = \text{const} \tag{8.68}$$

$$R(\tau) = 2\pi S_0 \delta(\tau) \tag{8.69}$$

where $\delta(\tau)$ ($= 1$ for $\tau = 0$, $= 0$ for $\tau \neq 0$) is the *Dirac function.*

Equation (8.55) implies that the autocorrelation is an even function. From equation (8.67), substituting $-\tau$ for τ, one can see that $S(\omega)$ is also an even function. Further,

$$S_x(\omega) = \int_{-\infty}^{\infty} R_x(\tau) e^{-i\omega\tau} \, d\tau = \int_{-\infty}^{\infty} R_x(\tau)(\cos \omega\tau - i \sin \omega\tau) \, d\tau \tag{8.70}$$

Because $R_x(\tau)$ is an even function, the integral of the sin term is zero. Therefore,

$$S_x(\omega) = 2 \int_{0}^{\infty} R_x(\tau) \cos \omega\tau \, d\tau \tag{8.71}$$

Similarly,

$$R_x(\tau) = \frac{1}{\pi} \int_0^\infty S_x(\omega) \cos \omega\tau \, d\omega \tag{8.72}$$

An important observation from equations (8.71) and (8.72) (Wiener–Khintchine formulas) is that, since that the autocorrelation function is real, so is the spectral density function $S_x(\omega)$.

The use of the amplitude, time-domain, and frequency-domain descriptions might look superfluous because $\bar{x}^2$, $R_x(\tau)$, and $S_x(\omega)$ are related to one another. The reason for their utilization is, mainly, the fact that the method of measurements facilitates monitoring of one parameter or another. For example, when amplitude-sensitive counters are used, the probability function is measured directly and the other parameters can be estimated from it. If analog filters are used, the spectral density function is measured. Finally, digital measuring techniques usually facilitate the measurement of the autocorrelation function.

☐ **Example 8.11   Frequency Spectrum of a Single Rectangular Pulse**

Determine the continuous frequency spectrum of the function shown in Figure E8.11a.

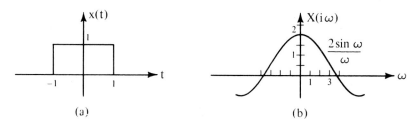

(a)   (b)

Figure E8.11

Solution   Equation (8.38) yields

$$X(i\omega) = \int_{-1}^{1} e^{i\omega t} \, dt = \frac{i}{\omega}(e^{-i\omega} - e^{i\omega}) = 2\frac{\sin \omega}{\omega}$$

This spectrum has only a real component and it is shown in Figure E8.11b.

(This example demonstrates analytical computation of the frequency spectrum of a single pulse.)

☐

☐ **Example 8.12   Spectral Density Function from the Autocorrelation Function**

From the autocorrelation function of (a) a harmonic vibration record, and (b) a periodic vibration record, find the respective spectral density functions.

Solution   (a) From Example 8.9, for the harmonic oscillation $R_x(\tau) = A^2 \cos \omega_0\tau$, and equation (8.67) gives

$$S_x(\omega) = \int_{-\infty}^{\infty} R_x(\tau)e^{-i\omega\tau} \, d\tau = \frac{1}{4\pi} \int_{-\infty}^{\infty} A^2 \cos \omega_0\tau e^{-i\omega\tau} \, d\tau$$

$$= \tfrac{1}{2} \int_{-\infty}^{\infty} A^2 \cos \omega_0\tau(\cos \omega\tau + i \sin \omega\tau) \, d\tau$$

$$= \tfrac{1}{4} \int_{-\infty}^{\infty} A^2[\cos(\omega_0 + \omega)\tau + \cos(\omega_0 - \omega)\tau] \, d\tau$$

The integral $\int_{-\omega}^{\omega} \cos \lambda t \, dt$ is different from zero only if $\lambda = 0$ and then it has value $2\pi\delta(\lambda)$, where $\delta(x)$ is the Dirac function. Therefore,

$$S_x(\omega) = \frac{\pi A^2}{2} [\delta(\omega_0 + \omega) + \delta(\omega_0 - \omega)]$$

(b) For the periodic function, using $R_x(\tau)$ from Example 8.9,

$$S_x(\omega) = \pi A_0^2\delta(0) + \sum_{n=1}^{\infty} \pi A_n^2[\delta(n\omega_0 + \omega) + \delta(n\omega_0 - \omega)]$$

The results are shown in Figures E8.12a and b, respectively.

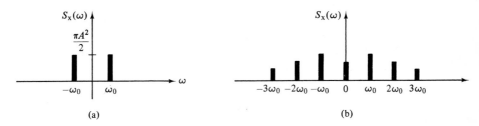

(a)   (b)

Figure E8.12

(This example demonstrates computation of the spectral density function from the autocorrelation function.)

☐

8.6 RESPONSE OF A SINGLE-DEGREE-OF-FREEDOM SYSTEM TO RANDOM EXCITATION

Response Parameters

The response of a linear, single-degree-of-freedom system (m, c, k) to a harmonic excitation $F_0 e^{i\omega t}$ was found (Chapter 3) to be

$$x(t) = \frac{F_0}{k} H(i\omega)e^{i\omega t} = X(i\omega)e^{i\omega t} \tag{8.73}$$

where

$$H(i\omega) = \frac{1}{1 - (\omega/\omega_n)^2 + i\zeta\omega/\omega_n}$$

is a function that multiplies the static deflection F_0/k to yield the dynamic amplitude, called the *amplification factor* or *gain*.

Due to the linearity and the principle of superposition, it was shown in Chapter 5 that the response to any force $f(t)$, periodic or not, given in the form of a Fourier series or Fourier integral transform, will be for a periodic excitation

$$x(t) = \sum_n \frac{F_n}{k} H(in\omega)e^{in\omega t} \tag{8.74}$$

or for a nonperiodic excitation,

$$x(t) = \frac{1}{2\pi} \int_{-\infty}^{\infty} F(i\omega)H(i\omega)e^{i\omega t}\, dt \qquad (8.75)$$

where, respectively,

$$F_n = \frac{1}{T} \int_0^T f(t)e^{-in\omega t}\, dt \qquad (8.76)$$

$$F(i\omega) = \int_{-\infty}^{\infty} f(t)e^{-i\omega t}\, dt \qquad (8.77)$$

From equation (8.75), for periodic excitation,

$$R_x(\tau) = \lim_{T \to \infty} \frac{1}{T} \int_{-T}^{T} \left[\sum_{n=1}^{\infty} \frac{F_n}{k} H(in\omega)e^{in\omega t} \right]\left[\sum_{m=1}^{\infty} \frac{F_m}{k} H(im\omega)e^{im\omega(t+\tau)} \right] dt \qquad (8.78)$$

Because of the orthogonality property of the harmonic functions, the products of the terms of the series will be nonzero upon integration only for $m = n$ (see also Example 8.12). Therefore,

$$R_x(\tau) = \lim_{T \to \infty} \frac{1}{T} \int_{-T}^{T} \left[\sum_{n=1}^{\infty} \left(\frac{F_n}{k}\right)^2 H^2(in\omega)e^{in\omega(2t+\tau)} \right] dt \qquad (8.79)$$

Equation (8.67) yields

$$S_x(\omega) = \int_{-\infty}^{\infty} \left[\lim_{T \to \infty} \frac{1}{T} \int_{-T}^{T} \left[\sum_{n=1}^{\infty} \left(\frac{F_n}{k}\right)^2 H^2(in\omega)e^{in\omega(2t+\tau)} \right] dt \right] d\tau \qquad (8.80)$$

$$S_x(\omega) = H^2(in\omega) \int_{-\infty}^{\infty} \left[\lim_{T \to \infty} \frac{1}{T} \int_{-T}^{T} \left[\sum_{n=1}^{\infty} \left(\frac{F_n}{k}\right)^2 e^{in\omega(2t+\tau)} \right] dt \right] e^{-i\omega\tau} d\tau \qquad (8.81)$$

$$S_x(\omega) = H^2(in\omega) \left[\lim_{T \to \infty} \frac{1}{T} \int_{-T}^{T} \left[\sum_{n=1}^{\infty} \left(\frac{F_n}{k}\right)^2 e^{in\omega 2t} \right] dt \right] \qquad (8.82)$$

$$S_x(\omega) = S_F(\omega)|H(i\omega)|^2 \qquad (8.83)$$

Equation (8.83) applies for nonperiodic excitation, as it was developed above as a limit case for $T \to \infty$; therefore, it also applies for random excitation. The autocorrelation function of the response will be

$$R_x(\tau) = \frac{1}{2\pi} \int_{-\infty}^{\infty} S_F H^2(\omega)e^{i\omega\tau}\, d\omega \qquad (8.84)$$

It is apparent that the response autocorrelation function is insensitive to the beginning of time; therefore, if the excitation is stationary, so is the response. Moreover, if the excitation is Gaussian, so is the response [Crandall and Mark, 1963]. The mean square will be, due to equation (8.82),

$$\overline{x^2} = R_x(0) = \frac{1}{2\pi} \int_{\infty}^{\infty} S_F(\omega)H^2(\omega)\, d\omega \qquad (8.85)$$

Random vibrations are responsible for many fatigue failures of machine and structures, turbine blades, and so on. One important parameter for this situation is the average number of zero crossings of the excitation. The number of zero crossings of a time function with spectral density function $S_x(\omega)$ is given by the formula[7]

$$N_0 = \frac{1}{\pi} \left[\frac{\int_{-\infty}^{\infty} \omega^2 S_x(\omega)\, d\omega}{\int_{-\infty}^{\infty} S_x(\omega)\, d\omega} \right]^{1/2} \tag{8.86}$$

and the number of times that the function exceeds a specified level x_0 is

$$N_x = N_0 e^{-x_0^2/\sigma^2} \tag{8.87}$$

The random records are often divided into two categories: narrowband and wideband (Figure 8.8). Narrowband records are dominated by a certain frequency and their frequency spectrum has a rather narrow band of frequencies. Wideband records have a widely distributed frequency spectrum. Some of their characteristics have been tabulated in Figure 8.8 for identification purposes, together with a harmonic function for comparison.

☐ **Example 8.13 Earthquake Response of a Nuclear Reactor Containment Vessel**

A nuclear reactor containment structure has a mass of 1.55×10^5 kg, a natural frequency $\omega_n = 20$ rad/s for horizontal vibration, and a damping constant 5% of the critical damping constant. Earthquakes have been modeled as random vibration with Rayleigh distribution. Records of the four largest earthquakes in the United States were analyzed and found that the spectral density function of the ground acceleration was $S_0 = 0.0063$ m²/s³ in the frequency range 0 to 35 Hz and zero outside. Determine the probability that acceleration of the reactor will exceed $0.65g$.

Solution The spring constant of the vessel for horizontal vibration is $k = m\omega_n^2 = 1.55 \times 10^5 \times 20^2 = 6.2 \times 10^7$ N/m. The critical damping constant is then

$$C_c = 2(mk)^{1/2} = 2(1.55 \times 10^5 \times 6.2 \times 10^7)^{1/2} = 6.20 \times 10^6 \text{ N·s/m}$$

Therefore, the damping constant is $C = 0.05C_c = 3.1 \times 10^5$ N·s/m.

If the base has a motion $y = Ye^{i\omega t}$, the relative motion x will be (Chapter 3)

$$|X| = |Y| \left[\frac{k^2 + (c\omega)^2}{(k - m\omega^2)^2 + (c\omega)^2} \right]^{1/2}$$

$$|X| = |Y|H(\omega), \qquad H(\omega) = \left[\frac{k^2 + (c\omega)^2}{(k - m\omega^2)^2 + (c\omega)^2} \right]^{1/2}$$

The same $H(\omega)$ will relate the accelerations. From equation (8.49) we obtain the mean square of the accelerations,

$$\overline{a^2} = \int_0^\infty S_y''(\omega)|H(\omega)|^2\, d\omega$$

[7]Ibid.

But $S_y''(\omega) = 0.0063$ m$^2 \cdot$rad/s^3 between frequencies 0 and $35 \times 2\pi = 219.9$ rad/s. Therefore,

$$\overline{a^2} = 0.0063 \int_0^{219.9} \frac{k^2 + (c\omega)^2}{(k - m\omega^2)^2 + (c\omega)^2} \, d\omega$$

With numerical integration we find that

$$\overline{a^2} = 0.0063 \times 1050 = 6.615 \ (\text{m/s}^2)^2$$

$$\sigma = (\overline{a^2})^{1/2} = 2.57 \ \text{m/s}^2$$

Entering Figure 8.7b with $a/\sigma = 0.65 \times 9.81/2.57 = 2.5$, we find from Rayleigh's cumulative probability that $P_R(a/\sigma) = 0.95$. Therefore, the probability that the acceleration will exceed 0.2g will be

$$1 - \frac{1}{\sigma} P_R\left(\frac{a}{\sigma}\right) = 1 - \left(\frac{1}{2.57}\right)0.95 = 0.63 \text{ or } 63\%$$

(This example demonstrates the probabilistic analysis of the response of a structure to an earthquake.)

□

Derivative Processes

For many technical problems, the derivatives of the response are of importance, notably velocity and acceleration. Differentiation of the autocorrelation function yields

$$R_x'(\tau) = \lim_{T \to \infty} \frac{1}{T} \int_{-T}^{T} x(t) \frac{d}{d\tau} x(t + \tau) \, dt \tag{8.88}$$

For a stationary process

$$R_x'(\tau) = \lim_{T \to \infty} \frac{1}{T} \int_{-T}^{T} x(t - \tau) \dot{x}(t) \, dt \tag{8.89}$$

One more differentiation yields for a stationary process

$$R_x''(\tau) = -\lim_{T \to \infty} \frac{1}{T} \int_{-T}^{T} \dot{x}(t - \tau) \dot{x}(t) \, dt = -\lim_{T \to \infty} \frac{1}{T} \int_{-T}^{T} \dot{x}(t) \dot{x}(t + \tau) \, dt \tag{8.90}$$

Therefore, the autocorrelation functions of the velocity $\dot{x}(t)$ and the acceleration $\ddot{x}(t)$ are

$$R_{\dot{x}}(\tau) = -R_x''(\tau), \qquad R_{\ddot{x}}(\tau) = -R^{(\text{iv})}(\tau) \tag{8.91}$$

Using equation (8.66) yields

$$R_{\dot{x}}(\tau) = \int_{-\infty}^{\infty} S_{\dot{x}}(\omega) e^{i\omega\tau} \, d\omega$$

$$R_{\ddot{x}}(\tau) = \int_{-\infty}^{\infty} S_{\ddot{x}}(\omega) e^{i\omega\tau} \, d\omega \tag{8.92}$$

and with equations (8.91),

$$S_{\dot{x}}(\omega) = \omega^2 S_x(\omega), \qquad S_{\ddot{x}}(\omega) = \omega^4 S_x(\omega) \tag{8.93}$$

□ **Example 8.14   Probabilistic Wind Loading of a Nuclear Reactor Containment Vessel**

The nuclear reactor of Example 8.13 will be subject to wind loading. Interpretation of meteorological data has concluded that the expected wind force on the structure follows Rayleigh's distribution and the spectral density function of the wind load will be $S_0 = 683 \times 10^5$ N$^2 \cdot$s in the frequency range 0 to 20 Hz and zero outside. Determine the probability that the acceleration of the reactor due to this wind loading will exceed $0.2g$.

Solution   From Example 8.13, $k = m\omega_n^2 = 1.55 \times 10^5 \times 20^2 = 6.2 \times 10^7$ N/m, $C = 3.1 \times 10^5$ N$\cdot$s/m. A harmonic force $f(t) = Fe^{i\omega t}$ will result in the response (Chapter 3)

$$x = Xe^{i\omega t}, \quad |X| = \frac{|F|}{[(k - m\omega^2)^2 + (c\omega)^2]^{1/2}}$$

$$|X| = |F|H(\omega), \quad H(\omega) = \frac{1}{[(k - m\omega^2)^2 + (c\omega)^2]^{1/2}}$$

From equation (8.93) we obtain the mean square of the accelerations,

$$\overline{a^2} = \int_0^\infty S_x''(\omega) \, d\omega = \int_0^\infty \omega^4 S_x(\omega) \, d\omega = \int_0^\infty \omega^4 S_F(\omega) |H(\omega)|^2 \, d\omega$$

But $S_F(\omega) = 9 \times 10^5$ N$^2 \cdot$s between frequencies 0 and $20 \times 2\pi = 125.6$ rad/s. Therefore,

$$\overline{a^2} = 683 \times 10^5 \int_0^{125.5} \frac{\omega^4}{(k - m\omega^2)^2 + (c\omega)^2} \, d\omega$$

With numerical integration we find that

$$\overline{a^2} = 683 \times 10^5 \times 1.79 \times 10^{-8} = 1.22 \ (\text{m/s}^2)^2$$

$$\sigma = (\overline{a^2})^{1/2} = 1.1 \ \text{m/s}^2$$

Entering Figure 8.7b with $a/\sigma = 0.2 \times 9.81/1.1 = 1.78$, we find from Rayleigh's cumulative probability that $P_R(a/\sigma) = 0.75$. Therefore, the probability that the acceleration will exceed $0.2g$ will be

$$1 - \frac{1}{\sigma} P_R\left(\frac{a}{\sigma}\right) = 1 - \left(\frac{1}{1.1}\right)0.75 = 0.32 \ \text{or} \ 32\%$$

(This example demonstrates the computation of the response of a structure to probabilistic wind loading.)

□

8.7 RESPONSE OF A LINEAR SYSTEM TO RANDOM EXCITATION

Superposition of Stationary Processes

Let a process $q(t)$ be the sum of n separate stationary processes $x_1(t), x_2(t), \ldots, x_n(t)$ with zero mean values. The autocorrelation function for the process is

$$R_q(\tau) = E[q(t)q(t + \tau)] \tag{8.94}$$

Substitution of $q(t) = x_1(t) + x_2(t) + \cdots + x_n(t)$ in (8.94) will yield integrals of the cross products of $x_1, x_2, \ldots, x_n$, called *cross-correlation functions* or *co-variances*, which are zero if the processes $x_1, x_2, \ldots, x_n$ are *statistically independent*. The nonzero terms will then give

$$R_q(\tau) = E[x_1(t)x_1(t + \tau)] + E[x_2(t)x_2(t + \tau)] + \cdots + E[x_n(t)x_n(t + \tau)]$$

$$= R_{x1}(\tau) + R_{x2}(\tau) + \cdots + R_{xn}(\tau) \tag{8.95}$$

$$\overline{q^2} = R_q(0) = \overline{x_1^2} + \overline{x_2^2} + \cdots + \overline{x_n^2} \tag{8.96}$$

since

$$S_q(\omega) = \int_{-\infty}^{\infty} R_q(\tau)e^{-i\omega\tau}\, d\tau$$

$$= \int_{-\infty}^{\infty} [R_{x1}(\tau) + R_{x2}(\tau) + \cdots + R_{xn}(\tau)]e^{-i\omega\tau}\, d\tau \tag{8.97}$$

$$S_q(\omega) = S_{x1}(\omega) + S_{x2}(\omega) + \cdots + S_{xn}(\omega)$$

Direct Method

The response of a linear system

$$\mathbf{M\ddot{x}} + \mathbf{C\dot{x}} + \mathbf{Kx} = \mathbf{f}(t) \tag{8.98}$$

was found to be

$$\mathbf{X} = \mathbf{H}(i\omega)\mathbf{f}_0 \tag{8.99}$$

if $\mathbf{f}(t) = \mathbf{f}_0 e^{i\omega t}$ and $\mathbf{H}(i\omega) = [-\omega^2\mathbf{M} + i\omega\mathbf{C} + \mathbf{K}]^{-1}$.
If $\mathbf{H} = \{\mathbf{h}_1 \quad \mathbf{h}_2 \quad \cdots \quad \mathbf{h}_n\} = [h_{ij}]$,

$$x_k(t) = \mathbf{h}_k \mathbf{f}_0 e^{i\omega t} = \sum_{j=1}^{n} h_{kj} f_{0j} e^{i\omega t} \tag{8.100}$$

where $\mathbf{h}_k$ is the kth row vector of the amplification matrix $\mathbf{H}(i\omega)$. Using the reasoning that yielded equations (8.95) to (8.97), and assuming that the *excitations are independent*,

$$R_{xk}(\tau) = E[x_k(t)x_k(t + \tau)] = E\left[\sum_{j=1}^{n} h_{kj}^2 f_{0j} e^{i\omega t} f_{0j} e^{i\omega(t+\tau)}\right]$$

$$= \sum_{j=1}^{n} h_{kj}^2 E[f_{0j} e^{i\omega t} f_{0j} e^{i\omega(t+\tau)}] = \sum h_{kj}^2 R_{fj}(\tau) \tag{8.101}$$

$$\overline{x_k^2} = R_{xk}(0) = \sum_{j=1}^{n} h_{kj}^2 \overline{f_j^2} \tag{8.102}$$

Since

$$S_{xk}(\omega) = \int_{-\infty}^{\infty} R_{xk}(\tau) e^{-i\omega\tau} \, d\tau = \sum_{j=1}^{n} \int_{-\infty}^{\infty} [R_{xj}(\tau)] e^{-i\omega\tau} \, d\tau$$

$$(8.103)$$

$$S_{xk}(\omega) = \sum_{j=1}^{n} h_{kj}^2 S_{ff}(\omega)$$

☐ **Example 8.15 Torsional Vibration of a Rotor Due to Probabilistic Torque**

The gas turbine rotor of Example 7.1 has on the third rotor (the generator) an unbalanced electric torque $T_0 \cos \omega t$, where T_0 has mean value 100 N·m and standard deviation 10 N·m and $\omega = 2 \times 377$ rad/s. Determine the mean value and the standard deviation of the vibration amplitudes.

Solution   Equation (8.102) gives

$$\overline{x_k^2} = R_{xk}(0) = \sum_{j=1}^{n} h_{kj}^2 \overline{f_j^2}$$

Here, $\overline{f_3^2} = 10{,}000$ N, $f_1 = f_2 = 0$. From Example 7.1,

$$h_{13} = -\frac{21.3 \times 10^{-6}}{100}, \qquad h_{23} = -\frac{9.19 \times 10^{-6}}{100}, \qquad h_{33} = -\frac{5.5 \times 10^{-6}}{100}$$

$$\overline{x_1^2} = h_{13}^2 f_3^2 = \left(\frac{21.3 \times 10^{-6}}{100}\right)^2 12^2 = 25.5^2 \times 10^{-14} \text{ m}^2$$

$$\overline{x_2^2} = h_{23}^2 f_3^2 = \left(\frac{9.19 \times 10^{-6}}{100}\right)^2 12^2 = 11.0^2 \times 10^{-14} \text{ m}^2$$

$$\overline{x_3^2} = h_{33}^2 f_3^2 = \left(\frac{5.5 \times 10^{-6}}{100}\right)^2 12^2 = 6.6^2 \times 10^{-14} \text{ m}^2$$

(This example demonstrates the computation of the torsional vibration of a rotor due to probabilistic torque.)

☐

Modal Analysis Method

Using normal coordinates and assuming that the damping matrix can be diagonalized, the uncoupled system of equations of motion, in terms of the normal coordinates q_k, is

$$\ddot{q}_k + 2\zeta_k \omega_k \dot{q}_k + \omega_k^2 q_k = \phi_k^T \mathbf{f}(t), \qquad k = 1, 2, \ldots, n \qquad (8.104)$$

Defining $H_k = 1/[(1 - \omega/\omega_k)^2 + (2\zeta_k\omega/\omega_k)^2]$, the transfer function of the mode k and independent forces,

$$R_{qk}(\tau) = \sum_{j=1}^{n} H_j^2 \phi_{kj}^2 R_{Fj}(\tau) \qquad (8.105)$$

$$\overline{q_k^2} = R_{qk}(0) = \sum_{j=1}^{n} [H_j^2 \phi_{kj}^2 \overline{f_j^2}] \qquad (8.106)$$

since

$$S_{qk}(\omega) = \int_{-\infty}^{\infty} R_{qk}(\tau) e^{-i\omega\tau} \, d\tau = \sum_{j=1}^{n} \int_{-\infty}^{\infty} [R_{qj}(\tau)] e^{-i\omega\tau} \, d\tau$$

$$= \sum_{j=1}^{n} H_{kj}^2 \phi_{kj}^2 S_{fj}(\omega) \tag{8.107}$$

□ **Example 8.16 Probabilistic Loading of a Motor Packaging**

The packaging of two electric motors in a container is modeled as shown in Figure E8.5. Determine the response to a harmonic horizontal acceleration on the container with amplitude that has mean value 1.2 m/s² and standard deviation 0.2 m/s².

Solution   From Example 8.5 the normalized modal matrix is

$$\mathbf{\Phi} = \begin{bmatrix} 1 & 1 & 0 & 0 \\ 1 & -1 & 0 & 0 \\ 0 & 0 & 1 & 0 \\ 0 & 0 & 0 & 1 \end{bmatrix} m^{-1/2}$$

The principal coordinates are given by the transformation

$$\mathbf{x} = \begin{bmatrix} 1 & 1 & 0 & 0 \\ 1 & -1 & 0 & 0 \\ 0 & 0 & 1 & 0 \\ 0 & 0 & 0 & 1 \end{bmatrix} m^{-1/2} \mathbf{q}$$

The modal forces are

$$f_1 = -m\phi_{11}\ddot{u}_1 - m\phi_{12}\ddot{u}_2 - m\phi_{13}\ddot{u}_3 - m\phi_{14}\ddot{u}_4$$
$$= -m(1 \times 1.2 + 1 \times 1.2 + 0 \times 0 + 0 \times 0) = -2.4m \text{ N}$$

$$f_2 = -m\phi_{21}\ddot{u}_1 - m\phi_{22}\ddot{u}_2 - m\phi_{23}\ddot{u}_3 - m\phi_{24}\ddot{u}_4$$
$$= -m(1 \times 1.2 - 1 \times 1.2 + 0 \times 0 + 0 \times 0) = 0$$

$$f_3 = -m\phi_{31}\ddot{u}_1 - m\phi_{32}\ddot{u}_2 - m\phi_{33}\ddot{u}_3 - m\phi_{34}\ddot{u}_4$$
$$= -m(0 \times 1 + 0 \times 1 + 1 \times 1.2 + 0 \times 0) = 0$$

$$f_4 = -m\phi_{41}\ddot{u}_1 - m\phi_{42}\ddot{u}_2 - m\phi_{43}\ddot{u}_3 - m\phi_{44}\ddot{u}_4$$
$$= -m(0 \times 1 + 0 \times 1 + 0 \times 0 + 1 \times 0) = 0$$

Therefore,

$$\ddot{q}_1 + \omega_1^2 q_1 = -2.4m \cos \omega t$$

$$\ddot{q}_2 + \omega_2^2 q_2 = 0$$

$$\ddot{q}_3 + \omega_3^2 q_3 = 0$$

$$\ddot{q}_4 + \omega_4^2 q_4 = 0$$

Only q_1 will be nonzero, and have the value

$$q_1 = -2.4m \frac{1}{\omega^2 - \omega_n^2} \cos \omega t$$

In terms of the physical coordinates, the response will be

$$x_1 = \overline{x_1} = (q_1 + q_2)m^{-1/2} = -2.4m^{1/2} \frac{1}{\omega^2 - \omega_n^2} \cos \omega t$$

$$x_2 = \overline{x_2} = (q_1 - q_2)m^{-1/2} = -2.4m^{1/2} \frac{1}{\omega^2 - \omega_n^2} \cos \omega t$$

$$\overline{X_1} = -2.4m^{1/2} \frac{1}{\omega^2 - \omega_n^2}$$

$$\overline{X_2} = -2.4m^{1/2} \frac{1}{\omega^2 - \omega_n^2}$$

$$\overline{X_3} = 0, \qquad \overline{X_4} = 0$$

Therefore,

$$h_{11} = h_{12} = -2.4m^{1/2} \frac{1/(\omega^2 - \omega_n^2)}{1.2}, \qquad h_{23} = 0, \qquad h_{14} = 0$$

$$h_{21} = h_{22} = -2.4m^{1/2} \frac{1/(\omega^2 - \omega_n^2)}{1.2}, \qquad h_{23} = 0, \qquad h_{24} = 0$$

$$\overline{X_1^2} = \left[2.4m^{1/2} \frac{1/(\omega^2 - \omega_n^2)}{1.2} \right]^2 \times 0.2^2$$

$$+ \left[2.4m^{1/2} \frac{1/(\omega^2 - \omega_n^2)}{1.2} \right]^2 \times 0.2^2$$

$$= 2 \left[0.4m^{1/2} \frac{1}{(\omega^2 - \omega_n^2)} \right]^2$$

$$\overline{X_2^2} = \left[2.4m^{1/2} \frac{1/(\omega^2 - \omega_n^2)}{1.2} \right]^2 \times 0.2^2$$

$$+ \left[2.4m^{1/2} \frac{1/(\omega^2 - \omega_n^2)}{1.2} \right]^2 \times 0.2^2$$

$$= 2 \left[0.4m^{1/2} \left(\frac{1}{\omega^2 - \omega_n^2} \right) \right]^2$$

$$\overline{X_3^2} = 0$$

$$\overline{X_4^2} = 0$$

(This example demonstrates computation of the response of a motor packaging to probabilistic wind loading.) ☐

□ Example 8.17   Modal Analysis of a Building for Earthquake Loading

As an example for the use of the vibration analysis routines package that accompanies this book, here is an example of modal analysis with the subroutine MODAL of a building with earthquake loading. The natural frequencies and modes have already been computed.

The input is given with data statements:

NHARM is the number of modes to be used.

NMAX is the number of masses in the lumped mass system (10-story building).

NPOINTS is the number of forward integration points.

TMAX is the maximum integration time.

IPRINT = 1 requires printing of results.

XM is the diagonal mass matrix.

OMEG is the natural frequencies vector.

ZHTA is the modal damping ratio.

Z is the natural modes vector.

The acceleration signal (Figure E8.17b) of the El Centro earthquake is used as input (see Chapter 3): file EL-CENT.SW in the data diskette. It is shown with the displacement signal (Figure E8.17a). The response at the first, fifth, and tenth floors is shown in Figure E8.17c, d, and e, respectively. The control program follows:

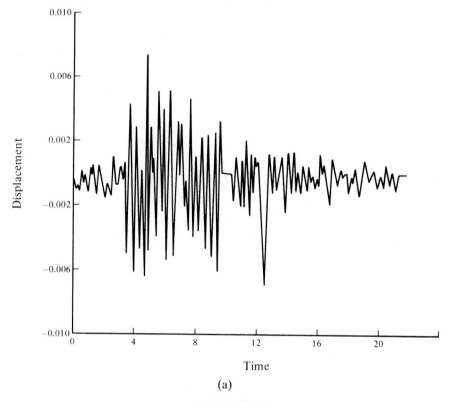

(a)

Figure E8.17

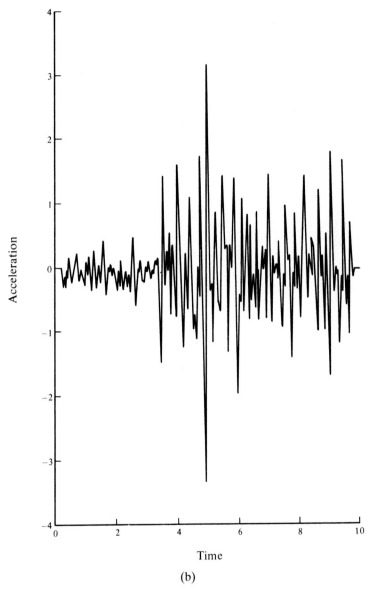

Time

(b)

Figure E8.17 (cont.)

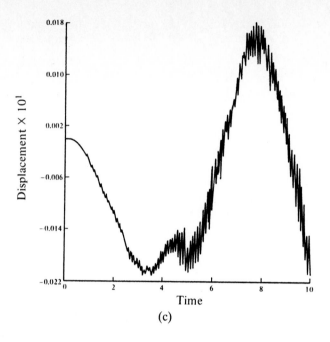

(c)

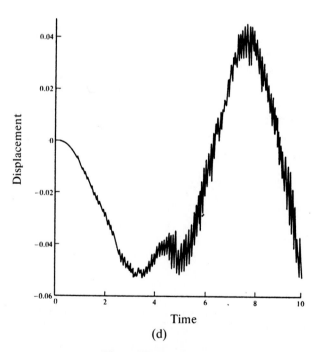

(d)

Figure E8.17 (cont.)

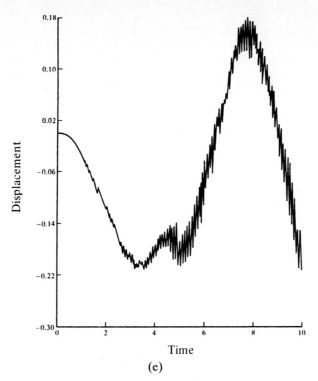

Figure E8.17 (cont.)

```
PROGRAM EX8_17
C       ......External: subroutines MODAL, PLOT, file EL_CENT.SW
C       ......System data are ficticious
        DIMENSION XM(10),F(1),Z(1,10),ZETA(1),OMEGA(1)
C       ......No of: modes, degrees of freedom, plotting points
        DATA NHARM,NMAX,NPOINTS/1,10,500/
C       ......Integration time, print results index
        DATA TMAX,IPRINT/10.,1/
C       .......Lumped masses
        DATA XM/1.,2.,3.,4.,5.,6.,7.,8.,9.,10./
C       .......Natural frequencies
        DATA OMEGA/1./
C       .......Modal damping ratios
        DATA ZETA/.05/
C       Natural modes stored row-by-row
        DATA Z/1,4,9,16,25,36,49,64,81,100/
        CALL MODAL(NHARM,NMAX,TMAX,IPRINT,XM,OMEGA,PHI,F,ZETA,
        NPOINTS)
        STOP
        END
```

(This example demonstrates computation of the response of a structure to earth-quake loading by modal analysis.)

□

TABLE 8.1 SUMMARY OF EQUATIONS FOR CHAPTER 8

$$\mathbf{M\ddot{x}} + \mathbf{C\dot{x}} + \mathbf{Kx} = \mathbf{F}_0 e^{i\omega t}$$

$$\mathbf{x} = \mathbf{X} e^{i\omega t}$$

$$\mathbf{X} = [-\omega^2 \mathbf{M} + i\omega \mathbf{C} + \mathbf{K}]^{-1} \mathbf{F}_0$$

General excitation
By Duhamel's integral

$$\mathbf{E} = \text{diag}[e^{p_1 t}, e^{p_2 t}, \ldots, e^{p_{2n} t}]$$

$$\mathbf{G}(t) = \mathbf{Z}^* \mathbf{E} \mathbf{Z}^{*-1}$$

$$\mathbf{x}(t) = \mathbf{M}^{-1} \int_0^t \mathbf{G}(t - t^*) \mathbf{f}(t^*) \, dt^*$$

$$\mathbf{f}(t) = \{\underbrace{0 \quad 0 \quad 0 \quad \cdots \quad 0}_{n} \quad f_i(t) \quad \cdots \quad f_n(t)\}$$

By modal analysis

$$\mathbf{\Phi} = [\mathbf{\phi}^{(1)} \quad \mathbf{\phi}^{(2)} \quad \cdots \quad \mathbf{\phi}^{(n)}]$$

$$\mathbf{\Phi}^T \mathbf{M} \mathbf{\Phi} = \mathbf{I} = \text{diag}[11 \quad \cdots \quad 1]$$

$$\mathbf{\Phi}^T \mathbf{K} \mathbf{\Phi} = \text{diag}[\omega_1^2 \, \omega_2^2 \ldots \omega_n^2]$$

$$\mathbf{x} = \mathbf{\Phi} \mathbf{q}$$

$$\mathbf{q} = \mathbf{\Phi}^{-1} \mathbf{x} = \mathbf{\Phi}^T \mathbf{M} \mathbf{x}$$

$$\mathbf{M\ddot{x}} + \mathbf{C\dot{x}} + \mathbf{Kx} = \mathbf{f}(t)$$

$$\ddot{q}_i + 2\zeta_i \omega_i \dot{q}_i + \omega_i^2 q = \mathbf{\phi}_i^T \mathbf{f}(t)$$

$$\mathbf{\Phi}^T \mathbf{C} \mathbf{\Phi} = \text{diag}[\lambda_1 \quad \lambda_2 \quad \cdots \quad \lambda_n]$$

$$\lambda_i = \mathbf{\phi}_i^T \mathbf{C} \mathbf{\phi}_i = 2\zeta_i \omega_i$$

Rayleigh damping

$$\mathbf{C} = a\mathbf{M} + b\mathbf{K}$$

$$\mathbf{\Phi}^T \mathbf{C} \mathbf{\Phi} = \mathbf{\Phi}^T (a\mathbf{M} + b\mathbf{K}) \mathbf{\Phi} = a\mathbf{\Phi}^T \mathbf{M} \mathbf{\Phi} + b\mathbf{\Phi}^T \mathbf{K} \mathbf{\Phi} = \text{diag}[a + \omega_i^2 b]$$

$$\mathbf{C} = (\gamma/\omega)\mathbf{K}, \qquad a = 0, \qquad b = \gamma/\omega$$

$$\mathbf{\Phi}^T \mathbf{C} \mathbf{\Phi} = \text{diag}[\gamma \omega_i^2 / \omega]$$

$$q_i(t) = e^{-\zeta_i \omega_i t} \left(q_i(0) \cos \omega_{di} t + \left[2\zeta_i \omega_i q_i(0) + \frac{\dot{q}_i(0)}{\omega_{di}} \right] \sin \omega_{di} t \right) + \int_0^t F_i(t^*) g_i(t - t^*) \, dt^*$$

$$\omega_{di} = \omega_i (1 - \zeta_i^2)^{1/2},$$

$$g_i(t) = \frac{1}{\omega_{di}} e^{-\zeta_i \omega_i t} \sin \omega_{di} t,$$

$$F_i(t) = \mathbf{\phi}_i^T \mathbf{f}(t)$$

TABLE 8.1 *(cont.)*

Initial conditions

$$\mathbf{q}(0) = \mathbf{\Phi}^T \mathbf{M} \mathbf{x}(0)$$

$$\dot{\mathbf{q}}(0) = \mathbf{\Phi}^T \mathbf{M} \dot{\mathbf{x}}(0), \ \zeta = 0:$$

$$q_i(t) = q_i(0) \cos \omega_{ni} t + \left[\frac{\dot{q}_i(0)}{\omega_{ni}} \right] \sin \omega_{ni} t + \int_0^t F_i(t^*) g_i(t - t^*) \, dt^*$$

$$g_i(t) = \frac{1}{\omega_{ni}} \sin \omega_{ni} t$$

$$F_i(t) = \mathbf{\phi}_i^T \mathbf{f}(t)$$

Random vibration
Probabilities

$$h(x_k) = n(x_k)/N$$

$$h(x_k) = \sum \Delta t_i / T$$

$$P_{-\infty}^x = \int_{-\infty}^x p(x) \, dx$$

$$P_x^\infty = \int_x^\infty p(x) \, dx$$

$$P_{x_1}^{x_2} = \int_{x_1}^{x_2} p(x) \, dx$$

$$P_{-\infty}^\infty = 1$$

$$P_{-\infty}^x + P_x^\infty = 1$$

Amplitude description
(Mean and standard deviation)

$$\overline{x(t)} = \lim_{T \to \infty} \frac{1}{T} \int_0^t x(t) \, dt$$

$$\overline{x^2(t)} = \lim_{T \to \infty} \frac{1}{T} \int_0^t x^2(t) \, dt$$

$$\overline{x} = E[x] = \int_{-\infty}^\infty x p(x) \, dx$$

$$\overline{x^2} = E[x^2] = \int_{-\infty}^\infty x^2 p(x) \, dx$$

$$\overline{\sigma}^2 = \int_{-\infty}^\infty (x - \overline{x})^2 p(x) \, dx$$

$$= \overline{x^2} - \overline{x}^2$$

(continues)

TABLE 8.1 *(cont.)*

The normal distribution

$$p_g(x) = \left[\frac{1}{\sigma (2\pi)^{1/2}}\right] e^{(x-\bar{x})^2/2\sigma^2}$$

$$P^x_{-\infty}(x) = \frac{1}{\sigma (2\pi)^{1/2}} \int_{-\infty}^x e^{(x-\bar{x})^2/2\sigma^2} \, dx$$

The Rayleigh distribution

$$p_r(x) = x/\alpha^2 e^{-(x/\alpha)^2/2}$$

$$\bar{x}_r = \int_0^\infty (x/\alpha)^2 e^{-(x/\alpha)^2/2} \, dx = 1.253\alpha$$

$$\bar{x}_r^2 = \alpha \int_0^\infty (x/\alpha)^3 e^{-(x/\alpha)^2/2} \, dx = 2\alpha^2$$

$$P^x_{r0} = \int_0^\infty (x/\alpha^2) e^{-(x/\alpha)^2/2} \, dx$$

Time domain description

$$R(\tau) = \lim_{T\to\infty} \frac{1}{T} \int_0^T x(t) x(t+\tau) \, dt$$

$$R_x(0) = \sigma_x^2$$

$$R_x(\tau) = R_x(-\tau)$$

$$R_x(\tau) \leq R_x(0)$$

Frequency domain description
Discrete Fourier transform

$$x(t) = \sum_{n=-\infty}^{\infty} C_n e^{in\omega_0 t}$$

$$C_n = \frac{1}{T} \int_{-T/2}^{T/2} x(t) e^{-in\omega_0 t} \, dt$$

Integral Fourier transform

$$x(t) = 1/2\pi \int_{-\infty}^{\infty} X(i\omega) e^{i\omega t} \, d\omega$$

$$X(i\omega) = \int_{-\infty}^{\infty} x(t) e^{-i\omega t} \, dt$$

Spectral density

$$G(n\omega_0) = \frac{T}{4\pi} C_n \overline{C}_n$$

$$\overline{x^2} = \frac{\sum_n C_n \overline{C}_n}{2 \sum_n G(n\omega_0)\Delta\omega}$$

TABLE 8.1 *(cont.)*

$$\overline{x^2} = \lim_{T \to \infty} \frac{1}{T} \int_{-T/2}^{T/2} |x(t)|^2 \, dt = \int_{-\infty}^{\infty} G_x(\omega) \, d\omega$$

$$G_x(\omega) = \lim_{\Delta\omega \to 0} \frac{|X(i\omega)|^2}{\Delta\omega}$$

$$R_x(\tau) = \frac{1}{2\pi} \int_{-\infty}^{\infty} S_x(\omega) e^{i\omega t} \, dt$$

$$S_x(\omega) = \int_{-\infty}^{\infty} R_x(\tau) e^{-i\omega t} \, d\tau = G_x(\omega)/2\pi$$

$$S_x(\omega) = 2 \int_{0}^{\infty} R_x(\tau) \cos \omega\tau \, d\tau$$

$$R_x(\tau) = \frac{1}{\pi} \int_{0}^{\infty} S_x(\omega) \cos \omega\tau \, d\omega$$

White noise

$$S(\omega) = S_0 = \text{const}$$
$$R(\tau) = 2\pi S_0 \delta(\tau)$$

Random response

$$S_x(\omega) = S_F(\omega) |H(\omega)|^2$$

$$R_x(\tau) = \frac{1}{2\pi} \int_{-\infty}^{\infty} S_F H^2(\omega) e^{i\omega\tau} \, d\omega$$

$$\overline{x^2} = R_x(0) = \frac{1}{2\pi} \int_{-\infty}^{\infty} S_F(\omega) H^2(\omega) \, d\omega$$

Number of crossings

$$N_0 = \frac{1}{\pi} \left[\frac{\int_{-\infty}^{\infty} \omega^2 S_x(\omega) \, d\omega}{\int_{-\infty}^{\infty} S_x(\omega) \, d\omega} \right]^{1/2}$$

$$N_x = N_0 e^{-x_0^2/\sigma^2}$$

Derivative processes

$$R_{\dot{x}}(\tau) = -R_x''(\tau), \qquad R_{\ddot{x}}(\tau) = -R^{IV}(\tau)$$

$$S_{\dot{x}}(\omega) = \omega^2 S_x(\omega), \qquad S_{\ddot{x}}(\omega) = \omega^4 S_x(\omega)$$

Superposition of stationary processes

$$R_q(\tau) = R_{x1}(\tau) + R_{x2}(\tau) + \cdots + R_{xn}(\tau)$$
$$\overline{q^2} = R_q(0) = \overline{x_1^2} + \overline{x_2^2} + \cdots + \overline{x_n^2}$$
$$S_q(\omega) = S_{x1}(\omega) + S_{x2}(\omega) + \cdots + S_{xn}(\omega)$$

(continues)

TABLE 8.1 *(cont.)*

Multi-degree of freedom systems

$$\mathbf{M\ddot{x}} + \mathbf{C\dot{x}} + \mathbf{Kx} = \mathbf{f}(t)$$

$$\mathbf{X} = \mathbf{H}(\omega)\mathbf{f}_0$$

$$\mathbf{f}(t) = \mathbf{f}_0 e^{i\omega t}, \qquad \mathbf{H}(\omega) = [-\omega^2 \mathbf{M} + i\omega \mathbf{C} + \mathbf{K}]^{-1}$$

$$\mathbf{H} = \{\mathbf{h}_1 \quad \mathbf{h}_2 \quad \cdots \quad \mathbf{h}_n\} = [h_{ij}]$$

$$x_k(t) = \sum_{j=1}^{n} h_{kj} f_{0j} e^{i\omega t}$$

$$R_{xk}(\tau) = \sum_{j=1}^{n} h_{kj}^2 R_{fj}(\tau)$$

$$\overline{x_k^2} = R_{xk}(0) = \sum_{j=1}^{n} [h_{kj}^2 \overline{f_j^2}]$$

$$S_{xk}(\omega) = \sum_{j=1}^{n} h_{kj}^2 S_{Fj}(\omega)$$

Modal analysis

$$H_k = \frac{1}{[(1 - \omega/\omega_k)^2 + (2\zeta_k \omega/\omega_k)^2]^{1/2}}$$

$$R_{qk}(\tau) = \sum_{j=1}^{n} H_j^2 \phi_{kj}^2 R_{Fj}(\tau)$$

$$\overline{q_k^2} = R_{qk}(0) = \sum_{j=1}^{n} [H_j^2 \phi_{kj}^2 \overline{f_j^2}]$$

$$S_{qk}(\omega) = \sum_{j=1}^{n} H_{kj}^2 \phi_{kj}^2 S_{Fj}(\omega)$$

REFERENCES AND FURTHER READING

BENDAT, J. S., and PIERSOL, A. G. 1971. *Random Data: Analysis and Measurement Procedures*. New York: Wiley.

BISHOP, R. E. D., and JOHNSON, C. D. 1979. *Mechanics of Vibration*, 2nd ed. Cambridge: Cambridge University Press.

CLOUGH, R. W., and PENZIEN, J. 1975. *Dynamics of Structures*. New York: McGraw-Hill.

CRANDALL, S. H., and MARK, W. D. 1963. *Random Vibration in Mechanical Systems*. New York: Academic Press.

HOHENEMSER, K., and PRAGER, W. 1933. *Dynamik der Stabwerke*. Berlin: Julius Springer.

NEWLAND, D. E. 1975. *Random Vibrations and Spectral Analysis*. London: Longman.

RAYLEIGH, J. W. S. 1894. *Theory of Sound*. New York: Dover Publications, 1946.

RICE, S. O. 1944, 1945. Mathematical analysis of random noise. *Bell Syst. Tech. J.*, *17, 18*.

SEIREG, A. 1966. *Mechanical Systems Analysis*. Scranton, Pa.: International Textbook Co.

PROBLEMS

Sections 8.1 and 8.2

8.1. The gas turbine rotor of Problem 7.1 has a harmonic unbalanced torque at the generator of frequency 120 Hz and amplitude 150 N·m. Find the resulting vibration.

8.2. At the propeller of the system of Problem 7.2, there is a harmonic torque at the blade passing frequency of 600 rpm and amplitude 600 N·m. Determine the resulting vibration at the propeller.

8.3. At the middisk of the compressor rotor of Problem 7.3, there is a constant torque 200 N·m acting during one-half of each rotation with speed 3600 rpm. Find the response at the two lower-frequency harmonics at the left disk.

8.4. At the pump of Problem 7.4, there is friction which applies on the pump rotor a constant torque of 20 N·m of duration one-tenth of the cycle at a speed of 3000 rpm. Find the motor torsional vibration at the synchronous frequency.

8.5. At the electric motor of Problem 7.5, there is a second harmonic of the voltage producing a harmonic torque at a frequency of 60 Hz. Determine the resulting torsional vibration of the blower.

8.6. On the top floor of the building of Problem 7.8, there is a reciprocating compressor that generates a harmonic horizontal force of amplitude 8 kN and a frequency of 157 rad/s. Find the dynamic response at the first floor.

8.7. On the machine of Problem 7.9, there is an unbalanced mass of 0.1 kg at a distance of 0.25 m from the axis of rotation, the axis of rotation is 0.3 m above the ground, and angular velocity 1750 rpm. Find the resulting vibration.

8.8. The rotor of a gas turbine is shown in Figure P8.8. On the center mass there is an unbalance of 80 g at a distance of 0.20 m, and the speed of rotation is 3600 rpm. The damping factor of the shaft material is 0.01. Determine the vibration at the midspan if $m_1 = 300$ kg, $m_2 = 350$ kg, $m_3 = 400$ kg, $L = 0.8$ m, and $d = 0.15$ m.

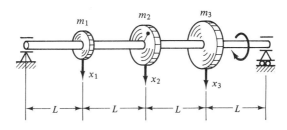

Figure P8.8

8.9. The rotor of a steam turbine is shown in Figure P8.9. On the center mass there is an unbalance of 80 g at a distance of 0.20 m, and the speed of rotation is 3600 rpm. Determine the vibration at the midspan if $m_1 = 300$ kg, $m_2 = 500$ kg, $m_3 = 400$ kg, $L = 0.8$ m, $d = 0.22$ m, $k_1 = k_2 = 50,000$ N/m, and $c_1 = c_2 = 800$ N·s/m.

8.10. The rotor of Problem 8.9 is balanced by placing two equal masses m_1 and m_2 of 40 g at the same distance 0.20 m but at a 180° phase angle with respect to the unbalance on mass m_2. Determine the vibration at midmass.

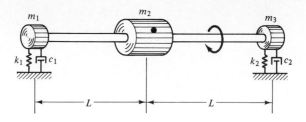

Figure P8.9

Sections 8.3 and 8.4

8.11–8.20. The forces in Problems 8.1 to 8.10, respectively, are applied suddenly. Find the resulting vibration.

8.21. The probability density function for the remaining unbalance after the balancing of rotors of small electric motors, on the basis of numerous production measurements, is nearly constant from 0 to 0.1 g·m (gram-meter). Determine the mean value and standard deviation of the unbalance force if the running speed is 3500 rpm.

8.22. On the basis of numerous production measurements, the probability density function for the remaining unbalance after the balancing of rotors of small electric motors varies linearly from 0 to 0.1 g·m. Determine the mean value and standard deviation of the unbalance force if the running speed is 3500 rpm.

8.23. On the basis of numerous soil tests, the probability density function for the soil elasticity in the foundation of a large compressor is nearly constant from 3×10^6 to 5×10^6 N/m³. Determine the mean value and standard deviation of the natural frequency of a concrete cube of 1 m edge length and 2600 kg mass for vertical vibration.

8.24. On the basis of numerous production measurements, the probability density function for the fraction of critical damping for an electronic instrument of 6 kg mass on an elastic mount of stiffness 2000 N/m is nearly constant from 0.2 to 0.4. Determine the mean value and standard deviation of the damped natural frequency.

8.25. On the basis of numerous measurements, the probability density function for the mass of a passenger car going over a wooden bridge of vertical stiffness 10^6 N/m is nearly constant from 1000 to 2000 kg. Determine the mean value and standard deviation of the natural frequency of a car on the bridge.

8.26–8.30. In Problems 8.21 to 8.25, respectively, assume normal distribution with mean the middle of the range of the variable and standard deviation one-fourth of the same range. Compute the percentage of the cases that will be above the high limit.

8.31–8.35. In Problems 8.21 to 8.25, respectively, assume Rayleigh distribution with mean the middle of the range of the variable and parameter α equal to one-fourth of the same range. Compute the percentage of the cases that will be above the high limit.

8.36–8.40. For the periodic waveforms shown in Figures P8.36 to P8.40, respectively (units: abscissa, mm, and ordinate, s), determine the Fourier transform, the autocorrelation function, and the spectral density function.

8.41. The mass of the rotor of a small electric motor is 20 kg and the bearing stiffness is 2×10^5 N/m. The mean value of the unbalance is 0.1 g·m and the standard deviation is 0.02 g·m. Determine the mean value and standard deviation of the dynamic response if the speed of rotation is 3500 rpm.

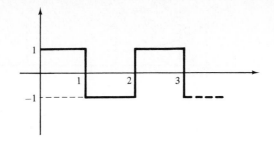

Figure P8.36

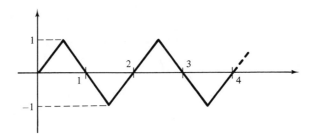

Figure P8.37

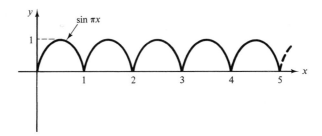

Figure P8.38

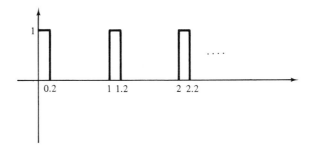

Figure P8.39

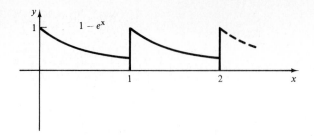

Figure P8.40

8.42. A refrigeration compressor is mounted on four springs of stiffness $k = 20,000$ N/m and has mass $m = 55$ kg. The mounts are made of spring steel with very low damping. Due to the design of the compressor, there is a vertical harmonic force of mean amplitude 12 N and standard deviation 3N, oscillating at the operating frequency of 1750 rpm. Determine the mean value and standard deviation of the vertical vibration of the compressor.

8.43. A piston-type air compressor of mass $m = 80$ kg is mounted on a platform of mass $M = 50$ kg that could vibrate horizontally without friction and had an elastic support with stiffness in the horizontal direction $k = 3500$ N/m. At the running frequency of 1150 rpm there is an unbalanced force of mean 40 N and a standard deviation of 10 N. Determine the mean value and standard deviation of the amplitude of the unbalanced horizontal force, neglecting damping.

8.44. An electric motor of mass $m = 22$ kg is mounted at the midspan of a simply supported steel beam of rectangular cross section, length $L = 1$ m, width $b = 0.2$ m, and thickness $t = 10$ mm. The amplitude of the unbalanced harmonic vertical force of the motor in the vertical direction is known to have mean 55 N and standard deviation 10 N at 58 Hz. Determine the mean value and standard deviation of the resulting vibration, neglecting damping.

8.45. The moving core of an electromagnetic relay has mass $m = 12$ g and it is supported at one end with a coil spring with $k = 3000$ N/m and at the other end, at the closed contact position, by the elastic strips of the electric contacts, which have stiffness 1200 N/m in the direction of coil motion (Figure P2.6). There is an unbalanced oscillating force due to the electric field along the core axis of mean value 1.3 N and $\sigma = 0.2$ N at the synchronous electric frequency of 60 Hz. Determine the mean and variance of the core vibration neglecting damping.

8.46–8.55. For the systems of Problems 8.1 to 8.10, find the mean value and standard deviation of the response if the value of the force given is the mean value and the standard deviation is 20% of the mean value.

9
Continuous Systems

9.1 INTRODUCTION

In previous chapters we studied the vibration of discrete systems, those that can be described with a finite number of degrees of freedom. It is almost impossible to find systems that can be described *exactly* this way. For many practical situations, however, an approximate description with one or more degrees of freedom is adequate for engineering applications. Consider, for example, the system in Figure 9.1.

In Figure 9.1a, a beam that weighs 2 kg supports a 100-kg mass. The error we commit by neglecting the weight of the beam is probably much less than other errors in the formulation of the problem, such as linearity of the material properties, rigidity of the supports, and small deflections in the beam theory. In Figure 9.1b half of the mass has been distributed on the beam. Here the error of neglecting the mass of the beam will be considerable. In Figure 9.1c all but 2 kg of the mass has been distributed over the beam. In this case the concentrated mass of 2 kg is

*The above photo portrays the cellist Guilhermina Suggia exciting standing waves on the cello strings by stick-slip. Portrait (1923) by Augustus John.

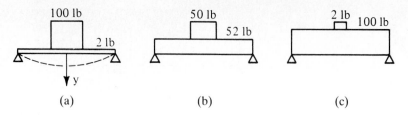

Figure 9.1 Beam with a concentrated mass.

probably so insignificant that it does not influence the vibration of the beam considerably. In this case we can only study the vibration of the beam with the distributed mass. In the case of a one-degree-of-freedom system model, the shape of the beam is assumed to be the *static* deflection due to a displacement y under the mass m. If the mass is distributed, we need an infinite number of coordinates to describe where the beam is at time t. In other words, the displacement is a function of the coordinate x and the time.

The study of the vibration of systems with distributed mass is of higher complexity than the one-degree-of-freedom system. Closed-form solutions exist only for some special cases. Study of these systems has didactic appeal primarily. In this chapter we provide only a brief introduction.[1]

9.2 THE ONE-DIMENSIONAL STRUCTURAL MEMBER WAVE EQUATION

Systems with continuous mass extend naturally in three dimensions. However, for many practical applications, we can make the approximation that all the points along a section perpendicular to one direction move together; in other words, the motion of this section is much greater than the relative motion of any two points on the section. A cylindrical member, for example a string, a beam, a rod, moving either along its axis or perpendicular to it can be considered as a one-dimensional structure, or *line structure*. We shall first consider one-dimensional problems.

Consider the mass of a line structure distributed in a certain way along the coordinate axis x. The line density (mass per unit length) is $\mu(x)$ (Figure 9.2). If $w(x, t)$ is the displacement of the axis of symmetry of this member along a given direction, application of Newton's law for a slice of the mass of length dx will be

$$F_w = \mu(x) \, dx \, \frac{\partial^2 w}{\partial t^2} \tag{9.1}$$

where F_w is the force acting on this mass due to the elasticity of the system. The following circumstances can give rise to a force F_w due to the displacement $w(x, t)$ of the section at the point x and time t.

[1]For a more complete discussion, see (a) [Rayleigh, 1894]; (b) Barre de Saint-Venant. 1883. *Théorie de l'élasticité des corps solides*. Paris; (c) [Timoshenko and Young, 1955].

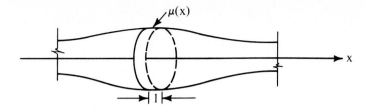

Figure 9.2 Mass distribution for a bar.

Taut String

We consider a string that is taut initially with a tension T. We assume that deflection of the string does not affect the tension of the string, so that tension remains constant at all times. We shall find the force F_y due to the lateral displacement of the string. The situation is shown in Figure 9.3.

We consider an element of length dx displaced by $y(x, t)$. Here we shall make use of the fundamental theorem of calculus. The value of a function $y = f(x)$ at $x + \Delta x$ is

$$f(x + \Delta x) = f(x) + \frac{df(x + \xi \Delta x)}{dx} \Delta x \tag{9.2}$$

where $0 \le \xi \le 1$. For $\Delta x \to 0$, we have

$$f(x + dx) = f(x) + \frac{df}{dx} dx \tag{9.3}$$

If f is also a function of additional variables, we have

$$f(x + dx, x_1, x_2, \ldots) = f(x, x_1, x_2, \ldots) + \frac{\partial f}{\partial x} dx \tag{9.4}$$

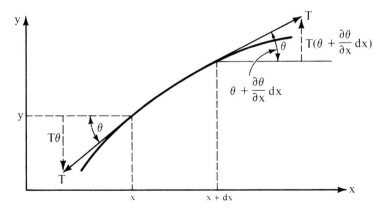

Figure 9.3 Deflected element of a taut string.

Application of equation (9.4) to the slope $\theta(x, t)$ yields

$$\theta(x + dx, t) = \theta(x) + \frac{\partial \theta}{\partial x} dx \tag{9.5}$$

Therefore, the sum of the components of T in the y direction (Figure 9.3) is

$$F_y = T\left(\theta + \frac{\partial \theta}{\partial x} dx\right) - T\theta = T\frac{\partial \theta}{\partial x} \tag{9.6}$$

or, in view of the fact that $\theta = \partial y/\partial x$,

$$F_y = T\frac{\partial^2 y}{\partial x^2} dx \tag{9.7}$$

Equations (9.1) and (9.7) yield for the taut string the *wave equation*[2]

$$\frac{\partial^2 y}{\partial x^2} = \frac{1}{c^2}\frac{\partial^2 y}{\partial t^2} \tag{9.8}$$

where $c = (T/\mu)^{1/2}$.

Longitudinal Vibrations of Rods

Along the coordinate x we assume a motion $u(x, t)$ of the rod in the x direction. Let us consider an element of length dx, at x. The motion and force at $x + dx$ will be, per equation (9.4),

$$u(x + dx, t) = u(x, t) + \frac{\partial u}{\partial x} dx \tag{9.9}$$

$$P(x + dx, t) = P(x, t) + \frac{\partial P}{\partial x} dx \tag{9.10}$$

According to Hooke's law, the change in the length of a prismatic bar of length dx is related to the transmitted force P is

$$\Delta(dx) = \left\{u(x, t) + \frac{\partial u}{\partial x} dx\right\} - u(x, t) = P\frac{dx}{AE} \tag{9.11}$$

or

$$\frac{\partial u}{\partial x} = \frac{P}{AE} \tag{9.12}$$

where A is the cross section and E the modulus of elasticity. The net force on the element is, from equation (9.10),

$$F_x = \left\{P(x, t) + \frac{\partial P}{\partial x} dx\right\} - P(x, t) = \frac{\partial P}{\partial x} dx \tag{9.13}$$

---

[2]Introduced by D'Alembert in a memoir to the Berlin Academy of 1750.

Differentiating (9.12) and using (9.13), we obtain from (9.1),

$$\frac{\partial}{\partial x}\left(EA\frac{\partial u}{\partial x}\right) = \mu\frac{\partial^2 u}{\partial t^2} \tag{9.14}$$

Equation (9.14) is the differential equation of motion for any form of E, A, and μ along the rod. For a rod of constant properties along its length, we obtain[3]

$$\frac{\partial^2 u}{\partial x^2} = \frac{1}{c^2}\frac{\partial^2 u}{\partial t^2} \tag{9.15}$$

where $c = (E/\rho)^{1/2}$, and $\rho = \mu/A$ is the density of the rod material.

Lateral Vibration: The Euler–Bernoulli Beam

Figure 9.4 shows an element of a beam of length dx. From strength of materials [Beer and Johnston, 1982] we know that for lateral deflection of the beam $y(x, t)$,

$$M = EI\frac{\partial^2 y}{\partial x^2} \tag{9.16}$$

$$V(x, t) = \frac{\partial M(x, t)}{\partial x} \tag{9.17}$$

In the y direction, the net force F_y will be

$$F_y = -\left(V + \frac{\partial V}{\partial x}\,dx\right) + V = -\frac{\partial V}{\partial x}\,dx \tag{9.18}$$

Differentiating (9.17) and using (9.16) and (9.18), we obtain

$$F_y = -\frac{\partial^2 M}{\partial x^2}\,dx = \frac{-\partial^2}{\partial x^2}\left(EI\frac{\partial^2 y}{\partial x^2}\right)dx \tag{9.19}$$

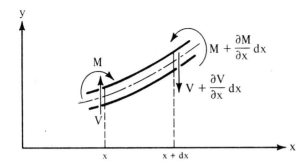

Figure 9.4 Deflected element of a beam in bending.

[3]Used by D'Alembert in his memoir to the Berlin Academy in 1750, also for longitudinal vibration of air columns in pipe organs. Experimental results for the same problem were obtained by Pythagoras.

Therefore, equation (9.1) becomes

$$-\frac{\partial^2}{\partial x^2}\left(EI\frac{\partial^2 y}{\partial x^2}\right) = \mu\frac{\partial^2 y}{\partial t^2} \tag{9.20}$$

If the properties of the beam are constant along its length, equation (9.20) becomes[4]

$$-\frac{\partial^4 y}{\partial x^4} = \frac{1}{c^2}\frac{\partial^2 y}{\partial t^2} \tag{9.21}$$

where $c = (EI/\mu)^{1/2}$.

Torsional Vibration of Rods

In Figure 9.5 a rod of circular cross section is shown. The twist $d\theta$ of an element of length dx will be, as we know from strength of materials [Beer and Johnston, 1982],

$$d\theta = \frac{T}{I_p G}\,dx \tag{9.22}$$

where I_p is the polar moment of inertia and G the material shear modulus. The net torque ΔT_θ on the element is $(\partial T/\partial x)\,dx$ or, using (9.22),

$$\Delta T_\theta = \frac{\partial}{\partial x}\left(I_p G\frac{\partial\theta}{\partial x}\right)dx \tag{9.23}$$

The moment of inertia of the element is $I_p\rho\,dx$, where ρ is the material density and the angular acceleration is $\partial^2\theta/\partial t^2$. Application of Newton's law for rotation in the θ direction yields

$$\frac{\partial}{\partial x}\left(I_p G\frac{\partial\theta}{\partial x}\right) = \rho I_p\frac{\partial^2\theta}{\partial t^2} \tag{9.24}$$

For constant properties along the length of the rod, we have[5]

$$\frac{\partial^2\theta}{\partial x^2} = \frac{1}{c^2}\frac{\partial^2\theta}{\partial t^2} \tag{9.25}$$

where

$$c = \left(\frac{G}{\rho}\right)^{1/2}$$

We observe that all cases discussed above can be expressed in the form

$$\mathbf{L}(w) = \frac{1}{c^2}\frac{\partial^2 w}{\partial t^2} \tag{9.26}$$

---

[4]Lateral vibration of beams was discussed by Euler in 1744 and Daniel Bernoulli in 1751. Equation (9.21) is called the Euler–Bernoulli equation, but in this form the equation appears to have been introduced by Jakob Bernoulli. 1789. *Nova Acta*. St. Petersburg.

[5][Timoshenko, 1928]; see also Timoshenko, S. 1911. Erzwungene Schwingungen Prismatischer Stäbe. Z. *Math. Phys.*, 59: 163.

Suppose that at $t = 0$ we impose a displacement $F_1 = F_2 = f(x)$, say to the vibrating string, where $f(x)$ is an even function, as shown in Figure 9.6. Because $t = 0$ we shall have

$$w(x, 0) = f(x) + f(-x) = 2f(x) \tag{9.28}$$

because we assumed that $f(x) = f(-x)$.

At time $t = 1$, we shall have

$$w(x, 1) = f(c - x) + f(c + x) \tag{9.29}$$

This function is equal to the function $f(x)$ displaced by c to both the left and right of $x = 0$. One can see immediately that the wave $f(x)$ traveled in time $t = 1$ a distance $x = c$ in both directions. Therefore, c has a simple physical interpretation; it is the velocity of the propagation of the wave along the one-dimensional structural member.

For this reason the second-order form of equation (9.26) is called a wave equation.

□ **Example 9.1 Alternative Form of the Wave Equation**

Find a linear transformation to transform the wave equation into the form $\partial^2 y/\partial\xi\, \partial\eta = 0$ and find its solution.

Solution Let a linear transformation $\xi = \alpha x + \beta t$ and $\eta = \alpha x - \beta t$, where α and β are constants to be determined. Since $y = y[x(\xi, \eta), t(\xi, \eta)]$,

$$\frac{\partial y}{\partial x} = \frac{\partial y}{\partial \xi}\frac{\partial \xi}{\partial x} + \frac{\partial y}{\partial \eta}\frac{\partial \eta}{\partial x} = \alpha\,\frac{\partial y}{\partial \xi} + \alpha\,\frac{\partial y}{\partial \eta}\frac{\partial^2 y}{\partial x^2}$$

$$= \frac{\partial}{\partial \xi}\left(\alpha\,\frac{\partial y}{\partial \xi} + \alpha\,\frac{\partial y}{\partial \eta}\right)\frac{\partial \xi}{\partial x} + \frac{\partial}{\partial \eta}\left(\alpha\,\frac{\partial y}{\partial \xi} + \alpha\,\frac{\partial y}{\partial \eta}\right)\frac{\partial \eta}{\partial x}$$

$$= \alpha^2\left(\frac{\partial^2 y}{\partial \xi^2} + \frac{\partial^2 y}{\partial \eta^2} + 2\,\frac{\partial^2 y}{\partial \xi\, \partial \eta}\right)$$

Similarly,

$$\frac{\partial^2 y}{\partial t^2} = \beta^2\left(\frac{\partial^2 y}{\partial \xi^2} + \frac{\partial^2 y}{\partial \eta^2} - 2\,\frac{\partial^2 y}{\partial \xi\, \partial \eta}\right)$$

The wave equation becomes

$$\alpha^2\left(\frac{\partial^2 y}{\partial \xi^2} + \frac{\partial^2 y}{\partial \eta^2} + 2\,\frac{\partial^2 y}{\partial \xi\, \partial \eta}\right) = \frac{\beta^2}{c^2}\left(\frac{\partial^2 y}{\partial \xi^2} + \frac{\partial^2 y}{\partial \eta^2} - \frac{\partial^2 y}{\partial \xi\, \partial \eta}\right)$$

If we assign $\alpha = 1$, $\beta = c$ we have

$$\frac{\partial^2 y}{\partial \xi\, \partial \eta} = 0$$

Thus $\xi = x + ct$ and $\eta = x - ct$ is the transformation.

The solution of the equation above is a sum of an arbitrary function of ξ and an arbitrary function of η. Thus the solution is[6]

$$w(x, t) = F_1(\eta) + F_2(\xi) = F_1(x - ct) + F_2(x + ct)$$

[6]Love, A. E. H. 1927. *Mathematical Theory of Elasticity*. London.

(In this example, the wave equation was transformed in a form that allows direct integration. Moreover, in this form of the solution, the wave character of the motion can be demonstrated.)

☐

☐ **Example 9.2 Speed of Sound in Solid and Air**

Miners trapped in a mine signal for help by knocking on the compressed air pipe. Determine the distance of the miners from the measured time difference of 0.2 s between the signal arriving through the steel pipe and through the air in the pipe. Assume longitudinal motion of the pipe and normal environmental conditions. Densities: air, $\rho = 1.29$ kg/m³; steel, $\rho = 7800$ kg/m³.

Solution The velocity of wave propagation for longitudinal vibration is (Table 9.1)

$$c = \left(\frac{E}{\rho}\right)^{1/2}$$

where, in general, $E = \sigma/\epsilon$. For isentropic compression of air, $pV^k = p_0 V_0^k$. For an air column of cross section A, free length x_0, and compressed length $x = x_0 - \Delta x$,

$$V = Ax, \qquad \sigma = p - p_0, \qquad \epsilon = \frac{\Delta x}{x_0}, \qquad E = \frac{(p - p_0)x_0}{\Delta x}$$

But

$$p[A(x_0 - \Delta x)]^k = p_0[Ax_0]^k, \qquad p = p_0\left(\frac{x_0}{x_0 - \Delta x}\right)^k$$

$$p = p_0 \frac{1}{[1 - (\Delta x/x_0)]^k} \approx 1 + k\frac{\Delta x}{x_0}, \qquad p - p_0 = p_0 k\frac{\Delta x}{x_0}, \qquad E \approx kp_0$$

assuming small Δx. Therefore, we have for the air at normal environment (temperature 20°C, pressure 1 bar) $E_a \approx kp_0 = 1.4 \times 9.81 \times 10^4 = 1.37 \times 10^5$ N/m². Therefore,

$$C_{\text{air}} = \left(\frac{E_a}{\rho}\right)^{1/2} = \left(\frac{1.37 \times 10^5}{1.29}\right)^{1/2} = 326 \text{ m/s}$$

$$C_{\text{pipe}} = \left(\frac{E}{\rho}\right)^{1/2} = \left(\frac{2.1 \times 10^{11}}{7800}\right)^{1/2} = 5189 \text{ m/s}$$

Due to the approximations used, the computed speed of sound in the air is somewhat smaller than the known speed of 343 m/s for the given conditions.

The time delay is $\Delta t = l/\Delta C$; therefore,

$$l = \Delta C \, \Delta t = (C_{\text{pipe}} - C_{\text{air}}) \, \Delta t = 4863 \times 0.2 = 972.6 \text{ m}$$

(This example demonstrates computation of the speed of sound in air and in metals.)

☐

☐ **Example 9.3 Speed of Sound in a Stretched Wire**

Determine the velocity of the wave propagation on an electric transmission line stretched by a constant tension $T = 20,000$ N and having mass per unit length 1.5 kg/m.

Solution For the taut spring, the speed of sound is computed with Table 9.1 as

$$c = \left(\frac{T}{\mu}\right)^{1/2} = \left(\frac{20,000}{1.5}\right)^{1/2} = 115 \text{ m/s}$$

(This example demonstrates computation of the speed of sound in a taut string.)

□

9.3 INITIAL AND BOUNDARY CONDITIONS

We have seen that the solution (9.27) satisfies the wave equation for arbitrary functions F_1 and F_2. The range of the selection of these functions is considerably narrowed in view of certain additional restraints, the initial and boundary conditions. In fact, it can be proved that only then is the solution *unique*, that is, there is only one function that satisfies the differential equation and all conditions.

The function $w(x, t)$ gives the displacement along the distributed mass. At time $t = 0$, as for the one-degree-of-freedom system, the displacement and velocity are usually known. This time, they are functions of the coordinate x along the distributed mass:

$$w(x, 0) = w_0(x) \tag{9.30}$$

$$\left. \frac{\partial w}{\partial t} \right|_{t=0} = v_0(x) \tag{9.31}$$

In addition, at certain points along x there are restrictions called *boundary conditions*, the most usual of which are listed below.

Taut Spring

Along the x-axis one or more points are fixed. In this case the boundary conditions are

$$y(l_1, t) = 0 \tag{9.32}$$
$$y(l_2, t) = 0$$

where l_1 and l_2 are the distances of the fixed points from the origin. Usually, we take $l_1 = 0$ and $l_2 = l$. Boundary conditions on the geometric configuration (displacement, slope) are *geometric boundary conditions*. Boundary conditions on the forces and moments are *natural boundary conditions*. Combinations of the two types are *mixed boundary conditions*.

Longitudinal Vibration of Rods

Due to equation (9.12), one can compile Table 9.2 with the usual boundary conditions for longitudinal vibration of rods.

TABLE 9.2 BOUNDARY CONDITIONS FOR LONGITUDINAL VIBRATION OF RODS

Case	Boundary condition left, $x = 0$	Boundary condition right, $x = l$
Free end	$\dfrac{\partial u}{\partial x} = 0$	$\dfrac{\partial u}{\partial x} = 0$
Fixed end	$u(0, t) = 0$	$u(l, t) = 0$
End spring	$AE\dfrac{\partial u}{\partial x} = ku$	$AE\dfrac{\partial u}{\partial x} = -ku$
End mass	$AE\dfrac{\partial u}{\partial x} = m\dfrac{\partial^2 u}{\partial t^2}$	$AE\dfrac{\partial u}{\partial x} = -m\dfrac{\partial^2 u}{\partial t^2}$
End damper	$AE\dfrac{\partial u}{\partial x} = c\dfrac{\partial u}{\partial t}$	$AE\dfrac{\partial u}{\partial x} = -c\dfrac{\partial u}{\partial t}$

TABLE 9.3 BOUNDARY CONDITIONS FOR TORSIONAL VIBRATION OF RODS

Case	Boundary condition left, $x = 0$	Boundary condition right, $x = l$
Fixed end	$\theta(0, t) = 0$	$\theta(l, t) = 0$
Free end	$\dfrac{\partial \theta}{\partial x} = 0$	$\dfrac{\partial \theta}{\partial x} = 0$
Torsional spring	$I_p G\dfrac{\partial \theta}{\partial x} = k\theta$	$I_p G\dfrac{\partial \theta}{\partial x} = -k\theta$
Inertia J_p	$I_p G\dfrac{\partial \theta}{\partial x} = J_p\dfrac{\partial^2 \theta}{\partial t^2}$	$I_p G\dfrac{\partial \theta}{\partial x} = -J_p\dfrac{\partial^2 \theta}{\partial t^2}$
Torsional damper	$I_p G\dfrac{\partial \theta}{\partial x} = c\dfrac{\partial \theta}{\partial t}$	$I_p G\dfrac{\partial \theta}{\partial x} = -c\dfrac{\partial \theta}{\partial t}$

Torsional Vibration of Rods

Using equation (9.22), the most usual boundary conditions for torsional vibration of rods have been compiled in Table 9.3.

Lateral Vibration of Beams

A beam of length l has two boundary conditions at every end. The *state* of every section of a beam is completely specified if we specify position (y), slope ($\partial y/\partial x$), moment ($EI\,\partial^2 y/\partial x^2$) and shear ($EI\,\partial^3 y/\partial x^3$). Two conditions relating the four are usually known at every boundary section. Therefore, we can compile Table 9.4.

These tables are by no means complete. One can develop boundary conditions for combinations of the given conditions or other ones.

TABLE 9.4 BOUNDARY CONDITIONS FOR LATERAL VIBRATION OF BEAMS

Case	Boundary condition left, at $x = 0$	Boundary condition right, at $x = l$
Clamped (deflection, slope = 0)	$y(0, t) = 0$ $\quad \dfrac{\partial y}{\partial x} = 0$	$y(l, t) = 0$ $\quad \dfrac{\partial y}{\partial x} = 0$
Pinned (deflection, moment = 0)	$y(0, t) = 0$ $\quad \dfrac{\partial^2 y}{\partial x^2} = 0$	$y(l, t) = 0$ $\quad \dfrac{\partial^2 y}{\partial x^2} = 0$
Sliding (slope, shear = 0)	$\dfrac{\partial y}{\partial x} = 0$ $\quad \dfrac{\partial^3 y}{\partial x^3} = 0$	$\dfrac{\partial y}{\partial x} = 0$ $\quad \dfrac{\partial^3 y}{\partial x^3} = 0$
Free (moment, shear = 0)	$\dfrac{\partial^2 y}{\partial x^2} = 0$ $\quad \dfrac{\partial^3 y}{\partial x^3} = 0$	$\dfrac{\partial^2 y}{\partial x^2} = 0$ $\quad \dfrac{\partial^3 y}{\partial x^3} = 0$
Mass m and moment of inertia J_p	$EI\,\dfrac{\partial^2 y}{\partial x^2} = -J_p\,\dfrac{\partial^3 y}{\partial x\,\partial t^2}$ $\quad EI\,\dfrac{\partial^3 y}{\partial x^3} = -m\,\dfrac{\partial^2 y}{\partial t^2}$	$EI\,\dfrac{\partial^2 y}{\partial x^2} = J_p\,\dfrac{\partial^3 y}{\partial x\,\partial t^2}$ $\quad EI\,\dfrac{\partial^3 y}{\partial x^3} = m\,\dfrac{\partial^2 y}{\partial t^2}$
Damper c and spring k	$EI\,\dfrac{\partial^3 y}{\partial x^3} = -ky - c\,\dfrac{\partial y}{\partial t}$ $\quad \dfrac{\partial^2 y}{\partial x^2} = 0$	$EI\,\dfrac{\partial^3 y}{\partial x^3} = ky + c\,\dfrac{\partial y}{\partial t}$ $\quad \dfrac{\partial^2 y}{\partial x^2} = 0$

□ **Example 9.4   Boundary Conditions of a Cantilever Beam[7]**

List the boundary conditions for lateral vibration of the two cantilever beams shown in Figure E9.4a and b of length L and flexural rigidity EI.

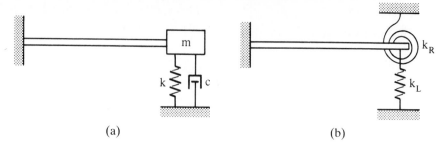

(a)   (b)

Figure E9.4

Solution   (a) The displacement and slope are equal to zero at the built-in end, whereas at the free end inertia force is equal to the shearing force:

$$y\big|_{x=0} = 0, \qquad \left(\frac{\partial y}{\partial x}\right)_{x=0} = 0$$

$$EI\frac{\partial^3 y}{\partial x^3} + ky\bigg|_{x=L} = m\frac{\partial^2 y}{\partial t^2} + c\frac{\partial y}{\partial t}\bigg|_{x=L}$$

$$EI\frac{\partial^2 y}{\partial x^2}\bigg|_{x=L} = 0$$

(b) Similarly, the displacement and slope are zero at the built-in end whereas the shearing force and bending moment are balanced by spring forces:

$$y\big|_{x=0} = 0, \qquad \frac{\partial y}{\partial x}\bigg|_{x=0} = 0$$

$$EI\frac{\partial^2 y}{\partial x^2}\bigg|_{x=L} = k_R\frac{\partial y}{\partial x}\bigg|_{x=L}$$

$$EI\frac{\partial^3 y}{\partial x^3}\bigg|_{x=L} = k_L(y)\big|_{x=L}$$

(This example demonstrates different boundary conditions of a cantilever beam.)

□

9.4 MODES OF NATURAL VIBRATION: THE EIGENVALUE PROBLEM

In Chapter 2 we have seen that the one-degree-of-freedom system has the property that any disturbance from equilibrium results in vibration at a single frequency ω_n, which does not depend on the disturbance itself but only on the properties m, c, k of the system. We called the resulting motion *natural vibration*.

[7]Small cantilever beams of varying length are used to measure vibration frequencies. The device is called a Frahm tachometer [Timoshenko, 1928].

It is natural to expect that continuous systems will exhibit similar behavior. Thus we pose the question: Can a continuous system undergo harmonic vibration; in other words, is it possible that every point of the system can have harmonic vibration $y = A \cos \omega t$ at some unknown yet circular frequency ω? Of course, we do not expect that every point x will have the same amplitude A. For example, for a simply supported beam, A must be zero at the ends of the beam, and we know intuitively that between supports it will take, in general, finite values. It is natural to search for a solution with A varying with x or $A = Y(x)$. Therefore, we shall investigate the possibility that the system can undergo vibration of the type $y(x, t) = Y(x) \cos \omega t$. This expression would assume that every point of the system has harmonic vibration of circular frequency ω and amplitude $Y(x)$. To make our investigation more general, we shall search for solutions of the type $y(x, t) = Y(x)T(t)$, where $T(t)$ is a function of time.

In partial differential equations, this is known as the *method of separation of variables*. We shall look for functions that satisfy the differential equation

$$\mathbf{L}(w) = \frac{1}{c^2} \frac{\partial^2 w}{\partial t^2} \tag{9.33}$$

where $\mathbf{L}(w)$ is a linear partial differential operator $\partial^2/\partial x^2$ or $-\partial^4/\partial x^4$ depending on the case, as in Table 9.1. We shall denote by $\mathbf{L}$ the partial differential operator and by L the operator

$$L(W) = \frac{W''}{W} \quad \text{or} \quad -\frac{W^{(iv)}}{W} \tag{9.34}$$

depending on the order of the operator $\mathbf{L}$. We shall investigate the possibility of existence of solutions of (9.33) of the form

$$w(x, t) = W(x)T(t) \tag{9.35}$$

We observe that

$$\frac{\partial^2 w}{\partial t^2} = W(x)T''(t)$$

$$\frac{\partial^2 y}{\partial x^2} = W''(x)T(t) \tag{9.36}$$

$$\frac{\partial^4 y}{\partial x^4} = W^{(iv)}(x)T(t)$$

Therefore, equation (9.33) yields

$$L(W) = \frac{1}{c^2} \frac{T''}{T} \tag{9.37}$$

We observe that the left-hand side of equation (9.37) is a function of x only, while the right is a function of t only. The equation is satisfied for any values of x and t only if both sides are equal to a constant. Let this (yet unknown) constant be b. We shall have

$$L(W) = b \qquad (9.38)$$

$$\frac{T''}{T} = c^2 b \qquad (9.39)$$

Equation (9.39) becomes

$$T'' - c^2 bT = 0 \qquad (9.40)$$

which has the general solution

$$T = B_1 e^{c\sqrt{b}t} + B_2 e^{-c\sqrt{b}t} \qquad (9.41)$$

Now we impose the first restriction on b: It has to be negative; otherwise, the function T goes to infinity for $t \to \infty$. Such a solution is not admissible as it is not physically realizable. Therefore, we set up b in the form

$$b = -\frac{\omega^2}{c^2} \qquad (9.42)$$

where ω is another, yet unknown constant and the solution (9.41) becomes the harmonic function

$$T = A_1 \cos \omega t + A_2 \sin \omega t \qquad (9.43)$$

where A_1, A_2, and ω as yet undetermined constants. The solution of equation (9.38) depends on the form of the operator L.

The constant ω is still undetermined. It has to be chosen such that equation (9.38) has a nontrivial solution satisfying the boundary conditions of system. The problem of finding such values of ω is called *eigenvalue problem* (or *characteristic value* or *proper value* or *latent roots* problem).

Second-Order Operator L

Equation (9.38) yields for the case of the second-order operator

$$W'' + \left(\frac{\omega}{c}\right)^2 W = 0 \qquad (9.44)$$

and the general solution of this differential equation is,[8] as obtained in Chapter 2,

$$W(x) = C_1 \cos \frac{\omega}{c} x + C_2 \sin \frac{\omega}{c} x \qquad (9.45)$$

where C_1 and C_2 are as yet undetermined constants.

The homogeneous boundary conditions on $y(x, t)$ apply on $W(x)T(t)$. For example, $y(0, t) = 0$ implies that $W(0)T(t) = 0$. Because $T(t)$ is not 0 for all values of time, we have $W(0) = 0$. Therefore, we have two boundary conditions on $W(x)$. This will yield two algebraic, homogeneous equations for C_1 and C_2. For example, if $W(0) = W(l) = 0$, we obtain, from (9.45),

---

[8]The solution of the string equation is due to Daniel Bernoulli, D'Alembert, and Euler, although Joseph Sauveur (1653–1716) and Brook Taylor (1685–1731) have previously obtained approximate solutions.

$$C_1 = 0$$

$$\tag{9.46}$$

$$C_1 \cos \frac{\omega}{c} l + C_2 \sin \frac{\omega}{c} l = 0$$

We conclude that $C_1 = 0$ and $C_2 \sin (\omega/c)l = 0$. The second equation implies either that $C_2 = 0$ or that $\sin (\omega/c)l = 0$. $C_2 = 0$ is not admissible because we are interested in nonzero solutions. Therefore,

$$\sin \frac{\omega l}{c} = 0 \tag{9.47}$$

This is true only if

$$\omega = \frac{\pi nc}{l} \tag{9.48}$$

where n is any positive integer. Equation (9.47) is called a *characteristic equation* and permits the determination of the unknown ω, which due to equation (9.43) now has an obvious physical meaning: it is the frequency of natural vibration.[9] Therefore, the boundary conditions yielded the constants C_1, C_2, C_3, C_4, and ω. Equation (9.48) indicates that there are infinite values of ω that are admissible, in other words, compatible with the boundary conditions. The value of ω for each value of integer n is called an *eigenvalue* or *natural frequency* of the nth order and is written as ω_n. The corresponding functions that satisfy equation (9.38) and the boundary conditions, such as the one defined in equation (9.45), signify the spatial form of the natural vibration and are called *eigenfunctions* or *natural modes of vibration*. Moreover, they represent waves which do not move along the length, called *standing waves*. The analogy of natural frequencies and natural modes with those for lumped mass systems is apparent.

□ **Example 9.5 Natural Frequencies of a Taut String**

The cord of a musical instrument of length $l = 1$ m, diameter $d = 1$ mm, and density 8000 kg/m³ is taut with 200 N tension. Determine the first two natural frequencies, the vibration modes, and the corresponding nodal points, those with zero vibration amplitudes.

Solution   For the taut string, $\omega_n = \pi nc/l$ are the natural frequencies [equation (9.48)] and the vibration modes are $W(x) = C_2 \sin(\omega/c)x$ [equation (9.44), $C_1 = 0$]. The velocity of wave propagation $c = (T/\mu)^{1/2}$. But $\mu = m/l =$

$$\frac{\pi \rho d^2}{4} = \frac{\pi \times 8000 \times 0.001^2}{4} = 0.00628 \text{ kg/m}$$

Thus

$$c = \left(\frac{200}{0.00628} \right)^{1/2} = 178 \text{ m/s}$$

$$\omega_n = \frac{\pi nc}{l} = \frac{\pi \times 178n}{1} = 558n \text{ rad/s}$$

[9]Pythagoras determined experimentally that ω is inversely proportional to the length and diameter, and depends on tension with "other analogies."

$$\omega_1 = 558, \qquad \omega_2 = 1116, \qquad \omega_3 = 1674 \text{ rad/s}$$

$$W_1 = \sin\frac{\omega_1}{c}x = \sin\frac{558x}{178} = \sin 3.14x$$

$$W_2 = \sin\frac{\omega_2}{c}x = \sin 6.28x$$

$$W_3 = \sin\frac{\omega_3}{c}x \sin 9.42x$$

Mode	1	2	2
Nodal points	$0, l$	$0, l/2, l$	$0, l/3, 2l/3, l$

The first 3 modes are shown in Figure E9.5.

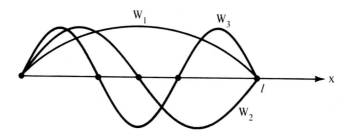

Figure E9.5

(This example demonstrates the determination of natural frequencies and modes of a taut string.)

☐

☐ **Example 9.6   Mass Supported by Heavy Helical Spring**

A mass m is supported by a helical spring of unstretched length l, spring constant k, and mass m (Figure E9.6). Determine the natural frequencies if the spring mass is nonnegligible.

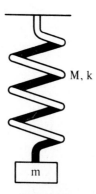

Figure E9.6

Solution The differential equation is

$$\frac{\partial^2 y}{\partial x^2} = \frac{1}{c^2}\frac{\partial^2 y}{\partial t^2}$$

with boundary conditions

$$y(0, t) = 0 \qquad\qquad\qquad\qquad\qquad\text{(a)}$$

$$AE\frac{\partial y}{\partial x} = -m\frac{\partial^2 y}{\partial t^2} \quad \text{at } x = l \qquad\qquad\text{(b)}$$

Separation of variables gives one solution,

$$y_n(x, t) = (A \cos \omega_n t + B \sin \omega_n t)\left(C \cos \frac{\omega_n}{c}x + D \sin \frac{\omega_n}{c}x\right)$$

The boundary conditions will now be applied: (a) yields $C = 0$; (b) yields

$$AE(A \cos \omega_n t + B \sin \omega_n t)\left(-C\frac{\omega_n}{c}\sin\frac{\omega_n l}{c} + D\frac{\omega_n}{c}\cos\frac{\omega_n l}{c}\right)$$

$$= -m(-A\omega_n^2 \cos \omega_n t - B\omega_n^2 \sin \omega_n t)\left(c \cos \frac{\omega_n l}{c} + D \sin \frac{\omega_n l}{c}\right)$$

Thus

$$AE\frac{\omega_n}{c}\cos\frac{\omega_n}{c}l = m\omega_n^2 \sin\frac{\omega_n l}{c}$$

or

$$\tan\frac{\omega_n l}{c} = \frac{AE}{mc\omega_n}$$

But for an elastic column of equal stiffness,

$$k = \frac{AE}{l}$$

Therefore, the frequency equation is

$$\frac{\omega_n l}{c}\tan\frac{\omega_n l}{c} = \frac{kl^2}{mc^2} = \frac{M}{m}$$

because $c = (E/\rho)^{1/2} = (EAl/M)^{1/2}$.
If $M \approx 0$, $C \to \infty$, $(\omega_n/c)l \to 0$, and $\tan(\omega_n/c)l \approx (\omega_n/c)l$. Thus

$$\left(\frac{\omega_n l}{c}\right)^2 = \frac{kl^2}{mc^2}, \qquad \omega_n^2 = \frac{k}{m}$$

as expected. The frequency equation can be written as $\lambda_n \tan \lambda_n = M/m$, where $\lambda_n = \omega_n l/c$ and the roots λ_n can be found in tables or numerically. The natural frequencies will be then $\omega_n = \lambda_n c/l$.

(This example demonstrates the determination of natural frequency of a mass on a helical spring of nonnegligible mass.)

□

Example 9.7 Torsional Vibration of a Circular Cylinder

For a uniform circular shaft of diameter E, shear modulus G, density ρ, and length l, fixed at both ends, determine (a) the differential equation of motion for torsional vibration, (b) the boundary conditions, and (c) the frequency equation and the natural frequencies.

Solution (a) The differential equation of motion is

$$\frac{\partial^2 \theta}{\partial x^2} = \frac{1}{c^2} \frac{\partial^2 \theta}{\partial t^2}, \qquad c = \left(\frac{G}{\rho}\right)^{1/2}$$

(b) The boundary conditions, with ends fixed, are

$$\theta(0, t) = \theta(l, t) = 0$$

(c) The general solution is

$$\theta(x, t) = \sum_{n=1}^{\infty} (A_n \cos \omega_n t + B_n \sin \omega_n t) \left(C_n \cos \frac{\omega_n x}{c} + D_n \sin \frac{\omega_n x}{c} \right)$$

For $\theta(0, t) = 0$, we obtain $C_n = 0$. For $\theta(l, t) = 0$, $D_n \sin \omega_n(l/c) = 0$ or (for $D_n \neq 0$), $\sin \omega_n(l/c) = 0$, $n = 1, 2, 3, \ldots$. The natural frequencies are $\omega_n = n\pi c/l$.

(This example demonstrates the determination of torsional natural frequencies and modes of a circular cylinder.)

□

Example 9.8 Torsional Circular Cylinder with Two Disks at the Ends[10]

Derive the frequency equation for the torsional vibration of a uniform circular shaft with disks J_1 and J_2 attached rigidly at the ends as shown in Figure E9.8.

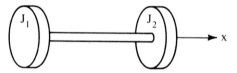

Figure E9.8

Solution The general solution for the torsional vibration of circular shafts can be expressed in the following form:

$$\theta(x, t) = \sum_{i=1,2,\ldots} (A_i \cos \omega_i t + B_i \sin \omega_i t) \left(C_i \cos \frac{\omega_i x}{c} + D_i \sin \frac{\omega_i x}{c} \right)$$

where $c^2 = (G/\rho)^{1/2}$ and ω_i = natural frequencies of the shaft.

The twisting of the shaft at both ends is produced by the inertia forces of the rotors; the boundary conditions are:

$$\text{At } x = 0: \qquad J_1 \frac{\partial^2 \theta}{\partial t^2} = GI_p \frac{\partial \theta}{\partial x} \qquad \text{(a)}$$

$$\text{At } x = L: \qquad J_2 \frac{\partial^2 \theta}{\partial t^2} = -GI_p \frac{\partial \theta}{\partial x} \qquad \text{(b)}$$

[10]Timoshenko, 1911, loc. cit.

where G is the shear modulus of elasticity and I_p is the polar moment of inertia. From (a) we have

$$\omega_i^2 J_i C_i + \frac{\omega_i G I_p}{c} D_i = 0$$

and from (b),

$$\left(\omega_i^2 J_2 \cos \frac{\omega_i L}{c} + \frac{\omega_i G I_p}{c} \sin \frac{\omega_i L}{c}\right) C_i + \left(\omega_i^2 J_2 \sin \frac{\omega_i L}{c} - \frac{\omega_i G I_p}{c} \cos \frac{\omega_i L}{c}\right) D_i = 0$$

This is a system of two homogeneous algebraic equations for C_i and D_i. It has a solution only if the determinant of the coefficients vanishes. This yields

$$\omega_i^2 \left(\cos \frac{\omega_i L}{c} - \frac{\omega_i c J_1}{G I_p} \sin \frac{\omega_i L}{c}\right) J_2 + \frac{\omega_i G I_p}{c} \left(\sin \frac{\omega_i L}{c} + \frac{\omega_i c J_1}{G I_p} \cos \frac{\omega_i L}{c}\right) = 0$$

This is the frequency equation. One can see immediately that $\omega_i = 0$ is a root of this equation. This corresponds to a rigid-body rotation of the system. $\omega_1 = 0$ is always found in cases where there is no constraint to rigid-body motion along one of the system coordinates.

(This example demonstrates the determination of the torsional natural frequencies and modes of circular cylinder with two disks at the ends.)

□

Principle of Orthogonality of the Natural Modes

The functions of the spatial coordinate that satisfy the wave equation and the boundary conditions were called vibration natural modes. We can repeat here that the natural modes, from all different configurations of the vibration of a continuous system, are the only ones at which the system can vibrate at a single frequency, and every point at x will then undergo harmonic motion at this frequency. It is also apparent that for all points the maximum value of the displacement y is reached simultaneously: at $t = 2\pi/\omega_n$, $\cos \omega_n t = 1$. The same is true for zero displacement: at $t = 2\pi/\omega_n + \pi/2\omega_n$, $\cos \omega_n t = 0$.

Another remarkable property of the natural modes is the orthogonality,[11] already known from the theory of Fourier series. That is, two different modes of order i and j satisfy the equation

$$\int_0^l W_i W_j \, dx = 0 \qquad \text{if } i \neq j$$

To prove this, we note that W_i and W_j must satisfy the differential equation

$$W_i'' + \left(\frac{\omega_i}{c}\right)^2 W_i = 0$$

$$W_j'' + \left(\frac{\omega_j}{c}\right)^2 W_j = 0$$

(9.49)

[11]For a rigorous proof, see von Kármán, T., and Biot, M. A. 1939. *Mathematical Methods in Engineering*. New York: McGraw-Hill.

Multiplying the two equations by W_j and W_i, respectively, and subtracting, we have

$$W_i''W_j + \left(\frac{\omega_i}{c}\right)^2 W_i W_j = 0$$

$$W_j''W_i + \left(\frac{\omega_j}{c}\right)^2 W_j W_i = 0 \qquad (9.50)$$

and

$$W_i W_j = -\frac{c^2}{\omega_i^2 - \omega_j^2} (W_i''W_j - W_j''W_i) \qquad (9.51)$$

Integrating over the length gives

$$\int_0^l W_i W_j \, dx = -\left(\frac{c^2}{\omega_i^2 - \omega_j^2}\right) \int_0^l (W_i''W_j - W_j''W_i) \, dx \qquad (9.52)$$

or

$$\int_0^l W_i W_j \, dx = -\left(\frac{c^2}{\omega_i^2 - \omega_j^2}\right) (W_i'W_j - W_i W_j')_0^l \qquad (9.53)$$

We note that the quantity in parentheses for the usual boundary conditions is zero. If $i = j$ we have $\omega_i^2 - \omega_j^2 = 0$, therefore, the right-hand side is not zero but indefinite.

It was said that the natural modes are computed to an arbitrary scale. The orthogonalization process can be used to assign them definite values. To this end, we set the scale factor of the natural modes to a value that satisfies the condition

$$\int_0^l W_i W_i \, dx = 1 \qquad (9.54)$$

The general solution of equation (9.33) can now be expressed as

$$w(x, t) = \sum_{i=1}^{\infty} u_i(t) W_i(x) \qquad (9.55)$$

Substituting in (9.33) yields

$$\sum_{i=1}^{\infty} u_i(t) W_i''(x) = \frac{1}{c^2} \sum_{i=1}^{\infty} \ddot{u}_i(t) W_i(x) \qquad (9.56)$$

Multiplying by $W_j(x)$ and integrating over the length and using the orthogonality properties of the natural modes, we obtain

$$-u_j \frac{\omega_j^2}{c^2} \int_0^l W_j^2 \, dx = \ddot{u}_j \frac{1}{c^2} \int_0^l W_j^2 \, dx \qquad (9.57)$$

Due to the normalization of the natural modes,

$$\ddot{u}_j(t) + \omega_j^2 u_j(t) = 0 \qquad (9.58)$$

where $j = 1, 2, 3, \ldots$. These are the modal analysis equations as obtained for the discrete systems.

Fourth-Order Operator L

For lateral vibration of beams, we have obtained in equation (9.38)

$$W^{(iv)} - \left(\frac{\omega}{c}\right)^2 W = 0 \qquad (9.59)$$

where ω is an undetermined constant. The substitution $W(x) = e^{kx}$ leads to the general solution, since $\cosh a = (e^a + e^{-a})/2$, $\sinh a = (e^a - e^{-a})/2$,

$$W(x) = C_1 \cosh kx + C_2 \sinh kx + C_3 \cos kx + C_4 \sin kx \qquad (9.60)$$

where $k = (\omega/c)^{1/2}$ and C_1, C_2, C_3, and C_4 are as yet undetermined constants. These constants will be determined by the boundary conditions. For lateral vibration of single-span beams, we have four boundary conditions. These conditions are homogeneous; in other words, they do not involve functions of t. Therefore, application of the boundary conditions to the solution (9.60) will yield four algebraic equations for C_1, C_2, C_3, and C_4. These algebraic equations are homogeneous. The condition for the existence of a solution is that the coefficient determinant is zero. This gives an equation for the determination of ω, which is the frequency equation. For example, for the simply supported beam, the boundary conditions are:

$$\text{Displacements:} \quad y(0) = y(l) = 0$$

$$\text{Bending moments:} \quad \left.\frac{d^2 y}{dx^2}\right|_{x=0} = 0, \quad \left.\frac{d^2 y}{dx^2}\right|_{x=l} = 0 \qquad (9.61)$$

Application of these conditions to equation (9.60) yields

$$
\begin{array}{llll}
C_1 & +C_3 & = 0 \\
C_1 \cosh kl + C_2 \sinh kl & +C_3 \cos kl + C_4 \sin kl & = 0 \\
C_1 & -C_3 & = 0 \\
C_1 k^2 \cosh kl + C_2 k^2 \sinh kl & -C_3 k^2 \cos kl - C_4 k^2 \sin kl & = 0
\end{array}
\qquad (9.62)
$$

This system has solution only if

$$
\begin{vmatrix}
1 & 0 & 1 & 0 \\
\cosh kl & \sin kl & \cos kl & \sin kl \\
1 & 0 & -1 & 0 \\
\cosh kl & \sin kl & -\cos kl & -\sin kl
\end{vmatrix} = 0 \qquad (9.63)
$$

Expansion of this determinant yields

$$\sin kl = 0 \qquad (9.64)$$

Therefore, the natural frequencies of the beam will be determined from the solution of (9.64), $kl = \pi n$, or

$$\omega_n = \frac{n^2 \pi^2 c}{l^2}, \qquad n = 1, 2, 3, \ldots \tag{9.65}$$

For these values of ω_n the system (9.62) has solutions. Three of C_1, C_2, C_3, and C_4 can then be determined as functions of the fourth. Indeed, we find that

$$C_1 = C_2 = C_3 = 0$$

Therefore, the natural modes will be

$$W_i(x) = C_4 \sin \frac{n \pi x}{l} \tag{9.66}$$

and the solution

$$y(x, t) = \sum_{n=1}^{\infty} [(A_1 C_4)_n \cos \omega_n t + (A_2 C_4)_n \sin \omega_n t] \sin \frac{n \pi x}{l} \tag{9.67}$$

The constants $(A_1 C_4)_n$ and $(A_2 C_4)_n$ are as yet undetermined and should be computed from the initial conditions. The natural frequencies and vibration modes for lateral vibration of beams for some common boundary conditions are tabulated in Table 9.5.

TABLE 9.5

$$\omega = \frac{(kl)^2}{l^2} \sqrt{\frac{EI}{\rho A}}, \quad Y(x) = A \cosh k x + B \sinh k x + C \cos k x + D \sin k x$$

Beam	Natural Modes	k	$(kl)^2$	Equation for k
		1.875	3.52	
		4.694	22.03	$1 + \cosh kl \cos kl = 0$
		7.855	61.70	
		3.141	$9.87 = \pi^2$	
		6.283	$39.5 = 4\pi^2$	$\sin kl = 0$
		9.425	$88.9 = 9\pi^2$	
		0	0	
		3.927	15.42	
		7.069	50.0	$(kl)^2 (\tan kl - \tanh kl) = 0$
		10.21	104.3	

TABLE 9.5 *(cont.)*

Beam	Natural Modes	k	$(kl)^2$	Equation for k
		0	0	
		2.362	5.59	
		5.943	30.2	$(\tan kl + \tanh kl) = 0$
		8.64	74.6	
		0	0	
		0	0	
		4.73	22.40	$(kl)^2 (1 - \cosh kl \cos kl) = 0$
		7.853	61.6	
		10.996	120.9	
		4.73	22.4	
		7.853	61.6	$1 - \cosh kl \cos kl = 0$
		10.996	120.9	
		3.927	15.42	
		7.069	50.0	$\tan kl - \tanh kl = 0$
		10.21	104.3	

☐ **Example 9.9   Proof or Orthogonality for Lateral Vibration of Beams**

Prove that the natural modes of lateral vibration of beams are orthogonal.
Solution   The differential equation for lateral vibration of beams is

$$\frac{1}{c^2} \frac{\partial^2 y}{\partial t^2} + \frac{\partial^4 y}{\partial x^4} = 0 \qquad (a)$$

and a particular solution is

$$y = Y(A \cos \omega t + B \sin \omega t) \qquad (b)$$

where Y is the natural mode, solution of

$$\frac{d^4 Y}{dx^4} = \frac{\omega^2}{c^2} Y \qquad (c)$$

If Y_i and Y_j are two eigenfunctions corresponding to two natural modes of vibration, we have the following two equations:

$$\frac{d^4 Y_i}{dx^4} = \frac{\omega_i^2}{c^2} Y_i \qquad \text{(d)}$$

$$\frac{d^4 Y_j}{dx^4} = \frac{\omega_j^2}{c^2} Y_j \qquad \text{(e)}$$

Multiplying equation (d) by Y_j and (e) by Y_i, subtracting (d) from (e), and integrating over the length, we have

$$\frac{\omega_j^2 - \omega_i^2}{c^2} \int_0^L Y_i Y_j \, dx = \int_0^L \left(\frac{Y_i \, d^4 Y_j}{dx^4} - Y_j \frac{d^4 Y_i}{dx^4} \right) dx \qquad \text{(f)}$$

from which

$$\frac{\omega_j^2 - \omega_i^2}{c^2} \int_0^L Y_i Y_j \, dx = \left(Y_i \frac{d^3 Y_j}{dx^3} - Y_j \frac{d^3 Y_i}{dx^3} + \frac{dY_j}{dx} \frac{d^2 Y_i}{dx^2} - \frac{dY_i}{dx} \frac{d^2 Y_j}{dx^2} \right)_0^L \qquad \text{(g)}$$

Some common boundary conditions are:

1. *Free end*: The bending moment and shear force are equal to zero:

$$\frac{d^2 Y}{dx^2} = 0, \qquad \frac{d^3 Y}{dx^3} = 0$$

2. *Simply supported end*: The bending moment and displacement are equal to zero:

$$\frac{d^2 Y}{dx^2} = 0, \qquad Y = 0$$

3. *Built-in end*: The displacement and slope are equal to zero:

$$Y = 0, \qquad \frac{dY}{dx} = 0$$

Substituting anyone of the boundary conditions above into equation (g), we obtain

$$\int_0^L Y_i Y_j \, dx = 0, \qquad i \neq j$$

(This example proves that the natural modes of lateral vibration for the above-mentioned cases of boundary conditions are orthogonal.)

□

□ **Example 9.10   Lateral Vibration for a Free–Free Beam**

Find the frequency equation of a beam with both ends free and having transverse vibration.

Solution   In this problem the bending moment and the shearing force are both zero at each end of the beam:

$$\left(\frac{d^2 X}{dx^2} \right)_{x=0} = 0, \qquad \left(\frac{d^3 X}{dx^3} \right)_{x=0} = 0, \qquad \left(\frac{d^2 X}{dx^2} \right)_{X=L} = 0, \qquad \left(\frac{d^3 X}{dx^3} \right)_{x=L} = 0$$

For this case it is convenient to take the general solution for the natural mode for transverse vibration of beams in the following form, with new arbitrary constants:

$$X = C_1(\cos kx + \cosh kx) + C_2(\cos kx - \cosh kx)$$
$$+ C_3(\sin kx + \sinh kx) + C_4(\sin kx - \sinh kx)$$

for $k = (\omega/c)^{1/2}$.

From the first two boundary conditions, $C_2 = C_4 = 0$, and hence

$$X = C_1(\cos kx + \cosh kx) + C_3(\sinh kx + \sinh kx)$$

From the other two boundary conditions,

$$(-\cos kL + \cosh kL)C_1 + (-\sin kL + \sinh kL)C_3 = 0$$

$$(\sin kL + \sinh kL)C_1 + (-\cos kL + \cos kL)C_3 = 0$$

Solving for the constants C_1 and C_3, we obtain the following determinant:

$$\begin{vmatrix} -\cos kL + \cosh kL & -\sin kL + \sinh kL \\ \sin kL + \sinh kL & -\cos kL + \cosh kL \end{vmatrix} = 0$$

Expanding the determinant, we obtain

$$- 2 \cos kL \cosh kL + \cos^2 kL + \cosh^2 kL - \sinh^2 kL + \sin^2 kL = 0$$

but

$$\cosh^2 kL - \sinh^2 kL = 1, \cos^2 kL + \sin^2 kL = 1$$

so the frequency equation is

$$\cos kL \cosh kL = 1$$

The lowest 3 roots are 0, 4.730 and 7.853.

(The frequency equation was found for lateral vibration of a beam with both ends free.)

□

□ **Example 9.11   Two-Span Beam Lateral Vibration**[12]

Determine the frequency equation of the two-span beam, clamped at the ends, simply supported at midspan, as shown in Figure E9.11.

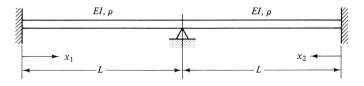

Figure E9.11

Solution   Two solutions will be developed for the two spans, and the compatibility at the interface will provide additional equations to compute the integration constants.

Let $X_1(x_1)$ and $X_2(x_2)$ be the natural modes of beam for the left and the right span, respectively. These have the forms

[12]Darnley, E. R. 1921. *Philos. Mag.*, *41*: 81. Also, [Hohenemser, and Prager, 1933].

$$X_1 = A_1 \cos kx_1 + B_1 \cosh kx_1 + C_1 \sin kx_1 + D_1 \sinh kx_1$$

$$X_2 = A_2 \cos kx_2 + B_2 \cosh kx_2 + C_2 \sin kx_2 + D_2 \sinh kx_2$$

The boundary conditions are:
At $x_1 = 0$:

$$X_1 = 0 \tag{a}$$

$$\frac{dX_1}{dx_1} = 0 \tag{b}$$

At $x_2 = 0$:

$$X_2 = 0 \tag{c}$$

$$\frac{dX_2}{dx_2} = 0 \tag{d}$$

At $x_1 = L, x_2 = L$:

$$X_1 = 0 \tag{e}$$

$$\frac{dX_1}{dx_1} = -\frac{dX_2}{dx_2} \tag{f}$$

$$X_2 = 0 \tag{g}$$

$$\frac{d^2X_2}{dx_2^2} = -\frac{d^2X_1}{dx_1^2} \tag{h}$$

From conditions (a) and (c), $A_1 = -B_1$ and $A_2 = -B_2$. Substituting into (b) and (d), we obtain $C_1 = -D_1$ and $C_2 = -D_2$.

Using (e) and (g), we have

$$A_1(\cos kL - \cosh kL) + C_1(\sin kL - \sinh kL) = 0$$

$$A_2(\cos kL - \cosh kL) + C_2(\sin kL - \sinh kL) = 0$$

from which A_1 and A_2 can be expressed in terms of C_1 and C_2, respectively:

$$A_1 = -\frac{C_1(\sin kL - \sinh kL)}{\cos kL - \cos kL}$$

$$A_2 = -\frac{C_2(\sin kL - \sinh kL)}{\cos kL - \cosh kL}$$

From conditions (f) and (h), we have

$$-A_1(\sin kL + \sinh kL) + C_1(\cos kL - \cosh kL)$$

$$= A_2(\sin kL + \sinh kL) - C_2(\cos kL - \cosh kL)$$

$$-A_1(\cos kL + \cos kL) - C_1(\sin kl + \sinh kL)$$

$$= A_2(\cos kL + \cos kL) + C_2(\sin kL + \sinh kL)$$

Substituting the expressions for A_1 and A_2 into the equations above and solving for C_1 and C_2, we find that

$$(\sin kL \cosh kL - \cos kL \sinh kL)(1 - \cos kL \cosh kL) = 0$$

and hence there are two families of natural frequencies given by the following frequency equations:

$$\cos kL \cosh kL = 1, \qquad \tan kL = \tanh kL$$

(The frequency equation was found for lateral vibration of a beam with two spans with both ends clamped and a simple support at the interface.)

□

9.5 NATURAL VIBRATION: THE INITIAL VALUE PROBLEM

The solution of equation (9.33) in the form (9.66) is only a particular solution; in fact there are an infinite number of such solutions, corresponding to the infinite number of eigenvales ω_n. It is known that the general solution of (9.33) is the sum of all the particular solutions:

$$w(x, t) \sum_{n=0}^{\infty} [\alpha_{1n} \cos \omega_n t + \alpha_{2n} \sin \omega_n t)] \sin \frac{\pi n x}{l} \qquad (9.68)$$

where $\alpha = A_1 C_2$ and $\alpha_2 = A_2 C_2$ will be different with every n. Obviously, only two constants, the products $\alpha_1 = A_1 C_2$ and $\alpha_2 = A_2 C_2$ for every n, are to be determined. Such vibration is called *transient*, although it does not disappear with time as the name suggests. In reality, however, some damping is expected, which will diminish the transient vibration over a sufficiently long period.

Let us now suppose that the initial conditions are

$$w(x, 0) = A \sin \frac{\pi n_0 x}{l}$$

$$\left. \frac{\partial w}{\partial t} \right|_{t=0} = 0 \qquad (9.69)$$

where n_0 is a given integer. In other words, it is assumed that initially the system has a half-sine configuration. Equations (9.68) and (9.69) yield

$$\sum_{n=0}^{\infty} \alpha_{1n} \sin \frac{\pi n x}{l} = A \sin \frac{\pi n_0 x}{l}$$

$$\omega_n \sum_{n=0}^{\infty} \alpha_{2n} \sin \frac{\pi n x}{l} = 0 \qquad (9.70)$$

These equations can be satisfied for every x if and only if

$$\alpha_{1n} = \begin{cases} 0 & \text{for } n \neq n_0 \\ A & \text{for } n = n_0 \end{cases}$$

$$\alpha_{2n} = 0$$

Therefore, only the term α_1 of order n_0 has nonzero coefficient, and we have

$$w(x, t) = A \cos \omega_{n0} t \sin \frac{\pi n_0 x}{l} \qquad (9.71)$$

Equation (9.71) is a solution of the problem. It represents vibration of one frequency only, ω_{n0}, and the configuration of the displacement is always the same; only the amplitude at every point changes, as a harmonic function of time. Therefore, in general, the system will not vibrate with many frequencies simultaneously, as equation (9.63) indicates, but it is capable, for appropriate initial conditions, of vibrating at a single frequency, the *natural frequency*, the *eigenvalue* of the wave equation. For every eigenfrequency, there corresponds a system configuration $\sin(\pi n_0 x/l)$, the *eigenfunction* or *natural vibration mode*.

In Section 9.4 it was shown that the system can execute simultaneous harmonic vibration in any one of an infinite number of distinct frequencies that depend only on the characteristics of the system. We have also seen that such vibrations can be inhibited only with specific forms of the initial conditions. In general, the initial conditions will have a general form, the system will vibrate at all eigenfrequencies, and its configuration will be the sum of all eigenfunctions. This will happen when the initial conditions do not coincide with any of the eigenfunctions. Consider, for example, a taut string displaced initially as shown in Figure 9.7.

As in Section 1.3, we shall analyze this triangular shape in a Fourier series. Let

$$
y(x, 0) = \begin{cases} 2A\,\dfrac{x}{l} & \text{for } 0 \le x < l/2 \\[2mm] 2A\left(1 - \dfrac{x}{l}\right) & \text{for } \dfrac{l}{2} \le x < l \end{cases}
$$

$$
\left.\frac{\partial y}{\partial t}\right|_{t=0} = 0
\tag{9.72}
$$

as the initial conditions. Equations (1.15) yield, for $n = 0, 1, 2, \ldots$,

$$
a_n = 0
$$

$$
b_n = (-1)^{(n-1)/2}\,\frac{8A}{\pi^2 n^2}, \qquad b_0 = 0
\tag{9.73}
$$

We obtain from (9.68),

$$
\alpha_{2n} = 0
\tag{9.74}
$$

$$
\sum_{n=1}^{\infty} \alpha_{1n} \sin \frac{n\pi x}{l} = \sum_{n=1}^{\infty} b_n \sin \frac{n\pi x}{l}
\tag{9.75}
$$

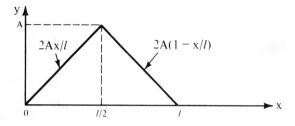

Figure 9.7 Initial displacement of a taut string.

This equation is true for every x only if $\alpha_{1n} = b_n$. Therefore, the solution is

$$w(x, t) = \sum_{n=1}^{\infty} (-1)^{(n-1)/2} \frac{8A}{\pi^2 n^2} \cos \omega_n t \sin \frac{n\pi x}{l} \qquad (9.76)$$

☐ **Example 9.12   Natural Longitudinal Vibration of a Bar[13]**

A uniform bar of length l is fixed at one and free at the other end is stretched uniformly to l_0 and released at $t = 0$ (Figure E9.12). Find the resulting longitudinal motion. This is a typical case of a bar welded while hot between two rigid walls. When it is cooled, tension thermal stresses are developing in the bar. If they exceed the material strength, the bar will brake, usually near one of the two ends, due to stress concentration at the weld.

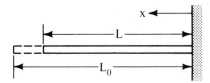

Figure E9.12

Solution   The general solution for bars with one end fixed and the other end free is given by

$$u(x, t) = \sum_{i=1,3,\ldots}^{\infty} \sin \frac{\omega_i x}{c} (A_i \cos \omega_i t + B_i \sin \omega_i t) \qquad (a)$$

$$\omega_i = \frac{\pi i c}{2l}$$

where A_i and B_i are constants determined by the initial conditions of the problem:

$$u(x, 0) = \frac{(l_0 - l)x}{l} \left. \frac{\partial u}{\partial t} \right|_{t=0} = 0 \qquad (b)$$

Using the initial conditions (b), we have

$$\left. \frac{\partial u}{\partial t} \right|_{t=0} = \sum_{i=1,3,\ldots}^{\infty} \omega_i B_i \sin \frac{\omega_i x}{c} = 0 \quad \text{or} \quad B_i = 0 \qquad (c)$$

From the initial condition (a),

$$u(x, 0) = \sum_{i=1,3,\ldots}^{\infty} A_i \sin \frac{\omega_i x}{c} = \frac{(l_0 - l)x}{l} \qquad (d)$$

Multiplying both sides of the equation above by $\sin (\omega_i x/c)$ and integrating over the length yields

$$\int_0^l A_i \sin \frac{\omega_i x}{c} \sin \frac{\omega_i x}{c} \, dx = \frac{l_0 - l}{l} \int_0^l x \sin \frac{\omega_i x}{c} \, dx$$

[13]The problem is related with the lift of a long oil drill stem. Langer, B. F., and Lamberger, E. H. 1943. *J. Appl. Mech.*, *10*: 1.

or

$$A_i = \frac{2(l_0 - l)}{l^2} \int_0^l x \sin \frac{\omega_i x}{c} \, dx = 8(-1)^{(i-1)/2} \frac{l_0 - l}{i^2 \pi^2}$$

where $i = 1, 3, 5, \ldots$. Therefore, the solution is

$$u(x, t) = 8 \frac{l_0 - l}{\pi^2} \sum_{i=1,3,\ldots}^{\infty} (-1)^{(i-1)/2} \frac{1}{i^2} \sin \frac{\omega_i x}{c} \cos \omega_i t$$

(The natural longitudinal vibration of a clamped-free column due to prescribed initial displacement is computed.)

□

□ **Example 9.13   Response of a Beam to Local Impact[14]**

A simple supported beam of length L is as shown in Figure E9.13. At time $t = 0$, a very short portion Δl of the beam is hit by a hammer, giving an initial velocity V_0 to that portion of the beam. Find the resulting lateral vibrations of the beam.

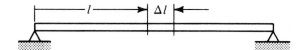

Figure E9.13

Solution   The initial velocity of the beam will be computed first and then the initial value problem will be solved.

The general expression for free lateral vibration of simply supported beams is given by

$$y(x, t) = \sum_{i=1,2,\ldots}^{\infty} \sin k_i x (A_i \cos \omega_i + B_i \sin \omega_i t), \quad k_i = \left(\frac{\omega_i}{c}\right)^{1/2} \tag{a}$$

where the constants A_i and B_i are to be evaluated by the initial conditions, and

$$\omega_i = \frac{c i^2 \pi^2}{L^2} \tag{b}$$

Substituting $t = 0$ in (a) and in the derivative of (a) with respect to t, we obtain

$$(y)_{t=0} = \sum_{i=1,2,\ldots} A_i \sin k_i x, \quad (\dot{y})_{t=0} = \sum_{i=1,2,\ldots} \omega_i B_i \sin k_i x \tag{c}$$

Now the initial conditions are, for the part of the beam of length l,

$$y = 0 \text{ at } t = 0, \quad \dot{y} = V_0, \quad l - \frac{\Delta l}{2} \leq x \leq l + \frac{\Delta l}{2} \tag{d}$$

and $y = 0$, $\dot{y} = 0$ elsewhere, and hence $A_i = 0$ and

$$V_0 \Big|_{c-l/2}^{c+l/2} = \sum_{i=1,2,\ldots} \omega_i B_i \sin k_i x \tag{e}$$

[14]Saint-Venant, loc. cit., and [Timoshenko and Young, 1955].

Multiplying (e) by $\sin k_i x$ and integrating, we obtain

$$\int_0^L \omega_i B_i \sin k_i x \sin k_i x \, dx = \int_{c-l/2}^{c+l/2} V_0 \sin k_i x \, dx \tag{f}$$

and

$$B_i = \frac{2V_0 L}{i\pi l \omega_i} \sin \frac{i\pi l}{L} 2 \sin \frac{i\pi \, \Delta l}{2L} \tag{g}$$

Therefore, the vibration of the beam is

$$y(x, t) = 4V_0 \sum_{i=1,2,\dots} \frac{l}{i\pi \omega_i} \sin \frac{i\pi l}{L} \sin \frac{i\pi x}{L} \sin \omega_i t \sin \frac{i\pi \, \Delta l}{2L} \tag{h}$$

(The natural lateral vibration of a simply supported beam due to prescribed initial impact is computed.)

□

9.6 FORCED VIBRATION

A continuous system will vibrate due to time-varying forces. In addition, system motion might be produced by motion of the boundaries or forces applied on the boundaries.[15]

Distributed Harmonic Excitation

Referring to Figure 9.2, if an external force of density $p(x, t)$ in the direction of positive y is distributed along the one-dimensional structural member (for a taut spring and beams a lateral force per unit length, for longitudinal vibration of bars a longitudinal force per unit length, for torsional vibration a torque per unit length), equilibrium will demand that

$$F_y = \mu(x) \, dx \frac{\partial^2 y}{\partial t^2} + p(x, t) \, dx \tag{9.77}$$

The differential equation of motion will then be, for the taut spring

$$\frac{\partial^2 y}{\partial x^2} = \frac{1}{c^2} \frac{\partial^2 y}{\partial t^2} + \frac{1}{T} p(x, t) \tag{9.78}$$

for longitudinal vibration of rods,

$$\frac{\partial^2 u}{\partial x^2} = \frac{1}{c^2} \frac{\partial^2 u}{\partial t^2} + \frac{1}{EA} p(x, t) \tag{9.79}$$

---

[15]Undamped elastic solids are considered only. For the effect of damping, see Holzer, H. 1928. *Z. Angew. Math. Phys.*, 8: 272. See also [Timoshenko and Young, 1955].

for lateral vibration of beams,

$$-\frac{\partial^4 u}{\partial x^4} = \frac{1}{c^2}\frac{\partial^2 y}{\partial t^2} + \frac{1}{EI}p(x, t) \tag{9.80}$$

and for torsional vibration,

$$\frac{\partial^2 \theta}{\partial x^2} = \frac{1}{c^2}\frac{\partial^2 \theta}{\partial t^2} + \frac{1}{GI_p}p(x, t) \tag{9.81}$$

For steady-state vibration, a closed-form solution can be obtained only for special forms of the exciting force $p(x, t)$, as in the case of the harmonic oscillator.

First, $p(x, t)$ has to be separable in time and space; in other words, $p(x, t) = P(x)Q(t)$.

Consider, for example, a beam loaded with a distributed force $P(x) \sin \omega t$. One can easily verify that the equation of motion is

$$c^2 \frac{\partial^4 y}{\partial x^4} = -\frac{\partial^2 y}{\partial t^2} - \frac{P(x)}{\rho A}\sin \omega t \tag{9.82}$$

We shall look here for a particular solution of the form

$$y(x, t) = Y(x) \sin \omega t \tag{9.83}$$

which leads to

$$Y^{(iv)} - \left(\frac{\omega}{c}\right)^2 Y = -\frac{P(x)}{A\rho c^2} \tag{9.84}$$

This is an ordinary differential equation and can be solved by the methods discussed in Chapter 5. For $P(x) = F_0 = $ constant, we have the solution, for $k^2 = \omega/c$,

$$Y(x) = C_1 \cosh kx + C_2 \sinh kx + C_3 \cos kx + C_4 \sin kx + \frac{F_0}{A\rho\omega^2} \tag{9.85}$$

For a simply supported beam of length l, we obtain

$$C_1 = C_2 = \frac{F_0}{2A\rho\omega^2}$$

$$C_3 = \frac{F_0}{2A\rho\omega^2}\tan kl$$

$$C_4 = -\frac{F_0}{2A\rho\omega^2}\tanh kl$$

and the steady-state vibration is

$$y(x, t) = \frac{F_0}{A^2}\left[\cos\frac{k(l - x)}{2}\cos kl\right.$$

$$\left. + \cosh\frac{k(l - x)}{2}\cosh (kl - 1)\right]\sin \omega t \tag{9.86}$$

Harmonic Motion of the Boundaries

In the case of nonhomogeneous (time-varying) boundary conditions, we shall look again for steady-state vibration in the case that the boundary conditions vary harmonically. Consider, for example, a cantilever beam with a free end at $x = l$ that moves as

$$y(l, t) = y_0 \sin \omega t \qquad (9.87)$$

We look for a steady-state solution,

$$y(x, t) = Y(x) \sin \omega t \qquad (9.88)$$

which yields

$$Y(x) = C_1 \cosh kx + C_2 \sinh kx + C_3 \cos kx + C_4 \sin kx \qquad (9.89)$$

and the boundary condition at $x = l$ will be

$$Y(l) = y_0 \qquad (9.90)$$

Therefore, we obtain

$$
\begin{aligned}
C_1 & & + C_3 & & & = 0 \\
C_1 \cosh kl & + C_2 \sinh kl & + C_3 \cos kl & + C_4 \sin kl & & = y_0 \\
& C_2 & & - C_4 & & = 0 \\
C_1 k^2 \cosh kl & + C_2 k^2 \sinh kl & - C_3 k^2 \cos kl & - C_4 k^2 \sin kl & & = 0
\end{aligned}
\qquad (9.91)
$$

The system (9.91) is not homogeneous and from it we can determine the constants C_1, C_2, C_3, and C_4. The steady-state response then will be given by equations (9.87) and (9.88).

☐ **Example 9.14   Forced Longitudinal Vibration of a Column**

A column of section A, Young's modulus E, density ρ, and length l supports a machine M through a slab S (Figure E9.14). The slab was observed vibrating with a harmonic

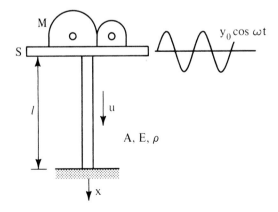

Figure E9.14

motion of amplitude y_0 and frequency ω. If the floor is to be assumed rigid, determine the longitudinal vibration of the column.

Solution   The differential equation for longitudinal vibration of bars is

$$\frac{\partial^2 u}{\partial x^2} = \frac{1}{c^2}\frac{\partial^2 u}{\partial t^2}, \qquad c = \left(\frac{E}{\rho}\right)^{1/2}$$

The boundary conditions are:

$$\text{At } x = 0: \qquad u(0, t) = y_0 \cos \omega t \tag{a}$$

$$\text{At } x = l: \qquad u(0, t) = 0 \tag{b}$$

Since the excitation is harmonic, one expects a harmonic response,

$$u(x, t) = U(x) \cos \omega t$$

Substitution in the differential equation yields

$$U'' + \left(\frac{\omega}{c}\right)^2 U = 0$$

The general solution is

$$U(x) = a \cos \frac{\omega x}{c} + b \sin \frac{\omega x}{c}$$

where a and b are constants to be determined by the boundary conditions. For condition (a),

$$u(0, t) = \left[a \cos \frac{\omega x}{c} + b \sin \frac{\omega x}{c}\right]_{x=0} \cos \omega t$$

$$= a \cos \omega t = y_0 \cos \omega t$$

Thus $a = y_0$. For (b),

$$u(l, t) = \left(a \cos \frac{\omega l}{c} + b \sin \frac{\omega l}{c}\right) \cos \omega t = 0$$

Since $\cos \omega t$ is not zero for all t,

$$b = -y_0 \frac{\cos (\omega l/c)}{\sin (\omega l/c)}$$

Therefore, the dynamic response of the bar will be

$$u(x, t) = y_0 \left[\cos \frac{\omega x}{c} - \frac{\cos (\omega l/c) \sin (\omega x/c)}{\sin (\omega l/c)}\right] \cos \omega t$$

(The longitudinal vibration of a column fixed at one end is computed due to a harmonic motion of the other end.)

□

Harmonic Force on the Boundary

A boundary force can be treated as a nonhomogeneous boundary condition, and then the methods of Section 9.5 applied. For example, a torque at the left end of a shaft $T_0 \cos \omega t$ will lead to the boundary condition

$$I_p G \, \frac{\partial \theta(0, t)}{\partial x} = T_0 \cos \omega t \tag{9.92}$$

and for $\theta(x, t) = \Theta(x) \cos \omega t$ will yield

$$I_p G \Theta'(x) = T_0 \tag{9.93}$$

☐ **Example 9.15 Forced Longitudinal Vibration of a Column**[16]

A column of section A, Young modulus E, density ρ, and length l supports a machine M through a slab S (Figure E7.14). There is an unbalanced vertical harmonic force on the machine $F \cos \omega t$. Determine the longitudinal vibration of the column.
Solution   The differential equation for longitudinal vibration of bars is

$$\frac{\partial^2 u}{\partial x^2} = \frac{1}{c^2} \frac{\partial^2 u}{\partial t^2}, \qquad c = \left(\frac{E}{\rho} \right)^{1/2}$$

The boundary condition at $x = 0$ is $F = M(\partial^2 u/\partial t^2) + EA(\partial u/\partial x)$. Therefore:

$$\text{At } x = 0: \qquad F \cos \omega t = m \frac{\partial^2 u}{\partial t^2} + EA \frac{\partial u}{\partial x} \tag{a}$$

$$\text{At } x = l: \qquad u(0, t) = 0 \tag{b}$$

Since the excitation is harmonic, one expects a harmonic response,

$$u(x, t) = U(x) \cos \omega t$$

Substitution in the differential equation yields

$$U'' + \left(\frac{\omega}{c} \right)^2 U = 0$$

The general solution is

$$U(x) = a \cos \frac{\omega}{c} x + b \sin \frac{\omega}{c} x$$

where a and b are constants to be determined by the boundary conditions. For (a),

$$F \cos \omega t = -M \left(a \cos \frac{\omega}{c} x + b \sin \frac{\omega}{c} x \right)_{x=0} \omega^2 \cos \omega t$$

$$+ \frac{\omega}{c} EA \left(-a \sin \frac{\omega}{c} x + b \cos \frac{\omega}{c} x \right)_{x=0} \cos \omega t$$

$$F = -Ma\omega^2 + \frac{\omega}{c} EAb$$

and for (b),

$$u(l, t) = \left(a \cos \frac{\omega}{c} l + b \sin \frac{\omega}{c} l \right) \cos \omega t = 0$$

Since $\cos \omega t \neq 0$, in general,

$$a \cos \frac{\omega}{c} l + b \sin \frac{\omega}{c} l = 0$$

[16]Timoshenko, 1911, loc. cit.

Therefore, equations (a) and (b) yield

$$a = - \frac{F}{M\omega^2 + (\omega/c)/\tan(\omega l/c)}$$

$$b = - \frac{a}{\tan(\omega l/c)} = \frac{F}{M\omega^2/\tan(\omega l/c) + \omega/c}$$

Finally, the dynamic response of the bar is

$$u(x, t) = F\left[-\frac{\cos(\omega x/c)}{M\omega^2 + (\omega/c)/\tan(\omega l/c)} + \frac{\sin(\omega x/c)}{M\omega^2/\tan(\omega l/c) + \omega/c} \right] \cos \omega t$$

(The longitudinal vibration of a column, fixed at one end and with a mass at the other end is computed due to a harmonic force of the end mass.)

□

General Excitation: Modal Analysis

Since the natural modes are orthogonal, one can apply the methods developed in Chapter 8 for continuous systems, since they are a natural extension of the multidegree-of-freedom systems. To illustrate this procedure, let us return to the problem of the simply supported beam with a distributed force $f(x, t)$. The differential equation of motion was found to be

$$c^2 \frac{\partial^4 y}{\partial x^4} = -\frac{\partial^2 y}{\partial t^2} - \frac{f(x, t)}{\rho A} \tag{9.94}$$

The natural modes are, for $\lambda^2 = \omega/c$,

$$Y_i(x) = C_4 \sin \frac{i\pi x}{l} \tag{9.95}$$

They are normalized as $\phi_i(x) = m_i Y_i(x)$, so that

$$\int_0^l m_i Y_i^2(x)\, dx = 1, \qquad m_i = \frac{1}{\int_0^l Y_i^2(x)\, dx} \tag{9.96}$$

We set up the solution in the form

$$y(x, t) = \sum_{k=1}^{\infty} \phi_k q_k(t) \tag{9.97}$$

where $q_k(t)$ are unknown functions of time.

Equation (9.97) is inserterd into (9.94) and the latter is then integrated over the length of the beam. Due to orthogonality and using (9.96),

$$\ddot{q}_k + \omega_k^2 q_k = Q_k(t), \qquad k = 1, 2, 3, \ldots \tag{9.98}$$

where $Q_k(t) = \int_0^l f(x, t)\phi_k(x)\, dx$.

Equations (9.98) are uncoupled and can be solved using the methods of Chapters 2, 3, 5, and 8.

□ **Example 9.16   Pipe in a Pulsating Cross-Flow**

A simply supported beam with flexural rigidity EI, mass per unit length μ constant, and length l, modeling a water pipe, is in a pulsating stream of water that acts on the pipe with a distributed lateral harmonic force of amplitude F constant along the beam and circular frequency ω. Determine its dynamic response.

Solution   The natural frequencies for lateral vibration of the simply supported beam are $\omega_k^2 = k^2\pi^2 c/l^2$, $k = 1, 2, 3, \ldots$ [equation (9.65)] where $c = (EI/\mu)^{1/2}$. From equation (9.66), the natural modes are for $k^2 = \omega/c$, $Y_n = \sin n\pi x/l$.

The normal modes are [equation (9.96)]

$$\phi_k = \frac{Y_k(x)}{\int_0^l Y_k^2(x)\,dx} = \frac{1}{2l}\sin\frac{k\pi x}{l}$$

The uncoupled equations (9.98) become

$$\ddot{q}_k + \omega_k^2 q_k = \int_0^l f(x, t)\phi_k(x)\,dx = \left(\int_0^l \sin\frac{k\pi x}{l}\,dx\right) F\cos\omega t$$

$$= \begin{cases} k\pi F\cos\omega t, & k = 1, 3, \ldots \\ 0 & k = 2, 4, \ldots \end{cases}$$

The modal force is harmonic; therefore, the response will be

$$q_k = \left(\int_0^l \sin\frac{k\pi x}{l}\,dx\right)\frac{F\cos\omega t}{\omega_k^2 - \omega^2} = \begin{cases} k\pi F\cos\omega t/(\omega_k^2 - \omega^2), & k = 1, 3, \ldots \\ 0, & k = 2, 4, \ldots \end{cases}$$

With equation (9.97), the pipe response will be

$$y(x, t) = \sum_{k=1}^\infty \phi_k q_k(t) = F\cos\omega t\sum_{k=1}^\infty \phi_k(x)\frac{\int_0^l \phi_k(x)\,dx}{\omega_k^2 - \omega^2}$$

$$= F\cos\omega t\left(\frac{1}{\omega_1^2 - \omega^2}\frac{\phi_1(x)}{\pi} + \frac{1}{\omega_3^2 - \omega^2}\frac{\phi_3(x)}{3\pi} + \cdots\right)$$

The response is a sum of the contributions of all natural modes. One can observe that due to the term $1/(\omega_k^2 - \omega^2)$, the contribution of modes that correspond to higher natural frequencies $\omega_k \gg \omega$ is diminished.

(The lateral vibration of a beam, simply supported at the ends, is computed due to a harmonic distributed force.)

□

9.7 SPECIAL TOPICS

Effects of the Rotatory Inertia and Shear Deformation: The Rayleigh–Timoshenko Beam[17]

Lateral vibration of beams are associated with the time-varying slope of the beam, which should resist rotatory acceleration with a rotatory inertia force. At high speeds (i.e., high angular accelerations) this resistance is considerable. More-

[17][Rayleigh, 1894]; Timoshenko, S. 1921. *Philos. Mag.*, Ser. 6. (41): 744.

Portrait of Timoshenko. (Courtesy ASME.)

STEPHEN P. TIMOSHENKO (1878–1972)

Born and educated in Russia, where he held professorships in Petrograd and Kiev, Timoshenko left Russia and came to the United States after he taught for a short time at Belgrade and studied for short periods with Prandtl. He had a profound influence in the United States, where, after a short tenure with Westinghouse, he became a professor at Michigan and at Stanford. Having contributed to the theory of elasticity and vibration, he was instrumental in the development of mechanics in the United States. He published several books in mechanics.

over, a beam carrying lateral forces has shear deformation, which might be substantial for shorter beams. Both factors can be considered together if we consider a beam element (Figure 9.8). Shear will cause the rectangular element to deform to a romboidal shape due to the relative parallel movement of the two opposite faces. If y is the displacement of the centerline of the beam, $\partial y/\partial x$ the total slope, ψ the slope due to bending, the slope due to shear will be $\psi - \partial y/\partial x$, equal to the shear angle $\gamma = V/kAG$ [Beer and Johnston, 1982]. Therefore,

$$\psi - \frac{\partial y}{\partial x} = \frac{V}{kAG} \tag{9.99}$$

and due to bending

$$\frac{\partial \psi}{\partial x} = \frac{M}{EI} \tag{9.100}$$

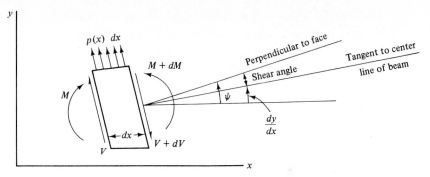

Figure 9.8 Shear deflection of a beam element.

where A is the cross section and k is a section shear constant, $\frac{5}{6}$ for a rectangular cross section, $\frac{9}{10}$ for a circular cross section. Newton's law for lateral and rotational motions yields

$$\mu \ddot{y} = -\frac{\partial V}{\partial x} + f(x, t) \tag{9.101}$$

$$\Gamma \ddot{\psi} = \frac{\partial M}{\partial x} - V \tag{9.102}$$

where μ and Γ are the mass and moment of inertia [Rayleigh, 1894] per unit length of the beam, respectively. Substituting (9.99) and (9.100) into (9.101) and (9.102) yields

$$\frac{\partial}{\partial x}\left(EI \frac{\partial \psi}{\partial x} \right) + kAG\left(\frac{\partial y}{\partial x} - \psi \right) - \Gamma \ddot{\psi} = 0 \tag{9.103}$$

$$\frac{\partial}{\partial x}\left(kAG \frac{\partial y}{\partial x} \right) - \mu \ddot{y} = f(x, t) \tag{9.104}$$

Eliminating ψ and for constant cross section,[18]

$$EI \frac{\partial^4 y}{\partial x^4} + \mu \frac{\partial^2 y}{\partial t^2} - \left(\Gamma + \frac{EI\mu}{kAG} \right) \frac{\partial^4 y}{\partial x^2 \partial t^2} + \frac{\Gamma\mu}{kAG} \frac{\partial^4 y}{\partial t^4}$$
$$= f(x, t) + \frac{\Gamma}{kAG} \frac{\partial^2 f}{\partial t^2} - \frac{EI}{kAG} \frac{\partial^2 f}{\partial x^2} \tag{9.105}$$

Equation (9.105) can be solved analytically only for very simple cases[19] but lends itself for numerical computation.

---

[18]Timoshenko, S. 1921; 1923. *Philos. Mag.*, 6(41): 744; 6(43): 125. The effect of shear deformation was recognized and discussed by Coulomb, 1773, in *Memoires par divers Savans*, Acad. Sci. Paris.

[19]Ward, P. F. 1913. *Philos. Mag.*, (25): 85.

Example 9.17   Effect of Rotatory Inertia on Lateral Vibration

Find the effect of rotatory inertia on the natural frequencies of a simply supported
beam of length l, flexural rigidity EI, and mass per unit length μ.
Solution   Equation (9.105) is satisfied with

$$Y_n(x) = \sin \frac{n\pi x}{l}$$

Disregarding the shear deformation terms (with k) and the forcing terms, we
obtain

$$EI \frac{\partial^4 y}{\partial x^4} + \mu \frac{\partial^2 y}{\partial t^2} - \Gamma \frac{\partial^4 y}{\partial x^2 \, \partial t^2} = 0$$

Substituting $y(x, t) = Y_n(x) \cos \omega t$,

$$\frac{EIn^4\pi^4}{l^4} + \omega^2 - \frac{\Gamma n^2 \pi^2 \omega^4}{l^2} = 0$$

Solving for ω^2 gives

$$\omega^2 = \frac{EIn^4\pi^4}{l^4 - \Gamma n^2 \pi^2}$$

It can be concluded that due to the term in the denominator, the rotatory inertia
Γ increases the natural frequencies, since it is a constraint. Moreover, the increase is
greater at higher modes, due to n^2 in the denominator.

An approximate solution with the shear deformation terms [Timoshenko and
Young, 1955] is

$$\omega^2 = \frac{EIn^4\pi^4}{l^4 - \Gamma n^2 \pi^2 (1 + A/kG)}$$

(The effect of the rotatory inertia on the lateral vibration natural frequencies
of a beam, simply supported at the ends, is computed.)

□

The Moving Taut String

A taut string that moves along its length with a constant linear velocity is an
interesting engineering problem. It is encountered, for example, in wire production
and in belt transmissions. Figure 9.9 shows a taut string with a force T moving with
a constant speed U. An element of mass dm is moving in the direction of the
y-axis with velocity $u = \partial y/\partial t + U(\partial y/\partial x)$. Newton's law in the y direction can be
written in the form

$$F = dm \frac{d(u)}{dt} \tag{9.106}$$

$$F = dm \frac{d}{dt} \left(\frac{\partial y}{\partial t} + U \frac{\partial y}{\partial x} \right) \tag{9.107}$$

Since u is a function of x and t,

$$\frac{du}{dt} = \frac{\partial u}{\partial x} \frac{\partial x}{\partial t} + \frac{\partial u}{\partial t} = \frac{\partial u}{\partial x} U + \frac{\partial u}{\partial t} \tag{9.108}$$

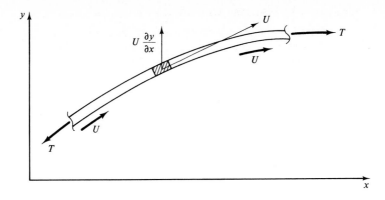

Figure 9.9 Traveling taut string.

Therefore, equation (9.107) gives

$$F = dm \left[\frac{\partial}{\partial x} \left(\frac{\partial y}{\partial t} + U \frac{\partial y}{\partial x} \right) U + \frac{\partial}{\partial t} \left(\frac{\partial y}{\partial t} + U \frac{\partial y}{\partial t} \right) \right] \tag{9.109}$$

$$F = dm \left(U^2 \frac{\partial^2 y}{\partial x^2} + 2U \frac{\partial^2 y}{\partial x \, \partial t} + \frac{\partial^2 y}{\partial t^2} \right) \tag{9.110}$$

Using equation (9.7) for F, for $dm = \mu \, dx$,

$$(T - \mu U^2) \frac{\partial^2 y}{\partial x^2} - 2\mu U \frac{\partial^2 y}{\partial x \, \partial t} = \mu \frac{\partial^2 y}{\partial t^2} \tag{9.111}$$

It is interesting to investigate what happens if $T = \mu U^2$. The first term on the right-hand side of equation (9.111) is eliminated and

$$-2\mu U \frac{\partial^2 y}{\partial x \, \partial t} = \mu \frac{\partial^2 y}{\partial t^2} \tag{9.112}$$

$$\frac{\partial}{\partial t} \left[2\mu U \frac{\partial y}{\partial x} + \mu \frac{\partial y}{\partial t} \right] = 0 \tag{9.113}$$

This implies that

$$2\mu U \frac{\partial y}{\partial x} + \mu \frac{\partial y}{\partial t} = f(x) \tag{9.114}$$

where $f(x)$ is an arbitrary function of x. This is possible for any x only if $\partial y/\partial t$ is constant. This means that $y = Y_1(x)t + Y_2(x)$, where $Y_1(x)$ and $Y_2(x)$ are arbitrary functions of x. The solution increases continuously in time, which implies instability. The threshold of instability is evidently at the point when

$$U = \left(\frac{T}{\mu} \right)^{1/2} \tag{9.115}$$

Therefore, when the string speed equals the speed of sound on the string, the system becomes unstable.

☐ **Example 9.18 Unstable Vibration of Belt Drives**

A belt drive operates between two pulleys of diameters 200 and 300 mm. Its cross section is $A = 250 \text{ mm}^2$, density $\rho = 1200 \text{ kg/m}^3$,[20] and the prestress is 600 N. Determine the angular velocity of the first pulley that will result in unstable lateral vibration of the belt.

Solution From equation (9.115), the unstable linear speed is

$$U = \left(\frac{T}{\mu}\right)^{1/2} = \left(\frac{T}{A\rho}\right)^{1/2}$$

$$= \left(\frac{600}{250 \times 10^{-6} \times 1200}\right)^{1/2} = 44.7 \text{ m/s}$$

Since $U = \omega R$, $\omega = U/R = 44.7/0.100 = 447$ rad/s or 4270 rpm. At 4270 rpm the belt will become unstable.

(The unstable speed of a moving belt is computed.)

☐

Vibration of Pipes at High Flow Velocities

A fluid pipe can develop unstable vibration at high fluid velocities. An approximate solution was obtained in Chapter 4, assuming the pipe and fluid mass to be lumped in the middle of the pipe length. For a more accurate solution, Figure 9.10 is used, where the pipe is considered as a beam of flexural rigidity EI, mass per unit length μ_p, carrying fluid of mass per unit length μ_f at a constant velocity U.

An element of length dx is considered. The fluid moves in the direction of the y-axis with velocity $u = \partial y/\partial t + U(\partial y/\partial x)$ while the pipe itself moves in the same direction with velocity $\partial y/\partial t$. Newton's law in the y direction can be written in the form of equation (9.106). Therefore, equation (9.107) gives

$$F = \mu_f \, dx \left(U^2 \frac{\partial^2 y}{\partial x^2} + 2U \frac{\partial^2 y}{\partial x \, \partial t} + \frac{\partial^2 y}{\partial t^2} \right) + \mu_p \, dx \frac{\partial^2 y}{\partial t^2} \tag{9.116}$$

Using equation (9.19) for F, we obtain

$$-\frac{\partial^2}{\partial x^2} \left(EI \frac{\partial^2 y}{\partial x^2} \right) + \mu_f U^2 \frac{\partial^2 y}{\partial x^2} + 2\mu_f U \frac{\partial^2 y}{\partial x \, \partial t} + (\mu_f + \mu_p) \frac{\partial^2 y}{\partial t^2} = 0 \tag{9.117}$$

Figure 9.10 Bending vibration of a fluid-carrying pipe.

[20]Dimarogonas, A. D. 1989. *Computer Aided Machine Design*. Englewood Cliffs, N.J.: Prentice Hall, p. 565.

Vibration of Membranes

Membrane is a thin elastic sheet that is initially taut with a uniform tension p per unit area through its boundary. It is further assumed that the applied deformation is much less than the initial stretch due to the uniform tension, which thus remains constant. The similarity with the taut string is apparent.

In Figure 9.11 a membrane initially on the (x, y) plane is stretched by uniform tension p and deflected by w along the z direction. A differential $dx\,dy$ of the membrane, of thickness h, is acted upon by elastic forces

$$F = ph\,dy\,\frac{\partial\theta_x}{\partial x}\,dx + ph\,dx\,\frac{\partial\theta_y}{\partial y}\,dy \qquad (9.118)$$

Because $\theta_x = \partial w/\partial x$, $\theta_y = \partial w/\partial y$, Newton's second law in the z direction yields

$$ph\,dy\,\frac{\partial\theta_x}{\partial x}\,dx + ph\,dx\,\frac{\partial\theta_y}{\partial y}\,dy = \rho\,dx\,dy \qquad (9.119)$$

$$\frac{\partial^2 w}{\partial x^2} + \frac{\partial^2 w}{\partial y^2} = \frac{1}{c^2}\frac{\partial^2 w}{\partial t^2} \qquad (9.120)$$

where $c = (ph/\rho)^{1/2}$, the sound velocity in the membrane.

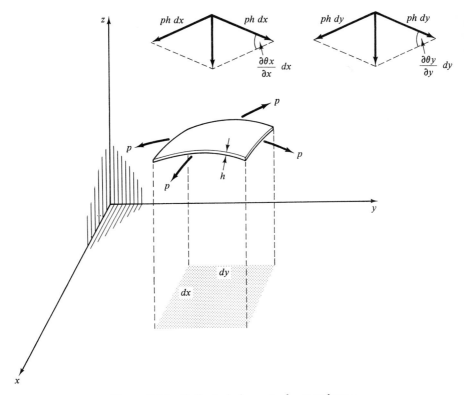

Figure 9.11 Deflected element of a membrane.

☐ **Example 9.19 Vibration of a Rectangular Membrane**

Find the natural frequencies of a rectangular membrane of sides a and b fixed at the perimeter.[21]

Solution We will use the method of separation of variables, as in the case of the beams. Let $w(x, y, t) = X(x)Y(y)T(t)$. Equation (9.120) gives

$$X''YT + XY''T = c^{-2}XYT''$$

Dividing through by XYT yields

$$c^2 \frac{X''}{X} + c^2 \frac{Y''}{Y} = \frac{T''}{T}$$

For the equation above to hold for any value of x, y, and t, we must have

$$c^2 \frac{X''}{X} = -\lambda_1^2, \qquad c^2 \frac{Y''}{Y} = -\lambda_2^2, \qquad \frac{T''}{T} = -\lambda_3^2$$

where λ_1, λ_2, and λ_3 are arbitrary constants, $\lambda_3^2 = \lambda_1^2 + \lambda_2^2$. We recognize λ_3 as the natural frequency ω. The solutions of the ordinary differential equations can be written in the form

$$X(x) = c_1 \cos \frac{\lambda_1}{c} x + c_2 \sin \frac{\lambda_1}{c} x$$

$$Y(y) = c_3 \cos \frac{\lambda_2}{c} y + c_4 \sin \frac{\lambda_2}{c} y$$

$$X(x) = c_5 \cos \omega t + c_6 \sin \omega t$$

The boundary conditions are

$$w(x, y, t) = 0 \qquad \text{for } x = 0, \quad x = a, \quad y = 0, y = a$$

Substituting, we have

$$c_1 = 0$$

$$c_1 \cos \frac{\lambda_1}{c} a + c_2 \sin \frac{\lambda_1}{c} a$$

$$c_3 = 0$$

$$c_3 \cos \frac{\lambda_2}{c} b + c_4 \sin \frac{\lambda_2}{c} b$$

The second and fourth, due to the first and third, give

$$\sin \frac{\lambda_1}{c} a = 0, \qquad \sin \frac{\lambda_2}{c} b = 0$$

$$\lambda_1 = \frac{\pi n c}{a}, \qquad \lambda_1 = \frac{\pi m c}{b}$$

Because $\lambda_1^2 + \lambda_2^2 = \omega^2$,

$$\omega^2 = \left(\frac{\pi n c}{a}\right)^2 + \left(\frac{\pi m c}{b}\right)^2$$

[21]Lamé, G. 1852. *Leçons sur élasticité des corps solides.* Paris.

where m and n are arbitrary integers. Each pair of m and n yields a different natural frequency and the associated natural modes

$$X(x)\,Y(y) = \sin\frac{\pi nx}{a}\,\sin\frac{\pi my}{b}$$

(The natural frequencies and modes of a rectangular membrane fixed at the boundary are computed.)

□

TABLE 9.6 SUMMARY OF EQUATIONS FOR CHAPTER 9

Lateral vibration of taut string

$$\frac{\partial^2 y}{\partial x^2} = \frac{1}{c^2}\frac{\partial^2 y}{\partial t^2}$$

$$c = \left(\frac{T}{\mu}\right)^{1/2}$$

Longitudinal vibrations of rods

$$\frac{\partial^2 u}{\partial x^2} = \frac{1}{c^2}\frac{\partial^2 u}{\partial t^2}$$

$$c = \left(\frac{E}{\rho}\right)^{1/2}$$

The Euler–Bernoulli beam

$$-\frac{\partial^4 y}{\partial x^4} = \frac{1}{c^2}\frac{\partial^2 y}{\partial t^2}$$

$$c = \left(\frac{EI}{\mu}\right)^{1/2}$$

Torsional vibration of rods

$$\frac{\partial^2\theta}{\partial x^2} = \frac{1}{c^2}\frac{\partial^2\theta}{\partial t^2}$$

$$c = \left(\frac{G}{\rho}\right)^{1/2}$$

General solution of wave equation

$$w(x,\,t) = F_1(ct - x) + F_2(ct + x)$$

$$w(x,\,t) = \sum_{n=0}^{\infty}[a_n\cos\omega_n t + b_n\sin\omega_n t]\,Y_n(x)$$

(continues)

TABLE 9.6 *(cont.)*

Forced vibration

Taut spring

$$\frac{\partial^2 y}{\partial x^2} = \frac{1}{c^2}\frac{\partial^2 y}{\partial t^2} + \frac{1}{T}p(x, t)$$

Longitudinal vibration of rods

$$\frac{\partial^2 u}{\partial x^2} = \frac{1}{c^2}\frac{\partial^2 u}{\partial t^2} + \frac{1}{EA}p(x, t)$$

Lateral vibration of beams

$$\frac{-\partial^4 u}{\partial x^4} = \frac{1}{c^2}\frac{\partial^2 y}{\partial t^2} + \frac{1}{EI}p(x, t)$$

Torsional vibration

$$\frac{\partial^2 \theta}{\partial x^2} = \frac{1}{c^2}\frac{\partial^2 \theta}{\partial t^2} + \frac{1}{GI_p}p(x, t)$$

Modal analysis

$$y(x, t) = \sum_{k=1}^{\infty} \phi_k q_k(t)$$

$$\ddot{q}_k + \omega_k^2 q_k = Q_k(t), \ k = 1, 2, 3, \dots$$

$$Q_k(t) = \int_0^l f(x, t)\phi_k(x)\, dx.$$

Rayleigh–Timoshenko beam

$$\frac{EI\partial^4}{\partial x^4} + \mu\frac{\partial^2 y}{\partial t^2} - \left(\Gamma + \frac{EI\mu}{kAG}\right)\frac{\partial^4 y}{\partial x^2\partial t^2} + \frac{\Gamma\mu}{kAG}\frac{\partial^4 y}{\partial t^4} = f(x, t) + \frac{\Gamma}{kAG}\frac{\partial^2 f}{\partial t^2} - \frac{EI}{kAG}\frac{\partial^2 f}{\partial x^2}$$

Moving string

$$(T - \mu U^2)\frac{\partial^2 y}{\partial x^2} - 2\mu U\frac{\partial^2 y}{\partial x\partial t} = \mu\frac{\partial^2 y}{\partial t^2}$$

Fluid carrying pipe

$$-\frac{\partial^2}{\partial x^2}\left(EI\frac{\partial^2 y}{\partial x^2}\right) + \mu_f U^2\frac{\partial^2 y}{\partial x^2} + 2\mu_f U\frac{\partial^2 y}{\partial x\partial t} + (\mu_f + \mu_p)\frac{\partial^2 y}{\partial t^2} = 0$$

Vibration of membranes

$$\frac{\partial^2 w}{\partial x^2} + \frac{\partial^2 w}{\partial y^2} = \frac{1}{c^2}\frac{\partial^2 w}{\partial t^2} \qquad c = (ph/\rho)^{1/2}$$

REFERENCES AND FURTHER READING

BEER, F. P., and JOHNSTON, W. 1982. *Mechanics of Materials*. New York: McGraw-Hill.

BISHOP, R. E. D., and JOHNSON, C. D. 1979. *Mechanics of Vibration*, 2nd ed. Cambridge: Cambridge University Press.

DEN HARTOG, J. P. 1952. *Mechanical Vibration*, 4th ed. New York: McGraw-Hill.

HOHENEMSER, K., and PRAGER, W. 1933. *Dynamik der Stabwerke*. Berlin: Julius Springer.

HORT, W. 1922. *Technische Schwingungslehre*, 2nd Ed. Berlin: Julius Springer.

RAYLEIGH, J. W. S. 1894. *Theory of Sound*. New York: Dover Publications, 1946.

TIMOSHENKO, S. P. 1928. *Vibration Problems in Engineering*. New York: D. Van Nostrand.

TIMOSHENKO, S. P. 1953. *History of the Strength of Materials*. New York: McGraw-Hill.

TIMOSHENKO, S. P., and YOUNG, D. H. 1955. *Theory of Structures*. New York: McGraw-Hill.

PROBLEMS

Sections 9.1 to 9.4

9.1. Miners trapped in a mine signal for help by knocking somewhere along a compressed air pipe. Determine the distance at which the miners are from the time difference between the signal coming to the outer end directly and that reflected at the inner end of the pipe if the total length of the pipe $L = 800$ m is known. Assume an air of 6 bar pressure at 20°C temperature.

9.2. Determine the velocity of the wave propagation on an underwater cable stretched by a tension $T = 8000$ N and a mass per unit length $\mu = 3$ kg/m.

9.3. Determine the velocity of the torsional wave propagation on the propeller shaft of a ship of diameter $d = 0.6$ m, density $\rho = 7800$ kg/m³, and shear modulus $G = 0.96 \times 10^{11}$ N/m².

9.4. Determine the velocity of the longitudinal wave propagation on a pipe of inner diameter $d_1 = 110$ mm and outer diameter $d_2 = 120$ mm, density 7800 kg/m³, and Young's modulus $E = 2 \times 10^{11}$ N/m².

9.5. A pipe AB of diameter d_1 and length l_1 is connected to a larger pipe BC of diameter d_2 and length l_2. At the end A, we knock at the pipe and measure the time difference between the returns of the reflections at B and C. Determine the density of the material.

9.6. Determine the tension needed in order that a cord of a musical instrument of diameter $d = 1$ mm and density 7800 kg/m³ will have a fundamental frequency of lateral vibration $f = 250$ Hz. The length of the cord is $L = 0.8$ m and it is fixed at the ends.

9.7. The cord of a musical instrument of length $l = 1$ m, diameter $d = 0.5$ mm, and density 7800 kg/m³ is taut with a tension $T = 250$ N. Determine the first three natural frequencies, the vibration modes, and the nodal points corresponding to these modes.

9.8. A spring of constant $k = 1000$ N/m and mass $m = 15$ kg is placed between two walls at a distance $L = 0.5$ m (Figure P9.8). Determine the natural frequencies of longitudinal vibration of the spring.

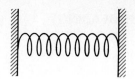

Figure P9.8

9.9. An oil drill (Figure P9.9) is made of steel with $E = 2.1 \times 10^{11}$ N/m^2, diameter $d = 0.30$ m, and density 7800 kg/m^3. Find the fundamental natural frequency for longitudinal vibration if it operates at a depth of 1500 m and while it is fixed at the upper end if **(a)** the lower end is free, and **(b)** the lower end rests on the bottom of the well.

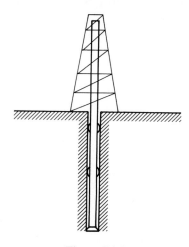

Figure P9.9

9.10–9.15. Determine the frequency equation for longitudinal vibration of the systems shown in Figures P9.10 to P9.15.

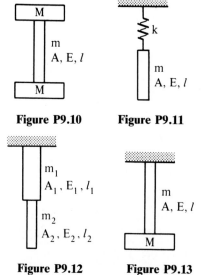

Figure P9.10   **Figure P9.11**

Figure P9.12   **Figure P9.13**

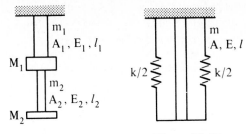

Figure P9.14

Figure P9.15

9.16–9.20. For the systems shown in Figures P9.16 to P9.20, determine the frequency equation for torsional natural vibration and the natural modes.

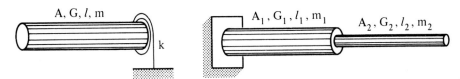

Figure P9.16

Figure P9.17

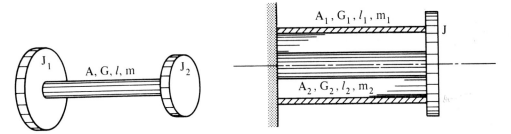

Figure P9.18

Figure P9.19

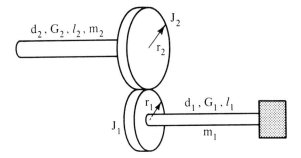

Figure P9.20

9.21. The propeller shaft of a ship has length between bearings $L = 12$ m, diameter $d = 0.30$ m, and shear modulus $G = 0.9 \times 10^{11}$ N/m². Determine the three lowest natural frequencies for torsional vibration, free-free, $\rho = 7800$ kg/m³.

9.22. For the drill of Problem 9.9, determine the lowest torsional natural frequency if the shear modulus of the drill is $G = 0.95 \times 10^{11}$ N/m^2.

9.23. The crankshaft of a diesel engine has diameter 50 mm and length 1.6 m. It is made of steel with $G = 0.85 \times 10^{11}$ Pa and $\rho = 7800$ kg/m^3. The left end is free while the right end is connected to a generator rotor with polar moment of inertia $J_p = 50$ kg/m^2. Find the lowest three torsional natural frequencies.

9.24. A 60-mm-diameter stub shaft, $\rho = 7800$ kg/m^3, $G = 0.95 \times 10^{11}$ N/m^2, connects an electric motor to a dc electric generator. The polar moments of inertia of the rotors are, respectively, 2 and 3 kg·m^2. Determine the lowest torsional natural frequency and compare it with the one found by neglecting the mass of the shaft.

9.25. For a free shaft of constant diameter, we assume that the shaft mass is lumped into two end disks. Determine the diameter and thickness of two equal disks that will give the same natural frequency as a two-degree-of-freedom massless shaft system with the fundamental torsional natural frequency of the free-free, distributed mass shaft.

9.26. A short turbine blade has length L, cross section A, radius of gyration r, and is made out of steel with $E = 2.1 \times 10^{11}$ Pa and $\rho = 7800$ kg/m^3. The blade is mounted at one end on a turbine wheel and the other end is free (Figure P9.26). The mounting is not rigid but has a rotational spring constant k. Determine the frequency equation for lateral vibrations.

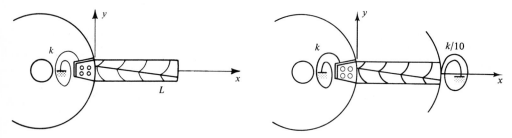

Figure P9.26   **Figure P9.27**

9.27. In Problem 9.26, the free end of the blade is reinforced by a blade cover that has a rotational spring constant $k/10$ (Figure P9.27). Determine the frequency equation.

9.28. The propeller shaft of a ship has length between bearings $L = 12$ m, diameter $d = 0.30$ m, $\rho = 7800$ kg/m^3, and Young's modulus $E = 2.1 \times 10^{11}$ N/m^2. Determine the three lowest natural frequencies for lateral vibration of the shaft, assumed simply supported.

9.29. The low-pressure rotor of a turbine can be considered as a shaft of constant diameter $d = 0.3$ m supported as a simply supported beam at two bearings at distance $L = 2.5$ m. Determine the three lower natural frequencies if $\rho = 7800$ kg/m^3 and $E = 2.1 \times 10^{11}$ N/m^2.

9.30. The shaft of Problem 9.29 is supported at the ends on two slider bearings that have stiffness $k = 2.5 \times 10^7$ N/m each. Determine the three lowest natural frequencies.

Section 9.5

9.31. A steel shaft of length $L = 200$ in. and diameter $d = 180$ mm rotating at 3600 rpm is suddenly stopped at its right end due to a malfunction of the driving motor. Determine the stresses on the shaft if $\rho = 7800$ kg/m^3 and $E = 2.1 \times 10^{11}$ N/m^2.

9.32. Solve Problem 9.31 assuming that at the right end the stopping mechanism is not rigid but has a torsional spring constant $k = 0.5 \times 10^7$ N·m/rad.

9.33. A steel stub shaft of diameter $d = 50$ mm and length $l = 0.75$ m connects a 3600-rpm electric motor whose rotor has a moment of inertia $J_1 = 5$ kg·m² to a compressor whose rotor has a moment of inertia $J_2 = 3$ kg·m². While the shaft transmits a power of 60 kW, suddenly the motor and compressor are unloaded. Determine the resulting vibration of the system.

9.34. Solve Problem 9.33 if the unloading was due to a shaft failure at the compressor end.

9.35. The generator of the system of Problem 9.24 is suddenly stopped due to an electrical fault. Determine the resulting vibration.

Section 9.6

9.36. A uniform rod of section $A = 200$ mm², density 7800 kg/m³, and length $L = 0.80$ m, placed on a shaker table, vibrates at 60 Hz with an amplitude of $y = 0.0010$ m in the direction of the rod. Determine the motion of the other end of the beam.

9.37–9.40. In Problem 9.36, the shaker is used with a function generator so that the vibration has the same amplitude and period but instead of being harmonic, it has the form shown in Figures P9.37 and P9.40. Determine the amplitude at the other end of the rod for each form.

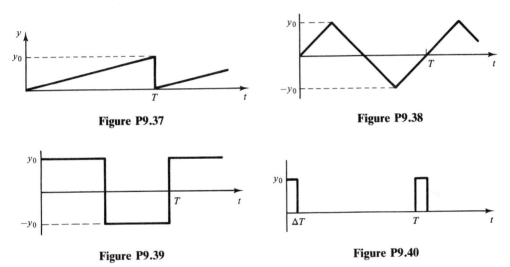

Figure P9.37   **Figure P9.38**

Figure P9.39   **Figure P9.40**

9.41. In Problem 9.33, determine the torsional vibration of the compressor if the induction motor has a torsional excitation of frequency $\frac{2}{10}$ of running and torque 4% of the transmitted steady torque at full load.

9.42. A cutting mill is attached to the end of a shaft of diameter $d = 20$ mm and length $l = 0.5$ m, and during a particular job it takes 4 hp at 90 rpm. If the diameter of the cutting tool is $D = 0.1$ m and one of its six teeth is broken, determine the resulting torsional vibration of the system assuming the other end of the shaft fixed on a driving system of very large mass.

9.43. Solve Problem 9.42 if the driving system has a torsional flexibility of $k = 0.8 \times 10^7$ N·m/rad.

9.44. The cord of Problem 9.5 has one end flexible in the direction perpendicular to the string with a spring constant $k = 2800$ N/m. Determine the frequency equation.

9.45. At the left end of the shaft of Problem 9.30, there is an unbalanced mass of 0.1 kg at radius $R = 0.1$ m. Determine the shaft vibration if it rotates at 3600 rpm.

10

Approximate and Numerical Methods

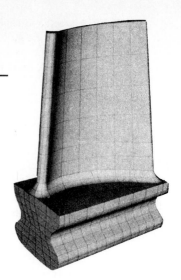

10.1 INTRODUCTION

In previous chapters we have modeled mathematically dynamic systems in a more or less exact form. Of course, masses concentrated at a point, rigid bodies, elastic massless members, and linear properties do not exist in nature. But if one can conceive such systems, they can have exact solutions. Furthermore, we have studied systems with distributed mass and elasticity. There again, idealizations of beams with constant properties, linearity, and boundary conditions of the rigid support type also do not exist. If one can conceive them, "exact" solutions, in the form of infinite series, exist and in certain cases, can be found.

This may have left us with a feeling of false security, however, because our solutions are, in any event, approximate. The first approximation was introduced in the modeling of the system itself, as we mentioned above. Second, only for few-degrees-of-freedom systems with special forms of excitation, if any, can an exact solution be found. For higher-order systems, the very first difficulty in obtaining

*The above photo portrays the solid model for vibration analysis of a turbine blade, from Dimarogonas, A. D., *Computer Aided Machine Design*, © 1988, p. 81. Reprinted by permission of Prentice Hall, Englewood Cliffs, New Jersey.

an exact solution is the finding of the roots of the characteristic equation. In general, this has to be done numerically. The same is true for the determination of the normal modes and the harmonic steady-state response. Therefore, these approaches do eventually resort to numerical methods. Moreover, in certain systems, one either cannot model it as a multidegree-of-freedom system or one cannot solve the equations describing the continuous system. The alternative is the application of certain approximate methods that often give results satisfying the engineering requirements of accuracy. Also, these methods often allow the correct answer to be bracketed between two limit values. This fact gives us assurance as to the accuracy of the computations, assurance that can rarely be obtained by all-computer solutions of the differential equations of the continuous system or by arbitrary lumping the mass of the system at specific points. This is very important for engineering computations, where "exact" solutions are not needed, but on the other hand, the approximation error should be within acceptable limits.

In Chapter 9, we have seen that continuous systems have infinite number of degrees of freedom. With modal analysis, the forced vibration response consists of summing up the contribution of the infinite number of natural modes, each multiplied by a function of time:

$$y(x, t) = \sum_{n=0}^{\infty} (a_n \cos \omega t + b_n \sin n\omega t)\phi_n(x) \tag{10.1}$$

where $\phi_n(x)$ are normal modes, all known in general and functions of the system properties and the boundary conditions only. The coefficients a_n and b_n depend in general on the initial conditions or the forcing functions. Thus we write

$$y(x, t) = \sum_{n=0}^{\infty} f_n(t)\phi_n(x) \tag{10.2}$$

where $f_n(t)$ are time functions, which depend on the initial conditions or the forcing functions.

In analyzing a continuous system, one can apply the principles of Chapters 7 to 9 and obtain the response to any form of initial conditions of forcing functions. However, there are certain difficulties that limit the application of the formal methods of analysis of continua to very simple geometries only:

1. The infinite series sometimes converge very slowly and it is difficult to estimate how many terms are needed for engineering accuracy.
2. The formulation and computation efforts are prohibitive for systems of engineering complexity.

The special methods that are presented in this chapter treat the continuous systems, for vibration analysis purposes, as discrete systems. This can done to a different extent, depending on the particular method used, with one of the following methods:

1. Retaining n natural modes only and considering them as generalized coordinates and computing the n weighing functions $f_n(t)$ to best fit the initial conditions or the forcing functions.

2. Substituting for the n modes an equal number of known functions $\phi_n(x)$ that satisfy the geometric conditions of the problem, and then compute the functions $f_n(t)$ to best fit the differential equation, the remaining boundary conditions, and the initial conditions or the forcing functions.

3. Taking as generalized coordinates the n physical coordinates of a certain number n of points of the system $q_1, q_2, \ldots, q_n$, considering them as functions of time, and computing them to fit the differential equation and the initial and boundary conditions.

The main advantage of all these methods is that instead of dealing with one or more partial differential equations, we deal with a larger number of ordinary differential equations (usually linear with constant coefficients) which are particularly suitable for solution in fast computing machines.

10.2 RAYLEIGH'S METHOD

For lumped mass and continuous systems, we discussed formal methods for the determination of natural frequencies. Sometimes, we are interested only in a quick determination of the lower natural frequency of the system, called *fundamental*. Rayleigh's method gives a fast and rather accurate computation of the fundamental frequency of the system and we discuss it in some detail, also in view of its implications on the subsequent discussions. Moreover, it applies for both discrete and continuous systems.

Consider a discrete, conservative system described by way of the matrix equation

$$\mathbf{M\ddot{x}} + \mathbf{Kx} = 0 \tag{10.3}$$

The equation above is satisfied by a set of n eigenvalues ω_i^2 and normalized eigenvectors $\boldsymbol{\phi}^{(i)}$, which satisfy the equation

$$\mathbf{K}\boldsymbol{\phi}^{(i)} = \omega_i^2 \mathbf{M}\boldsymbol{\phi}^{(i)}, \qquad i = 1, 2, \ldots, n \tag{10.4}$$

Multiplying both sides of (10.4) by $\boldsymbol{\phi}^{(i)T}$ and dividing by the scalar $\boldsymbol{\phi}^{(i)T}\mathbf{M}\boldsymbol{\phi}^{(i)}$, which is a quadratic form, we obtain

$$\omega_i^2 = \frac{\boldsymbol{\phi}^{(i)T}\mathbf{K}\boldsymbol{\phi}^{(i)}}{\boldsymbol{\phi}^{(i)T}\mathbf{M}\boldsymbol{\phi}^{(i)}} \tag{10.5}$$

If we know the eigenvector $\boldsymbol{\phi}^{(i)}$, we can obtain the corresponding eigenvalue ω_i^2 by equation (10.5). However, in general, we do not know the eigenvectors in advance. Suppose that we consider an arbitrary vector $\mathbf{z}$ in equation (10.5):

$$\omega^2 = R(\mathbf{z}) = \frac{\mathbf{z}^T\mathbf{Kz}}{\mathbf{z}^T\mathbf{Mz}} \tag{10.6}$$

where $R(\mathbf{z})$ depends on the vector $\mathbf{z}$ and is called *Rayleigh's quotient*. When the vector $\mathbf{z}$ coincides with an eigenvector $\boldsymbol{\phi}$, Rayleigh's quotient coincides with the corresponding eigenvalue.

From vector algebra we know that we can express any vector z by a linear combination of the n linearly independent vectors (expansion theorem):

$$z = \sum_{j=1}^{n} c_i z^{(i)} = Zc \qquad (10.7)$$

where Z is a square modal matrix $[z^{(1)} \cdots z^{(n)}]$ and $c = \{c_1 \ c_2 \ \cdots \ c_n\}$. If the vectors $z^{(i)}$ have been normalized so that $Z^T M Z = I$, then

$$Z^T K Z = \text{diag}[\omega_1^2, \omega_2^2, \ldots, \omega_n^2] = P \qquad (10.8)$$

Introducing (10.7) into (10.6) and using the orthogonality property, we obtain

$$R(z) = \frac{c^T Z^T K Z c}{c^T Z^T M Z c} = \frac{c^T P c}{c^T I c} = \frac{\sum_{i=1}^{n} c_i^2 \omega_i^2}{\sum_{i=1}^{n} c_i^2} \qquad (10.9)$$

We expect that if the selection of z differs very little from a certain eigenvector $\phi^{(r)}$, the corresponding c_r is much larger then the other c's and we can write

$$\left| \frac{c_i}{c_r} \right| = \epsilon_i \ll 1 \qquad \text{if } i \neq r \qquad (10.10)$$

where ϵ_i is a small number. Dividing the fraction in (10.9) by c_r^2, we obtain

$$R_r(z) = \frac{\omega_r^2 + \sum_{i=1}^{n} \epsilon_i^2 \omega_i^2}{1 + \sum_{j=1}^{n} \epsilon_j^2}, \qquad i \neq r \qquad (10.11)$$

We can also write

$$R_r(z) = \frac{\omega_r^2}{1 + \epsilon^2} \qquad (10.12)$$

Equation (10.12) indicates that the Rayleigh's quotient differs from the eigenvalue ω_r^2 by second-order terms of the error of the eigenvector. In other words, if the error we committed by selecting the nonexact eigenvector z instead of $\phi^{(r)}$ was ϵ, the error in the computation of the eigenvalue is ϵ^2. A 10% error in the form of the eigenvector (natural mode) will result in only 1% error in the eigenvalue (natural frequency). From equation (10.11) we also conclude that

$$\omega_n^2 \geq R_r(z) \geq \omega_1^2 \qquad (10.13)$$

In words, the Rayleigh's quotient is never lower than the smallest eigenvalue or higher than the greatest eigenvalue.

In view of the preceding discussion, one can state that in a conservative system the frequency of vibration has a stationary value in the neighborhood of a natural mode. This is called *Rayleigh's principle*. This principle enables us to calculate the fundamental frequency of a system as follows.

For a conservative system, the total energy remains constant. When the system vibrates at a natural mode it has harmonic motion at the corresponding natural frequency. When it passes through the equilibrium position, the potential energy is zero (no deformation) and thus the kinetic energy is maximum. When the system reaches the most extreme position, the kinetic energy becomes zero (no velocity) and the potential energy has a maximum. Both maxima must equal the total energy of the system:

$$E = T + V$$

$$E = T_{max} + 0 = 0 + V_{max} \qquad (10.14)$$

$$T_{max} = V_{max}$$

To calculate the fundamental frequency one can assume an approximate natural mode and from equation (10.6) obtain the natural frequency. However, one can do it an easier way, using the principle of equations (10.14).

We assume a certain mode $\mathbf{z} = \{z_1 \quad z_2 \quad \cdots \quad z_n\}$. If the system has harmonic vibration at a frequency ω (unknown yet), the maximum kinetic energy will be, for a discrete system,

$$T_{max} = \tfrac{1}{2}\mathbf{z}^T\mathbf{Mz}\omega^2 \qquad (10.15)$$

or for a diagonal mass matrix,

$$T_{max} = \tfrac{1}{2}[m_1(\omega z_1)^2 + m_2(\omega z_2)^2 + \cdots + m_n(\omega z_n)^2] \qquad (10.16)$$

As a first guess, we assume that the fundamental mode $\mathbf{z} = \{z_1, z_2, \ldots, z_n\}$ resembles the static deflection of the system due to the weights $m_1g, m_2g, \ldots, m_ng$. At static deflection, the potential energy of elastic deformation of the system must be equal to the work of the weights or body forces in general from the straight (1) to the deflected position (2):

$$V_{max} = W_{1\rightarrow 2} = \tfrac{1}{2}(m_1gz_1 + m_2gz_2 + \cdots + m_ngz_n) \qquad (10.17)$$

Equating T_{max} and V_{max}, we obtain

$$\omega^2 = \frac{g \sum\limits_{i=1}^{n} m_i z_i^2}{\sum\limits_{i=1}^{n} m_i z_i} \qquad (10.18)$$

One can extend Rayleigh's method for continuous systems observing that if $y(x)$ designates the mode, we shall have for lateral vibration of a beam of length L,

$$U_{max} = \frac{1}{2} \int_0^L EI\left(\frac{d^2y}{dx^2}\right)^2 dx \qquad (10.19)$$

$$T_{max} = \tfrac{1}{2}\omega^2 \int_0^L \dot{y}^2 \, dm \qquad (10.20)$$

Therefore,

$$\omega^2 = \frac{\int_0^L EI(d^2y/dx^2)^2 \, dx}{\int_0^L \dot{y}^2 \, dm} \tag{10.21}$$

Owing to the stationarity of Rayleigh's quotient, remarkably good estimates of the fundamental frequency can be obtained even if the trial eigenvector or eigenfunction does not resemble the first mode too closely. Hence the quality of the estimate obtained depends on the experience and skill in selecting a trial vector. Due to the inequality (10.13), one could try many different trial modes and find the minimum of the resulting frequencies. The method can be used for any frequency, but acceptable accuracy is obtained only for the fundamental, due to the fact that in most cases, the static deflection curve resembles closely the first mode. In cases when the static deflection is not relevant, torsional vibration, for example, one finds other approximations for the mode, for example by applying body forces proportional to the masses or moments of inertia.

☐ **Example 10.1   Natural Frequencies with Rayleigh's Method**

Determine the natural frequencies of the system shown in Figure E10.1a.

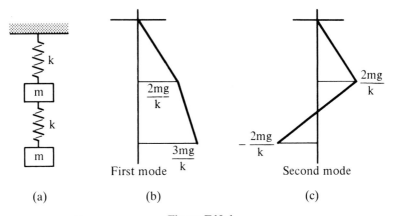

Figure E10.1

Solution   For comparison, one can easily find that the natural frequencies are, using methods of Chapter 7,

$$\omega_1^2 = 0.36 \frac{k}{m}, \qquad \omega_2^2 = 2.64 \frac{k}{m}$$

We shall try to find them with Rayleigh's method. The energies are

$$T_{max} = \tfrac{1}{2}m\dot{x}_1^2 + \tfrac{1}{2}m\dot{x}_2^2 = \tfrac{1}{2}m\omega^2(A_1^2 + A_2^2)$$

$$V_{max} = \tfrac{1}{2}kx_1^2 + \tfrac{1}{2}k(x_2 - x_1)^2 = \tfrac{1}{2}k[A_1^2 + (A_2 - A_1)^2]$$

Therefore,

$$\omega^2 = \frac{k}{m} \frac{A_1^2 + (A_2 - A_1)^2}{A_1^2 + A_2^2}$$

The first mode is assumed to be the static deflection

$$A_1 = \frac{2mg}{k}, \qquad A_2 = \frac{3mg}{k}$$

Therefore,

$$\omega_1^2 = 0.38 \frac{k}{m}$$

(compare with the exact value $0.36\ k/m$). The second mode is assumed to be

$$A_1 = \frac{2mg}{k}, \qquad A_2 = -\frac{2mg}{k}$$

We obtain

$$\omega_2^2 = 2.5 \frac{k}{m}$$

which, owing to a good selection of the mode, does not differ considerably from the exact value of $2.64\ k/m$.

(In this example, the natural frequencies of a two-degree-of-freedom system were computed with Rayleigh's method, assuming forms of the vibration modes.)

□

□ **Example 10.2   Natural Frequency of a Cantilever Beam Using Rayleigh's Method**

At the end to a cantilever beam of negligible mass, there is an attached disk of mass m and polar moment of inertia J_p (Figure E10.2). The cantilever has flexural rigidity EI and length l. Find the natural frequency for vertical vibration of the disk using Rayleigh's method and assuming one degree of freedom.

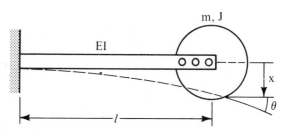

Figure E10.2

Solution   We assume that the inertia of the disk does not result in considerable bending moment and the beam bents only by the lateral force. The kinetic energy is

$$T = \tfrac{1}{2}m\dot{x}^2 + \tfrac{1}{2}J_p\dot{\theta}^2$$

But

$$\theta = \frac{3}{2}\frac{x}{l}, \qquad \dot{\theta} = \frac{3}{2l}\dot{x}$$

For $x = A \cos \omega t$,

$$T_{\max} = \frac{1}{2} mA^2\omega_n^2 + \frac{1}{2} J_p \left(\frac{3}{2l}\right)^2 A^2\omega_n^2$$

The potential energy of elastic deformation (we neglect the change of elevation of the weight)

$$V = \tfrac{1}{2} kx^2, \qquad V_{\max} = \tfrac{1}{2} kA^2$$

For $T_{\max} = V_{\max}$ we obtain

$$\omega_n^2 \left(\frac{1}{2} mA^2 + \frac{1}{2} \frac{9J_p}{4l^2} A^2\right) = \frac{1}{2} kA^2$$

or

$$\omega_n^2 = \frac{k}{m} + \left(1 + \frac{9J_p}{4ml^2}\right)$$

$$\omega_n = \left(\frac{k}{m}\right)^{1/2} \left(1 + \frac{9J_p}{4ml^2}\right)^{1/2}$$

(In this example, the natural frequency and vibration modes of a cantilever beam were computed with Rayleigh's method, assuming a form of the vibration mode.) $\square$

☐ **Example 10.3   Natural Frequency of a Cantilever Beam with End Mass**

A cantilever beam of length l carries at the free end a mass m (Figure E10.3). Determine which portion λ of the mass of the beam M should be added to the mass m in order that the natural frequency can be computed with a massless beam bearing at the end a mass $m + \lambda M$.

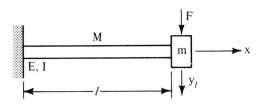

Figure E10.3

Solution   From strength of materials, we know that the deflection due to a load F at the end of a cantilever beam is

$$y(x) = \frac{1}{2}\left[3\left(\frac{x}{l}\right)^2 - \left(\frac{x}{l}\right)^3\right]\frac{Fl^3}{3EI}$$

But $Fl^3/3EI$ is the deflection of the end y_l. Therefore, the shape of the elastic line, due to a static end deflection y_l is

$$y(x) = \frac{1}{2}\left[3\left(\frac{x}{l}\right)^2 - \left(\frac{x}{l}\right)^3\right]y_l$$

Selecting y_l as the coordinate, we have

$$\dot{y}(x) = \frac{1}{2}\left[3\left(\frac{x}{l}\right)^2 - \left(\frac{x}{l}\right)^3 2\right]\dot{y}_l$$

The kinetic energy (beam + end mass) is

$$T = \frac{1}{2} \int_0^l \dot{y}^2 \, dM + \frac{1}{2} m\dot{y}_l^2 = \frac{1}{2} \left(\frac{33M}{140} + m \right) \dot{y}_l^2$$

The potential energy

$$V = \frac{1}{2} ky_l^2 = \frac{1}{2} \frac{3EI}{l^3} y_l^2$$

for

$$y_l = A \cos \omega t, \qquad \dot{y}_l = -\omega A \sin \omega t, \qquad (y_l)_{max} = A, \qquad (\dot{y}_l)_{max} = \omega A$$

Thus

$$T_{max} = \frac{1}{2} \left(\frac{33}{140} M + m \right) \omega^2 A^2$$

$$V_{max} = \frac{1}{2} \frac{2EI}{l^3} A^2$$

$$\omega = \left(\frac{3EI/l^3}{0.235M + m} \right)^{1/2}$$

Finally,

$$m_{eq} = m + \lambda M, \qquad \lambda = 0.235$$

(In this example, the natural frequency of a cantilever beam with a mass at the end was computed using Rayleigh's method, assuming a form of the vibration mode. A simple formula was developed for estimation of the beam mass effect on the end mass with a one-degree-of-freedom system.)

□

□ **Example 10.4 Heavy Spring with End Mass**

A spring of not negligible mass M supports a mass m (Figure E10.4). Determine the fundamental natural frequency by Rayleigh's method and the effective mass $m + \lambda M$ that would yield the same natural frequency with a massless spring.

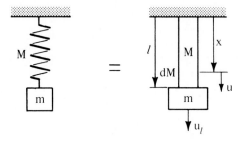

Figure E10.4

Solution   Let l be the length of the spring at static equilibrium. If u_l, the vertical static deflection of the mass m, a point at distance x from the top will have vertical displacement,

$$u = \frac{x}{l} u_l, \qquad \dot{u} = \frac{x}{l} \dot{u}_l, \qquad dM = \frac{M}{l} dx$$

The kinetic energy

$$T = \frac{1}{2} m \dot{u}_l^2 + \frac{1}{2} \int_0^l \dot{u}^2 \, dM = \frac{1}{2} m \dot{u}_l^2 + \frac{1}{2} \frac{M}{l} \int_0^l \frac{x^2}{l^2} \dot{u}_l^2 \, dx$$

$$= \frac{1}{2} m \dot{u}_l^2 + \frac{1}{2} \frac{M \dot{u}_l}{l^3} \int_0^l x^2 \, dx = \frac{1}{2} \left(m + \frac{M}{3} \right) \dot{u}_l^2$$

$$V = \tfrac{1}{2} k u_1^2$$

For

$$u_1 = A \cos \omega t, \qquad (u_1)_{\max} = A, \qquad (\dot{u}_1)_{\max} = \omega A$$

Therefore,

$$T_{\max} = \frac{1}{2} \left(m + \frac{M}{3} \right) \omega^2 A^2 = \frac{1}{2} k A^2 = V_{\max}$$

$$\omega = \left(\frac{k}{m + M/3} \right)^{1/2}$$

Thus

$$m_{\text{eq}} = m + \lambda M, \qquad \lambda = \tfrac{1}{3}$$

(In this example, the natural frequency of a helical spring with a mass at the end was computed with Rayleigh's method, assuming a form of the vibration mode. A simple formula was developed for estimation of the spring mass effect on the end mass with a one-degree-of-freedom system.)

□

□ **Example 10.5   Torsional Natural Frequency of a Shaft with the Rayleigh Method**

Find the fundamental torsional natural frequency of the shaft of Figure E10.5 using Rayleigh's method.

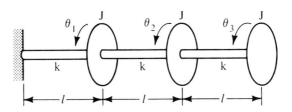

Figure E10.5

Solution   The shaft segments have torsional spring constant

$$k = \frac{J_p G}{l} = \frac{\pi d^4 G}{32 l}$$

$$T = \tfrac{1}{2} J \dot{\theta}_1^2 + \tfrac{1}{2} J \dot{\theta}_2^2 + \tfrac{1}{2} J \dot{\theta}_3^2 = \tfrac{1}{2} J (\dot{\theta}_1^2 + \dot{\theta}_2^2 + \dot{\theta}_3^2)$$

$$V = \tfrac{1}{2} k \theta_1^2 + \tfrac{1}{2} k (\theta_2 - \theta_1)^2 + \tfrac{1}{2} k (\theta_3 - \theta_2)^2$$

Assume that $\theta_1 = A_1 \cos \omega t$, $\theta_2 = A_2 \cos \omega t$, and $\theta_3 = A_3 \cos \omega t$, to obtain

$$T_{max} = \tfrac{1}{2}J\omega^2(A_1^2 + A_2^2 + A_3^2)$$

$$V_{max} = \tfrac{1}{2}k[A_1^2 + (A_2 - A_1)^2 + (A_3 - A_2)^2]$$

A "static deflection" is used for the mode, or A_1, A_2, and A_3 are calculated as the static rotations if we apply static moments, proportional to the moments of inertia. One can see that because k and J are common, we shall have $A_1 = 1$, $A_2 = 2$, and $A_3 = 3$ (scale does not matter). Taking $T_{max} = V_{max}$, we have

$$\omega^2 = \frac{k}{J}\frac{1 + 1 + 1}{1 + 4 + 9}, \qquad \omega = 0.463\left(\frac{k}{J}\right)^{1/2}$$

This is the fundamental natural frequency, 4% greater than the exact one.

(In this example the torsional natural frequency of a shaft was computed with Rayleigh's method, assuming a form of the vibration mode.)

$\square$

10.3 THE METHOD OF CONSTRAINTS

The *method of constraints*, also known as an *assumed modes method*, leads to the determination of a system of differential equations approximately describing the system.

We assume a continuous system and we express its configuration by way of n coordinates, $q_1, q_2, \ldots, q_n$. We assume that the response will be of the form

$$W_n = \sum_{i=1}^{n} z_i q_i(t) \tag{10.22}$$

which in essence treats the continuous system as an n-degree-of-freedom system. The functions z_i are functions of the spatial coordinates and satisfy the geometric constraints of the problem. Since the system has an infinite number of degrees of freedom, the expansion theorem requires that the system response should be expressed as a weighted sum of an infinite number of appropriate functions. In fact, in modal analysis we have seen that these functions are the natural modes of the system. It is apparent that there are some assumptions in describing the system with equations (10.22). Beyond the form of the functions z_i, it is assumed that only n functions are sufficient to describe the system to engineering accuracy, and consequently, the functions beyond order n are zero; that is, $z_{n+1} = z_{n+2} = \cdots = 0$. These conditions are known as *constraints*, and one can expect that they will have a stiffening effect on the system.

On the basis of equations (10.22), the kinetic energy is

$$T(t) = \tfrac{1}{2}\sum_{i=1}^{n}\sum_{j=1}^{n} m_{ij}\dot{q}_i(t)\dot{q}_j(t) \tag{10.23}$$

where m_{ij} are coefficients that will be determined from the mass distribution and the functions z_i. The potential energy will be

$$V(t) = \tfrac{1}{2}\sum_{i=1}^{n}\sum_{j=1}^{n} k_{ij}q_i(t)q_j(t) \tag{10.24}$$

where k_{ij} will be determined from the properties of the system and the functions z_i.

To apply Lagrange's equations we note that for the coordinate r,

$$\frac{\partial T}{\partial \dot{q}_r} = \frac{1}{2} \sum_{i=1}^{n} m_{ir} \dot{q}_i + \frac{1}{2} \sum_{j=1}^{n} m_{rj} \dot{q}_j \qquad (10.25)$$

For a symmetric matrix $[m_{ij}]$, we have

$$\frac{\partial T}{\partial \dot{q}_r} = \sum_{i=1}^{n} m_{ir} \dot{q}_i, \qquad \frac{d}{dt}\left(\frac{\partial T}{\partial \dot{q}_r}\right) = \sum_{i=1}^{n} m_{ir} \ddot{q}_i \qquad (10.26)$$

Similarly, we obtain

$$\frac{\partial V}{\partial q_r} = \sum_{i=1}^{n} k_{ir} q_i, \qquad \frac{\partial T}{\partial q_r} = 0 \qquad (10.27)$$

Therefore, Lagrange's equations yield

$$\sum_{i=1}^{n} m_{ir} \ddot{q}_i + \sum_{i=1}^{n} k_{ir} q_i = f_r, \qquad r = 1, 2, \ldots, n \qquad (10.28)$$

or in matrix form

$$\mathbf{M}\ddot{\mathbf{q}} + \mathbf{K}\mathbf{q} = \mathbf{f} \qquad (10.29)$$

For a lumped mass system, this is the system of equations that we have already obtained, where $\mathbf{M}$ is usually diagonal. Here the matrix $\mathbf{M}$ will be, in general, fully populated. The vector $\mathbf{f}$ contains all external forces for which we cannot write a potential function. It does not contain forces of constraint, gravity, or internal forces. Typical forces in the vector $\mathbf{f}$ include damping, friction, bearing reactions, fluid forces, any nonlinear forces, and so on.

After we have obtained equations (10.29), we continue with the usual methods, as in the case of the multidegree-of-freedom systems. The selection of the functions z and the coordinates q is very important. A usually good selection is a number of node displacements q and the static displacements of the system between nodes, usually in polynomial form. If the functions z are the normal modes of the system, the method of constraints is identical with modal analysis.

□ **Example 10.6 Vibrations of a Cantilever Beam with the Method of Constraints**

Write the differential equation with one degree of freedom that approximates the dynamic response of the cantilever beam of the Figure E10.6 with constant section properties EI and mass μ per unit length along its length l. Then find the natural

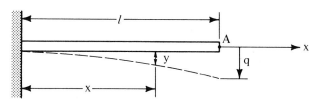

Figure E10.6

frequency and compare it with the one found with continuous beam analysis (Chapter 9).

Solution We select only one degree of freedom, the vertical deflection q of the end A of the beam. For a static deflection q of the end of the beam, the static deflection at distance x will be, assuming a polynomial shape of order 3 with zero displacement and slope at $x = 0$ and displacement q at $x = l$,

$$Y(x) = \frac{1}{2}\left[3\left(\frac{x}{l}\right)^2 - \left(\frac{x}{l}\right)^3\right]q$$

We then select

$$z_1 = z = \frac{1}{2}\left[3\left(\frac{x}{l}\right)^2 - \left(\frac{x}{l}\right)^3\right]$$

and we assume that the motion of the beam will be

$$y(x, t) = z(x)q(t)$$

where q is a function of time, not yet determined. The kinetic energy will be

$$T = \frac{1}{2}\int_0^l \dot{y}^2 \, dm = \frac{A\rho}{2}\int_0^l \frac{1}{4}\left[3\left(\frac{x}{l}\right)^2 - \left(\frac{x}{l}\right)^3\right]^2 \dot{q}^2 \, dx$$

$$= \frac{33A\rho l}{280}\dot{q}^2$$

The potential energy of elastic deformation is

$$V = \frac{1}{2}\int_0^l EI\left(\frac{\partial^2 y}{\partial x^2}\right)^2 dx$$

$$= \frac{1}{2}EI\int_0^l \frac{3}{2l^3}\left(1 - \frac{x}{l}\right)^2 d\left(\frac{x}{l}\right) = \frac{3EI}{2l^3}q^2$$

The Lagrangian function is

$$L = T - V = \frac{33A\rho l}{280}\dot{q}^2 - \frac{3EI}{2l^3}q^2$$

$$\frac{d}{dt}\left(\frac{\partial L}{\partial \dot{q}}\right) - \frac{\partial L}{\partial q} = \frac{33A\rho l}{140}\ddot{q} + \frac{3EI}{l^3}q = 0$$

$$\ddot{q} + \frac{140EI}{11A\rho l^4}q = 0$$

The natural frequency is then

$$\omega_n = 3.56\left(\frac{EI}{A\rho l^4}\right)^{1/2}$$

The exact theory for a continuous beam gives

$$\omega_1 = 3.52\left(\frac{EI}{A\rho l^4}\right)^{1/2}$$

which implies that the approximate frequency is only a little over 1% greater than the exact frequency.

The differential equation can also be used for further analysis, such as forced response. Equally good results are obtained in comparison with exact methods, provided that the forcing frequency is not much higher than the fundamental natural frequency ω_n.

(In this example the method of constraints was used to obtain an approximate differential equation of a single-degree-of-freedom system for the lateral vibration of a cantilever beam.)

☐

☐ **Example 10.7   Vibrations of a Building with the Method of Constraints**

On the top of the tall building of Figure E10.7, there is a motor that operates at angular velocity ω rad/s and has a vertical unbalance force $F_0 \cos \omega t$. Calculate the response of the building in the vertical direction.

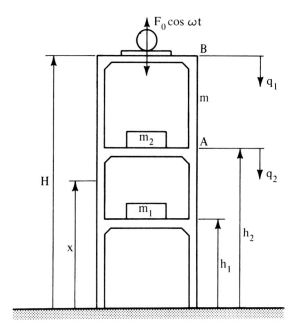

Figure E10.7

Solution   We select only two generalized coordinates, the vertical motion of the points A and B. For static deflections q_1 and q_2 at A and B, respectively, the vertical static deflection at height x will be, between A and B,

$$y_{AB}(x) = q_2 + \frac{q_1 - q_2}{H - h_2}(x - h_2)$$

$$= \frac{x - h_2}{H - h_2}q_1 + \left(1 - \frac{x - h_2}{H - h_2}\right)q_2 \tag{a}$$

and between A and the ground will be

$$y_{A0}(x) = \frac{x}{h_2}q_2 \tag{b}$$

Therefore, we shall seek solutions in the form

$$q(x, t) = z_1(x)q_1(t) + z_2(x)q_2(t) \qquad \text{(c)}$$

where

$$z_1(x) = \begin{cases} \dfrac{x - h_2}{H - h_2}, & h_2 \le x \le H \\[2mm] 0, & \text{elsewhere} \end{cases} \qquad \text{(d)}$$

$$z_2(x) = \begin{cases} 1 - \dfrac{x - h_2}{H - h_2}, & h_2 \le x \le H \\[2mm] \dfrac{x}{h_2}, & 0 \le x \le h_2 \end{cases} \qquad \text{(e)}$$

Our task is now to find $q_1(t)$ and $q_2(t)$. Then $q(x, t)$ will immediately follow from equation (c). The velocity will be

$$\dot{q} = z_1(x)\dot{q}_1 + z_2(x)\dot{q}_2 \qquad \text{(f)}$$

The kinetic energy

$$\begin{aligned} T &= \frac{1}{2} \int_0^H \dot{q}^2 \, dm = \frac{1}{2} \left[m_1 \dot{q}_1^2(h_1) + m_2 \dot{q}_2^2(h_2) + \frac{m}{H} \int_0^H \dot{q}^2 \, dx \right] \\ &= \frac{1}{2} \left\{ m_1 \dot{q}_1^2 + m_2 \dot{q}_2^2 + \frac{m}{H} \int_0^H [z_1(x)\dot{q}_1 + z_2(x)\dot{q}_2]^2 \, dx \right\} \end{aligned} \qquad \text{(g)}$$

We see that lumped masses m_1 and m_2 appear as independent terms, whereas distributed masses m/H appear in an integral. This integral over x ($\dot{q}_1$ and $\dot{q}_2$ are independent of x) can be carried out and yield the kinetic energy T in terms of $\dot{q}_1^2$, $\dot{q}_2^2$, and $q_1 q_2$. Let

$$T = \tfrac{1}{2}(a_1 \dot{q}_1^2 + a_2 \dot{q}_2^2 + 2a_3 \dot{q}_1 \dot{q}_2) \qquad \text{(h)}$$

where

$$a_1 = \frac{m}{H} \int_0^H z_1^2(x) \, dx$$

$$a_2 = m_1 \frac{h_1^2}{h_2^2} + m_2 + \frac{m}{H} \int_0^H z_2^2(x) \, dx$$

$$a_3 = \frac{m}{H} \int_0^H z_1(x)z_2(x) \, dx$$

To compute the potential energy, we note that for a longitudinal load on rod of length L, the spring constant is $k = EA/L$ and the potential energy is

$$V_e = \tfrac{1}{2}ku^2$$

where u is the displacement; therefore,

$$V_e = \frac{1}{2} \frac{EA}{H} q_2^2 + \frac{1}{2} \frac{EA}{H} (q_1 - q_2)^2$$

$$= \frac{1}{2}\left(\frac{EA}{H} q_1^2 + 2 \frac{EA}{H} q_2^2 - 2 \frac{EA}{H} q_1 q_2 \right)$$

The Lagrangian is $L = T - V_e$. Applications of Lagrange's equations in q_1 and q_2 yields

$$\frac{d}{dt}\left(\frac{\partial L}{\partial \dot{q}_1}\right) - \frac{\partial L}{\partial q_1} = a_1 \ddot{q}_1 + a_3 \ddot{q}_2 + \frac{EA}{H} q_1 - \frac{EA}{H} q_2 = F_0 \cos \omega t$$

$$\frac{d}{dt}\left(\frac{\partial}{\partial \dot{q}_2}\right) - \frac{\partial L}{\partial q_2} = a_2 \ddot{q}^2 + a_3 q_1 + \frac{EA}{H} q_2 - \frac{EA}{H} q_1 = 0$$

In matrix form,

$$\begin{bmatrix} a_1 & a_3 \\ a_3 & a_2 \end{bmatrix}\begin{bmatrix} \ddot{q}_1 \\ \ddot{q}_2 \end{bmatrix} + \begin{bmatrix} \dfrac{EA}{H} \end{bmatrix}\begin{bmatrix} 1 & -1 \\ -1 & 1 \end{bmatrix}\begin{bmatrix} q_1 \\ q_2 \end{bmatrix} = \begin{bmatrix} 1 \\ 0 \end{bmatrix} F_0 \cos \omega$$

The solution of this system of equations is very easy. If the mass distribution along the vertical axis is irregular, one has to take it into account in the formulation of the kinetic energy. Very complicated systems can thus be modeled by a small number of degrees of freedom; modeling with continuous systems or with lumped masses would be more complicated.

(In this example, the method of constraints was used to obtain an approximate differential equation of a two-degree-of-freedom system for the vertical vibration of a building.)

□

10.4 THE HOLZER–VAN DEN DUNGEN METHOD: TRANSFER MATRICES

The *Holzer–van den Dungen* method, also known in matrix form as the *transfer matrix method,* was introduced by Holzer, Güembel, and Tolle[1] for calculation of torsional natural frequencies of shafts. It makes use of the fact that in a large class of engineering problems, the vibrating system is arranged in a line and the behavior at every point in the system is influenced by the behavior at neighboring points only. Typical examples are torsional, longitudinal, and lateral vibrations of beams and shafts. In this case, the matrices **M**, **C**, and **K** that describe the system are not fully populated, but they have a narrow band around the diagonal. This method allows computation of static and dynamic response without finding the equations of motion first. Further, the method substitutes the system with an assembly of standard elements that have geometries and properties simple enough to be tabulated.

[1]Holzer, H. 1907. *Schifbau,* 8: 823, 866, 904. Gümbel, E. 1912. *Z. VDI, 56:* 1025. Tolle, M. 1921. *Regelung der Kraftmachinen.* Berlin.

Static Deflection of Beams

We shall start with the *static* lateral deflection of beams.[2] We assume a beam, say simply supported, which consists of $n - 1$ *beam elements* of different but constant section moment of inertia (Figure 10.1). Thus the beam has $n - 1$ elements with constant cross section and n *nodes,* points (or planes) that define the beginning or the end of a beam element.

To fully describe the situation at each node, we need to know four quantities: the deflection y, the slope θ, the moment M, and the shear force V. The section j of a beam between nodes j and $j + 1$ is shown in Figure 10.2 with the usual in statics sign conventions. These four quantities can be arranged in a vector $\mathbf{s} = \{y \quad \theta \quad M \quad V\}$ which, because it describes the state of the system at node j, is called a *state vector,* and to designate the node j we shall use it with a subscript j.

Let us suppose that at the node 1 the state vector is

$$\mathbf{z}_1 = \{x_1 \quad \theta_1 \quad M_1 \quad V_1\}$$

yet unknown. If no force is acting between nodes 1 and 2, the deflection, slope, moment, and shear at node 2, from simple beam theory, will be

$$y_2 = y_1 + l_1\theta_1 + \frac{l_1^2}{2EI_1}M_1 + \frac{l_1^3}{6EI_1}V_1$$

$$\theta_2 = \theta_1 + \frac{l_1}{EI_1}M_1 + \frac{l_1^2}{2EI_1}V_1 \qquad (10.30)$$

$$M_2 = M_1 + l_1V_1$$

$$V_2 = V_1$$

The last two equations express the equilibrium of the forces and moments on the beam, and the first two give the deflections due to these moments and forces.

Equation (10.30) can be written in the matrix form

$$\mathbf{z}_2 = \mathbf{L}_1\mathbf{z}_1 \qquad (10.31)$$

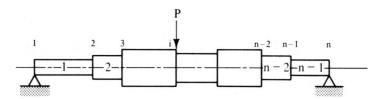

Figure 10.1 Stepped, simply supported beam.

[2]The method, generally known as the Myklestadt–Prohl method, was developed by Van Den Dungen, M. F.-H. 1928. Les problèmes généreux de la technique des vibrations. *Mem. Sci. Phys. Acad. Sci.* Paris: Gauthier-Villars. It was used by Myklestadt and Prohl later (see the Historical Introduction). It was further developed in the transfer matrix form by W. T. Thomson. 1950. *J. Appl. Mech., 17:* 337–339. For an extensive discussion, see [Pestel and Leckie, 1963].

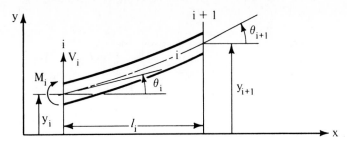

Figure 10.2 Prismatic beam element connecting two nodes.

where

$$
\mathbf{L}_1 =
\begin{bmatrix}
1 & l & \dfrac{l^2}{2EI} & \dfrac{l^3}{6EI} & 0 \\[2mm]
0 & 1 & \dfrac{l}{EI} & \dfrac{l^2}{2EI} & 0 \\[2mm]
0 & 0 & 1 & l & 0 \\[2mm]
0 & 0 & 0 & 1 & 0 \\[2mm]
0 & 0 & 0 & 0 & 1
\end{bmatrix}_1,
\qquad
\mathbf{z}_1 =
\begin{bmatrix}
y \\ \theta \\ M \\ V \\ 1
\end{bmatrix}_1,
\qquad
\mathbf{z}_2 =
\begin{bmatrix}
y \\ \theta \\ M \\ V \\ 1
\end{bmatrix}_2
$$

and the subscript 1 of the matrix indicates that the quantities l, E, and I are properties of the element number 1. The upper left 4×4 part of the matrix $\mathbf{L}$ will be used in some analyses. The fifth column and row are added here for computational convenience.

Equation (10.31) tells us that the state vector at node 2 is the state vector at node 1 multiplied by a square 5×5 matrix $\mathbf{L}$ which depends on the element properties only and it is well known. This matrix transferred the state from node 1 to node 2, and therefore is called a *transfer matrix*. For every element of the beam there exists one, known, transfer matrix $\mathbf{L}$. We can repeat the procedure for elements 2, 3, . . . , to obtain, also using the previous relations,

$$
\mathbf{z}_2 = \mathbf{L}_1 \mathbf{z}_1
$$

$$
\mathbf{z}_3 = \mathbf{L}_2 \mathbf{z}_2 = \mathbf{L}_2 \mathbf{L}_1 \mathbf{z}_1
$$

(10.32)

$$
\mathbf{z}_4 = \mathbf{L}_3 \mathbf{z}_3 = \mathbf{L}_3 \mathbf{L}_2 \mathbf{L}_1 \mathbf{z}_1
$$

$$
\vdots
$$

At the nodes the state vector as we approach the node from left and right is the same. However, if at the node we have a static force F, this is not true. In Figure 10.3 we show the situation for a small length about the node; the deflection, slope, and moment remain unchanged, but to have equilibrium, we must have $V^R = V^L + F$, where with superscript L we designate the situation at the left of the node and R refers to the situation at the right of the node.

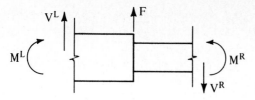

Figure 10.3 Node force equilibrium.

We can write

$$y^R = y^L$$
$$\theta^R = M^L$$
$$M^R = M^L$$
$$V^R = V^L + F$$

(10.33)

We can write this in matrix form as

$$\mathbf{z}_j^R = \mathbf{P}\mathbf{z}_j^L$$

(10.34)

where

$$\mathbf{P}_j = \begin{bmatrix} 1 & 0 & 0 & 0 & 0 \\ 0 & 1 & 0 & 0 & 0 \\ 0 & 0 & 1 & 0 & F \\ 0 & 0 & 0 & 1 & 0 \\ 0 & 0 & 0 & 0 & 1 \end{bmatrix}_j$$

$\mathbf{L}$ is called the *field matrix* (and the element i, j is called a *field*), and $\mathbf{P}$, a *point matrix*. Therefore, for transferring from left to right of a loaded node, we have to multiply by the point matrix. Then the last of the equations (10.32) will be

$$\mathbf{z}_n^L = \mathbf{L}_{n-1}\mathbf{L}_{n-2} \cdots \mathbf{L}_j\mathbf{P}_j\mathbf{L}_{j-1} \cdots \mathbf{L}_2\mathbf{L}_1\mathbf{z}_1^R$$

(10.35)

In this product one can take into account any number of loads at the nodes by multiplying with all the point matrices. Let $\mathbf{A} = \mathbf{P}_n\mathbf{L}_{n-1}\mathbf{P}_{n-1}\mathbf{L}_{n-2} \cdots \mathbf{P}_3\mathbf{L}_2\mathbf{P}_2\mathbf{L}_1\mathbf{P}_1$, a square 5×5 matrix that can be computed easily by multiplication of a chain of 5×5 matrices. Then equations (10.35) can be written in the form

$$\mathbf{z}_n^L = \mathbf{A}\mathbf{z}_1^R$$

(10.36)

The first four of the matrix equations (10.36) can be written as

$$y_n = a_{11}y_1 + a_{12}\theta_1 + a_{13}M_1 + a_{14}V_1 + b_{14}F$$
$$\theta_n = a_{21}y_1 + a_{22}\theta_1 + a_{23}M_1 + a_{24}V_1 + b_{24}F$$
$$M_n = a_{31}y_1 + a_{32}\theta_1 + a_{33}M_1 + a_{34}V_1 + b_{34}F$$
$$V_n = a_{41}y_1 + a_{42}\theta_1 + a_{43}M_1 + a_{44}V_1 + b_{44}F$$

(10.37)

We have four equations with eight unknowns, the end conditions y_1, θ_1, M_1, y_n, θ_n, V_1, M_n, V_n. However, because of the boundary conditions, we know four of these quantities. For example, for a simply supported beam we shall have $y_1 = y_n = 0$ and $M_1 = M_n = 0$. Therefore, equations (10.37) have four unknowns, θ_1, θ_n, V_1, and V_n. After computation of these unknowns, we can obtain the state vectors at the nodes, thus the deflection of the beam, from equations (10.36).

What did we accomplish at this point? We have been able to compute the static deflection of the beam with only chain multiplications of 5×5 matrices and solution of a system of four algebraic equations. With the usual matrix inversion methods we would have to invert a large matrix, and in addition to the much greater computation effort, one would have to use a digital computer for this purpose. Although the line solution method is very suitable for machine computations, it can also be used for hand computations.

Natural Vibration

Consider a beam with $n - 1$ massless elements (fields) and n nodes (Figure 10.4a). A free-body diagram (Figure 10.4b) shows the shear forces on the left and right. For harmonic motion, the acceleration of the mass m_j will be $\ddot{y}_j = -\omega^2 y_j$. Therefore, Newton's law gives

$$-V_j^R + V_j^L = -\omega^2 m_j y_j \tag{10.38}$$

Therefore, we can write

$$
\begin{aligned}
y_j^R &= y_j^L, & M_j^R &= M_j^L \\
\theta_j^R &= \theta_j^L, & V_j^R &= V_j^L + \omega^2 m_j y_j^L
\end{aligned}
\tag{10.39}
$$

In matrix form,

$$\mathbf{Z}_j^R = \mathbf{P}_j^M \mathbf{z}_j^L \tag{10.40}$$

where

$$
\mathbf{P}_j = \begin{bmatrix}
1 & 0 & 0 & 0 & 0 \\
0 & 1 & 0 & 0 & 0 \\
0 & 0 & 1 & 0 & 0 \\
\omega^2 m & 0 & 0 & 1 & 0 \\
0 & 0 & 0 & 0 & 1
\end{bmatrix}_j
$$

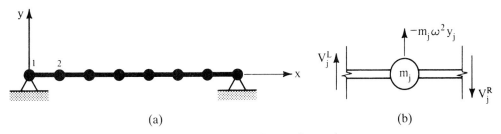

(a)   (b)

Figure 10.4 Lumped mass beam.

We shall obtain

$$\mathbf{z}_n^L = \mathbf{L}_{n-1}\mathbf{P}_{n-1}\mathbf{L}_{n-2} \cdots \mathbf{P}_3\mathbf{L}_2\mathbf{P}_2\mathbf{L}_1\mathbf{z}_1 = \mathbf{A}\mathbf{z}_1^R \tag{10.41}$$

Application of the boundary conditions will yield a system of four homogeneous algebraic equations. Because the last equation will be the identity $1 = 1$, the condition for existence of solution will yield a 4×4 determinant equal to zero, thus the frequency equation.

For a large number of masses, if a computer is used, one cannot proceed with explicit evaluation of the frequency equation. Instead, the chain multiplication of matrices will yield a relation of the form (10.41). In explicit form,

$$\begin{bmatrix} y \\ \theta \\ M \\ V \end{bmatrix}_n = \begin{bmatrix} a_{11} & a_{12} & a_{13} & a_{14} \\ a_{21} & a_{22} & a_{23} & a_{24} \\ a_{31} & a_{32} & a_{33} & a_{34} \\ a_{41} & a_{42} & a_{43} & a_{44} \end{bmatrix} \begin{bmatrix} y \\ \theta \\ M \\ V \end{bmatrix}_1 \tag{10.42}$$

We shall have four boundary conditions, in addition. For example, for a simply supported beam, we have $y_1 = y_n = 0$, $M_1 = M_n = 0$, and the system (10.42) becomes

$$0 = a_{12}\theta_1 + a_{14}V_1$$

$$\theta_n = a_{22}\theta_1 + a_{24}V_1$$

$$0 = a_{32}\theta_1 + a_{34}V_1 \tag{10.43}$$

$$V_n = a_{42}\theta_1 + a_{44}V_1$$

or, in terms of the unknowns θ_1, θ_n, V_1, and V_n:

$$a_{12}\theta_1 \qquad + a_{14}V_1 \qquad\qquad = 0$$

$$a_{22}\theta_1 - \theta_n + a_{24}V_1 \qquad\qquad = 0$$

$$a_{32}\theta_1 \qquad + a_{34}V_1 \qquad\qquad = 0 \tag{10.44}$$

$$a_{42}\theta_1 \qquad + a_{44}V_1 - V_n = 0$$

The condition for existence of solution for this homogeneous system of linear algebraic equations is

$$D(\omega) = \begin{vmatrix} a_{12} & 0 & a_{14} & 0 \\ a_{22} & -1 & a_{24} & 0 \\ a_{32} & 0 & a_{34} & 0 \\ a_{42} & 0 & a_{44} & -1 \end{vmatrix} = 0 \tag{10.45}$$

The value of ω satisfying equation (10.45) are the natural frequencies. The general computation procedure is as follows:

1. Select a frequency ω arbitrarily smaller than the expected fundamental (see Chapter 7).

2. Form the transfer matrices of field and masses.

3. Find matrix **A** by multiplying the transfer matrices from the left to the right end [equation (10.41)].
4. Applying the boundary conditions, write down the system of equations for the unknown end forces and displacements (10.43).
5. Rearrange to form a homogeneous system such as (10.39).
6. Derive the frequency determinant (10.45).
7. Plot the value of the determinant on a (D, ω) plot (Figure 10.5).
8. Give an increment $\Delta\omega$ to ω and repeat the procedure. This way you obtain a curve $D = D(\omega)$. The points at which this curve crosses the ω-axis correspond to the natural frequencies.
9. For every frequency calculated in this manner, equations (10.44) yield values for the end conditions. Starting from the left end, we find the state vectors at the nodes on the right by chain matrix multiplications. This yields the natural mode at this frequency. The values of ω_1, ω_2, . . . can be found on the graph. However, there are computational methods to find the roots of the nonlinear algebraic equation $D(\omega) = 0$ with any desired degree of accuracy, as discussed in Chapter 7.

The matrix A does not depend on the boundary conditions, but only on ω and the properties of the beam. The boundary conditions show up in the structure of the determinant (10.45). A usual situation is when at the end of the beam there is a supporting spring of constant k. The boundary conditions will be this end

$$M = 0, \qquad V = ky \tag{10.46}$$

Therefore, if the spring is at the left end, equations (10.43) will be

$$0 = a_{12}\theta_1 + \left(a_{14} + \frac{1}{k}\right)V_1$$

$$\theta_n = a_{22}\theta_1 + \left(a_{24} + \frac{1}{k}\right)V_1 \tag{10.47}$$

$$0 = a_{32}\theta_1 + \left(a_{34} + \frac{1}{k}\right)V_1$$

$$V_n = a_{42}\theta_1 + \left(a_{44} + \frac{1}{k}\right)V_1 - V_n$$

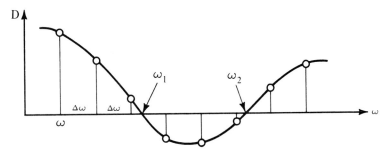

Figure 10.5 Numerical evaluation of natural frequencies.

and the frequency determinant

$$
D(\omega) = \begin{vmatrix}
a_{12} & 0 & a_{14} + \dfrac{1}{k} & 0 \\[2ex]
a_{22} & -1 & a_{24} + \dfrac{1}{k} & 0 \\[2ex]
a_{32} & 0 & a_{34} + \dfrac{1}{k} & 0 \\[2ex]
a_{42} & 0 & a_{44} + \dfrac{1}{k} & -1
\end{vmatrix} \tag{10.48}
$$

A list of transfer matrices for common beam elements and nodes is given in Appendix V.

Forced Harmonic Vibration

As in the static case, we assume that at the node j there is a harmonic force $F_0 \cos \omega t$. Therefore, all the components of the state vectors will be of the form $y \cos \omega t$, $\theta \cos \omega t$, $M \cos \omega t$, and $V \cos \omega t$. In all equations we shall divide throughout with $\cos \omega t$ and thus deal with amplitudes.

Multiplying the transfer matrices from left to right, we obtain

$$
\mathbf{z}_n^R = \mathbf{A}\mathbf{z}_n^L \tag{10.49}
$$

The last of equations (10.49) is the identity $1 = 1$. The first four, supplemented by boundary conditions, have four unknowns. Because they are nonhomogeneous (they include F explicitly) they can be solved to yield the end conditions. The state vectors at the several nodes can then be calculated as usual from the first node [equation (10.41)].

The method can be used for other kinds of vibration, such as torsional vibration, shells of revolution, disks, branched systems, and so on. The technique is the same, but the transfer matrices are different. Depending on the application, the state vector can have more or fewer than five components. For example, for torsional vibration the state vector is $\{\theta \quad T \quad 1\}$. For coupled vertical/horizontal vibration of beams, we have 10×10 matrices.

For natural vibration problems one can omit the last column and last row of the matrices, since they lead to identity and use 4×4 matrices for lateral vibration, 2×2 for torsional vibration.

The line solution method has been used extensively for beam vibration and for vibration of rotating shafts.

☐ **Example 10.8   Point Transfer Matrix for Mass–Spring Node**

Determine the point transfer matrix for the mass–spring station shown in Figure E10.8 for lateral vibration of shafts. Then determine the transfer matrix for the mass m with spring constant k.

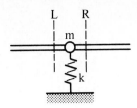

Figure E10.8

Solution Let the state vector be $s = \{x \quad \theta \quad M \quad v \quad 1\}$.

  (a) For the spring only, equilibrium demands

$$x^R = x^L$$

$$\theta^R = \theta^L$$

$$M^R = M^L$$

$$V^R = -kx^L + V^L$$

Thus

$$\mathbf{P}_{\text{spring}} = \begin{bmatrix} 1 & 0 & 0 & 0 \\ 0 & 1 & 0 & 0 \\ 0 & 0 & 1 & 0 \\ -k & 0 & 0 & 1 \end{bmatrix}$$

  (b) For mass only, for $x = A \cos \omega t$ and $\ddot{x} = -\omega^2 x$, application of Newton's law yields

$$m\ddot{x} = -\omega^2 mx = V^L - V^R$$

Thus the equilibrium equations are

$$x^R = x^L$$

$$\theta^R = \theta^L$$

$$M^R = M^L$$

$$V^R = m\omega^2 x^L + V^L$$

Thus

$$\mathbf{L}_{\text{mass}} = \begin{bmatrix} 1 & 0 & 0 & 0 \\ 0 & 1 & 0 & 0 \\ 0 & 0 & 1 & 0 \\ m\omega^2 & 0 & 0 & 1 \end{bmatrix}$$

and

$$
\mathbf{L}_{mass}\mathbf{L}_{spring} =
\begin{bmatrix}
1 & 0 & 0 & 0 \\
0 & 1 & 0 & 0 \\
0 & 0 & 1 & 0 \\
m\omega^2 & 0 & 0 & 1
\end{bmatrix}
\begin{bmatrix}
1 & 0 & 0 & 0 \\
0 & 1 & 0 & 0 \\
0 & 0 & 1 & 0 \\
-k & 0 & 0 & 1
\end{bmatrix}
=
\begin{bmatrix}
1 & 0 & 0 & 0 \\
0 & 1 & 0 & 0 \\
0 & 0 & 1 & 0 \\
m\omega^2 - k & 0 & 0 & 1
\end{bmatrix}
$$

(Using Newton's law, the point transfer matrix was developed for a node with elastically restrained mass.)

□

10.5 THE FINITE ELEMENTS METHOD[3]

Direct Method

With the transfer matrix method the reader has been introduced to the idea of an elastic member whose state can be approximated at some points on its boundary, if the state is known. For the static deformation of a prismatic beam, for example (Figure 10.6), if the conditions at the ends are known (deflection, slopes), the shape of the beam can be determined for static conditions.

This can be done with methods from strength of materials. For static loading the deflection of the beam will be v and

$$
\frac{d^2v}{dx^2} = \frac{1}{EI}M(x), \qquad V(x) = \frac{dM}{dx}, \qquad q(x) = \frac{d^2M}{dx^2}
\tag{10.50}
$$

If no load is assumed on the beam, $q(x) = 0$, differentiating twice the first equation and using the second, we obtain $d^4v/dx^4 = 0$. Its general solution is

$$
v(x) = c_1x^3 + c_2x^2 + c_3x + c_4
\tag{10.51}
$$

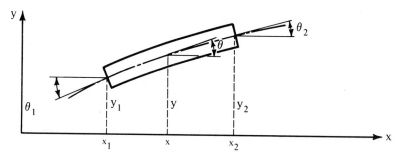

Figure 10.6 Geometry of deflected prismatic beam element.

[3]The basis of the idea as it applies to beams is Holzer's method and the transfer matrix methods. For a general continuum it was suggested by Courant. 1943. Variational methods for the solution of problems of equilibrium and vibrations. *Bull. Am. Math. Soc., 49:* 1–23. It was developed for structures by M. J. Turner, R. W. Clough, H. H. Martin, and L. J. Topp. 1956. Stiffness and deflection analysis of complex structures, *J. Aeronaut. Sci., 23:* 805–824.

where c_1, c_2, c_3, and c_4 are constants to be determined from the boundary conditions v_1, θ_1, v_2, and θ_2 ($\theta = dv/dx$). Therefore, if the end conditions are known, the deflections in between are functions only of these conditions and independent of the beam properties E and I. In fact, it can be shown very easily that

$$v(x) = h_1(x)v_1 + h_2(x)\theta_1 + h_3(x)v_2 + h_4(x)\theta_2 = \mathbf{hw} \qquad (10.52)$$

where for $s = x/l$ and l the length of the element,

$$h_1(x) = 1 - 3s^2 + 2s^3, \qquad h_2(x) = l(s - 2s^2 + s^3)$$

$$h_3(x) = 3s^2 - 2s^3, \qquad h_4(x) = l(-s^2 + s^3) \qquad (10.53)$$

$$\mathbf{h} = [h_1 \quad h_2 \quad h_3 \quad h_4], \qquad \mathbf{w} = \{v_1 \quad \theta_1 \quad v_2 \quad \theta_2\}$$

The slope $\theta(x) = dv/dx$ will be found by differentiation of equation (10.52). Equation (10.52) can also give the moments and shear forces along the beam, using equations (10.50). Therefore, the state at a distance x can be calculated once the end conditions are known. Therefore, instead of dealing with this beam segment along its length, we can, in principle, deal only with the conditions at the two ends, called *nodes*. Every piece of a continuum that can be described by way of the state at a finite number of points along its boundary will be called a *finite element*.

The displacements at the nodes (deflections and slopes) are related to the generalized forces at these nodes (forces and moments). In the sequel we shall use the terms *displacements* and *forces*, respectively. For a prismatic bar in flexure, we have already seen that the end conditions are related by way of a transfer matrix L. As we see from equations (10.30), we can solve them for v_1, θ_1, v_2, and θ_2 in terms of M_1, V_1, M_2, and V_2, and reverse. In general, we shall have the relations

$$\mathbf{f}^{(e)} = \mathbf{K}^{(e)}\mathbf{w}^{(e)}, \qquad \mathbf{w}^{(e)} = \mathbf{K}^{(e)-1}\mathbf{f}^{(e)} \qquad (10.54)$$

where superscript (e) designates an element property,

$$\mathbf{f}^{(e)} = \{V_1 \quad M_1 \quad V_2 \quad M_2\}^{(e)}, \qquad \mathbf{w}^{(e)} = \{v_1 \quad \theta_1 \quad v_2 \quad \theta_2\}^{(e)}$$

The matrix $\mathbf{K}^{(e)}$ is the element *stiffness matrix* and its inverse $\mathbf{A}^{(e)} = \mathbf{K}^{(e)-1}$ is the element *flexibility matrix*.

The state of displacements at each node of a beam element can have up to six components: three deflections and three slopes. Consequently, the forces at each node can have six components: three forces and three moments. Therefore, the stiffness and flexibility matrices can be of dimension 12×12, at the most, for two nodes. In general, for m nodes per element, the dimension of these matrices will be $6m \times 6m$. In practical situations, we do not always use all the possible coordinates. In flexure of beams, for example, we use only two displacements per node: the deflection and slope if we confine the motion in one plane and along a direction perpendicular to the beam.

For vibration analysis, the finite element method is used in conjunction with the method presented in Section 10.3. The functions h are special forms of the assumed modes ϕ, called *interpolation functions*. The stiffness matrix obtained in equation (10.54) from the transfer matrix is in tune with the basic assumption in the transfer matrix analysis, namely the lumped masses. The procedure is called a *direct method*.

Consistent Mass Method

To account for the distributed mass, although it can also be done with transfer matrix methods, we shall use the *consistent mass matrix* approach. To this end we use the equation for the potential energy along the prismatic element with equations (7.48) and (7.49),

$$V^{(e)} = \frac{1}{2}\int_0^l EI\left(\frac{d^2v}{dx^2}\right)^2 dx = \frac{1}{2}\int_0^l EI\mathbf{w}^T\mathbf{h}''^T\mathbf{h}''\mathbf{w}\ dx$$

$$= \frac{1}{2}\mathbf{w}^{(e)T}\mathbf{K}^{(e)}\mathbf{w}^{(e)} = \frac{1}{2}\sum_{i=1}^{4}\sum_{j=1}^{4} k_{ij}^{(e)} w_i w_j \qquad (10.55)$$

where

$$k_{ij}^{(e)} = \int_0^l EI\mathbf{h}''^{(e)T}\mathbf{h}''^{(e)}\ dx$$

To find the kinetic energy, we observe that the displacements are functions of time; therefore, we can differentiate equation (10.52):

$$\frac{\partial v(x,\ t)}{\partial t} = h_1(x)\dot{v}_1 + h_2(x)\dot{\theta}_1 + h_3(x)\dot{v}_2 + h_4(x)\dot{\theta}_2 \qquad (10.56)$$

Therefore, the kinetic energy is

$$T^{(e)} = \frac{1}{2}\int_0^l \frac{\rho A}{l}\dot{v}^{(e)2}\ dx = \frac{1}{2}\int_0^l \frac{\rho A}{l}\dot{\mathbf{w}}^{(e)T}\mathbf{h}^{(e)T}\mathbf{h}^{(e)}\dot{\mathbf{w}}^{(e)}\ dx$$

$$= \frac{1}{2}\dot{\mathbf{w}}^{(e)T}\mathbf{M}^{(e)}\ \dot{\mathbf{w}}^{(e)} = \frac{1}{2}\sum_{i=1}^{4}\sum_{j=1}^{4} m_{ij}^{(e)}\dot{w}_i\dot{w}_j \qquad (10.57)$$

where $\mathbf{M}^{(e)} = \int_0^l (\rho A/l)\mathbf{h}^{(e)T}\mathbf{h}^{(e)}\ dx$ is the mass matrix of the element.
For straight prismatic beams, the integrations can be carried out easily. For constant properties EI along the beam,

$$\mathbf{K}_v^{(e)} = EI\begin{bmatrix} \dfrac{12}{l^3} & & \text{symmetric} & \\ \dfrac{6}{l^2} & \dfrac{4}{l} & & \\ -\dfrac{12}{l^3} & -\dfrac{6}{l^2} & \dfrac{12}{l^3} & \\ \dfrac{6}{l^2} & \dfrac{2}{l} & -\dfrac{6}{l^2} & \dfrac{4}{l} \end{bmatrix} \qquad (10.58)$$

$$\mathbf{M}_v^{(e)} = \frac{\rho A}{420} \begin{bmatrix} 156l & & & \text{symmetric} \\ 22l^2 & 4l^3 & & \\ 54l & 13l^2 & 156l & \\ -13l^2 & -3l^3 & -22l^2 & 4l^3 \end{bmatrix} \tag{10.59}$$

For variable E, I, ρ, and A one has to integrate numerically.

Using the mass and stiffness matrices above, the differential equations of motion for the element can be written in the form

$$\mathbf{M}_v^{(e)}\ddot{\mathbf{w}}^{(e)} + \mathbf{K}_v^{(e)}\mathbf{w}^{(e)} = \mathbf{f}_v^{(e)} \tag{10.60}$$

where $\mathbf{f}_v^{(e)}$ is the nodal force vector.

Finally, the element mass and stiffness matrices will be assembled to yield the system equations, as indicated in Figures 10.7 and 10.8 for lateral vibration of

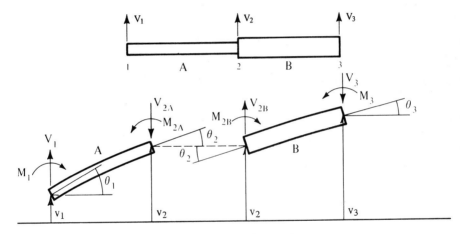

Figure 10.7 Two-element beam.

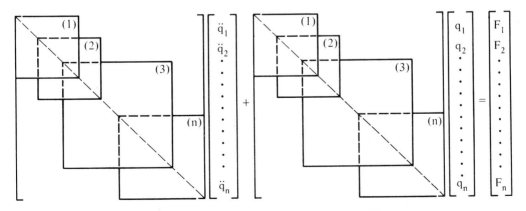

Figure 10.8 System matrix assembly.

a beam. We observe that nodes connect elements; therefore, for the displacements of the node that is common to two adjacent elements, we can add the equations for all elements, as indicated in Figure 10.8. Each element equation is added to the system equation with which it has common diagonal index. The system displacements and forces are defined as

$$\mathbf{q} = \{\mathbf{w}_1 \mid \mathbf{w}_2 \mid \cdots \mid \mathbf{w}_n\}, \qquad \mathbf{f}_q = \{\mathbf{f}_{q1} \mid \mathbf{f}_{q2} \mid \cdots \mid \mathbf{f}_{qn}\}$$

The elements of the nodal force vector $\mathbf{f}_q$ *are not* the forces transmitted through the node but the external forces at the node. If we add the respective equations during the assembly process, only the sum of the elements nodal forces will appear, which is the known external force, since the interaction forces are equal in magnitude and opposite in direction. For example, if two elements A and B have one node in common, say the number 2 of the first and the number 1 of the second element, the third and fourth equations (10.60) for the first element and the first and second for the second element, will be, for the nomenclature of Figure 10.7, omitting the subscript v:

$$m_{A31}\ddot{w}_1 + m_{A32}\ddot{w}_2 + m_{A33}\ddot{\theta}_1 + m_{A34}\ddot{\theta}_2 + k_{A31}w_1 + k_{A32}w_2$$
$$+ k_{A33}\theta_2 + k_{A34}\theta_2 = V_{2A}$$

$$m_{A41}\ddot{w}_1 + m_{A42}\ddot{w}_2 + m_{A43}\ddot{\theta}_1 + m_{A44}\ddot{\theta}_2 + k_{A41}w_1 + k_{A42}w_2$$
$$+ k_{A43}\theta_1 + k_{A44}\theta_2 = M_{2A} \qquad (10.61)$$

$$m_{B11}\ddot{w}_3 + k_{B12}\ddot{w}_4 + k_{B13}\ddot{\theta}_3 + k_{B14}\ddot{\theta}_4 + k_{B11}w_3 + k_{B12}w_4$$
$$+ k_{B13}\theta_3 + k_{B14}\theta_4 = M_{2B}$$

$$m_{B21}\ddot{w}_3 + m_{B22}\ddot{w}_4 + m_{B23}\ddot{\theta}_3 + m_{B24}\ddot{\theta}_4 + m_{B21}w_3 + m_{B22}w_4$$
$$+ m_{B23}\theta_3 + m_{B24}\theta_4 = M_{2B}$$

If we add the first to the third and the second to the fourth, we shall have

$$m_{A31}\ddot{w}_1 + \cdots + k_{B14}\theta_4 = V_{2A} + V_{2B} = V_2$$
$$m_{A41}\ddot{w}_1 + \cdots + k_{B24}\theta_4 = M_{2A} + M_{2B} = M_2 \qquad (10.62)$$

Therefore, in the equations of motion we have the external forces V_2 and M_2. In a general system, this is called *assembly* of the elements and is shown in Figure 10.8. When, in Figure 10.8, stiffness or matrices overlap each other, this means addition of the respective equations. We can see that the elements (1, 2), (1, 2, 3), and (3, n) have common nodes.

The system differential equations are obtained directly:

$$\mathbf{M}\ddot{\mathbf{q}} + \mathbf{C}\dot{\mathbf{q}} + \mathbf{K}\mathbf{q} = \mathbf{f}_q(t) \qquad (10.63)$$

where

$$\mathbf{M} = \sum_{i=1}^{n} \mathbf{M}_i^{(e)}$$

$$\mathbf{C} = \sum_{i=1}^{n} \mathbf{C}_i^{(e)} \tag{10.64}$$

$$\mathbf{K} = \sum_{i=1}^{n} \mathbf{K}_i^{(e)}$$

The addition of the element matrices, equations (10.64), is done in the sense explained above and n is the number of elements.

☐ **Example 10.9   Three-Node Beam FE Longitudinal Vibration Analysis**

Using three nodes, derive the equations of motion for the longitudinal vibration of the beam in Figure E10.9.

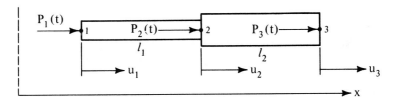

Figure E10.9

Solution   First, we have to find the mass and flexibility matrices for the element of a prismatic bar of length l, area A, and density ρ. To this end we observe that for this element, there are only two displacements, u_1 and u_2. At distance x from the undeflected position of the first node, the *static* deflection will be $u = (u_2 - u_1)x/l$. The potential energy will be

$$V = \frac{1}{2} AE \int_0^l \left(\frac{\partial u}{\partial x}\right)^2 dx = \frac{1}{2} AE \int_0^l \left(u_2 - \frac{u_1}{l}\right)^2 dx$$

$$= \frac{1}{2}\frac{AE}{l}(u_1^2 + u_2^2 - 2u_1 u_2)$$

Therefore,

$$k_{11} = \frac{AE}{l}, \qquad k_{22} = \frac{AE}{l}, \qquad k_{12} = k_{21} = -\frac{AE}{l}$$

The kinetic energy

$$T = \frac{1}{2} \rho A \int_0^l \dot{u}^2 \, dx = \frac{1}{2} \rho \int_0^l \left[\frac{(\dot{u}_2 - \dot{u}_1)x}{l} + \dot{u}_1\right]^2$$

$$= \frac{1}{2}\frac{\rho A l}{3}(\dot{u}_1^2 + \dot{u}_2^2 - \dot{u}_1 \dot{u}_2)$$

Therefore,

$$m_{11} = m_{22} = \frac{m}{3}, \qquad m_{12} = m_{21} = \frac{m}{6}, \qquad m = \rho A l$$

The equations of motion will be

$$\frac{m}{3} \begin{bmatrix} 2 & 1 \\ 1 & 2 \end{bmatrix} \begin{bmatrix} \ddot{u}_1 \\ \ddot{u}_2 \end{bmatrix} + \frac{AE}{l} \begin{bmatrix} 1 & -1 \\ -1 & 1 \end{bmatrix} \begin{bmatrix} u_1 \\ u_2 \end{bmatrix} = \begin{bmatrix} F_1 \\ F_2 \end{bmatrix} \qquad \text{(a)}$$

where $m = \rho A l$, or

$$\frac{m}{3} \ddot{u}_1 + \frac{m}{6} \ddot{u}_2 + \frac{AE}{l} u_1 - \frac{AE}{l} u_2 = F_1$$

$$\frac{m}{6} \ddot{u}_1 + \frac{m}{3} \ddot{u}_2 + \frac{AE}{l} u_1 - \frac{AE}{l} u_2 = F_2 \qquad \text{(b)}$$

If we divide the bar in two elements by way of three nodes 1, 2, and 3, application of equations (b) yields, with indices indicating elements,

$$\left.\begin{array}{c} \dfrac{m_1}{3} \ddot{u}_1 + \dfrac{m_1}{6} \ddot{u}_2 + \dfrac{AE}{l} u_1 - \left(\dfrac{AE}{l}\right)_1 u_2 = p_1 \\[3mm] \dfrac{m_1}{6} \ddot{u}_1 + \dfrac{m_1}{3} \ddot{u}_2 + \left(\dfrac{AE}{l}\right)_1 u_1 + \left(\dfrac{AE}{l}\right)_1 u_2 = (F_2)_1 \end{array}\right\} \text{element 1}$$

$$\left.\begin{array}{c} \dfrac{m_2}{3} \ddot{u}_2 + \dfrac{m_2}{6} \ddot{u}_3 + \left(\dfrac{AE}{l}\right)_2 u_2 - \left(\dfrac{AE}{l}\right)_2 u_3 = (F_2)_2 \\[3mm] \dfrac{m_2}{6} \ddot{u}_2 + \dfrac{m_2}{3} \ddot{u}_3 - \left(\dfrac{AE}{l}\right)_2 u_2 + \left(\dfrac{AE}{l}\right)_2 u_3 = p_2 \end{array}\right\} \text{element 2}$$

If we add the second and third equations, we obtain (together with the first and fourth equations)

$$\frac{1}{6}\begin{bmatrix} 2m_1 & m_1 & \vdots & 0 \\ \cdots & \cdots & \vdots & \cdots \\ m_1 & 2m_1 + 2m_2 & \vdots & m_2 \\ \cdots & \cdots & \vdots & \cdots \\ 0 & \vdots & m_2 & 2m_2 \end{bmatrix} \begin{bmatrix} \ddot{u}_1 \\ \ddot{u}_2 \\ \ddot{u}_3 \end{bmatrix} + \begin{bmatrix} k_1 & -k_1 & \vdots & 0 \\ \cdots & \cdots & \vdots & \cdots \\ -k_1 & k_1 + k_2 & \vdots & -k_2 \\ \cdots & \cdots & \vdots & \cdots \\ 0 & \vdots & -k_2 & k_2 \end{bmatrix} \begin{bmatrix} u_1 \\ u_2 \\ u_3 \end{bmatrix} = \begin{bmatrix} p_1 \\ p_2 \\ p_3 \end{bmatrix}$$

where $k = AE/l$. The assembly of the elements is indicated by the dotted lines.

(The consistent mass and stiffness matrices have been computed for prismatic, two-node elements for longitudinal vibration, and the assembly procedure has been performed to obtain the system differential equations.) □

Vibration of Framed Structures

In developing the element mass and stiffness matrices the coordinate system used was selected so that the x-axis would coincide with the geometric axis of the element. In framed structures it is not always so. Consider, for example, the plane prismatic element in Figure 10.7a, inclined in respect to the x-axis of the global coordinate system x, y, z by angle ϕ. A coordinate system $x'y'z'$ is attached to the beam element, $xOx' = \theta$. The nodal displacements are u', v', and θ'. The element displacement and force vectors will be

$$
\mathbf{w} = \{u'_1 \quad u'_2 \quad v'_1 \quad \theta'_1 \quad v'_2 \quad \theta'_2\}
$$
$$
\mathbf{f} = \{U'_1 \quad U'_2 \quad V'_1 \quad M'_1 \quad V'_2 \quad M'_2
$$

(10.65)

For longitudinal vibration, the element equations were developed in Example 10.9:

$$
\mathbf{M}_u^{(e)} = \frac{m}{3}\begin{bmatrix} 2 & 1 \\ 1 & 2 \end{bmatrix}, \qquad \mathbf{K}_u^{(e)} = \frac{AE}{l}\begin{bmatrix} 1 & -1 \\ -1 & 1 \end{bmatrix}, \qquad \mathbf{f}_u^{(e)} = \begin{bmatrix} U_1 \\ U_2 \end{bmatrix}
$$

(10.66)

while the element equations for bending are (10.58) and (10.59). The element mass and stiffness matrices and the force vector for the element motion defined by equations (10.63) will be

$$
\mathbf{M}^{(e)} = \begin{bmatrix} \mathbf{M}_u & 0 \\ \hline 0 & \mathbf{M}_v \end{bmatrix}, \qquad \mathbf{K}^{(e)} = \begin{bmatrix} \mathbf{K}_u & 0 \\ \hline 0 & \mathbf{K}_v \end{bmatrix}, \qquad \mathbf{f}^{(e)} = \begin{bmatrix} f_u \\ f_v \end{bmatrix}
$$

(10.67)

The nodal displacements and forces are now defined in respect to the global coordinate system:

$$
\mathbf{x} = \{u_1 \quad u_2 \quad v_1 \quad \theta_1 \quad v_2 \quad \theta_2\}
$$
$$
\mathbf{f} = \{U_1 \quad U_2 \quad V_1 \quad M_1 \quad V_2 \quad M_2\}
$$

(10.68)

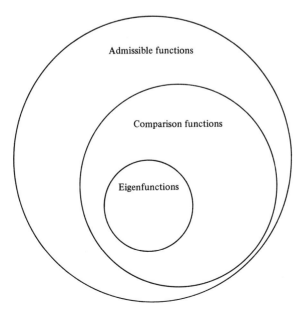

Figure 10.9 Classes of functions used in Rayleigh's quotient.

We observe that the slopes and moments remain the same in both coordinate systems. The linear displacements u and v change, in general, upon rotation by angle ϕ,

$$
\begin{bmatrix} u \\ v \end{bmatrix} = \begin{bmatrix} \cos\phi & -\sin\phi \\ \sin\phi & \cos\phi \end{bmatrix} \begin{bmatrix} u' \\ v' \end{bmatrix} \tag{10.69}
$$

Therefore, the displacement vector components in the two coordinate systems are related as

$$
\mathbf{w}' = \mathbf{R}\mathbf{w} \tag{10.70}
$$

where the transformation matrix is

$$
\mathbf{R} = \begin{bmatrix}
\cos\phi & -\sin\phi & 0 & 0 & 0 & 0 \\
\sin\phi & \cos\phi & 0 & 0 & 0 & 0 \\
0 & 0 & \cos\phi & 0 & -\sin\phi & 0 \\
0 & 0 & 0 & 1 & 0 & 0 \\
0 & 0 & \sin\phi & 0 & \cos\phi & 0 \\
0 & 0 & 0 & 0 & 0 & 1
\end{bmatrix}
$$

The same transformation matrix applies for the force vectors. The mass, damping, and stiffness matrices in the two coordinate systems are different, in general. To find the applicable transformation, we note that the kinetic and potential energies and the dissipation function are the same in the two coordinate systems. This yields

$$
T = \tfrac{1}{2}\dot{\mathbf{w}}'^{(e)T}\mathbf{M}'^{(e)}\dot{\mathbf{w}}'^{(e)} = \tfrac{1}{2}\dot{\mathbf{w}}^{T(e)}\mathbf{R}^{T(e)}\mathbf{M}'^{(e)}\mathbf{R}^{(e)}\dot{\mathbf{w}}^{(e)}
$$

$$
U = \tfrac{1}{2}\dot{\mathbf{w}}'^{T(e)}\mathbf{C}'^{(e)}\dot{\mathbf{w}}'^{(e)} = \tfrac{1}{2}\dot{\mathbf{w}}^{T(e)}\mathbf{R}^{T(e)}\mathbf{C}'^{(e)}\mathbf{R}^{(e)}\dot{\mathbf{w}}^{(e)} \tag{10.71}
$$

$$
V = \tfrac{1}{2}\mathbf{w}'^{T(e)}\mathbf{K}'^{(e)}\mathbf{w}'^{(e)} = \tfrac{1}{2}\mathbf{w}^{T(e)}\mathbf{R}^{T(e)}\mathbf{K}'^{(e)}\mathbf{R}^{(e)}\mathbf{w}^{(e)}
$$

Therefore, the transformation equations for the mass, damping, and stiffness matrices are, respectively,

$$
\mathbf{M}^{(e)} = \mathbf{R}^{T(e)}\mathbf{M}'^{(e)}\mathbf{R}^{(e)}
$$

$$
\mathbf{C}^{(e)} = \mathbf{R}^{T(e)}\mathbf{C}'^{(e)}\mathbf{R}^{(e)} \tag{10.72}
$$

$$
\mathbf{K}^{(e)} = \mathbf{R}^{T(e)}\mathbf{K}'^{(e)}\mathbf{R}^{(e)}
$$

Now the system matrices can be assembled using equations (10.64). Equations (10.63) are obtained here using Lagrange's equations with equations (10.71).

Node Restraints

In many cases the structures are constrained against rigid-body motion. This is done by restraining the motion of some nodes. For plane motion we need, in general, three restraints for a rigid body to be fully restrained. More restraints will simply make it redundant. There are several ways to incorporate the restraint of a node into the differential equations of motion:

1. *Directly,* by removing the corresponding coordinate. Assume, for example, that coordinate j is restrained; therefore, $q_j \equiv 0$. Row and column j of the system matrices $\mathbf{M}$, $\mathbf{C}$, and $\mathbf{K}$ and element j of the vectors $\mathbf{q}$ and $\mathbf{f}_q$ are removed and the dimensions are reduced by 1.
2. *Indirectly,* by making all the above row and column j vector and matrix elements zero except k_{jj}, which is set equal to 1. This yields a larger system but is easier to implement.

Nonrigid restraints can be incorporated in the same way. Assume, for example, that along the coordinate j there is a mass, a spring, and a damper with values m, k, and c, respectively, the spring and damper connected with the ground. The elements jj of the mass, damping, and stiffness matrices will be, respectively, $m_{jj} + m$, $c_{jj} + c$, and $k_{jj} + k$.

10.6 THE RAYLEIGH–RITZ METHOD

In Section 10.2 it was shown that if the natural mode of a system is known, the corresponding natural frequency can be found by Rayleigh's quotient. Moreover, an approximation to the natural mode yields the natural frequency to a better approximation. For the fundamental natural frequency, this method is applicable because it is possible in most problems to obtain a good approximation of the fundamental natural mode intuitively, assuming that at the lowest natural frequency the vibration mode approaches the static deflection. It was shown in Example 10.2 that higher modes can be found with Rayleigh's method, provided that a good intuitive knowledge exists for the form of the second mode. This becomes progressively more difficult to have as the nodes become higher. To overcome this, the natural modes are substituted by a set of geometric forms that connect the physical nodes of the system, and thus a complicated high mode can be approximated. Up to this point, the accuracy of this method and its relation to modeling decisions (i.e., number of nodes, shape of the assumed functions, etc.) has not been established but has been left more or less to the intuition and experience of the engineer. Therefore, there is a need to obtain approximations with a certain error control built-in.

The eigenfunctions or eigenvectors (the natural modes) have to satisfy the differential equations and the boundary conditions. They yield limit value for the Rayleigh's quotient. Since they are not known in advance, application of the Rayleigh's principle needs the selection of approximate functions or vectors.

The *trial* functions used to approximate the system behavior in the Rayleigh's principle application have to approximate the natural nodes; therefore, they have to satisfy the problem boundary conditions. Such functions we call *comparison functions.* Moreover, such functions should be differentiable as many times as the order of the system, as for example, computation of forces involve differentiation. One could intuitively understand that it is much easier to find functions that satisfy

WALTER RITZ (1878–1909)

Walter Ritz was born in Switzerland, son of a well-known landscape painter. He initially entered the Federal Institute of Technology to study engineering, but an accident forced him to abandon the idea. Instead, he joined a group of students, including Einstein, to study pure mathematics. His poor health forced him to transfer to Götingen for a drier climate. There he studied under Voigt, Hilbert, and Klein, and Voigt approved his dissertation in spectroscopy in 1902. Until 1907, he was occupied with health problems again. He moved to Tübingen, and also having financial problems, he worked on a *habilitation* to qualify for a faculty position. The last two years of his life were very productive. The related work was in spectroscopy, electrodynamics, and numerical solution of boundary value problems, areas essentially unrelated. In 1904, the Paris Academy of Science announced the 4000-franc Prix Vaillant for the solution of the plate equation with a rigid rectangular boundary. In his professional and financial situation, the opportunity was attractive and Ritz decided to apply himself to the problem and soon developed a practicable procedure to develop the solution by successive approximations. In a striking demonstration, he produced 13 harmonics of the square plate in the form of Chladni figures. His entry did not win even honorable mention. Shortly afterward Poincaré, one of the referees, apologized to Ritz for this, and the method was adopted quickly. The intensive effort, however, caused the deterioration of his health and he died in 1909 from tuberculosis. For a fistful of francs. . . .

only the geometric boundary conditions (such as displacements and slopes) and not the natural boundary conditions (such as forces and moments) and are differentiable. Such functions we call *admissible*. The eigenfunctions or vectors are a subset of the comparison functions or vectors and they, in turn, are a subset of the admissible functions or vectors, Figure 10.9. The first satisfy the differential equations and all boundary conditions, the second satisfy all the boundary conditions and the third satisfy only the geometric boundary conditions. Since in the expressions for the kinetic and potential energy one can take into account inertias and springs, we can have the natural boundary conditions accounted for while the admissible functions do not satisfy them. In fact, this can be proved for all physically realizable natural boundary conditions. Therefore, we set up the system response in the form[4]

$$z(x) = \sum_{i=1}^{n} c_i z_i(x) \tag{10.73}$$

for continuous systems where $z_i(x)$ are admissible functions and

$$\mathbf{z} = \sum_{i=1}^{n} c_i \mathbf{z}^{(i)} = \mathbf{Z}\mathbf{c} \tag{10.74}$$

[4]Ritz, W. 1909. *Ann. Phys.*, 28: 737.

for discrete systems, where $\mathbf{Z}$ is a square matrix of the admissible vectors $[\mathbf{z}^{(1)} \cdots \mathbf{z}^{(n)}]$ and $\mathbf{c} = \{c_1 \ c_2 \ \cdots \ c_n\}$, where c_i are constants to be determined and n is the number of admissible functions, which can be substantially smaller than the number of degrees of freedom of the system.

If the vectors $\mathbf{z}^{(i)}$ have been normalized so that $\mathbf{Z}^T\mathbf{M}\mathbf{Z} = \mathbf{I}$, then, for discrete systems,

$$R(\mathbf{z}) = \frac{\mathbf{c}^T\mathbf{Z}^T\mathbf{K}\mathbf{Z}\mathbf{c}}{\mathbf{c}^T\mathbf{Z}^T\mathbf{M}\mathbf{Z}\mathbf{c}} = \frac{\mathbf{c}^T\mathbf{P}\mathbf{c}}{\mathbf{c}^T\mathbf{I}\mathbf{c}} \tag{10.75}$$

$$\mathbf{Z}^T\mathbf{K}\mathbf{Z} = \text{diag}[\omega_1^2, \omega_2^2, \ldots, \omega_n^2]\mathbf{c}^T\mathbf{Z}^T\mathbf{M}\mathbf{Z}\mathbf{c} \tag{10.76}$$

and for a continuous beam,

$$\omega^2 = \frac{\displaystyle\int_0^L EI(z'')^2 \, dx}{\displaystyle\int_0^L m(x)\dot{z}^2 \, dx} \tag{10.77}$$

The Rayleigh–Ritz method finds the proper selection of c_i by making the value of the Rayleigh quotient minimum. This is accomplished by making the derivatives of this quotient, in respect to the parameters c_i, zero. For example, equation (10.77) gives

$$\frac{\partial \omega^2}{\partial c_i} = \frac{B(\partial A/\partial c_i) - A(\partial B/\partial c_i)}{B^2} = 0, \qquad i = 1, 2, \ldots, n \tag{10.78}$$

where

$$\omega^2 = \frac{A}{B}, \ A = \int_0^L EI(z'')^2 \, dx, \qquad B = \int_0^L m(x)\dot{z}^2 \, dx$$

Therefore, since B is not zero everywhere,

$$\frac{\partial A}{\partial c_i} - \omega^2 \frac{\partial B}{\partial c_i} = 0, \qquad i = 1, 2, \ldots, n \tag{10.79}$$

This is a system of n equations in n unknowns $c_1, c_2, \ldots, c_n$. The solution will yield the lowest quotient for the selected trial functions. Considering equations (10.23) and (10.24), A and B can be written in the form

$$A = \sum_{i=1}^{n} k_{ij}c_ic_j, \qquad B = \sum_{i=1}^{n} m_{ij}c_ic_j \tag{10.80}$$

Therefore, using the reciprocity relations,

$$\frac{\partial A}{\partial c_i} = 2(k_{i1}c_1 + k_{i2}c_2 + \cdots + k_{in}c_n) \tag{10.81}$$

$$\frac{\partial B}{\partial c_i} = 2(m_{i1}c_1 + m_{i2}c_2 + \cdots + m_{in}c_n) \tag{10.82}$$

and equations (10.79) take the form

$$\sum_{i=1}^{n} (k_{ij} - \omega^2 m_{ij})c_j = 0, \quad i = 1, 2, \ldots, n \quad (10.83)$$

There are n equations of this form, which can be written in the matrix form

$$(\mathbf{K} - \mathbf{M}\omega^2)\mathbf{c} = 0 \quad (10.84)$$

The eigenvalue problem (10.84) gives an estimate of the natural frequencies and modes of the system. The latter will be obtained by inserting the eigenvectors of the system (10.84) into (10.83) or (10.84). For the discrete system, from (10.76), let $\mathbf{K}^* = \mathbf{Z}^T\mathbf{K}\mathbf{Z}$, $\mathbf{M}^* = \mathbf{Z}^T\mathbf{M}\mathbf{Z}$ and equation (10.84) becomes

$$(\mathbf{K}^* - \mathbf{M}^*\omega^2)\mathbf{c} = 0 \quad (10.85)$$

The system of equations (10.85) is smaller than the original system; its dimension equals the number of the trial functions used. The Rayleigh–Ritz method was thus used to either discretize a continuous system [equations (10.84)] or to reduce the size of a discrete system [equation (10.85)].

☐ **Example 10.10   Natural Frequency of a Clamped Beam Using the Rayleigh–Ritz Method**

Determine the natural frequencies of a beam of length l, mass per unit length μ, and flexural rigidity EI, both constant, clamped at both ends, using the Rayleigh–Ritz method.

Solution   The functions $z_1(x) = 1 - \cos 2\pi x/l$ and $z_2(x) = 1 - \cos 4\pi x/l$ are admissible because at $x = 0$ and l they are zero and have zero slope. Then, from equation (10.78)

$$A = 8\pi EIl(c_1^2 + c_2^2), \quad B = \frac{\mu l^3}{2}(3c_1^2 + 3c_2^2 + 4c_1 c_2)$$

Equation (10.79) gives

$$\frac{16\pi^4 EI}{\mu l^4}\begin{bmatrix} 1 & 0 \\ 0 & 16 \end{bmatrix}\begin{bmatrix} c_1 \\ c_2 \end{bmatrix} = \omega^2 \begin{bmatrix} 3 & 2 \\ 2 & 3 \end{bmatrix}\begin{bmatrix} c_1 \\ c_2 \end{bmatrix}$$

which yield the solution

$$\omega_1 = \frac{22.35}{l^2}\left(\frac{EI}{\mu}\right)^{1/2}, \quad \mathbf{c}_1 = \{1.0 \quad 0.575\}$$

$$\omega_2 = \frac{124.0}{l^2}\left(\frac{EI}{\mu}\right)^{1/2}, \quad \mathbf{c}_2 = \{1.0 \quad -1.4477\}$$

The exact values from Table 9.5 for $(kl)^2$ are 22.4 and 61.6. The value $(kl)^2 = 124.0$ is close to the third mode, $(kl)^2 = 120.9$. Apparently, the second mode was missed due to the selection of the admissible functions.

(The natural frequencies of a beam with clamped ends were computed with the Rayleigh–Ritz method using approximate admissible functions.)

☐

10.7 THE GALERKIN METHOD

The Galerkin method[5] differs from the Rayleigh–Ritz method in that it achieves the best approximation by minimizing the error in satisfying a differential equation over the system range. The expansion

$$z(x) = \sum_{j=1}^{n} c_i z_i(x) \qquad (10.86)$$

approximates a function $z(x)$ that is the solution of the differential equation

$$L(z) = 0 \qquad (10.87)$$

where L is a differential operator. For example, for lateral vibration of a beam,

$$L(z) = EIz(x)^{(iv)} + \mu\omega^2 z(x) = 0 \qquad (10.88)$$

Substitution of (10.86) into (10.88) will not, in general, yield zero on the left-hand side. A measure of the deviation from zero over the length of the beam l is the integral $\int_0^l z_i L(x)\, dx$ (residual) for each of the trial functions $z_i(x)$. Setting this deviation equal to 0 for each trial function,

$$R_i = \int_0^l z_i L(z)\, dx = 0, \qquad i = 1, 2, \ldots, n \qquad (10.89)$$

This is a homogeneous system of n linear equations in $c_1, c_2, \ldots, c_n$. It will yield the natural frequencies and natural modes.

☐ **Example 10.11   Beam Natural Frequencies with the Galerkin Method**

Determine the natural frequencies of the beam of Example 10.10 with the Galerkin method.

Solution   The functions $z_1(x) = 1 - \cos 2\pi x/l$ and $z_2(x) = 1 - \cos 4\pi x/l$ are admissible functions because at $x = 0$ and l they are zero and have zero slope. The residual is

$$R_i = \int_0^l z_i L(x)\, dx$$

$$L(z) = c_1[(\lambda^4 - \beta^4)\cos \lambda x + \beta^4] + c_2[(16\lambda^4 - \beta^4)\cos 2\lambda x + \beta^4]$$

where $\beta = \mu\omega^2/EI$ and $\lambda = 2\pi/l$.

For the two trial functions,

$$R_1 = \int_0^l \left(1 - \cos\frac{2\pi x}{l}\right)\{c_1[(\lambda^4 - \beta^4)\cos\lambda x + \beta^4] + c_2[(16\lambda^4 - \beta^4)\cos 2\lambda x + \beta^4]\}\, dx$$

$$R_2 = \int_0^l \left(1 - \cos\frac{4\pi x}{l}\right)\{c_1[(\lambda^4 - \beta^4)\cos\lambda x + \beta^4] + c_2[(16\lambda^4 - \beta^4)\cos 2\lambda x + \beta^4]\}\, dx$$

Performing the integrations,

$$R_1 = \left(\frac{\lambda^4}{2} - \beta^4\right)c_1 - \beta^4 c_2 = 0$$

$$R_2 = -\beta^4 c_1 + (8\lambda^4 - \beta^4)c_2 = 0$$

[5]Galerkin, B. G. 1915. Sterzhni plastinki. *Vestn. Inzh.*, *1*(19): 897–908. For an extensive account of the original Galerkin paper in Russian, see Hensky, H. 1927. *Z. Angew. Math. Mech.*, 7: 80.

Portrait of Galerkin (Courtesy of the USSR Academy of Science.)

BORIS GRIGOR'EVICH GALERKIN (1871–1945)

Born in Polotsk, Galerkin studied engineering at the St. Petersburg Technological Institute. He became a Lieutenant General in the Corps of Engineers and later a member of the Academy of Sciences in 1935. He consulted on the design and construction of large public works. His main research work was in the theory of elasticity and the theory of plates. He formulated the solution of the general elasticity problem in the form of three functions (the Galerkin functions) and developed the general approximate solution of problems described by partial differential equations.

The frequency determinant must be zero:

$$\begin{vmatrix} \dfrac{\lambda^4}{2} - \beta^4 & -\beta^4 \\[2ex] -\beta^4 & 8\lambda^4 - \beta^4 \end{vmatrix} = 0$$

The solutions of the frequency equation are $\beta l = 4.741$ and 11.140. The corresponding natural frequencies and constants c_i are

$$\omega_1 = \frac{22.48}{l^2} \left(\frac{EI}{\mu} \right)^{1/2}, \qquad c_1 = \{23.0 \quad 1\}$$

$$\omega_2 = \frac{124.1}{l^2} \left(\frac{EI}{\mu} \right)^{1/2}, \qquad c_2 = \{-0.69 \quad 1\}$$

Therefore, the two modes are

$$\phi_1(x) = 23\left(1 - \cos \frac{2\pi x}{l} \right) + 1 - \cos \frac{4\pi x}{l}$$

$$\phi_2(x) = -0.69\left(1 - \cos \frac{2\pi x}{l} \right) + 1 - \cos \frac{4\pi x}{l}$$

(The natural frequencies of a beam with clamped ends were computed with the Galerkin method using approximate comparison functions.)

$\square$

10.8 THE STODOLA METHOD: MATRIX ITERATION

Fundamental Natural Frequency

The determinant expansion method discussed in Chapter 7 can in principle be used to solve the eigenvalue problem of any size, since a determinant of even large size can be evaluated numerically. However, only for a small system the method of the frequency determinant is used. For larger systems, the computation effort is prohibitive and the accuracy progressively lower as the system becomes larger.

Stodola had to compute critical speeds[6] of turbine shafts of variable cross section.

1. He started with Rayleigh's method of determining the fundamental natural frequency $\omega_{0(0)}$ using as a trial function the static deflection of the shaft due to its own weight $y_{(0)}$.

2. He then lumped the mass of the shaft at a number of nodes and computed the inertia forces $m_i y_{(0)} \omega_{0(0)}^2$. Then he found the static deflection $y_{(1)}$ due to the inertia forces.

3. He used this deflection $y_{(1)}$ as a trial function with Rayleigh's method to determine a new value for the fundamental natural frequency $\omega_{0(1)}$.

[6]In a later chapter we explain the difference between the critical speed and the natural frequency. At present, we consider them equal.

Portrait of Stodola. (Courtesy of the Swiss Federal Institute of Technology, Zurich, Switzerland.)

AREL BORESLAV STODOLA (1859–1942)

Stodola was born in a small Slovakian town. He studied at the Budapest Technical University, University of Zurich, and Swiss Federal Institute of Technology. He was self-educated, taking up work and studies in Hungary, Berlin, Paris, and Czechoslovakia. The chief engineer of Ruston & Co. in Prague for six years, he was given in 1892 the newly created chair of thermal machinery at the Federal Institute of Technology in Zurich, where he remained until his retirement in 1929. He combined high mathematical competence with devotion to practical utility. His first work was with automatic control, but his attention was shifted to the steam turbines. Although he published many articles, the body of his work is incorporated in his *Dampf- und Gasturbinen,* first published in 1903. Beyond his fundamental works in thermodynamics, he was a pioneer in rotor dynamics and approximate methods.

---

4. He repeated the procedure and computed a sequence of values for the fundamental natural frequency $\omega_{0(1)}$, $\omega_{0(2)}$, $\omega_{0(3)}$, $\omega_{0(4)}$, . . . until the value of the frequency would not change with the iteration.[7]

[7]Stodola, A. 1910. *Dampf und Gasturbinen,* 4th ed. Berlin: Julius Springer. The theoretical basis was developed by R. von Mises. 1911. *Zeit. Math. Phys., 22:* 33. Similar procedures were used by other authors between 1900 and 1920.

Convergence[8] of the procedure to the lowest natural frequency follows equations (10.10) and (10.11).

In matrix form, this procedure can be formulated as follows.

1. The eigenvalue problem is set up in the form

$$\mathbf{Kx} = -\omega^2 \mathbf{Mx} \qquad (10.90)$$

An initial guess for the vector $\mathbf{x}$ is assumed, usually the unit vector $\mathbf{x}_{(0)} = \{1 \quad 1 \quad \cdots \quad 1\}$. This value is used in the right-hand side of equation (10.81). This yields a system of linear equations in a vector $\mathbf{x}_{(1)}^*$:

$$\mathbf{Kx}_{(1)}^* = -\omega^2 \mathbf{Mx}_{(0)} \qquad (10.91)$$

2. The vector $\mathbf{x}_{(1)}^*$ is normalized so that the first element is 1:

$$\mathbf{x}_{(1)} = \lambda \mathbf{x}_{(1)}^* \qquad (10.92)$$

3. The procedure is repeated to yield a sequence $\lambda_1, \lambda_2, \lambda_3, \lambda_4, \ldots$.
4. λ_n converges to $\omega_{(0)}^2$ and the vector $\mathbf{x}_{(n)}$ converges to the corresponding lowest natural mode.

The highest mode can be computed by iteration on the inverted system

$$\mathbf{x} = \frac{1}{\omega^2} \mathbf{M}^{-1} \mathbf{Kx} \qquad (10.93)$$

☐ **Example 10.12   Fundamental Natural Frequency of a Three-Degree-of-Freedom System with Matrix Iteration**

Determine the fundamental natural frequency of the three-degree-of-freedom system shown in Figure E10.12.

Solution   Considering coordinates $\theta_1, \theta_2, \theta_3$, the flexibility influence coefficients α_{ij} are the rotations of the disk i due to unit torque at disk j. Since the rotation of a circular cylinder of length l, shear modulus G, and section moment of inertia I_p due to a torque T at its ends is $\Delta\theta = Tl/GI_p = T/k_T$, the influence coefficient matrix is

$$\mathbf{A} = \frac{1}{k_T}\begin{bmatrix} 1 & 1 & 1 \\ 1 & 4 & 4 \\ 1 & 4 & 7 \end{bmatrix}, \qquad \mathbf{M} = J_p\begin{bmatrix} 4 & 0 & 0 \\ 0 & 2 & 0 \\ 0 & 0 & 1 \end{bmatrix}, \qquad \mathbf{K} = \mathbf{A}^{-1}$$

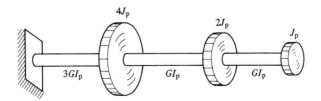

Figure E10.12

[8]Grammel, R. 1922. *Ergeb. Exakten Naturwiss.*, *1*: 93–119.

This eigenvalue problem can then be stated as

$$(-\mathbf{M}\omega^2 + \mathbf{K})\mathbf{\theta} = 0 \quad \text{or} \quad (-\mathbf{K}^{-1}\mathbf{M}\omega^2 + \mathbf{I})\mathbf{\theta}$$

$$= 0 \quad \text{or} \quad (-\mathbf{A}\mathbf{M}\omega^2 + \mathbf{I})\mathbf{\theta} = 0 \quad \text{or} \quad \mathbf{\theta} = \mathbf{A}\mathbf{M}\omega^2\mathbf{\theta}$$

or

$$\begin{bmatrix} \theta_1 \\ \theta_1 \\ \theta_1 \end{bmatrix} = \frac{\omega^2 J_p}{3k_T} \begin{bmatrix} 4 & 2 & 1 \\ 4 & 8 & 4 \\ 4 & 8 & 7 \end{bmatrix} \begin{bmatrix} \theta_1 \\ \theta_1 \\ \theta_1 \end{bmatrix}$$

To obtain a guess of the starting vector of iteration, we select $\mathbf{\theta}_{(0)} = \{1 \quad 2 \quad 4\}$. The first iteration gives

$$\begin{bmatrix} \theta_1 \\ \theta_1 \\ \theta_1 \end{bmatrix} = \frac{\omega^2 J_p}{3k_T} \begin{bmatrix} 4 & 2 & 1 \\ 4 & 8 & 4 \\ 4 & 8 & 7 \end{bmatrix} \begin{bmatrix} 1 \\ 2 \\ 4 \end{bmatrix} = \frac{\omega^2 J_p}{3k_T} \begin{bmatrix} 12 \\ 36 \\ 48 \end{bmatrix} = \frac{\omega^2 J_p}{3k_T} (12) \begin{bmatrix} 1 \\ 3 \\ 4 \end{bmatrix}$$

The second iteration gives

$$\begin{bmatrix} \theta_1 \\ \theta_1 \\ \theta_1 \end{bmatrix} = \frac{\omega^2 J_p}{3k_T} \begin{bmatrix} 4 & 2 & 1 \\ 4 & 8 & 4 \\ 4 & 8 & 7 \end{bmatrix} \begin{bmatrix} 1 \\ 3 \\ 4 \end{bmatrix} = \frac{\omega^2 J_p}{3k_T} \begin{bmatrix} 14 \\ 44 \\ 56 \end{bmatrix} = \frac{\omega^2 J_p}{3k_T} (14) \begin{bmatrix} 1 \\ 3.2 \\ 4.0 \end{bmatrix}$$

The third iteration

$$\begin{bmatrix} \theta_1 \\ \theta_1 \\ \theta_1 \end{bmatrix} = \frac{\omega^2 J_p}{3k_T} \begin{bmatrix} 4 & 2 & 1 \\ 4 & 8 & 4 \\ 4 & 8 & 7 \end{bmatrix} \begin{bmatrix} 1 \\ 3.2 \\ 4.0 \end{bmatrix} = \frac{\omega^2 J_p}{3k_T} \begin{bmatrix} 14.4 \\ 45.6 \\ 57.6 \end{bmatrix} = \frac{\omega^2 J_p}{3k_T} (14.4) \begin{bmatrix} 1 \\ 3.18 \\ 4.00 \end{bmatrix}$$

The value of the mode obtained is close to the one assumed and we stop the iteration here. The first equation gives

$$1 = \frac{\omega^2 J_p}{3k_T} (14.4) \times 1 \quad \text{or} \quad \omega^2 = \frac{3k_T}{14.4 J_p}, \quad \omega = \omega_1 = 0.46 \frac{k_T}{J_p}$$

(The Stodola iteration was used to find the lower natural frequency of a three-degree-of-freedom system.)

□

Higher Modes: Matrix Deflation

Computation of the higher modes follows the observation that since the iteration $\mathbf{x} = (1/\omega^2) \mathbf{D}\mathbf{x}$, where $\mathbf{D} = \mathbf{M}^{-1}\mathbf{K}$ always converges to the lowest natural frequency, iteration on $\mathbf{x} = (1/\omega^2)\mathbf{D}_{(2)}\mathbf{x}$, where $\mathbf{D}_{(2)} = \mathbf{D} - (1/\omega_1^2)\mathbf{x}^T\mathbf{x}\mathbf{M}$ converges to the second lower natural frequency.[9] Indeed, multiplying the matrix $\mathbf{D}_2$ by the normal mode $\mathbf{\phi}_1$, obtained by normalization of $\mathbf{z}_1$ so that $\mathbf{\phi}_1^T\mathbf{M}\mathbf{\phi}_1 = 1$, we obtain

$$\mathbf{D}_2\mathbf{\phi}_1 = \sum_{i=1}^{n} c_i\mathbf{D}_2\mathbf{\phi}_i$$

$$= \sum_{i=1}^{n} c_i\mathbf{D}\mathbf{\phi}_i - \frac{1}{\omega^2} \mathbf{\phi}_1 \sum_{i=1}^{n} c_i\mathbf{\phi}_1^T\mathbf{M}\mathbf{\phi}_i \qquad (10.94)$$

[9]Wielandt, H., 1944. Bestimung Höherer Eigenwerte durch gebrochene Iteration, Proc. B44/J/37, Aerodynamischen Versuchsanstalt, Götingen.

$$= \sum_{i=1}^{n} c_i \phi_i - c_1 \phi_1 = \sum_{i=2}^{n} c_i \phi_i$$

This means that the first mode is not included in the expansion of matrix $\mathbf{D}_2$ and the iteration

$$\mathbf{x} = \frac{1}{\omega^2} \mathbf{D}_2 \mathbf{x} \qquad (10.95)$$

will indeed yield the second natural frequency and the second natural mode. The process, called *matrix deflation,* can be repeated with a sequence of matrices $\mathbf{D}_3$, $\mathbf{D}_4$, . . . to yield the desired number of modes.

☐ **Example 10.13 Higher Natural Frequencies with the Matrix Deflation Method**

Determine the second and third natural frequencies of the three-degree-of-freedom system of Example 10.12.

Solution The deflated matrix without the first mode is

$$\mathbf{D}_{(2)} = \mathbf{D} - \frac{1}{\omega_1^2} \mathbf{x}_1^T \mathbf{x}_1 \mathbf{M}$$

$$\mathbf{D}_{(2)} = \frac{\omega^2 J_p}{3k_T} \begin{bmatrix} 4 & 2 & 1 \\ 4 & 8 & 4 \\ 4 & 8 & 7 \end{bmatrix} - \frac{1}{\omega_1^2} [1 \quad 3.18 \quad 4] \begin{bmatrix} 1 \\ 3.18 \\ 4 \end{bmatrix} \begin{bmatrix} 4 & 0 & 0 \\ 0 & 2 & 0 \\ 0 & 0 & 1 \end{bmatrix}$$

$$\mathbf{D}_{(2)} = \frac{\omega^2 J_p}{3k_T} \begin{bmatrix} 0 & -4.4 & -3 \\ 0 & 1.6 & 0 \\ 0 & 1.6 & 3 \end{bmatrix}$$

In analogy with observed results in Chapter 7, we assume a second mode $x(2)$ = $\{1 \quad 0 \quad -1\}$. The iteration now is repeated. The first iteration gives

$$\begin{bmatrix} \theta_1 \\ \theta_1 \\ \theta_1 \end{bmatrix} = \frac{\omega^2 j_p}{3k_T} \begin{bmatrix} 0 & -4.4 & -3 \\ 0 & 1.6 & 0 \\ 0 & 1.6 & 3 \end{bmatrix} \begin{bmatrix} 1 \\ 0 \\ -1 \end{bmatrix} = \frac{\omega^2 J_p}{3k_T} \begin{bmatrix} 3 \\ 0 \\ -3 \end{bmatrix} = \frac{\omega^2 j_p}{3k_T} (3) \begin{bmatrix} 1 \\ 0 \\ -1 \end{bmatrix}$$

The iteration does not need to continue because the mode repeats itself. The second natural frequency is obtained from the first equation

$$1 = \frac{\omega^2 J_p}{3k_T} (3), \qquad \omega^2 = \frac{3k_T}{3} \qquad \omega = \left[\frac{k_T}{J_p} \right]^{1/2}$$

The deflated matrix without the first and second modes is

$$\mathbf{D}_{(3)} = \mathbf{D}_{(2)} - \frac{1}{\omega_2^2} \mathbf{x}_2^T \mathbf{x}_2 \mathbf{M}$$

$$\mathbf{D}_{(3)} = \frac{\omega^2 j_p}{3k_T} \begin{bmatrix} 0 & -4.4 & -3 \\ 0 & 1.6 & 0 \\ 0 & 1.6 & 3 \end{bmatrix} - \frac{1}{\omega_2^2} [1 \quad 0 \quad -1] \begin{bmatrix} 1 \\ 0 \\ -1 \end{bmatrix} \begin{bmatrix} 4 & 0 & 0 \\ 0 & 2 & 0 \\ 0 & 0 & 1 \end{bmatrix}$$

$$\mathbf{D}_{(3)} = \frac{\omega^2 J_p}{3k_T} \begin{bmatrix} 0 & 0 & 0.437 \\ 0 & 0 & -1.36 \\ 0 & 0 & 1.75 \end{bmatrix}$$

We assume a second mode $\mathbf{x}_{(3)} = \{2 \quad 0 \quad 1\}$. The iteration is now repeated. The first iteration gives

$$\begin{bmatrix} \theta_1 \\ \theta_1 \\ \theta_1 \end{bmatrix} = \frac{\omega^2 J_p}{3k_T} \begin{bmatrix} 0 & 0 & 0.437 \\ 0 & 0 & -1.36 \\ 0 & 0 & 1.75 \end{bmatrix} \begin{bmatrix} 2 \\ 0 \\ 1 \end{bmatrix} = \frac{\omega^2 J_p}{3k_T} \begin{bmatrix} 3 \\ 0 \\ -3 \end{bmatrix} = \frac{\omega^2 J_p}{3k_T} (1.75) \begin{bmatrix} 0.25 \\ -0.72 \\ 1.00 \end{bmatrix}$$

and the second gives

$$\begin{bmatrix} \theta_1 \\ \theta_1 \\ \theta_1 \end{bmatrix} = \frac{\omega^2 J_p}{3k_T} \begin{bmatrix} 0 & 0 & 0.437 \\ 0 & 0 & -1.36 \\ 0 & 0 & 1.75 \end{bmatrix} \begin{bmatrix} 0.25 \\ -0.72 \\ 1.00 \end{bmatrix} = \frac{\omega^2 J_p}{3k_T} (1.75) \begin{bmatrix} 0.25 \\ -0.72 \\ 1.00 \end{bmatrix}$$

The iteration does not need to continue because the mode repeats itself. The third natural frequency is obtained from the first equation,

$$0.25 = \frac{\omega^2 J_p}{3k_T} (1.75 \times 0.25), \qquad \omega^2 = \frac{3k_T}{1.75 J_p}, \qquad \omega = \omega_3 = 1.31 \left(\frac{k_T}{J_p} \right)^{1/2}$$

(The matrix deflation method was used to compute the higher modes of a three-degree-of-freedom system with matrix iteration.)

□

10.9 CONCURRENT COMPUTATION METHODS

Jacobi Method of Matrix Rotation

With the matrix iteration method the natural frequencies and modes are computed one at a time. Moreover, the error in the computation of any mode is carried over to computation of the next mode. This might lead to a propagation of large errors at higher modes. The matrix rotation method computes concurrently all modes.

If we rotate a planar vector $\mathbf{x}(x_1, x_2)$ in a Cartesian coordinate system (x, y) by an angle θ, the new vector $\mathbf{x}^*(x_1^*, x_2^*)$ will be, with elementary trigonometry,

$$\begin{aligned} x_1^* &= (\cos \theta) x_1 - (\sin \theta) x_2 \\ x_2^* &= (\sin \theta) x_1 + (\cos \theta) x_2 \end{aligned} \tag{10.96}$$

In matrix notation $\mathbf{x}^* = \mathbf{R}\mathbf{x}$, or

$$\begin{bmatrix} x_1^* \\ x_2^* \end{bmatrix} = \begin{bmatrix} \cos \theta & -\sin \theta \\ \sin \theta & \cos \theta \end{bmatrix} \begin{bmatrix} x_1 \\ x_2 \end{bmatrix} \tag{10.97}$$

With a proper selection of the angle θ, one can diagonalize a 2×2 symmetric matrix $\mathbf{A}$ by rotation of its columns or rows, considered as vectors. To this end we form the product

$$\mathbf{A}^* = \mathbf{R}^T \mathbf{A} \mathbf{R} \tag{10.98}$$

The elements of the matrix $\mathbf{A}^*$ will be

$$\begin{aligned} a_{11}^* &= a_{11} \cos^2\theta + 2a_{12} \sin \theta \cos \theta + a_{22} \sin^2\theta \\ a_{12}^* &= a_{21}^* = (a_{22} - a_{11}) \sin \theta \cos \theta + a_{12}(\cos^2\theta - \sin^2\theta) \\ a_{22}^* &= a_{11} \sin^2\theta - 2a_{12} \sin \theta \cos \theta + a_{22} \cos^2\theta \end{aligned} \tag{10.99}$$

If $a_{12}^* = 0$, matrix $\mathbf{A}^*$ will be diagonal. This requires that

$$\tan 2\theta = \frac{2a_{12}}{a_{11} - a_{22}} \tag{10.100}$$

Let us now assume a real symmetric matrix $\mathbf{A}$ of dimension $n \times n$. We multiply it with a matrix $\mathbf{R}$ according to equation (10.98), where $\mathbf{R}$ consists of the 2×2 rotation matrix added to the ith and jth row and column of an $n \times n$ identity matrix:

Column: 1 2 3 · i · · · · j · · · · n Row:

$$\mathbf{R} = \begin{bmatrix} 1 & 0 & 0 & \cdot & \cdot & \cdot & \cdot & \cdot & \cdot & \cdot & \cdot & \cdot & \cdot & 0 & 0 \\ 0 & 1 & 0 & \cdot & \cdot & \cdot & \cdot & \cdot & \cdot & \cdot & \cdot & \cdot & \cdot & 0 & 0 \\ \cdot & \cdot & \cdot & \cdot & \cdot & \cdot & \cdot & \cdot & \cdot & \cdot & \cdot & \cdot & \cdot & \cdot & \cdot \\ & & & & \cos\theta & & & & \sin\theta & & & & & & \\ \cdot & \cdot & \cdot & \cdot & \cdot & \cdot & \cdot & \cdot & \cdot & \cdot & \cdot & \cdot & \cdot & \cdot & \cdot \\ 0 & 0 & \cdot & \cdot & \cdot & 0 & 1 & 0 & \cdot & \cdot & \cdot & \cdot & 0 & 0 \\ \cdot & \cdot & \cdot & \cdot & \cdot & \cdot & \cdot & \cdot & \cdot & \cdot & \cdot & \cdot & \cdot & \cdot & \cdot \\ & & & -\sin\theta & & & & \cos\theta & & & & & & & \\ \cdot & \cdot & \cdot & \cdot & \cdot & \cdot & \cdot & \cdot & \cdot & \cdot & \cdot & \cdot & \cdot & \cdot & \cdot \\ \cdot & \cdot & \cdot & \cdot & \cdot & \cdot & \cdot & \cdot & \cdot & \cdot & \cdot & \cdot & \cdot & \cdot & \cdot \\ 0 & 0 & 0 & \cdot & \cdot & \cdot & \cdot & \cdot & \cdot & \cdot & \cdot & \cdot & \cdot & 0 & 1 \\ 0 & 0 & 0 & \cdot & \cdot & \cdot & \cdot & \cdot & \cdot & \cdot & \cdot & \cdot & \cdot & 0 & 1 \end{bmatrix} \begin{matrix} 1 \\ 2 \\ \cdot \\ i \\ \cdot \\ \cdot \\ \cdot \\ j \\ \cdot \\ \cdot \\ n-1 \\ n \end{matrix} \tag{10.101}$$

If θ is computed by (10.100), the elements a_{ij}^* and a_{ji}^* of the matrix $\mathbf{A}^* = \mathbf{R}^T\mathbf{A}\mathbf{R}$ will be zero. Repeating this procedure for $i, j = 1, 2, \ldots, n$, it seems that the matrix $\mathbf{A}^*$ will become diagonal. This is not quite true because operation with new values of i and j will make nonzero the values of the off-diagonal elements previously made zero. They become progressively smaller, however, so that if we repeat the procedure another time, the off-diagonal terms of the matrix become smaller compared with the diagonal terms. The procedure is terminated when some ratio of the maximum off-diagonal term to the diagonal terms become smaller than a preset limit. Finally, since the identity matrix will not change under rotation, the multiplications with the rotation matrix are applied to the equation $(-\omega^2\mathbf{I} + \mathbf{A})\mathbf{x} = 0$. The result will be

$$[-\omega^2\mathbf{I} + \text{diag}(a_{11}^* \quad a_{22}^* \quad \cdots \quad a_{nn}^*)]\mathbf{x} = 0 \tag{10.102}$$

or

$$\omega^2 = a_{11}^*$$
$$\omega^2 = a_{22}^*$$
$$\vdots$$
$$\omega^2 = a_{nn}^* \tag{10.103}$$

Therefore, the diagonal elements of the matrix $\mathbf{A}^*$ are the natural frequencies squared ω^2. Moreover, the product $\mathbf{R}$ of the rotation matrices $\mathbf{R} = \mathbf{R}_1\mathbf{R}_2\mathbf{R}_3\mathbf{R}_4\cdots$ has the natural modes in the respective columns.

Cholesky Method of Matrix Decomposition

It seems that the standard eigenvalue problem encountered in vibration analysis

$$(-\omega^2\mathbf{M} + \mathbf{K})\mathbf{x} = 0 \tag{10.104}$$

can be solved by Jacobi's method in the form

$$(-\omega^2\mathbf{I} + \mathbf{D})\mathbf{x} = 0 \tag{10.105}$$

But the matrix $\mathbf{D} = \mathbf{M}^{-1}\mathbf{K}$ is not, in general, symmetric although $\mathbf{M}$ and $\mathbf{K}$ are symmetric. To transform problem (10.104) into a form that can be handled with Jacobi's method, the Cholesky method of matrix decomposition[10] can be used. To this end, the matrix $\mathbf{D}$ is set in the form

$$\mathbf{D} = \mathbf{U}^T\mathbf{U} \tag{10.106}$$

where

$$u_{11} = (d_{11})^{1/2}$$

$$u_{1j} = \frac{d_{1j}}{u_{11}}, \qquad j = 2, 3, \ldots, n$$

$$u_{ii} = \left(d_{ii} - \sum_{j=1}^{i-1} u_{ji}^2\right)^{1/2}, \qquad i = 2, 3, \ldots, n$$

$$u_{ij} = \left(d_{ij} - \sum_{k=1}^{i-1} u_{ki}u_{kj}\right), \qquad i = 2, 3, \ldots, n, \tag{10.107}$$

$$j = i + 1, \quad i + 2, \ldots, n$$

$$u_{ij} = 0, \qquad i > j$$

Obviously, because of the last relation, $\mathbf{U}$ is an upper triangular matrix.

This decomposition is applicable to the symmetric matrix $\mathbf{K} = \mathbf{U}^T\mathbf{U}$. Equation (10.104) becomes

$$(-\omega^2\mathbf{M} + \mathbf{U}^T\mathbf{U})\mathbf{x} = 0 \tag{10.108}$$

---

[10]Benoit, J., 1924. Sur une methode de resolution des equations normales (Procedé du Commandant Cholesky.) Bull. Geodesique.

Multiplying from the left by $(\mathbf{U}^T)^{-1}$, we obtain

$$(-\omega^2(\mathbf{U}^T)^{-1}\mathbf{M} + (\mathbf{U}^T)^{-1}\mathbf{U}^T\mathbf{U})\mathbf{x} = 0$$

$$\omega^2(\mathbf{U}^T)^{-1}\mathbf{M}\mathbf{x} = \mathbf{U}\mathbf{x} \qquad (10.109)$$

$$\omega^2(\mathbf{U}^T)^{-1}\mathbf{M}\mathbf{U}^{-1}\mathbf{U}\mathbf{x} = \mathbf{U}\mathbf{x}$$

Let $\mathbf{y} = \mathbf{U}\mathbf{x}$. Then equation (10.99) yields

$$\omega^2\mathbf{B}\mathbf{y} = \mathbf{y} \qquad (10.110)$$

where $\mathbf{B} = (\mathbf{U}^T)^{-1}\mathbf{M}\mathbf{U}^{-1}$ is a symmetric matrix. Equation (10.110) defines an eigenvalue problem that can be solved with the Jacobi or the power iteration method. From the eigenvectors $\mathbf{y}$ we can compute the natural modes as

$$\mathbf{x} = \mathbf{U}^{-1}\mathbf{y} \qquad (10.111)$$

Due to the specific form of the matrix $\mathbf{U}$, it is $\mathbf{U}\mathbf{U}^{-1} = \mathbf{I}$ and its inverse $\mathbf{U}^{-1}$ can be found from the relations

$$\mathbf{U}^{-1} = [v_{ij}]$$

$$v_{ii} = \frac{1}{u_{ii}}$$

$$v_{ij} = \frac{\displaystyle\sum_{k=i+1}^{j} u_{ik}v_{kj}}{u_{ii}}, \qquad i < j \qquad (10.112)$$

$$v_{ij} = 0, \qquad i > j$$

Due to the last equation, matrix $\mathbf{U}^{-1}$ is upper triangular, too.

☐ **Example 10.14 Torsional Natural Frequencies of a Shaft with the Jacobi Method**

For the shaft shown in Figure E10.14, determine the torsional natural frequencies using the Jacobi method and lumping the mass at the nodes.

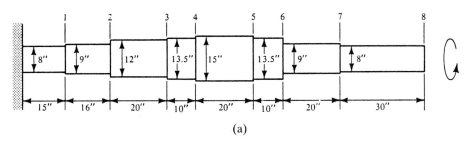

(a)

Figure E10.14

```
      PROGRAM TEST(INPUT,OUTPUT,TAPE6=OUTPUT,TAPE5=INPUT)
      DIMENSION D(10),XL(10),S(10),XK(10,10),A(10,10)
      DIMENSION XM(10,10),W(10,10),T(10)
      DATA PI/3.14159/
C       READ INPUT DATA
C       READ PROBLEM IDENTIFICATION
      READ 5,W1,W2,W3,W4,W5,W6,W7
 5    FORMAT(7A10)
C    A. NUMBER OF SECTIONS,SHEAR MODULUS
      READ 10,N,GSHEAR
 10   FORMAT(I10,E10.1)
C    B.   SECTION PROPERTIES (FROM    LEFT)  LENGTH,DIAMETER
      DO 30 I=1,N
      READ 20,XLEN,DIAM
 20   FORMAT(2F10.4)
      D(I)=DIAM
      XL(I)=XLEN
 30   CONTINUE
C
C       PRINTOUT OF INPUT DATA
C
C       IDENTIFICATION
C
      PRINT 5,W1,W2,W3,W4,W5,W6,W7
      PRINT 100
 100  FORMAT(//////,* INPUT DATA *,/,*---------------------------*,/)
      PRINT 110,GSHEAR,N
 110  FORMAT(* SHEAR MODULUS = *,E15.1,/,* NUMBER OF SECTIONS = *,I4,//)
      PRINT 105
 105  FORMAT(///,* NO       DIAMETER       LENGTH  *,/,
     1*---------------------------------------------------*,/)
      DO 120 I=1,N
      PRINT 130,I,D(I),XL(I)
 130  FORMAT(I4,2F18.2)
C       ELEMENT STIFFNESSES S(I) , MOMENT=S(I)*ANGLE OF TWIST
      DO 40 I=1,N
      XINERT=3.14159*D(I)**4/32.
      S(I)=XINERT*GSHEAR/XL(I)
C       NESS MATRIX
        1,N
        1,N
 50   0.
      S(1)*S(2)
      -S(1)
```

```
      XK(N,N)=S(N)
      XK(N,NM1)=-S(N)
      DO 60 I=2,NM1
      IM1=I-1
      IP1=I+1
      XK(I,I)=S(I)*S(IP1)
      XK(I,IM1)=-S(I)
 60   XK(I,IP1)=-S(IP1)
      CALL PMAT(XK,1,N,1,N,10H  STIFFNESS)
C     FLEXIBILITY MATRIX A = INVERSE OF XK
      CALL INVE(XK,A,N)
      CALL PMAT(A,1,N,1,N,10H  FLEXIBL  )
C     COMPUTATION OF THE MASS MATRIX XM
C     MASSES WILL BE LUMPED AT NODAL POINTS
C
      DO 190 I=1,N
      DO 190 J=1,N
 190  XM(I,J)=0.
      DENS=62.2*8.5/386.4
      XM1 =PI*D(1)**2/4.*XL(1)*DENS+PI*D(2)**2/8.*XL(2)*DENS
      XM(1,1)=XM1
      XMN =PI*D(N)**2/8.*XL(N)*DENS
      XM(N,N)=XMN
      DO 200 I=2,NM1
      XM I =PI/8.*(D(I)**2*XL(I)*D(I*1)**2*XL(I+1))*DENS
      XM(I,I)=XMI
 200  CONTINUE
      CALL PMAT(XM,1,N,1,N,10H  MASS MATR)
      CALL EIGENC(XK,XM,W,T,N)
      CALL PMAT(XK,1,N,1,N,10H EIGENVALU)
      CALL PMAT(XM,1,N,1,N,10H EIGENVECT)
      STOP
      END
```

STIFFNESS AND FLEXIBILITY MATRICES OF A STEPPED SHAFT IN TORSION

INPUT DATA

SHEAR MODULUS =	1.5E+07	
NUMBER OF SECTIONS =	8	

NO	DIAMETER	LENGTH
1	8.00	15.00
2	9.00	16.00
3	12.00	20.00
4	13.50	10.00
5	15.00	20.00
6	13.50	10.00
7	9.00	20.00
8	8.00	30.00

(b)

Figure E10.14   (cont.)

504

PRINTOUT OF MATRIX   STIFFNESS

J=	1	2	3	4	5	6	7	8
I=								
1	1.01E+09	-4.02E+08	0.	0.	0.	0.	0.	0.
2	-6.04E+08	2.13E+09	-1.53E+09	0.	0.	0.	0.	0.
3	0.	-1.53E+09	6.42E+09	-4.89E+09	0.	0.	0.	0.
4	0.	0.	-4.89E+09	8.62E+09	-3.73E+09	0.	0.	0.
5	0.	0.	0.	-3.73E+09	8.62E+09	-4.89E+09	0.	0.
6	0.	0.	0.	0.	-4.89E+09	5.37E+09	-4.83E+08	0.
7	0.	0.	0.	0.	0.	-4.83E+08	6.84E+08	-2.01E+08
8	0.	0.	0.	0.	0.	0.	-2.01E+08	2.01E+08

PRINTOUT OF MATRIX   FLEXIBL

J=	1	2	3	4	5	6	7	8
I=								
1	1.66E-09	1.10E-09	1.10E-09	1.10E-09	1.10E-09	1.10E-09	1.10E-09	1.10E-09
2	1.66E-09	2.76E-09	2.76E-09	2.76E-09	2.76E-09	2.76E-09	2.76E-09	2.76E-09
3	1.66E-09	2.76E-09	3.41E-09	3.41E-09	3.41E-09	3.41E-09	3.41E-09	3.41E-09
4	1.66E-09	2.76E-09	3.41E-09	3.62E-09	3.62E-09	3.62E-09	3.62E-09	3.62E-09
5	1.66E-09	2.76E-09	3.41E-09	3.62E-09	3.89E-09	3.89E-09	3.89E-09	3.89E-09
6	1.66E-09	2.76E-09	3.41E-09	3.62E-09	3.89E-09	4.09E-09	4.09E-09	4.09E-09
7	1.66E-09	2.76E-09	3.41E-09	3.62E-09	3.89E-09	4.09E-09	6.16E-09	6.16E-09
8	1.66E-09	2.76E-09	3.41E-09	3.62E-09	3.89E-09	4.09E-09	6.16E-09	1.11E-08

PRINTOUT OF MATRIX   MASS MATR

J=	1	2	3	4	5	6	7	8
I=								
1	1.73E+03	0.	0.	0.	0.	0.	0.	0.
2	0.	2.24E+03	0.	0.	0.	0.	0.	0.
3	0.	0.	2.53E+03	0.	0.	0.	0.	0.
4	0.	0.	0.	3.40E+03	0.	0.	0.	0.
5	0.	0.	0.	0.	3.40E+03	0.	0.	0.
6	0.	0.	0.	0.	0.	1.85E+03	0.	0.
7	0.	0.	0.	0.	0.	0.	1.90E+03	0.
8	0.	0.	0.	0.	0.	0.	0.	1.03E+03

PRINTOUT OF MATRIX   EIGENVALU

J=	1	2	3	4	5	6	7	8
I=								
1	5.11E+05	-1.10E-07	-4.73E-28	-2.49E-08	-5.56E-13	-2.87E-16	2.04E-18	-2.08E-17
2	-1.10E-07	8.49E+05	-3.06E-10	1.89E-18	-8.13E-10	7.37E-22	4.38E-08	-1.79E-21
3	-4.73E-28	-3.06E-10	5.11E+06	-5.72E-11	-1.36E-19	0.	2.19E-11	4.40E-24
4	-2.49E-08	1.89E-18	-5.72E-11	1.97E+04	-2.40E-15	8.24E-12	0.	-8.57E-08
5	-5.56E-13	-8.13E-10	-1.36E-19	-2.40E-15	1.57E+06	-6.08E-08	1.95E-25	-2.59E-10
6	-2.07E-16	7.37E-22	0.	8.24E-12	-6.08E-08	4.92E+06	3.39E-08	1.92E-12
7	2.04E-18	4.38E-08	2.19E-11	0.	1.95E-25	3.39E-06	4.05E+05	-5.93E-12
8	-2.08E-17	-1.79E-21	4.40E-24	-8.57E-08	-2.59E-10	1.92E-12	-5.93E-12	1.29E+05

PRINTOUT OF MATRIX   EIGENVECT

J=	1	2	3	4	5	6	7	8
I=								
1	1.85E-02	-1.18E-02	5.29E-05	2.34E-03	-2.61E-03	1.90E-04	-8.97E-03	-2.28E-03
2	5.69E-03	1.36E-02	-1.03E-02	5.66E-03	1.11E-02	-2.80E-03	-6.84E-03	-4.43E-03
3	-1.19E-02	5.12E-03	6.28E-03	7.12E-03	-9.36E-03	1.26E-02	-3.11E-03	-4.74E-03
4	-3.03E-03	2.39E-04	-8.02E-03	7.51E-03	-8.15E-03	-8.72E-03	-1.30E-03	-4.52E-03
5	-4.02E-03	-6.35E-03	1.06E-02	7.87E-03	5.07E-03	-4.76E-03	1.56E-03	-3.70E-03
6	-3.36E-03	-7.62E-03	-1.27E-02	8.05E-03	9.63E-03	1.15E-02	3.30E-03	-2.74E-03
7	9.96E-03	4.23E-03	6.82E-04	9.19E-03	-2.05E03	-8.02E-04	1.58E-02	8.31E-03
8	-6.15E-03	-1.26E-03	-2.70E-05	1.02E-02	2.91E-04	4.09E-05	-1.47E-02	2.47E-02

(c)

Figure E10.14 *(cont.)*

Solution First, the stiffness matrix is computed (see Example 8.1).

$$
\begin{bmatrix}
s_1 + s_2 & -s_2 & 0 & 0 & 0 & 0 & 0 & 0 \\
-s_2 & s_2 + s_3 & -s_3 & 0 & 0 & 0 & 0 & 0 \\
0 & -s_3 & s_3 + s_4 & -s_4 & 0 & 0 & 0 & 0 \\
\cdot & \cdot & \cdot & & \cdot & \cdot & \cdot & \cdot \\
0 & 0 & 0 & & & & -s_{n-1} & s_{n-1} + s_n
\end{bmatrix}
$$

$$
s_i = \frac{GI_p}{l}, \qquad I_p = \frac{\pi d^4}{32}
$$

The mass moment of inertia at each node will be the sum of one-half of each of the two neighboring elements:

$$
J_{pi} = \frac{1}{2} \frac{\pi \rho d_i^2 (l_{i-1} + l_i)}{8}, \qquad i = 2 \text{ to } 7
$$

$$
J_{p1} = \frac{1}{2} \frac{\pi \rho d_i^2 l_1}{16}, \qquad J_{p8} = \frac{1}{2} \frac{\pi \rho d_7^2 l_7}{16}
$$

The mass matrix will be diagonal.

The subroutine package will be used. The control program and the results are shown in Figure E10.14b.

$\square$

10.10 THE DUNKERLEY PROCEDURE

When dealing with the vibration of continuous shafts carrying masses along their length, S. Dunkerley[11] intuitively noticed that if the natural frequencies Ω_1, Ω_2, . . . , Ω_n for the continuous shaft without masses and the massless shaft carrying each mass separately were combined with the formula

$$
\frac{1}{\omega^2} = \frac{1}{\Omega_1^2} + \frac{1}{\Omega_2^2} + \cdots + \frac{1}{\Omega_n^2} \tag{10.113}
$$

the resulting frequency ω agreed very closely with the system natural frequency determined experimentally. Equation (10.113), known as the Dunkerley formula, was later proved by H. H. Jeffcott.[12]

Consider the problem of equation (10.93) in the form

$$
(-\omega^2 \mathbf{A} + \mathbf{I})\mathbf{x} = 0 \tag{10.114}
$$

The frequency determinant will be

$$
|-\omega^2 \mathbf{A} + \mathbf{I}| = 0 \tag{10.115}
$$

[11]1894. On the whirling and vibration of shafts. *Philos. Trans. R. Soc. London*, Ser. A, *185*, part I: 279–360. Dunkerley acknowledged in his paper that most of the analysis was done by Osborne Reynolds.

[12]1919. The periods of natural vibration of loaded shafts: the rational derivation of Dunkerley's empirical rule for determining the whirling speeds. *Proc. R. Soc. London*, Ser. A, *95*(A666): 106–115.

This determinant can be expanded into a polynomial equation of order $2n$ of the form

$$c_n\omega^{2n} - c_{n-1}\omega^{2n-2} + \cdots + (-1)^{n+1}c_1\omega^2 + (-1)^n = 0 \qquad (10.116)$$

where

$$c_1 = b_{11} + b_{22} + \cdots + b_{nn}$$

$$c_2 = \begin{vmatrix} b_{11} & b_{12} \\ b_{21} & b_{22} \end{vmatrix} + \cdots + \begin{vmatrix} b_{n-1,n-1} & b_{n-1,n} \\ b_{n,n-1} & b_{nn} \end{vmatrix}$$

$$c_3 = (\text{sum of terms } b_{ij}b_{kl}b_{mn}) \qquad (10.117)$$

$$\vdots$$

$$c_n = (\text{sum of products of } n \; b_{ij} \text{ terms}).$$

Further, $\qquad\qquad b_{ij} = a_{ij}m_j$

also assuming a diagonal $\mathbf{M}$ matrix. From equations (10.117) it is evident that c_n is on the order of $b_{ij}^n = a_{ij}^n m_{ij}^n$, which is on the order of $1/\Omega^{2n}$, where $\Omega = a_{ij}m_j$. Therefore, if we are seeking the lowest natural frequency ω_1, the higher-order terms in the polynomial of equation (10.116) are of the order $(\omega_1/\Omega)^2 \ll (\omega_1/\Omega)^2 = c_1\omega_1^2 < 1$. Therefore, a good approximation is

$$c_1\omega_1^2 - 1 = 0 \qquad (10.118)$$

In view of the first of equations (10.117), equation (10.118) is identical with the Dunkerley formula, equation (10.113).

This procedure can be extended to higher modes [Dimarogonas and Paipetis, 1983] by dividing the polynomial in equation (10.116) by $\omega^2 - \omega_1^2$ and retaining the lowest-order term, to obtain

$$\omega_1^2\omega_2^2 c_2 - 1 = 0$$

$$\omega_1^2\omega_2^2\omega_3^2 c_3 - 1 = 0$$

$$\omega_1^2\omega_2^2\omega_3^2\omega_4^2 c_4 - 1 = 0 \qquad (10.119)$$

$$\vdots$$

$$\omega_1^2\omega_2^2\omega_3^2 \cdots \omega_n^2 c_n - 1 = 0$$

The last result was expected from the properties of polynomials, since

$$c_n = \frac{1}{\omega_1^2\omega_2^2\omega_3^2 \cdots \omega_n^2}$$

A similar procedure, based on the distribution of elasticity rather than distribution of mass, was developed by Southwell [Bishop and Johnson, 1979].

Example 10.15   Natural Frequencies of a Two-Degree-of-Freedom System with the Dunkerley Formula

Using the Dunkerley formula, compute the lowest natural frequencies of the system of Example 10.1.

Solution   The natural frequency with the upper mass only is $\omega_1^2 = k/m$. With the lower mass only, the spring constant of the two springs in series is $kk/(k + k) = k/2$ and the natural frequency is $\omega_2^2 = k/2m$. Therefore, the lower natural frequency of the system is

$$\frac{1}{\omega^2} = \frac{1}{\omega_1^2} + \frac{1}{\omega_2^2} = \frac{m}{k} + \frac{2m}{k} = 3\frac{m}{k}$$

and $\omega^2 = 0.333(k/m)$, while the exact value is $\omega^2 = 0.36(k/m)$. Evidently, the error of Dunkerley's formula in this problem is about 7.5%.

(The Dunkerley formula was used to compute the lowest natural frequency of a two-degree-of-freedom system.)

□

□ **Example 10.16   Natural Frequencies of a Beam with Two Masses Using the Dunkerley Formula**

Using the Dunkerley formula, compute the natural frequencies of a simply supported shaft of uniform flexural rigidity EI and length l carrying two equal masses m at $l/3$ and $2l/3$, respectively.

Solution   From Appendix II, the influence coefficients are

$$a_{ij} = \frac{l^3}{6EI} [q_j(1 - q_i)[1 - q_j^2 - (1 - q_i)^2]], \qquad q_j < q_i$$

$$a_{ij} = \frac{l^3}{6EI} [q_i(1 - q_j)[1 - q_i^2 - (1 - q_j)^2]], \qquad q_j > q_i$$

where $q = x/l$ is the dimensionless coordinate along the beam, here $q_1 = 1/3$, $q_2 = 2/3$.

Using the first two of equations (10.117), we obtain

$$c_1 = 329 \times 10^{-4} \frac{EI}{ml^3}, \qquad c_2 = 638 \times 10^{-7} \left(\frac{EI}{ml^3}\right)^2$$

Using equation (10.118), $\omega_1(EI/ml^3)^{1/2} = 5.5169$. Using the first of equations (10.119), $\omega_2(EI/ml^3)^{1/2} = 22.72$.

The respective values using the two-degree-of-freedom system determinant expansion would be 5.69 and 21.99, respectively. Evidently, the error of Dunkerley's formula in this problem is about 3%.

(The Dunkerley formula was used to compute within 3% error the two lowest natural frequencies of a uniform beam carrying two masses.)

□

10.11 VIBRATION OF PLATES

Consider a plate extending on the $x - y$ plane of a Cartesian coordinate system, assuming small thickness in the z direction and with the middle surface at $z = 0$.

The strains in the plate are,[13] assuming that the middle surface is not stretched,

$$\epsilon_x = -z \frac{\partial^2 w}{\partial x^2}, \qquad \epsilon_y = -z \frac{\partial^2 w}{\partial y^2}, \qquad \gamma_{xy} = -z \frac{\partial^2 w}{\partial x \, \partial y} \qquad (10.120)$$

where w is the transverse displacement of the plate. Using the well-known stress–strain relationships [Timoshenko and Goodier, 1970], we obtain

$$\sigma_x = \frac{E}{(1 - v^2)} (\epsilon_x + v\epsilon_y) = -\frac{Ez}{(1 - v^2)} \left(\frac{\partial^2 w}{\partial x^2} + v \frac{\partial^2 w}{\partial y^2} \right)$$

$$\sigma_y = \frac{E}{(1 - v^2)} (\epsilon_y + v\epsilon_x) = -\frac{Ez}{(1 - v^2)} \left(\frac{\partial^2 w}{\partial y^2} + v \frac{\partial^2 w}{\partial x^2} \right) \qquad (10.121)$$

$$\tau_{xy} = G\gamma_{xy} = -\frac{Ez}{(1 + v)} \frac{\partial^2 w}{\partial x \, \partial y}$$

The potential energy of elastic deformation of an element $dx \, dy \, dz$ is

$$dV = \left(\frac{\epsilon_x}{2} + \frac{\epsilon_y}{2} + \frac{\gamma_{xy}\tau_{xy}}{2} \right) dx \, dy \, dz \qquad (10.122)$$

Integrating (10.122) over the volume V or the surface S of the plate and using (10.121) yields

$$V = \iiint_V dV = \frac{D}{2} \iint_S \left[\left(\frac{\partial^2 w}{\partial x^2} \right)^2 + \left(\frac{\partial^2 w}{\partial y^2} \right)^2 \right.$$
$$\left. + 2v \frac{\partial^2 w}{\partial x^2} \frac{\partial^2 w}{\partial y^2} + 2(1 - v) \left(\frac{\partial^2 w}{\partial x \, \partial y} \right)^2 \right] dx \, dy \qquad (10.123)$$

where $D = Eh^3/12(1 - v^2)$ is the flexural rigidity of the plate of thickness h. The kinetic energy is

$$T = \frac{\rho h}{s} \int_S \dot{w}^2 \, dx \, dy \qquad (10.124)$$

For circular plates, use of the cylindrical coordinates r and θ with $x = r \cos \theta$ and $y = r \sin \theta$ yields

$$V = \iiint_V dV = \frac{D}{2} \iint_S \left\{ \left(\frac{\partial^2 w}{\partial r^2} + \frac{1}{r} \frac{\partial w}{\partial r} + \frac{1}{r^2} \frac{\partial^2 w}{\partial r^2} \right)^2 \right.$$
$$\left. + 2(1 - v) \left[\frac{\partial}{\partial r} \left(\frac{1}{r} \frac{w}{\partial \theta} \right) \right]^2 \right\} dr \, d\theta \qquad (10.125)$$

$$T = \frac{\rho h}{s} \int_S \dot{w}^2 r \, dr \, d\theta \qquad (10.126)$$

---

[13]Kirchhoff, G. R. 1876. *Vorlesungen über mathematische Physik, Mechanik.* Leipzig. See also [Timoshenko and Young, 1972].

(a)

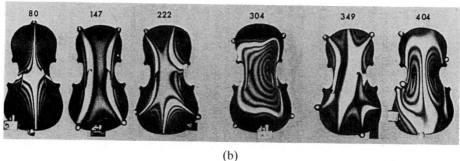

80   147   222   304   349   404

(b)

Figure 10.10 Modes of natural vibration of plates. (a) Rectangular and circular plates. (b) Backplate of a violin.

□ **Example 10.17 Vibration of a Rectangular Plate Simply Supported at the Edges**

Determine the natural frequencies of transverse vibration for a rectangular plate $0 < x < a$, $0 < y < b$, of thickness h, simply supported at the four edges.

Solution We will use Lagrange's equations expanding the natural modes as triginometric expansions in terms of admissible functions.

The functions

$$w(x, y, t) = \sum_{m=1}^{\infty} \sum_{n=1}^{\infty} q_{mn}(t) \sin \frac{m\pi x}{a} \sin \frac{n\pi y}{b}$$

where q_{mn} are yet unknown functions of time, satisfy the geometric boundary conditions at $x = 0$, a and $y = 0$, b. Therefore, they are admissible. Equations (10.123) and (10.124) give, also due to the orthogonality,

$$V = \frac{\pi^4 abD}{8} \sum_{m=1}^{\infty} \sum_{n=1}^{\infty} (m^2/a^2 + n^2/b^2)^2 q_{mn}^2$$

$$T = \frac{abh\rho}{8} \sum_{m=1}^{\infty} \sum_{n=1}^{\infty} q_{mn}^2$$

Application of Lagrange's equation, considering q_{mn} as the principal coordinates, yields

$$\rho h \ddot{q}_{mn}(t) + \pi^4 D \left(\frac{m^2}{a^2} + \frac{n^2}{b^2} \right)^2 q_{mn}(t) = 0, \qquad m, n = 1, 2, 3, \ldots$$

The natural frequencies are

$$\omega_{mn} = \pi^2 \left(\frac{m^2}{a^2} + \frac{n^2}{b^2} \right) \left(\frac{D}{\rho h} \right)^{1/2}, \qquad m, n = 1, 2, 3, \ldots$$

☐

☐ **Example 10.18   Vibration of a Circular Plate Fixed at the Boundary**

Determine the natural frequencies of transverse vibration for a circular plate $0 < r < R$, of thickness h, fixed at the boundary, using the Rayleigh–Ritz method.

Solution   We will use admissible functions satisfying the geometric boundary conditions. Considering only modes that are symmetric in respect to the origin $w(r, t) = W(r)T(t)$, we select functions

$$W(r) = \sum_{n=1}^{\infty} c_n \left(1 - \frac{r^2}{R^2} \right)^{n+1}$$

where c_n are yet unknown constants, which have $W(r)$ and $dW(r)/dr$ zero at $r = R$; therefore, they satisfy the geometric boundary conditions. Therefore, they are admissible. Equations (10.78), (10.125), and (10.126) yield

$$\frac{\partial}{\partial q_n} \int_0^R \left[\left(\frac{d^2 W}{dr^2} + \frac{1}{r} \frac{dW}{dr} \right)^2 - \frac{\omega^2 \rho h}{D} W^2 \right] r \, dr = 0$$

which gives sufficient number of homogeneous equations for the determination of the natural frequencies. Taking only one term in the series, we obtain

$$\frac{96}{9R^2} - \frac{\omega^2 \rho h}{D} \frac{R^2}{10} = 0$$

$$\omega = \frac{10.33}{R^2} \left(\frac{D}{\rho h} \right)^{1/2}$$

Taking two terms gives

$$c_1 \left(\frac{192}{2} - \frac{\lambda}{5} \right) + c_2 \left(\frac{144}{9} - \frac{\lambda}{6} \right) = 0$$

$$c_1 \left(\frac{144}{9} - \frac{\lambda}{6} \right) + c_2 \left(\frac{96}{5} - \frac{\lambda}{7} \right) = 0$$

where $\lambda = R^4 \omega^2 \rho h / D$. The frequency equation is then

$$\begin{vmatrix} \dfrac{192}{2} - \dfrac{\lambda}{5} & \dfrac{144}{9} - \dfrac{\lambda}{6} \\[2mm] \dfrac{144}{9} - \dfrac{\lambda}{6} & \dfrac{96}{5} - \dfrac{\lambda}{7} \end{vmatrix} = 0$$

from which

$$\lambda_1 = 104.3, \qquad \lambda_2 = 1854,$$

$$\omega_1 = \frac{10.21}{R^2}\left(\frac{D}{\rho h}\right)^{1/2}, \qquad \omega_2 = \frac{43.06}{R^2}\left(\frac{D}{\rho h}\right)^{1/2}$$

In the general case, w is function of r and θ and the natural modes are functions of both r and θ. The nodal lines are circles of order n and diameters of order m. The natural frequencies are then

$$\omega_{mn} = \frac{a_{mn}}{R^2}\left(\frac{D}{\rho h}\right)^{1/2}$$

where a_{mn} is given in the following table [Timoshenko, 1928]:

n	$m = 0$	$m = 1$	$m = 2$
0	10.21	21.22	34.84
1	39.78		
2	88.90		

For extensive tables for different boundary conditions, see [Szilard, 1974].

(The Rayleigh–Ritz method was used with symmetric circular admissible functions with one- and two-term approximations. This problem is very important in acoustics because the results are applicable for microphones and speakers.) □

TABLE 10.1 SUMMARY OF EQUATIONS OF CHAPTER 10

$$y(x, t) = \sum_{n=0}^{\infty} f_n(t) Y_n(x)$$

Rayleigh quotient

$$R(\mathbf{z}) = \frac{\omega^2}{(1 + \epsilon^2)}$$

$$\omega_i^2 = \frac{[\boldsymbol{\phi}^{(i)T}\mathbf{K}\boldsymbol{\phi}^{(i)}]}{[\boldsymbol{\phi}^{(i)T}\mathbf{M}\boldsymbol{\phi}^{(i)}]}$$

$$R(\mathbf{z}) = \mathbf{z}^T\mathbf{K}\mathbf{z}/\mathbf{z}^T\mathbf{M}\mathbf{z}$$

$$\omega_n^2 \geq R_r(\mathbf{z}) \geq \omega_1^2$$

Continuous systems

$$U_{max} = \frac{1}{2}\int_0^L EI\left(\frac{d^2 y}{dx^2}\right)^2 dx$$

$$T_{max} = \frac{1}{2}\omega^2 \int_0^L \dot{y}^2 dm$$

TABLE 10.1 *(cont.)*

$$\omega^2 = \frac{\int_0^L EI\left(\frac{d^2y}{dx^2}\right)^2 dx}{\int_0^L \dot{y}^2 \, dm}$$

Method of constraints

$$W_n = \sum_{i=1}^n z_i q_i(t)$$

$$T(t) = \frac{1}{2} \sum_{i=1}^n \sum_{j=1}^n m_{ij} \dot{q}_i(t)\dot{q}_j(t)$$

$$V(t) = \frac{1}{2} \sum_{i=1}^n \sum_{j=1}^n k_{ij} q_i(t)q_j(t)$$

$$\frac{\partial T}{\partial \dot{q}_r} = \frac{1}{2} \sum_{i=1}^n m_{ir}\dot{q}_i + \frac{1}{2} \sum_{j=1}^n m_{rj}\dot{q}_j$$

$$\frac{\partial T}{\partial \dot{q}_r} = \sum_{i=1}^n m_{ir}\dot{q}_i$$

$$\frac{d}{dt}\left(\frac{\partial T}{\partial \dot{q}_r}\right) = \sum_{i=1}^n m_{ir}\ddot{q}_i$$

$$\frac{\partial V}{\partial q_r} = \sum_{i=1}^n k_{ir}q_i, \qquad \frac{\partial T}{\partial q_r} = 0$$

$$\sum_{i=1}^n m_{ir}\ddot{q}_i + \sum_{i=1}^n k_{ir}q_i = f_r$$

$$r = 1, 2, \ldots, n$$

$$\mathbf{M}\ddot{\mathbf{q}} + \mathbf{K}\mathbf{q} = \mathbf{f}$$

Transfer matrix method

$$\mathbf{z}_1 = \{x_1 \quad \theta_1 \quad M_1 \quad V_1\}$$

$$\mathbf{z}_n^L = \mathbf{L}_{n-1}\mathbf{L}_{n-2} \cdots \mathbf{L}_2\mathbf{L}_2\mathbf{z}_1^R$$

$$\mathbf{z}_n^L = \mathbf{A}\mathbf{z}_1^R$$

Finite element method

$$\mathbf{K}^{(e)} = \int_0^l EI \, \mathbf{h}''^{(e)T}\mathbf{h}''^{(e)} \, dx$$

$$\mathbf{M}^{(e)} = \int_0^l \frac{\rho A}{l} \, \dot{\mathbf{h}}^{(e)T}\dot{\mathbf{h}}^{(e)} \, dx$$

(continues)

TABLE 10.1 *(cont.)*

Rayleigh–Ritz method

$$z(z) = \sum_{j=1}^{n} c_i z_i(x)$$

$$\mathbf{z} = \sum_{j=1}^{n} c_i \mathbf{z}^{(i)} = \mathbf{Zc}$$

$$A = \int_0^L EI(z'')^2 dx, \qquad B = \int_0^L m(x)z^2 dx$$

$$\frac{\partial A}{\partial c_i} - \omega^2 \frac{\partial B}{\partial c_i} = 0, \qquad i = 1, 2, \ldots, n$$

$$A = \sum_{i=1}^{n} k_{ij} c_i c_j, \qquad B = \sum_{i=1}^{n} m_{ij} c_i c_j$$

$$\frac{\partial A}{\partial c_i} = 2(k_{i1}c_1 + k_{i2}c_2 + \cdots + k_{in}c_n)$$

$$\frac{\partial B}{\partial c_i} = 2(m_{i1}c_1 + m_{i2}c_2 + \cdots + m_{in}c_n)$$

$$\sum_{i=1}^{n} (k_{ij} - \omega^2 m_{ij})c_j = 0, \qquad i = 1, 2, \ldots, n$$

$$(\mathbf{K} - \mathbf{M}\omega^2)\mathbf{c} = 0$$

$$(\mathbf{K}^* - \mathbf{M}^*\omega^2)\mathbf{c} = 0$$

Galerkin method

$$z(x) = \sum_{j=1}^{n} c_i z_i(x)$$

$$L(z) = 0$$

$$R_i = \int_0^l z_i L(z) dx = 0, \qquad i = 1, 2, \ldots, n$$

Dunkerley procedure

$$\frac{1}{\omega^2} = \frac{1}{\omega_1^2} + \frac{1}{\omega_2^2} + \cdots + \frac{1}{\omega_n^2}$$

Vibration of plates
Rectangular plates

$$\epsilon_x = -z \frac{\partial^2 w}{\partial x^2}, \qquad \epsilon_y = -z \frac{\partial^2 w}{\partial y^2}, \qquad \gamma_{xy} = -z \frac{\partial^2 w}{\partial x \partial y}$$

$$\sigma_x = \frac{E}{(1 - v^2)} (\epsilon_x + v\epsilon_y) = -\frac{Ez}{(1 - v^2)} \left(\frac{\partial^2 w}{\partial x^2} + \frac{v \partial^2 w}{\partial y^2} \right)$$

$$\sigma_y = \frac{E}{(1 - v^2)} (\epsilon_y + v\epsilon_x) = -\frac{Ez}{(1 - v^2)} \left(\frac{\partial^2 w}{\partial y^2} + \frac{v \partial^2 w}{\partial x^2} \right)$$

TABLE 10.1 *(cont.)*

$$\tau_{xy} = G\gamma_{xy} = -\frac{Ez}{(1+v)}\frac{\partial^2 w}{\partial x\partial y}$$

$$dV = \left(\frac{\epsilon_x}{2} + \frac{\epsilon_y}{2} + \gamma_{xy}\frac{\tau_{xy}}{2}\right)dx\,dy\,dz$$

$$V = \iiint_v dV = \frac{D}{2}\iint_s\left[\left(\frac{\partial^2 w}{\partial x^2}\right)^2 + \left(\frac{\partial^2 w}{\partial y^2}\right)^2 + 2v\frac{\partial^2 w}{\partial x^2}\frac{\partial^2 w}{\partial y^2} + 2(1-v)\left(\frac{\partial^2 w}{\partial x\partial y}\right)^2\right]dx\,dy$$

$$D = \frac{Eh^3}{12(1-v^2)}$$

$$T = \frac{\rho h}{s}\int_s \dot{w}^2\,dx\,dy$$

Circular plates

$$V = \iiint_v dV = \frac{D}{2}\iint_s\left\{\left(\frac{\partial^2 w}{\partial r^2} + \frac{1}{r}\frac{\partial w}{\partial r} + \frac{1}{r^2}\frac{\partial^2 w}{\partial r^2}\right)^2 + 2(1-v)\left[\frac{\partial}{\partial r}\left(\frac{1}{r}\frac{w}{\partial\theta}\right)\right]^2\right\}dr\,d\theta$$

$$T = \frac{\rho h}{s}\int_s \dot{w}^2 rd\,rd\theta$$

REFERENCES AND FURTHER READING

BISHOP, R. E. D., and JOHNSON, D. C. 1979. *Mechanics of Vibrations*, 2nd ed. Cambridge: Cambridge University Press.

CLOUGH, R. W., and PENZIEN, J. 1975. *Dynamics of Structures*. New York: McGraw-Hill.

DIMAROGONAS, A. D., and PAIPETIS, S. A. 1983. *Analytical Methods in Rotor Dynamics*. London: Elsevier Science.

HOHENEMSER, K., and PRAGER, W. 1933. *Dynamik der Stabwerke*. Berlin: Julius Springer.

HOLZER, H. 1921. *Die Berechnung der Drehschwingungen*. Berlin: Julius Springer.

KLOTTER, K. 1960. *Technische Schwingungslehre*. Berlin: Springer-Verlag.

MEIROVITCH, L. 1980. *Computational Methods in Structural Dynamics*. Alphen aan den Rijn, The Netherlands: Sijthoff en Noordhoff.

MYKLESTAD, N. O. 1944. *Vibration Analysis*. New York: McGraw-Hill.

PESTEL, E. C., and LECKIE, F. A. 1963. *Matrix Methods in Elastomechanics*. New York: McGraw-Hill.

STODOLA, A. 1927. *Steam and Gas Turbines*, 2 vol. New York: McGraw-Hill.

SZILARD, R. 1974. *Theory and Analysis of Plates*. Englewood Cliffs, N.J.: Prentice Hall.

TIMOSHENKO, S. P. 1928. *Vibration Problems in Engineering*. New York: D. Van Nostrand.

TIMOSHENKO, S. P., and GOODIER, J. N. 1972. *Theory of Elasticity*, 3rd ed. New York: McGraw-Hill.

TIMOSHENKO, S. P., and YOUNG, D. H. 1937. *Vibration Problems in Engineering*, 2nd ed. New York: D. Van Nostrand.

VAN DEN DUNGEN, M. F.-H. 1928. *Les problèmes généreux de la technique des vibrations*. Paris: Gauthier-Villars.

VON KÁRMÁN, T., and BIOT, M. A. 1939. *Mathematical Methods in Engineering*. New York: McGraw-Hill.

PROBLEMS

Section 10.2

10.1. Using Rayleigh's method, determine the first natural frequency of a cantilever beam of negligible mass, bearing two equal masses at equal distances (Figure P10.1).

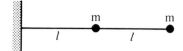

Figure P10.1

10.2. In Problem 10.1, calculate the natural frequency assuming that the mass M of the beam is not negligible.

10.3. A beam fixed at both ends carries at the midspan a mass m. Determine which portion λ of the mass of the beam M should be added to the mass m in order that the natural frequency can be computed with a massless beam bearing at the midspan a mass $m + \lambda M$.

10.4. Solve Problem 10.3 for a simply supported beam of mass M supporting a mass m at the middle of the beam.

10.5. Find the fundamental torsional natural frequency of the constant-cross-section shaft of Figure P10.5 using Rayleigh's method.

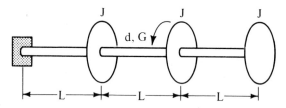

Figure P10.5

10.6. Find the torsional natural frequency of the branched system shown in Figure P10.6.

10.7. Using the Rayleigh method, determine the lateral natural frequency of a beam of mass m, fixed at the one end, of shape shown in Figure P10.7, if $a = 25$ mm, $b = 300$ mm, $L = 900$ mm, and $E = 1.2 \times 10^{11}$ Pa.

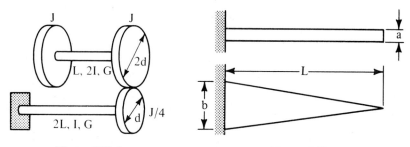

Figure P10.6   **Figure P10.7**

10.8–10.40. Using Rayleigh's method, determine the natural frequency of the systems shown in Figures P10.8 to P10.40, respectively.

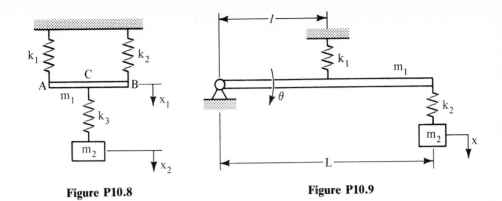

Figure P10.8

Figure P10.9

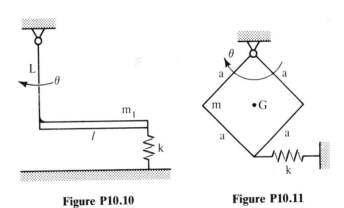

Figure P10.10

Figure P10.11

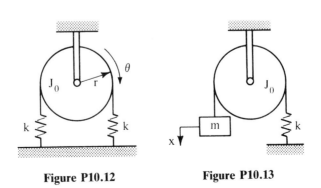

Figure P10.12

Figure P10.13

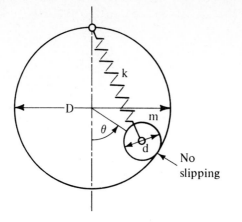

Figure P10.14

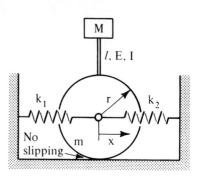

Figure P10.15

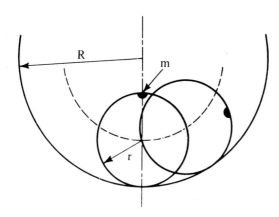

Figure P10.16

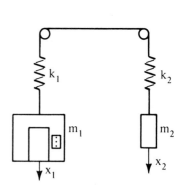

Figure P10.17

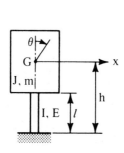

Figure P10.18

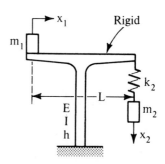

Figure P10.19

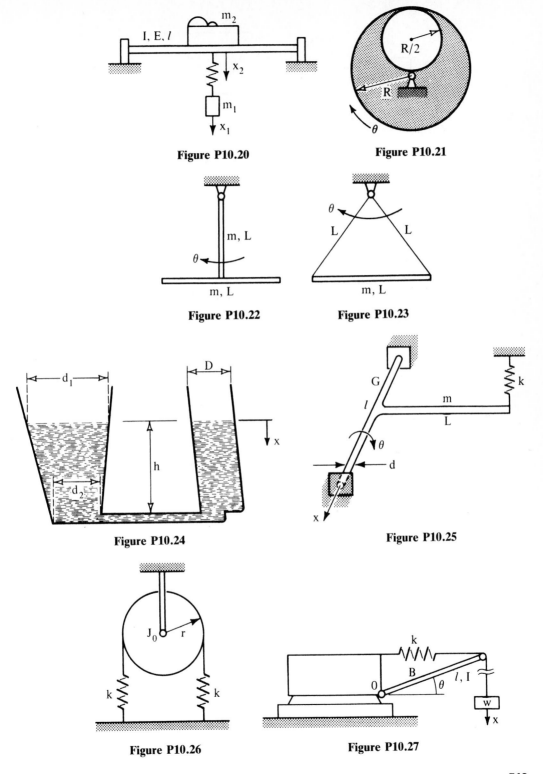

Figure P10.20

Figure P10.21

Figure P10.22

Figure P10.23

Figure P10.24

Figure P10.25

Figure P10.26

Figure P10.27

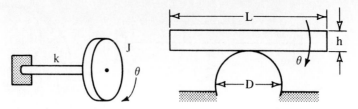

Figure P10.28   **Figure P10.29**

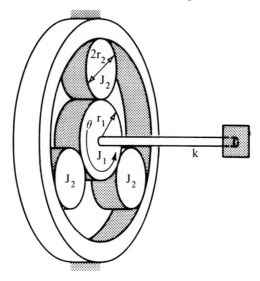

Figure P10.30

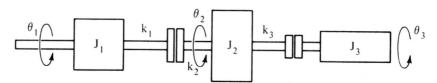

Figure P10.31

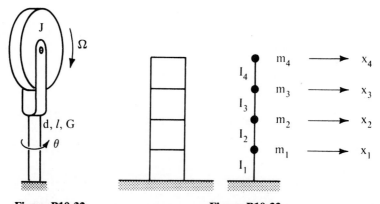

Figure P10.32   **Figure P10.33**

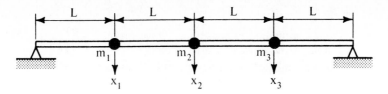

Figure P10.34

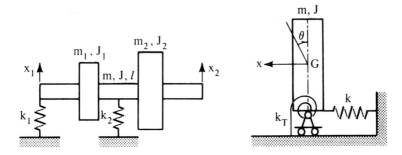

Figure P10.35

Figure P10.36

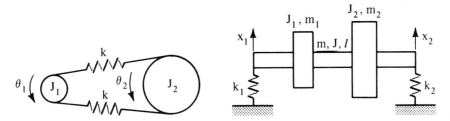

Figure P10.37

Figure P10.38

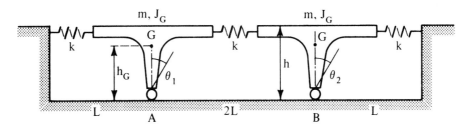

Figure P10.39

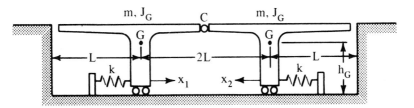

Figure P10.40

Section 10.3

10.41. Using the method of constraints, solve Problem 10.5 assuming coordinates θ_1, θ_2, and θ_3, the rotations of the three disks.

10.42. Using the method of constraints, solve Problem 10.6 with coordinates θ_1 and θ_2, the rotations at the ends of the upper shaft.

10.43. Using the method of constraints, determine the free vibration of a spring with constant k and mass m using as coordinates **(a)** the end deflection, and **(b)** the end and middle deflections.

10.44. Using the method of constraints, solve Problem 10.33 using only one coordinate, the deflection of the top of the building.

10.45. Using the method of constraints, solve Problem 10.34, with only one coordinate, the deflection at midspan.

Section 10.4

10.46. Derive the transfer matrix for the mass in Figure P10.46 if the state vector is $[x, F]^T$.
10.47. Derive the transfer matrix for the spring k_1 in Figure P10.47.

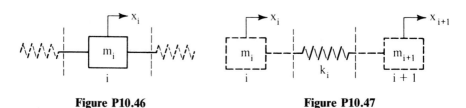

Figure P10.46   **Figure P10.47**

10.48. Determine the frequency determinant for the system of Figures P10.5 and P10.6 for free–free end conditions.

10.49. Determine the frequency determinant for the system of Figures P10.5 and P10.6 for free–clamped end conditions.

10.50. For the element in Figure P10.50, determine the transfer matrix, for the state vector $[x \quad F]^T$.

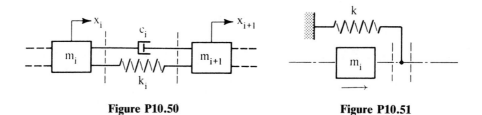

Figure P10.50   **Figure P10.51**

10.51. Determine the transfer matrix for a spring k as in Figure P10.51.

10.52. Determine the transfer matrix for the element of Figure P10.52 for lateral beam vibration, taking as state vector $[x \quad \theta \quad M \quad V]^T$.

10.53. Determine the transfer matrix for the disk of Figure P10.53 for lateral vibration.

10.54–10.57. Determine the mass and stiffness matrices for the elements shown in Figures P10.54 to P10.57.

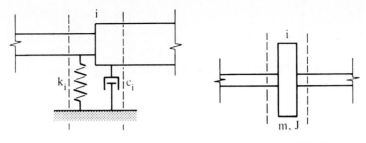

Figure P10.52　　　　　　　　**Figure P10.53**

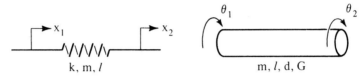

Figure P10.54　　　　　　　　**Figure P10.55**

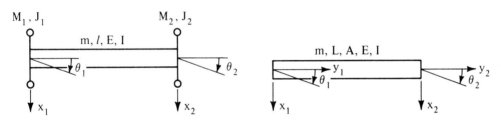

Figure P10.56　　　　　　　　**Figure P10.57**

Sections 10.5 to 10.9

10.58. Using the finite element method, find the natural frequency of the beam in Figure P10.1 if the beam has constant flexural rigidity EI and mass m.

10.59. Using the finite element method, find the torsional natural frequency of the shaft in Figure P10.5 if it has constant flexural rigidity EI and mass m.

10.60. Using the finite element method, find the torsional natural frequency of the system in Figure P10.6.

10.61. Using the finite element method, find the natural frequency of the tapered beam in Figure P10.7. Assume three equally spaced nodes and a constant cross section in between: **(a)** the maximum along the element, and **(b)** the section at the midlength of the element.

10.62. Determine the two lower longitudinal natural frequencies of a bar of constant cross section with the finite element method, taking four equidistant nodes along the bar.

10.63. Using the Rayleigh–Ritz method, determine the natural frequencies of the systems of Problems 10.1 to 10.7 and 10.62, selecting proper admissible functions.

10.71–10.76. Using the Dunkerley method, determine the natural frequencies of the systems of Problems 10.1 to 10.6, respectively.

10.77–10.78. Using the Dunkerley method, determine the natural frequencies of the systems of Problems 10.8 and 10.9, respectively.

11

Rotor
Dynamics

11.1 DYNAMICS OF A ROTOR ON A FLEXIBLE SHAFT

The De Laval Rotor

Up to this point, we have considered only vibration about a static equilibrium position. There are instances where a system vibrates about a steady motion. Typical examples include the vibration of rotating shafts, disks, blades, and so on. In this chapter we discuss a very common case of vibration about steady motion: the vibration of rotating shafts.

We have seen on many occasions that in some problems the equations of motion are uncoupled. In other words, the motion along one coordinate is unaffected by the motion along another coordinate. Consider, for example, the vibration of a horizontal cantilever beam. It can vibrate in the vertical or horizontal lateral direction. It can also have longitudinal or torsional vibration. However, we have

*The above photo portrays a Turbine Rotor on a balance stand. (Courtesy of Bruel & Kjaer Co., Denmark). Reprinted by permission.

treated these problems separately because in most cases motion in one direction does not affect considerably the motion in another direction. We have seen other examples where this is not the case. Therefore, before we go into details, we can state that the steady motion might not affect the vibration about it.

Let us consider the simplest case of a long and light shaft rotating with an angular velocity ω and carrying on its midspan a disk of mass m and mass polar moment of inertia J with respect to the axis of rotation (Figure 11.1a)—the *de Laval rotor*. Due to unbalance, the mass center is not at the geometric center of the disk O, but at G, and the distance e of these two points is the disk eccentricity. If x and y are the coordinates of the disk center O, the coordinates of the mass center G will be $x + e \cos \Omega t$ and $y + e \sin \Omega t$, respectively, if the time is measured when OG is in the horizontal position and Ω is the angular velocity of rotation of the disk.

If the shaft is flexible (with a lateral spring constant k), the disk is capable of movement perpendicular to its axis in the horizontal and vertical directions x and y (Figure 11.1b), but also rotations about the three axes, such as in Figure 11.1c. The rotation about the z-axis is the steady motion.

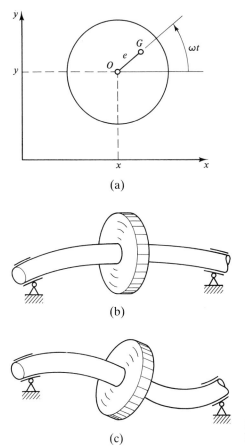

(a)

(b)

(c)

Figure 11.1 Whirling of a de Laval rotor.

Portrait of de Laval.
(Courtesy of de Laval Turbines, Inc.)

CARL GUSTAF PATRIC DE LAVAL (1845–1913)

De Laval was born in Sweden. He studied in the Uppsala Technical University, where he was awarded a Ph.D. in 1872 in mathematics and natural sciences. His first contributions were in metallurgy: the extraction of copper and the iron–nickel–phosphorus alloys. He further contributed substantial developments in metallurgy, such as the arc melting and lead and zinc metallurgy. Then he developed the centrifugal separator for the production of butter. He developed a working steam turbine and experimented at high speeds, up to 30,000 rpm. This won him the title "Man of High Speeds." More important, he observed and found engineering solutions for most problems of rotor dynamics, such as whirling, critical speeds, and accelerating beyond the critical speed.

We assume at this point that the disk moves only about z and along x and y. In other words, it remains parallel to itself and perpendicular to z-axis. General motion of the disk will be considered later. Applying Newton's law with respect to the motion of the mass center and assuming small motions, we can see immediately that the equations of motion will be[1]

[1]Whirling of shafts was anticipated by Rankine, W. A. 1869. On the centrifugal force of rotating shafts. *Engineer (London)*, 27: 249. Extensive analytical investigations were done by Dunkerley, S., and Reynolds, O. 1883. On the whirling and vibration of shafts. *Philos. Trans. A*, 279–359. The whirling problem was solved by Föppl, A. 1895. Das Problem der DeLaval'schen Turbinenwelle. *Civilingenieur*, 61: 333–342. Stodola, A. 1916. Neuere Beobachtungen über die kritischen Umlaufzahlen von Wellen. *Schweiz. Bauztg.*, 68: 210–214. This analysis is sometimes inaccurately credited to Jeffcott (*Philos. Mag*, Mar. 1919, pp. 304–314) and the de Laval rotor is sometimes misnamed the "Jeffcott rotor."

$$m \frac{d_2(x + e \cos \Omega t)}{dt^2} + kx = 0$$

$$m \frac{d_2(y + e \sin \Omega t)}{dt^2} + ky = 0 \qquad (11.1)$$

$$J\ddot{\phi} = 0$$

We can see that equation (11.1) is uncoupled and not dependent explicitly on the angular coordinate ϕ if $\dot{\phi} = \Omega$ is constant. We could also deduct this result from the energy equations. Indeed, the kinetic and potential energies are

$$T = \frac{1}{2} m \left[\frac{d(x + e \cos \Omega t)}{dt} \right]^2 + \frac{1}{2} m \left[\frac{d(y + e \sin \Omega t)}{dt} \right]^2 + \frac{1}{2} J\omega^2 \qquad (11.2)$$

$$V = \frac{1}{2} kx^2 + \frac{1}{2} ky^2 \qquad (11.3)$$

Application of Lagrange's equation with respect to the coordinates x, y, and ϕ yields (for $\Omega = \dot{\phi}$)

$$\frac{d}{dt} \left[\frac{\partial}{\partial \dot{x}} (T - V) \right] - \frac{\partial}{\partial x} (T - V) = m \frac{d_2(x + e \cos \Omega t)}{dt^2} + kx = 0$$

$$\frac{d}{dt} \left[\frac{\partial}{\partial \dot{y}} (T - V) \right] - \frac{\partial}{\partial x} (T - V) = m \frac{d_2(y + e \cos \Omega t)}{dt^2} + ky = 0 \qquad (11.4)$$

$$\frac{d}{dt} \left[\frac{\partial}{\partial \dot{\phi}} (T - V) \right] - \frac{\partial}{\partial \phi} (T - V) = \frac{d}{dt} [J\omega] = 0$$

The first two are identical with equation (11.1). The last one simply implies that $\omega = \Omega$, the constant speed of rotation, which was assumed. Therefore, the lateral motion of the disk is independent of the steady rotation about the z-axis, under the assumptions stated.

Polar Representation, Whirl, and Precession

We multiply the second of equations (11.1) by $i = (-1)^{1/2}$ and add to the first one, to obtain

$$m(\ddot{x} + i\ddot{y}) + k(x + iy) = me\Omega^2 e^{i\Omega t} \qquad (11.5)$$

Introducing $r = x + iy$, we can write

$$m\ddot{r} + kr = me\Omega^2 e^{i\Omega t} \qquad (11.6)$$

To study the physical significance of the quantity r we consider a coordinate system (x, iy) on the plane of the disk with its origin on the center of the disk at the static equilibrium (Figure 11.2). If the center of the disk is displaced from equilibrium by x and y, $r = x + iy$ represents the vector $\mathbf{OO''}$ if we consider the

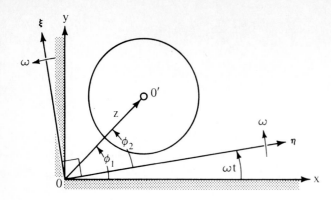

Figure 11.2 Stationary and moving coordinate systems.

plane (x, y) as a complex plane (x, iy). One advantage of equation (11.6) is that we have halved the number of coordinates and equations, but in turn, we deal with complex numbers.

From this point on, the real part of any resulting vector quantity (force, displacement, etc.) will be meant in the x direction and any imaginary part in the y direction. Readers should note the similarity with the method of complex vector representation of the harmonic motion presented in Chapter 1, but should remember that the physical interpretation is not exactly the same.

Furthermore, let us assume a coordinate system (η, ξ) that has the same origin O but rotates about O with angular velocity Ω. We introduce

$$\zeta = \eta + i\xi \tag{11.7}$$

The complex number ζ represents the same vector **OO'**, but in the rotating coordinate system. Between ζ and r, the following relations exist:

$$r = \zeta e^{i\Omega t}, \qquad \zeta = r e^{-i\Omega t} \tag{11.8}$$

because if R is the length of this vector, its polar form in the coordinate system (x, iy) will be

$$r = R e^{i\phi_1}, \qquad \zeta = R e^{i\phi_2} \tag{11.9}$$

Dividing through, we obtain

$$\frac{r}{\zeta} = e^{i(\phi_1 - \phi_2)} = e^{i\Omega t} \tag{11.10}$$

In terms of the moving coordinates, equation (11.6) can now be written as

$$m(\ddot{\zeta} + i\Omega\dot{\zeta} - \Omega^2\zeta)e^{i\Omega t} + k\zeta e^{i\Omega t} = me\Omega^2 e^{i\Omega t} \tag{11.11}$$

or, dividing through with $me^{i\Omega t}$,

$$\ddot{\zeta} + 2i\Omega\dot{\zeta} + (\omega_n^2 - \Omega^2)\zeta = e\Omega^2 \tag{11.12}$$

where $\omega_n = (k/m)^{1/2}$ is the natural frequency of the system. Equations (11.6) and (11.12) express the motion of the rotor on the stationary and the moving coordinate system, respectively.

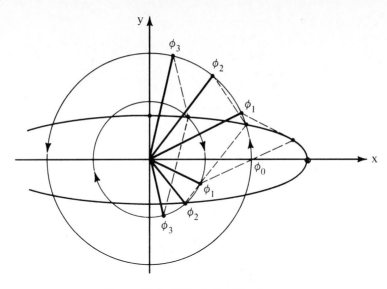

Figure 11.3 Elliptical orbit of rotor.

The solution of the homogeneous part of equation (11.6) is, as we know,

$$r(t) = Ae^{i\omega_n t} + Be^{-i\omega_n t} \tag{11.13}$$

where A and B are arbitrary complex constants that can be determined from the initial conditions. The term $Ae^{i\omega_n t}$ represents a counterclockwise circular orbit of radius A and angular velocity ω_n, and the term $Be^{-i\omega_n t}$, a clockwise orbit of radius B and angular velocity ω_n. The motion of the point O' will be the vector sum of the two orbits. This motion of the shaft superposed to the unbalance response is called *whirling*, the frequency $\omega_n/2\pi$ is called *whirl frequency*, the component $Ae^{i\omega_n t}$ is called *forward whirl* (in the direction of Ω) or *forward precession*, and the component $Be^{-i\omega_n t}$ is called *backward whirl* or *backward precession*.

We can plot the two precessions on the *x-y* plane, as in Figure 11.3.

The orbit of the shaft center is, in general, an ellipse, if $A \neq B \neq 0$. If $B = 0$, the orbit is a circle. If $A = B$, the orbit is a straight line. A and B are complex numbers, in general. One should not deduce that the shaft has forward and backward whirl at the same time. This is merely a calculation aid. The whirl is elliptical, in general, and the direction of whirling depends on A and B. If $|A| > |B|$, the direction is counterclockwise (forward). Otherwise, it is clockwise (backward). The reader might want to verify that by way of a graphical construction similar to Figure 11.3 with $|B| > |A|$.

☐ **Example 11.1**

A shaft of diameter $d = 50$ mm is carrying a rotor of 200 kg at one-fourth of its length of 1.6 m. Determine its natural frequency for lateral vibration assuming simple supports. The material is steel with $E = 2.1 \times 10^{11}$ Pa.

Solution The spring constant, from Appendix II, is

$$k = \frac{16^2 EI}{3l^3} = \frac{16^2 \times 2.1 \times 10^{11} \times \pi \times 0.050^4}{3 \times 64 \times 1.6^3} = 1.34 \times 10^6 \text{ N/m}$$

The natural frequency is

$$\omega_c = \omega_n = \left(\frac{k}{m}\right)^{1/2} = \left(\frac{1.34 \times 10^6}{200}\right)^{1/2} = 81.8 \text{ rad/s}$$

□

Unbalance and Critical Speeds

Suppose that the disk in Figure 11.1a has a small unbalanced mass m_e at a distance u from the center. From equation (11.6), the differential equation of motion will be[2]

$$\ddot{r} + \omega_n^2 r = \frac{m_e}{m} u\Omega^2 e^{i\Omega t} \tag{11.14}$$

Portrait of Föppl. (From Timoshenko, S. P. 1953. *History of the Strength of Materials.* New York: McGraw-Hill.)

AUGUST FÖPPL (1854–1924)

Born in Gross-Umstadt in the family of a doctor, Föppl studied structural engineering in Darmstadt, Stuttgart, and Karlsruhe, mostly under Mohr. For many years he taught at trade schools and did research on structures and later on Maxwell's theory of electricity. He succeeded Bauschinger in the chair of mechanics at Munich. His textbook *Lectures in Mechanics* had a lasting impact in the field. He did experimental work on fatigue and stress concentration. He was the first to give a satisfactory theory for shaft whirling and did extensive experimental work on the dynamics of the single-disk rotor, which he called the "de Laval rotor."

[2]Föppl, A., 1895, loc. cit.

The same result can be obtained from equation (11.6) observing that the eccentricity of the mass center due to the unbalance is $e = u(m_e/m)$. The steady-state solution is

$$r = Re^{i\Omega t}, \qquad R = \frac{m_e u}{m} \frac{(\Omega/\omega_n)^2}{1 - (\Omega/\omega_n)^2} \qquad (11.15)$$

We observe, as we expected, that for $\Omega = \omega_n$, the amplitude becomes infinite. This corresponds to an unwanted situation in rotating machinery, where for certain rotating speeds the whirl amplitude becomes large. These speeds have traditionally been called *critical speeds*. In the simple model we have considered here, there is only one critical speed, equal to the natural frequency ω_n. In most rotors, there are several significant critical speeds, indicated by the peaks of the vibration amplitude versus speed of rotation diagram (Figure 11.4). We will exemplify this point later.

In the case of unbalance, equation (11.15) indicates that the orbit will be a *circle* of radius R. This orbit is *synchronous*: in other words, moves with the angular velocity of the shaft and its direction. From equation (11.15) we see that for $\Omega < \omega_n$, called *subcritical operation*, the amplitude R is positive: in other words, is in phase with the unbalance. If $\omega_n < \Omega$, called *supercritical operation*, the amplitude R is negative, which means that the unbalance has a 180° phase angle from the steady-state whirl. If we observe a line on the disk, we shall see at sequential times the situation of Figure 11.5a and b, corresponding to synchronous orbit and preces-

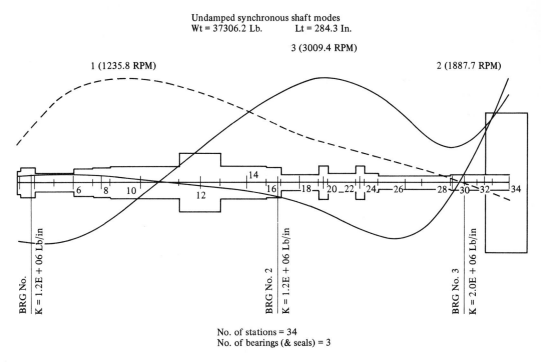

Figure 11.4 Critical speeds of a turbine rotor.

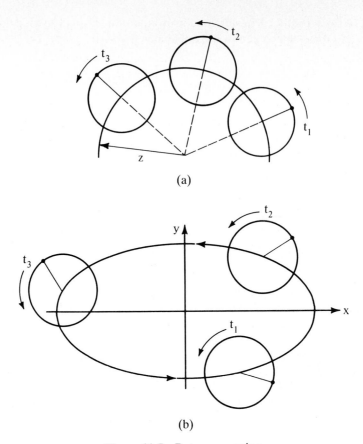

(a)

(b)

Figure 11.5 Rotor precession.

sion respectively, the latter due to superposition of the steady-state and the transient responses.

☐ **Example 11.2**

The shaft of Example 11.1 operates at 1000 rpm and has unbalance $m_e = 100$ g at radius $u = 200$ mm. Determine the vibration amplitude.

Solution From Example 11.1 the critical speed is $\omega_n = 81.8$ rad/s. Since $\Omega = 2\pi \times 1000/60 = 105$ rad/s, the response will be, from equation (11.15),

$$R = \frac{m_e u}{m} \frac{(\Omega/\omega_n)^2}{1 - (\Omega/\omega_n)^2}$$

$$= \frac{0.100 \times 0.200}{200} \frac{(105/81.8)^2}{1 - (105/81.8)^2} = -2.54 \times 10^{-4} \text{ m}$$

☐

Damping

In Chapter 3 we have introduced viscous damping forces proportional to the velocity. We mentioned earlier that this kind of damping is common when viscous flow is involved. We have seen that, in general, damping of this kind results in vibration suppression or gradual elimination. For this reason the name *damping* was originally used. We have called this *viscous damping*, to distinguish it from other kinds of energy dissipative forces that originate from deformation and its time rate of change, which we shall call *general damping*.

Viscous damping.   Considering first the case of viscous damping, we can find that the energy dissipated per cycle in a harmonic oscillator vibrating at frequency Ω and amplitude X, with damping constant c, is (for $P = c\dot{x}$, $x = X \cos \Omega t$)

$$D_v = \int_0^{2\pi/\Omega} P\dot{x}\, dt = \pi X^2 c\Omega \qquad (11.16)$$

We see that the energy dissipated per cycle is proportional to the frequency of the oscillation Ω. The damping constant c is unchanged with frequency.

Hysteretic damping.   In solid materials, especially metals, experiments show that if we impose a deformation at a certain rate, there is a phase angle between force and deformation. If we plot them on a Cartesian diagram, we shall see a loop as in Figure 3.14, called a *hysteresis loop*.

If the applied force (or stress) is of an oscillatory nature, the material will go through one loop every period of vibration. Since the loop area gives *force* × *displacement* = *work*, the loop area can be considered as energy D_h dissipated per cycle. Experiments show that for a wide range of frequencies the hysteretic energy per cycle is roughly independent of frequency and proportional to the square of amplitude and the stiffness. It has also been found that for every material there is a certain coefficient of proportionality γ such that

$$D_h = \pi X^2 k\gamma \qquad (11.17)$$

Working backward we can show that this is compatible with a force P in equation (11.16) which is inversely proportional to Ω, in fact,

$$P = c(\Omega)\dot{x} = \frac{\gamma}{\Omega} k\dot{x} = \frac{\gamma}{2\pi f} k\dot{x} \qquad (11.18)$$

where f is the frequency. Therefore, the hysteretic damping constant is $\gamma k/\Omega$, where γ is a material property.

For a nonrotating system, the equation of motion will be

$$m\ddot{x} + \frac{\gamma k}{\Omega} \dot{x} + kx = 0 \qquad (11.19)$$

where Ω signifies the frequency of harmonic oscillation. In a rotating shaft, the situation is more complicated. Assume, for example, that the vibration amplitude is $Xe^{i\Omega t}$, which means that the shaft is deflected by X, but this deflection rotates with the shaft. Therefore, the shaft does not change shape; the strain at any particular point is constant. Furthermore, an observer on the rotating coordinate system (η, ξ), will see a deflected shaft rotating with him or her, and not changing shape. Since there is no change of strain, there is no hysteresis loop and therefore there is no hysteretic damping force. This damping starts appearing when there is a change $\dot{\zeta}$ in respect to the observer on the rotating coordinate system. According to equation (11.18), it will be

$$P = \frac{\gamma}{2\pi f} k\dot{\zeta} \qquad (11.20)$$

The frequency f has to be equal to the number of reversals of the sign of the strain in a unit of time. Suppose that we have a forward whirl of frequency $f_n = \omega_n/2\pi$ and a rotating speed $f_\Omega = \Omega/2\pi$. Any one fiber of the shaft material will change the sign of strain (from tension to compression) $f_\Omega - f_n$ times per unit of time. Therefore,

$$f = f_n - f_\Omega \qquad (11.21)$$

and equation (11.20) gives

$$P = \frac{\gamma}{\omega_n - \Omega} k\dot{\zeta} \qquad (11.22)$$

Therefore, the equation of motion in the rotating coordinate system will be

$$m\ddot{\zeta} + \left(\frac{\gamma}{\omega_n - \Omega} k + 2i\Omega m\right)\dot{\zeta} + (k - m\Omega^2)\zeta = e\Omega^2 \qquad (11.23)$$

In the stationary coordinate system, the damping force rotates with the speed of the moving coordinate system; therefore, the force in (11.22) should be multiplied by $e^{-i\Omega t}$. Therefore,

$$-P = \frac{\gamma}{\Omega - \omega_n} k\left(-i\Omega r e^{i\Omega t} + \dot{r}e^{i\Omega t}\right)e^{-i\Omega t} = \frac{\gamma}{\omega_n - \Omega} k(-i\Omega r + \dot{r}) \qquad (11.24)$$

The equation of motion will be

$$m\ddot{r} + \frac{\gamma}{\omega_n - \Omega} k\dot{r} + k\left(1 - \frac{\gamma\Omega i}{\omega_n - \Omega}\right)r = e\Omega^2 e^{i\Omega t} \qquad (11.25)$$

If $\Omega = 0$ (no rotation), the equation reduces to

$$m\ddot{r} + \frac{\gamma}{\omega_n} k\dot{r} + kr = 0 \qquad (11.26)$$

Substituting $r = Re^{i\omega_n t}$, we obtain

$$[-m\omega_n^2 + k(1 + \gamma i)]R = 0 \qquad (11.27)$$

The same equation would have been obtained for an undamped system with a modified spring constant $k^* = k(1 + i\gamma)$. This is accomplished if instead of shear or Young's moduli we had used the complex moduli

$$G^* = G(1 + i\gamma) \qquad (11.28)$$

$$E^* = E(1 + i\gamma) \qquad (11.29)$$

provided that $\gamma \ll 1$, as often happens. Indeed, the spring constant will be, in general, an expression in the form $k = E/\lambda$, where the parameter λ includes the geometry of the spring. Substituting for E the complex value, we obtain,

$$k^* = E^*/\lambda = E(1 + i\gamma)/\lambda = k(1 + i\gamma) \tag{11.30}$$

The equation of motion will be

$$m\ddot{r} + k^*r = 0 \tag{11.31}$$

For $r = Re^{i\Omega t}$ we obtain

$$[-m\Omega^2 + k(1 + i\gamma)]R = 0 \tag{11.32}$$

which is identical with equation (11.27).

The complex modulus of elasticity is a very convenient way of taking hysteretic damping into account in free or forced vibrations, provided that *the response is harmonic and damping is due to material hysteresis*. The damping associated with the material of the shaft is sometimes called *internal damping*.[3]

☐ **Example 11.3**

Determine the vibration amplitude for the shaft of Example 11.2 if it operates at 800 rpm with an unbalance of 100 g at radius 200 mm, assuming that there is hysteretic damping in the shaft material with a material damping constant $\gamma = 0.01$.

Solution From Example 11.1, the critical speed is $\omega_n = 81.8$ rad/s and $\Omega = 800 \times 2\pi/60 = 84$ rad/s. The differential equation will be, from equation (11.26),

$$m\ddot{r} + \frac{\gamma}{\omega_n} k\dot{r} + kr = m_e u\Omega^2 \cos \Omega t$$

The unbalance response amplitude will be, for $r = Re^{i\Omega t}$, from Chapter 3,

$$R = \frac{\dfrac{m_e u}{m}\left(\dfrac{\Omega}{\omega_n}\right)^2}{\{[1 - (\Omega/\omega_n)^2]^2 + \gamma^2\}^{1/2}}$$

$$= \frac{\dfrac{0.100 \times 0.200}{200}\left(\dfrac{84}{81.8}\right)^2}{\left\{\left[1 - \left(\dfrac{84}{81.8}\right)^2\right]^2 + 0.05^2\right\}^{1/2}} = 0.0013 \text{ m.}$$

☐

Vector Methods

Another convenient way to deal with vibration about rotary motion is with vector methods. In the same stationary (x, y) and rotating (η, ξ) coordinate systems we assign unit vectors $(\mathbf{i}, \mathbf{j})$ and $(\mathbf{l}, \mathbf{m})$, respectively. Furthermore, we assume the unbalance in the form of a small eccentricity e of the mass of the disk. The situation

[3]Stodola, A. 1916, loc. cit., p. 197. Newkirk, B. L. 1924. Shaft whipping. *Gen. Electr. Rev.*, 27(3): 169–178. Kimball, A. T., Jr. 1924. Internal friction theory of shaft whirling, *Gen. Electr. Rev.*, 27(4): 244–251. Kimball, A. T. 1925. Internal friction as a cause of shaft whirling. *Philos. Mag.*, Ser. 6, 49: 724–727.

is shown in Figure 11.6, where the origin O is at the static equilibrium position of the center of the disk.

We have, in the two coordinate systems, if $t = 0$ when the mass center is on the x-axis,

$$\mathbf{r} = (e + \eta)\mathbf{l} + \xi\mathbf{m} \tag{11.33}$$

$$\mathbf{r} = (x + e \cos \Omega t)\mathbf{i} + (y + e \sin \Omega t)\mathbf{j} \tag{11.34}$$

where Ω is the angular velocity of the disk. The restoring forces are, for elastic force,

$$\mathbf{f}_s = -k(\eta\mathbf{l} + \xi\mathbf{m}) = -k(x\mathbf{i} + y\mathbf{j}) \tag{11.35}$$

for internal damping,

$$\begin{aligned} \mathbf{f}_i &= -c_i(\dot{\eta}\mathbf{l} + \dot{\xi}\mathbf{m}) \\ &= -c_i[(\dot{x} + \Omega y)\mathbf{i} + (\dot{y} - \Omega x)\mathbf{j}] \end{aligned} \tag{11.36}$$

and for viscous damping,

$$\begin{aligned} \mathbf{f}_v &= -c_v[(\dot{\eta} - \Omega\xi)\mathbf{l} + (\dot{\xi} + \Omega\eta)\mathbf{m}] \\ &= -c_v(\dot{x}\mathbf{i} + \dot{y}\mathbf{k}) \end{aligned} \tag{11.37}$$

The equations of motion will be

$$m\ddot{\mathbf{r}} = \mathbf{f}_s + \mathbf{f}_i + \mathbf{f}_v \tag{11.38}$$

Differentiating equations (11.33) and (11.34) and using equations (11.35) to (11.38) yields

$$\begin{aligned} m\ddot{x} + (c_i + c_v)\dot{x} + kx + c_i\dot{y} &= m\Omega^2 e \cos \Omega t \\ m\ddot{y} + (c_i + c_v)\dot{y} + ky - c_i\dot{x} &= m\Omega^2 e \sin \Omega t \end{aligned} \tag{11.39}$$

$$\begin{aligned} m\ddot{\eta} + (c_i + c_v)\dot{\eta} + (k - m\Omega^2)\eta - 2m\Omega\dot{\xi} - c_v\Omega\xi &= m\Omega^2 e \\ m\ddot{\xi} + (c_i + c_v)\dot{\xi} + (k - m\Omega^2)\xi + 2m\Omega\dot{\eta} + c_v\Omega\eta &= 0 \end{aligned} \tag{11.40}$$

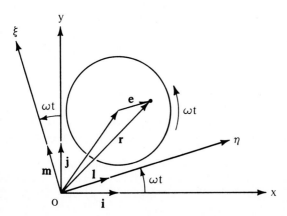

Figure 11.6 Vector representation of rotor motion.

Rotor Dynamics Chap. 11

Again, the equations for the lateral vibration are coupled and they may be dependent on the speed of rotation.

Acceleration through the Critical Speed

Often, the operating speed has to be higher than the rotor critical speed, especially in large machines such as steam and gas turbines, turbocompressors, and so on. In this case, it is obvious that the rotor has to pass through the critical speed when it accelerates from rest to reach the operating speed. It seems that when it passes through the critical speed, the amplitude will be the resonance amplitude, generally very high. This is not necessarily the case.[4] To illustrate this, we assume an undamped rotor with mass m and stiffness k with a mass eccentricity e which appears suddenly at $t = 0$. The vibration amplitude will be given by the general solution of the first of equations (11.39), neglecting damping,

$$x(t) = x_0 \cos \omega_n t + \frac{v_0}{\omega_n} \sin \omega_n t + \frac{me\Omega^2 \cos \Omega t}{(k - m\Omega^2)} \qquad (11.41)$$

At $t = 0$, the amplitude has to be zero; therefore,

$$x_0 = -\frac{me\Omega^2}{k - m\Omega^2}, \qquad v_0 = 0 \qquad (11.42)$$

and the response will be

$$x(t) = \frac{me\Omega^2}{k - m\Omega^2} (\cos \Omega t - \cos \omega_n t) \qquad (11.43)$$

At resonance, $k = m\Omega^2$ and $\Omega = \omega_n$ and the vibration response has the form 0/0. Application of L'Hospital's rule to the fraction with respect to Ω yields, for $\Omega \to \omega_n$,

$$x(t) = \frac{e\Omega}{2} t \sin \Omega t \qquad (11.44)$$

We observe that the response increases with time beyond bound, but it takes time for the amplitude to increase. Therefore, if the passage through the critical speed is rapid, the amplitude will not have time to reach very high values. The maximum amplitude will depend on the rate of acceleration of the rotor. This is shown in Figure 11.7 for values of increasing angular acceleration $\dot{\Omega} = \alpha = 0, \alpha_1, \alpha_2, \alpha_3, \ldots$.

☐ **Example 11.4   Acceleration of Rotor through Critical Speed**

The shaft of Example 11.2 is quickly accelerated to the critical speed and then run at this speed. The shaft will fail when the amplitude reaches 12 mm, due to interference with the casing. Determine the time of failure.

---

[4]G. de Laval proved that experimentally in 1889. The analytical solution is due to Föppl, O. 1918. *Z. Gesamte Turbinenwes.*

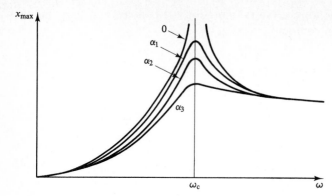

Figure 11.7 Acceleration of a rotor through the critical speed.

Solution From Example 11.1 the critical speed is $\omega_n = 81.8$ rad/s. The amplitude is given by equation (11.44):

$$X = \frac{e\Omega}{2} t$$

The equivalent eccentricity of the mass is

$$e_{eq} = \frac{um_e}{m} = \frac{0.200 \times 0.100}{200} = 0.0001 \text{ m.}$$

Solving for t gives us

$$t = \frac{2X}{e\omega_n} = \frac{2 \times 0.012}{0.0001 \times 81.8} = 2.93 \text{ s}$$

$\square$

Gyroscopic Effects on the Critical Speeds

If the rotor has a substantial moment of inertia about its diameter and a general rigid-body motion, changing the orientation of its axis requires application of considerable moments at high speeds. This is known to change the critical speeds of high-speed rotors, rendering them functions of the rotating speed, and to generate new ones.[5]

It is known from dynamics that the position of the disk can be defined by way of the coordinates of its mass center, plus the three angles ϕ, χ, and ψ, called *Euler angles* (Figure 11.8). The reader can find the resulting equations of motion in any standard textbook on advanced dynamics or mechanics. From a vibration engineering standpoint, the Euler angle description of the motion is not a very convenient one because these angles are difficult to measure or interpret. Instead, to conform with practices from statics, we shall describe the orientation of the disk by way of the vertical and horizontal slopes of its axis of rotation, θ and ϕ: in other words, the angles θ and ϕ of the axis of rotation of the disk with respect to the planes (y, z) and (x, z), respectively (Figure 11.9).

[5]Stodola, A. 1918. Neue kritische Drehzahlen als Folge der Kreiselwirkung der Läufräder. *Z. Gesamte Turbinennwes. 15*: 269–275.

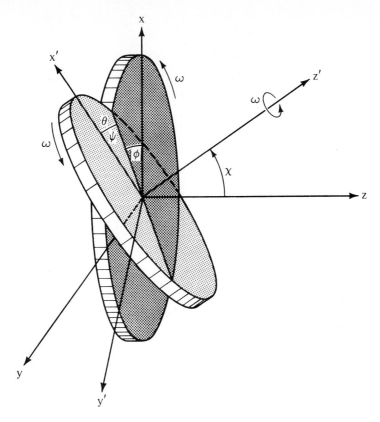

Figure 11.8 Euler angles.

It is known from kinematics that finite rotations are not commutative; that is, they do not obey the vector law of addition. However, infinitesimal rotations are commutative and can be added as vectors. Since we consider only small displacements and rotations, we can consider the angles θ and ϕ as infinitesimal. A consequence of this is that the order of application of the rotations θ and ϕ has no effect on the final position. Thus we consider a coordinate system (x'', y'', z'') affixed to the disk but not rotating with it and oriented along the principal axis of inertia of the disk. We also consider an inertial coordinate system (X, Y, Z). Let the two systems initially coincide and rotate the system (x'', y'', z'') about the z'' axis by an angle ϕ to a new position (x', y', z'). Then we rotate it by an angle θ about the axis y' to the final position (x, y, z). If the angles θ and ϕ change with time, the disk will have, in addition to its angular velocity of rotation Ω about the z axis, angular velocities $\dot\theta$ and $\dot\phi$ about the axes Y and Z, respectively.

We select the coordinate system (x, y, z) to coincide with the principal axes of inertia of the disk but not to rotate with it about the z axis.[6] Thus the disk rotates

[6]For asymmetric disks, the coordinate system (x, y, z) does not always coincide with the principal axes of inertia of the disk and the problem needs special treatment. A disk or rotor that is symmetric with respect to the z-axis is assumed in the sequel.

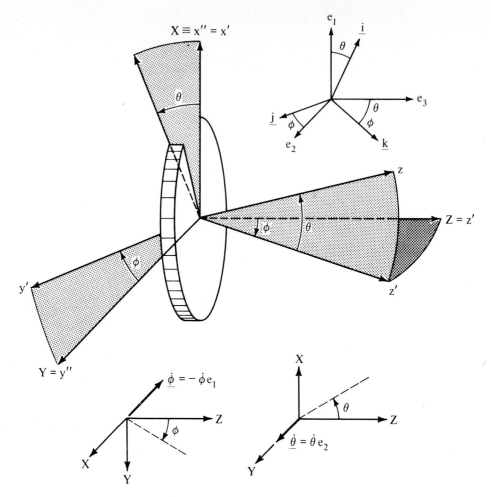

Figure 11.9 Coordinates for rotor motion.

with respect to this coordinate system with angular velocity $\boldsymbol{\omega}$ about the z-axis, which is called *spin*. Then Newton's second law yields[7]

$$\mathbf{M}_G = (\dot{\mathbf{H}}_G)_{x,y,z} + \boldsymbol{\omega} \times \mathbf{H}_G \qquad (11.45)$$

where $\mathbf{M}_G$ is the sum of the moment vectors about the mass center G, $\mathbf{H}_G$ the momentum vector with respect to a coordinate system (X', Y', Z') through G of fixed orientation and always parallel to (X, Y, Z), and $(\dot{\mathbf{H}}_G)_{x,y,z}$ is the rate of change of the angular momentum vector with respect to the moving coordinate system (x, y, z). If J_x, J_y, and J_z are the principal moments of inertia of the disk, $\mathbf{J}$ is the diagonal inertia tensor $= \text{diag}[J_x \quad J_y \quad J_z]$, $\mathbf{H}_G = \mathbf{J}\boldsymbol{\Omega}$, $(\dot{\mathbf{H}}_G)_{x,y,z} = \mathbf{J}\dot{\boldsymbol{\omega}}$, $\boldsymbol{\omega}$ is the disk

---

[7]For an extensive treatment, see Beer, F. P., and Johnston, E. R. 1984. *Vector Mechanics for Engineers*, 4th ed. New York: McGraw-Hill, Chapter 4.

instant angular velocity, and $\boldsymbol{\Omega}$ is the constant angular velocity of the disk vector, we obtain the Euler equations

$$\Sigma M_x = J_x \dot{\omega}_x - J_y \Omega_y \omega_z + J_z \Omega_z \omega_y \tag{11.46}$$

$$\Sigma M_y = J_y \dot{\omega}_y - J_z \Omega_z \omega_x + J_x \Omega_x \omega_z \tag{11.47}$$

$$\Sigma M_z = J_z \dot{\omega}_z - J_x \Omega_x \omega_y + J_y \Omega_y \omega_x \tag{11.48}$$

Let $\mathbf{e}_1$, $\mathbf{e}_2$, and $\mathbf{e}_3$ be the unit vectors of the coordinate system (X, Y, Z) and $\mathbf{i}$, $\mathbf{j}$, and $\mathbf{k}$ be the unit vectors of (x, y, z). We set up the instant angular velocity of the nonrotating disk plane with angular velocities $\dot{\phi}$ about the X-axis and $\dot{\theta}$ about the Y-axis, while the constant angular velocity of rotation has component Ω about the z-axis:

$$\boldsymbol{\omega} = -\dot{\phi}\mathbf{e}_1 + \dot{\theta}\mathbf{e}_2, \qquad \boldsymbol{\Omega} = \Omega\mathbf{k} \tag{11.49}$$

We note that from the geometry of Figure 11.9,

$$\mathbf{e}_1 = \mathbf{i}\cos\theta + \mathbf{k}\sin\theta \tag{11.50}$$
$$\mathbf{e}_2 = \mathbf{i}\sin\theta\sin\phi + \mathbf{j}\cos\phi - \mathbf{k}\cos\theta\sin\phi$$

Thus $\boldsymbol{\omega} = -(\dot{\phi}\cos\theta)\mathbf{i} + \dot{\theta}(\cos\phi + \sin\theta\sin\phi)\mathbf{j} - (\dot{\phi}\sin\theta + \dot{\theta}\cos\theta\sin\phi)\mathbf{k}$. We assume that $\Omega = $ const, which means that the shaft is infinitely stiff in torsion and attached to a prime mover with infinite inertia. Moreover, in tune with the general assumptions, we assume small angles θ and ϕ, so that $\sin\theta \approx \theta$, $\cos\theta \approx 1$, and we neglect products of θ and ϕ and their time derivatives to obtain

$$\boldsymbol{\omega} = -\dot{\phi}\mathbf{i} + \dot{\theta}\mathbf{j}, \qquad \omega_x = -\dot{\phi}, \qquad \omega_y = \dot{\theta}, \qquad \omega_z = 0 \tag{11.51}$$

$$\boldsymbol{\Omega} = \Omega\mathbf{k}, \qquad \Omega_x = 0, \qquad \Omega_y = 0, \qquad \Omega_z = \Omega \tag{11.52}$$

Furthermore, for a circular disk we have polar moment of inertia $J_z = J_p$ and moment of inertia about a diameter $J_x = J_y = J$. Therefore,

$$\Sigma M_x = -J\ddot{\phi} + J_p\Omega\dot{\theta} \tag{11.53}$$

$$\Sigma M_y = J\ddot{\theta} + J_p\Omega\dot{\phi} \tag{11.54}$$

If the mass center of the disk has, in addition, deflections x and y, considering no motion along the z axis, we have

$$m\ddot{x} + k_{xx}x - k_{x\theta}\theta = 0 \tag{11.55}$$

$$m\ddot{y} + k_{yy}y + k_{y\phi}\phi = 0 \tag{11.56}$$

$$J\ddot{\theta} + J_p\Omega\dot{\phi} + k_{\theta\theta}\theta - k_{\theta x}x = 0 \tag{11.57}$$

$$J\ddot{\phi} - J_p\Omega\dot{\theta} + k_{\phi\phi}\phi + k_{\phi y}y = 0 \tag{11.58}$$

where k_{ij}, $i, j = x, y, \theta, \phi$, are stiffness influence coefficients.

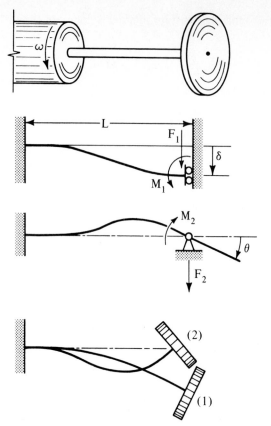

Figure 11.10 Gyroscopic effects on an overhang rotor.

To demonstrate the implications of the system of equations (11.55) to (11.58), we consider an overhang rotating shaft (Figure 11.10), which carries at the end a thin disk of mass m and radius r.

From strength of materials we know that unit deflection at the free end of a cantilever $\delta = 1$ generates at the clamped end force and moment

$$F_1 = \frac{12EI}{L^3}, \qquad M_1 = \frac{6EI}{L^2} \tag{11.59}$$

Similarly, unit rotation at the free end $\theta = 1$ rad generates force and moment

$$F_2 = \frac{6EI}{L^2}, \qquad M_2 = \frac{4EI}{L} \tag{11.60}$$

where the flexural rigidity of the shaft is EI. These forces and moments are, by definition, the stiffness influence coefficients (the reciprocity should be noted)

$$k_{xx} = F_1, \qquad k_{\theta x} = M_1, \qquad k_{x\theta} = F_2, \qquad k_{\theta\theta} = M_2 \tag{11.61}$$

Therefore, the equations of motion become for a symmetric shaft

$$m\ddot{x} + 6\lambda x - 3\lambda L\theta = 0 \tag{11.62}$$

$$m\ddot{y} + 6\lambda y + 3\lambda L\phi = 0 \tag{11.63}$$

$$J\ddot{\theta} + J_p\Omega\dot{\phi} - 3\lambda Lx + 2\lambda L^2\theta = 0 \qquad (11.64)$$

$$J\ddot{\phi} - J_p\Omega\dot{\theta} + 3\lambda Ly + 2\lambda L^2\phi = 0 \qquad (11.65)$$

where $\lambda = 2EI/L^3 = 2k/3$, $k = 3EI/L^3$, the cantilever beam spring constant for end loading. The frequency equation is, for $x = x_0 e^{ipt}$, and so on,

$$\begin{vmatrix} -mp^2 + 6\lambda & 0 & -3\lambda L & 0 \\ 0 & -mp^2 + 6\lambda & 0 & 3\lambda L \\ -3\lambda L & 0 & -Jp^2 + 2\lambda L^2 & +J_p\Omega ip \\ 0 & 3\lambda L & -J_p\Omega ip & -Jp^2 + 2\lambda L^2 \end{vmatrix} = 0 \qquad (11.66)$$

The solutions of this equation are the system natural frequencies at the running speed Ω. When the running speed is increasing, it might at some point coincide with one of the natural frequencies. This speed is called the *critical speed*. If we plot the frequencies derived from equation (11.66) versus the running speed Ω and also the straight line $p = \Omega$, the points at which they coincide give the critical speeds (Figure 11.11). To compute them, we introduce into equation (11.66) $p = \Omega = \omega_c$ and we obtain

$$\begin{vmatrix} -m\omega_c^2 + 6\lambda & 0 & -3\lambda L & 0 \\ 0 & -m\omega_c^2 + 6\lambda & 0 & 3\lambda L \\ -3\lambda L & 0 & -J\omega_c^2 + 2\lambda L^2 & +iJ_p\omega_c^2 \\ 0 & 3\lambda L & -iJ_p\omega_c^2 & -J\omega_c^2 + 2\lambda L^2 \end{vmatrix} = 0 \qquad (11.67)$$

The roots ω_c of this equation are the critical speeds.

For an isotropic shaft, the frequency equation can be simplified considerably. Introducing complex notation,

$$r = x + iy, \qquad -\psi = \phi + i\theta \qquad (11.68)$$

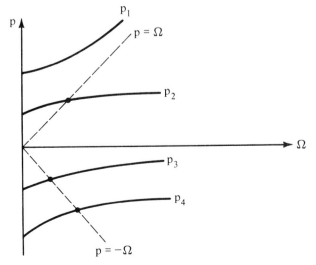

Figure 11.11 Campbell diagram for critical speeds.

equations 11.62 to 11.65 become

$$m\ddot{r} \quad + \quad 6\lambda r \quad - \quad i3\lambda L\psi = 0 \tag{11.69}$$

$$J\ddot{\psi} - iJ_p\Omega\dot{\psi} + 3i\lambda Lr + 2\lambda L^2\psi = 0 \tag{11.70}$$

The frequency equation is then, for $r = r_0 e^{i\omega_n t}$,

$$mJ\omega_c^4 - mJ_p\Omega\omega_c^3 - (2\lambda L^2 m - 6\lambda J)\omega_c^2 + 6\lambda J_p\Omega\omega_c + 3\lambda^2 L^2 = 0 \tag{11.71}$$

The critical speeds can be computed from the condition $\omega_c = \Omega_n$ which yields

$$m(J_p - J)\omega_c^4 + 2\lambda[L^2 m - 3(J_p - J)]\omega_c^2 - 3\lambda^2 L^2 = 0 \tag{11.72}$$

Two more critical speeds can be computed from the condition $\Omega = -\omega_c$, which yields

$$m(J_p + J)\omega_c^4 - 2\lambda[L^2 m - 3(J_p + J)]\omega_c^2 + 3\lambda^2 L^2 = 0 \tag{11.73}$$

The four roots of these two biquadratic equations are the four critical speeds.

A more natural way of searching for critical speeds is by observing the resonance of the forced response. To this end we consider a disk having eccentricity e: in other words, the mass is located at radial distance e from the axis of rotation. Assuming also a thin disk, we have $J_p = 2J = 1/2mr^2$ and the equations of motion are

$$m\ddot{x} + 6\lambda x + 3\lambda L\theta = me\omega^2 \cos \omega t$$

$$m\ddot{y} + 6\lambda y + 3\lambda L\phi = me\omega^2 \sin \omega t$$

$$\frac{1}{4} mr^2\ddot{\theta} + \frac{1}{2} mr^2\Omega\dot{\phi} + 3\lambda Lx + 2\lambda L^2\theta = 0 \tag{11.74}$$

$$\frac{1}{4} mr^2\ddot{\phi} - \frac{1}{2} mr^2\Omega\dot{\theta} + 3\lambda Ly + 2\lambda L^2\phi = 0$$

The steady-state solution is,

$$x = G(\omega)me\omega^2(3\lambda L^2 + 2J\omega^2) \cos \omega t$$

$$y = G(\omega)me\omega^2(3\lambda L^2 + 2J\omega^2) \sin \omega t$$

$$\theta = -3\lambda LG(\omega)me\omega^2 \cos \omega t \tag{11.75}$$

$$\phi = -3\lambda LG(\omega)me\omega^2 \sin \omega t$$

where,

$$G(\omega) \simeq \frac{1}{(6\lambda - m\omega^2)(2\lambda L^2 + J\omega^2) - 9(\lambda L)^2} \tag{11.76}$$

If the denominator of $G(\omega)$ becomes zero, the amplitudes tend to infinity and we have resonance. Therefore, the critical speeds are roots of the equation

$$(6\lambda - m\omega^2)(2\lambda L^2 + J\omega^2) - 9(\lambda L)^2 = 0 \tag{11.77}$$

the solution of which is

$$\omega_c^2 = \frac{\omega_n^2[1 - \mu^2 \pm [(1 - \mu^2)^2 - 4v^2 + 4\mu^2]^{1/2}}{2} \tag{11.78}$$

where $\omega_n = 6\lambda/m = 12EI/mL^3$, the natural frequency for a disk at the end of the cantilever shaft which remains parallel to itself, and

$$\mu^2 = \frac{2\lambda L^2/J}{6\lambda/m} = \frac{mL^2}{3J} = \frac{m}{J + J_p}\frac{k_{\theta\theta}}{k_{xx}}$$

$$v = \frac{3\lambda L/(mJ)^{1/2}}{6\lambda/m} = \frac{L}{2(m/J)^{1/2}} = \frac{m}{J + J_p}\frac{k_{x\theta}^2}{k_{xx}^3}$$

But for a thin disk of mass m and radius R, $J = mR^2/4$. Therefore,

$$\mu^2 = \frac{4}{3}\xi^2, \qquad v = \frac{L}{R}$$

$$\omega_c^2 = \frac{\omega_n^2[1 - \frac{4}{3}\xi^2 + [(1 - \frac{4}{3}\xi^2)^2 + \frac{4}{3}\xi^2]^{1/2}}{2} = \frac{\omega_n^2}{f(\xi)} \tag{11.79}$$

If we plot $f(\xi)$, we observe that $f(\xi)$ approaches 1 for $\xi \to 0$ and that the critical speed is $\omega_c^2 \to 12EI/mL^3$ or $\omega_c = 2\omega_n$ (Figure 11.12), the natural frequency of the mass m at the end of the cantilever shaft if the disk always remains in the vertical plane. For $\xi \to \infty$, $R \to 0$, $f(\xi) \to \frac{1}{4}$, and $\omega_c^2 \to 3EI/mL^3 = \omega_n$ and there is no gyroscopic effect, as expected. It can be concluded that for the cantilever shaft, the gyroscopic effects can increase the critical speed of the overhang rotor by 100% at sufficiently high disk inertia.

Here we have obtained only one critical speed, due to the approximations involved in deriving equation (11.76). If we use the frequency equation (11.67) instead and then plot the roots p (Figure 11.11), we observe higher critical speeds. The intersection of $p = \Omega$ with the curves p_1 and p_2 gives the two critical speeds at which the whirl has the same direction (and angular velocity) with the rotation of the shaft. It corresponds to *forward precession*. We observe that the line $p =$

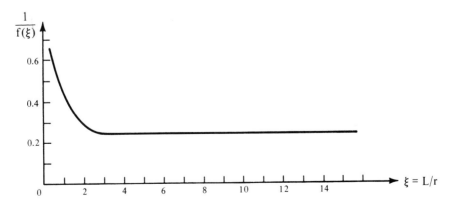

Figure 11.12 Gyroscopic effects function.

$-\Omega$ intersects the two negative roots p_3 and p_4. This indicates that critical speed is possible with whirling angular velocity and opposite in direction to the shaft's rotation. This corresponds to *backward precession*. In Figure 11.11, called a *Campbell diagram*, the abscissa is in rpm and the mantissa in Hz.

If at these critical speeds we solve for the vibration modes, we shall observe modes as in Figure 11.10. The first (and fundamental) critical speed of forward precession is of type (1), and it is of the most dangerous. The second critical speed of forward precession is of type (2) and it is very high. Generally, type (2) modes are not as dangerous because they result in high strain in the shaft and thus high structural damping. Similarly, the critical speeds of backward precession are the first very low and at mode (1), and the second higher, at mode (2). In general, the relative magnitude of the critical speeds is

$$(\omega_{c1})_{bw} < (\omega_{c1})_{fw} < (\omega_{c2})_{bw} < (\omega_{c2})_{fw} \qquad (11.80)$$

□ **Example 11.5 Gyroscopic Effects on the Critical Speed of a Simply Supported de Laval Rotor**

A disk of 0.4 m diameter, mass $m = 200$ kg, and mass polar moment of inertia $J = 4$ kg·m² is carried by a slender steel shaft of diameter $d = 80$ mm at one-third of its length $L = 1.50$ m and $E = 2.11 \times 10^{11}$ N/m² (Figure E11.5). Calculate the critical speed of the shaft.

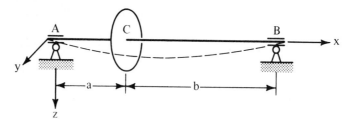

Figure E11.5

Solution The influence coefficients are

$$\alpha = k_{xx} = \frac{3EIl(a^2 - ab + b^2)}{a^3 b^3}, \qquad \gamma = k_{x\theta} = \frac{3EIl(\alpha - b)}{a^2 b^2},$$

$$\delta = k_{\theta\theta} = \frac{3EIl}{ab}$$

$$I = \frac{\pi d^4}{64} = \frac{\pi \times 0.08^4}{64} = 2.011 \times 10^{-6} \text{ m}^2$$

$$\alpha = 3 \times 2.1 \times 10^{11} \times 2.011 \times 10^{-6} \times \frac{1.5(1^2 - 0.5 + 0.5^2)}{1^3 \times 0.5^3}$$

$$= 11.68 \times 10^6 \text{ N/m}$$

Equation (11.21) gives

$$\omega_c^2 = \frac{\alpha}{m} \frac{1 - \mu^2 \pm [(1 - \mu^2)2 - 4\nu^2 + 4\mu^2]^{1/2}}{2}$$

with

$$\mu^2 = \frac{a^2 b^2 m}{(a^2 - ab + b^2)J} = \frac{4a^2 b^2}{(a^2 - ab + b^2)r^2}$$

$$\nu = \frac{(\alpha - b)mab}{(a^2 - ab + b^2)(mJ)^{1/2}}$$

$$= \frac{2(a - b)(ab)}{(a^2 - ab + b^2)r}$$

Therefore,

$$\mu^2 = \left(\frac{b}{r}\right)^2 \frac{4}{1 - b/a + (b/a)^2} = \frac{4}{7}\xi^2, \qquad \xi = \frac{b}{r}$$

$$b = \frac{L}{3} = \frac{1.5}{3} = 0.5, \qquad \xi = \frac{b}{r} = \frac{0.5}{0.2} = 0.25, \qquad \mu = [\tfrac{4}{7}(0.25^2)]^{1/2} = 0.189$$

$$\nu = \frac{2(1 - b/a)}{1 - (b/a) + (b/a)^2}\frac{b}{r} = -\tfrac{4}{7}\xi = -\tfrac{4}{7}(0.25) = -0.143$$

The critical speed is

$$\omega_c^2 = \frac{\alpha}{m}\frac{\{1 - \tfrac{4}{7}\xi^2 + [1 - \tfrac{4}{7}\xi^2)^2 + 0.978\xi^2]^{1/2}\}}{2}$$

$$= (11.68 \times 10^6/200)\{0.964^2 + 0.964^4 + (0.978 \times 0.25^2)^{1/2}\}/2$$

$$\omega_c = 291 \text{ rad/s}$$

$$\omega_n = \left(\frac{\alpha}{m}\right)^{1/2} = \left(\frac{11.68 \times 10^6}{200}\right)^{1/2} = 241 \text{ rad/s}$$

The difference in the critical speed from the natural frequency is substantial (20.7% higher).

(The critical speeds of a de Laval rotor on rigid bearings have been computed, including gyroscopic effects.) □

11.2 ROTORS ON FLEXIBLE BEARINGS

Bearing Properties

In almost any machine, there are elements that are used as means of transferring forces between other members having relative motion. An interesting case in vibration is journal bearings. Some types of journal bearings are shown in Figure 11.13.

Due to rotation, fluid pressure is developed between journal and bearing which has resultant equal to the static load of the bearing. Deviations of the journal from its equilibrium position result in additional forces that, in general, oppose this deviation. The oil film has a spring action.

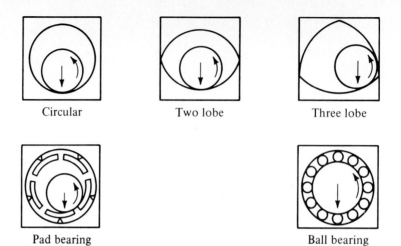

Figure 11.13 Different bearing types.

To quantify this, we consider vertical and horizontal components of these forces:

$$F_x = F_x(x, \dot{x}, y, \dot{y}) \tag{11.81}$$

$$F_y = F_y(x, \dot{x}, y, \dot{y}) \tag{11.82}$$

The functions F_x and F_y depend on the bearing design, the static force, the angular velocity Ω, and so on. Expanding them in a Taylor series, we obtain[8]

$$-F_x = -F_{x0} + K_{xx}\Delta x + K_{xy}\Delta y + C_{xx}\Delta \dot{x} + C_{xy}\Delta \dot{y}$$
$$+ o(\Delta x^2, \Delta y^2, \Delta \dot{x}^2, \Delta \dot{y}^2) \tag{11.83}$$

$$-F_y = -F_{y0} + k_{yy}\Delta y + k_{yx}\Delta x + C_{yy}\Delta \dot{y} + C_{yx}\Delta \dot{x}$$
$$+ o(\Delta x^2, \Delta \dot{y}^2, \Delta x^2, \Delta \dot{y}^2) \tag{11.84}$$

where the K's and C's are called *spring* and *damping coefficients*, respectively, and are functions of Ω, static load, and bearing design but independent of the perturbations Δx, Δy, $\Delta \dot{x}$ and $\Delta \dot{y}$. As a first approximation we neglect the higher-other terms in equations (11.83) and (11.84), and thus the bearing can be represented as in Figure 11.14. The terms K_{xy}, K_{yx}, C_{xy}, and C_{yx}, which relate forces to the displacements perpendicular to them, are called *cross-coupling terms*. These terms couple the vertical and horizontal vibrations of the shaft. Their dependence on the

[8]Stodola, A. 1927. *Steam and Gas Turbines*. New York: McGraw-Hill. Hagg, A. C., and Sankey, G. O. 1958. Oil film properties for unbalance vibration calculations. *Trans. ASME J. Appl. Mech*, 25: 141–143. Sternlicht, B. 1959. Elastic and damping properties of cylindrical journal bearings. *Trans. ASME*, Ser. D, *81*: 101–108.

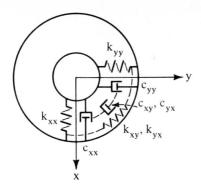

Figure 11.14 Linear model for a slider bearing.

rotating speed Ω makes the lateral vibration of the system dependent on the steady-state motion.

The bearing spring and damping coefficients are usually given in dimensionless form as $\beta = cK/W$ and $\gamma = c\Omega C/W$ respectively, as functions of the *Sommerfeld* and *Reynolds numbers*, defined as

$$S = \frac{\eta NLD}{W}\left(\frac{R}{c}\right)^2 \tag{11.85}$$

$$Re = \frac{\Omega R^2}{\nu} \tag{11.86}$$

where

η = viscosity of the lubricant $(N \cdot s/m^2)$
ν = kinematic viscosity = η/ρ (m^2/s)
N = revolutions per second $\Omega/2\pi$ (s^{-1})
W = bearing static load (N)
L = bearing length (m)
R = bearing radius (m)
D = bearing diameter (m)
c = bearing radial clearance (m)
ρ = lubricant density (kg/m^3)

The radial clearance is the difference between bearing radius and journal radius.

The Sommerfeld number is defined by many authors as $So = W/\eta\Omega LD(c/R)^2$. We shall call this quantity the *Sommerfeld parameter* or *dimensionless load*. Obviously, $So = 1/2\pi S$.

In terms of the Sommerfeld number, Figures 11.15 and 11.16 give the dimensionless linear stiffness and damping coefficients β and γ, respectively, for various types of bearings.[9]

[9]An extensive compilation of such properties for bearings can be found in Lund, J. W., et al. 1965. *Design Handbook for Fluid Film Type Bearings, Rotor-Bearing Dynamics Design Technology*, Vol. III. Mechanical Technology, Inc. Technical Report AFAPL-TR-65-45 to Wright-Paterson Air Force Base, Ohio.

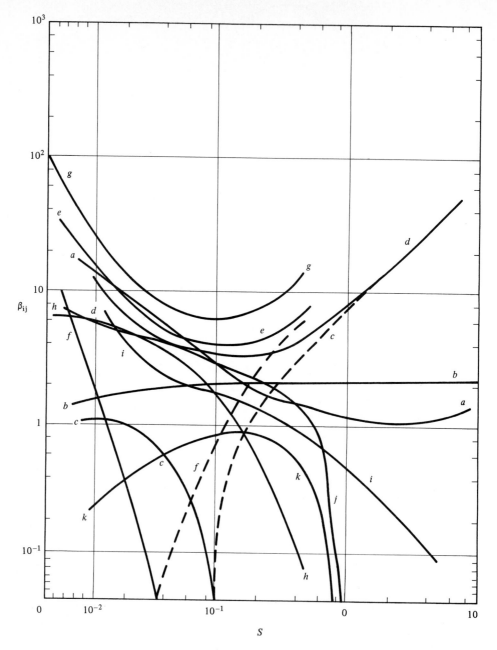

Full Circular Bearing, L/D = 1: $a = \beta_{xx}$, $b = \beta_{yy}$, $c = \beta_{yx}$, $d = \beta_{xy}$. *Elliptical Bearing*, L/D = 1 (clearances, $c_x = 0.5c$, $c_y = c$): $e = \beta_{xy}$, $f = \beta_{yx}$, $g = \beta_{xx}$, $h = \beta_{yy}$. *4-Pad Bearing*, L/D = 0.75, Load on pad: $i = \beta_{xx} = \beta_{yy}$. *6-Pad Bearing*, L/D = 0.5, Load on pad: $j = \beta_{xx}$, $k = \beta_{yy}$. Dashed lines designate negative values.

Figure 11.15 Spring constants for slider bearings.

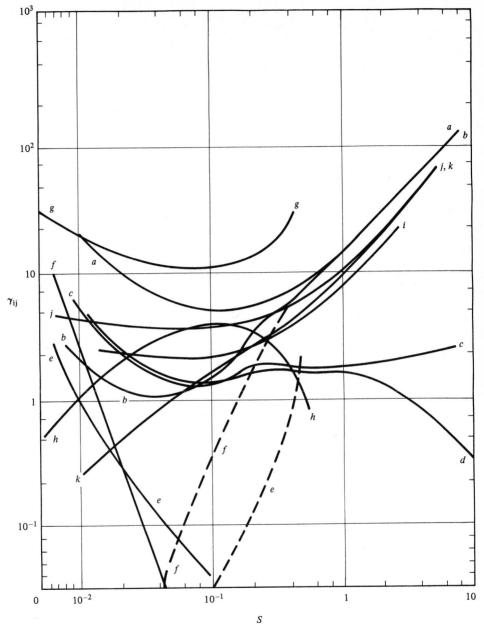

Full Circular Bearing, L/D = 1: $a = \gamma_{xx}$, $b = \gamma_{yy}$, $c = \gamma_{yx}$, $d = \gamma_{xy}$. *Elliptical Bearing*, L/D = 1 (clearances, $c_x = 0.5c$, $c_y = c$): $e = \gamma_{xy}$, $f = \gamma_{yx}$, $g = \gamma_{xx}$, $h = \gamma_{yy}$. *4-Pad Bearing*, L/D = 0.75, Load on pad: $i = \gamma_{xx} = \gamma_{yy}$. *6-Pad Bearing*, L/D = 0.5, Load on pad: $j = \gamma_{xx}$, $k = \gamma_{yy}$. Dashed lines designate negative values.

Figure 11.16 Damping constants for slider bearings.

□ Example 11.6 Critical Speed of Rigid Rotor in Fluid Bearings

A drive turbine rotor weighs $W = 10,000$ N and the shaft is of the welded drum type with very high stiffness and can be considered rigid. The bearings have $L/D = 0.75$, $c/R = 0.0015$, $D = 60$ mm, and oil viscosity $\eta = 0.020$ Pa·s. Find the critical speed.

Solution The mass of the rotor is $m = 10,000/g = 1019$ kg. The bearing length is $L = 0.8D = 48$ mm and the clearance $c = 0.0015 \times (0.060/2) = 45 \times 10^{-6}$ m.

We start with the assumption that $\Omega = 1000$ rad/s, $W = 10,000/2$ and

$$S = \eta \Omega L D (R/c)^2 / 2\pi W = 0.020 \times 1000 \times 0.045$$

$$\times 0.060 \ (1/0.0015)^2 / 2\pi (10,000/2) = 0.764$$

Entering Figure 11.15, curve i, with $S = 0.764$ we obtain $k_{xx}(c/W) = 0.65$. Therefore

$$k_{xx} = 0.65 \times \frac{(10,000/2)}{45 \times 10^{-6}} = 72.2 \times 10^6 \text{ N/m}$$

The critical speed is then

$$\Omega_c = \left(\frac{k}{m}\right)^{1/2} = \left[\frac{72.2 \times 106}{(1019/2)}\right]^{1/2} = 376 \text{ rad/s}$$

We repeat the computation with $\Omega = 500$ rad/s. $S = 0.764 \times 500/1000 = 0.382$. Entering Figure 11.15 (i) with $S = 0.382$ we obtain $k_{xx}(c/W) = 1$. Therefore $k_{xx} = 1 \times (10,000/2)/45 \times 10^{-6} = 111 \times 10^6$ N/m. The critical speed is then

$$\Omega_c = \left(\frac{k}{m}\right)^{1/2} = [111 \times 106/(1019/2)]^{1/2} = 466 \text{ rad/s},$$

close to the assumed value of 500 rad/s. The iteration can be repeated for more accuracy.

(The critical speed of a rigid rotor in fluid bearings was computed by iteration.)

□

Fluid or Magnetic Forces on the Shaft

If the shaft works in a fluid or magnetic environment, force can be exerted by the magnetic field[10] (Figure 11.17). The radial force changes with the gap, and the sum of the two pole radial forces is a nonlinear force as a function of the radial displacement from the concentric position. There is a tangential component, due to power generation or dissipation, which is a similar function of the radial displacement.

A similar situation occurs in a power stage on a turbine or turbocompressor rotor (Figure 11.18). Part of the fluid flows past the blades as loss. A radial displacement reduces the flow past the blade on the side of the smaller gap and increases the flow past the blade on the side of the larger gap. The tangential force produced in the blade is thus greater at the small-gap side, and the net result is a

---

[10]Freise, W., and Jordan, H. 1962. *Einseitige magnetische Zugkräfte in Drehstrommmaschinen.* ETZ-A83, pp. 299–303.

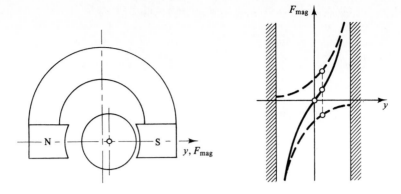

Figure 11.17 Electromagnetic effects on rotors (From E. Krämer, 1984, Maschinen-dynamik. By permission of Springer-Verlag.)

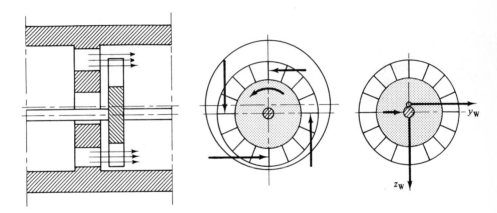

Figure 11.18 Fluid flow effects on rotors.

tangential force[11] perpendicular to the direction of the radial displacement. This situation, known as *stream whirl*, can induce rotor instability.[12]

For the magnetic and fluid forces, we can develop a linearized expansion similar to the one for the bearings:

$$-P_x = -P_{x0} + L_{xx} \Delta x + L_{xy} \Delta y + D_{xx} \Delta \dot{x} + D_{xy} \Delta \dot{y} \qquad (11.87)$$

$$-P_y = -P_{y0} + L_{yy} \Delta y + L_{yx} \Delta x + D_{yy} \Delta \dot{y} + D_{yx} \Delta \dot{x} \qquad (11.88)$$

---

[11]Often called "Alford's force"; actually introduced by Thomas, H. J. 1958. Instabile Eigen-schwingungen von Turbinenläufern, angefächt durch Spalströmungen. AEG-Sonderdruck.

[12]Thomas, loc. cit. Lanzberg, A. H. 1960. Stability of a turbine-generator rotor: types of steam and bearing excitations. *Trans. ASME J. Appl. Mech.*, Ser. E, *27*: 410–416. Dimarogonas, A. D. 1970. *Analysis of Steam Whirl*. General Electric Technical Information Series, DF70LS48, Schenectady, N.Y.

The coefficients L and D are external field force gradients. The bearing and external field force gradients are not symmetrical, which makes it necessary to write the equations of motion explicitly in terms of x and y rather than in the compact r form.

11.3 STABILITY OF ROTORS

The stability of a dynamic system will depend on the behavior of the homogeneous equation. Going back to the complex representation, we can write, in general,

$$m\ddot{r} + (c_1 + ic_2)\dot{r} + (k_1 + ik_2)r = 0 \tag{11.89}$$

Although we can deal with the solution of equation (11.89) directly, we can do it with a generalization of the Routh–Hurwitz criteria of complex equations. To this end we seek a solution $r = Re^{pt}$, to obtain

$$mp^2 + (c_1 + ic_2)p + (k_1 + ik_2) = 0 \tag{11.90}$$

In general, p is a complex number. For stability, its real part should be negative (Chapter 4, also see, Hurwitz, A., Math. Werke, vol. II, p. 533.)

The Routh–Hurwitz criteria that a polynomial equation of the form

$$(a_0 + ib_0)p^n + (a_1 + ib_1)p^{n-1} + \cdots + (a_n + ib_n) = 0 \tag{11.91}$$

has all roots with negative real parts (stability) are

a. All coefficients of the polynomial are real, $b_1 = b_2 = \cdots = b_n = 0$. It must be $a_0 > 0$ and all coefficients $a_j \neq 0$, $j = 1, 2, \ldots, n$. In addition, all the partial determinants of the following matrix, indicated with dotted lines, have to be positive:

$$\begin{bmatrix} a_1 & a_0 & 0 & 0 & 0 & 0 & 0 \\ a_3 & a_2 & a_1 & a_0 & 0 & 0 & 0 \\ a_5 & a_4 & a_3 & a_2 & a_1 & a_0 & 0 \\ & & & & & & \\ a_{2n-1} & a_{2n-2} & a_{2n-3} & & & & a_n \end{bmatrix}$$

i.e., $a_1 > 0$, $a_1 a_2 - a_3 a_0 > 0$, $-a_1(a_1 a_4 - a_5 a_0) + a_3(a_1 a_2 - a_3 a_0) > 0$, etc.

b. All coefficients $a_j, b_j \neq 0$, $j = 1, 2, \ldots, n$. In addition, all the following signed determinants have to be positive:

$$(-1)\begin{vmatrix} a_0 & a_1 \\ b_0 & b_1 \end{vmatrix} > 0 \qquad \begin{vmatrix} a_0 & a_1 & a_2 & 0 \\ b_0 & b_1 & b_2 & 0 \\ 0 & a_0 & a_1 & a_2 \\ 0 & b_0 & b_1 & b_2 \end{vmatrix} > 0 \cdots \tag{11.92}$$

$$(-1)^n \begin{vmatrix} a_0 & a_1 & \cdot & \cdot & \cdot & a_n & 0 & 0 & \cdot & \cdot & & \cdot \\ b_0 & b_1 & \cdot & \cdot & \cdot & b_n & 0 & 0 & \cdot & \cdot & & \cdot \\ 0 & a_0 & \cdot & \cdot & \cdot & a_{n-1} & a_n & 0 & \cdot & \cdot & & \cdot \\ 0 & b_0 & \cdot & \cdot & \cdot & b_{n-1} & b_n & 0 & \cdot & \cdot & & \cdot \\ \cdot & \cdot & \cdot & \cdot & \cdot & \cdot & & & \cdot & \cdot & & \cdot \\ \cdot & \cdot & \cdot & \cdot & \cdot & 0 & a_0 & \cdot & \cdot & \cdot & a_{n-1} & a_n \\ \cdot & \cdot & \cdot & \cdot & \cdot & 0 & b_0 & \cdot & \cdot & \cdot & b_{n-1} & b_n \end{vmatrix} > 0 \qquad (11.93)$$

Applied to equation (11.90), it yields for $a_0 = m$, $b_0 = 0$, $a_1 = c_1$, $b_1 = c_2$, $a_2 = k_1$, and $b_2 = k_2$:

$$\begin{vmatrix} m & c_1 \\ 0 & c_2 \end{vmatrix} < 0 \qquad \text{(a)}$$

$$\begin{vmatrix} m & c_1 & k_1 & 0 \\ 0 & c_2 & k_2 & 0 \\ 0 & m & c_1 & k_1 \\ 0 & 0 & c_2 & k_2 \end{vmatrix} > 0 \qquad \text{(b)} \qquad\qquad (11.94)$$

Condition (a) implies that $mc_2 < 0$ or $c_2 < 0$. Condition (b) gives

$$\begin{vmatrix} c_2 & k_2 & 0 \\ m & c_1 & k_1 \\ 0 & c_2 & k_2 \end{vmatrix} > 0 \qquad (11.95)$$

or

$$c_1 c_2 k_2 - c_2^2 k_1 - m k_2^2 > 0 \qquad (11.96)$$

In engineering problems almost all the factors contributing to the equations of motion are present simultaneously. However, depending on the particular application, some factors are dominant and the associated instability phenomena are labeled accordingly. Readers are warned again to be alert for oversimplifications that result from omitting factors considered less important, very common in the contemporary literature. We present below some cases of instability, with the warning that they are altogether qualitative.

Internal Friction

Internal friction, the internal damping[13] we discussed previously, plus some other factors, contribute to the same effect. The most common is dry friction in shrink-fit joints, such as disks, gears, and pulleys on shaft. Figure 11.19 illustrates this point. A gear G is mounted on the shaft by way of a shrink-fit. Due to this, there

---

[13]Newkirk, B. L. 1924. Shaft whipping. *Gen. Electr. Rev.*, *27*(3): 169–178. Kimball, A. T., Jr. 1924. Internal friction theory of shaft whirling. *Gen. Electr. Rev.*, *27*(4): 244–251. Kimball, A. T. 1925. Internal friction as a cause of shaft whirling. *Philos. Mag.*, Ser. 6, *49*: 724–727.

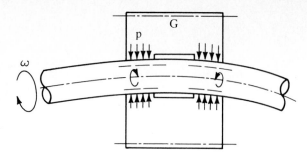

Figure 11.19 Shrink-fit effects on rotors. [Adapted from Newkirk, B. L. 1924. Shaft whipping. *Gen. Electr. Rev.*, 27(3).]

is always a pressure p, almost uniform, between shaft and hub. If the shaft bends, as in the figure, the upper fibers are elongated and the lower are contracted. There is a friction force associated with this if the fit slips, which produces a bending moment, which is a function of the deflection of the shaft. From experiments, we can incorporate this effect into the internal damping coefficient. We shall assume, in general, that c_i in equations (11.39) includes all similar effects. These equations, in complex form (multiplying the second by i, adding, and substituting $\zeta = \eta + i\xi$) are, neglecting unbalance,

$$m\ddot{\zeta} + (c_i + c_e)\dot{\zeta} + (k - m\Omega^2)\zeta + 2m\Omega i\dot{\zeta} + c_e\Omega i\zeta = 0 \qquad (11.97)$$

Substituting $\zeta = \zeta_0 e^{pt}$, we obtain the characteristic equation

$$p^2 + \left(\frac{c_i + c_e}{m} + 2\Omega i\right)p + \left(\frac{k}{m} - \Omega^2\right) + \frac{c_e}{m}\Omega i = 0 \qquad (11.98)$$

Applying the criterion (11.96) we obtain, for stability,

$$\Omega < \left(1 + \frac{c_e}{c_i}\right)\omega_n, \qquad \omega_n = \left(\frac{k}{m}\right)^{1/2} \qquad (11.99)$$

In the absence of external damping ($c_e = 0$) we observe that for rotating speed smaller than the critical ω_n, we have stability. Furthermore, equation (11.99) implies that there is always a rotating speed $\Omega = (1 + c_e/c_i)\omega_n$ above which the system is unstable. Also, for any given rotating speed $\Omega > \omega_n$, we always make the system stable by providing adequate external damping.

Bearing Instabilities: Orthotropic Bearings

Fluid flow in hydrodynamic bearings can inhibit rotor instability.[14] The bearing forces in equations (11.73) and (11.74) can be written for orthotropic bearings ($K_{xx} = K_{yy}$, etc.) for $F_B = F_x + iF_y$ and $r = x + iy$,

$$-F_B = Kr - iK_c r + C\dot{r} - iC_c\dot{r} \qquad (11.100)$$

---

[14]Newkirk, B. L., and Taylor, H. D. 1925. Shaft whipping due to oil action in journal bearings. *Gen. Electr. Rev.*, 25(8): 559–568. Stodola, A. 1925. Kritische Wellenstörung infolge der Nachgiebkeit des Ölposters im Lager, *Schweiz. Bauzt.*, 85: 265.

where

$$K = K_{xx} = K_{yy}, \qquad K_c = K_{xy} = -K_{yx}$$

$$C = C_{xx} = C_{yy}, \qquad C_c = C_{xy} = -C_{yx}$$

are functions of the speed of rotation.

For a rigid rotor, the equation $m\ddot{r} = F_B$ yields

$$m\ddot{r} + (C - iC_c)\dot{r} + (K - iK_c)r = 0 \qquad (11.101)$$

Stability conditions (11.94) to (11.96) yield

$$-C_c < 0, \ CC_cK_c - C_c^2K - mK_c^2 > 0 \qquad (11.102)$$

The first condition indicates that negative bearing damping always means instability. The second condition, upon division by m, yields

$$\omega_n^2 < \frac{C}{C_c}\frac{K_c}{m} - \left(\frac{K_c}{C_c}\right)^2 \qquad (11.103)$$

The right-hand side of this equation is a function of the rotating speed Ω, because so are the bearing properties. If this expression, at some value of Ω, becomes smaller than ω_n^2, this will imply instability.

Bearing Instabilities, Rigid Rotor

In general, journal bearings are not orthotropic and the equations of motion have to be developed explicitly in terms of x and y. For a mass m of the journal with eccentricity e rotating at an angular velocity Ω,

$$m\ddot{x} + C_{xx}\dot{x} + C_{xy}\dot{y} + K_{xx}x + K_{xy}y = me\Omega^2 \cos \Omega t \qquad (11.104)$$

$$m\ddot{y} + C_{yy}\dot{y} + C_{yx}\dot{x} + K_{yy}y + K_{yx}x = me\Omega^2 \sin \Omega t \qquad (11.105)$$

The dimensionless stiffness $\beta_{ij} = K_{ij}c/W$, $\gamma_{ij} = C_{ij}\Omega c/W$, $i, j = x, y$, are introduced where $W = mg$ is the static load on the bearing, c the radial clearance, and Ω the angular velocity. Equations (11.104) and (11.105) can be rewritten in the form

$$\frac{c}{2\pi gS}\ddot{x} + \frac{\gamma_{xx}}{\Omega}\dot{x} + \frac{\gamma_{xy}}{\Omega}\dot{y} + \beta_{xx}x + \beta_{xy}y = \frac{e\Omega^2c}{2\pi gS}\cos \Omega t \qquad (11.106)$$

$$\frac{c}{2\pi gS}\ddot{y} + \frac{\gamma_{yy}}{\Omega}\dot{y} + \frac{\gamma_{yx}}{\Omega}\dot{x} + \beta_{yy}y + \beta_{yx}x = \frac{e\Omega^2c}{2\pi gS}\sin \Omega t \qquad (11.107)$$

Stability of the system will depend on the homogeneous system of equations. Substitution of $x = x_0 e^{pt}$ and $y = y_0 e^{pt}$ yields the characteristic equation

$$\alpha^2 \left(\frac{p}{\Omega}\right)^4 + \alpha A_3 \left(\frac{p}{\Omega}\right)^3 + (\alpha A_4 + A_2)\left(\frac{p}{\Omega}\right)^2 + A_1 \frac{p}{\Omega} + A_0 = 0 \qquad (11.108)$$

where

$$A_0 = \gamma_{xx}\gamma_{yy} - \gamma_{yx}\gamma_{xy}$$

$$A_1 = \beta_{xx}\beta_{yy} + \beta_{yx}\beta_{xy} - (\beta_{xy}\gamma_{yx} + \beta_{yx}\gamma_{xy})$$

$$A_2 = \beta_{xx}\beta_{yy} - \beta_{yx}\beta_{xy}$$

$$A_3 = \beta_{xx} + \beta_{yy}$$

$$A_4 = \gamma_{xx} + \gamma_{yy}$$

$$\alpha = \frac{c\Omega^2}{2\pi g S}$$

Applying the Routh–Hurwitz stability criterion to the polynomial of equation (11.108) yields the stability condition

$$\alpha^2 A_1^2 + A_1(\alpha A_4 + A_2)\alpha A_3 + \alpha^2 A_3^2 A_0 = 0 \tag{11.109}$$

Substituting α and solving for Ω, we obtain the threshold speed for instability,

$$\frac{\Omega_t^2 c}{g} = 2\pi S \frac{A_1 A_2 A_3}{A_1^2 - A_1 A_2 A_3 + A_0 A_3^2} \tag{11.110}$$

Above this speed, the bearing becomes unstable. Since the bearing properties depend on the Sommerfeld number and the bearing design, the threshold speed can be plotted as function of S for different bearing designs. Equation (11.110) has to be solved by iteration because the angular velocity in the Sommerfeld number here will be the threshold speed. In rotating machinery design this is not usually necessary, because the designer needs to know if the system is stable at the operating speed. Therefore, the Sommerfeld number is computed using the operating speed. If the threshold speed is higher than the operating speed, the system is stable; otherwise, it is unstable.

Bearing Instabilities: Elastic Shaft

An elastic symmetric shaft with a rotor at midspan on two identical bearings is shown in Figure 11.20. Equilibrium of forces on the bearings and Newton's law on the rotor mass yield

$$K_{xx}x_1 + K_{xy}y_1 + C_{xx}\dot{x}_1 + C_{xy}\dot{y}_1 - \frac{k}{2}(x_2 - x_1) = 0$$

$$K_{yx}x_1 + K_{yy}y_1 + C_{yx}\dot{x}_1 + C_{yy}\dot{y}_1 - \frac{k}{2}(x_2 - x_1) = 0 \tag{11.111}$$

$$m\ddot{x}_2 + k(x_2 - x_1) = em\Omega^2 \cos \Omega t$$

$$m\ddot{y}_2 + k(y_2 - y_1) = em\Omega^2 \sin \Omega t$$

To study the stability conditions for this system, the homogeneous system of equations is used with $x = x_0 e^{pt}$ and $y = y_0 e^{pt}$ to yield the characteristic determinant, which is expanded to a polynomial in p:

$$r_6\left(\frac{p}{\omega_n}\right)^6 + r_5\left(\frac{p}{\omega_n}\right)^5 + r_4\left(\frac{p}{\omega_n}\right)^4 + r_3\left(\frac{p}{\omega_n}\right)^3$$

$$+ r_2\left(\frac{p}{\omega_n}\right)^2 + r_1\frac{p}{\omega_n} + r_0 = 0 \qquad (11.112)$$

where

$$r_0 = A_0$$

$$r_1 = A_1\frac{\omega_n}{\Omega}$$

$$r_2 = 2A_0 + A_2\left(\frac{\omega_n}{\Omega}\right)^2 + \frac{ck}{2\pi mgS}A_4$$

$$r_3 = \frac{\omega_n}{\Omega}\left(2A_1 + \frac{ck}{2\pi mgS}A_3\right)$$

$$r_4 = A_0 + 2A_2\left(\frac{\omega_n}{\Omega}\right)^2 + \frac{ck}{2\pi mgS}A_4 + \left(\frac{ck}{2\pi mgS}\right)^2$$

$$r_5 = \frac{\omega_n}{\Omega}\left(A_1 + \frac{ck}{2\pi mgS}A_3\right)$$

$$r_6 = A_2\left(\frac{\omega_n}{\Omega}\right)^2$$

$$\omega_n^2 = \frac{k}{m}$$

Application of the Routh–Hurwitz stability criterion on the polynomial of equation (11.112), yields the stability condition as for the rigid rotor. Solving for Ω, we obtain the threshold speed for instability:

$$\left(\frac{\Omega_t}{\omega_n}\right)^2 = \frac{A_2 A_3^2}{1 + (ck/2\pi Smg)(A_3/A_1)}\frac{1}{A_1^2 - A_1 A_3 A_4 + A_0 A_3^2} \qquad (11.113)$$

As for the rigid rotor, equation (11.113) has to be solved by iteration because the angular velocity in the Sommerfeld number will be equal to the threshold speed.

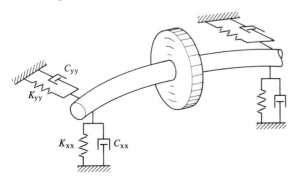

Figure 11.20 De Laval rotor on linear bearings.

In rotating machinery design this is not usually necessary, because the designer needs to know if the system is stable at the operating speed. Therefore, the Sommerfeld number is computed using the operating speed. If the threshold speed is higher than the operating speed, the system is stable; otherwise, it is unstable.

To devise design charts for the threshold of instability, the Sommerfeld number based on the rigid bearings natural frequency of the shaft ω_n is used in the form

$$S_n = \frac{\eta NLD(c/R)^2}{W}\frac{\omega_n}{N} = \eta\omega_n LD\left(\frac{c}{R}\right)^2/W \qquad (11.114)$$

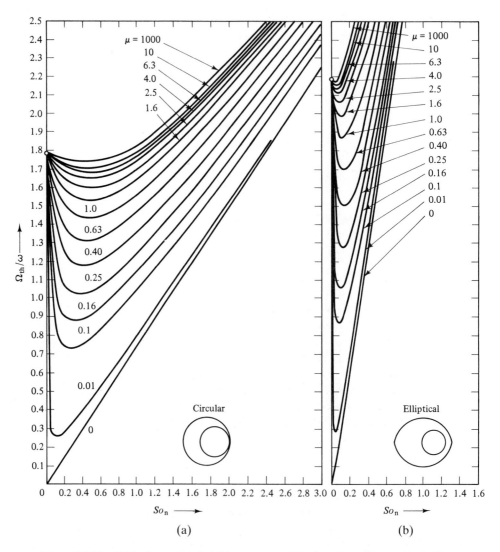

Figure 11.21 Critical speeds of rigid rotors on slider bearings. (From Gash, R., and Pfützner, H. 1975. *Rotordynamik*. By permission of Springer-Verlag.)

On the basis of this definition of the Sommerfeld number, Figure 11.21 shows the threshold of instability, computed by iteration, for a circular and an elliptical bearing (two-lobe, Figure 11.13), the latter with a ratio of radial clearance in the horizontal direction three times that in the vertical direction. The absissa is the Sommerfeld parameter $So_n = W/\eta\omega_n LD(c/R)^2$, with parameter the rotor flexibility ratio $\mu = mg/kc$, static deflection of the shaft over the radial clearance. The straight line ($\mu = 0$) represents a rigid rotor on elastic bearings, discussed in the preceding section.

For the same types of bearings, Figure 11.22 gives the maximum amplitude at resonance over the equivalent mass eccentricity e due to unbalance at the two lower critical speeds versus the dimensionless load So_n, with parameter the rotor flexibility ratio μ. The corresponding critical speeds are shown in Figure 11.23.

□ **Example 11.7** **Critical Speeds and Stability of an Elastic Rotor in Fluid Bearings**

A drive turbine rotor weighs $W = 10,000$ N and the shaft stiffness on rigid bearings is 2.55×10^8 N/m. The bearings have $L/D = 0.8$, $c/R = 0.0015$, $D = 60$ mm, and oil viscosity $\eta = 0.020$ Pa·s. Find (a) the critical speed, (b) the threshold of instability,

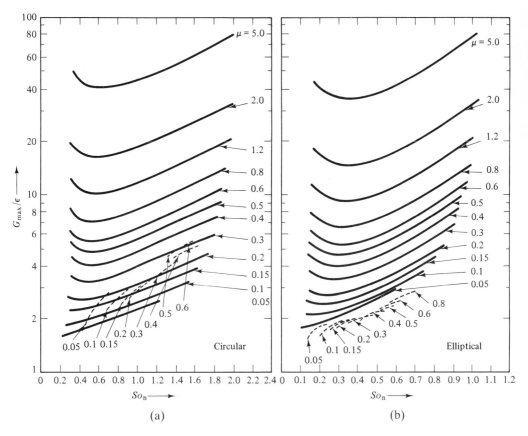

Figure 11.22 Maximum vibration of rigid rotors on slider bearings. (From Gash, R., Pfützner, H. 1975. *Rotordynamik*. By permission of Springer-Verlag.)

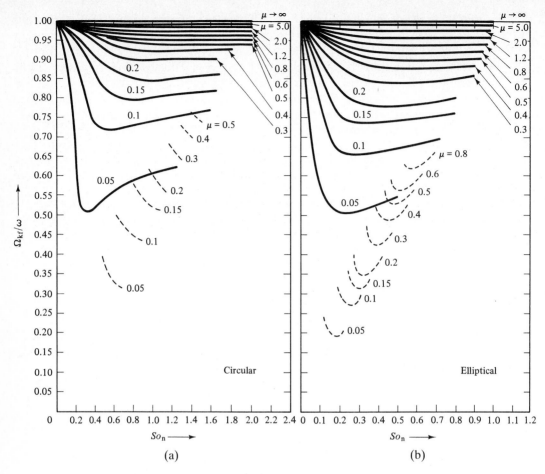

Figure 11.23 Critical speeds of elastic rotors on slider bearings. (From Gash, R., Pfützner, H. 1975, *Rotordynamik*. By permission of Springer-Verlag.)

and (c) the vibration amplitude at resonance if the equivalent mass eccentricity due to unbalance is $e = 0.12$ mm.

Solution The mass of the rotor is $m = 10,000/g = 1019$ kg. The natural frequency of the shaft on rigid bearings is $\omega_n = (k/m)^{1/2} = (2.55 \times 10^8/1019)^{1/2} = 500$ rad/s. The bearing length is $L = 0.8D = 48$ mm and the clearance $c = 0.0015 \times (0.060/2) = 45 \times 10^{-6}$. The parameters are

$$So_n = \frac{W}{LD}\frac{(c/R)^2}{\eta\omega_n} = \frac{10,000}{2 \times 0.048 \times 0.060}\frac{(1.5 \times 10^{-3})^2}{0.020 \times 500} = 0.39$$

$$\mu = \frac{mg}{kc} = \frac{10,000}{2.55 \times 10^8 \times 45 \times 10^{-6}} = 0.87$$

(a) Entering Figure 11.21a with these values, we obtain $\Omega_c/\omega_n = 0.98$. The critical speed is then $\Omega_c = 0.98\omega_n = 0.98 \times 500 = 490$ rad/s, very near the rotor natural frequency on rigid bearings.

(b) Entering Figure 11.22a with these values, we obtain $\Omega_{th}/\omega_n = 1.4$. The threshold speed is then $\Omega_{th} = 1.4\omega_n = 1.4 \times 500 = 700$ rad/s.

(c) Entering Figure 11.23a with these values, we obtain $Y_{max}/e = 4.5$. Therefore, the maximum amplitude is $Y_{max} = 4.5e = 4.5 \times 0.12 = 0.54$ mm.

(An elastic rotor on fluid bearings was used to find the critical speed, threshold of instability due to fluid bearing, and unbalance response.)

□

□ Example 11.8   Instability of an Elastic Rotor in Fluid Four-Pad Bearings Due to Internal Damping

A drive turbine rotor weighs $W = 10,000$ N and the shaft stiffness on rigid bearings is 2.55×10^8 N/m. The four-pad bearings have $L/D = 0.75$, $c/R = 0.0015$, $D = 60$ mm, and oil viscosity $\eta = 0.020$ Pa·s. At the frequency of rotation $\Omega = 1000$ rad/s, find the value of the rotor material loss factor γ that will cause instability due to internal damping, assuming that orthotropic bearings having horizontal stiffness and damping equal to the vertical ones.

Solution   The mass of the rotor is $m = 10,000/g = 1019$ kg. The clearance $c = 0.0015 \times 30 = 0.045$ mm. The natural frequency of the shaft on rigid bearings is $\omega_{rn} = (k/m)^{1/2} = (2.55 \times 10^8/1019)^{1/2} = 500$ rad/s.

The Sommerfeld number

$$S = \frac{\eta NLD}{W}\left(\frac{R}{c}\right)^2 = 2 \times 0.060 \times 0.048 \times (1/0.0015)^2/10,000 = 0.901$$

Entering Figures 11.15 and 11.16 with $S = 0.901$ we obtain $k_{xx}(c/W) = 0.7$ and $c_{xx}(c\Omega/W) = 6.0$. Therefore,

$$k_{xx} = \frac{0.7 \times 10,000}{45 \times 10^{-6}} = 1.55 \times 10^8 \text{ N/m}$$

$$c_{xx} = \frac{6.0 \times 10,000}{1000 \times 45 \times 10^{-6}} = 13.3 \times 10^5 \text{ N·s/m}$$

The combined stiffness of rotor and bearing is

$$k = \frac{k_r k_b}{k_r + k_b} = \frac{2.55 \times 1.55 \times 10^8}{2.55 + 1.55} = 1.03 \times 10^6 \text{ N/m}$$

The critical speed is then $\omega_n = \Omega_c = (k/m)^{1/2} = [1.03 \times 10^6/(1019)/2)]^{1/2} = 44.9$ rad/s. The damping constant due to internal damping is $c_i = \gamma k_r/(\Omega - \omega_n)$. The threshold speed is $\Omega = (1 + c_e/c_i)\omega_n$. Therefore,

$$\Omega = \left(1 + \frac{c_e}{\gamma k_r/(\Omega - \omega_n)}\right)\omega_n$$

Solving for γ yields

$$\gamma = \frac{c_e \omega_n}{k_r} = \frac{13.3 \times 10^5 \times 44.9}{2.55 \times 10^8} = 0.23$$

It is obvious that the rotor is safe against internal damping-induced instability since γ is less than 0.05 for most shaft steels.

(An elastic rotor on fluid bearings was used to find the threshold of instability due to internal damping.)

□

Asymmetry of Rotating Parts

We have confined our discussion up to this point to cylindrical shafts with similar properties in both directions. If the shaft has bending spring constants k_η and k_ξ different along two perpendicular directions (a situation common in electric generators, Figure 11.24a), the restoring forces in the rotating coordinate system will be $k_\eta\eta$ and $k_\xi\xi$, respectively. Substituting into equations (11.39) and using

$$k_1 = \frac{k_\eta + k_\xi}{2}, \qquad k_2 = \frac{k_\eta - k_\xi}{2} \tag{11.115}$$

we obtain, for an unbalance eccentricity e along η,

$$m\ddot{\eta} + (c_e + c_i)\dot{\eta} + (k_1 - m\Omega^2 + k_2)\eta - 2m\Omega\dot{\xi} - c_e\Omega\xi$$
$$= e\Omega^2 + mg\cos\Omega t \tag{11.116}$$

$$m\ddot{\xi} + (c_e + c_i)\dot{\xi} + (k_1 - m\Omega^2 - k_2)\xi$$
$$+ 2m\Omega\dot{\eta} + c_e\Omega\eta = mg\sin\Omega t \tag{11.117}$$

Substituting $\eta = \eta_0 e^{pt}$ and $\xi = \xi_0 e^{pt}$, we obtain the frequency equation, for the system of homogeneous equations,

$$\left(p^2 + \frac{c_e + c_i}{m} + \frac{k_\eta}{m} - \Omega^2\right)\left(p^2 + \frac{c_e + c_i}{m} + \frac{k_\xi}{m} - \Omega^2\right)$$
$$+ \left(2\Omega - \frac{c_e}{m}\right)p^2 = 0 \tag{11.118}$$

Neglecting damping, the frequency equation becomes

$$p^4 + 2(\omega_n^2 + \Omega^2)p^2 + (\omega_n^2 - \Omega^2)^2 - \lambda^2\omega_n^2 = 0 \tag{11.119}$$

where $\omega_n^2 = k_1/m$ and $\lambda = (k_\eta - k_\xi)/(k_\eta + k_\xi)$. The roots are

$$p = \pm[-(\omega_n^2 + \Omega^2) \pm (4\omega_n^2\Omega^2 + \lambda^2\omega_n^4)^{1/2}]^{1/2} \tag{11.120}$$

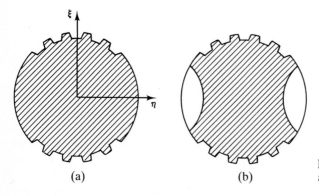

Figure 11.24 Rotor with dissimilar area moments of inertia.

(a)　　　　　(b)

One root is real and positive, implying instability, for $k_\eta < k_\xi$ if $4\omega_n^2\Omega^2 + \lambda\omega_n^4 > (\omega_n^2 + \Omega^2)^2$, or

$$\left(\frac{k_\eta}{m}\right)^{1/2} < \Omega < \left(\frac{k_\xi}{m}\right)^{1/2} \qquad (11.121)$$

In other words, the operation between the critical speeds by the two spring constants of the shaft is unstable. The instability range is reduced due to damping. In fact, for a value of the damping factor, for $\mu = \omega_n/\Omega$,

$$\zeta^2 = \frac{\mu^2}{4}[\lambda^2 - (1 - \mu^2)^2 \qquad (11.122)$$

the instability region disappears.

The influence of the rotor's own weight is important in this case since the static deflection has one full cycle of change with the rotation every half-rotation, because of the changing lateral stiffness in the vertical direction. The forced response is obtained substituting $\eta = \eta_0 e^{\Omega t}$ and $\xi = \xi_0 e^{\Omega t}$ in equations (11.116) and (11.117).

The displacement due to the constant unbalance term is, neglecting damping,

$$A_\epsilon = \frac{\Omega^2\epsilon}{k_i/m - \Omega^2}, \qquad i = 1 \text{ or } 2 \qquad (11.123)$$

The vibration amplitude due to the weight term mg, neglecting damping, for $\omega_i^2 = k_i/m$, $i = 1$ or 2, is

$$A_g = \frac{g(\omega_2 - \omega_1)}{2[\omega_2\omega_1 - 2\Omega^2(\omega_2 + \omega_1)]} \qquad (11.124)$$

The denominator becomes zero for

$$\Omega_g = \frac{\omega_n}{2}(1 - \lambda^2)^{1/2} \qquad (11.125)$$

This indicates a new critical speed due to the rotor's weight of a little less than one-half of the rotor critical speed.[15] It is to be noted that even for perfectly symmetric rotor, the critical speed Ω_g exists because then $\lambda = 0$ and $\Omega_g = \omega_n/2$.

Using the transformation to the stationary coordinate system, the forced response is

$$A_f = A_\epsilon e^{i\Omega t} + A_g e^{2i\Omega t} \qquad (11.126)$$

The first term is the synchronous unbalance response. The second term, due to the rotor weight, has angular velocity 2Ω. In other words, the rotor weight results in a response with frequency that is twice the rotating speed. This phenomenon is used to detect conditions of dissimilar moment of inertia of the shaft, first observed in two-pole electric generator rotors (Figure 11.24a). To correct the problem, additional slots are made on the rotor to equalize the rotor stiffness in two perpendicular directions (Figure 11.24b).

---

[15]Stodola, 1916, loc. cit.

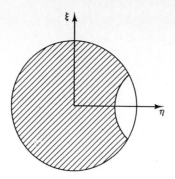

Figure 11.25 Cracked rotor.

An edge crack on the rotor results in dissimilar stiffness along the direction of the crack tip and perpendicular to it (Figure 11.25). The dynamic response is similar to the response of the rotor, with dissimilar moments of inertia along the two directions [Dimarogonas and Paipetis, 1983].

It must be noted that the response due to the rotor weight, such as the additional critical speed at $\omega_n/2$ and the vibration at frequency 2Ω, are observed in horizontal rotors only, not in vertical rotors.

□ **Example 11.9   Dynamic Response of a Cracked Shaft**

A drive turbine rotor weighs $W = 10,000$ N and the shaft stiffness on rigid bearings is 2.55×10^8 N/m. Due to a crack on the shaft, the stiffness along the tip of the crack was reduced by 3% and in a direction perpendicular to it by 10%. Determine the vibration amplitudes at the frequency of rotation $\Omega = 650$ rad/s and the second harmonic of 1300 rad/s if the equivalent eccentricity due to the unbalance is $e = 0.01$ mm.

Solution   The mass of the rotor is $m = 10,000/g = 1019$ kg. The natural frequencies of the shaft on rigid bearings in the two directions related with the crack are

$$\omega_{n1} = \left(\frac{0.97k}{m}\right)^{1/2} = \left(\frac{0.97 \times 2.55 \times 10^8}{1019}\right)^{1/2} = 492 \text{ rad/s}$$

$$\omega_{n2} = \left(\frac{0.9k}{m}\right)^{1/2} = \left(\frac{0.9 \times 2.55 \times 10^8}{1019}\right)^{1/2} = 474 \text{ rad/s}$$

The synchronous vibration amplitude is [equation (11.123)]

$$A_\epsilon = \frac{\Omega^2\epsilon}{k_1/m - \Omega^2} = \frac{650^2 \times 0.01}{492^2 - 650^2} = 0.023 \text{ mm}$$

The amplitude at the frequency 2×650 rad/s is found from equation (11.124):

$$A_g = \frac{g(\omega_2 - \omega_1)}{2[\omega_2\omega_1 - 2\Omega^2(\omega_2 + \omega_1)]}$$

$$= \frac{9.81(474^2 - 492^2)}{2 \times [474^2492^2 - 2 \times 650^2(474^2 + 492^2)]} = 1.36 \times 10^{-5} \text{ m}$$

$$= 0.0136 \text{ mm}$$

It should be noted that the amplitude at twice the speed of rotation does not depend on the unbalance.

(This example demonstrates the existence of measurable vibration of frequency twice the running frequency, due to gravity and edge cracks on a rotating shaft.)

□

Thermal Unbalance: The Newkirk Effect

If about the shaft there are stationary components on which the shaft might rub, there will be a heat input in the rotor due to rubbing which is in proportion and in the direction of the amplitude. Therefore, there will be an unbalance in the rotating coordinate system of

$$e = \rho \eta \tag{11.127}$$

where ρ is a complex constant depending on the geometric and thermoelastic characteristics of the system, $\rho = u + iv$. Neglecting damping, equation (11.77) becomes

$$m\ddot{\eta} + (k - m\Omega^2)\eta + 2m\Omega i\dot{\eta} - m\rho\Omega^2\eta = 0 \tag{11.128}$$

The characteristic equation, for synchronous response $\eta = He^{ipt}$,

$$p^2 + 2\Omega p + (-\omega_n^2 + \Omega^2 + \rho\Omega^2) = 0 \tag{11.129}$$

Stability condition (11.96) yields, with $\rho = u + iv$,

$$(2\Omega)^2(-\omega_n^2 - \Omega^2 - u\Omega^2) - (v\Omega^2)^2 > 0 \tag{11.130}$$

Finally,

$$\Omega^2 > \frac{\omega_n^2}{1 - u + v^2/4} \tag{11.131}$$

In most practical situations, $v \approx 0$ and $u \ll 1$; therefore, we find that stability occurs for rotating speed greater than the critical speed. The main characteristic of this type of vibration, known as Newkirk effect,[16] is the very slowly varying amplitude at the frequency of rotation (typical times are in minutes) and the phase angle change, together with the amplitude.

11.4 THE TRANSFER MATRIX METHOD

Up to this point we have discussed special shafts of very simple geometry. Such models are useful in extracting qualitative results and general features. For many engineering systems, one has to take into account the realistic geometries of such systems, and this leads to the application of methods similar to those discussed in Chapters 7 to 10. Since the most commonly used method for such systems is the transfer matrix method, we discuss here its application on rotating shafts.

---

[16]Newkirk, B. L. 1926. Shaft rubbing, *Mech. Engi.* 48: 830–834. Dimarogonas, A. D. 1973. *Newkirk Effect: Thermally Induced Dynamic Instability of High Speed Rotors*. ASME Gas Turbine Conference, Wasington, D.C. Paper 73-GT-26.

Portrait of Newkirk. (Courtesy of
Dr. John B. Newkirk, Evergreen, Co.)

BURT LEROY NEWKIRK (1876–1964)

Born in Ellenville, New York, Newkirk received A.B. and A.M. degrees from
the University of Minnesota before transferring to Munich, where he received a Ph.D. in astronomy in 1902. He taught until 1921, when he joined
General Electric's Research Laboratories in Schenectady. In his 15 years
with GE, he advanced rotor-bearing dynamics to its present position. He
was the first to study bearing effects on stability, bearing whirl and whip,
internal damping (with Kimball), rubbing (the Newkirk effect), the effect
of shrink-fits, inductive vibration transducers, and the effect of bearing
stiffness. It is interesting to note that Newkirk, whose formal education
was in mathematics, was more concerned with the physics and engineering aspects of rotor dynamics problems than with the mathematical
ones, did little mathematical modeling. From 1937 to 1950 he was professor of vibration practice and theory at Rensselaer Polytechnic Institute
and was consultant to industry and the National Advisory Command on
Aeronautics (now NASA) on machinery vibration and stability.

Natural Vibration: Critical Speeds

The first consideration in such application is the coupling between horizontal and vertical vibration. If there is an element in the system, such a fluid bearing, which couples the vertical and horizontal vibration, the state vector should include the state parameters in both directions. Referring to Figure 11.26, we set up the state vector

$$\mathbf{s} = \{x \quad \theta \quad M \quad V \quad y \quad \phi \quad N \quad W\}$$

For cylindrical elements the transfer matrices will have the form, due to symmetry,[17]

$$\begin{bmatrix} \mathbf{L} & \vdots & 0 \\ \cdots\cdots & + & \cdots\cdots \\ 0 & \vdots & \mathbf{L} \end{bmatrix} \tag{11.132}$$

where $\mathbf{L}$ is the 4×4 field matrix for a prismatic element already introduced in Chapter 10 with the addition of the Timoshenko shear term[18] (see Chapter 9)

$$(l_{14})_s = l_{14} - \frac{\kappa L}{AG} \tag{11.133}$$

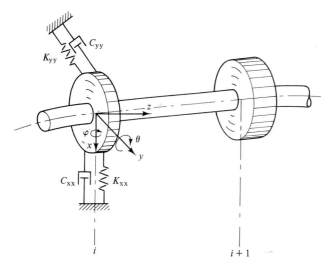

i $i + 1$ **Figure 11.26** General rotor node.

[17]Van den Dungen, M. F.-H. 1928. *Les Problèmes Généraux des la technique des vibrations*. Paris: Gauthier-Villars. Misnamed the Myklestad–Prohl method: Myklestad, N. O. 1944. A new method of calculating natural modes of uncoupled bending vibration of airplane wings and other types of beams. *J. Aeronaut. Sci.*, II, pp. 153–162. Rotating shaft: Prohl, M. 1945. A general method for calculating critical speeds of rotating shafts. *Trans. ASME J. Appl. Mech.*, Ser. A, 67: 142, 148. (See Historical Introduction.)

[18]Corrections due to attached hubs, stress concentrations due to diameter changes are discussed in Stodola, 1927, loc. cit.

where κ is the shear coefficient for the section (1.33 for a circular section), L the element length, A the element area of cross section, and G the element shear modulus.

The bearing reactions can be written as

$$-F_x = K_{xx}x + K_{xy}y + C_{xx}\dot{x} + C_{xy}\dot{y}$$

$$-F_y = K_{yx}x + K_{yy}y + C_{yx}\dot{x} + C_{yy}\dot{y} \tag{11.134}$$

Since free vibrations analyses normally do not consider damping, the damping terms may be omitted. The point or concentrated field matrix for a node with mass m, rotatory inertia J, polar moment of inertia J_p for the gyroscopic effects [equations (11.53 and 11.54)], rotating speed Ω, and a linear bearing with constants K_{xx} and K_{yy}, functions of Ω, has the form

$$\mathbf{P}_{xy} = \left[\begin{array}{cccc:cccc} 1 & 0 & 0 & 0 & 0 & 0 & 0 & 0 \\ 0 & 1 & 0 & 0 & 0 & 0 & 0 & 0 \\ 0 & -J_p\Omega^2 + J\omega^2 & 1 & 0 & 0 & 0 & 0 & 0 \\ -K_{xx} + m\omega^2 & 0 & 0 & 1 & -K_{yx} & 0 & 0 & 0 \\ \hdashline 0 & 0 & 0 & 0 & 1 & 0 & 0 & 0 \\ 0 & 0 & 0 & 0 & 0 & 1 & 0 & 0 \\ 0 & 0 & 0 & 0 & 0 & J_p\Omega^2 + J\omega^2 & 1 & 0 \\ -K_{xy} & 0 & 0 & 0 & K_{yy} + m\omega^2 & 0 & 0 & 1 \end{array}\right] \tag{11.135}$$

The elastic constants K_{xx}, K_{xy}, K_{yx}, and K_{yy} for the bearing may be constant or functions of the angular velocity of rotation Ω. For a rotating speed Ω we can compute the natural frequencies of the shaft using the method of Chapter 10. Critical speeds can be found by equating $\Omega = \omega$ in the expressions for the bearing spring constants and solving for the frequencies ω, which now will be the critical speeds.

Special attention must be given to the coupling terms K_{xy} and K_{yx}. If these terms do not exist or are very small, there will be numerical problems because the frequency equation plot will be as shown in Figure 11.27 because it will be of the form, for a symmetric system,

$$D(\omega) = \left|\begin{array}{c:c} D_x(\omega) & 0 \\ \hdashline 0 & D_x(\omega) \end{array}\right| = |D_x(\omega)|^2 = 0 \tag{11.136}$$

which is tangent to the ω axis and remains nonnegative. Determination of the natural modes presents similar problems. In such a case, the critical speeds are

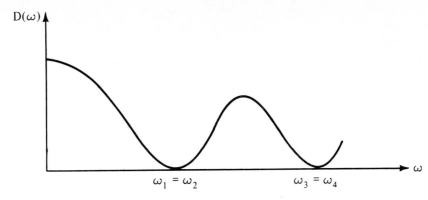

Figure 11.27 Double critical speeds.

computed separately in the vertical and horizontal directions. The transfer matrix method can also be used for rotor stability analysis.[19]

Forced Vibration

We assume harmonic forces and moments $F_j e^{i\Omega t}$ and $T_j e^{i\Omega t}$ at node j, where F_j and T_j are complex force amplitudes. Let the dynamic motion of the biplane system be defined by the expressions

$$x_j = X_j e^{i\Omega t}$$
$$y_j = Y_j e^{i\Omega t}$$
$$\theta_j = \Theta_j e^{i\Omega t}$$
$$\phi_j = \Phi_j e^{i\Omega t}$$

(11.137)

where X_j, Y_j, Θ_j, and Φ_j are the amplitudes of the displacements and slopes. If, in addition, V, W, M, and N are the respective nodal force and moment amplitudes, the state vector will be

$$s_j = \{X \quad \Theta \quad M \quad V \quad Y \quad \Phi \quad N \quad W\}_j$$

The typical 8×8 transfer matrix for a section takes the form of

$$\mathbf{L}_{jxy} = \begin{bmatrix} \mathbf{L} & \vdots & 0 \\ \text{---} & \text{+} & \text{---} \\ 0 & \vdots & \mathbf{L} \end{bmatrix}$$

(11.138)

where $\mathbf{L}$ is the 4×4 field matrix of the rotor section, assumed circular cylinder. The point matrix for a node has the form

[19]Lanzberg, loc. cit.

$$\mathbf{P}_{jxy} = \begin{bmatrix}
1 & 0 & 0 & 0 & \vline & 0 & 0 & 0 & 0 & \vline & 0 \\
0 & 1 & 0 & 0 & \vline & 0 & 0 & 0 & 0 & \vline & 0 \\
0 & -J_p\Omega^2 + J\Omega^2 & 1 & 0 & \vline & 0 & 0 & 0 & 0 & \vline & T_x \\
\begin{matrix} -K_{xx} + m\Omega \\ -iC_{xx}\Omega \end{matrix} & 0 & 0 & 1 & \vline & \begin{matrix} -K_{yx} \\ -iC_{yx}\Omega \end{matrix} & 0 & 0 & 0 & \vline & F_x \\
\hline
0 & 0 & 0 & 0 & \vline & 1 & 0 & 0 & 0 & \vline & 0 \\
0 & 0 & 0 & 0 & \vline & 0 & 1 & 0 & 0 & \vline & 0 \\
0 & 0 & 0 & 0 & \vline & 0 & J_p\Omega^2 + J\Omega^2 & 1 & 0 & \vline & T_y \\
\begin{matrix} -K_{xy} \\ -iC_{xx}\Omega \end{matrix} & 0 & 0 & 0 & \vline & \begin{matrix} -K_{yy} + m\Omega^2 \\ -iC_{yy}\Omega \end{matrix} & 0 & 0 & 1 & \vline & F_y \\
\hline
0 & 0 & 0 & 0 & \vline & 0 & 0 & 0 & 0 & \vline & 1
\end{bmatrix}$$

$$(11.139)$$

where K_{xx}, K_{xy}, K_{yx}, K_{yy}, C_{xx}, C_{xy}, C_{yx}, and C_{yy} are the bearing linearized stiffness and damping coefficients, defined in equation (11.33).

The solution methodology is described in Chapter 10. The forward sweep

$$\mathbf{s}_n = \mathbf{P}_n \mathbf{L}_{(n-1)xy} \cdots \mathbf{P}_3 \mathbf{L}_{2xy} \mathbf{P}_2 \mathbf{L}_{1xy} \mathbf{P}_1 \mathbf{s}_1 \qquad (11.140)$$

This is a system of eight equations, plus the identity $1 = 1$, in 16 unknowns, the elements of the end state vectors. Supplemented by eight boundary conditions, equation (11.140) will yield the response, since the system of equations is non-homogeneous, due to nodal loads.

11.5 THE FINITE ELEMENT METHOD

Finite element modeling of rotor systems follows the methodology for beams developed in Chapter 10, or it can be derived directly from the transfer matrix analysis of Section 11.4 if only lumped mass analysis is adequate. The consistent mass formulation, however, is usually employed. To this end, the displacement vector is defined with the notation of Figure 11.26 as

$$\delta^{(e)} = \{x_1 \quad y_1 \quad \theta_1 \quad \phi_1 \quad x_2 \quad y_2 \quad \theta_2 \quad \phi_2\}$$

and the corresponding force vector

$$\mathbf{F}^{(e)} = \{V_1 \quad W_1 \quad M_1 \quad N_1 \quad V_2 \quad W_2 \quad M_2 \quad N_2\}$$

where x, θ, and y, ϕ are the vertical and horizontal displacements and slopes, respectively. The consistent mass and stiffness matrices follow directly from equations (10.58) and (10.59) for a circular-cross-section rotor, since the relations between forces and displacements are the same in both the horizontal and vertical lateral directions[20]:

---

[20]Rotatory inertia and shear deformation effects can be found in Davis, R., Henshell, R. D., and Warburton, G. B. 1972. A Timoshenko beam element. *J. Sound Vibrat.* 22(4): 475–487. A general element including distributed gyroscopic effects and variable cross section along the element can be found in Dimarogonas, A. D. 1975. A general method for stability analysis of rotating shafts. *Ing. Arch.*, 44: 9–20.

$$\mathbf{K}_d^{(e)} = EI \begin{bmatrix} \dfrac{12}{l^3} & & & & & & & \\ 0 & \dfrac{12}{l^3} & & & & \text{symmetric} & & \\ \dfrac{6}{l^2} & 0 & \dfrac{4}{l} & & & & & \\ 0 & \dfrac{6}{l^2} & 0 & \dfrac{4}{l} & & & & \\ -\dfrac{12}{l^3} & 0 & -\dfrac{6}{l^2} & 0 & \dfrac{12}{l^3} & & & \\ 0 & -\dfrac{12}{l^3} & 0 & -\dfrac{6}{l^2} & 0 & \dfrac{12}{l^3} & & \\ \dfrac{6}{l^2} & 0 & \dfrac{2}{l} & 0 & \dfrac{6}{l^2} & 0 & \dfrac{4}{l} & \\ 0 & \dfrac{6}{l^2} & 0 & \dfrac{2}{l} & 0 & \dfrac{6}{l^2} & 0 & \dfrac{4}{l} \end{bmatrix} \qquad (11.141)$$

$$\mathbf{M}_d^{(e)} = \frac{\rho A}{420} \begin{bmatrix} 156l & & & & & & & \\ 0 & 156l & & & & & & \\ 22l^2 & 0 & 4l^3 & & & \text{symmetric} & & \\ 0 & 22l^2 & 0 & 4l^3 & & & & \\ 54l & 0 & 13l^2 & 0 & 156l & & & \\ 0 & 54l & 0 & 13l^2 & 0 & 156l & & \\ -13l^2 & 0 & -3l^3 & 0 & -22l^2 & 0 & 4l^3 & \\ 0 & -13l^2 & 0 & -3l^3 & 0 & -22l^2 & 0 & 4l^3 \end{bmatrix} \qquad (11.142)$$

where ρ is the material density, A the element cross section, I the element section moment of inertia about a diameter, l the element length, and E is Young's modulus.

To the element mass matrix, the effects of concentrated masses m_i and J_i, m_{i+1} and J_{i+1}, and inertias should be added:

$$\mathbf{M}_c^{(e)} = \tfrac{1}{2} \operatorname{diag}[m_i \quad m_i \quad J_i \quad J_i \quad m_{i+1} \quad m_{i+1} \quad J_{i+1} \quad J_{i+1}] \qquad (11.143)$$

To the element stiffness matrix, the effects of concentrated bearing properties at the end nodes should be added:

$$\mathbf{K}_c^{(e)} = \frac{1}{2} \begin{bmatrix} K_{1xx} & K_{1xy} & 0 & 0 & 0 & 0 & 0 & 0 \\ K_{1yx} & K_{1yy} & 0 & 0 & 0 & 0 & 0 & 0 \\ 0 & 0 & 0 & 0 & 0 & 0 & 0 & 0 \\ 0 & 0 & 0 & 0 & 0 & 0 & 0 & 0 \\ 0 & 0 & 0 & 0 & K_{2xx} & K_{2xy} & 0 & 0 \\ 0 & 0 & 0 & 0 & K_{2yx} & K_{2yy} & 0 & 0 \\ 0 & 0 & 0 & 0 & 0 & 0 & 0 & 0 \\ 0 & 0 & 0 & 0 & 0 & 0 & 0 & 0 \end{bmatrix} \tag{11.144}$$

The element damping matrix due to bearing damping at the nodes and the gyroscopic effects of the inertias at the nodes will be

$$\mathbf{C}_c^{(e)} = \frac{1}{2} \begin{bmatrix} C_{1xx} & C_{1xy} & 0 & 0 & 0 & 0 & 0 & 0 \\ C_{1yx} & C_{1yy} & 0 & 0 & 0 & 0 & 0 & 0 \\ 0 & 0 & 0 & J_1\omega & 0 & 0 & 0 & 0 \\ 0 & 0 & -J_1\omega & 0 & 0 & 0 & 0 & 0 \\ 0 & 0 & 0 & 0 & C_{2xx} & C_{2xy} & 0 & 0 \\ 0 & 0 & 0 & 0 & C_{2yx} & C_{2yy} & 0 & 0 \\ 0 & 0 & 0 & 0 & 0 & 0 & 0 & J_2\omega \\ 0 & 0 & 0 & 0 & 0 & 0 & -J_2\omega & 0 \end{bmatrix} \tag{11.145}$$

The element mass, damping, and stiffness matrices will then be

$$\mathbf{M}^{(e)} = \mathbf{M}_d^{(e)} + \mathbf{M}_c^{(e)}$$

$$\mathbf{C}^{(e)} = \mathbf{C}_c^{(e)} \tag{11.146}$$

$$\mathbf{K}^{(e)} = \mathbf{K}_d^{(e)} + \mathbf{K}_c^{(e)}$$

The system equations are assembled, with the procedures discussed in Chapter 10, in the form

$$\mathbf{M}\ddot{\mathbf{x}} + \mathbf{C}\dot{\mathbf{x}} + \mathbf{K}\mathbf{x} = f(t) \tag{11.147}$$

In this form, further analysis can be performed with the methods discussed in Chapters 7 to 10.

Analysis in two planes needs to be used with great care and only if the cross-coupling terms due to the bearings and the gyroscopic effects are of substantial magnitude. Otherwise, not only will the computation effort be unnecessarily high, but numerical problems are likely to be encountered, due to the weak coupling of vertical and horizontal vibration.

All rotor dynamic analyses above are coded in the program RODYNA. Its use is demonstrated by way of the following example.

☐ **Example 11.10**

Using RODYNA, compute the natural frequencies of the high-pressure turbine rotor shown.

Solution   RODYNA runs from the current drive, say C, with the command C>ROTOR
After the sign-in page, the main menu appears in the screen with pull-down sub-menus from the top. Selection is performed with the keypad arrows and ENTER. Figure E11.10a is the main menu. From the <File> menu, <load> is selected.

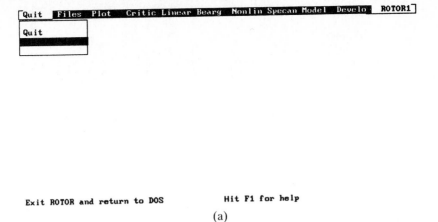

Exit ROTOR and return to DOS   Hit F1 for help

(a)

Data file to be loaded   ? gen.rot█

File manipulations   Hit F1 for help

(b)

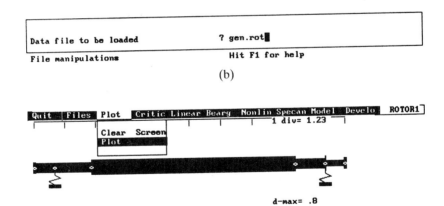

1 div= 1.23

Clear Screen
Plot

d-max= .8

Data file to be loaded   ? gen.rot

Screen operations   Hit F1 for help

(c)

Figure E11.10a, b, c, d, e, f, g, h, i

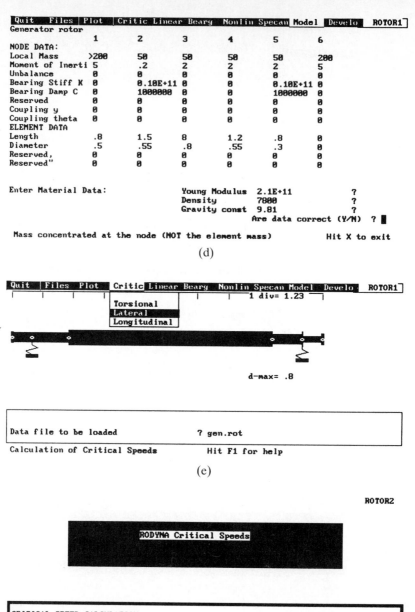

Quit   Files Plot   Critic Linear Beary   Nonlin Specan Model Develo   ROTOR1
Generator rotor

NODE DATA:

	1	2	3	4	5	6
Local Mass	>200	50	50	50	50	200
Moment of Inerti	5	.2	2	2	2	5
Unbalance	0	0	0	0	0	0
Bearing Stiff K	0	0.10E+11	0	0	0.10E+11	0
Bearing Damp C	0	1000000	0	0	1000000	0
Reserved	0	0	0	0	0	0
Coupling y	0	0	0	0	0	0
Coupling theta	0	0	0	0	0	0

ELEMENT DATA

Length	.8	1.5	8	1.2	.8	0
Diameter	.5	.55	.8	.55	.3	0
Reserved,	0	0	0	0	0	0
Reserved"	0	0	0	0	0	0

Enter Material Data: Young Modulus 2.1E+11   ?
                              Density   7800   ?
                              Gravity const 9.81   ?
                                        Are data correct (Y/N) ? ▮

Mass concentrated at the node (NOT the element mass)   Hit X to exit

(d)

Quit   Files Plot   Critic Linear Beary   Nonlin Specan Model Develo   ROTOR1

                                              1 div= 1.23

                    Torsional
                    Lateral
                    Longitudinal

                                        d-max= .8

Data file to be loaded   ? gen.rot

Calculation of Critical Speeds   Hit F1 for help

(e)

                                                        ROTOR2

RODYNA Critical Speeds

CRITICAL SPEED CALCULATION   ....Jacobi rotation method....
Distributed or lumped mass (D/L)? d   Jacobi or Power Iteration (J/P)? J

Expanding matrix into factors...

(f)

Figure E11.10 *(cont.)*

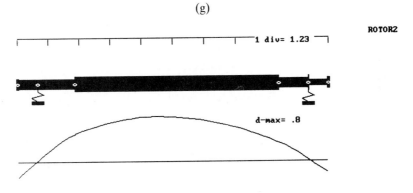

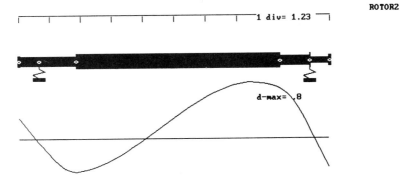

Figure E11.10 *(cont.)*

Figure E11.10b is the file loading selection menu. File GEN.ROT is selected, depicting a generator rotor.

Figure E11.10c is the file plotting menu. *Plot* is selected and the rotor geometry appears on the screen.

In Figure E11.10d, the critical speed menu is selected and <Lateral> is selected from the submenu.

The critical speed menu appears on the screen, Figure E11.10e. Distributed system and Jacobi iteration are selected. The computed critical speeds and vibration modes (up-to-eight) are printed, Figure E11.10f.

Then, the vibration modes are animated, one-by-one, in slow motion. The first critical speed is shown in Figure E11.10g and the second one in Figure E11.10h. We terminate the plotting of critical speeds and return to the main menu. <Model> is selected and a spreadsheet with the rotor data appears on the screen, Figure E11.10i. We can alter the data and repeat the computation or invoke <Model> from the beginning and enter data for a new rotor.

Other rotor dynamic analyses, such as torsional critical speeds, dynamic response, nonlinear response, cracked or misaligned rotor response, nonlinear analysis, bearing analysis, vibration analysis, balancing, can be selected from the menu. The bottom line is a help line.

□

TABLE 11.1 SUMMARY OF EQUATIONS OF CHAPTER 11

Whirling shaft

$$m(\ddot{x} + i\ddot{y}) + k(x + iy) = e\Omega^2 e^{i\Omega t}$$

$$r = x + iy,$$

$$m\ddot{r} + kr = e\Omega^2 e^{i\Omega t}$$

Rotating coordinates

$$\zeta = \eta + i\xi$$

$$r = \zeta e^{i\Omega t}, \qquad \zeta = re^{-\Omega t}$$

$$r = Re^{i\phi_1}, \qquad \zeta = Re^{i\phi_2}$$

$$r/\zeta = e^{i(\phi_1 - \phi_2)} = e^{i\Omega t}$$

$$\ddot{\zeta} + 2i\Omega\dot{\zeta} + (\omega_n^2 - \Omega^2)\zeta = e\Omega^2/m$$

$$\omega_n = (k/m)^{1/2}$$

$$r(t) = Ae^{i\omega_n t} + Be^{-i\omega_n t}$$

$$\ddot{r} + \omega_n^2 r = \frac{m_e}{m} u\Omega^2 e^{i\Omega t}$$

$$r = Re^{i\Omega t},$$

$$R = \frac{m_e u}{m} \left(\frac{\Omega}{\omega_n}\right)^2 \Big/ \left[1 - \left(\frac{\Omega}{\omega_n}\right)^2\right]$$

TABLE 11.1 *(cont.)*

Hysteretic damping

$$D_h = \pi X^2 k \gamma$$

$$m\ddot{x} + \frac{\gamma k}{\Omega} \dot{x} + kx = 0$$

$$-P = \frac{\gamma}{2\pi f} k \dot{\zeta} 20$$

$$m\ddot{\zeta} + \left[\frac{\gamma}{(\Omega - \omega_n)} k + 2i\Omega m \right] \dot{\zeta} + (k - m\Omega^2)\zeta = e\Omega^2$$

$$m\ddot{r} + \frac{\gamma}{(\Omega - \omega_n)} k\dot{r} + k\left[1 - \frac{\gamma\Omega i}{(\Omega - \omega_n)} \right] r = e\Omega^2 e^{i\Omega t}$$

$$G^* = G(1 + i\gamma)$$

$$E^* = E(1 + i\gamma)$$

$$k^* = k(1 + i\gamma)$$

Vector representation

$$\mathbf{r} = (e + \eta)\mathbf{1} + \zeta\mathbf{m}$$

$$\mathbf{r} = (x + e\cos\Omega t)\mathbf{i} + (y + e\sin\Omega t)\mathbf{j}$$

Elastic force

$$\mathbf{f}_s = -k(\eta\mathbf{1} + \xi\mathbf{m}) = -k(x\mathbf{i} + y\mathbf{j})$$

Internal damping

$$\mathbf{f}_i = -c_i(\dot{\eta}\mathbf{1} + \dot{\xi}\mathbf{m}) = -c_i[(\dot{x} + \Omega y)\mathbf{i} + (\dot{y} - \Omega x)\mathbf{j}]$$

Viscous damping

$$\mathbf{f}_v = -c_v[(\dot{\eta} - \Omega\xi)\mathbf{1} + (\dot{\xi} + \Omega\eta)\mathbf{m}] = -c_v(\dot{x}\mathbf{i} + \dot{y}\mathbf{k})$$

$$m\ddot{\mathbf{r}} = \mathbf{f}_s + \mathbf{f}_i + \mathbf{f}_v$$

$$m\ddot{x} + (c_i + c_v)\dot{x} + kx + c_i\dot{y} = m\Omega^2 e \cos\Omega t$$

$$m\ddot{y} + (c_i + c_v)\dot{y} + ky - c_i\dot{x} = m\Omega^2 e \sin\Omega t$$

$$m\ddot{\eta} + (c_i + c_v)\dot{\eta} + (k - m\Omega^2)\eta - 2m\Omega\dot{\xi} - c_v\Omega\xi = m\Omega^2 e$$

$$m\ddot{\xi} + (c_i + c_v)\dot{\xi} + (k - m\Omega^2)\xi + 2m\Omega\dot{\eta} + c_v\Omega\eta = 0$$

Acceleration through critical speed

$$x(t) = \frac{e\Omega}{2} t \sin\Omega t$$

(continues)

TABLE 11.1 *(cont.)*

Gyroscopic effects

$$\mathbf{M}_G = (\dot{\mathbf{H}}_G)_{x,y,z} + \omega \times \mathbf{H}_G$$

$$m\ddot{x} + k_{xx}x + k_{x\theta}\theta = 0$$

$$m\ddot{y} + k_{yy}y + k_{y\phi}\phi = 0$$

$$J\ddot{\theta} - J_p\Omega\dot{\phi} + k_{\theta\theta}\theta + k_{\theta x}x = 0$$

$$J\ddot{\phi} + J_p\Omega\dot{\theta} + k_{\phi\phi}\phi + k_{\phi y}y = 0$$

Elastic bearings

$$F_x = F_x(x, \dot{x}, y, \dot{y})$$

$$F_y = F_y(x, \dot{x}, y, \dot{y})$$

$$S = \frac{\eta NLD}{W}\left(\frac{R}{c}\right)^2, \; So = \frac{W}{\eta\omega LD}\left(\frac{c}{R}\right)^2 = \frac{1}{2\pi S}$$

$$Re = \frac{\Omega R^2}{\nu}$$

Routh-Hurwitz stability criteria

$$(a_0 + ib_0)p^n + (a_1 + ib_1)p^{n-1} + \cdots + (a_n + ib_n) = 0$$

Real coefficients

$$\begin{bmatrix} a_1 & \cdot & a_0 & \cdot & 0 & \cdot & 0 & & 0 & & 0 & \cdot & \cdot & 0 \\ \cdot & \cdot & \cdot & & \cdot & & \cdot & & & & & & & \\ a_3 & & a_2 & \cdot & a_1 & \cdot & a_0 & & 0 & & 0 & \cdot & \cdot & 0 \\ \cdot & \cdot & \cdot & & \cdot & \cdot & & & & & & & & \\ a_5 & & a_4 & & a_3 & \cdot & a_2 & & a_1 & & a_0 & \cdot & \cdot & 0 \\ \cdot & \cdot & \cdot & & \cdot & \cdot & \cdot & & \cdot & & & & & \\ \hline a_{2n-1} & & a_{2n-2} & & a_{2n-3} & \cdot & \cdot & \cdot & \cdot & \cdot & \cdot & \cdot & \cdot & a_n \end{bmatrix}$$

$$a_1 > 0, \; a_1a_2 - a_3a_0 > 0, \; -a_1(a_1a_4 - a_5a_0) + a_3(a_1a_2 - a_3a_0) > 0, \text{ etc.}$$

Complex coefficients

$$(-1)\begin{vmatrix} a_0 & a_1 \\ b_0 & b_1 \end{vmatrix} > 0, \quad \begin{vmatrix} a_0 & a_1 & a_2 & 0 \\ b_0 & b_1 & b_2 & 0 \\ 0 & a_0 & a_1 & a_2 \\ 0 & b_0 & b_1 & b_2 \end{vmatrix} > 0, \cdots,$$

$$(-1)^n \begin{vmatrix} \cdot & \cdot & & \cdot & & & & & \\ \cdot & \cdot & & \cdot & & & & & \\ a_0 & a_1 & \cdots & a_n & 0 & 0 & \cdot & \cdot \\ b_0 & b_1 & \cdots & b_n & 0 & 0 & \cdot & \cdot \\ 0 & a_0 & \cdots & a_{n-1} & a_n & 0 & \cdot & \cdot \\ \cdot & \cdot & & b_{n-1} & b_n & \cdot & \cdot \\ \cdot & \cdot & 0 & a_0 & \cdot & \cdot & a_{n-1} & a_n \\ \cdot & \cdot & 0 & b_0 & \cdot & \cdot & b_{n-1} & b_n \end{vmatrix} > 0$$

TABLE 11.1 *(cont.)*

Newkirk effect

$$e = \rho\eta$$

$$m\ddot{\eta} + (k - m\Omega^2)\eta + 2m\Omega i\dot{\eta} - m\rho\Omega^2\eta = 0$$

$$\Omega^2 > \frac{\omega_n^2}{\left[1 - u + \dfrac{v^2}{4}\right]}$$

Electromagnetic effects

$$-P_x = -P_{x0} + L_{xx}\Delta x + L_{xy}\Delta y + D_{xx}\Delta\dot{x} + D_{xy}\Delta\dot{y}$$

$$-P_y = -P_{y0} + L_{yy}\Delta y + L_{yx}\Delta x + D_{yy}\Delta\dot{y} + D_{yx}\Delta\dot{x}$$

Internal damping

$$m\ddot{\zeta} + (c_i + c_e)\dot{\zeta} + (k - m\Omega^2)\zeta + 2m\Omega i\dot{\zeta} + c_e\Omega i\zeta = 0$$

Stability condition

$$\Omega < \left(1 + \frac{c_e}{c_i}\right)\omega_n, \qquad \omega_n = \left(\frac{k}{m}\right)^{1/2}$$

Bearings

$$m\ddot{x} + C_{xx}\dot{x} + C_{xy}\dot{y} + K_{xx}x + K_{xy}y = me\Omega^2\cos\Omega t$$

$$m\ddot{y} + C_{yy}\dot{y} + C_{yx}\dot{x} + K_{yy}y + K_{yx}x = me\Omega^2\sin\Omega t$$

$$\beta_{ij} = K_{ij}\frac{c}{W}, \qquad \gamma_{ij} = C_{ij}\frac{\Omega c}{W}, \qquad i, j = x, y$$

$$\frac{c}{2\pi gS}\ddot{x} + \frac{\gamma_{xx}}{\Omega}\dot{x} + \frac{\gamma_{xy}}{\Omega}\dot{y} + \beta_{xx}x + \beta_{xy}y = \frac{e\Omega^2 c}{2\pi gS}\cos\Omega t$$

$$\frac{c}{2\pi gS}\ddot{y} + \frac{\gamma_{yy}}{\Omega}\dot{y} + \frac{\gamma_{yx}}{\Omega}\dot{x} + \beta_{yy}y + \beta_{yx}x = \frac{e\Omega^2 c}{2\pi gS}\sin\Omega t$$

where

$$A_0 = \gamma_{xx}\gamma_{yy} - \gamma_{yx}\gamma_{xy}$$

$$A_1 = \beta_{xx}\beta_{yy} + \beta_{yx}\beta_{xy} - (\beta_{xy}\gamma_{yx} + \beta_{yx}\gamma_{xy})$$

$$A_2 = \beta_{xx}\beta_{yy} - \beta_{yx}\beta_{xy}$$

$$A_3 = \beta_{xx} + \beta_{yy}$$

$$A_4 = \gamma_{xx} + \gamma_{yy}$$

$$\alpha = \frac{c\Omega^2}{2gS}$$

(continues)

TABLE 11.1 *(cont.)*

Stability condition

$$\alpha^2 A_1^2 + A_1(\alpha A_4 + A_2)\alpha A_3 + \alpha^2 A_3^2 A_0 = 0$$

Threshold speed

$$\Omega_t^2 \frac{c}{g} = \frac{1}{2\pi S} A_1 A_2 A_3 / (A_1^2 - A_1 A_2 A_3 + A_0 A_3^2)$$

Elastic, symmetric shaft

$$K_{xx} x_1 + K_{xy} y_1 + C_{xx} \dot{x}_1 + C_{xy} \dot{y}_1 - \frac{k}{2}(x_2 - x_1) = 0$$

$$K_{yx} x_1 + K_{yy} y_1 + C_{yx} \dot{x}_1 + C_{yy} \dot{y}_1 - \frac{k}{2}(x_2 - x_1) = 0$$

$$m\ddot{x}_2 + k(x_2 - x_1) = em\Omega^2 \cos \Omega t$$

$$m\ddot{y}_2 + k(y_2 - y_1) = em\Omega^2 \sin \Omega t$$

$$r_0 = A_0$$

$$r_1 = A_1 \left(\frac{\omega_n}{\Omega} \right)$$

$$r_2 = 2A_0 + A_2 \left(\frac{\omega_n}{\Omega} \right)^2 + \frac{ck}{2\pi mgS} A_4$$

$$r_3 = \left(\frac{\omega_n}{\Omega} \right) \left(2A_1 + \frac{ck}{2\pi mgS} A_4 \right)$$

$$r_4 = A_0 + 2A_2 \left(\frac{\omega_n}{\Omega} \right)^2 + \frac{ck}{2\pi mgS} A_4 + \left(\frac{ck}{2\pi mgS} \right)^2$$

$$r_5 = \left(\frac{\omega_n}{\Omega} \right) \left(A_1 + \frac{ck}{2\pi mgS} A_3 \right)$$

$$r_6 = A_2 \left(\frac{\omega_n}{\Omega} \right)^2$$

$$\omega_n^2 = \frac{k}{m}$$

$$\left(\frac{\Omega_t}{\omega_n} \right)^2 = \frac{A_2 A_3^2}{\left(1 + \dfrac{ck}{2\pi mgS} \right) \dfrac{A_3}{A_1}} \frac{1}{(A_1^2 - A_1 A_3 A_4 + A_0 A_3^2)}$$

Asymmetric rotor

$$k_1 = \frac{(k_\eta + k_\xi)}{2}, \qquad k_2 = \frac{(k_\eta - k_\xi)}{2}$$

$$m\ddot{\eta} + (c_e + c_i)\dot{\eta} + (k_1 - m\Omega^2 + k_2)\eta - 2m\Omega\dot{\xi} - c_e\Omega\xi = e\Omega^2 + mg \cos \Omega t$$

$$m\ddot{\xi} + (c_e + c_i)\dot{\xi} + (k_1 - m\Omega^2 - k_2)\xi + 2m\Omega\dot{\eta} + c_e\Omega\dot{\eta} = mg \sin \Omega t$$

TABLE 11.1 *(cont.)*

Instability range

$$\left(\frac{k_\eta}{m}\right)^{1/2} < \Omega < \left(\frac{k_\xi}{m}\right)^{1/2}$$

Secondary critical speed

$$\Omega_g = \frac{\omega_n}{2}(1 - \lambda^2)^{1/2}$$

REFERENCES AND FURTHER READING

BIEZENO, C. B., and GRAMMEL, R. 1953. *Technische Dynamik*. Berlin: Springer-Verlag.

DEN HARTOG, J. P. 1952. *Mechanical Vibration* 4th ed. New York: McGraw-Hill.

DIMAROGONAS, A. D. 1976. *Vibration Engineering*. St. Paul, Minn.: West Publishing.

DIMAROGONAS, A. D., and PAIPETIS, S. A. 1983. *Analytical Methods in Rotor Dynamics*. London: Elsevier–Applied Science.

DIMENTBERG, F. 1961. *Flexural Vibrations of Rotating Shafts*, English Translation. London: Butterworth.

FÖPPL, A. 1933. *Vorlesungen über Technische Mechanik*, Vol. 4: *Dynamik*. Leipzig: Teubner.

GASH, R., and PFÜTZNER, H. 1975. *Rotordynamik*. Berlin: Springer-Verlag.

GOODWIN, M. J. 1989. *Dynamics of Rotor-Bearing Systems*. London: Unwin Hyman.

HOUSNER, G. W., and HUDSON, D. E. 1950. *Applied Mechanics: Dynamics*. Princeton, N.J.: D. Van Nostrand.

HORT, W. 1910. *Technische Schwingungslehre*. Berlin: Julius Springer.

KRÄMER, E. 1984. *Maschinendynamik*. Berlin: Springer-Verlag.

LALANE, M., and FERRARIS, G. 1990. *Rotordynamics Prediction in Engineering*. New York: Wiley.

LAZAN, B. J. 1968. *Damping of Materials and Members in Structural Mechanics*. Oxford: Pergamon Press.

PESTEL, E. C., and LECKIE, F. A. 1963. *Matrix Methods in Elastomechanics*. New York: McGraw-Hill.

PONTRYAGIN, L. S. 1962. *Ordinary Differential Equations*. Reading, Mass.: Addison-Wesley.

RAYLEIGH, J. W. S. 1894. *Theory of Sound*. New York: Dover Publications, 1946.

SPECTRAL DYNAMICS CORP. 1990. *Vibration Handbook*.

STODOLA, A. 1924. *Dampf- und Gasturbinen*. Berlin: Springer-Verlag.

TONDL, A. 1965. *Some Problems of Rotor Dynamics*, English translation. London: Chapman & Hall.

PROBLEMS

Section 11.1

11.1. Calculate the critical speed of a shaft of spring constant $k = 10^7$ N/m and length $L = 6$ m carrying a disk of mass 1,000 kg and diameter $D = 1.30$ m, at one-third of its length. Compare it with the natural frequency at zero and a running speed of 3600 rpm.

11.2. In Problem 11.1, calculate the response to an unbalance of 200 g at the outer diameter, at the running speed. Determine the influence of the gyroscopic effect.

11.3. In Problem 11.2, compute the influence of a hysteretic damping $\gamma = 0.001$.

11.4. An electric motor has a massive rotor of 150 kg on two bearings, each of spring constant 1500 N/cm. The magnetic field has a spring constant $k_m = -250\omega$ (N/m). Compute the response at 3600 rpm to an unbalance of 50 g located at a radius of 150 mm.

11.5. A cantilever beam of length 4 m, width 25 mm, and thickness 12 mm has a small motor at the end running at 2000 rpm and having an unbalance of 40 g at a 50-mm radius, rotor $m = 50$ kg, $J = 3$ kg m². Find the steady-state response if the modulus of elasticity is $E^* = 2 \times 10^5 \times (1 + 0.01i)$ N/mm².

11.6. In Problem 11.5, determine the logarithmic decrement of the lateral vibration.

11.7. A simply supported beam has on its middle a heavy load, as in Figure P11.7. The beam is vibrating at a frequency of 100 rad/s due to a harmonic force at the mass m with an amplitude 2.5 mm and $L = 2.5$ m, $l = 250$ mm, $m = 40$ kg, $d = 50$ mm, width $b = 50$ mm. If the coefficient of friction is 0.1, calculate the rate of the energy dissipation.

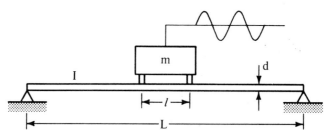

Figure P11.7

11.8. A shaft carrying a disk of mass $m = 100$ kg at midspan has a spring constant of 1000 N/cm. If the damping factor $\gamma = 0.001$, calculate the external damping coefficient in order that the shaft will be stable at a running speed three times the critical speed.

11.9. A shaft carrying a disk of mass $m = 100$ kg at one-third of its length of $l = 0.80$ m has a spring constant of 10,000 N/cm. If the damping factor is $\gamma = 0.001$, calculate the external damping coefficient in order that the shaft will be stable at a running speed three times the critical speed.

11.10. A windmill operates at the top of a steel tube of diameter $d = 200$ mm, thickness $s = 12$ mm, and height $h = 40$ m. The mill consists of a housing and moving parts having a moment of inertia in respect to the vertical axis $J_z = 250$ kg·m². The rotor and blades have a moment of inertia about the horizontal axis of rotation $J_x = 12$ kg·m². If the mill is rotating at 200 rpm, determine the natural frequency of the torsional vibration about the z axis.

11.11. In Problem 11.10, determine the torsional vibration of the mill housing if one of the four blades of length $l = 8$ m and mass 6 kg breaks (Figure P11.11).

11.12. An electric motor of mass $m = 55$ kg rotates at 3500 rpm at the end of the cantilever beam of Problem 11.5 (Figure P11.12). If the moment of inertia of the rotor and fan are $J_x = 0.9$ kg·m² and the moment of inertia of the cantilever beam can be neglected, determine **(a)** the natural frequency at running speed, and **(b)** the critical speed.

11.13. An inertia-propelled bus (Figure P11.13) has a heavy horizontal wheel of mass $m = 700$ kg and radius of gyration $r = 0.6$ m which rotates at 6000 rpm maximum. The

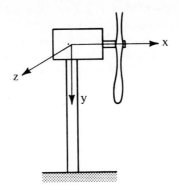

Figure P11.11

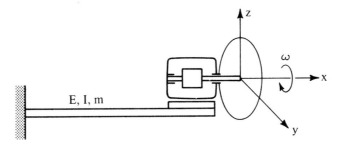

Figure P11.12

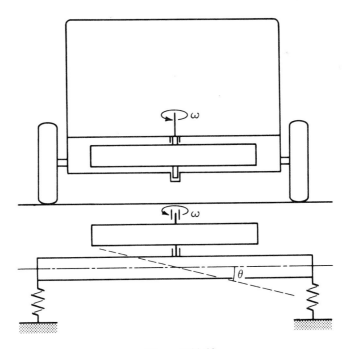

Figure P11.13

mass of the bus is $M = 2000$ kg and the radius of gyration in respect to the longitudinal axis of the car, without the wheel, is $R = 0.75$ m. The bus is mounted on four springs which result in an undamped vertical natural frequency 8 rad/s. For rocking vibration about the longitudinal axis, determine **(a)** the undamped natural frequency if the wheel speed is $\omega = 0$, **(b)** the same frequency if $\omega = \omega_{max}$, and **(c)** the critical speed.

11.14. In Problem 11.13, determine the size of the shock absorbers in order that the system will be critically damped **(a)** for vertical vibration, **(b)** for rocking vibration about the longitudinal axis with $\omega = 0$, and **(c)** the same for $\omega = \omega_{max}$.

11.15. A crane is lifting a load $w = 800$ kg with a speed $v_0 = 1$ m/s (Figure P11.15). Determine the frequency of oscillation of the load at cable length $l = 80$ m, 30 m, and 10 ft.

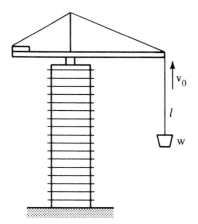

Figure P11.15

11.16. A heavy circular disk of radius r and mass m is attached to a light shaft of length l (Figure P11.16). The other end is attached to a vertical shaft rotating at angular velocity ω by way of a ball joint that allows vertical oscillations of the shaft and disk on any vertical plane. Assuming uniform rotation of the disk, determine **(a)** the natural frequency of oscillation of the disk, and, **(b)** the angular velocity at which the system becomes unstable.

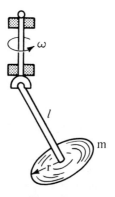

Figure P11.16

11.17. A ventilation fan consisting of mass $m = 6$ kg and radius of gyration $r = 0.5$ m is operated by a vertical motor of mass $M = 8$ kg that is supported from the ceiling by way of a vertical bracket of length $l = 0.25$ m (Figure P11.17). If the ceiling joint of the bracket is loose such that it allows oscillations of the fan–motor system, determine the natural frequency of oscillations if the fan runs at 500 rpm, and the maximum operating speed that the fan can have and still maintain the vertical position.

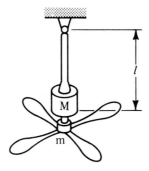

Figure P11.17

11.18. A circular steel shaft of length $l = 1.2$ m and diameter $d = 25$ mm is carrying at its midspan a disk of mass $m = 120$ kg and radius of gyration $r = 0.5$ m (Figure P11.18). Determine the speed range at which the shaft is safe both from internal damping instability and from the possibility that the disk will be unstable if the bearing B becomes loose.

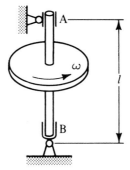

Figure P11.18

11.19. A heavy disk has mass m and moments of inertia J_x and J_y in respect to the coordinates x' and y' of a coordinate system $(x'y'z')$ attached to the disk and coinciding with its principal axis of inertia (Figure P11.19). The disk is balanced and rotates about the z'-axis with a constant angular velocity. Prove that if F_x and F_y are restoring forces due to the deflections x and y of the mass center from equilibrium and M_x and M_y the restoring moments for rotations ϕ_x and ϕ_y about the axes x and y, respectively, the differential equations of motion are

$$m\ddot{x} = -F_x$$

$$m\ddot{y} = -F_y$$

$$J_y\dot{\phi}_y - J\omega\dot{\phi}\dot{x} = -M_y$$

$$J_x\dot{\phi}\ddot{x} + J_p\omega\dot{\phi}\dot{y} = -M_x$$

where J_p is the polar moment of inertia $J_p^2 = J_x^2 + J_y^2$.

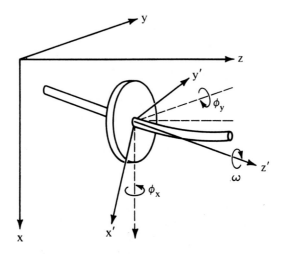

Figure P11.19

11.20. Prove that for shaft having an overhang disk as shown in Figure P11.20, the differential equations of motion are

$$m\ddot{z} + c_{11}z + c_{12}\phi = 0$$

$$J\ddot{\phi} + c_{11}z + c_{22}\phi - i\phi J = 0$$

where $z = x + iy$, $\phi = -\phi_y + i\phi_x$, $J = J_x = J_y$,

$$c_{11} = \frac{12EI}{a^3l}\frac{3a + l}{3a + 4l}, \qquad c_{12} = \frac{GEI}{a^2}\frac{3a + 2l}{3a + 4l} = c_{21}, \qquad c_{22} = \frac{4EI}{a}\frac{3a + 3l}{3a + 4l}$$

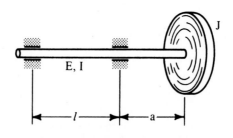

Figure P11.20

11.21. Prove that the natural frequencies at running speed for the system of Problem 11.20 are the solutions of the equation

$$p^4 - 2\omega p^3 - \left(\frac{c_{11}}{m} + \frac{c_{22}}{J}\right)p^2 + 2\frac{c_{11}}{m}\frac{\omega p + c_{11}c_{22} - c_{12}^2}{Jm} = 0$$

11.22. An electric generator rotor has mass $m = 600$ kg and can be considered rigid with respect to the two identical circular bearings of length 40 mm, diameter 50 mm, radial and clearance ratio $c/R = 0.001$, lubricated with oil of viscosity $\eta = 0.020$ N·s/m². Determine the stiffness and damping coefficients of the bearings and the lowest critical speed.

11.23. A 3600 RPM turbocompressor rotor has mass $m = 200$ kg and can be considered rigid in respect to the two identical elliptical bearings of length 40 mm, diameter 50 mm, radial clearance ratio $c/R = 0.002$, ellipticity $\frac{1}{3}$, and lubricated with oil of viscosity $\eta = 0.020$ N·s/m². Determine the stiffness and damping coefficients of the bearings and the lowest critical speed.

11.24. An electric generator rotor has mass $m = 600$ kg and can be considered rigid with respect to the two identical circular bearings of length 40 mm, diameter 50 mm, and radial clearance ratio $c/R = 0.001$, lubricated with oil of viscosity $\eta = 0.020$ N·s/m². Determine the stiffness and damping coefficients of the bearings and investigate the stability at angular velocity of rotation $\Omega = 377$ rad/s.

11.25. A turbocompressor rotor has mass $m = 200$ kg and can be considered rigid with respect to the two identical elliptical bearings of length 40 mm, diameter 50 mm, radial clearance $c/R = 0.002$, ellipticity $\frac{1}{3}$, and lubricated with oil of viscosity $\eta = 0.020$ N·s/m². Determine the stiffness and damping coefficients of the bearings and investigate the stability at angular velocity of rotation $\Omega = 377$ rad/s.

11.26. Solve Problem 11.22 assuming an elastic rotor with natural frequency on rigid bearings $\omega_n = 180$ rad/s. Then compute the unbalance response with angular velocity of rotation $\Omega = 377$ rad/s.

11.27. Solve Problem 11.23 assuming an elastic rotor with natural frequency on rigid bearings $\omega_n = 160$ rad/s. Then compute the unbalance response with angular velocity of rotation $\Omega = 377$ rad/s for an equivalent eccentricity due to unbalance of $e = 0.005$ mm.

11.28. Solve Problem 11.24 assuming an elastic rotor with natural frequency on rigid bearings $\omega_n = 140$ rad/s. Then compute the unbalance response for an equivalent eccentricity due to unbalance of $e = 0.005$ mm.

11.29. Solve Problem 11.25 assuming an elastic rotor with natural frequency on rigid bearings $\omega_n = 120$ rad/s. Then compute the unbalance response for an equivalent eccentricity due to unbalance of $e = 0.005$ mm.

12

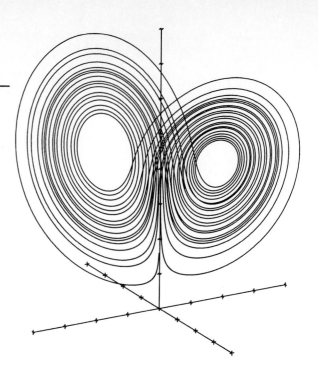

Vibration of Nonlinear Systems

12.1 GEOMETRIC METHODS: THE PHASE PLANE

Nonlinear Elements

In most considerations of physical systems, certain idealizing assumptions or limitations are introduced. In many problems one of the critical assumptions is that of linearity between two of the variables (force and displacement, current and voltage, stress and strain, etc.). There is almost no real physical system that is linear under all conditions.

Nonlinear behavior can be the result of:

1. *Material nonlinearities,* for example, nonlinear relation of stress and strain in rubberlike materials, and deformation of materials beyond the elastic limit or saturation of magnetic materials.
2. *Geometric nonlinearities,* due to nonlinear relation between force and deformation, even without exceeding the elasticity limit of the materials, due to large deflections, contact stresses, closing gaps, and backlash.

3. *Kinematic nonlinearities* due to large motions which change continuously the system configuration, such as in linkages

The real question then is: Are these important effects or is the linearization approximation good enough? The answer is that sometimes they are very important and sometimes they are not. The unfortunate aspect is that one cannot tell whether or not a nonlinear effect is large or small until considerable experience exists and careful thought has been given to the problem. Perhaps the point that causes many engineering difficulties is the fact that effects which are small in one case are frequently dominant in another, very similar situation. This is particularly true for instability phenomena.

To illustrate this point, we consider the case of a simple pendulum. The differential equation of motion was obtained in Chapter 1,

$$\ddot{\theta} + \frac{g}{l} \sin \theta = 0 \tag{12.1}$$

An equation like this cannot be solved with the methods we used for linear systems. The problem can be simplified if we look for solutions near $\theta = 0$, where with a "good" approximation we can state that $\sin \theta \simeq \theta$. Then the problem assumes the usual linear form and can be solved. The first question that arises is: To what extent is this approximation good enough? Unfortunately, most of the time one must obtain the solution of the nonlinear problem in some way or rely on experimental observations in order to check the fidelity of the linear approximation. Perhaps, one can get an idea by considering the series expansion of the $\sin z$ function: $\sin z = z - z^3/3! + \cdots$. Our approximation was that for small z, the terms of order z^3 and higher are very small compared with z. This approximation is good for some purposes and bad for others. If, for example, we are concerned with the period of oscillations, the error for oscillations, say between $+10°$ and $-10°$, is quite negligible. What about stability? The approximation $\theta \simeq \sin \theta$ yields $\ddot{\theta} + (g/l)\theta = 0$. This equation has an equilibrium at $\theta = 0$. Equation (12.1) has an equilibrium at $\theta = 0$ and at $\theta = \pi$. The solution of the linearized equations at $\theta = 0$ is $\theta = \theta_0 \cos \omega_n t$, which is stable; in other words, it stays bounded and system damping will eventually bring the solution to $\theta = 0$. At the other equilibrium point (inverted pendulum) a Taylor expansion, $\sin(\pi + \theta) = -\theta + \theta^3/3! - \theta^5/5! + \theta^7/7! - \cdots$, will be used. Neglecting higher-order terms, the differential equation of motion will be

$$\ddot{\theta} - \frac{g}{l} \theta = 0 \tag{12.2}$$

The solution of this equation is $\theta = \theta_0 e^{\omega_n t}$, where $\omega_n^2 = g/l$, which increases beyond bound and implies instability. Similar situations in which the linearized solution is stable and the nonlinearities result in instabilities appear quite often.

The Phase Plane

In the study of vibration of nonlinear systems we shall make use of a method of plotting the solutions known as *phase portrait*, introduced in Chapter 2. We shall extend its application to the study of nonlinear systems.

The *phase plane* is simply a plane in which the velocity $\dot{x}$ and the displacement x are the coordinates. The parameter t (time) does not appear explicitly. The phase portrait is a graphical construction that retains more information about system behavior than a graphical solution for $x(t)$ would give. In the following we discuss how some of the oscillations we have studied appear on the phase plane.

The harmonic oscillator is expressed with the differential equation $m\ddot{x} + kx = 0$. The velocity of the oscillator is defined as $v = \dot{x}$. We already know that the solution is $x(t) = x_0 \cos \omega_n t + (v_0/\omega_n) \sin \omega_n t$. If we differentiate, we obtain the velocity $\dot{x}(t) = -x_0\omega_n \sin \omega_n t + v_0 \cos \omega_n t$. Eliminating the time, we can obtain a relation between the displacement $x(t)$ and the velocity $\dot{x}(t)$. Alternatively, we can rewrite the differential equation of motion as

$$m\frac{dv}{dx} + \frac{kx}{v} = 0 \tag{12.3}$$

The solution of this equation is

$$mv^2 + kx^2 = R_0^2 = mv_0^2 + kx_0^2 \tag{12.4}$$

where R_0^2 is a constant determined by the initial conditions, twice the sum of the maximum kinetic energy $mv_0^2/2$ and the maximum potential energy $\int kx\,dx$. Equation (12.4) is an ellipse in the $x - v$ plane. By using the normalized variable $y = v(k/m)^{-1/2}$ instead of v, the solution becomes a circle. The arc length from the initial position (x_0, v_0) gives the time t. From any point (x_0, v_0) there pass the ellipse (12.4). The situation is shown in Figure 12.1.

When damping is considered in the oscillator, the governing equation is

$$\ddot{x} + 2\mu\dot{x} + \omega_n^2 x = 0, \qquad \mu = \omega_n\zeta \tag{12.5}$$

The solution is

$$x = e^{-\mu t} A \cos(\omega_d t + \phi) \tag{12.6}$$

and the velocity is

$$\dot{x} = -Ae^{-\mu t}[\mu \cos(\omega_d t + \phi) + \omega_d \sin(\omega_d t + \phi)] \tag{12.7}$$

These are the equations of a spiral, and the phase plane solution shows that as time goes to infinity, the particle comes to rest at the origin. The corresponding phase-plane equation is

$$\frac{dv}{dx} + 2\mu + \frac{\omega_n^2 x}{v} = 0 \tag{12.8}$$

Analytical solution of this equation is possible, but cumbersome. One of the approximate methods of sketching the solution is the *phase-plane delta method*.

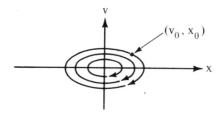

Figure 12.1 Phase-plane portrait.

The Delta Method

The governing differential equation of a nonlinear system can be written as

$$\ddot{x} + \omega^2 x = -\omega^2 \delta(x, \dot{x}, t) \tag{12.9}$$

In general, if the equation of the system is $\ddot{x} + f(x, x, t) = 0$, we shall have

$$\delta(x, x, t) = -x + \frac{f(x, \dot{x}, t)}{\omega^2}$$

It is also convenient to use a normal form in which $y = \dot{x}/\omega$ (rather than $v = \dot{x}$) is the ordinate. When this is done, equation (12.9) becomes

$$\frac{dy}{dx} = -\frac{x + \delta}{y} \tag{12.10}$$

If δ is slowly changing, it can be treated as a constant for small increments. The solution (with constant δ) is

$$(x + \delta)^2 + y^2 = c^2 \tag{12.11}$$

where c is a constant. The instantaneous phase trajectory is therefore a circle centered at $(-\delta, 0)$. This makes it easy to construct the solution graphically with a compass as a sequence of circular segments.

☐ **Example 12.1**

For the case of the damped oscillator [equation (12.5)] with $\omega = 1.0$ rad/s, $\mu = 0.1$, $x(0) = 1$ mm, and $\dot{x}(0) = 3$ mm/s, construct a graph of the solution using the delta method.

Solution For the damped linear oscillator,

$$\delta(x, x, t) = -x + \frac{f(x, \dot{x}, t)}{\omega^2} = -x + \frac{(2\zeta\omega\dot{x} + \omega^2 x)}{\omega^2}$$

$$= \frac{2\mu\dot{x}}{\omega^2} = \frac{2 \times 0.1 \times 3}{1^2} = 0.6$$

The solution is starting by drawing a circle of radius $c = [(x + \delta)^2 + (x/\omega)^2]^{1/2} = [(1 + 0.6)^2 + (3/1)^2]^{1/2} = (11.56)^{1/2} = 3.4$ mm and centered at $x = -\delta = -0.6$. The first arc passes through the point (2.0, 2.2), which serves as a new initial condition for another arc of radius $(2.2^2 + 2.44^2)^{1/2}$ and centered $x = -0.44$. This arc passes through the point (2.5, 1.4) and the process is repeated. The first five steps are shown below and in Figure E12.1.

Step	x	y	New δ
0	1.0	3.0	0.6
1	2.0	2.2	0.44
2	2.5	1.4	0.28
3	2.85	0	0
4	2.0	−2.0	−0.45
5	1.0	−2.5	−0.50

☐

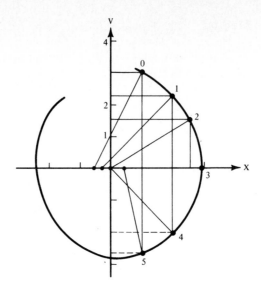

Figure E12.1

Phase-Plane Transient Solutions

In many problems, there are situations where the purely graphical phase-plane analysis must be supplemented by an analytical solution valid for small time intervals. Such situation arises, for example, when the differential equation has the form

$$\ddot{x} + 2\left(\frac{\dot{x}}{t}\right) + f(x) = 0 \tag{12.12}$$

and

$$x(0) = 1.0, \qquad \dot{x}(0) = 0 \tag{12.13}$$

in the vicinity of $t = 0$. The procedure starts assuming a Taylor series solution in the form

$$x = A + Bt + Ct^2 + Dt^3 + Et^4 \tag{12.14}$$

In view of the initial conditions equation (12.14) requires $A = 1.0$ and $B = 0$. Noting that

$$\frac{\partial f}{\partial t} = \frac{\partial f}{\partial x}\,\dot{x} \qquad \text{and} \qquad \frac{\partial^2 f}{\partial t^2} = \frac{\partial^2 f}{\partial x^2}\,(\dot{x})^2 + \frac{\partial f}{\partial x}\,\ddot{x} \tag{12.15}$$

and at $t = 0$

$$\frac{\partial f}{\partial t} = 0 \qquad \text{and} \qquad \frac{\partial^2 f}{\partial t^2} = -f'\left[2\left(\frac{x}{t}\right) + f(x)\right] \tag{12.16}$$

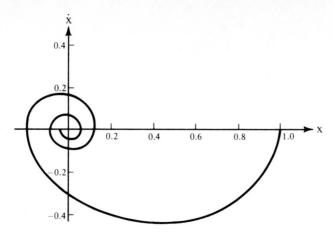

Figure 12.2 Phase-plane transient solution.

the solution is obtained by taking time derivatives of equation (12.12) and using (12.16) at time $t = 0$. This computes C, D, E and then,

$$x = 1.0 - \frac{f(x_0)}{6} t^2 + f(x_0) \frac{f'(x_0)}{120} t^4 \tag{12.17}$$

For the special case $f(x) = \sin x$, equation (12.17) becomes

$$x = 1.0 - 0.140t^2 + 0.00378t^4 \tag{12.18}$$

Thus $x(1) = 0.864$ and $\dot{x}(1) = -0.265$ serve as initial conditions for the phase-plane construction. The results by the phase-plane delta method are shown in Figure 12.2. The first five numerical values are

x	$\dot{x}$	t	δ
1.0	0	0	—
0.864	−0.265	1.0	−0.634
0.770	−0.320	1.32	−0.559
0.700	−0.359	1.53	−0.526
0.600	−0.390	1.79	−0.481
0.510	−0.410	2.01	−0.430

☐ **Example 12.2 Bilinear Spring Systems with Backlash**

Determine the response of a block of mass m oscillating between two linear springs of equal spring constant k having a backlash d between the block and the springs if the block is given an initial velocity v_0 at $x_0 = 0$ [Hort, 1922].

Solution The equation of motion is $m\ddot{x} + f(x) = 0$, where $f(x)$ is as shown in Figure E12.2b. In the δ form,

$$\ddot{x} + \omega_n^2 \left[x + \frac{f(x)}{k} - x \right] = 0$$

$$x + \omega_n^2(x + \delta_x) = 0, \qquad \delta_x = \frac{f(x)}{k} - x$$

and the function δ_x is shown in Figure E12.2c.

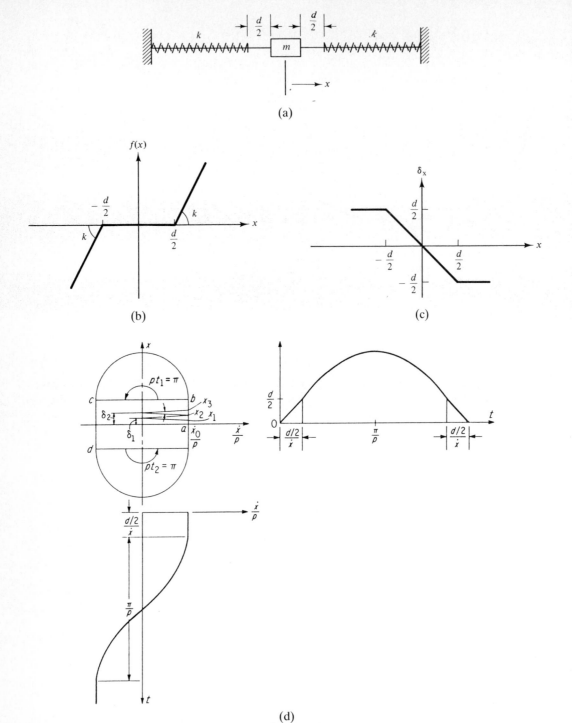

Figure E12.2 (From Seireg, A. 1961, Mechanical Systems Analysis. Reprinted by permission of the author.)

The δ procedure is shown in Figure E12.2d. From the center position a to the point of contact on the right b, the velocity is constant and the displacement increases. From b to the rightmost position and then back to point of separation c, the radius δ is constant $-d/2$ and the angle $\omega t = \pi$. The block then continues to the left with constant velocity $-v_0$ until it hits the spring on the left at d. The continuation is obvious.

The time plot gives the period of vibration:

$$T = 2\left(\frac{d}{v_0} + \frac{\pi}{\omega_n}\right)$$

with $\omega_n^2 = k/m$.

(The delta method is used to plot the response of an oscillator with a backlash-type bilinearity.)

□

□ **Example 12.3   Bouncing Pendulum**

Determine the response of a pendulum of mass m and length l free to oscillate on the right of a vertical wall through its pivot if the impact has a coefficient of restitution e [Seireg, 1969] (Figure E12.3).

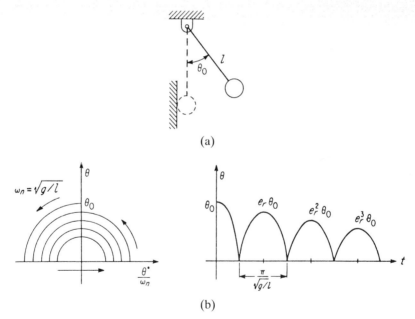

(a)

(b)

Figure E12.3

Solution   The solution can be mapped on the phase plane. The first half-swing, from the wall up and then down until the bob hits the wall, is represented by the circle. At the time it hits the wall it has zero displacement and a certain velocity ω. When it leaves the wall the velocity is $e\omega$ and the path is a circle with smaller radius. The process is continued with diminishing radius.

(The bouncing of a pendulum on a vertical wall through its pivot is plotted on the phase plane.)

□

Classification of Nonlinear Systems

The main difficulty in analyzing nonlinear problems is that superposition does not apply. This is easily seen in the case where a spring force F is related to the displacement by $F = \beta x^3$. If the displacement x_1 is added to an existing one, x_0, the force is

$$F = \beta x_1^3 + 3\beta x_1^2 x_0 + 3\beta x_1 x_0^2 + \beta_0^3 \tag{12.19}$$

and not $\beta x_1^3 + \beta x_0^3$.

There are a few nonlinear differential equations for which exact or approximate solutions are known. Therefore, rational approximate methods or purely numerical solutions are usually employed. The modern computer provides solutions for most nonlinear problems at moderate computation effort.

Many nonlinear systems are governed by differential equations of the general form

$$\ddot{x} + \phi(\dot{x}) + f(x) = F(t) \tag{12.20}$$

where $\phi(x)$, $f(x)$, and $F(t)$ are general functions, the latter usually harmonic $F(t) = F_0 \cos \omega t$. If x is not a scalar function but a vector $\mathbf{x} = \{x_1 \quad x_2 \quad \cdots \quad x_n\}$, $\boldsymbol{\phi}$, $\mathbf{f}$, and $\mathbf{F}$ are vector functions of $\mathbf{x}$ and t.

The function $f(x)$ is referred to as a *nonlinear restoring force*, and if $\phi(\dot{x}) = 0$ or $c\dot{x}$, equation (12.20) is known as a *Duffing equation*.[1] In particular, if $f'(x)$ increases with x, the restoring spring is called a *hard spring*, while if $f'(x)$ decreases with x, the restoring spring is called a *soft spring*.

The case $f(x) = x$ leads, for certain forms of $\phi(\dot{x})$, to a class of oscillations called *self-excited oscillations*. In particular, $\phi(\dot{x}) = -\dot{x} + \frac{1}{3}\dot{x}^3$ leads to the *van der Pol equation*.[2]

The substitution $v = \dot{x}$ in equation (12.20) yields, in the absence of $F(t)$,

$$\frac{dv}{dx} = -\frac{\phi(v) + f(x)}{v} \tag{12.21}$$

This equation defines the velocity field and yields the phase portraits of the vibration, as defined in Chapter 2 for linear systems, and for certain forms of the nonlinear functions, equation (12.21) can be integrated and used to define phase portraits as in Figure 12.3.

At certain points it is possible that $\phi(v) + f(x)$ and v are both zero and the velocity field is indefinite. Such points are called *singular*. Equilibrium points are typical singular points. For the pendulum, equation (12.21) for $x = l\theta$, $\alpha = gl$ becomes

$$v \frac{dv}{dx} = -\frac{\alpha}{l} \sin \frac{x}{l} \tag{12.22}$$

---

[1]Duffing, G. 1918. *Enzwungene Schwingungen bei veränderlicher Eigenfrequenz*. Braunschweig: F. Vieweg.

[2]van der Pol, B. 1927. Forced oscillations in a system with nonlinear resistance. *Philos. Mag.*

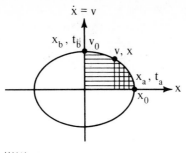

|||||| First Integration

≡≡≡ Second Integration

Figure 12.3 Integration procedure for the vibration period of a nonlinear system.

and the trajectories on the (x, v) plane are obtained with direct integration:

$$\frac{v^2}{2} = \alpha \cos \frac{x}{l} + \frac{h}{2} \tag{12.23}$$

where h is an integration constant. The dimensions of α and h is m^2/s^2 or $N \cdot m/kg$, energy per unit mass. In fact, one can observe that $-\alpha \cos \theta$ is the potential energy $= \int \alpha \sin \theta \, dx$, $v^2/2$ is the kinetic energy, and $h/2$ is the sum of the two energies at $x = 0$, the initial energy of the bob of the pendulum. For $|h| < 2\alpha$, equation (12.23) represents a closed curve, while for $|h| > 2\alpha$, the velocity is always > 0 and thus the pendulum moves in one direction continuously while the velocity oscillates (Figure 12.4). There is an obvious physical interpretation: If the initial velocity is low, the pendulum oscillates about the lower equilibrium position. If the initial velocity is sufficiently large to overcome the potential energy $2gl$, the pendulum rotates continuously in the same direction.

The points $\theta = 0, \mp\pi, \mp2\pi, \ldots$ are singular points, points of equilibrium. For $\theta = 0, \mp2\pi, \mp4\pi, \ldots$ the closed trajectories enclose the equilibrium points,

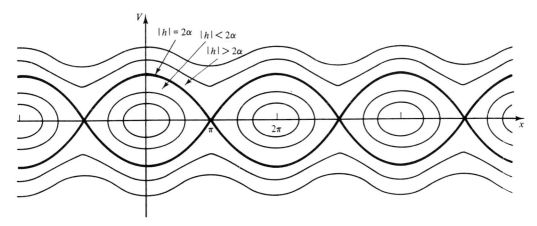

Figure 12.4 Phase portrait of a pendulum.

which are stable. Physically, these points are at the lower equilibrium point of the pendulum. They represent the lower energy levels, $h = 0$.

For $\theta = \mp\pi, \mp 3\pi, \ldots$ the open trajectories cross each other at the equilibrium points, which are then unstable. Physically, these points are at the higher equilibrium point of the pendulum. They represent the higher energy levels, $h = 2\alpha$.

Autonomous Systems

Autonomous systems, for $F(t) = 0$, can be effectively represented in the (x, v) plane, as in the case of the pendulum. Figure 12.5 shows different phase portraits of equation (12.21), depending on the functions ϕ and f. Poincaré [1892] has shown

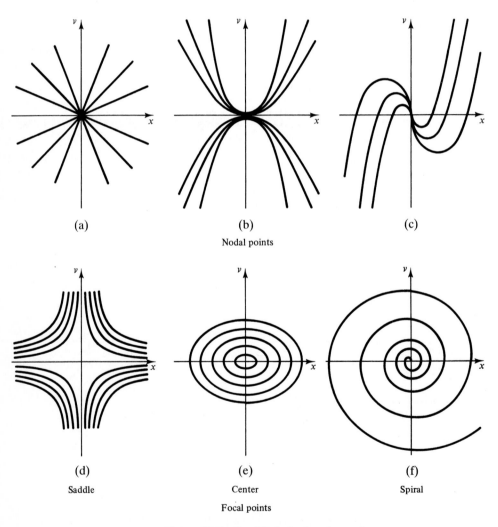

Figure 12.5 Equilibria in the phase plane.

that the behavior of any differential equation of the form (12.21) is equivalent to the equation

$$\frac{dv}{dx} = \frac{ax + bv}{\alpha x + \beta v} \qquad (12.24)$$

where $[\phi(v) + f(x)]/v = [ax + bv]/[\alpha x + \beta v + o(x^2, v^2)]$. Figure 12.5 shows all possible types of singular points of equation (12.21):

(a) $a = 1, b = 0, \alpha = 1, \beta = 0$, nodal point
(b) $0 < a < 1, b = 0, \alpha = 1, \beta = 0$, nodal point
(c) $a = 1, b = 1, \alpha = 1, \beta = 0$, nodal point
(d) $a < 0, b = 0, \alpha = 1, \beta = 0$, saddle point
(e) $a = -\lambda^2, b = 0, \alpha = 0, \beta = 1$, center point
(f) $a = 1, b = \alpha, \beta = -1$, spiral point

Nonautonomous Systems

Equation (12.21) can be set in the form of the Duffing equation

$$\ddot{x} + c\dot{x} + \alpha x + \beta x^3 = F_0 \cos \omega t \qquad (12.25)$$

For small levels of excitation, the amplitude A is expected to be low and then $\alpha A \gg \beta A^3$. Therefore, the system will behave more or less as a linear system, as shown in Figure 12.6a for $\beta = 0$. The resonant speed is constant with increasing amplitude.

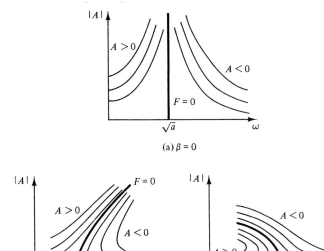

(a) $\beta = 0$

(b) $\beta > 0$

(c) $\beta < 0$

Figure 12.6 Natural vibration of a Duffing oscillator. $\beta = 0$, linear spring; $\beta > 0$, hard spring; $\beta < 0$, soft spring.

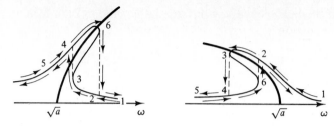

Figure 12.7 Forced response and jumping phenomena of a Duffing oscillator. From Seireg, A. 1961, Mechanical Systems Analysis. By permission of the author.)

At higher amplitudes, for $\beta > 0$, increasing amplitude will result in harder spring and thus increasing resonant speed as the amplitude becomes greater. The result is a response spectrum distorted for high amplitudes to the right (Figure 12.6b). For $\beta < 0$, increasing amplitude will result in softer spring and thus decreasing resonant speed as the amplitude becomes greater. The result is a response spectrum distorted for high amplitudes to the left (Figure 12.6c).

An interesting jump phenomenon is observed in such nonlinear systems as they change speed near resonance with constant-amplitude excitation. At the speed where a vertical is tangent to the distorted response curve, the amplitude jumps from one branch of the curve to the other. The situation is shown in Figure 12.7 for hard spring $\beta > 0$ and soft spring $\beta < 0$.

12.2 EXACT SOLUTIONS: PERIOD OF VIBRATION OF UNDAMPED SYSTEMS

The period of vibration of conservative nonlinear systems expressed in the form

$$\ddot{x} + f(x) = 0 \tag{12.26}$$

can easily be obtained on the phase plane. The substitution $\ddot{x} = v(dv/dx)$ yields

$$v \frac{dv}{dx} + f(x) = 0 \tag{12.27}$$

Integrating from $(v = 0, x = x_0)$ to (v, x), we obtain

$$\tfrac{1}{2}v^2 = \int_{x_0}^{0} f(x)\, dx \tag{12.28}$$

Because $v = dx/dt$, we have

$$dt = \frac{dx}{\left[2 \int_{x_0}^{x} f(x)\, dx\right]^{1/2}} \tag{12.29}$$

Integrating from (x_a, t_a) to the point (x_b, t_b) and observing that $t_a = 0$, $x_a = x_0$, $t_b = T/4$, $x_b = 0$, we obtain the period of natural vibration

$$T = 4 \int_0^{x_0} \frac{dx}{\left[2 \int_0^{x_0} f(x)\, dx \right]^{1/2}} \tag{12.30}$$

Certain forms of the function $f(x)$ permit direct integration and thus a closed-form expression for the period T.

For example, for the simple pendulum, $f(\theta) = (g/l) \sin \theta$ and we obtain

$$T = 4 \left(\frac{l}{g} \right)^{1/2} \int_0^{\pi/2} \frac{d\phi}{(1 + u^2 \sin^2 \phi)^{1/2}} \tag{12.31}$$

where $u = \sin \theta_0/2$. The integral in equation (12.31) is a tabulated function $K(u)$ called the *complete elliptic integral of the first kind*.[3] The significant point here is the amplitude dependence of the period of vibration. In general, the period of a conservative system can be expressed as a definite integral which can also be evaluated with numerical methods and tabulated.

☐ **Example 12.4 Period of Pendulum as a Function of Amplitude**

Determine the period of vibration of a pendulum as a function of the amplitude.

Solution Equation (12.31) will be used with values of the elliptic functions obtained from tables.[4] Alternatively, one can use in equation (12.31) the expansion

$$\frac{1}{(1 + u^2 \sin^2 \phi)^{1/2}} = 1 + \frac{1}{2} u^2 \sin^2 \phi + \frac{3}{8} u^4 \sin^4 \phi + \frac{5}{16} u^6 \sin^6 \phi + \cdots$$

Integration gives us

$$T = \left(\frac{l}{g} \right)^{1/2} \left(1 + \frac{1}{4} u^2 + \frac{9}{64} u^4 + \frac{25}{256} u^6 + \cdots \right)$$

The numerical ratios of the nonlinear system period of vibration to the one of the linearized system are

θ_1 (deg)	Nonlinear period/linear period
2	1.0000
22.5	1.0097
45	1.0390
90	1.1831
135	1.5279
157.5	1.9360
180.0	∞

(A series expansion is used to find the dependence of the period of pendulum oscillation on the amplitude on the phase plane.)

☐

[3] Abramovits, M. and Stegun, I., 1965. *Handbook of Mathematical Functions*. New York: Dover Publications.

[4] Abramovits & Stegun, *loc. cit.*

Portrait of Poincaré.
(Library of Congress.)

JULES-HENRI POINCARÉ (1854–1912)

Poincaré was born at Nancy to a prominent family. He graduated from the Ecole Polytechnique with a degree in mining engineering and was employed by the Department of Mines for the remainder of his life. He earned a doctorate from the University of Paris, where he held several professorships. He wrote very extensively, more than any mathematician of our century, in a variety of areas, including differential equations, probability, and celestial mechanics. He is considered the founder of topology. He did not stay long in one area: "He was a conqueror, not a colonist." He died at the age of 58.

□ **Example 12.5** **Forced Vibration of Bilinear Systems**

Determine the response of a bilinear spring system, with spring constant k_1 if $-x_0 < x < x_0$ and k_2 otherwise, to harmonic excitation [Klotter, 1951] and [Seireg, 1969]).
Solution This is one of the very limited number of nonlinear problems with analytical solution. The motion is represented by two equations,

$$m\ddot{x}_1 + k_1 x_1 = \mp P \cos \omega t \qquad \text{for } |x| < x_0$$

$$m\ddot{x}_2 + (k_1 - k_2)x_0 - k_2 x_2 = \mp P \cos \omega t \qquad \text{for } |x| > x_0$$

The conditions of compatibility are

$$t = 0, \qquad x_1 = 0$$

$$t = t_0, \qquad x_1 = x_2 = x_0$$

$$t = t_0, \qquad \dot{x}_1 = \dot{x}_2$$

$$t = \frac{\pi}{2\omega}, \qquad \dot{x}_2 = 0$$

The solution is

$$x_1 = \frac{\mp P/k_1}{1 - (\omega/\omega_1)^2} \cos \omega t + A_1 \sin \omega_1 t + A_2 \cos \omega_1 t$$

$$x_2 = \frac{\mp P/k_2}{1 - (\omega/\omega_2)^2} \cos \omega t + A_1 \sin \omega_2 t + A_2 \cos \omega_2 t + \left(1 - \frac{k_1}{k_2}\right)x_0$$

The $+$ in the $\mp$ sign represents in-phase motion and the $-$ sign, out-of-phase motion. The constants A_1, A_2, A_3, and A_4 can be determined from the compatibility conditions.

(A piecewise-linear solution was used for an oscillator with a bilinear spring.)
□

12.3 APPROXIMATE ANALYTICAL METHODS

Duffing Method of Successive Approximations

Duffing[5] tried, as a first approximation to the solution of equation (12.31) without damping and assuming that $m = 1$ in the form

$$\ddot{x} = -\alpha x - \beta x^3 + F_0 \cos \omega t, \tag{12.32}$$

the function $x_0 = A \cos \omega t$, to obtain

$$\ddot{x}_1 = -(\alpha A + \tfrac{3}{4}\beta A^3 - F) \cos \omega t - \tfrac{1}{4}\beta A^3 \cos^3 \omega t \tag{12.33}$$

using $\cos^3 \omega t = \tfrac{3}{4}\cos \omega t + \tfrac{1}{4}\cos 3\omega t$. Integrating (12.33) twice yields

$$x_1 = \frac{1}{\omega^2}\left(\alpha A + \frac{3}{4}\beta A^3 - F\right) \cos \omega t - \frac{1}{36}\frac{\beta A^3}{\omega^2}\cos^3 \omega t \tag{12.34}$$

[5]Loc. cit.

having taken the integration constants as zero to assure a periodic solution. The iteration can continue in the same manner, and it converges provided that β is sufficiently small; that is, the nonlinearity is not very strong. Higher-order approximations yield progressively higher harmonics, known as *ultraharmonics*.

Duffing made the assumption that A_1 should not differ appreciably from A if β is small. If $\beta \to 0$, equation (12.34) with $A = A_1$ gives

$$\omega^2 = \omega_n^2 + \frac{3}{4}\beta A^2 - \frac{F}{A} \tag{12.35}$$

because $\alpha = k/m = \omega_n^2$. Equation (12.35) yields the relationship of the resonant speed as a function of the amplitude, which results in the curves shown in Figure 12.6 for the Duffing oscillator for a hard ($\beta > 0$) and a soft ($\beta < 0$) spring.

☐ **Example 12.6 Resonant Frequency of an Elastically Restrained Column**

Determine the resonant frequency as a function of the amplitude of a restrained column with a mass m, as shown in Figure E12.6 [Stoker, 1950].

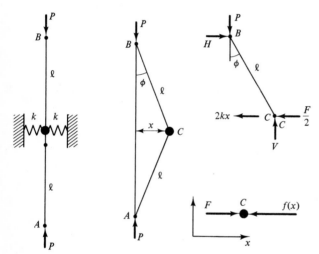

Figure E12.6 (From Seireg, A. 1961, Mechanical Systems Analysis. By permission of the author.)

Solution   In the static equilibrium equations, we assume that $P \gg mg$. Then equilibrium of vertical forces and moments about B gives

$$P = V, \qquad -\frac{2kl\theta}{2} + Vl\sin\theta - \frac{Fl}{2}\cos\theta = 0$$

The equation of motion in the horizontal direction for the mass m is, with a horizontal force $F_0\cos\omega t$,

$$m\ddot{x} + kx = F = \frac{2(Px - kl\arcsin x/l)}{l^2 - x^2} + F_0\cos\omega t$$

Expanding the fraction in power series in x and retaining terms up to x^3, we obtain

$$m\ddot{x} + \left(3k - \frac{2P}{l}\right)x + \left(\frac{4k}{3l^2} - \frac{P}{l^3}\right)x^3 = F_0\cos\omega t$$

It is obvious that for the linearized case, omitting the x^3 term, the system is unstable for $3k - 2P/l < 0$ or $P_{cr} = 3kl/2$. For a larger deformation, the problem has to be treated as nonlinear.

The system is of the hard sprint type if $4k/3l^2 - P/l^3 > 0$, that is, when $P < 4kl/3$; otherwise, it is of the soft spring type.

The resonant frequency is given by equation (12.35):

$$\omega^2 = \frac{3k - 2P/l}{m} + \frac{3}{4}\left(\frac{4k}{3ml^2} - \frac{P}{ml^3}\right)A^2 - \frac{F_0}{A}$$

The first term on the right represents the linear system response and the second the effect of the nonlinearity.

(Expansion of the nonlinear element in a power series is used to evaluate the resonant frequency of an elastically restrained column on the vibration amplitude.)

□

Linstedt Perturbation Method

Working on celestial mechanics problems,[6] Linstedt developed the solution of equation (12.32) in a power series with respect to a small parameter ϵ to obtain harmonic solutions:

$$x = x_0 + \epsilon x_1 + \epsilon^2 x_2 + \epsilon^3 x_3 + \cdots \tag{12.36}$$

where $x, x_0, x_1, x_2, \ldots$ are harmonic functions of time. In particular, $x(t)$ should be harmonic with frequency ω.

If $\theta = \omega t$ is used as an independent variable, let

$$\omega^2 x'' + \alpha x + \beta x^3 = F_0 \cos\theta \tag{12.37}$$

where a prime denotes derivative in respect to θ. Assuming a periodic solution, $x(0) = A$, $\dot{x}(0) = 0$. Using β as the perturbation parameter, assuming that it is small,

$$x = x_0 + \beta x_1 + \beta^2 x_2 + \beta^3 x_3 + \cdots \tag{12.38}$$

$$\omega = \omega_0 + \beta\omega_1 + \beta^2\omega_2 + \beta^3\omega_3 + \cdots \tag{12.39}$$

Substituting in (12.37) will yield a power in β that must vanish identically in β. Therefore, all coefficients of the successive powers of β must vanish. The resulting differential equations, together with the initial conditions that apply also to $x_1, x_2, x_3, \ldots$, yield the unknown functions $x_0, x_1, x_2, x_3, \ldots$. We obtain [Stoker, 1950]

$$x_0 = A\cos\omega t, \qquad \omega_0^2 = \alpha \tag{12.40}$$

$$x_1 = \frac{A^3}{32\alpha}(-\cos\theta + \cos 3\theta), \qquad \omega_1 = \frac{1}{2\omega_0}\left(\frac{3}{4}A^2 - \frac{F_0}{A}\right)$$

[6]Linstedt, A. 1883. Differentialgleichungen der Störungtheorie. *Comment. Acad. Sci. Imperialis Petropolitanae, 31.*

Higher harmonics can be obtained if the procedure is continued. We observe again the superharmonics. Subharmonics can also be found in nonlinear systems [Stoker, 1950].

The Ritz–Galerkin Method

It was shown in Chapter 10 that the Galerkin method can be used[7] to obtain the best approximation by minimizing the error in satisfying the differential equation over the system range. The expansion

$$x(t) = \sum_{j=1}^{n} c_i x_i(t) \tag{12.41}$$

approximates a function $x(t)$ that is a solution of the differential equation

$$L(x; t) = 0 \tag{12.42}$$

where L is a differential operator, linear or nonlinear. For example, for a nonlinear system,

$$L(x; t) = \ddot{x} + \phi(\dot{x}) + f(x) - F(t) = 0 \tag{12.43}$$

Substitution of (12.41) into (12.43) will not, in general, yield zero on the left-hand side. A measure of the deviation from zero over a vibration cycle is the integral $\int_0^T x_i(t)L(x; t)\, dt$ for each of the trial functions $x_i(t)$. Setting this deviation equal to 0 for each trial function gives

$$R_i = \int_0^T x_i(t)L(x; t)\, dt = 0, \qquad i = 1, 2, \ldots, n \tag{12.44}$$

This is a homogeneous system of n linear algebraic equations in $c_1, c_2, \ldots, c_n$. Its solution will yield the natural frequencies and the constants c_i, which, in turn, will yield the natural modes with equation (12.41).

☐ **Example 12.7 Application of Ritz–Galerkin Method to a Nonlinear System**

Find an approximate solution to a general second-order nonlinear system under harmonic excitation using the Ritz–Galerkin method ([Klotter, 1951] and [Seireg, 1969]). **Solution** A general form of the differential equation for a nonlinear system with harmonic excitation is

$$L(x; t) = \ddot{x} + f_1(\dot{x}) + f_2(x) - F \cos \omega t = 0 \tag{a}$$

A two-term Galerkin approximation is used in the form

$$x = a_1 \sin \omega t + a_2 \cos \omega t = X \sin(\omega t - \phi) \tag{b}$$

---

[7]Galerkin, B. G. 1915. Reihenentwinklungen für einige Fäle des Gleichgewichts von Platten und Balken, *Wjestn. Ing. Petrograd, 19.* Biezeno, C. B., and Grammel, R. 1939. *Technische Dynamik.* Berlin: Springer-Verlag. Introduced for vibration problems by [Klotter, 1951].

Galerkin conditions are

$$\int_0^{2\pi/\omega} L(x; t) \sin \omega t \, dt = 0 \tag{c}$$

$$\int_0^{2\pi/\omega} L(x; t) \cos \omega t \, dt = 0 \tag{d}$$

Substituting $x = X \sin(\omega t - \phi)$ into the differential equation yields

$$L(x; t) = -\omega^2 X \cos(\omega t - \phi) + f_1[X\omega \cos(\omega t - \phi)]$$
$$+ f_2[X \sin(\omega t - \phi)] - F \cos \omega t \tag{e}$$

The Galerkin conditions give

$$-\omega^2 \cos \phi + \sin \phi \, G(X, \omega) + \cos \phi \, H(X) = \frac{F}{X} \tag{f}$$

$$-\omega^2 \sin \phi - \cos \phi \, G(X, \omega) + \sin \phi \, H(X) = 0 \tag{g}$$

where

$$G(X, \omega) = \frac{1}{\pi X} \int_0^{2\pi} f_1(X\omega \cos \lambda) \cos \lambda \, d\lambda \tag{h}$$

$$H(X) = \frac{1}{\pi X} \int_0^{2\pi} f_2(X \sin \lambda) \sin \lambda \, d\lambda \tag{i}$$

Squaring equations (f) and (g) and adding yields

$$[H(X) - \omega^2]^2 + G^2(X, \omega) = \left(\frac{F}{X}\right)^2 \tag{j}$$

which gives the response implicitly. From equation (g),

$$\tan \phi = \frac{G(X, \omega)}{H(X) - \omega^2} \tag{k}$$

(A two-term Galerkin approximation is used to obtain the solution of a second-order nonlinear system.)

□

12.4 NUMERICAL METHODS

Point Methods

Nonlinear and, sometimes, linear problems can conveniently be solved with numerical methods in a digital computer. We consider here systems of ordinary differential equations in the normal form[8]

$$\dot{\mathbf{x}} = \mathbf{f}(\mathbf{x}, t) \tag{12.45}$$

---

[8] Euler, L. 1728. *Comment. Acad. Sci. Imperialis Petropolitanae, 3:* 124–137.

$$\mathbf{x} = \{x_1 \quad x_2 \quad \cdots \quad x_n\}, \qquad \mathbf{f} = \{f_1 \quad f_2 \quad \cdots \quad f_n\}$$

A second-order system can generally be transformed in normal form. For example, for a single-degree-of-freedom system, we have, in general

$$\ddot{x} + f(x, \dot{x}, t) = 0 \tag{12.46}$$

where $f(x, \dot{x}, t)$ is a nonlinear function. We can define $x_1 = \dot{x}$, $x_2 = x$ and equation (12.46) becomes

$$\dot{x}_1 = f_1(x_1, x_2, t) \tag{12.47}$$
$$\dot{x}_2 = f_2(x_1, x_2, t)$$

where

$$f_2(x_1, x_2, t) = x_1, \qquad f_1(x_1, x_2, t) = -f(x_2, x_1, t) \tag{12.48}$$

The system (12.47) can be written in matrix form as in equation (12.45). The initial conditions are $\mathbf{x} = \mathbf{x}_0$ at time t.

The solution of equation (12.45) is to be computed numerically at time $t + h$, where h is a small increment of time. Using the computed values of $\mathbf{x}$ as initial conditions, we can repeat the procedure for times $t + 2\Delta t$, $t + 3\Delta t$, and so on. In this way we compute the response at distinct points in time and construct the response function as a sampled function (see Chapter 1).

The value of a function $x(t)$ at an incremental value of the independent variable $t + h$ is given by the Taylor expansion[9]:

$$x(t + h) = x(t) + \dot{x}(t)h + \ddot{x}(t)\frac{h^2}{2!} + \dddot{x}(t)\frac{h^3}{3!} \cdots \tag{12.49}$$

If the time derivative of the function $\dot{x}(t)$ is $f(x, t) = \dot{x}(t)$, for small enough h, one can retain the first derivative only:

$$x(t + h) \simeq x(t) + f(x, t)h \tag{12.50}$$

and the *truncation error* is on the order of $\ddot{x}(t)h^2$, the next term in the Taylor expansion. Incidentally, this estimate of the error shows that it is related to the time step h and the derivative at this point. This means that the higher the value of the derivative, the smaller the time step should be to keep the truncation error below a certain value. The method of numerical integration expressed by equation (12.50) is known as the *Euler method*.

The estimation of the function at time $t + h$ with equation (12.50) is shown in Figure 12.8, corresponding to point B. It can be seen that this point is below the true function, because the value of the derivative changes in the interval h. An improvement of the estimate given by equation (12.50) can be achieved if the value of the derivative is computed at time $t + h$, then using equation (12.50) with the average value of the two derivatives at t and $t + h$, point C in Figure 12.8,

---

[9]Known in some form in India a century before, was discovered in the Western world by Gregory, J. 1668. *Geometriae pars universalis,* Padua, and *Exersitationes geometricae,* London. Traditionally related to Taylor, B. 1715. *Methodus incrementorum directa et inversa.* London.

$$x_0 = x(t)$$

$$x_1(t + h) \simeq x_0 + f(x, t)h \qquad \text{(a)} \quad (12.51)$$

$$x_2(t + h) = x_0 + \frac{h[f(x_0, t) + f(x_1, t + h)]}{2} \qquad \text{(b)}$$

Equation (12.51b) is the *corrector,* while the one computed by equation (12.51a) is the *predictor,* and they express the *improved Euler method.* The error is on the order h^3.

Equations (12.51) can be written in an alternative form,

$$k_1(t + h) \simeq f(x_0, t)h \qquad (12.52)$$

$$k_2(t + h) \simeq f(x_1, t + h)h \qquad (12.53)$$

$$x(t + h) \simeq 0.5k_1(t + h) + 0.5k_2(t + h) + x_0 \qquad (12.54)$$

Equations (12.52) to (12.54) constitute the first-order *Runge–Kutta method.*[10] In a more general form, Runge's method consists of setting up the forward integration scheme, equation (12.54), in the form

$$x(t + h) = x(t) + R_1 K_1 + R_2 K_2 + R_3 K_3 + R_4 K_4 \qquad (12.55)$$

where

$$K_1 = hf(x_0, t)$$

$$K_2 = hf(x_0 + \beta K_1, t + \alpha h)$$

$$K_3 = hf(x_0 + \beta_1 K_1 + \gamma_1 K_2, t + \alpha_1 h)$$

$$K_4 = hf(x_0 + \beta_2 K_1 + \gamma_2 K_2 + \delta K_3, t + \alpha_2 h)$$

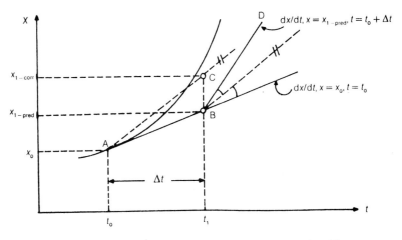

Figure 12.8 Euler and modified Euler integration algorithms.

[10]Runge, K. 1895. *Math. Ann., 46:* 167. Later modifications are due, among others, to Kutta, W. 1901. *Z. Math. Phys., 46:* 435.

and the four constants $R_1, \ldots, R_4$ and the nine constants $\alpha, \beta, \ldots, \delta$ should be determined so that equation (12.55) matches the Taylor series, equation (12.47), up to and including the term in h^4. Runge's solution is[11]

$$R_1 = R_4 = \tfrac{1}{6}, \qquad R_2 = R_3 = \tfrac{1}{3}$$

$$\alpha = \beta = \alpha_1 = \beta_1 = \gamma_1 = \tfrac{1}{2} \tag{12.56}$$

$$\alpha_2 = \delta = 1, \qquad \beta_2 = \gamma_2 = 0$$

Extrapolation Methods

The methods discussed in the preceding section utilize only information at the last step x_0. Every forward step, one has to compute two values of the function $f(x, t)$ with the modified Euler method, and four values with the fourth-order Runge–Kutta method. It is possible to use previous values of the function $x(t)$ and its derivative to improve accuracy or reduce the computation effort.

We assume that the values of the function $x(t)$ and the derivative $\dot{x} = f(x, t)$ are available at times $\ldots, t_{n-2}, t_{n-1}, t_n$ and that the value at the forward step t_{n+1} is to be computed. This can be done by integrating the derivative between times $t - h$ and $t + h$:

$$x_{n+1} = x_{n-1} + \int_{t-h}^{t+h} f(x, t)\, dt \tag{12.57}$$

The integral in equation (12.57) can be approximated with a cubic polynomial in t that coincides with the values of $f(x, t)$ at four successive points, $t_{n-3}, t_{n-2}, t_{n-1}$, and t_n, and integrate it from t_n to t_{n+1} to yield the *predictor* of the value of the function $u_{n+1} = x(t_{n+1})$:

$$u_{n+1} = x_n + \frac{h}{24} \left[-9f(x_{n-3}, t_{n-3}) + 37f(x_{n-2}, t_{n-2}) \right.$$

$$\left. - 59f(x_{n-1}, t_{n-1}) + 55f(x_n, t_n) \right] \tag{12.58}$$

Then a cubic polynomial is computed to coincide with the function $f(x, t)$ at four successive points, t_{n-2}, t_{n-1}, t_n, and t_{n+1}, and integrate it from t_n to t_{n+1} to yield the *corrector* of the value of the function $x_{n+1} = x(t_{n+1})$:

$$x_{n+1} = x_n + \frac{h}{24} \left[f(x_{n-2}, t_{n-2}) - 5f(x_{n-1}, t_{n-1}) \right.$$

$$\left. + 19f(x_n, t_n) + 9f(u_{n+1}, t_{n+1}) \right] \tag{12.59}$$

where u_{n+1} on the right-hand side is the predictor value from equation (12.58). This method, called the *Adams–Moulton method,* needs at the start the value of the function at three points, which can be obtained with a point method. A new start is necessary at points of discontinuity of the function $f(x, t)$, because the extrapolation is not valid. Moreover, the method needs two evaluations of the function at every integration step.

[11]The lengthy derivation can be found in [Ince, 1926].

Integration Errors

Inherent questions in any numerical integration scheme relate to numerical accuracy. Truncation error was discussed above and was related to the order of approximation of the Taylor expansion of the function. Thus the order of the *local* (for every integration step) truncation error is h^2 for the Euler method, h^3 for the modified Euler method, h^5 for the fourth-order Runge–Kutta method, and h^5 for the Adams–Moulton method.

It seems that when the integration step decreases, the computation effort increases, but the error becomes smaller to a certain power. Beyond the increase in the computation effort, there is a low limit to the integration step because of the limited accuracy of the numerical computation. If n digits are used in the computation, the *round-off error* is on the order of 10^{-n}. If the integration is performed over time T, the maximum accumulation of the round-off error is $10^{-n}T/h$. The ratio $T/10^n h$ is the estimate of the error and it can be quite large for small enough h and long integration time T.

A very important feature for an intelligent numerical integration scheme for numerical vibration analysis is the error control. To this end, an estimate of the error is performed at every step. If found to be greater than a preset higher value, the integration step is reduced (say, to one-half), to improve accuracy. If found to be less than a preset lower value, the integration step is increased (say, is doubled), to improve computational efficiency.

To estimate the integration error, Merson has modified the Runge–Kutta method in the following form [Gerald and Wheatley, 1984]:

$$x(t + h) = x(t) + \tfrac{1}{6}(K_1 + 4K_4 + K_5) \tag{12.60}$$

where

$$K_1 = hf(x_0, t)$$

$$K_2 = hf\left(x_0 + \frac{K_1}{3}, t + \frac{h}{3}\right)$$

$$K_3 = hf\left(x_0 + \frac{K_1}{6} + \frac{K_2}{6}, t + \frac{h}{3}\right)$$

$$K_4 = hf\left(x_0 + \frac{K_1}{8} + \frac{3K_3}{8}, t + \frac{h}{2}\right)$$

$$K_5 = hf\left(x_0 + \frac{K_1}{2} - \frac{3K_3}{8} + 2K_4, t + h\right)$$

The error is then

$$E = \tfrac{1}{30}(2K_1 - 9K_3 + 8K_4 - K_5) \tag{12.61}$$

The truncation error is proportional to some power of h and the time derivative of the function. For a harmonic function, the amplitude of its derivative is ωA, where ω is the angular velocity and A the vibration amplitude. Therefore, the higher the frequencies in the response spectrum, the higher the integration error expected. To overcome this, the integration step has to be small enough, subject to the limitations due to the round-off error. A rule of thumb is that the integration step should be one to two orders of magnitude smaller than the period corresponding to the maximum frequency expected in the response spectrum. It is apparent that high frequencies in the spectrum require very small integration steps for adequate error control. In addition to requiring substantial integration effort, they might lead to accumulation of high round-off error. This type of system, having a mixture of very low and very high frequencies (i.e., very high and very low eigenvalues), is called a *stiff system* and requires that special methods for *stiff differential equations* be employed, such as *Gear's algorithm* ([Gear, 1971] and [Gerald and Wheatley, 1984]).

ALGORITHM RKM($x,n,t,E_{max},E_{min},h,f$)

```
/* Runge-Kutta-Merson Method for a system of n Ordinary Differential
Equations with error control*/
External Derivative(x,f,n,t)
Global Real Array x(n),f(n)
Global Real Variable x,E_max,E_min,h
Global Integer Variable n
Local Real Array x1(n),k1(n),k2(n),k3(n),k4(n),k5(n)
Local Real Variable t0
begin
  For i=1 to n, do ; x1(i)=x(i) ; end do
  t0=t ; DERIVATIVE(x,f,n,t)
  For i=1 to n, do ; k1(i)=hf(i) ; end do
  t=t0+h/3 ; For i=1 to n, do ; x(i)=x1(i)+k1(i)/3; end do
  DERIVATIVE(x,f,n,t)
  For i=1 to n, do ; k2(i)=hf(i) ; end do
  For i=1 to n, do ; x(i)=x1(i)+k1(i)/3+k2/6 ; end do
  DERIVATIVE(x,f,n,t)
  For i=1 to n, do ; k3(i)=hf(i) ; end do
  t=t0+h/2 ; For i=1 to n, do ; x(i)=x1(i)+k1(i)/8+3k3/8; end do
  DERIVATIVE(x,f,n,t)
  For i=1 to n, do ; k4(i)=hf(i) ; end do
  t=t0+h ; For i=1 to n, do ; x(i)=x1(i)+k1(i)/2-3k3/2+2k4; end do
  DERIVATIVE(x,f,n,t)
  For i=1 to n, do ; k5(i)=hf(i) ; end do
  Error=0
  For i=1 to n, do
     x(i) = x1(i) + [k1(i)+4k4(i)+k5(i)]/6
     E = [2k1(i) - 9k3(i) + 8k4(i) - k5(i)] /30
     if abs(E)>error then error=abs(E)
  end do
  if error>E_max then         /* half time step and repeat step */
     t=t0 ; h=h/2
     For i=1 to n, do ; x(i)=x1(i) ; end do
     go to begin
```

```
    else
        if error‹Emin then          /* double time step and repeat step */
        t=t₀ ; h=h/2
        For i=1 to n, do ; x(i)=x₁(i) ; end do
        go to begin
        else exit
        end if
    end if
end
```

☐ **Example 12.8 Numerical Solution of the Torsional Vibration of a Single-Cylinder Engine**

The differential equation for torsional vibration of a single-cylinder piston engine (Figure E12.8a) is

$$(1 - \epsilon \cos 2\tau)\gamma'' + 2\epsilon \sin 2\tau\gamma' + \left(\frac{1}{r^2} + 2\epsilon \cos 2\tau\right)\gamma = -\epsilon \sin 2\tau \qquad (a)$$

where $\tau = \omega t$, ω is the angular velocity, γ is the displacement of torsional motion, $\epsilon = 1/2Ma^2/(I + 1/2Ma^2)$, M is the mass of reciprocating parts, I is the moment of inertia of the rotating parts, a is the crank radius, and r is the ratio of the angular velocity ω to the natural frequency ω_n of the system. Calculate the response using the Runge–Kutta method.

Solution   In the routine RUKUTA, of the software package, the fourth-order Runge–Kutta method of numerical integration has been programmed. To this end, we rewrite equation (a) in the normal form (phase plane)

$$\dot{u}_1 = f_1(u_1, u_2, t)$$

$$\dot{u}_2 = f_2(u_1, u_2, t)$$

or

$$\dot{\mathbf{u}} = \mathbf{f}(\mathbf{u}, t) \qquad (b)$$

where

$$u_1 = v = \gamma', \qquad u_2 = \gamma, \qquad \mathbf{u} = \{u_1 \quad u_2\}, \qquad \mathbf{f} = \{f_1 \quad f_2\}$$

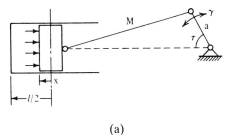

(a)

Figure E12.8

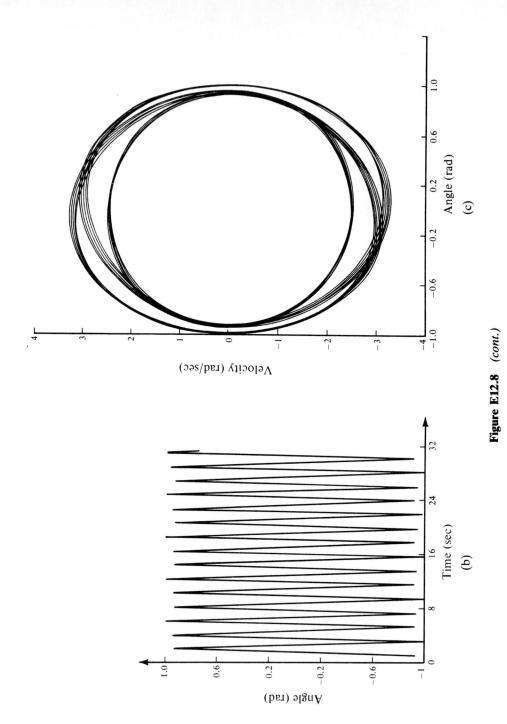

Figure E12.8 *(cont.)*

Starting from any initial value of $u_1 = v_0$, $u_2 = \gamma_0$, we can compute the values of u_1 and u_2 after time h with the Runge formulas, from equations (12.56):

$$K_1 = hf(u(t), t)$$

$$K_2 = hf\left(u(t) + \left(\frac{K_2}{2}, t + \frac{h}{2}\right)\right)$$

$$K_3 = hf\left(u(t) + \frac{K_2}{2}, t + \frac{h}{2}\right)$$

$$K_4 = hf(u(t) + K_3, t + h)$$

$$u(t + h) = u(t) + \tfrac{1}{6}(K_1 + 2K_2 + 2K_3 + K_4)$$

This can be repeated for many time steps h and thus the solution can be constructed step by step. This process has been programmed in a digital computer and the results plotted in a CALCOMP plotter (Figure E12.8b and c). The FORTRAN computer program and results are shown below together with the input parameters.

```
      PROGRAM SIMULTN
      DIMENSION TIME(1500),X(1500),V(1500)
      DIMENSION U(2),f(2)
      U(1)=1.
      U(2)=0.
      NPOINT=1500
      NOVAR=2
      DT=6.28/100.
      T=0.
      DO 40 I=1,NPOINT
      CALL RUKUTA(T,NOVAR,U,F)
      TIME(1)=T
      X(I)=U(2)
      V(I)=U=(1)
   40 CONTINUE
      CALL NAMPLT
      NP=NPOINT
 CALL QIKPLT(TIME,X, NP,6H*TIME*,7H*ANGLE*,20H*TIME PLOT*)
      CALL QIKPLT(X,V, NP,7H*ANGLE*,15H* VELOCITY      *)
      140H PHASE PLANE PORTRAIT OF THE MOTION      *)
      CALL ENDPLT
      STOP
      END

      SUBROUTINE DERIV(U,F,NOVAR,T)
      DIMENSION U(NOVAF),F(NOVAR)
      R=SORT(1./9. )
      EPS=.80
      DIV=1.-EPS*COS(2.*T)
      F(1)=-2.*EPS*SIN(2.*T)/DIV*U(1)-(1./R**2+2.*EPS*COS(2.*T))
      1/DIV*U(2)
    C 1/DIV*U(2)-EPS*SIN(2.*T)/DIV
```

```
C   FORCING TERM OUT
      F(2)=U(1)
      RETURN
      END
```

(Numerical solution of the torsional vibration of a single-cylinder engine was obtained using the Runge–Kutta method.)

☐

12.5 STABILITY

Up to this point, stability has been discussed on many occasions but without a general definition. Perhaps the term *stability* is so expressive that tells its own story. Simply stated, when a system or device of some sort operates under certain conditions, if we slightly vary these conditions, this can have small or considerable effect on the device. The first case is called *stability,* the second *instability*.

For example, if the simple pendulum is at equilibrium at the low point $\theta = 0$ and we displace it from this point by an angle θ, it will move in a harmonic oscillation of the order of the initial displacement. In fact, due to the presence of damping, after some time it is going to return to equilibrium. The system is stable *at the equilibrium* $\theta = 0$. At the higher equilibrium position, $\theta = \pi$, a slight displacement from equilibrium will make the pendulum move downward and the amplitude of the oscillation will be unrelated to the initial displacement but will be π. After any length of time, the system will not return to the equilibrium $\theta = \pi$, but in the presence of damping, it will return to another equilibrium position $\theta = 0$. The system is unstable *at the equilibrium position* $\theta = \pi$.

We shall restrict our discussion to a system:

1. That can be described by ordinary differential equations

$$\dot{\mathbf{x}} = \mathbf{f}(\mathbf{x}, t) \tag{12.62}$$

   where

$$\mathbf{x} = \{x_1 \quad x_2 \quad \cdots \quad x_n\} \quad \text{and} \quad \mathbf{f} = \{f_1 \quad f_2 \quad \cdots \quad f_n\}$$

   The functions $f_1, f_2 \ldots$ are nonlinear functions of the state variables $x_1, x_2,$ . . . . If a system has another form, it can usually be transformed to the form (12.62).

2. In which the functions are *autonomous,* in other words, $f_1, f_2, \ldots, f_n$ do not have the independent variable t appearing explicitly. This also means that there is no external force acting on the system.

In linear systems generally there is one *equilibrium point,* or *singular point,* at a value of the coordinate vector $\mathbf{x}$ that satisfies the system of equations $\mathbf{f}(\mathbf{x}) = 0$. For such circumstances of linear systems we have, in general, a unique solution.

Without loss in generality, we can assume that $\mathbf{x}(0) = \mathbf{0}$ is the equilibrium, for if it is not, we can redefine the coordinate system as $\mathbf{x}^* = \mathbf{x} - \mathbf{x}(0)$.

For a nonlinear system, the equilibrium is given by the equation $\mathbf{f}(\mathbf{x}) = \mathbf{0}$. This is a system of nonlinear algebraic equations and can have many solutions corresponding to the several equilibria of the system. For example, the equation for the simple pendulum can be written as $\ddot{\theta} + (g/l) \sin \theta = 0$. In the normal form of equation (12.62) we can write, for $x_1 = \dot{\theta}$, $x_2 = \theta$,

$$\dot{x}_1 = -\frac{g}{l} \sin x_2$$

$$\dot{x}_2 = x_1 \tag{12.63}$$

The equilibrium will be at $-(g/l) \sin x_2(0) = 0$ and $x_1(0) = 0$. There are an infinite number of solutions: $x_1(0) = 0$ and $x_2(0) = 0$, π, 2π, 3π, ... In reality, only two solutions exist: $x_1(0) = 0$ and $x_1(0) = \pi$. The others are repetitions of the two. Defining $x_1^* = x_1$, $x_2^* = x_2 - \pi$, the system is equivalent to the following two systems:

$$
\begin{array}{cc}
\text{(a)} & \text{(b)} \\
\dot{x}_1 = -\dfrac{g}{l} \sin x_2 & \dot{x}_1^* = -\dfrac{g}{l} \sin(x_2^* + \pi) \\
\dot{x}_2 = x_1 & \dot{x}_2^* = x_1^* \\
x_1(0) = 0, \quad x_2(0) = 0 & x_1(0) = 0, \quad x_2^*(0) = 0
\end{array}
\tag{12.64}
$$

We know already that system (a) is stable and that system (b) is unstable.

For an autonomous system, we shall define stability as follows. The equilibrium state $\mathbf{x}(0) = \mathbf{0}$ is:

1. *Stable* if, given an initial deviation from the equilibrium, every element of $\mathbf{x}$ remains bounded for any length of time and is of the same order of magnitude as the initial deviation.
2. *Asymptotically* stable if, in addition, $\mathbf{x}$ tends to the equilibrium $\mathbf{x}(0) = \mathbf{0}$ as time increases indefinitely.
3. *Unstable* if, no matter how small the initial perturbation is, at least some of the elements of $\mathbf{x}$ increase far beyond the initial deviation as time increases indefinitely.

These definitions are not free of ambiguity, and from a practical standpoint they are not always acceptable. For example, if a system is stable in the sense of definition 1 but it has a stable solution $\mathbf{x}(t)$ with high amplitudes, the design engineer would call it unstable. However, these definitions will serve as a starting point. Moreover, the definitions above have to be amended to include equilibrium states associated with periodic responses, such as the response to harmonic excitation. On the phase plane, they appear as closed trajectories, also called *limit cycles*. Stability then has to be defined with respect to deviation from the trajectory instead of the equilibrium point, and is known as stability *in the Poincaré sense*.

The stability of differential equations was advanced in its present form by the Russian mathematician Liapounov.[12] His methods will be outlined here.

The First Method of Liapounov

A *normal autonomous system* will be defined as an autonomous system expressed by way of a normal system of ordinary differential equations of the form of equations (12.62). Let an equilibrium state of the system be $\mathbf{a} = \{a_1 \ a_2 \ \cdots \ a_n\}$. Let

$$x_i = a_i + \Delta x_i, \qquad i = 1, 2, 3, \ldots, n \qquad (12.65)$$

Substituting (12.53) into (12.50) and expanding in Taylor series in respect to Δx_1, $\Delta x_n, \ldots, \Delta x_n$ yields

$$\Delta x_i = f_i(\mathbf{a}) + \sum_{j=1}^{n} \frac{\partial f_i(\mathbf{a})}{\partial x_j} \Delta x_i + R_i, \qquad i = 1, 2, 3, \ldots, n \qquad (12.66)$$

where R is a higher-other remainder. Because $\mathbf{a}$ is an equilibrium state, $f_i(\mathbf{a}) = 0$. Defining $a_j^i = \partial f_i(\mathbf{a})/\partial x_j$, we obtain the *perturbed system equations*

$$\Delta \dot{x}_i = \sum_{j=1}^{n} a_j^i \Delta x_j + R_i \qquad (12.67)$$

Theorem 1. If all the eigenvalues of the matrix $\mathbf{A} = [a_j^i]$ have negative real parts, the equilibrium state $\mathbf{a}$ of the system (12.62) is asymptotically stable.

For linear systems, we have seen this already. Positive real parts will result in exponential solutions that grow beyond bound. In nonlinear systems we observe that in the vicinity of the equilibrium they behave like the linear system that results if we linearize the system as indicated in equation (12.67), geometrically as if we substitute the nonlinear functions f_i with their derivatives through the equilibrium state. Linearization is a very common method of dealing with nonlinear systems if we are interested in their behavior near the equilibrium state.

The validity of the first method of Liapounov extends only to a certain region about the equilibrium position. The obvious question is: How large is this region? This is the same equation as the one for the validity of linearization. Only application of more general methods or a complete solution of the differential equations or experimental verification could answer such questions. For practical problems, experience can usually assure validity of the method. However, as we have seen in several examples, stability of the linearized model is accompanied by instabilities at other equilibrium states or at a considerable distance from the equilibrium state. The first method of Liapounov, which can be used to investigate stability in the vicinity of equilibrium states, is referred to as *stability in the small,* and although it looks insecure, it is a very useful method for practical applications where repeated application can give the engineer an assurance of its range of validity.

However, the need is apparent for a method that will investigate stability in a specified large range, the range of operation of the device under investigation. This is called *stability in the large* and is dealt with by way of the second method of Liapounov.

---

[12]Liapounov. 1907. *Ann. Fac. Sci. Toulouse, 2:* 203–469. Originally published in Russian.

Portrait of Huygens.
(Relief by Cornelis de Viszcher.)

CHRISTIAN HUYGENS (1629–1695)

Huygens was born in the Hague and studied at Leyden with Frans van Schooten the Younger. In 1651, at 22, he published the first of a number of papers dealing with geometry. His brother devised a better lens and Huygens was able to settle many problems in observational astronomy. He further introduced the idea of "mathematical expectation" while working on probabilities. In 1667 he published his major work, *Horologium oscillatorium,* in which he discusses the pendulum clock he invented in 1656, falling bodies in vacuum or sliding on an inclined plane or curve, motion on an inverted cycloid, and the theory of clocks. He studied the nonlinear behavior of the pendulum and devised the constant-period pendulum. He did extensive work on the centrifugal force, developing the formula $F = mv^2/R$, and on the wave theory of light.

□ **Example 12.9 Stability of Journal Bearings**

A fluid film type of bearing is a device used to support rotating shafts (Figure E12.9). The rotating journal generates a pressure profile of the oil film between journal and bearing. This results in both a vertical and a horizontal force, which are functions of the bearing geometry, the rotating speed, and the motion of the journal, described by the vertical and horizontal displacements x and y and velocities $\dot{x}$ and $\dot{y}$. Apply the first method of Liapunov to study the stability of the system.

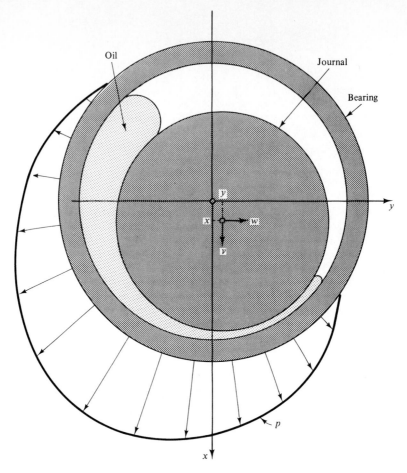

Oil

Journal

Bearing

p

Figure E12.9

Solution If we call F_x and F_y the vertical and horizontal force, respectively, and if m is the journal mass, assuming that it is supported by one journal bearing only, application of Newton's law yields

$$m\ddot{x} = -F_x(x, y, \dot{x}, \dot{y}) + mg$$

$$m\ddot{y} = -F_y(x, y, \dot{x}, \dot{y})$$

In normal form,

$$\dot{v} = \frac{-F_x(x, y, v, w)}{m} + g = f_1(x, y, v, w)$$

$$\dot{x} = v \qquad\qquad = f_2(x, y, v, w)$$

$$\dot{w} = \frac{-F_y(x, y, v, w)}{m} \qquad = f_3(x, y, v, w)$$

$$\dot{y} = w \qquad\qquad = f_4(x, y, v, w)$$

The equilibrium is at $f_1 = f_2 = f_3 = f_4 = 0$. This yields

$$F_x[x(0), y(0), 0, 0] = mg$$

$$F_y[x(0), y(0), 0, 0] = 0$$

Solution of this system of algebraic equations will yield the equilibrium state $x(0)$, $y(0)$, $v(0) = 0$, and $w(0) = 0$. If $x(0) = x_0$ and $y(0) = y_0$, we set up $x = x_0 + \Delta x$, $y = y_0 + \Delta y$, $v = 0 + \Delta v$, and $w = 0 + \Delta w$ and we obtain

$$\Delta \dot{v} = - \frac{K_{xx} \Delta x + K_{xy} \Delta y + C_{xx} \Delta v + C_{xy} \Delta w}{m} + o(x^2, \ldots)$$

$$\Delta \dot{x} = \Delta v$$

$$\Delta \dot{w} = - \frac{K_{yy} \Delta y + K_{yx} \Delta x + C_{yy} \Delta v + C_{yx} \Delta w}{m} + o(x^2, \ldots)$$

$$\Delta \dot{y} = \Delta w$$

where

$$K_{ij} = \frac{\partial F_i}{\partial j} \quad \text{with } i, j = x, y$$

$$C_{ij} = \frac{\partial F_i}{\partial \dot{j}} \quad \text{with } i, j = x, y$$

Therefore, the stability of the system will depend on the sign of the real part of the eigenvalues of the matrix

$$\frac{1}{m} \begin{bmatrix} C_{xx} & K_{xx} & C_{xy} & K_{xy} \\ 1 & 0 & 0 & 0 \\ C_{yx} & K_{yx} & C_{yy} & K_{yy} \\ 0 & 0 & 1 & 0 \end{bmatrix}$$

which are the roots p of the characteristic equation

$$\begin{vmatrix} C_{xx} + p & K_{xx} & C_{xy} & K_{xy} \\ 1 & p & 0 & 0 \\ C_{yx} & K_{yx} & C_{yy} + p & K_{yy} \\ 0 & 0 & 1 & p \end{vmatrix} = 0$$

(The condition for stability of the equilibrium for a fluid bearing was developed using the first method of Liapounov.)

□

□ **Example 12.10   The Simple Pendulum**

Investigate the stability of the pendulum by the first method of Liapounov.
Solution   The equation is, as we know, $\ddot{\theta} + (g/l) \sin \theta = 0$. In normal form we can write $\dot{\omega} = -(g/l) \sin \theta$, $\dot{\theta} = \omega$. The equilibrium is given by $(g/l) \sin \theta = 0$. Let us

consider the equilibrium position $\theta_0 = 0$. Writing the equations of motion in the form (12.66), we obtain

$$\Delta\dot{\omega} = -\frac{g}{l}\cos\theta_0\,\Delta\theta + R$$

$$\Delta\dot{\theta} = \omega \qquad A = \begin{bmatrix} 0 & -\dfrac{g}{l} \\ 1 & 0 \end{bmatrix}$$

The eigenvalues of the matrix A are calculated from the characteristic determinant

$$\begin{vmatrix} p & -\dfrac{g}{l} \\ 1 & p \end{vmatrix} = p^2 + \frac{g}{l} = 0 \qquad \text{or} \qquad p = \pm i\left(\frac{g}{l}\right)^{1/2}$$

The eigenvalues are complex, with zero real part. Therefore, the equilibrium $\theta = 0$ is stable.

For the equilibrium point $\theta_0 = \pi$ (equations 12.64)],

$$\Delta\dot{\omega} = -\frac{g}{l}\cos\theta_0\,\Delta\theta + R$$

$$\Delta\dot{\theta} = \omega \qquad A = \begin{bmatrix} 0 & \dfrac{g}{l} \\ 1 & 0 \end{bmatrix}$$

The eigenvalues of the matrix A are calculated from the characteristic determinant

$$\begin{vmatrix} p & \dfrac{g}{l} \\ 1 & p \end{vmatrix} = p^2 - \frac{g}{l} = 0 \qquad \text{or} \qquad p = \pm\left(\frac{g}{l}\right)^{1/2}$$

The eigenvalues are real; one is positive. Therefore, the equilibrium $\theta = \pi$ is unstable.

(The condition for stability of the equilibrium for the pendulum was developed using the first method of Liapounov.)

$\square$

The Second Method of Liapounov

We consider a normal autonomous system $\dot{\mathbf{x}} = \mathbf{f}(\mathbf{x})$, where $\mathbf{x}$ and $\mathbf{f}$ are vectors. The inner product of two vectors is defined as $\mathbf{x}\cdot\mathbf{y} = x_1 y_1 + x_2 y_2 + \cdots + x_n y_n$. We also consider a scalar function $V(\mathbf{x})$ for which $\mathbf{grad\ V} = [\partial V/\partial x_1 \quad \partial V/\partial x_2 \quad \cdots \quad \partial V/\partial x_n]$. If this function has the following properties, it is called a *Liapounov function:*

1. $V(\mathbf{x})$ is continuous with its first partial derivatives $\partial V/\partial x_1 \cdots \partial V/\partial x_n$ in the region of interest.
2. $V(\mathbf{0}) = 0$.
3. For $\mathbf{x} \neq \mathbf{0}$ but in the region of interest, $V(\mathbf{x}) > 0$.
4. $\dot{V} = \mathbf{f}\cdot\mathbf{grad\ V} = f_1(\partial V/\partial x_1) + f_2(\partial V/\partial x_2) + \cdots + f_n(\partial V/\partial x_n) < 0$.

Properties 1 to 3 depend only on the structure of the function V itself. For example, if $n = 2$, the function $V = x_1^2 + x_2^2$ fulfills requirements 1 to 3 in the entire plain (x_1, x_2). Whether it is a Liapounov function or not will then depend on requirement 4, which depends both on the structure of the function V and on the nonlinear functions f of the system.

The tests for the Liapounov function here are applied in the "region of interest," in other words, not for any value of x but in a certain region, for example in the circle $x_1^2 + x_2^2 < R^2$ or in the square $|x_1| < a$, $|x_2| < a$ or in any other region. Therefore, a Liapounov function is always defined with respect to a certain region and with respect to a system of ordinary differential equation. A function can be a Liapounov function with respect to a certain region or a certain system of ordinary differential equations, but the same function might not be a Liapounov function for another region or for another system of ordinary differential equations. We shall call this the *region of interest R*.

The following theorems are given without proofs.

Stability Theorem.   If for a system $\dot{\mathbf{x}} = \mathbf{f}(\mathbf{x})$ there exists a Liapounov function in a region R about an equilibrium state, this equilibrium state is stable.

Asymptotic Stability Theorem.   If, in addition, the function $-\dot{V}$ is positive definite in R, the stability is asymptotic ($-\dot{V} > 0$ if $\mathbf{x} \neq \mathbf{0}$).

First Instability Theorem.   If $\dot{V}$ is positive definite ($\dot{V} > 0$ for $x \neq 0$), and if V can assume positive values arbitrarily near the equilibrium state, this implies instability of the state.

In all theorems, continuity of V and its first partial derivatives are assumed.

These theorems should be considered as sufficient but not necessary conditions. Thus only partial answers can be offered to questions about stability. This is the reason for the existence of both stability and instability theorems. Application of both theorems with appropriate Liapounov functions will yield regions of stability and instability for the governing parameters. The area between is an area of uncertainty which narrows as the selection of the functions V is more successful. This point is illustrated in the following examples.

☐ **Example 12.11   Stability of the Duffing Oscillator in the Large.**

An oscillator has a nonlinear spring with a restoring force expressed in the form of a power series,

$$m\ddot{x} + k_1 x + k_2 x^2 + \cdots = 0$$

Investigate the stability of the system in the large.
Solution   In normal form the system equation is

$$\dot{v} = -\frac{k_1}{m} x - \frac{k_2}{m} x^2 - \cdots$$

$$\dot{x} = v$$

We select for W the total energy for the oscillator

$$W(x, v) = T + V = \frac{1}{2} mv^2 + \frac{k_1}{2} x^2 + \frac{k_2}{3} x^3 + \cdots$$

Then

$$\dot{W} = \frac{dW}{dt} = \frac{\partial W}{\partial v}\dot{v} + \frac{\partial W}{\partial t} + \frac{\partial W}{\partial x}\dot{x}$$

or

$$\dot{W} = (k_1 x + k_2 x^2 + \cdots)v + mv\left(-\frac{k_1}{m}x - \frac{k_2}{m}x^2 - \cdots\right) \equiv 0$$

If $k_1, k_2, \ldots > 0$, $V > 0$ and the system is stable according to the first stability theorem. However, asymptotic stability cannot be established because the second stability theorem is not satisfied. This does not necessarily mean that the system is not asymptotically stable. If $k_1 < 0$, we take $W = mv^3/3$. We find that

$$\dot{W} = mv^2\left(-\frac{k_1}{m}x - \frac{k_2}{m}x^2 - \cdots\right)$$

There is an area around the equilibrium state where the term $-k_1 v^2 x$ (which is positive) is greater than the rest terms (if x is small enough); therefore, in this region $\dot{W}$ is positive definite. Since W can take positive values (for any positive v), the conditions of the first instability theorem are met and the equilibrium state $x = 0$, $v = 0$ is unstable.

An interesting case is when $k_1 > 0$, $k_2 < 0$; we take $W = mv^3/3 + k_2 x^3/3 + k_3 x^4/4 + \cdots$ and obtain $\dot{W} = -k_2 v^2 x^2$. Therefore, $\dot{W}$ is positive definite and W takes positive values. Therefore, the equilibrium state is unstable. This is very important because the linearized system is stable. However, for $k_1 > 0$ it was proven that the system is stable for a small enough region: Therefore, if $k_1 > 0$ and $k_2 < 0$, there are regions of stability and instability. In such systems stability or instability depends on the initial conditions, because if the orbit comes through an unstable region, this might start the instability.

(The second method of Liapounov is used to investigate the stability of the general case of a second-order nonlinear system, where the nonlinear restoring force is expressed as power series.)

□

□ **Example 12.12 Stability of the Pendulum in the Large**

Investigate the stability of a pendulum in the large.
Solution The equation of motion for the simple pendulum is

$$\ddot{\theta} + \frac{g}{l}\sin\theta = 0 \qquad\qquad (a)$$

To study the stability at the origin, we take

$$V(\theta, \dot{\theta}) = \frac{g}{l}(1 - \cos\theta) + \frac{\dot{\theta}^2}{2} \qquad\qquad (b)$$

This function is positive definite for $|\theta| < \pi$. Further,

$$\dot{V} = \frac{g}{l}\sin\theta\,\dot{\theta} + \dot{\theta}\left(-\frac{g}{l}\sin\theta\right) \equiv 0 \qquad\qquad (c)$$

Therefore, in the region $|\theta| < \pi$ the system is stable, in accordance with the first stability theorem.

In the case that damping $c\dot\theta$ is added to equation (a), $\dot{V} = -c\dot\theta^2$, which is negative definite, because it is zero only at $\theta = 0$, $\dot\theta = 0$. Therefore, the system, again in region $|\theta| < \pi$, is asymptotically stable.

(The second method of Liapounov is used to investigate the stability of the pendulum in the large.)

$\square$

Liapounov Function for Linear Autonomous Systems

We return now to the linear system

$$\dot{x} = Ax \qquad (12.68)$$

The stability of such systems depends on the sign of the real parts of the eigenvalues of the matrix A, according to the first method of Liapounov. If any eigenvalue has a positive real part, the system is unstable. If all eigenvalues have negative real parts, the system is asymptotically stable. If any eigenvalue has zero real part, the system might have a closed cycle.

The latter case is when the second method of Liapounov becomes necessary. However, there are more reasons for this:

1. Determination of all eigenvalues of large systems presents insurmountable difficulties.
2. Application of Routh–Hurwitz criteria is practical only when the system is small enough to determine the characteristic polynomial by expanding the characteristic determinant.

To find a proper Liapounov function for the system (12.68), we try a quadratic form $x^T Q x$, where the matrix Q is symmetric and positive definite (see Chapter 7). Then

$$\dot{V}(x) = \dot{x}^T Q x + x^T Q \dot{x} = x^T A^T Q x + x^T Q A z = x^T (A^T Q + QA)x \qquad (12.69)$$

Let

$$A^T Q + QA = -C \qquad (12.70)$$

If C is positive definite, the system is asymptotically stable in the large. It can also be proved that C is symmetric.

Going backward, we begin with any positive definite matrix C, say $C = I$, and we solve for the $n(n + 1)/2$ components of Q using the systems of equations (12.70). If the resulting matrix Q is positive definite, the system (12.68) is asymptotically stable (Liapounov).

$\square$ **Example 12.13**

Investigate the stability of a harmonic oscillator in the large.

Solution We consider the harmonic oscillator $m\ddot{x} + c\dot{x} + kx = 0$. This can be written, for $m = 1$,

$$\dot{v} = -cv - kx \qquad (a)$$

$$\dot{x} = v$$

Let $\mathbf{C} = \text{diag}[1, 1]$. Equations (12.70) can be written as

$$\begin{bmatrix} -c & 1 \\ -k & 0 \end{bmatrix}\begin{bmatrix} q_{11} & q_{12} \\ q_{21} & q_{22} \end{bmatrix} + \begin{bmatrix} q_{11} & q_{12} \\ q_{21} & q_{22} \end{bmatrix}\begin{bmatrix} -c & -k \\ 1 & 0 \end{bmatrix} = \begin{bmatrix} 1 & 0 \\ 0 & 0 \end{bmatrix}$$

$$\begin{bmatrix} q_{21} - cq_{11} & q_{22} - cq_{12} \\ -kq_{11} & -kq_{12} \end{bmatrix} + \begin{bmatrix} q_{12} - cq_{11} & -kq_{11} \\ q_{22} - cq_{21} & -kq_{21} \end{bmatrix} = -\begin{bmatrix} 1 & 0 \\ 0 & 1 \end{bmatrix}$$

This results in the system of equations

$$q_{21} \qquad - cq_{11} = -\tfrac{1}{2}, \qquad q_{22} - cq_{12} - kq_{11} = 0$$

$$-kq_{11} + q_{22} - cq_{21} = 0, \qquad -kq_{12} - kq_{21} = -1$$

Since $q_{12} = q_{21}$, we obtain

$$q_{11} = 1 + \frac{k}{2ck}, \qquad q_{12} = \frac{1}{2k}, \qquad q_{22} = k + k^2 + \frac{c^2}{2ck}$$

Then $\mathbf{Q}$ is positive definite if $c > 0$, $k > 0$, which, as we already know, are the conditions for asymptotic stability.

$\square$

Limit Cycles of Autonomous Systems: Chaotic Vibration

It has been noted above that there are system states, called limit cycles, which consist of closed trajectories traced with a periodic motion. Some simple such states were:

1. Response of linear systems to harmonic excitation. The amplitude depends on the excitation.
2. Response of conservative systems to initial deviation from equilibrium. The amplitude depends on the initial deviation.
3. Certain parametrically excited systems with damping.

It is apparent that a closed cycle implies that the system is always in motion; therefore, some energy input mechanism is needed to set it in motion initially, and if the system is nonconservative, to maintain the motion.

Consider, for example, the motion of a clock pendulum (Figure 12.9a) of amplitude A. Due to damping, every cycle the amplitude is reduced by $A(1 - e^{-2\pi\zeta})$ (Chapter 2) while the maximum velocity will be reduced by $\omega_d A(1 - e^{-2\pi\zeta})$. Gradually, the motion is reduced to the equilibrium point at the origin. The equilibrium point is called an *attractor*.

If there is a mechanism which at the two positions of maximum velocity gives an impact that increases the velocity by $d/2 = \omega_d A(1 - e^{-2\pi\zeta})/2$, the pendulum will maintain the amplitude $A = d/[\omega_d(1 - e^{-2\pi\zeta})]$ (Figure 12.9b). If the pendulum is displaced further from equilibrium, at every cycle it will lose velocity more than the increase d and gradually will return to the amplitude A. If the initial displacement is less than A, at every cycle the increase in amplitude d will be greater than the decrease due to damping. Therefore, the amplitude will again become A. The

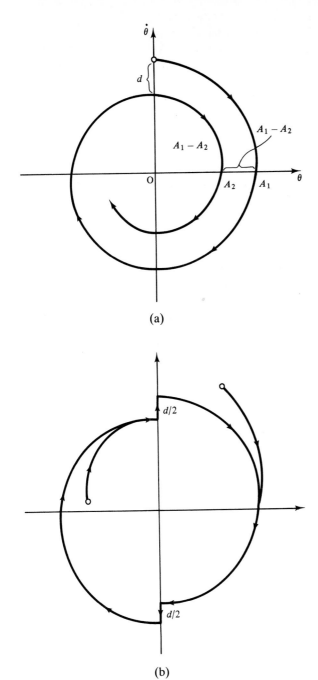

(a)

(b)

Figure 12.9 Limit cycle of an impulse-sustained pendulum.

pendulum is said to move in a closed cycle, which, in addition, is stable. The closed cycle is also an *attractor*.[13]

If the initial displacement of the pendulum and/or the velocity increase per cycle is increased so that the pendulum will move with full rotation, the closed cycle is lost (Figure 12.10). The pendulum moves in a trajectory that is never attracted to an equilibrium point or to a closed cycle. Such motion is called *chaotic*.

There are several other characteristics that contribute to the chaotic behavior. Nonlinearity is certainly a factor, and another factor can be identified by observing Figure 12.10. If the initial amplitude or the velocity increase are such that the maximum amplitude never exceeds the escape line, oscillation is confined within $-\pi < \theta < \pi$ and the motion is a stable limit cycle. At the slightest deviation of the initial displacement or d that can move the trajectory outside the escape line, the trajectory is open and the motion is chaotic. We identify here a mechanism by which a small change in initial conditions can result in disproportionally higher change in the response.[14]

Going back to Figure 12.10, we devise a phase portrait where the pendulum rotation angle θ is taken between $-\pi$ to π, and for $\theta > \pi$ we plot the response in the range $-\pi$ to π. This phase portrait of the response is then shown in Figure 12.11. If the plotting is continued, the figure is going to be covered with a great number of seemingly unrelated segments of the trajectory. There is very little to be concluded from this chaotic plot in this form, which is called a *strange attractor*.

Figure 12.12 shows several types of phase plots. The first system converges on a steady equilibrium point. The second converges to a limit cycle. The third is also a limit cycle with three discrete frequencies. The fourth is chaotic: the Lorenz

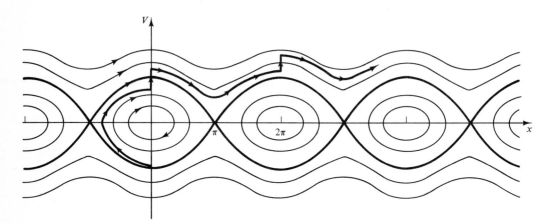

Figure 12.10 Phase portrait of chaotic motion of an impulse-sustained pendulum.

[13]Sarkovskii, A. N. 1964. Coexistence of cycles of a continuous map of a line into itself. *Ukranian Math. J., 16:* 61.

[14]Poincaré, H. 1921. *The Foundations of Science: Science and Method*, English translation. New York: Science Press.

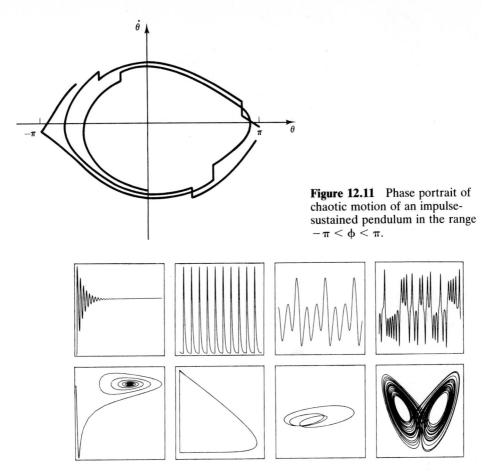

Figure 12.11 Phase portrait of chaotic motion of an impulse-sustained pendulum in the range $-\pi < \phi < \pi$.

Figure 12.12 Phase portraits of different attractors. (From Gleik, J. 1987. *Chaos.* New York: Penguin Books.)

attractor, the first reported strange attractor for thermally induced fluid convection in the atmosphere.[15]

Suppose now that we take a photograph of the phase portrait at time intervals equal to the period of oscillation of the pendulum. For an oscillating undamped pendulum that moves on a limit cycle, every photograph will show the same point on the phase portrait; therefore, the resulting plot will be a point in every frame (Figure 12.13a).

If we do the same with the same frequency but at different locations, say at angles $\pi/6$ apart, the trace of the response on the phase portraits will not be the same. If we assemble the frames as shown in Figure 12.13b with a common velocity axis, rotated with respect to one another at angles $\pi/6$ apart and connect the trace points of the different frames, the closed cycle will appear as a single three-dimensional curve which, in the phase space, in addition to phase portrait infor-

[15]Lorenz, E. N. 1963. Deterministic non-periodic flow. *J. Atmos. Sci., 20:* 130–141.

Portrait of Liapounov.
(Courtesy of the USSR Academy of Science.)

ALEKSANDR MIKHAILOVICH LIAPOUNOV (1857–1918)

Liapounov was born in Jaroslavl and graduated from the University of St. Petersburg in 1880, where he studied under P. L. Chebyshev. He became a professor of the University of Kharkov in 1892 and began working for the St. Petersburg Academy of Sciences in 1902. Liapounov made fundamental contributions to the theory of systems stability, and his dissertation, *General Problem of the Stability of Motion* (1892), had a lasting effect on the field. He worked on the Dirichlet problem and the probability theorem, for which he developed a very general proof of the central limit theorem.

mation, gives the dimension of time since $\phi = \omega t$. This is useful in itself because the phase portrait does not give information about time explicitly.

If the response is chaotic, at every frame there will be no single point but several, because the three-dimensional response curve is different at each cycle. The plot on each of the frames is called a *Poincaré map*. Such maps play a major role in the study of chaotic vibration because in many chaotic phenomena, Poincaré maps show an orderly arrangement of points, some of them of aesthetic quality. For example, the full three-dimensional plot for the chaotic pendulum is shown in Figure 12.14a. The Poincaré map (Figure 12.14b) shows remarkable order. Poincaré maps can be computed numerically using the program POINCARE.

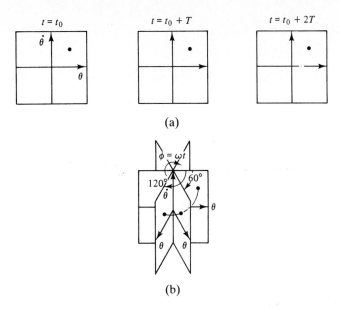

Figure 12.13 Construction of a Poincaré map.

Friction-Induced Vibration: Stick-Slip

The fact that friction force is a nonlinear function of the velocity, $F = \text{sgn}(\dot{x})\mu W$, while the coefficient of friction is itself a function of the velocity (Figure 12.15a), introduces interesting nonlinear vibration phenomena in a variety of applications, such as bearings, friction drives, and manufacturing processes. For example, we assume that the mass of a linear oscillator is placed on a moving belt (Figure 12.15b) moving with constant velocity U. The mass moves with the belt until the restoring force of the spring kx equals the friction force $\mu_s W$, where μ_s is the static coefficient of friction. This will happen at $x_s = \mu_s W/k$. The mass then slides on the belt and is slowed down; therefore, there is a static force to the right $\mu_k W$, where μ_k is the kinetic coefficient of friction, assumed constant. In the phase portrait (Figure 12.15c), the system will leave the line $\dot{x} = U$ at point A and will follow an elliptical trajectory ABC with center at $x_k = \mu_k W/k$ and maximum amplitude $x_s + U/\omega_n$, where $\omega_n^2 = k/m$. At point C, the velocity of the mass will equal the velocity of the belt and will stick on the belt and follow the straight trajectory CA. It is seen that the system follows a limit cycle. On CA the mass sticks on the belt, while on ABC it slips. For this reason the motion, called stick-slip, is very usual in engineering systems with sliding motion.

The surface of the belt is often not uniform and the coefficient of friction is not the same everywhere but is distributed in a more-or-less random manner. In this case, as the trajectory moves from C to A (Figure 12.15c), the slip point A' will be different and the trajectory will follow a different elliptical path, $A'B'C'$. This will continue in an unpredictable way and the motion will be chaotic, under certain assumptions for the surface properties of the belt. Most squeaking noises during sliding friction are due to stick-slip vibration.

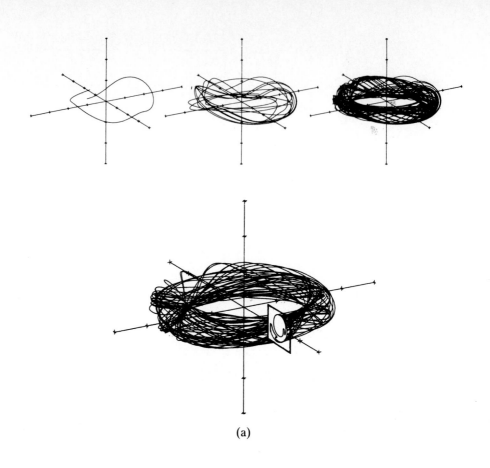

(a)

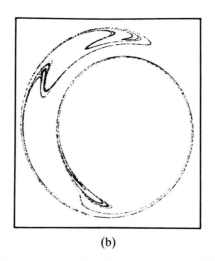

(b)

Figure 12.14 Poincaré map of a chaotic pendulum. (From Gleik, loc. cit.)

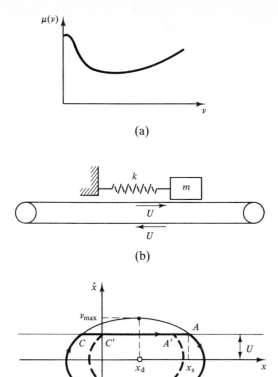

(a)

(b)

(c)

Figure 12.15 Stick-slip motion of a nonlinear slider.

Mechanical Systems with Backlash

Mechanical systems very often have backlash due to gaps, which can be functional or due to wear or failure. Some sources of backlash are gear gaps, bearing clearances, fatigue cracks, joint clearances, worn guides, and so on. Backlash introduces a strong nonlinearity, and if there is an energy input mechanism, limit cycles and chaotic vibration are often observed.

Consider, for example, the turning gear mechanism of turbine rotors. During heating or cooling the machine is set in slow rotation, to avoid creep deformations, by a motor and a gear couple (Figure 12.16a). The gear teeth have a gap b. The system can be modeled as in Figure 12.16b and the stiffness of the bilinear system with the gap in torsion is shown in Figure 12.16c. Assuming that ϕ is the angle of rotation and that the angular velocity of the motor is constant $R\Omega$, the speed of the turbine Ω, where R is the gear ratio, the equations of motion of the system are

$$J\ddot{\phi} + k(\phi - \Omega t) = -\,\text{sgn}(\dot{\phi})\mu mg \qquad (12.71)$$

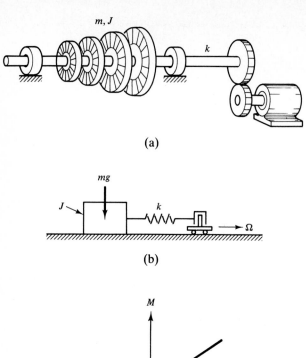

(a)

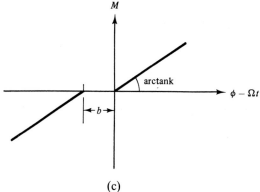

(b)

(c)

Figure 12.16 Nonlinear systems with backlash.

In normal form

$$\dot{x}_1 = -\frac{k(\phi - \Omega t) - \mathrm{sgn}(\dot{\phi})\mu mg}{J}$$

(12.72)

$$\dot{x}_2 = x_1$$

The program POINCARE will be used with $J = 10{,}000 \ \mathrm{kg \cdot m^2}$, $k = 10{,}000$ N·m/rad, $\mu = 0.1 \ e^{1-|\dot{\phi}|}$, $\Omega = 0.1$ rad/s, $c = 0.01$ N·m·s/rad, and $mg = 10{,}000$ N. Chaotic motion is observed (Figure 12.17a). Increasing the rotation frequency to 10 rad/s results in steady-state (Figure 12.17b). The parameters of the Poincaré Map will be angular displacement, $x_2 - \Omega t$, and angular velocity $x_1 - \Omega$. At every time step, the restoring torque $k(\phi - \Omega t)$ is tested against the friction torque. If greater, equation 17.72 holds. If smaller, $\dot{x}_1 = 0$ (stick).

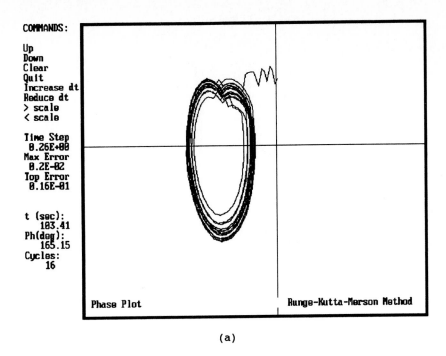

(a)

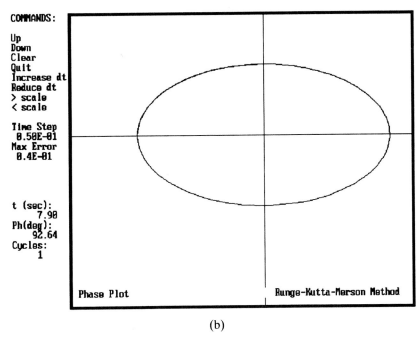

(b)

Figure 12.17 Chaotic motion of a geared system with backlash.

TABLE 12.1 SUMMARY OF EQUATIONS OF CHAPTER 12

Nonlinear pendulum

$$\ddot{\theta} + \left(\frac{g}{l}\right) \sin \theta = 0$$

Phase plane

$$m\frac{dv}{dx} + \frac{kx}{v} = 0$$

$$mv^2 + kx^2 = R_0^2 = mv_0^2 + kx_0^2$$

Delta method

$$\ddot{x} + \omega^2 x = -\omega^2 \delta(x, \dot{x}, t)$$

$$\delta(x, x, t) = -x + f(x, \dot{x}, t)/\omega^2$$

$$\frac{dy}{dx} = -(x + \delta)/y$$

$$(x + \delta)^2 + y^2 = c^2$$

Duffing method

$$\ddot{x} + \phi(\dot{x}) + f(x) = F(t)$$

$$\frac{dv}{dt} = -\frac{[\phi(v) + f(x)]}{v}$$

Duffing equation

$$m\ddot{x} + c\dot{x} + \alpha x + \beta x^3 = F_0 \cos \omega t$$

Exact methods

$$\ddot{x} + f(x) = 0$$

$$T = 4 \int_0^{x_0} \frac{dx}{\sqrt{\int_0^{x_0} f(x)dx}}$$

The Ritz–Galerkin Method

$$x(t) = \sum_{j=1}^{n} c_i x_i(t)$$

$$L(x; t) = \ddot{x} + \phi(\dot{x}) + f(x) - F(t) = 0$$

$$R_i = \int_0^T x_i(t)L(x; t)\, dt = 0, \quad i = 1, 2, \ldots, n$$

$$\frac{\partial R_i}{\partial c_i} = 0, \quad i = 1, 2, \ldots, n$$

TABLE 12.1   *(cont.)*

Euler method
  predictor

$$x(t + h) \simeq x(t) + f(x, t)h$$

  corrector

$$x_2(t + h) = x_0 + h[f(x_0, t) + f(x_1, t + h)]/2$$

Runge–Kutta method, second order

$$k_1(t + h) \simeq f(x_0, t)h$$

$$k_2(t + h) \simeq f(x_1, t + h)h$$

$$x(t + h) \simeq 0.5k_1(t + h) + 0.5k_2(t + h) + x_0$$

Runge–Kutta method, fourth order

$$x(t + h) = x(t) + R_1K_1 + R_2K_2 + R_3K_3 + R_4K_4$$

$$K_1 = hf(x_0, t)$$

$$K_2 = hf(x_0 + \beta K_1, t + \alpha h)$$

$$K_3 = hf(x_0 + \beta_1 K_1 + \gamma_1 K_2, t + \alpha_1 h)$$

$$K_4 = hf(x_0 + \beta_2 K_1 + \gamma_2 K_2 + \delta K_3, t + \alpha_2 h)$$

$$R_1 = R_4 = \frac{1}{6}, \qquad R_2 = R_3 = \frac{1}{3}$$

$$\alpha = \beta = \alpha_1 = \beta_1 = \gamma_1 = \frac{1}{2}$$

$$\alpha_2 = \delta = 1, \qquad \beta_2 = \gamma_2 = 0$$

Runge–Kutta–Merson method

$$x(t + h) = x(t) + \frac{1}{6}(K_1 + 4K_4 + K_5)$$

$$K_1 = hf(x_0, t)$$

$$K_2 = hf\left(x_0 + \frac{K_1}{3}, t + \frac{h}{3}\right)$$

$$K_3 = hf\left(x_0 + \frac{K_1}{6} + \frac{K_2}{6}, t + \frac{h}{3}\right)$$

$$K_4 = hf\left(x_0 + \frac{K_1}{8} + \frac{3K_3}{8}, t + \frac{h}{2}\right)$$

$$K_5 = hf\left(x_0 + \frac{K_1}{2} - \frac{3K_3}{8} + 2K_4, t + h\right)$$

$$E = \frac{1}{30}[2K_1 - 9K_3 + 8K_4 - K_5]$$

(continues)

Sec. 12.5   Stability

TABLE 12.1 *(cont.)*

Adams–Moulton Method
predictor

$$u_{n+1} = x_n + \left(\frac{h}{24}\right)[-9f(x_{n-3}, t_{n-3}) + 37f(x_{n-2}, t_{n-2}) - 59f(x_{n-1}, t_{n-1}) + 55f(x_n, t_n)]$$

corrector

$$x_{n+1} = x_n + \left(\frac{h}{24}\right)[f(x_{n-2}, t_{n-2}) - 5f(x_{n-1}, t_{n-1}) + 19f(x_n, t_n) + 9f(u_{n+1}, t_{n+1})]$$

REFERENCES AND FURTHER READING

BISHOP, R. E. D., and JOHNSON, D. C. 1979. *Mechanics of Vibration*, 2nd ed. Cambridge: Cambridge University Press.

GEAR, C. W. 1971. *Numerical Initial Value Problems in Ordinary Differential Equations*. Englewood Cliffs, N.J.: Prentice-Hall.

GERALD, C. F., and WHEATLEY, P. O. 1984. *Applied Numerical Analysis*, 3rd ed. Reading, Mass.: Addison-Wesley.

HORT, W. 1922. *Technische Schwingungslehre*. Berlin: Julius Springer.

INCE, E. L. 1926. *Ordinary Differential Equations*. London: Longmans, Green.

JORDAN, D. W., and SMITH, P. 1987. *Nonlinear Ordinary Differential Equations*, 2nd ed. Oxford: Clarendon Press.

KLOTTER, K. 1951. *Technische Schwingungslehre*. Berlin: Julius Springer.

KRYLOV, N., and BOGOLIUBOV, N. 1943. *Introduction to Nonlinear Mechanics*. Translated by S. Lefschetz. Princeton, N.J.: Princeton University Press.

MOON, F. C. 1987. *Chaotic Vibration*. New York: Wiley.

POINCARÉ, H. 1892. *Sur les courbes définies par une équation différentielle*, in *Oevres*. Paris: Gauthier-Villars.

SEIREG, A. 1969. *Mechanical Systems Analysis*. Scranton, Pa.: International Textbook Co.

STOKER, J. J. 1950. *Nonlinear Vibrations*. New York: Interscience.

VON KÁRMÁN, T., and BIOT, M. A. 1939. *Mathematical Methods in Engineering*. New York: McGraw-Hill.

PROBLEMS

Section 12.1

12.1. An undamped harmonic oscillator has a natural frequency $\omega_n = 400$ rad/s. Plot its phase-plane trajectory for initial conditions $x_0 = 0.25$ mm, $v_0 = 0$. What initial velocity v_0 with $x_0 = 0$ would yield the same trajectory?

12.2. An automobile of mass $m = 1364$ kg and undamped natural frequency $\omega_n = 5$ rad/s has shock absorbers with $\zeta = 0.8$. If the car is given an initial vertical displacement $x_0 = 25$ mm, determine its phase-plane trajectory.

12.3. Plot the phase-plane trajectory of the system of Problem 12.2 for $\zeta = 1$.

12.4. A small turbine stage of mass $m = 5.5$ kg completely balanced is mounted on a light shaft of stiffness $k = 535 \times 10^5$ N/m. While running at 3000 rpm, a blade of mass $m_e = 45$ g at a pitch radius $r_i = 0.024$ m brakes away. Plot the resulting phase-plane trajectory of the motion of the shaft.

12.5. Plot the phase-plane trajectory of Problem 12.4 if the system has a fraction of critical damping $\zeta = 0.1$.

12.6. Show that a spring–mass system with Coulomb friction $-F\,\mathrm{sgn}(\dot{x})$ consists of elliptical segments.

12.7. In Problem 12.6, plot the time–displacement response and show that the amplitude decreases linearly with time. Find an approximate expression for the time to stop.

12.8. A simple pendulum of length $l = 0.35$ m and mass $m = 0.7$ kg is pivoted by way of a pin of diameter $d = 12$ mm. Determine the coefficient of friction using the phase-plane trajectory if the pendulum starts from $\theta_0 = 15°$ and stops in 12 oscillations.

12.9. The phenomenon of stick-slip in systems with backlash and friction is as follows. A large block A moves to the right with a constant speed v_0 (Figure P12.9). A second, much heavier block B is resting on a horizontal plane and a horizontal spring of stiffness k attaches to it. When the block A meets the end c of the spring, it starts compressing it while the block B is kept in the plane by the friction force. When the spring is compressed by δ_0 and $k\delta_0 = \mu_s mg$, where m is the mass of B and μ_s the static coefficient of friction, the block B starts moving to the right. But the friction force is not $\mu_s mg$ because the coefficient of kinetic friction $\mu_k \ll \mu_s$. Therefore, the block B moves with a velocity v_0 until the spring is disengaged and at some point the block B stops again because of friction and the cycle is repeated. Plot the phase-plane trajectory of blocks A and B and write the differential equation of motion.

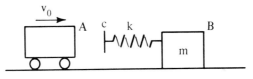

Figure P12.9

12.10. A turning gear is used to rotate a turbine while it is hot, to avoid permanent deformation. It consists of a motor M of constant angular velocity with a gear G_1 of diameter d_1 that is engaged with a gear G_2 of diameter d_2 (Figure P12.10). The gear G_2 is attached to a shaft of torsional stiffness k. At the bearing B_2 due to the rotor weight mg there is a friction moment in which the rotor is still until the torsion of the shaft k_0 equals the friction moment $\mu_s mg\, r$, where r is the bearing radius. Then

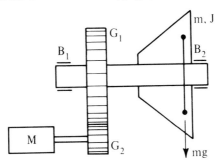

Figure P12.10

the rotor starts moving and the coefficient of friction becomes $\mu_k \ll \mu_s$. The rotor then moves faster than the motor, and somewhere it stops, to start the same cycle later.

(a) Interpret this system as a stick-slip couple such as that of Figure P12.9.

(b) Write the differential equation of motion.

(c) Plot the phase-plane trajectory of the system.

12.11. Using the delta method, complete one full trajectory of Example 12.1. Show by observation that the system is stable.

12.12. In Problem 12.11, if $\zeta = -0.1$, plot one full trajectory and comment on the stability of the system.

12.13. A vibrating table of mass $m = 68$ kg is supported by springs of constant $k = 3.5 \times 10^5$ N/m. To limit its motion, there are two stops of stiffness $k = 1.8 \times 10^6$ N/m on each side of the table. The free play of the table is 12 mm. A weight of 45,000 kg was placed on the table and suddenly removed. Using the delta method, determine the phase-plane trajectory of the system.

12.14. Use the delta method to plot a full phase-plane trajectory for the internal combustion engine of Example 12.8 if $a = 0$, $e = 0$, $b = 2000$, and $k = 1.0$.

12.15. Use the delta method to plot a full phase-plane trajectory for a simple pendulum of mass $m = 0.45$ kg and length $l = 0.5$ m if it oscillates due to a large initial displacement $\theta_0 = 45°$ from equilibrium.

Section 12.2

12.16. Determine the period of oscillation of an oscillator of mass m that has a nonlinear spring which reacts to a displacement x by a force $-kx^n$, where k and n are constants, as a function of maximum amplitude X.

12.17. An electronic instrument is packaged in a box with a polystyrene plate of thickness $d = 25$ mm. The polystyrene has nonlinear spring characteristics, as shown in Figure P12.17. If the mass of the instrument is $m = 20$ kg and the horizontal section is A 0.3×0.3 m, determine the natural frequency as a function of the amplitude, integrating using the trapezoidal rule.

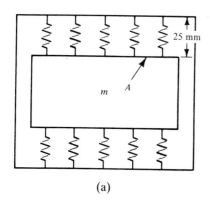

(a)

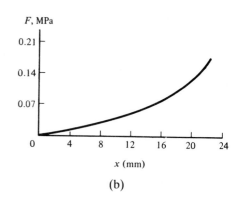

(b)

Figure P12.17

12.18. The foundation of a heavy compressor rests on soil that has nonlinear characteristics. These characteristics have been tested by loading the foundation with static loads of 1, 2, 3, 4 tons and measuring static deflections 20, 40, 50, 60 mm, respectively. If

Vibration of Nonlinear Systems   Chap. 12

the mass of the compressor and foundation is 4 tons and assuming the nonlinear reaction of the soil to consist of two linear segments, determine **(a)** the constants k_1 and k_2, and **(b)** the range of the natural frequencies for amplitudes 0 to 65 mm.

12.19. A mass $m = 4$ kg is attached at the middle of a horizontal spring of length $l = 1$ m pinned at the ends without initial tension. The mass is free to vibrate on a horizontal plane in a direction perpendicular to the spring without friction, the spring is of diameter $D = 10$ mm, wire $d = 0.25$ mm, and shear modulus $G = 1.05 \times 10^{11}$ N/m^2. If the mass is displaced by $x_0 = 0.1$ m and released, determine the frequency of oscillation.

12.20. A foundation for a reciprocating engine of mass $m = 1.5$ ton has a horizontal natural frequency $\omega = 150$ rad/s. An overhead crane moving a weight in the horizontal direction hits the engine with a large, unknown force. It is suspected that the impact left a gap between the foundation and the cork isolation in the horizontal direction. A test is made that shows a natural frequency $\omega_1 = 135$ rad/s. Determine the size of the gap, assuming that the cork was not damaged.

Section 12.3

12.21. Using the Duffing method of successive approximations, determine the dependence of the natural frequency of the pendulum on the oscillation amplitude, using up to the third-order term of the sin function in a series expansion.

12.22. A machine of mass $m = 2$ ton rests on a cork support for vibration isolation with a nonlinear force–displacement relationship $F = 2 \times 10^6 x + 5 \times 10^{12} x^3$, units of N and m. Determine the natural frequency for vibration amplitudes of 1, 10, and 20 mm, using the Duffing method of successive approximations.

12.23. A pendulum has length $l = 1$ with a bob of mass $m = 2$ kg and a torsional spring at the pivot of stiffness 30 N·m. Using the Duffing method of successive approximations, determine the dependence of the natural frequency on the oscillation amplitude using up to the third-order term of the sin function in a series expansion, if the pendulum is at equilibrium **(a)** at the lower position, and **(b)** at the upper position.

12.24–12.26. Solve Problems 12.21 to 12.23 using the Linstedt perturbation method.

12.27–12.29. Solve Problems 12.21 to 12.23 using the Ritz–Galerkin method.

12.30. A cantilever steel beam of $E = 2.1 \times 10^{11}$ N/m^2, length $l = 1$ m, negligible mass, and rectangular cross section 30 × 30 mm supports a motor of mass 300 kg at the free end. At the point where the beam is welded on a vertical rigid wall there is a crack which can be modeled as a bilinear spring with infinite rotational stiffness when closed and $K_T = 2 \times 10^4$ N·m/rad when open. Using the Ritz–Galerkin method, determine the dynamic response to a harmonic force lateral at a mass of magnitude $F = 2000$ N and frequency $\omega = 400$ rad/s.

Section 12.4

12.31. On the machine of Problem 12.22, there is a vertical harmonic force of magnitude $F = 8000$ N and frequency $\omega = 200$ rad/s. Find the response using numerical integration.

12.32. Solve Problem 12.23 for an initial deviation of 45° deviation from the lower equilibrium position.

12.33–12.35. Solve Problems 12.30, 12.2, and 12.13 using numerical integration.

12.36. An elliptical journal bearing has $K_{xx} = 5 \times 10^9$ N/m, $K_{xy} = 2 \times 10^8$ N/m, $K_{yx} = 2 \times 10^8$ N/m, $K_{yy} = 2 \times 10^9$ N/m, $C_{xx} = 4 \times 10^5$ N·s/m, $C_{xy} = 1 \times 10^5$ N·s/m, $C_{yx} = 1 \times 10^5$ N·s/m, and $C_{yy} = 2 \times 10^5$ N·s/m. Determine if it is stable using the first method of Liapounov.

12.37. Determine the stability of the system of Example 12.6 with the first method of Liapounov.

13

Vibrating Systems Design

13.1 ACCEPTABLE VIBRATION LEVELS

The acceptable level of vibration on a machine or structure is determined (1) by its acceptability to the operating personnel; and (2) by the ability of the machinery, equipment, and structure to carry the vibration without damage or failure.[1] The equation of the level of vibration that can be endured by human beings without annoyance, loss of efficiency, or physical damage has been the subject of many studies. The question has also been a subject of study by International Standards Organization Technical Committee ISO/TC 108, Mechanical Vibration and Shock. Figure 13.1 shows their proposed recommendations for acceptable levels of vertical vibration on standing and sitting human beings, expressed in terms of acceleration. Figure 13.2 gives their recommendations for acceptable levels of horizontal vibration. Similar methods of determining the vibration levels acceptable to people have been proposed in the literature. One thing that should be noted here is the de-

*The above photo portrays a balancing machine (Courtesy of Mechanalysis, Inc. Reprinted by permission.)

[1]Rathbone, T. C. 1939. Vibration tolerances. *Power Plant Eng.*, *43*: 721.

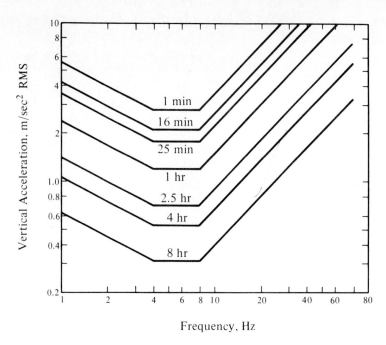

Figure 13.1 ISO—suggested acceptable vertical vibration levels.

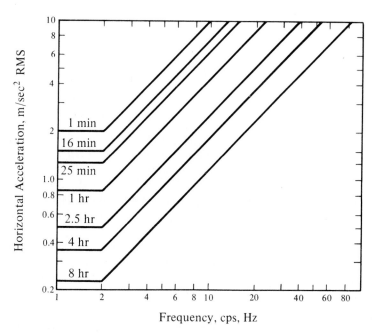

Figure 13.2 ISO—suggested acceptable horizontal vibration levels.

pendence of the annoyance levels on the duration of exposure and the frequency of vibration.[2]

The acceptability of a vibration that is not a pure single frequency is more complicated. The ISO recommendations are not definitive in this case. Apparently, the annoyance levels from two simultaneous vibrations at widely separated frequencies are not cumulative. The ISO suggests that in the case of narrowband vibration concentrated in a third-octave band or less, the rms value of the acceleration within the band is to be evaluated with reference to the appropriate limit at the center frequency of that band. For broadband distributed or random vibration, the rms acceleration is to be evaluated separately for all one-third octave bands with respect to the appropriate limit for each. However, the recommendation does consider weighing networks for broadband vibration (i.e., integration over the full frequency spectrum). It must be noted that if the vibration is measured as a velocity, there is no weighing required for rms measurements above 8 Hz in the vertical vibration and 2 Hz in the horizontal vibration.

Where vibrations occur in the vertical and horizontal directions simultaneously, the ISO recommends that the corresponding limit apply to each component. Since the movement, for vibrations of the same frequency, is the resultant of the three components of motion, it is suggested that the annoyance level be taken as the square root of the sum of the squares of the annoyance level associated with each component.

☐ **Example 13.1**

Determine the proficiency boundary of a worker who is standing on a machine foundation having a vertical vibration of 480 cpm and acceleration 2g.

Solution   The frequency is $480/60 = 8$ Hz and the acceleration is $a = 0.2g = 0.2 \times 9.81 = 1.97$ m/s^2. Entering these values in Figure 13.1, we obtain an endurance limit of 25 min.

(The ISO charts have been used to find the endurance limit of a person standing on a vibrating body.)

☐

13.2 EFFECTS OF VIBRATION ON MACHINERY DETERIORATION AND FAILURES

The failure of a machine is preceded by an increase in its vibration level in more than 90% of cases. Common causes of vibration are unbalance, misalignment, worn bearings, and so on. Many attempts have been made to develop a definite set of vibration level criteria which will indicate that failure is near. Most authorities appear to agree that a vibration velocity level of 12 mm/s or higher indicates that failure is near and repairs should be scheduled as soon as possible. If the vibration level exceeds 18 mm/s, immediate shutdown is advisable. Figures 13.3a and 13.3b and Table 13.1 are typical guides for evaluating the severity of vibration of rotating machinery. Figure 13.4 is a conversion nomogram, relating harmonic vibration

---

[2]Reiher, H., and Meister, E. J. 1931. Die Empfindichkeit des Menschen den Errechuterungen. *Forsch. Gebiete Ing.* 2(11): 381–386.

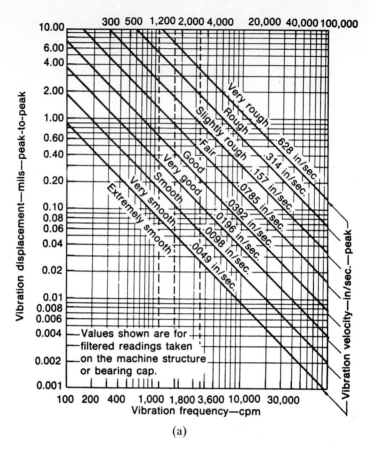

Figure 13.3 Acceptable machinery vibration levels. (Adopted a) from product documentation of IRD Mechanalysis, Inc., 6150 Huntley Road, Columbus, Ohio, b) from ANSI Standard 19–1975.)

frequency, acceleration, velocity and displacement. The range of operating conditions for a variety of systems is indicated on the same figure.

13.3 VIBRATING SYSTEMS SYNTHESIS

Up to this point we have discussed, almost exclusively, the *performance* of a *given* system whose properties are well known (or assumed). This process, called *analysis*, is a basic function in engineering. It comes second, however. The first task for design engineers is to embody their design concepts in a three-dimensional form: in other words, to determine the specific form and detailed dimensions of the system in order to perform the specified task under given constraints. This process is called *synthesis*. In Figure 13.5 we show an oversimplification of the procedure used for the development of a piece of machinery or structure.

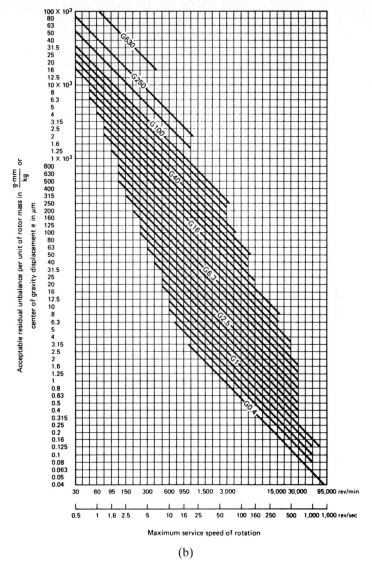

Figure 13.3 *(cont.)*

Someone has to conceive the idea and the requirements and plan its implementation. Then someone with familiarity in the general area synthesizes the subject, in other words, decides form, materials, speeds, and so on. Then someone should analyze it before it is manufactured or sold, to make sure that it comes up to the original expectations (design specifications).

Synthesis is a very complicated process. Contrary to analysis, where there are usually enough equations for determination of the problem unknowns (*performance*), in synthesis almost everything is unknown. Also, some of the known

TABLE 13.1 VIBRATION VELOCITY STANDARDS FOR SHIPBOARD MACHINERY

Balance quality grades G	$e\omega$ (mm/sec)	Rotor types—General examples
G 4 000	4 000	Crankshaft-drives[a] of rigidly mounted slow marine diesel engines with uneven number of cylinders.[b]
G 1 600	1 600	Crankshaft-drives of rigidly mounted large two-cycle engines.
G 630	630	Crankshaft-drives of rigidly mounted large four-cycle engines. Crankshaft-drives of elastically mounted marine diesel engines.
G 250	250	Crankshaft-drives of rigidly mounted fast four-cylinder diesel engines.[b]
G 100	100	Crankshaft-drives of fast diesel engines with six or more cylinders.[b] Complete engines (gasoline or diesel) for cars, trucks, and locomotives.[b]
G 40	40	Cars wheels, wheel rims, wheel sets, drive shafts Crankshaft-drives of elastically mounted fast four-cycle engines (gasoline or diesel) with six or more cylinders[b] Crankshaft-drives for engines of cars, trucks, and locomotives.
G 16	16	Drive shafts (propeller shafts, cardan shafts) with special requirements. Parts of crushing machinery. Parts of agricultural machinery. Individual components of engines (gasoline or diesel) for cars, trucks, and locomotives. Crankshaft-drives of engines with six or more cylinders under special requirements. Slurry or dredge pump impeller.
G 6.3	6.3	Parts or process plant machines. Marine main turbine gears (merchant service). Centrifuge drums. Fans. Assembled aircraft gas turbine rotors. Fly wheels. Pump impellers. Machine-tool and general machinery parts. Normal electrical armatures. Individual components of engines under special requirements.
G 2.5	2.5	Gas and steam turbines, including marine main turbines (merchant service). Rigid turbo-generator rotors. Rotors. Turbo-compressors. Machine-tool drives. Medium and large electrical armatures with special requirements. Small electrical armatures. Turbine-driven pumps.

TABLE 13.1 *(cont.)*

Balance quality grades G		eω (mm/sec)	Rotor types—General examples
G	1	1	Tape recorder and phonograph (gramophone) drives. Grinding-machine drives. Small electrical armatures with special requirements.
G	0.4	0.4	Spindles, disks, and armatures of precision grinders. Gyroscopes.

[a] A crankshaft-drive is an assembly which includes the crankshaft, a flywheel clutch, pulley, vibration damper, rotating portion of connecting rod, etc.

[b] For purposes of this Standard, slow diesel engines are those with a piston velocity of less than 9 m/sec; fast diesel engines are those with a piston velocity of greater than 9 m/sec.

In complete engines, the rotor mass comprises the sum of all masses belonging to the crankshaft-drive described in Note a above.

Excerpted from ANSI Standard 19-1975: Balance Quality of Rotating Rigid Bodies. Washington, D.C., American National Standards Institute.

facts are in forms very difficult to set in mathematical form—experience, for example.

To determine the size, form, properties, and so on, of a design subject, one must specify a number of parameters. For example, synthesis of a bolted connection ends up with a list of several items, including dimensions (length, diameter, thread, nut, head), materials, surface finish, color, and packaging. We shall call these *design parameters*.

For the bolt to satisfy the designer's expectations, it must be capable of taking a certain load and having a certain length (equality constraints). Also, some of its design parameters, or other quantities, must be within certain bounds, such as maximum diameter and materials available (inequality constraints). The general statement of *synthesis* will therefore be that synthesis is the determination of the design parameters of a problem subject to equality and inequality constraints.

In engineering problems there is often another, similar task: for a given system of unknown characteristics (something in a black box, for example), to perform enough number of (nondestructive) tests to determine the design parameters. For example, we already know that a velocity pickup type or vibration measuring transducer consists of a metal case with a simple oscillator inside, and we can do a number of tests on it without opening it. Its weight gives us the mass of the case plus the oscillator. Its volume has to be a bound for the volume of the oscillating mass. Then we can place it on a shaker table and by varying the frequency, find its natural frequency. From the natural frequency and the mass we can estimate the spring stiffness. This process is called *identification*. The general statement of the problem would be: *Perform enough tests on a black box to determine its contents*.

We shall call *input* whatever external influence makes a system move. The resulting motion of the system will be called *output*. There is always a relation between the input and the output. For example, for a harmonic oscillator (m, c,

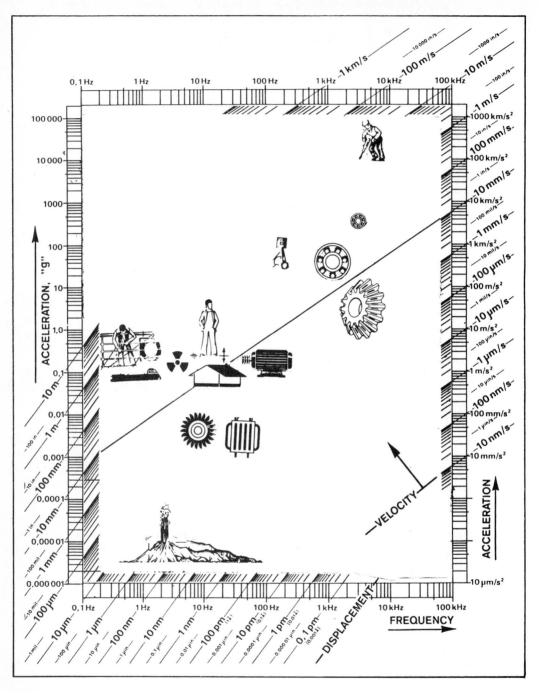

Figure 13.4 Vibration range nomogram levels.

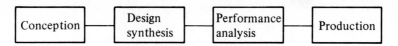

Figure 13.5 System design procedure.

k), if a harmonic exciting force $F_0 e^{i\omega t}$ is the input, the resulting motion $x(t) = F_0 G(m, c, k, \omega) e^{i\omega t - \phi}$ is the output. The function G has the value

$$G(m, c, k, \omega) = \frac{1}{[(k - m\omega^2) + (c\omega)^2]^{1/2}} \qquad (13.1)$$

and it is only a function of the system properties m, k, and c and the angular velocity ω. The function G is called a *transfer function* or *receptance*. We shall use the latter term in the sequel.

In multidegree-of-freedom linear systems we have a matrix relation instead:

$$\mathbf{x}(t) = (-\omega^2 \mathbf{M} + i\omega \mathbf{C} + \mathbf{K})^{-1} \mathbf{F}_0 e^{i\omega t} \qquad (13.2)$$

where $\mathbf{x}$ and $\mathbf{F}_0$ are $n \times 1$ vectors and $\mathbf{M}$, $\mathbf{C}$, and $\mathbf{K}$ are $n \times n$ matrices. The $n \times n$ matrix $\mathbf{R} = (-\omega^2 \mathbf{M} + i\omega \mathbf{C} + \mathbf{K})^{-1}$ is then the system *receptance* matrix.

Following these definitions, the synthesis and identification problems for linear multidegree-of-freedom systems will be restated as follows:

1. *Synthesis* is the determination of the receptance matrix $\mathbf{R}$ and its components $\mathbf{M}$, $\mathbf{C}$, and $\mathbf{K}$ in order that a specified input results in a specified output.
2. *Identification* is the determination of the receptance matrix $\mathbf{R}$ by way of testing by using several inputs $\mathbf{F}_0(t)$, measuring the resulting outputs $\mathbf{x}(t)$, and using equations (13.2).
3. The *design* problem consists of synthesis and identification.

We set an *a priori* restriction: that the system is discrete, linear, and has n degrees of freedom. In many systems this is quite arbitrary and ambiguous; in other cases there is some insight into the problem. For example, a transparent box will allow us to see how many masses are in there, but we cannot measure them. If the system in there is a flexible beam, we can always approximate it with a discrete system, as we have already done. Also, sometimes part of the system is known and part is unknown. For example, we can usually estimate masses from design drawings, and the problem is to compute the stiffness and damping coefficients.

13.4 THE GENERAL DESIGN PROBLEM

As we have mentioned, we shall consider the n-degree-of-freedom linear system. The general design problem, synthesis or identification, will consist of finding the receptance matrix $\mathbf{R}$. Furthermore, we shall assume that the system can be described by way of three matrices $\mathbf{M}$, $\mathbf{C}$, and $\mathbf{K}$ of dimension $n \times n$, which give the mass, damping, and stiffness influence coefficients, respectively. In practical problems one has to specify the components of these matrices directly instead of the matrix **R**, only on two counts: (1) The receptance matrix cannot be interpreted directly in terms of hardware, therefore is of little use to the designer, and (2) the receptance matrix is a function of ω, therefore again is of little use. The **M**, **C**, and **K** matrices are often independent of ω. The latter observation will naturally distinguish the problem in the two cases described below.

System Properties Dependent on Frequency

For an input $\mathbf{F}_0 e^{i\omega t}$ and an output $\mathbf{X} e^{i\omega t}$, we shall have

$$(-\omega^2 \mathbf{M} + i\omega \mathbf{C} + \mathbf{K})\mathbf{X} = \mathbf{F}_0 \tag{13.3}$$

Equations (13.3) are n complex linear algebraic equations on $3n(n + 1)/2$ real unknowns (the components of the symmetric matrices **M**, **C**, and **K**). To obtain a unique solution, we have to specify $3n(n + 1)/4$ sets of linearly independent inputs and outputs $\mathbf{F}_0$ and $\mathbf{X}$ (both vectors of dimension n) *at the same frequency* ω. This will result in the determination of the $3n(n + 1)/2$ unknown elements of the matrices **M**, **C**, and **K** for the frequency ω. This has to be repeated for a sufficient number of frequencies in order to determine these elements as functions of ω.[3] The problem is this general form is not well-posed, as one can see from equation (13.3); therefore it is not solvable in this general form. As stated above, however, many system elements are known. This allows the problem to have a unique solution in many practical applications.

Selection of the input and output vectors is not altogether arbitrary. The several sets of vectors should be linearly independent and consistent. Simply stated, if we select, for example, the first set of vectors $\mathbf{X}_1 = \{1 \quad 1 \quad 1\}$, $\mathbf{F}_1 = \{2 \quad 2 \quad 2\}$, for $n = 3$, we cannot select as another set $\mathbf{X}_2 = \{1 \quad 1 \quad 1\}$, $\mathbf{F}_2 = \{3 \quad 3 \quad 3\}$ because we ask that the same output results from different input—nor the vectors $\mathbf{X}_2 = \{2 \quad 2 \quad 2\}$, $\mathbf{F}_2 = \{4 \quad 4 \quad 4\}$ because it is simply the first set multiplied by 2, and because since the system is linear, if $\{\mathbf{X}_1, \quad \mathbf{F}_1\}$ is a solution, $\{2\mathbf{X}_1, \quad 2\mathbf{F}_1\}$ will also be a solution.

Generally, the problem is not so large because we know some of the matrix elements; therefore, we must find enough input–output sets to determine the unknown elements. In fact, this is always necessary to a certain extent. Moreover, inputs–outputs are known to a certain engineering accuracy and are statistical quantities, as they contain noise. For such cases, estimation methods can be employed, used extensively in control systems.

☐ **Example 13.2   Identification of Stiffness of a de Laval Rotor**

A rotating shaft of unknown stiffness k_r carries a disk of mass $M = 1.5$ kg and is supported symmetrically by two identical bearings of unknown stiffness, both a func-

---

[3]Haag, A. C., and Sankey, G. O. 1956. Some dynamic properties of oil-film journal bearings. *Trans. ASME J. Appl. Mech.*, *78*: 302–306. Woodcock, J. S., and Holmes, R. 1970. Determination and application of the dynamic properties of a turbo-rotor bearing oil film. *Proc. Inst. Mech. Eng.*, *184*: 31, 111–119. Morton, P. G. 1971. Measurement of the dynamic characteristics of large sleeve bearings. *ASME J. Lubr. Technol.*, pp. 143–150.

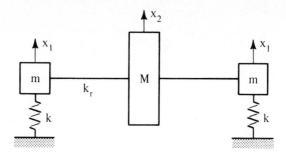

Figure E13.2

tion of ω, a mass for both of $m = 1$ kg, and negligible damping (Figure E13.2). A test is conducted with added unbalances. Find the bearing and shaft stiffness from the following test results:

Angular velocity ω	100 rad/s
Unbalance m_1 (bearings)	0.001 kg
Unbalance m_2 (rotor)	0.001 kg
Eccentricity e_1	0.25 m
Eccentricity e_2	0.25 m
Amplitude x_1	0.050 mm
Amplitude x_2	0.100 mm
Unbalance location $\phi_1 = \phi_2$	0

Solution If x_2 is the displacement of the rotor and x_1 that of the bearings the equation of motion will be

$$M\ddot{x}_2 + k_r(x_2 - x_1) = m_2 e_2 \omega^2 \sin \omega t$$

$$m\ddot{x}_1 + kx_1 - k_r(x_2 - x_1) = m_1 e_1 \omega^2 \sin \omega t$$

For $x_1 = X_1 \cos \omega t$, $x_2 = X_2 \cos \omega t$, we obtain

$$k_r(X_2 - X_1) = m_2 e_2 \omega^2 + M\omega^2 X_2$$

$$kX_1 - k_r(X_2 - X_1) = m_1 e_1 \omega^2 + m\omega^2 X_1$$

Substituting the numerical values, we have

$$k_r(0.1 - 0.05) \times 10^{-3} = 0.001 \times 0.25 \times 10^4$$

$$+ 0.005 \times 10^4 \times 0.1 \times 10^{-3} = 2.55$$

$$k \times 0.050 \times 10^{-3} - k_r(0.1 - 0.05) \times 10^{-3}$$

$$= 0.001 \times 0.25 \times 10^4 + 1 \times 10^4 \times 0.050 = 3.0$$

From the first equation

$$k_r = 51,000 \text{ N/m}$$

From the second equation,

$$0.050k \times 10^{-3} - 51{,}000 \times 0.05 \times 10^{-3} = 3, \qquad k = 111{,}000 \text{ N/m}$$

$$k_r = 51{,}000 \text{ N/m}$$

$$k = 111{,}000 \text{ N/m}$$

If we repeat the test at several values of ω, we will be able to plot the stiffness versus the angular velocity ω.

(The shaft and bearing stiffness of a de Laval rotor on flexible bearings was found using unbalance test results.)

☐

Constant System Properties

For systems with properties that are independent of the frequency, the system of equations (13.3) will be used again to yield the system matrices elements. The difference is that because they are unchanged with the angular velocity, one can give sets of vectors at different velocities. For the computation, this makes no difference. But this can facilitate the test because one has to run a very easy test, keeping the excitation constant and measuring the output at several speeds. If the number of unknown parameters is sufficiently small, they can be determined by static tests only.[4]

☐ **Example 13.3 Identification of a Velocity Transducer**

A velocity transducer, consisting of a cage with a harmonic oscillator inside, was tested for the determination of its characteristics. It was put on a shaker and between the table and the transducer there was a force-measuring device (Figure E13.3). For the following test results, determine the characteristics of c and k of the transducer inside the cage, provided that m is known (by weighing).

Angular velocity	100 rad/s
Base motion amplitude Y	0.25 mm
Force amplitude F	10 N
Phase angle between F and Y	60°
Mass m	0.386 kg

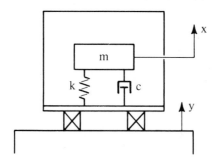

Figure E13.3

[4]Michell, J. R., Holmes, R., and van Bellegoyen, H. 1966. Experimental determination of a bearing oil film thickness. *Proc. Inst. Mech. Eng.*, *180*(31K): 90–96.

Solution The equation of motion is (Chapter 3)

$$m\ddot{x} + c(\dot{x} - \dot{y}) + k(x - y) = 0 \tag{a}$$

The force transmitted is

$$f = -c(\dot{x} - \dot{y}) - k(x - y) \tag{b}$$

For harmonic motion we set up $x = Xe^{i\omega t}$, $f = Fe^{i\omega t}$, and $y = Ye^{i\omega t}$, where Y and F are known and X is unknown. Equations (a) and (b) yield

$$-\omega^2 m\ddot{x} = f, \qquad X = \frac{F}{\omega^2 m} \tag{c}$$

$$F = k(X - Y) + i\omega c(X - Y)$$

$$k + i\omega c = \frac{F}{F/m\omega^2 - Y} \tag{d}$$

Equation (d) is complex in two real unknowns k and c; therefore, one test is enough for the determination of k and c, provided that the mass m is known. Of course, F is complex and has the value

$$F = |F|\,(\cos \phi + i \sin \phi) = 10(0.833 + 0.5i)$$

$$k + i\omega c = \frac{8.33 + 5i}{(8.33 + 5i)/0.386 \times 10^4 - 0.01}$$

$$= \frac{8.33 + 5i}{-0.08 + 0.0013i}$$

$$= \frac{(8.33 + 5i)(-0.08 + 0.0013i)}{0.08^2 + 0.05^2}$$

Therefore,

$$k = 1890 \text{ N/m}, \qquad \omega c = 4816 \text{ N/m}, \qquad c = 4816/\omega = 48.16 \text{ N·s/m}$$

(The spring and damping constants of a velocity transducer have been estimated from shaker table vibration tests results.)

$\square$

The Modal Analysis Method

Although, in principle, one can apply a sufficient number of tests on a lumped system to determine its linear elements, for reasons explained above this is not usually possible, except for small systems only. For larger systems, only certain parameters can be identified directly, if all other lumped elements are known. Bearing properties in rotor-bearing systems are such examples. For larger systems determination of the detailed system structure is extremely difficult. However, such knowledge is not always necessary. In the case of modal analysis, for example, the mass matrix with a number of natural frequencies and modes, together with the values of modal damping for these nodes, are all that is needed. The mass matrix can usually be derived if the modal coordinates are selected. Natural frequencies and vibration modes can be measured directly. The modal damping only needs to be estimated indirectly. To this end, a resonance curve can be used to yield the damping factor with the half-power bandwidth method (Chapter 3). For light damping an alternative method is used in the complex plane.

A modal equation for the mode j has the form (Chapter 7)

$$\ddot{q}_j + 2\zeta_j \omega_{nj} \dot{q}_j + \omega_{nj}^2 q_j = A_j e^{i\omega t} \tag{13.4}$$

The steady-state solution is $q_j = Q_j e^{i\omega t}$, where

$$Q_j = \frac{A_j\{[1 - (\omega/\omega_{nj})^2] - (2\zeta\omega/\omega_{nj})i\}}{\{[1 - (\omega/\omega_{nj})^2]^2 + (2\zeta\omega/\omega_n^2)\}^{1/2}} \tag{13.5}$$

Therefore, the real and imaginary parts of the response are

$$Q_{rj} = \frac{A_j[1 - (\omega/\omega_{nj})^2]}{\{[1 - (\omega/\omega_{nj})^2]^2 + (2\zeta\omega/\omega_n)^2\}^{1/2}} \tag{13.6}$$

$$Q_{imj} = \frac{A_j[-2\zeta(\omega/\omega_{nj})]}{\{[1 - (\omega/\omega_{nj})^2]^2 + (2\zeta\omega/\omega_n)^2\}^{1/2}} \tag{13.7}$$

Eliminating the frequency from equations (13.6) and (13.7) gives

$$\left(Q_{imj} + \frac{A_j}{4\zeta_j}\right)^2 + Q_{rj}^2 = \left(\frac{A_j}{4\zeta}\right)^2 \tag{13.8}$$

Equation (13.8) represents a circle on the (Q_r, Q_{im}) plane with radius $A_n/4\zeta_j$ (Figure 13.6), passing through the origin and having a diameter on the imaginary axis. The center is at $Q_r = 0$, $Q_{im} = -A_j/4\zeta_j$; point A corresponds to the undamped natural frequency $1 - (\omega/\omega_{nj})^2 = 0$, $\omega = \omega_{nj}$, and points B and C, to the half-power bandwidth points and corresponding frequencies ω_1 and ω_2. If the three frequencies can be measured, $\zeta = (\omega_2 - \omega_1)/2\omega_n$ (Chapter 3).

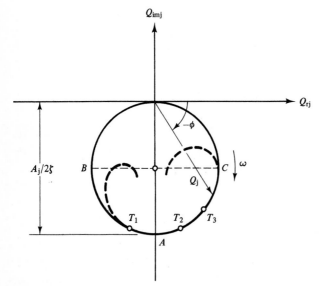

Figure 13.6 Modal analysis on Nyquist diagram.

If different natural frequencies are close to one another, there is no clear single resonance and the response curve is not a circle but a higher-order curve with several loops and many resonances. In this case, each mode is separated by taking three successive points near resonance and computing the circle that passes through these points. The diameter of this circle will be the ratio $A_j/2\zeta_j$. It should be pointed out that this circle might be displaced from the symmetric position and the origin [Buzdugan et al., 1986].

☐ **Example 13.4   Three-Point Formula for Damping Identification**

In a modal analysis experiment, three points of the response curve are measured. The real and imaginary part of the response are (Q_{rj1}, Q_{imj1}), (Q_{rj2}, Q_{imj2}), and (Q_{rj3}, Q_{imj3}), respectively. Determine the ratio $A_j/2\zeta_j$.

Solution   We assume that the circle is expressed as, with the subscript j dropped,

$$(Q_r - Q_{r0})^2 + (Q_{im} - Q_{0im})^2 = R^2$$

Application for points 1, 2, and 3 gives

$$(Q_{r1} - Q_{r0})^2 + (Q_{im1} + Q_{im0})^2 = R^2$$

$$(Q_{r2} - Q_{r0})^2 + (Q_{im2} - Q_{im0})^2 = R^2$$

$$(Q_{r3} - Q_{r0})^2 + (Q_{im3} - Q_{im0})^2 = R^2$$

Eliminating R^2 yields

$$(Q_{r1} - Q_{r0})^2 + (Q_{im1} - Q_{im0})^2 = (Q_{r2} - Q_{r0})^2 + (Q_{im2} - Q_{im0})^2$$

$$(Q_{r2} - Q_{r0})^2 + (Q_{im2} - Q_{im0})^2 = (Q_{r3} - Q_{r0})^2 + (Q_{im3} - Q_{im0})^2$$

Solving both for Q_{r0}, we obtain

$$Q_{r0} = \frac{Q_{r2}^2 + Q_{im2}^2 - Q_{r1}^2 + Q_{im1}^2 + 2Q_{im1}Q_{0im} - 2Q_{im2}Q_{0im}}{2(Q_{r2} - Q_{r1})}$$

$$Q_{r0} = \frac{Q_{r3}^2 + Q_{im3}^2 - Q_{r2}^2 + Q_{im2}^2 + 2Q_{im2}Q_{0im} - 2Q_{im3}Q_{0im}}{2(Q_{r3} - Q_{r2})}$$

Eliminating Q_{r0}, we have

$$Q_{im0} = \frac{A}{B}$$

where

$$B = \frac{Q_{im1} - Q_{im2}}{Q_{r2} - Q_{r1}} - \frac{Q_{im3} - Q_{im2}}{Q_{r2} - Q_{r3}}$$

$$A = \frac{Q_{r2}^2 - Q_{im2}^2 - Q_{r1}^2 - Q_{im1}^2}{2(Q_{r2} - Q_{r1})} - \frac{Q_{r3}^2 - Q_{im3}^2 - Q_{r2}^2 - Q_{im2}^2}{2(Q_{r3} - Q_{r2})}$$

Using the formulas backward, we compute Q_{r0} and finally, $R = A_j/2\zeta_j$.

(Three test points of the frequency response curve were used to determine the diameter of the modal circle and the value of the modal damping.)

☐

13.5 MASSES NEEDED TO ESTABLISH KNOWN MOTION BALANCING

In some cases the mass distribution in a system is not known. A general statement of the problem would be: Determine the mass matrix $\mathbf{M}$ of a system, if the damping and stiffness matrices $\mathbf{C}$ and $\mathbf{K}$ are known, from the measured response to known forces. One can write, for harmonic excitation,

$$\mathbf{MX} = \frac{1}{\omega^2}[(i\omega\mathbf{C} + \mathbf{K})\mathbf{X} - \mathbf{F}] \qquad (13.9)$$

If the matrix $\mathbf{M}$ is fully populated, this is a system of n equations for every ω in $n(n + 1)/2$ unknowns. If enough test data $(\mathbf{X}, \mathbf{F}, \omega)$ are available, one can compute, in principle, the mass matrix $\mathbf{M}$. However, this is an unusual situation in engineering. Generally, only some of the unbalance masses leading to forces $\mathbf{F}$ and their location are unknown. This is a very common problem in rotating machinery, known as *balancing*.

For a variety of reasons, rotating members have their mass distributed about the geometric axis (axis of rotation) without perfect polar symmetry. Along the rotating shaft, unbalance can exist in many places, continuous or lumped. The latter case is most common and we discuss it in some detail.

We suppose that the shaft has been modeled by way of n nodes (see Chapter 11) along the axis (Figure 13.7). Since we do not consider motion of these nodes along the z-axis, they are confined to move on planes parallel to the x-y plane. We confine the unbalanced masses to these planes. Next, we assign a reference point on the shaft. Suppose that at the right-hand end we attach a small disk that has a marked radius A which rotates with the shaft with respect to another marked line, B, fixed in space. We start measuring the time when the two marks are aligned with one another. At time t the angle between them will be ωt, where ω is the shaft's angular velocity, assumed constant.

The unbalanced masses along the shaft need not be in the same direction. Every one of them, assumed on a nodal plane, is characterized by the mass m_e at a distance e from the axis and in a direction that forms an angle ϕ from the moving mark A. Since the inertia forces have the form $m_e e\omega^2$, we shall use the product $u_i = m_e e$ at each nodal plane and call it *unbalance* related to the nodal plane i. The magnitude of the unbalanced mass is usually negligible compared with the system mass. Therefore, we shall neglect it in the mass matrix. The equations of motion will be, in the stationary coordinate system,

$$\mathbf{M\ddot{r}} + \mathbf{C\dot{r}} + \mathbf{Kr} = \omega^2\mathbf{U}e^{i\omega t} \qquad (13.10)$$

where

$$\mathbf{U} = \{u_1 e^{i\phi_1} \quad u_2 e^{i\phi_2} \quad \cdots \quad u_n e^{i\phi_n}\}$$

The elements of this vector are complex numbers expressed in a complex plane with a real axis along the direction of the moving mark A and the imaginary axis perpendicular to it.

We suppose that on every nodal plane we have instruments to measure the response r of the shaft. Consider, for example, a nodal plane (Figure 13.8). In the

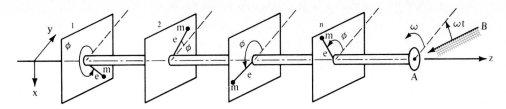

Figure 13.7 Multiplane balancing.

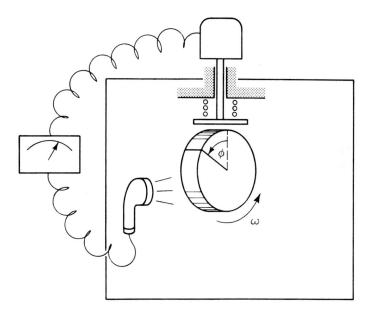

Figure 13.8 Stroboscopic balancing.

vertical direction, there is a vibration transducer riding the shaft by way of a spring. With electronic means, we light a stroboscope lamp for a very short time every time the shaft is at its higher position. Therefore, we "freeze" the reference mark A and we can observe the angle between the vertical (direction of measurement) and the reference mark. This angle is the phase angle between the response vector r and the *reference mark* (not the unbalance). Expressing the response on the same coordinate system as the unbalance U (in the moving coordinate system), we shall have

$$\zeta_i = A_i(\cos \phi_i + i \sin \phi_i) \tag{13.11}$$

and in the stationary coordinate system

$$r_i = A_i(\cos \phi_i + i \sin \phi_i)e^{i\omega t} = A_i e^{i\phi_i}e^{i\omega t} \tag{13.12}$$

Therefore, for $\mathbf{Z} = \{A_1 e^{i\phi_1} \quad \cdots \quad A_n e^{i\phi_n}\}$

$$[-\omega^2 \mathbf{M} + i\omega \mathbf{C} + \mathbf{K}]\mathbf{Z} = \mathbf{U} \tag{13.13}$$

Equation (13.13) yields the unbalance at every nodal plane in magnitude and direction. To correct the vibration, one has to add balance weights $-U$.

However, the situation is not as simple as it appears here, for the following reasons:

1. The system properties, especially $\mathbf{C}$ and $\mathbf{K}$, are not always known.
2. The system usually is not accessible for measurements at all the nodal planes.
3. Unbalance cannot be added on all nodal planes.

If the added unbalance is indeed $-\mathbf{U}$, where $\mathbf{U}$ is computed using equation (13.13), the response will be zero at all nodal planes. This is called *complete balancing*. However, for the reasons mentioned, complete balancing is beyond reach or very difficult in most engineering problems; therefore, we have to resort to approximate balancing.

A common approach is to try to eliminate the response at m selected nodal planes, $n > m$. This is called *m-plane balancing*. This can be accomplished by putting correction weights at p planes, where $m \leq p \leq n$. This is called *correction at p planes*. The next question is: What is the minimum number of balancing planes m to yield acceptable results?[5]

As we have said in Chapters 7 and 9, the system vibrates at all vibrating modes. However, the participation of every mode is higher the closer the corresponding critical speed is to the running speed. For example, if the running speed is between the first and second critical speed, the nodes way above the second will have very small amplitudes and can be neglected.

A single-span rotor on two flexible bearings has modes as shown in Figure 13.9. We see that each mode of order j can be eliminated by forcing the shaft to have amplitude zero at $j - 1$ planes: in other words, by a $j - 1$ plane balancing. Because of the difficulties mentioned, empirical methods have been developed for rotor balancing, based on trial and error or experience methods.

One-Plane Balancing

The one-plane balancing method is based on a test with a known unbalance and measurement of the response. The situation is shown in Figure 13.10. We assume a coordinate system (x, iy) attached to the shaft and having the real axis along the reference mark on the shaft. The strobe light shows the response in the direction of z_0, and the magnitude z_0 is measured by the vibration transducer. Therefore, we plot the initial response (due to the unknown unbalance) as the vector z_0. Next we add a trial mass m_1 at any convenient position and we run the rotor again. We observe the response vector in a new position z_1. This vector is the sum of the initial vector z_0 and the effect of the added unbalance $z_v = z_1 - z_0$. To have zero response z_1, the effect of the added unbalance ought to be $-z_0$. This can be accomplished if we rotate the trial weight m_1 by an angle ϕ and increase it by the

[5]Akimoff, B. 1918. Balancing apparatus. *Trans. ASME, 39*: 779. Bishop, R. E. D., and Gladwell, G. M. L. 1959. The vibration and balancing of an unbalanced flexible rotor. *J. Mech. Eng. Sci.*, 1: 66–67. Kellenberger, W. 1972. Should a flexible rotor be balanced in N or $N + 2$ planes? *ASME J. Eng. Ind.*, *94B*(2): 548–560.

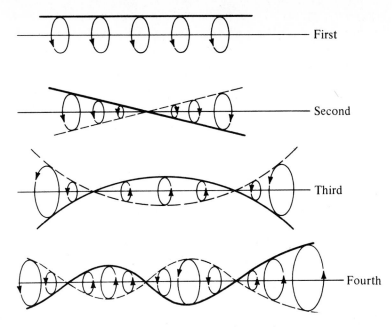

Figure 13.9 Rotor natural modes.

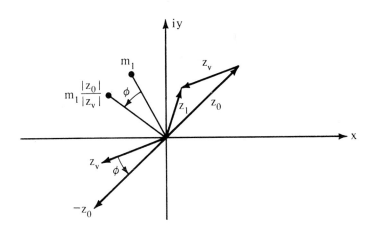

Figure 13.10 Vector diagram for one-plane balancing. (From Somervaile, I. J. 1954. Balancing a rotating disk: simple graphical construction. *Engineering,* Feb. 19.)

ratio $|z_0| \, / \, |z_v|$. Therefore, we determine the location of the required correction weight by magnitude and direction.

☐ **Example 13.5   One-Plane Balancing of a Rotor**

An axial fan rotor has 0.010-mm vibration and the reference mark appears 60° ahead of the vibration pickup location. A trial weight of 0.1 kg at 180° ahead of the reference mark resulted in 0.014-mm vibration 120° ahead of the reference mark. Compute the necessary balance correction in magnitude and direction.

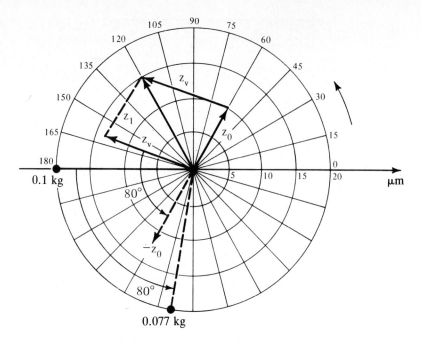

Figure E13.5

Solution We take a complex plane (x, iy) attached to the shaft and having the reference mark along the x-axis and the direction of rotation from x to iy. The vectors appear as in Figure E13.5. The effect of the added unbalance 0.1 kg is then the vector $z_u = z_1 - z_0$, because the vector z_1 is the result of the initial (unknown) unbalance and the added weight. Rotation of the added unbalance will result in equal rotation of its response z_u by the same angle. Therefore, in order that the effect of the added unbalance will correct the initial vibration z_0, the response z_n should appear opposite to z_0. Therefore, the added unbalance must be rotated by an angle 80°. To have complete balancing, the effect of the added unbalance should have magnitude z_0. Therefore, the added unbalance should be

$$U = 0.1 \times \frac{|z_0|}{|z_v|} = 0.1 \times \frac{10}{13} = 0.077 \text{ kg at } \phi = 180 + 80 = 260°$$

(The unit trial unbalance method was used to compute the required balance mass for a single plane rotor.)

☐

Two-Plane Balancing

In long rotors, unbalance might appear at different angles in different places. This will excite several modes, depending on the relation of the running speed to the critical speeds. To balance the first and second modes, we must balance in two planes.[6]

Suppose that the measured vibration is, at two balancing planes, as indicated in Figure 13.11. If the subscript indicates the plane (1 or 2) and the superscript

[6]Kellenberger, W. 1967. Balancing flexible rotors on two generally flexible bearings. *Brown Boveri Rev. 54*(9).

Portrait of Hunter. (Courtesy of Hunter Engineering.)

indicates the mode (1 or 2), we can analyze the measured vibration z_1 and z_2 in terms of components:

$$z_1 = z^{(1)} + z^{(2)}$$
$$z_2 = z^{(1)} - z^{(2)}$$

(13.14)

Solving for $z^{(1)}$ and $z^{(2)}$, we obtain

$$z^{(1)} = \tfrac{1}{2}(z_1 + z_2)$$
$$z^{(2)} = \tfrac{1}{2}(z_1 - z_2)$$

(13.15)

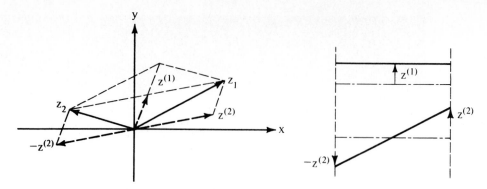

Figure 13.11 Vector diagram for two-plane balancing.

Graphically, these relations express the parallelogram law (Figure 13.11). We shall correct the two modes separately. The first mode can be treated as one plane balancing by adding, at both planes, weights of the same magnitude and direction. The second mode must be corrected by adding weights that are 180° apart in the two balancing planes. This correction is called a *balancing couple*. This method is applicable only if the system is symmetric; in other words, unbalance in plane 2 results in the same response as identical unbalance in plane 1. In general, this is not the case and we resort to a different method.

In two separate tests, we add unit unbalances in the two correction planes. Unit unbalance at correction plane 1 results in added unbalance effects $z_{11} = r_{11}e^{i\phi_{11}}$ and $z_{12} = r_{12}e^{i\phi_{12}}$ on balancing planes 1 and 2, respectively. Similarly, unit unbalance at correction plane 2 results in effects $z_{21} = r_{21}e^{i\phi_{21}}$ and $z_{22} = r_{22}e^{i\phi_{22}}$ on the two balancing planes. Without loss in generality, we assume that both trial unit unbalances were put on the reference mark. If the required balancing is u_1 and u_2, respectively, yet to be determined, their combined effects should add to the opposite of the initially measured response z_1 and z_2. Therefore,

$$u_1 r_{11}e^{i\phi_{11}} + u_2 r_{21}e^{i\phi_{21}} = -z_1$$
$$u_2 r_{12}e^{i\phi_{12}} + u_2 r_{22}e^{i\phi_{22}} = -z_2$$

(13.16)

In Cartesian form,

$$u_1 z_{11} + u_2 z_{21} = -z_1$$
$$u_1 z_{12} + u_2 z_{22} = -z_2$$

(13.17)

These are two complex equations in two complex unknowns, u_1 and u_2. Therefore, they can be used to determine the balancing required. Solving for u_1 and u_2 yields

$$u_1 = \frac{-z_1 z_{22} + z_2 z_{21}}{z_{11} z_{22} - z_{12} z_{21}}$$

$$u_2 = \frac{-z_2 z_{11} + z_1 z_{12}}{z_{11} z_{22} - z_{12} z_{21}}$$

(13.18)

If $z = x + iy$,

$$u_1 = \frac{(a_1 + ia_2)(b_1 + ib_2)}{D} \tag{13.19}$$

where

$$a_1 = -x_1 x_{22} + y_1 y_{22} + x_2 x_{21} - y_2 y_{21}$$

$$a_2 = -x_1 y_{22} - y_1 x_{22} + x_2 y_{21} + y_2 x_{21}$$

$$b_1 = x_{11} x_{22} - y_{11} y_{22} - x_{12} x_{21} + y_{12} y_{21}$$

$$b_2 = x_{11} y_{22} + y_{11} x_{22} - x_{12} y_{21} - y_{12} x_{21} \tag{13.20}$$

$$D = (x_{11} x_{22} - y_{11} y_{22} - x_{12} x_{21} + y_{12} y_{21})^2$$

$$+ (x_{11} y_{22} + y_{11} x_{22} - x_{12} y_{21} - y_{12} x_{21})^2$$

$$u_2 = \frac{(c_1 + ic_2)(b_1 + ib_2)}{D}$$

where

$$c_1 = -x_2 x_{11} + y_2 y_{11} + x_1 x_{12} - y_1 y_{12}$$

$$c_2 = -x_2 y_{11} - y_2 x_{11} + x_1 y_{12} + y_1 x_{12}$$

Finally,

$$u_1 = \frac{(a_1 b_1 - a_2 b_2) + i(a_1 b_2 + a_2 b_1)}{D}$$

$$u_2 = \frac{(c_1 b_1 - c_2 b_2) + i(c_1 b_2 + c_2 b_1)}{D} \tag{13.21}$$

The balance magnitudes will then be the moduli of the complex quantities

$$r_1 = \frac{[(a_1 b_1 - a_2 b_2)^2 + (a_1 b_2 + a_2 b_1)^2]^{1/2}}{D}$$

$$r_2 = \frac{[(c_1 b_1 - c_2 b_2)^2 + (c_1 b_2 + c_2 b_1)^2]^{1/2}}{D} \tag{13.22}$$

multiplied by the mass of the unit trial unbalance, and the balance angles ϕ_1 and ϕ_2 will be

$$\phi_1 = \arctan \frac{a_1 b_2 + a_2 b_1}{a_1 b_1 - a_2 b_2}$$

$$\phi_2 = \arctan \frac{c_1 b_2 + c_2 b_1}{c_1 b_1 - c_2 b_2} \tag{13.23}$$

Vector diagrams for balancing can be conveniently plotted in the polar diagram of Figure 13.12.

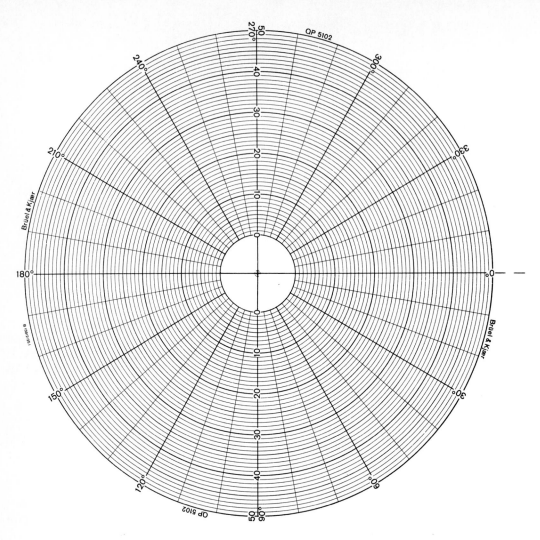

Figure 13.12 Polar diagram for balancing. (Courtesy of Bruel & Kjaer, Naerum, Denmark. Reprinted by permission.)

ALGORITHM A13.1

```
BALANCE2
Global variables x₁, x₂, x₁₁, x₂₂, x₂₁, x₁₂
Global variables y₁, y₂, y₁₁, y₁₂, y₂₁, y₂₂,
Global variables D, r₁, r₂, φ₁, φ₂
Global variables a₁, a₂, b₁, b₂, c₁, c₂
begin
    input x₁, x₂, y₁, y₂      /* rotor initial vibration  */
    input x₁₁, x₁₂, y₁₁, y₁₂   /* vibration with trial weight #1 */
```

```
input x₂₂, x₂₁, y₂₁, y₂₂   /* vibration with trial weight #2 */
' Computation of the added unbalance effects
x₁₁=x₁₁-x₁: x₁₂=x₁₂-x₂: x₂₁=x₂₁-x₁: x₂₂=x₂₂-x₂
y₁₁=y₁₁-y₁: y₁₂=y₁₂-y₂: y₂₁=y₂₁-y₁: y₂₂=y₂₂-y₂
' Computation of the balance shots
a₁=(-x₁x₂₂+y₁y₂₂+x₂x₂₁-y₂y₂₁)
a₂=-(-x₁y₂₂-y₁x₂₂+x₂y₂₁+y₂x₂₁)
b₁=(x₁₁x₂₂-y₁₁y₂₂-x₁₂x₂₁+y₁₂y₂₁)
b₂=-(x₁₁y₂₂+y₁₁x₂₂-x₁₂y₂₁-y₁₂x₂₁)
D=(x₁₁x₂₂-y₁₁y₂₂-x₁₂x₂₁+y₁₂y₂₁)²+(x₁₁y₂₂+y₁₁x₂₂-x₁₂y₂₁-y₁₂x₂₁)²
c₁=(-x₂x₁₁+y₂y₁₁+x₁x₁₂-y₁y₁₂)
c₂=-(-x₂y₁₁-y₂x₁₁+x₁y₁₂+y₁x₁₂)
r₁=[(a₁b₁-a₂b₂)²+(a₁b₂+a₂b₁)²]¹ᐟ²/D   /* magnitude of balance #1*/
r₂=[(c₁b₁-c₂b₂)²+(c₁b₂+c₂b₁)²]¹ᐟ²/D   /* magnitude of balance #2*/
φ₁ = arctan[(a₁b₂+a₂b₁)/(a₁b₁-a₂b₂)]   /* angle of balance #1    */
φ₂ = arctan[(c₁b₂+c₂b₁)/(c₁b₁-c₂b₂)]   /* angle of balance #2    */
Print r₁, r₂, φ₁, φ₂
end
```

□ **Example 13.6 Two-Plane Balancing**

A compressor rotor is to be balanced. A reference mark is marked on the rotor and vibration measurements are taken in the two bearings in the vertical direction. The amplitudes are 0.4 and 0.3 mm and the strobe light freezes the reference mark at angles 120° and 240°, respectively. Two tests were performed: one with a 0.1-kg unbalance at a radius of 150 mm on the reference mark in correction plane I, which resulted in vibrations of 0.5 mm at 140° and 0.35 mm at 150°, respectively, on the two bearings, and similar unbalance on correction plane II resulted in vibrations of 0.45 mm at 50° and 0.4 mm at 300°, respectively. Compute the required balance correction on the two correction planes (Figure E13.6).

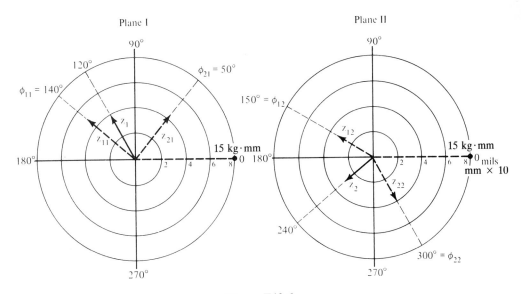

Figure E13.6

Solution   To balance the rotor, we have to use balance shots u_1 and u_2, yet unknown. The combined effect of the two balance shots should be equal and opposite to the rotor vibration. This is expressed by equations (13.19) and (13.23).

The solution was programmed in BASIC as follows:

```
pi = 3:14159
READ z1, ph1, z2, ph2: DATA 0.4, 120, 0.3, 240
x1 = z1*COS(ph1*pi / 180)
x2 = z2*COS(ph2*pi / 180)
y1 = z1*SIN(ph1*pi  / 180)
y1 = z1*SIN(ph2*pi  / 180)
READ z11, ph11, z12, ph12: DATA 0.5, 140, 0.35, 150
x11 = z11*COS(ph11*pi / 180)
x12 = z12*COS(ph12*pi / 180)
y11 = z11*SIN(ph11*pi / 180)
y11 = z11*SIN(ph12*pi / 180)
READ z21, ph21, z22, ph22: DATA 0.45, 50, 0.4, 300
x21 = z21*COS(ph21*pi / 180)
x22 = z23*COS(ph22*pi / 180)
y21 = z21*SIN(ph21*pi / 180)
y21 = z21*SIN(ph22*pi / 180)
' Computation of the added unbalance effects
x11=x11-x1: x12=x12-x2: x21=x21-x1: x22=x22-x2
y11=y11-y1: y12=y12-y2: y21=y21-y1: y22=y22-y2
' Computation of the balance shots
a1 = (-x1*x22+y1*y22+x2*x21-y2*y21)
a2 = -(-x1*y22-y1*x22+x2*y21+y2*x21)
b1 = (x11*x22-y11*y22-x12*x21+y12*y21)
b2 = -(x11*y22+y11*x22-x12*y21-y12*x21)
D = (x11*x22-y11*y22-x12*x21+y12*y21) ^ 2
D=D +(x11*y22+y11*x22-x12*y21-y12*x21) ^ 2
c1 = (-x2*x11+y2*y11+x1*x12-y1*y12)
c2 = -(-x2*y11-y2*x11+x1*y12+y1*x12)
r1 = SQR((a1*b1-a2*b2) ^ 2+(a1*b2+a2*b1) ^ 2) / D
r2 = SQR((c1*b1-c2*b2) ^ 2+(c1*b2+c2*b1) ^ 2) / D
phi1 = ATN((a1*b2+a2*b1) / (a1*b1-a2*b2))
phi2 = ATN((c1*b2+c2*b1) / (c1*b1-c2*b2))
IF phi1 < 0 THEN phi1 = phi1+pi
IF phi2 < 0 THEN phi2 = phi2+pi
PRINT r1, r2, phi1*180 / pi, phi2*180 / pi
END

RUN

    1.236931  0.8310182  141.1438  10.20661
```

Since the trial balance shots were 0.1 kg, the required balance weights are 0.124 and 0.083 kg·m at angles 141.1° and 10.20°, respectively.

□

TABLE 13.2 SUMMARY OF EQUATIONS OF CHAPTER 13

Receptance

$$G(m, c, k, \omega) = 1/[(k - m\omega^2) + (c\omega)^2]^{1/2}$$

$$R = (-\omega^2 M + i\omega C + K)^{-1}$$

Identification

$$(-\omega^2 M + i\omega C + K)X = F_0$$

Modal analysis

$$\ddot{q}_j + 2\zeta_j \omega_{nj} \dot{q}_j + \omega_{nj}^2 q_j = A_j e^{i\omega t}$$

$$\left(Q_{imj} + \frac{A_j}{4\zeta_j}\right)^2 + Q_{rj}^2 = \left(\frac{A_j}{4\zeta}\right)$$

Balancing

$$z_1 = z^{(1)} + z^{(2)}$$

$$z_2 = z^{(1)} - z^{(2)}$$

$$z^{(1)} = \frac{1}{2}(z_1 + z_2)$$

$$z^{(2)} = \frac{1}{2}(z_1 - z_2)$$

$$z = x + iy,$$

$$u_1 = \frac{(a_1 + ia_2)(b_1 + ib_2)}{D}$$

$$a_1 = (-x_1 x_{22} + y_1 y_{22} + x_2 x_{21} - y_2 y_{21})$$

$$a_2 = -(-x_1 y_{22} - y_1 x_{22} + x_2 y_{21} + y_2 x_{21})$$

$$b_1 = (x_{11} x_{22} - y_{11} y_{22} - x_{12} x_{21} + y_{12} y_{21})$$

$$b_2 = -(x_{11} y_{22} + y_{11} x_{22} - x_{12} y_{21} - y_{12} x_{21})$$

$$D = (x_{11} x_{22} - y_{11} y_{22} - x_{12} x_{21} + y_{12} y_{21})^2 + (x_{11} y_{22} + y_{11} x_{22} - x_{12} y_{21} - y_{12} x_{21})^2$$

$$u_2 = \frac{(c_1 + ic_2)(b_1 + ib_2)}{D}$$

$$c_1 = (-x_2 x_{11} + y_2 y_{11} + x_1 x_{12} - y_1 y_{12})$$

$$c_2 = -(-x_2 y_{11} - y_2 x_{11} + x_1 y_{12} + y_1 x_{12})$$

$$u_1 = \frac{(a_1 b_1 - a_2 b_2) + i(a_1 b_2 + a_2 b_1)}{D}$$

(continues)

Sec. 13.5   Masses Needed to Establish Known Motion Balancing   671

TABLE 13.2 *(cont.)*

$$u_2 = \frac{(c_1 b_1 - c_2 b_2) + i(c_1 b_2 + c_2 b_1)}{D}$$

$$r_1 = \frac{[(a_1 b_1 - a_2 b_2)^2 + (a_1 b_2 + a_2 b_1)^2]^{1/2}}{D}$$

$$r_2 = \frac{[(c_1 b_1 - c_2 b_2)^2 + (c_1 b_2 + c_2 b_1)^2]^{1/2}}{D}$$

$$\phi_1 = \arctan \frac{(a_1 b_2 + a_2 b_1)}{(a_1 b_1 - a_2 b_2)}$$

$$\phi_2 = \arctan \frac{(c_1 b_2 + c_2 b_1)}{(c_1 b_1 - c_2 b_2)}$$

REFERENCES AND FURTHER READING

BUZDUGAN, G., MIHAILESCU, E., and RADES, M. 1986. *Vibration Measurement*. Dordrecht, The Netherlands: Martinus Nijhoff.

DEN HARTOG, J. P. 1952. *Mechanical Vibration*, 4th ed. New York: McGraw-Hill.

DIMAROGONAS, A. D. 1976. *Vibration Engineering*. St. Paul, Minn.: West Publishing.

DIMAROGONAS, A. D., and PAIPETIS, S. A. 1983. *Analytical Methods in Rotor Dynamics*. London: Elsevier–Applied Science.

FEDERN, K. 1977. *Auswuchttechnik*. Berlin: Springer-Verlag.

GOODWIN, M. J. 1989. *Dynamics of Rotor-Bearing Systems*. London: Unwin Hyman. (ISO/DIS 1 940: Balance quality of rotating rigid bodies, TC 108.)

ISO/DIS 2 372: *Mechanical Vibration of Machines with operating speeds from 10 to 200 rps*. (Basis for specifying evaluation standards, TC 108.)

KLOTTER, K. 1960. *Technische Schwingungslehre*. Berlin: Springer-Verlag.

SPECTRAL DYNAMICS CORP. 1990. *Vibration Handbook*.

WILCOX, J. B. 1967. *Dynamic Balancing of Rotating Machinery*. London: Pitman.

PROBLEMS

Sections 13.1 and 13.2

13.1. Determine the maximum endurance of the worker in Example 13.1 and his time limit for reduced comfort.

13.2. You are an expert appointed by the court to study a union complaint that power plant operators feel discomfort working in the control room. You measured vertical vibration amplitude of 0.037 mm and horizontal amplitude 0.06 mm at a frequency of 1800 cpm. Comment on the validity of the complaint if the maximum interval between work breaks is 2 h.

13.3. You are now a company engineer in the case of Problem 13.2 assigned to design a vibration isolation system for the control console and the operator, together weighing 5350 N. Determine the stiffness of the isolation under the console to assure 2 h of work at full comfort if the critical damping factor is expected to be 4%, if only vertical vibrations are to be considered.

13.4. A compressor rotor weighs 5350 N, rotates at 5000 rpm, and has a natural frequency of 55 Hz. Determine the maximum allowed unbalance $(m_e e)$ in order that the operation be smooth, according to the chart of Figure 13.3, **(a)** for $\zeta = 0$, and **(b)** for $\zeta = 25\%$.

13.5. In Problem 13.4, determine the severity of vibration when the rotor runs through the natural frequency, for $\zeta = 25\%$.

13.6. An aircraft engine operates at 12,000 rpm and has a natural frequency of 140 Hz. The rotor, which weighs 2580 N, has a certain unbalance due to which there is vibration classified as "slightly rough" in Figure 13.3. What should be the system fraction of critical damping that a special damper should provide in order that the vibration be reduced to "smooth"?

Sections 13.3 and 13.4

13.7. An electric motor weighs 670 N and is mounted through springs on a base plate. If an unbalance of 0.0075 kg·m results in a vibration amplitude of 0.09 mm at 1800 rpm and 0.05 mm at 2600 rpm, determine the natural frequency and the fraction of critical damping of the system, assuming both to be invariable with the frequency of vibration.

13.8. Outline the tests necessary on a vibrating table to identify the mass, stiffness, and damping of a vibration transducer having one degree of freedom.

13.9. On the foundation of a gas turbine–compressor system, an engineer performed tests and found that there were two natural frequencies in the vertical direction at 43 and 95 Hz. She also computed the masses m_1 of the foundation and m_2 of the gas turbine and its support: $m_1 = 10,000$ kg and $m_2 = 5000$ kg. Determine the spring constants k_1 of the soil and k_2 of the flexible mounting (Figure P13.9).

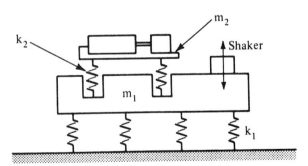

Figure P13.9

13.10. In Problem 13.9 it was determined that while a shaker delivering a constant harmonic force of 3850 N was at a frequency equal to the first natural frequency, the vibration amplitude at the foundation was 0.15 mm. Assuming that only the soil has significant damping, determine its damping constant.

13.11. To perform static balance of a large disk we put the shaft of the disk on light antifriction bearings and rotate it slowly (Figure P13.11). Then, before the disk stops, it oscillates for a while, indicating the existence of an unbalance. If the moment of inertia of the disk is 730 kg·m² and the period of oscillation is 1.3 s, determine the balance weight needed to balance the disk at a radius $r = 1.2$ m.

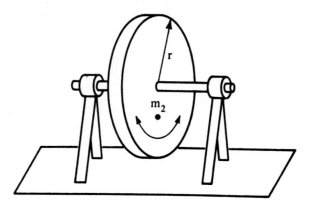

Figure P13.11

14

Machinery Vibration: Monitoring and Diagnosis

14.1 INTRODUCTION

Vibration and noise signals of any machine, engine, or structure contain a great deal of dynamic information related to the various exciting forces applied to them and the condition of the system. Therefore, changes in these response signals could be used to identify undesirable external loads or the onset of system faults before drastic failure occurs.

It has been found that noise is normally difficult to interpret unless measured in a good acoustic environment; otherwise, certain corrections must be applied to account for changing environmental factors, reflections, and interference. However, surface vibration measured at appropriate locations has been shown to present a reliable diagnostic tool by means of observing the overall vibration signal or by processing the vibration signal using certain techniques.

The mechanism of vibration generation in machines, engines, and common structures can be modeled and quantified in relation to changing operating or

*The above photo portrays predictive maintenance by vibration analysis. (Courtesy of Spectral Dynamics. Reprinted by permission.)

excitation conditions. For example, in piston engines the vibration excitation can be modeled as combustion and mechanical exciting forces (piston slap, timing gears, bearing impacts, fuel injection equipment, valve mechanism, etc.). These forces excite the engine structure and produce vibration response that is related both to the type of force and engine structure properties at the point of force application.

Another example is vibration generation in an off-shore structure. A structure placed in the sea vibrates continuously under the action of wind and waves. The movements of an off-shore platform occur mainly at specific frequencies, its natural frequencies. These frequencies are dependent upon the mass, stiffness (including foundation stiffness) and geometry of the structure, but are independent of the excitation. If the platform is damaged so that there is a change in stiffness, damping or mass, then some of its natural frequencies will be changed. The nature of the damage can be deducted from the pattern of these changes.

Similar models can be established for most engines, machines and structures where it would be possible to predict the trend of changing vibration corresponding to changing excitations or operating conditions, thus providing us with prior knowledge that can be used for monitoring purposes. Naturally these models can be perfected by correlations with simulated or real faults.

A vibration signal measured on a machine can be rendered very useful for detecting the onset of faults by processing it either in its time domain or frequency domain forms.[1]

14.2 TIME-DOMAIN VIBRATION ANALYSIS AND SIGNAL PROCESSING

Time-domain vibration signals can be processed in some of the following ways:

Weighting is the process where high-pass or/and low-pass filters are used to exclude certain parts of the vibration signal that are known to be unrelated to the development of a specific fault under investigation. It is a direct consequence of the orthogonality. Assume, for example, that the physical phenomenon being investigated is the unbalance response of a rotating machine and that the measured signal is of very broad spectrum due to noise or other vibration unrelated to unbalance. The frequency of the vibration due to unbalance is known to be the frequency of rotation. Therefore, if we multiply the signal by a harmonic function of unit amplitude and frequency equal to the frequency of rotation and integrate, the harmonics which are different from the frequency of rotation will give zero result and the nonzero result will be related only to the synchronous vibration.

High-pass filtering consists of using as a weighting function a function with a wide spectrum of high frequencies, to eliminate the low frequencies in the signal. *Low-pass filtering* consists of using as a weighting function a function with a wide spectrum of low frequencies, to eliminate the high frequencies in the signal. The procedure for a low-pass filter is illustrated in Figure 14.1. *Band pass filtering* consists of using as a weighting function a function with a wide spectrum of fre-

[1]Yates, H. G. 1949. Vibration diagnosis in marine geared turbines. *Trans. NE Coast Inst. Eng. Shipbuilders, 65:* 225. Sadowy, M. 1959. Elementary chatter in machine tools. *Tool Eng., 43.*

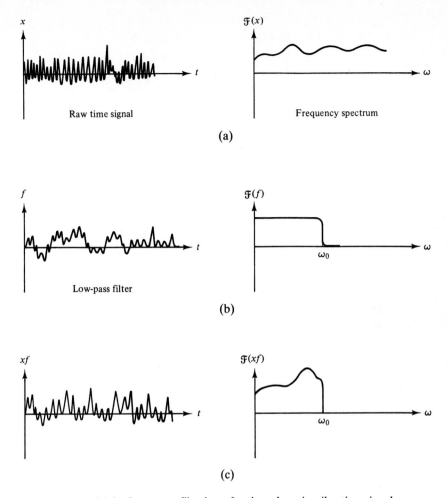

Figure 14.1 Low-pass filtering of a time-domain vibration signal.

quencies within a narrow range, to eliminate the frequencies in the signal that are outside the filter range.

Time-Series Averaging

Figure 14.2 illustrates how this method works for the case of an accelerometer-monitored gearbox. The output signal from the accelerometer mounted on the gearbox casing is sampled over a period equal to the time for one revolution of the particular gear being examined, and this sample is stored. Another sample of equal period is then taken and the "average" of this and the first sample is then stored. When this procedure is repeated many times (usually, several hundred), those frequency components of the signal related to the gear being examined are enhanced, while those related to other gears in the system or random causes tend to be averaged out. The net result is a *signal average plot,* which exhibits the same

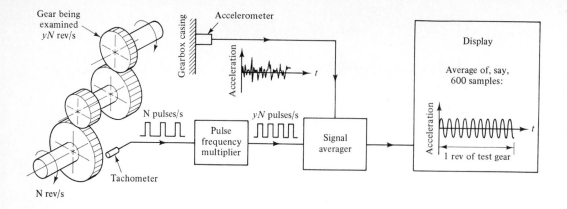

The type of gear defects which this method can
detect, and their associated signal average plots, are:

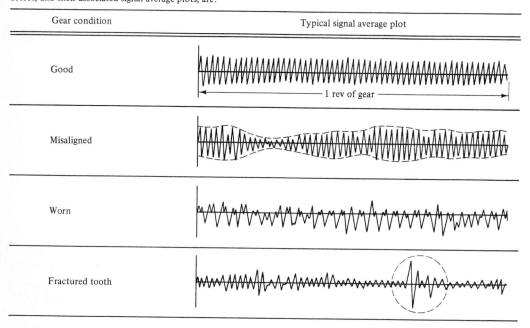

Gear condition	Typical signal average plot
Good	
Misaligned	
Worn	
Fractured tooth	

Figure 14.2 Time-series averaging of time-domain gearbox signals.

number of complete cycles as the number of teeth on the gear. A tachometer is
required to generate pulses at a frequency equal to the rotational frequency of
some reference shaft in the system, and a pulse frequency multiplier then adjusts
the pulse frequency by some preset amount to give a sample-length reference pulse
at the rotational frequency of the gear being examined.

Typical signal-average plots for an undamaged rolling element bearing and
for the same bearing with a small local defect in the outer face are shown in Figure
14.3. For comparison, the unprocessed transducer output signals for the two bear-

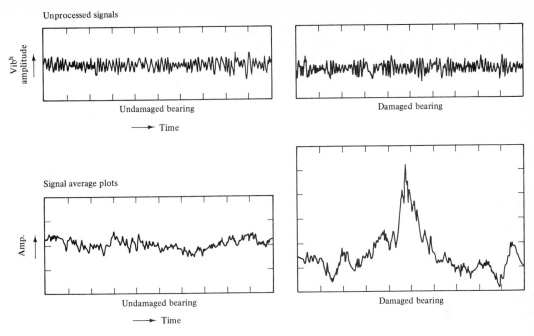

Figure 14.3 Typical signal-average plots for rolling element vibration.

ings are also shown, and it can be seen that there is no apparent difference in the "raw" signals for these two bearing conditions.

☐ **Example 14.1   Analysis of Gearbox Vibration**

Analysis of a gearbox's vibration signal generally starts with an assessment of its amplitude modulations. Generally speaking, these may be of two kinds: (1) synchronous with gear rotation, and (2) asynchronous with rotation. For the vast majority of gears, individual tooth faults emerge most clearly as the result of synchronous analysis. For stationary gears, as opposed to marine drives, so do most "whole" gear faults—in particular, misalignment. The signal processing technique to use for synchronous analysis is signal averaging, and the method of application is shown in Figure E14.1. Two signals are taken from the gearbox: casing vibration and rotational speed. A signal average is then derived for each gear in the system, which in turn involves the electronic manipulation of the available tacho signal to produce the necessary once-per-revolution synchronization pulse. The accuracy of this manipulation is obviously crucial, with phase errors of less than 1 part in 1000 being looked for.

The signatures that stem from signal averaging of the casing vibration tend to be of two general types. First, there are what may be called "regular" (Figure E14.1b), produced primarily by fixed-axis and low-speed epicyclic gears. The predominant component within them is gear meshing frequency (the average is performed across the frequency band from 0 Hz to greater than $1.5\times$ meshing frequency). Whole gear and external faults tend to modulate this frequency in a major way, whereas individual tooth faults may or may not.

Figure E14.1c, d, and e show the modulations produced by various common faults. Therefore, we should have a suite of pattern recognition techniques for the

Sec. 14.2   Time-Domain Vibration Analysis and Signal Processing   **679**

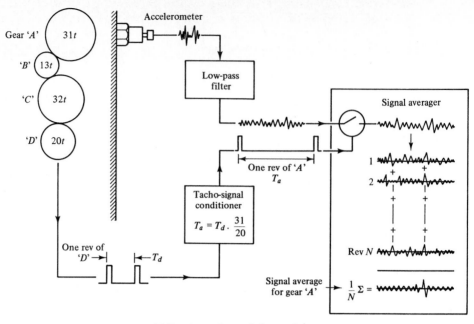

Gear 'A'   31t

'B'   13t

'C'   32t

'D'   20t

Accelerometer

Low-pass filter

Signal averager

One rev of 'A'
T_a

Tacho-signal conditioner

$T_a = T_d \cdot \dfrac{31}{20}$

One rev of 'D' $\longleftarrow T_d$

1

2

Rev N

Signal average for gear 'A' $\dfrac{1}{N} \Sigma =$

(a) Signal averaging applied to gearboxes

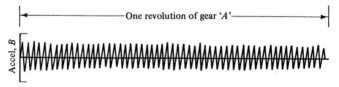

——————One revolution of gear 'A'——————

Accel, B

(b) Signal average for a perfect gear 'A'

(c) Signal average for a slightly misaligned gear 'A'

(d) Signal average for a heavily worn gear 'A'

(e) Signal average for a gear with fractured tooth

Figure E14.1

various modulations, and the more mutually exclusive these are the better. The other type of average has a much more random character. The randomness is most often the result of low dynamic isolation of the gear, as opposed to the "regular" average that stems from pure forced vibration.

□

Signal Enveloping

In this case we produce a time-domain envelope from the raw time-domain vibration signal such that the resulting envelope could be identified in relation to specific fault algorithms. Normally, a time-series averaging is applied first to produce a signal average, as shown in Figure 14.4a. Then we envelope the average signal using a process of full-wave rectification and low-pass filtration with a cutoff frequency set to an appropriate magnitude depending on the application (see Figure 14.4b). The low beat frequency associated with the beating phenomenon (see Chapter 1) is one form of signal enveloping.

Figure 14.5 shows an example of the technology applied to a high-speed epicyclic gearbox. The monitoring positions are marked P1 $\cdots$ P4 and the fault was chipping of two teeth on the sun gear. The signal average for the sun gear is shown in Figure 14.6a. The gear in question had 19 teeth, and there is little or no evidence of this frequency component in the average. The search for tooth damage then proceeds as follows. The time average is enveloped using a process of full-wave rectification and low-pass filtration, the cutoff frequency of the latter set to not less than 1.2 × meshing frequency (see Figure 14.6b). Then the dc component of the envelope is removed and calculation of its standard deviation is performed (Figure 14.6c).

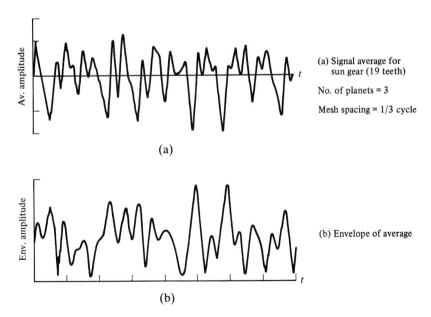

(a) Signal average for sun gear (19 teeth)

No. of planets = 3

Mesh spacing = 1/3 cycle

(a)

(b) Envelope of average

(b)

Figure 14.4 Process of signal averaging.

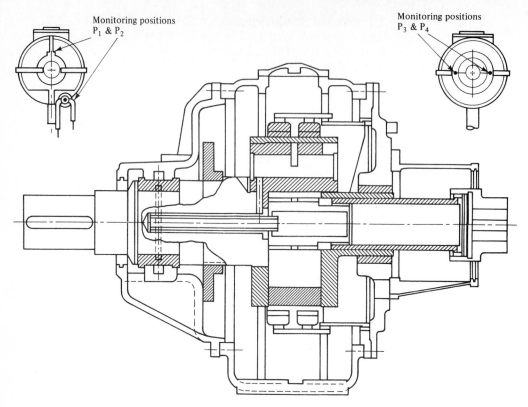

Figure 14.5 High-speed epicyclic gearbox showing the location of vibration measurements for monitoring and diagnosis.

The ac-coupled envelope is then clipped to preserve components about a threshold level of one standard deviation. This particular threshold level is not immutable, but experience has shown it to be the most satisfactory. To clip below this seems to allow too much signal through to the next stage of correlation, and to clip above it does not allow enough through. Much the same form of argument has been applied to certain pitch detection algorithms used in speech processing.

The clipped envelope is then correlated to ascertain whether or not any of the peaks occur at intervals of mesh spacing. The logic here is that tooth damage will excite the system only when it passes through the mesh (three times per revolution in the case shown), while under normal circumstances the gear in good condition will produce an envelope with Gaussian statistics. Experience has shown the latter to be true, and also that a correlation level at the mesh point of 0.2 represents a suitable division between "good" and "bad."

The correlogram of Figure 14.6d is a relatively simple signature to interpret. The high peak at the lag value corresponding to mesh spacing is quite obvious from the clipped envelope itself, and the two minor peaks beside it reflect the existence of two damaged teeth rather than one. In practice it is the robustness of the general method that is important, because to be economical, all such work must be performed by a computer.

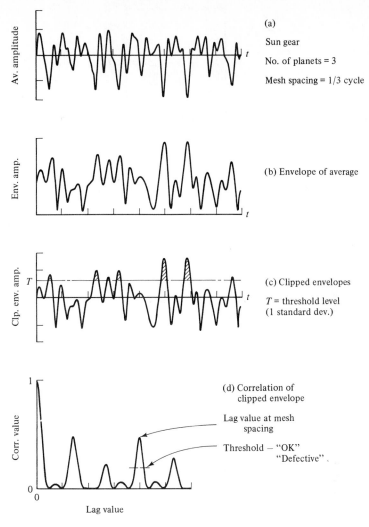

Figure 14.6 Diagnosis of faults in the gearbox system of Figure 14.5 using signal processing.

The Kurtosis Method

Of all components associated with monitoring of the rotating machine it is the rolling contact bearing that has received by far the most attention. The reasons for this are basically threefold: (1) it is simply the commonest component; (2) because it generally fails through fatigue, it possesses a definite catalog life with high scatter, which in turn forces the adoption of conservative periods for scheduled maintenance and confers the possibility of substantial economic advantages on the alternative strategy of condition-based maintenance, and (3) because of its high tolerance to abuse, it is then often abused, and so fails more frequently.

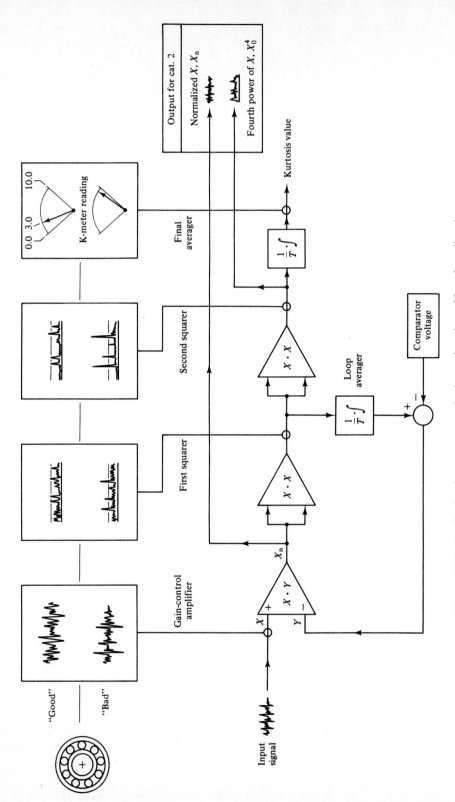

Figure 14.7 Basic circuit to compute the kurtosis value of bearing vibration signals.

There are a number of techniques that set out to detect the impulsive signals so characteristic of bearing track damage. One such technique involves the use of the signal's normalized fourth moment, otherwise known as its *kurtosis value*. The principle of operation is shown in Figure 14.7, and in one of its many possible implementations involves the use of three analog squaring devices. The simplest way to visualize the process is to think of the first two as acting to produce a signal normalized level equal to 0.11 V^2. So if the signal is "white," only the odd spike will go through the 1-V threshold, and in passage through the third quarter the vast bulk of the signal will be "squelched." The impulsive signal will, however, by definition, have many spikes that pass through the 1-V level, so far less of the normalized signal will be squelched by the final squarer, and a good deal in fact amplified. The kurtosis value can then be related directly to the integrated output of the third squarer and the mathematical formula (normalized fourth moment):

$$\text{kurtosis, } k = \frac{\int_{-\infty}^{\infty} (x - \bar{x})^4 p(x) \, dx}{\sigma^4} \qquad (14.1)$$

where x = signal level
$\bar{x}$ = signal mean
$p(x)$ = probability density distribution of x
σ = signal standard deviation

or in less precise terms by the more practical time-averaged estimator:

$$k = \frac{\int_{0}^{T} [x(t) - \bar{x}]^4 \, dt}{T\sigma^4} \qquad (14.2)$$

The Spike Energy Method

Spike energy, a unit of measure for judging bearing condition, is based on high-frequency peak acceleration. It is used for rolling element bearings, where damage is usually in the form of local spalling. Every time a rolling element passes over the local damage, there is a short impact, resulting in the release of a certain amount of energy in a short time (Figure 14.8). The vibration due to the short pulse is much smaller than the overall vibration and cannot be measured using conventional methods. However, the acceleration during the pulse is very high, due to the short duration of the energy release. The spike energy method consists of capturing the high-acceleration pulses and relating them to bearing damage.

The spike energy measuring circuit uses a high-frequency bandpass filter to reject low-frequency signals caused by unbalance, misalignment, looseness, and so on. It detects and holds the peak signals of vibration and displays them on a meter. The spike energy amplitude is affected by pulse duration and repetition frequency. Spike energy measurements differ from ordinary acceleration measurements by detecting only high-frequency vibrations and holding their peak amplitudes. Spike energy joins the list of vibration parameters—displacement, velocity, and acceleration—to provide an additional tool for predictive maintenance. This tool yields

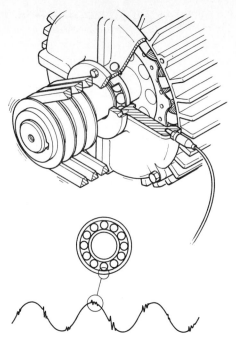

Figure 14.8 Local spalling in a ball bearing and its effect on the bearing vibration signal.

a single reading that provides an indication of rolling element bearing condition without having to interpret data from a vibration signature.

The spike energy of different signals is shown in Figure 14.9. A low-frequency signal (Figure 14.9a) is associated with high acceleration and low spike energy. A high-frequency signal (Figure 14.9b) is associated with high acceleration and high spike energy. A low-frequency pulse (Figure 14.9c), is associated with low acceleration and medium spike energy. A high-frequency pulse (Figure 14.9d) is associated with low acceleration and high spike energy.

14.3 FREQUENCY-DOMAIN VIBRATION ANALYSIS AND SIGNAL PROCESSING

The frequency-domain technique involves frequency analysis of the vibration signal and further processing of the resulting spectrum to obtain clearly defined diagnostic information. Figure 14.10 shows a typical use of straight frequency analysis to detect faults at blade passing and gear meshing frequencies. The top spectrum plot depicts a gearbox in good mechanical condition with reasonably low vibration acceleration levels and a normal mixture of events. A similar measurement made on a unit that had cracked teeth is presented in the bottom plot. We note the high amplitudes at gear mesh frequency, the pinion rotating speed ($1\times$), and the seventh harmonic ($7\times$) of pinion rotational speed.[2]

[2]Yates, loc. cit.; Shatoff, F. 1976. Using vibration analysis to determine the dynamic health of turbine/generators. *Power, 120*(5): 23–28.

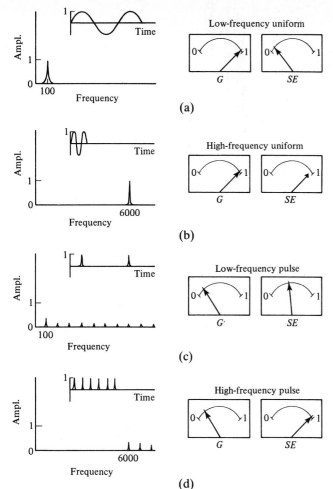

(a)

(b)

(c)

(d)

Figure 14.9 Spike energy values for various types of vibration signals.

In some cases, however, this normal frequency analysis may not clearly define the onset of a specific fault, and further processing may be required. Now let us look at some methods to process standard signals for fault diagnosis.

Signal Enhancement with Gating

Signal enhancement is often called *synchronous time averaging* or *history ensemble averaging* (Figure 14.11a), that is, the averaging takes place in the time domain (Figure 14.3), in contrast to spectrum averaging (Figure 14.11b), where the averaging is done in the frequency domain. This technique is very useful when combined with windowing or gating to detect specific events in portions of a machine cycle as shown in Figure 14.11. As explained before, time averaging should be initiated by a trigger, to assure that the averaged signals are in phase. If only a portion of the cycle is needed, to isolate a specific event, electronic triggers are used to effect the start and finish of sampling.

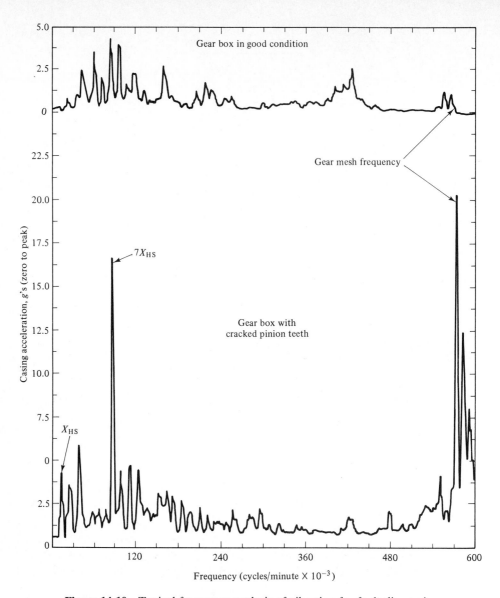

Figure 14.10 Typical frequency analysis of vibration for fault diagnosis.

RPM Spectral Maps (Waterfalls)

The spectrum for a specific vibration often changes with some parameter of the system operation. For a rotating machine, for example, the spectrum changes with the speed of rotation. Arranging the various spectra vertically in ascending value of this parameter (angular velocity in this case) results in a very useful diagram for fault identification. This display format allows the engineer to scan the data quickly to identify areas of interest. Such a plot is called a *spectral map* or *waterfall* or *cascade diagram*.

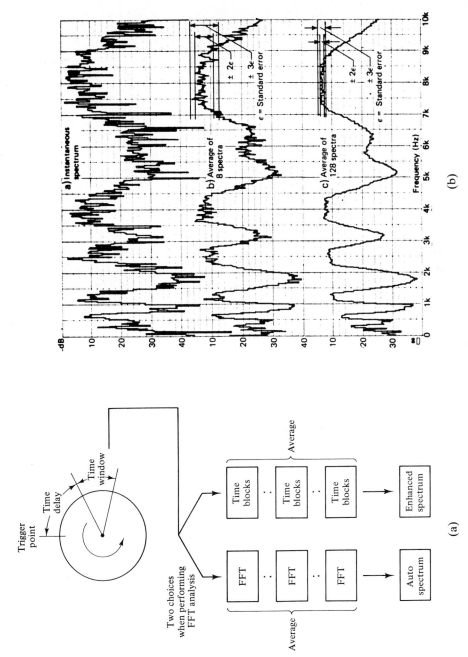

Figure 14.11 Spectrum averaging of a vibration signal.

The waterfall plot is very useful to distinguish between harmonic-order-related frequencies and those related to structural natural frequencies. Such an identification can be obtained when performing coast-down analysis, such as that shown in Figure 14.12, where the frequency spectrum is plotted for various speeds of rotation, in rpm. The synchronous vibration amplitude (called 1/rev) is then plotted along the straight line $\omega = 2\pi(rpm)/60$. Higher harmonics of order n, $n = 2, 3, 4, \ldots$, will be shown along the straight lines $\omega = 2\pi(rpm)/60n$ on the right of the synchronous vibration line. Subharmonics will appear on the left of the synchronous vibration line. Figure 14.12 shows that little harmonic activity is present. The peaks along the $1\times$ line indicate two critical speeds.

Figure 14.13 is another waterfall plot for an internal combustion engine, showing how this form of analysis clarifies the speed-related and constant natural frequencies. On the lower part, a 120-Herz constant frequency is shown, related to some engine resonance that does not depend on the exciting forces. At higher speeds, the different harmonics of the rotation are shown, of orders $1\times$, $2\times$, $3\times$, $5\times$, $6\times$, $8\times$, and so on. Moreover, at high speeds the valve gear dynamics are depicted at frequencies of about 600 Herz.

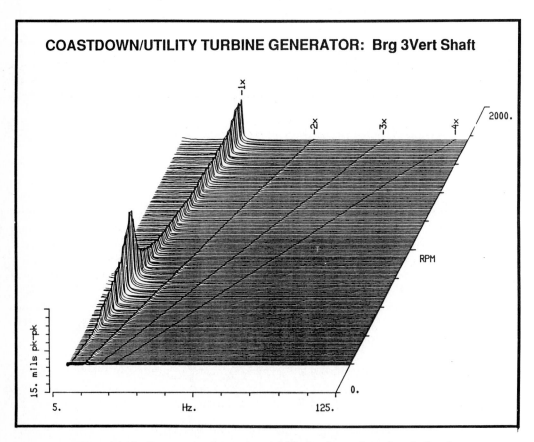

Figure 14.12 Rpm spectral map (waterfall) showing order-related vibration activity.

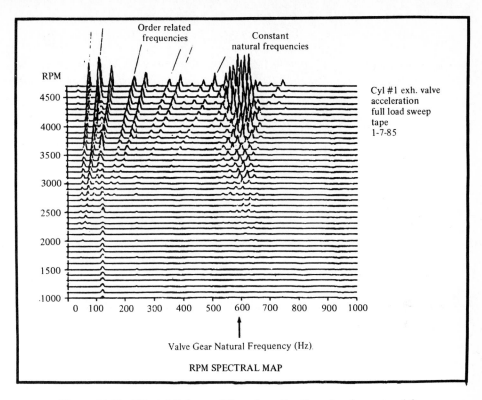

Figure 14.13 Waterfall for an IC engine vibration showing natural frequencies and speed-related events.

The Campbell Diagram

A Campbell diagram represents the same information as does a waterfall plot, but the information is plotted in a different way. The machine speed in rpm is plotted along the horizontal axis and frequency in hertz as the vertical axis. Orders of harmonics are shown as dashed lines originating at the lower left-hand corner. Resonances are then plotted using a code, for instance, as an intensity or color modulation, or as shown in Figure 14.14, as crosses with a circle, the diameter of the circle representing the peak amplitude at resonance.

The Campbell diagram shown in Figure 14.14 provides a very effective presentation for this type of data, wherein discrete vibration modes are excited by various harmonics of the running speed of the machine as the speed is varied. The Campbell diagram for a gas turbine blade vibration is shown in Figure 14.15. In general it is possible to distinguish resonant response on the blading from background noise by observing that the resonant modes reach their largest amplitudes on harmonic lines. It can be noted that certain harmonics, such as the twenty-fourth and fortieth, give rise to a much higher level of response than others. When this occurs in turbines at the higher orders, it generally can be explained by disturbances in the flow path upstream such as radial struts or stationary vanes.

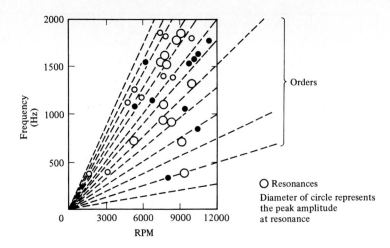

Figure 14.14 Construction of a vibration Campbell diagram.

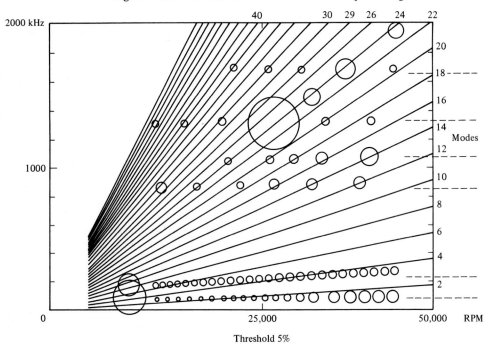

Figure 14.15 Campbell diagram for gas turbine blade vibration.

Cepstrum Analysis

The cepstrum was originally defined as the "power spectrum of the logarithm of the power spectrum," or mathematically,

$$C(\tau) = \left| F\{\log F_{xx}(f)\} \right|^2 \tag{14.3}$$

where $F_{xx}(f)$ is the power spectrum of the time signal $f_x(t)$ [i.e., $F_{xx}(f) = |F\{f_{x(t)}\}|^2$]

and $F\{\cdot\}$ represents the forward Fourier transform of quantity in braces. Later, a newer definition was coined as the "inverse transform of the logarithm of the power spectrum," or

$$C(\tau) = F^{-1}\{\log F_{xx}(f)\} \qquad (14.4)$$

One of the reasons for using this definition is that it highlights the connection between the cepstrum and the autocorrelation function, which can be obtained as the inverse transform of the power spectrum: that is,

$$R_{xx}(\tau) = F^{-1}\{F_{xx}(f)\} \qquad (14.5)$$

The definition in equation (14.4) is also closer to that of the complex cepstrum, which is the "inverse transform of the complex logarithm of the complex spectrum" and in fact is identical to it for spectra with zero phase.

The cepstrum can be considered as an aid to the interpretation of the spectrum in particular with respect to sideband families because it presents the information in a more efficient manner. Figure 14.16 shows a typical cepstrum for a truck axle system. Note the terminology used: *gamnitude* (*y*-axis), *quefrency* (*x*-axis), and *rahmonics* for multiples of the fundamental quefrency.

Figure 14.17 shows the results of this type of analysis (on a 1200-line spectrum extending from 150 to 1650 Hz) for a gearbox, both before and after repair. What is of interest here is that two main cepstrum components dominate, even over this

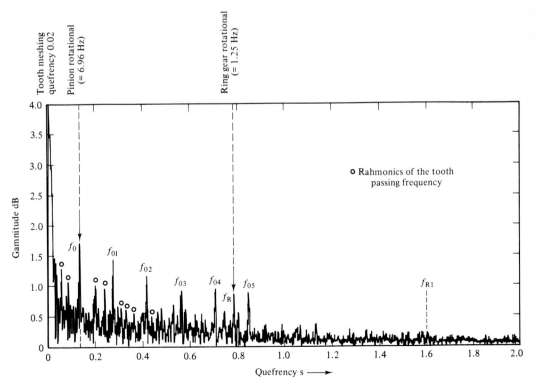

Figure 14.16 Typical vibration cepstrum for a truck axle system.

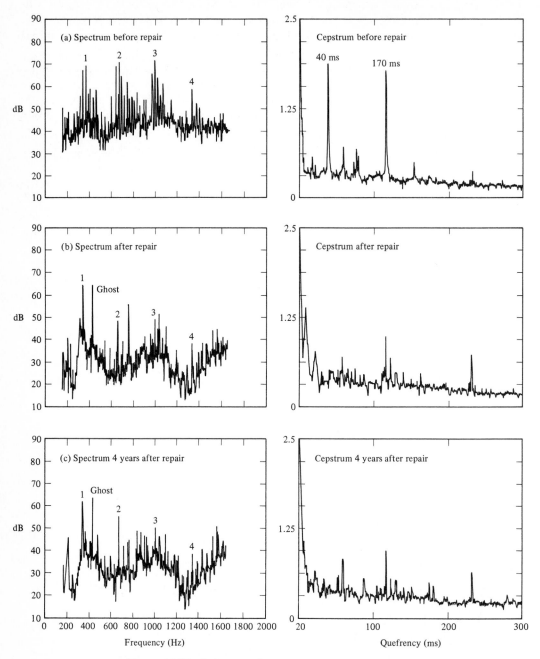

Figure 14.17 Spectra and cepstra of a gearbox system.

range, including the first four tooth-meshing harmonics (but eliminating frequencies below approximately half the tooth-meshing frequency). Both the 40- and 120-ms frequency components are significant before repair, indicating that there is modulation at both the shaft speed (8.3 Hz) and its third harmonic (25 Hz). The shaft speed modulation was probably due to misalignment that was corrected during repair, while the third harmonic was found to be due to a measurable "triangularity" of the gear. After repair the 40-ms component has disappeared completely and the shaft speed component has fallen drastically. It was found that the triangularity was due to excessive lapping during manufacture and the repair consisted in part of reversing the pinion on its shaft, utilizing the unused flanks. Figure 14.17 shows that 4 years later the problem had not redeveloped, although there is some indication of wear in the development of the tooth-mesh harmonics.

☐ **Example 14.2**

Find the Fourier spectrum and cepstrum of a periodic impulse of amplitude F_0, duration $\Delta t \to 0$, and period T.

Solution

$$F(t) = \frac{a_0}{2} + \sum_{n=1}^{\infty} a_n \cos n\omega t + b_n \sin n\omega t, \quad \omega = \frac{2\pi}{T}$$

is the Fourier series expansion of the function $F(t)$, where

$$a_n = \frac{2}{T} \int_0^T F(t) \cos n\omega t \, dt$$

$$= \frac{2}{T} \int_0^{\Delta t} F_0 \cos n\omega t \, dt + \frac{2}{T} \int_{\Delta t}^T (0) \cos n\omega t \, dt = \frac{2F_0 \, \Delta t}{T}$$

$$a_0 = \frac{2}{T} \int_0^T F(t) \, dt = \frac{2F_0}{T} \int_0^{\Delta t} dt = \frac{2F_0 \, \Delta t}{T}$$

$$b_n = \frac{2}{T} \int_0^{\Delta t} F(t) \sin n\omega t \, dt = 0$$

assuming that $\Delta t \to 0$. Therefore,

$$a_0 = a_1 = a_2 = a_3 = \cdots = \frac{2F_0 \, \Delta t}{T}, \quad b_n = 0, \quad n = 1, 2, 3, \ldots$$

The complex Fourier components are

$$C_n = \frac{2F_0 \, \Delta t}{T}, \quad n = 0, 1, 2, 3, \ldots$$

We observe that for $\Delta t \to 0$ the frequency spectrum has constant amplitude. The cepstrum is

$$C(\tau) = F^{-1}\{\log F_{xx}(f)\} = F^{-1}\left\{\log\left(2F_0 \frac{\Delta t}{T}\right)\right\} = F^{-1}(A)$$

The inverse of the Fourier transform is [equation (1.30)]

$$C(\tau) = \int_{-\infty}^{\infty} A e^{i2\pi ft} \, df = A \int_{-\infty}^{\infty} e^{i2\pi ft} \, df$$

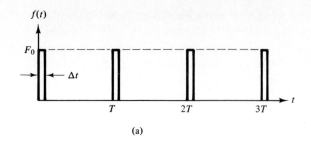

(a)

(b)

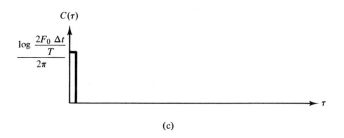

(c)

Figure E14.2

which is an impulse at $\tau = 0$ of amplitude $A/2\pi$. The situation is shown in Figure E14.2, which also illustrates the utility of the cepstrum. Indeed, only one spike exists in the cepstrum, which can easily be monitored and interpreted. The application of cepstrum analysis to rolling element bearings is rendered natural due to the impulsive character of the signal in the presence of a fault.

$\square$

14.4 FAULT DIAGNOSIS

Signal processing in the time and frequency domains has a major part to play in machinery health monitoring (MHM). However, its role is a subsidiary one to the task of connecting failure with vibration produced due to specific faults. Other digital signal processing techniques are being applied to MHM. These include modal analyses, digital filtering, convolution, inverse filtering, parametric methods, autoregressive-moving average (ARMA) processes, parametric estimation, and maximum entropy spectral analysis. These techniques require a great deal of understanding and relatively sophisticated measuring and analyzing equipment.

In most cases, however, basic frequency-domain analysis (FFT and narrow-band) can be used effectively for condition monitoring and fault diagnosis of machinery, provided that such analysis is accompanied by knowledge-based diagnostic charts or expert systems, such as in Table 14.1, with a typical diagnostic chart.

Unbalance

The most common machine fault is unbalance, since most machines have rotating members and complete balancing of any rotating member is impossible. On the spectral plot diagram (Figure 14.18), the signature has the form of amplitude along the $1\times$ line. High amplitudes along this line indicate critical speeds or system resonance to the harmonic excitation. Higher harmonics at higher vibration amplitudes indicate nonlinear effects due to high amplitudes: for example, in bearings and clearance annuli [Haddad and Corless, 1978]. In particular, high odd harmonics with very small even harmonics indicate loose parts or backlash, since they appear as square functions, which are known to have only odd harmonics (see Example 1.3).

☐ **Example 14.3 Vibration Spectrum of a Turbocompressor**

A turbocompressor shaft has a high level of vibration. A frequency spectrum at the running speed 3600 rpm shows a peak at 60 Hz and some smaller components at 180 and 300 Hz. The rotor critical speeds are known to be 2500 and 6000 rpm. Comment on the cause of vibration.

Solution   Resonance with the critical speed should be excluded, because the running speed is far from the critical speed. The cause must be unbalanced and loose parts in the system.

☐

Slow Transients: Rubbing

Temporary unbalance can be caused by local heating of shafts due to rubbing or water injection in steam turbines. Both are associated with slowly varying amplitude and continuously shifting phase angle, in a time of many seconds or minutes. Rubbing is associated with a repeating pattern, whereas water injection accompanies a single event, such as a sudden change in load.[3] A typical response to rubbing is shown in Figure 14.19

☐ **Example 14.4 Vibration Spectrum of a Turbocompressor**

A turbocompressor shaft has a high level of vibration. A frequency spectrum at the running speed 3600 rpm shows a peak at 60 Hz and some smaller components at 120 and 180 Hz. During the last 4 hours, the vibration amplitude fluctuated between 0.1 and 0.3 mm every hour, approximately. Comment on the probable cause.

[3]Newkirk, B. 1926. Shaft rubbing, *Mech. Eng.*, *48:* 830. Dimarogonas, A. 1970, *Packing Rub Effect.* General Electric Technical Information Series, DF70LS87. Dimarogonas, A. 1974. A study of the Newkirk effect in turbomachinery. *Wear, 28:* 369–382.

TABLE 14.1 DIAGNOSTIC CHART

Cause of vibration		0-40%	40-50%	50-100%	1x running frequency	2x RF	Higher multiples	½ RF	¼ RF	Lower multiples	Odd frequency	Very high frequency
1. Initial unbalance		..	..	..	90	5	5	..	..	..	..	..
2. Permanent bow or lost rotor parts (vanes)		..	..	..	90	5	5	..	..	..	..	..
3. Temporary rotor bow		..	..	..	90	5	5	..	..	..	..	..
4. Casing distortion — Temporary		←——10——→			80	5	5	..	..	..	..	..
Permanent		←——10——→			80	5	5	..	..	..	..	..
5. Foundation distortion		..	20	..	50	20	..	..	..	..	10	..
6. Seal rub		10	10	10	20	10	10	..	..	10	10	10
7. Rotor rub, axial		←——20——→			30	10	10	..	..	10	10	10
8. Misalignment		..	..	..	40	50	10	..	..	..	..	..
9. Piping forces		..	..	..	40	50	10	..	..	..	..	..
10. Journal & bearing eccentricity		..	..	..	80	20	..	..	..	..	..	..
11. Bearing damage		20——————→			40	20	..	..	..	..	..	20
12. Bearing & support excited vibration (oil whirls, etc.)		←—10—→	←—70—→	..	..	..	..	10	10	..	..	..
13. Unequal bearing stiffness horizontal-vertical		..	..	..	..	80	20	..	..	..	..	..
14. Thrust bearing damage		90————————————→			..	..	..	..	..	..	10
Insufficient tightness in assembly of:		Predominant frequency will show at lowest critical or resonant frequency									..	..
	15. Rotor (shrink fits)	40	40	10	..	..	..	..	..	..	10	..
	16. Bearing liner	90————→			..	..	..	..	..	..	10	..
	17. Bearing cases	90————→			..	..	..	..	..	..	10	..
	18. Casing & support	50————→			..	..	..	..	..	..	50	..
19. Gear inaccuracy		..	..	..	..	..	20	..	..	..	20	60
20. Coupling inaccuracy or damage		10	20	10	20	30	10	..	..	..	..	..
21. Rotor & bearing system critical		..	..	..	100	..	..	..	..	..	..	..
22. Coupling critical		..	..	..	100	Also make sure tooth fit is *tight!*						
23. Overhang critical		..	..	..	100	..	..	..	..	..	..	..
Structural resonance of: 24. Casing		..	10	..	70	10	..	10	..	..	..	..
25. Supports		..	10	..	70	10	..	10	..	..	..	..
26. Foundation		..	20	..	60	10	..	10	..	..	..	..
27. Pressure pulsations		Most troublesome if combined with resonance									100	..
28. Electrically excited vibration		..	..	..	..	..	..	..	..	..	..	..
29. Vibration transmission		..	..	..	..	..	..	..	..	..	90	..
30. Valve vibration		..	..	..	..	..	..	..	..	..	..	100
Problem		The section below is meant to identify basic mechanisms										
31. Sub-harmonic resonance		..	Rare — Look for aerodynamic origin (seals)					←————100————→			..	..
32. Harmonic resonance		..	..	..	..	←————100————→		..	..	..	..	..
33. Friction induced whirl		80	10	10	..	..	..	..	..	..	..	..
34. Critical speed		..	..	..	100	..	..	..	..	..	..	..
35. Resonant vibration		..	..	..	100	..	..	..	..	..	..	..
36. Oil whirl		..	100		Watch for aerodynamic rotor-lift (partial admission, etc.)							
37. Resonant whirl		..	100	..	..	..	..	..	..	..	..	..
38. Dry whirl		..	..	..	..	..	..	..	..	..	..	100
39. Clearance induced vibrations		10	80	10	..	..	..	..	..	..	..	..
40. Torsional resonance		..	..	..	40	20	20	..	..	..	20	..
41. Transient torsional		..	..	..	50	..	..	..	..	..	50	..

Numbers indicate percent of cases showing above symptoms, for causes listed in vertical column at left.

Source: Adapted from Sohre, G. S., 1968. Operating problems with high-speed turbomachinery, causes and corrections. *ASME Petroleum Mechanical Eng. Conf.,* Dallas.

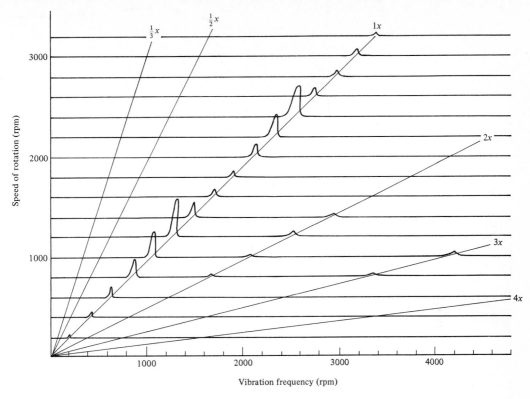

Figure 14.18 Spectral plot exhibiting large (1X) frequency due to unbalance.

> **Solution**   It is apparent that the problem is rotor rubbing. The higher harmonics are
> due to the high vibration amplitudes and nonlinearities, probably in the bearings.
>
> □

Misalignment

A very common machine fault is misalignment, since most machines consist of
compound rotating members, and exact alignment of all of them and with support
bearings is impossible. On the spectral plot diagram (Figure 14.20), the signature
has the form of amplitude along the $1\times$ line and higher harmonics, notably the
second. Higher harmonics are due primarily to nonlinear effects, and some of the
second harmonic is due to a dissimilar moment of inertia in couplings with lateral
misalignment. The latter increases continuously with speed; the former are stronger
when the rotating speed is near critical speeds. Misalignment is also associated with
substantial axial vibration.

Cracks

Structural cracks are associated with a shift of the natural frequencies to lower
values. Rotor cracks are associated (1) with high ($2\times$) amplitudes which remain
almost constant with the speed of rotation, and (2) with half-critical speed com-

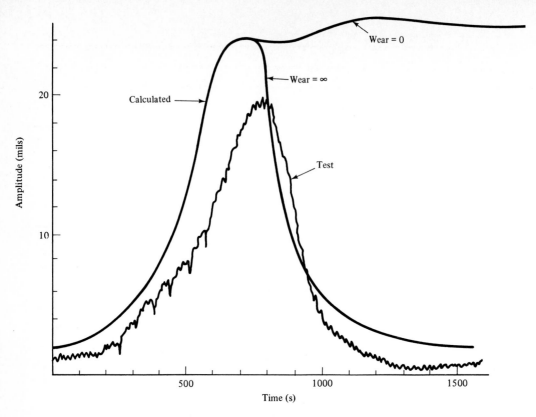

Figure 14.19 Rubbing response of a turbine rotor.

ponents, due to dissimilar moment of inertia and rotor weight, and higher harmonics due to the nonlinearity of closing cracks (Figure 14.21). In vertical rotors, the foregoing are not present or are very weak. In all cases, cracks are associated with vibration coupling; that is, axial and torsional resonances appear in the lateral vibration spectrum.[4]

☐ **Example 14.5 Vibration Spectrum of a Turbine Shaft**

A turbine shaft has a high level of vibration. A frequency spectrum at the running speed 3600 rpm shows a peak at 60 Hz, a larger one at 120 Hz, and substantial components at 180 and 240 Hz. The rotor critical speeds are known to be 2500 and 6000 rpm. Comment on the cause of vibration.

Solution Resonance with the critical speed should be excluded, because the running speed is far from the critical speed. The cause must be a crack or misalignment. Misalignment is most probable, due to the high ($3\times$) level of vibration. Additional observations must be made. For example, a 1250-rpm component will definitely indicate a crack.

☐

[4]Dimarogonas, A. D. 1970. *Dynamics of Cracked Rotors*. General Electric Report, Schenectady, N.Y. Dimarogonas, A. D. 1976. *Vibration Engineering*. St. Paul, Minn.: West Publishers. See also [Dimarogonas and Paipetis, 1983].

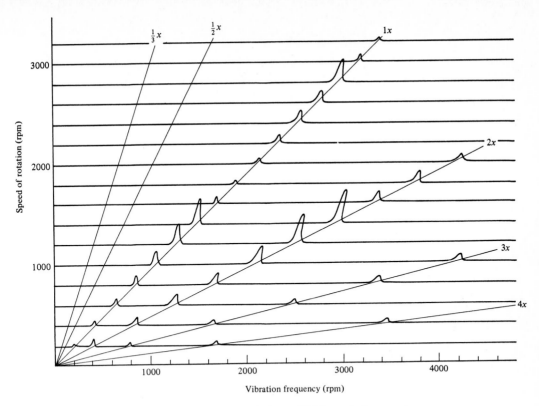

Figure 14.20 Spectral plot exhibiting misalignment.

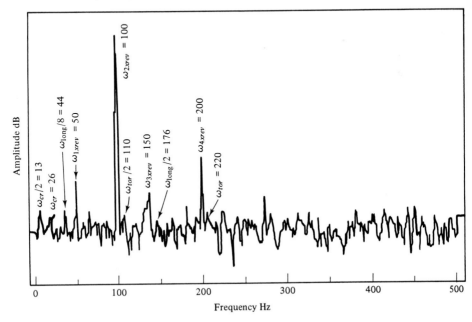

Figure 14.21 Vibration spectrum of a cracked shaft.

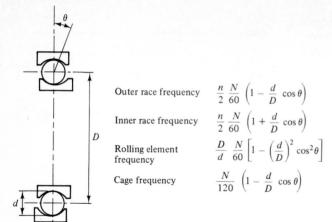

Outer race frequency $\quad \dfrac{n}{2}\dfrac{N}{60}\left(1-\dfrac{d}{D}\cos\theta\right)$

Inner race frequency $\quad \dfrac{n}{2}\dfrac{N}{60}\left(1+\dfrac{d}{D}\cos\theta\right)$

Rolling element
frequency $\qquad\quad \dfrac{D}{d}\dfrac{N}{60}\left[1-\left(\dfrac{d}{D}\right)^2\cos^2\theta\right]$

Cage frequency $\qquad \dfrac{N}{120}\left(1-\dfrac{d}{D}\cos\theta\right)$

Figure 14.22 Rolling element
bearing frequencies.

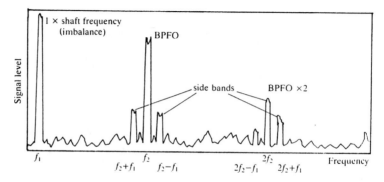

Figure 14.23 Typical rolling element bearing vibration spectrum.

Rolling Element Bearing Faults

A rolling element bearing is associated with several characteristic frequencies, beyond the frequency of rotation $N = \omega/2\pi$. These are shown in Figure 14.22. Since they are numerous, they can interfere with the harmonics of the fundamental frequency, and this is one of the reasons for the different signal processing methods developed for rolling element bearing fault diagnosis. Due to interaction of different harmonics, additional "sum and difference" sideband harmonics appear in the spectrum (Figure 14.23).

☐ **Example 14.6   Vibration Spectrum of a Rotor on Rolling Bearings**

The vibration spectrum of a 1000-rpm machine shows peaks at the following frequencies: 17, 95, 97, 114, and 132 Hz. The machine runs on rolling element bearings with 10 rollers of 16 mm diameter, 0 contact angle, and 112 mm pitch diameter. Investigate possible causes.

Solution   Using Figure 14.22, the bearing characteristic frequencies are:

Outer race frequency	71 Hz
Inner race frequency	95 Hz
Rolling element frequency	114 Hz
Cage frequency	7.1 Hz
Machine frequency	17 Hz

The 17-Hz component is probably caused by unbalance. The 114-Hz component indicates rolling element damage and the 97- and 132-Hz components are sidebands of the 114-Hz frequency $\mp$ 17 Hz. The 95-Hz signal indicates inner race wear. □

Gear Faults

The characteristic frequencies of gears are (1) the gear speeds N, and (2) the tooth passing frequencies nN, where n is the number of teeth and the sidebands are $(n - 1)N$ and $(n + 1)N$, plus the structural resonances of the gear shafts, gearbox, bearings, and so on. A typical gearbox vibration spectrum is shown in Figure 14.24.

□ **Example 14.7   Vibration Spectrum of a Gearbox**

A gearbox has a high level of vibration. There are two shafts connected with two gears with 20 and 35 teeth, respectively. The first shaft runs at 3600 rpm. A frequency spectrum at the running speed 3600 rpm shows a peak at 60 Hz, a larger peak at 35 Hz, and peaks at 1165, 1200, and 1235 Hz. Comment on the cause of vibration.
Solution   The characteristic frequencies are

Shaft 1: 60 Hz
Shaft 2: 60 × 20/35 = 34.3 Hz
Tooth passing frequency, gear 1: 60 × 20 = 1200 Hz
Tooth passing frequency, gear 2: 34.3 × 35 = 1200 Hz
Sidebands, gear 1: 1200 $\mp$ 60 = 1140 and 1260 Hz
Sidebands, gear 2: 1200 $\mp$ 34.3 = 1165.7 and 1234.3 Hz

The problem is probably a broken tooth on gear 2. □

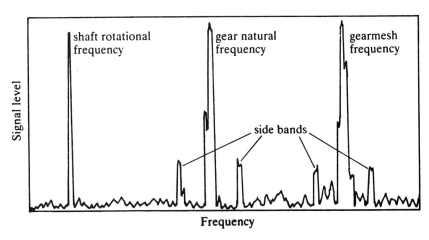

Figure 14.24   Typical gearbox vibration spectrum.

Fluid-Induced Instabilities

As discussed in Chapter 11, fluid bearings can lead to rotor-bearing system instabilities. They appear with subharmonics with frequency equal to the fundamental critical speed ω_{cr} when it is smaller than the operating speed Ω, in general $\omega_{cr} <$ 0.4Ω (Figure 14.25), called *oil whirl*. When the critical speed is nearly half the operating speed, the whirl becomes violent and stays at the critical speed for increasing operating speed. This condition is called *oil whip*. The vibration amplitude and frequency do not depend on the machine loading.

Flow of fluids or magnetic fields can induce violent instabilities, generally at the lower critical speed. Vibration amplitude is strongly dependent on the machine load, wind speed, and in general, excitation load.[5]

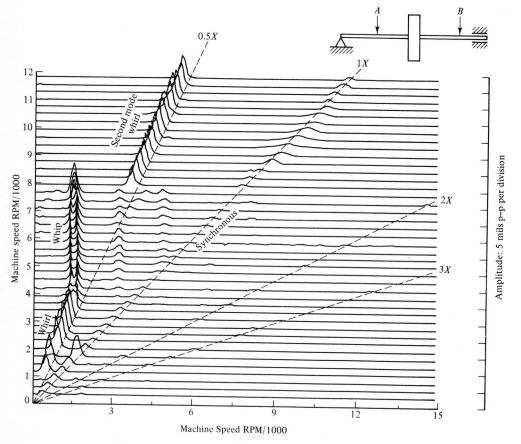

Figure 14.25 Spectral plot showing oil whirl and oil whip.

[5]Thomas, H. J. 1956. *Unstable Oscillations of Turbine Rotors Due to the Leakage in the Clearance of Rotor and Bucket Packings.* AEG Technical Publication 1150. Dimarogonas, A. D. 1971. *Analysis of Steam Whirl.* General Electric Technical Information Series, DF71LS48.

TABLE 14.2 SUMMARY OF EQUATIONS IN CHAPTER 14

Fourier Series of sampled functions

$$a_0 = \left(\frac{2}{N}\right) \sum_{i=1}^{N} g(t_i)$$

$$a_n = \left(\frac{2}{N}\right) \sum_{i=1}^{N} g(t_i) \cos \frac{2n\pi t_i}{T},$$

$$b_n = \left(\frac{2}{N}\right) \sum_{i=1}^{N} g(t_i) \sin \frac{2n\pi t_i}{T}$$

$$t_i = \frac{iT}{N}, \qquad T = \frac{2\pi}{\omega}$$

Complex Fourier Series of sampled functions

$$G_k = \sum_{n=0}^{N-1} g_n e^{-i(2\pi kn/N)}$$

$$g_n = \left(\frac{1}{N}\right) \sum_{k=0}^{N-1} G_k e^{i(2\pi kn/N)}$$

Kurtosis

$$k = \left[\int_{-\infty}^{\infty} (x - \bar{x})^4 \, p(x) \, dx \right] / \sigma^4$$

$$k = \frac{\left[\int_0^T [x(t) - \bar{x}]^4 \, dt \right]}{T\sigma^4}$$

Cepstrum

$$C(\tau) = \left| F\{\log F_{xx}(f)\} \right|^2$$

$$C(\tau) = F^{-1}\{\log F_{xx}(f)\}$$

Autocorrelation

$$R_{xx}(\tau) = F^{-1}\{F_{xx}(f)\}$$

REFERENCES AND FURTHER READING

BRIGHAM, E. O. 1974. *The Fast Fourier Transform*. Englewood Cliffs, N.J.: Prentice Hall.

COOLEY, J. W., and TUKEY, J. W. 1965. An algorithm for the machine calculation of complex Fourier series, *J. Math. Comp.*, 19(90): 297–301.

DIMAROGONAS, A. D., and PAIPETIS, S. A. 1983. *Analytical Methods in Rotor Dynamics*. London: Elsevier–Applied Science.

HADDAD, S. D., and CORLESS, M. J. 1978. Vibration measurements to monitor faults in rotating machines. *Proc. Int. Conf. Noise Control Eng.*, San Francisco, pp. 963–966.

KLOTTER, K. 1951. *Technische Schwingungslehre*. Berlin: Springer-Verlag.

KRUSE, R. L. 1987. *Data Structures and Program Design*. Englewood Cliffs, N.J.: Prentice Hall.

RAMIREZ, R. W. 1985. *The FFT Fundamentals and Concepts*. Englewood Cliffs, N.J.: Prentice Hall.

RANDALL, R. B. 1977. *Frequency Analysis*. Naerum, Denmark: Bruel & Kjaer.

RUNGE, C. 1904. *Theorie und Praxis der Reihen*. Leipzig.

PROBLEMS

Section 14.2

14.1. A vibration signal consists of 10 harmonics of fundamental frequency $\omega_0 = 377$ rad/s and constant amplitude $a_n = 1$ mm, $b_n = 0$, $n = 0, 1, 2, \ldots$.
  (a) Compute the time signal.
  (b) Form the product $x(t)w(t)$, where $w(t)$ has harmonics 4, 5, 6 of unit real amplitude.
  (c) Compute the frequency spectrum of the result.

14.2. A vibration signal consists of 10 harmonics of fundamental frequency $\omega_0 = 377$ rad/s and constant amplitude $a_n = 1$ mm, $b_n = 0$, $n = 0, 1, 2, \ldots$.
  (a) Compute the time signal of length $50T_0$.
  (b) Form the product $x(t)w(t)$, where $w(t) = \cos(\omega_0 t/25)$.
  (c) Compute the frequency spectrum of the result.

14.3. A vibration signal consists of 10 harmonics of fundamental frequency $\omega_0 = 377$ rad/s and constant amplitude $a_n = 1$ mm, $b_n = 0$, $n = 0, 1, 2, \ldots$.
  (a) Compute the time signal.
  (b) Form the product $x(t)w(t)$, where $w(t) = \cos^2(\omega_0 t/25)$.
  (c) Compute the frequency spectrum of the result.

14.4. A vibration signal consists of 10 harmonics of fundamental frequency $\omega_0 = 377$ rad/s and constant amplitude $a_n = 1$ mm, $b_n = 0$, $n = 0, 1, 2, \ldots$.
  (a) Compute the time signal.
  (b) Form the product $x(t)w(t)$, where $w(t) = \cos^3(\omega_0 t/25)$.
  (c) Compute the frequency spectrum of the result.

14.5. A vibration signal consists of 10 harmonics of fundamental frequency $\omega_0 = 377$ rad/s and constant amplitude $a_n = 1$ mm, $b_n = 0$, $n = 0, 1, 2, \ldots$.
  (a) Compute the time signal.
  (b) Form the product $x(t)w(t)$, where $w(t) = \cos^4(\omega_0 t/25)$.
  (c) Compute the frequency spectrum of the result.

14.6. Compute the kurtosis of a signal that is the $x(t) = x_1(t) + x_2(t)$, where $x_1(t) = 0.5 \cos 377t$ and $x_2(t)$ is an impulse of amplitude 1.5, duration 1 ms, and frequency 60 Hz.

14.7. Compute the kurtosis of a signal that is the $x(t) = x_1(t) + x_2(t)$, where $x_1(t) = 0.5 \cos 377t + 0.8 \sin(200t + 2.5)$ and $x_2(t)$ is an impulse of amplitude 1.5, duration 1 ms, and frequency 300 Hz.

14.8. Compute the spike energy of the signal in Problem 14.6.

14.9. Compute the spike energy of the signal in Problem 14.7.

14.10. Compute the envelope of the beating signal, which is the sum of two signals of amplitude 1 and frequencies 350 and 377 rad/s.

Section 14.3

14.11. A turbine rotor has two critical speeds at 1400 and 2700 rpm. Draw a spectral map illustrating the unbalance response.

14.12. A turbine rotor has two critical speeds at 1400 and 2700 rpm. In addition, there is a constant vibration of amplitude 0.2 and frequency 60 Hz due to excitation from neighboring machines. Draw a spectral map illustrating the response.

14.13. A turbine rotor has two critical speeds at 1400 and 2700 rpm. There is an edge crack on the rotor. The fundamental torsional and longitudinal natural frequencies are 3000 and 4000, respectively. Draw a spectral map illustrating the unbalance response.

14.14. A turbine rotor has two critical speeds at 1400 and 2700 rpm. It is supported by roller bearings with $\theta = 0$, pitch diameter 200 mm, roller diameter 20 mm, and number of rollers 12. Draw a spectral map illustrating the unbalance response.

14.15. A compressor rotor has two critical speeds at 1400 and 2700 rpm. It is connected to a motor with gears with 20 teeth (motor) and 45 teeth (rotor). Draw a spectral map illustrating the response to unbalance and broken teeth.

14.16–14.20. For the systems of Problems 14.11 to 14.15, draw a Campbell diagram.

Section 14.4

14.21. A motor rotor is supported by roller element bearings with 10 rollers, pitch diameter 60 mm, roller diameter 8 mm, and contact angle 0. Determine the characteristic frequencies.

14.22. An engine shaft is supported by roller element bearings. At the left end the bearing has 10 rollers, pitch diameter 48 mm, roller diameter 6 mm, and contact angle 0. At the right end the bearing has 10 rollers, pitch diameter 60 mm, roller diameter 7 mm, and contact angle 0. Determine the characteristic frequencies.

14.23. A gearbox has the following characteristics:

Shaft no.	Pinion teeth	Gear teeth
1	18	
2	20	40
3		35

Determine the characteristic frequencies.

14.24. An epicyclic gear drive has the following characteristics:

Gear	Number of teeth
Sun	18
Planets	20
Outer	58

Determine the characteristic frequencies for all speeds.

14.25. A rotating machine has the following values of the amplitude peaks in the vibration spectrum at the frequencies indicated:

$$\frac{\omega_{cr}}{2} = 0.2 \qquad \omega_{cr} = 0$$

$$1 \times \Omega = 1.1 \qquad 2 \times \Omega = 2.4 \qquad 3 \times \Omega = 0.1 \qquad 4 \times \Omega = 0.2$$

Determine the probable cause of the problem.

14.26. A rotating machine has the following values of the amplitude peaks in the vibration spectrum at the frequencies indicated:

$$\frac{\omega_{cr}}{2} = 0 \qquad \omega_{cr} = 0$$

$$1 \times \Omega = 2.2 \qquad 2 \times \Omega = 0.1 \qquad 3 \times \Omega = 0.1 \qquad 4 \times \Omega = 0.05$$

Determine the probable cause of the problem.

14.27. A rotating machine has the following values of the amplitude peaks in the vibration spectrum at the frequencies indicated:

$$\frac{\omega_{cr}}{2} = 0 \qquad \omega_{cr} = 2.4$$

$$1 \times \Omega = 1 \qquad 2 \times \Omega = 0 \qquad 3 \times \Omega = 0 \qquad 4 \times \Omega = 0$$

Increasing load has no effect. Determine the probable cause of the problem.

14.28. A rotating machine has the following values of the amplitude peaks in the vibration spectrum at the frequencies indicated:

$$\frac{\omega_{cr}}{2} = 0 \qquad \omega_{cr} = 0$$

$$1 \times \Omega = 3 \qquad 2 \times \Omega = 0.1 \qquad 3 \times \Omega = 0 \qquad 4 \times \Omega = 0.1$$

The $1\times$ amplitude fluctuates regularly between 3 and 1 mm at intervals of 18 min. Determine the probable cause of the problem.

14.29. A rotating machine has the following values of the amplitude peaks in the vibration spectrum at the frequencies indicated:

$$\frac{\omega_{cr}}{2} = 0 \qquad \omega_{cr} = 3$$

$$1 \times \Omega = 1 \qquad 2 \times \Omega = 0.1 \qquad 3 \times \Omega = 0.1 \qquad 4 \times \Omega = 0$$

The vibration at $\omega = \omega_{cr}$ appears at high loads only. Determine the probable cause of the problem.

Appendix

I

Algorithmic Notation

The algorithms given throughout the book are described in a pseudocode format with the following conventions:

1. Module name, any alphanumeric name: for example,

```
SUBNAME(A,B,x,n)
```

2. Declaration of external modules

```
EXTERNAL LinearEquations(A,b,x,n)
```

3. Declaration of global variables, communicated to other modules as arguments, as with SHARED, COMMON, and so on (Floating, Integer, Complex, . . . ,/Variable, Constant, . . ./Array, Scalar): for example,

```
GLOBAL FLOATING VARIABLE ARRAY A(3,5), B(20)
GLOBAL INTEGER CONSTANT SCALAR Pi, GraviCon
```

4. Comments: for example,

```
/* Any Comment .............................*/
```

5. Loops, For-do/End do, Do until, and so on: for example,

```
For index=1 to 20 step 4, do
  statement
  statement
End do
```

6. Assignment statements: for example,

```
Sum=Sum+2*Area(i) ; Volume=Volume+Area(i)*Xlength(i)
```

7. Branching: for example,

```
Go to Bypass /* Bypass is a line identifier */
```

8. Decisions: for example,

```
If a›b+3 then
  statements
else
  statements
end if
```

9. Markings: for example,

```
begin           /* beginning of module execution */
  statements
end             /* end of module execution */
```

10. Some other conventions are used, most of which are C-like.

Appendix

II

Discrete Elements

Figure	Description	Formula

A. Mass Elements

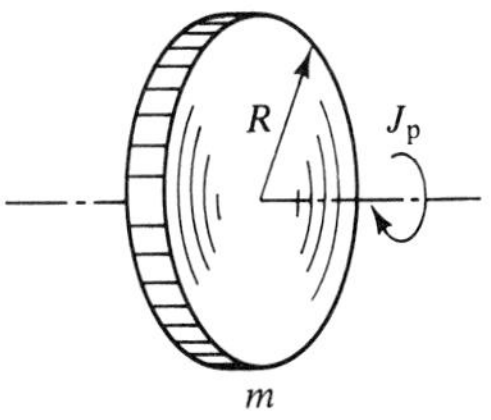

Mass moment of inertia of a disk about the axis of rotation

$J_p = \frac{1}{2}mR^2$

about a diameter

$J_{x-x} = \frac{1}{4}mR^2$

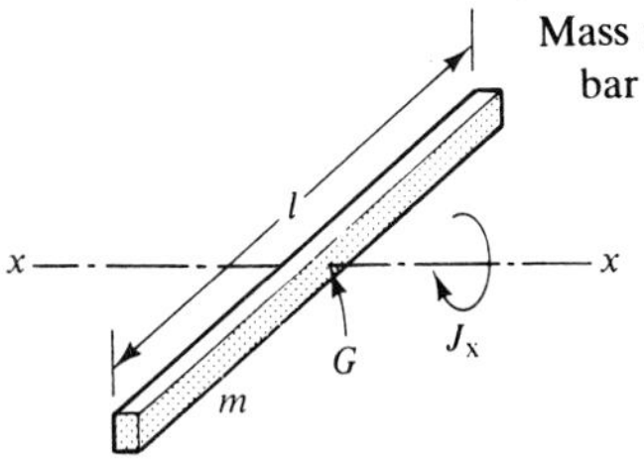

Mass moment of inertia of a bar about the mass center

$J_x = \frac{1}{12}ml^2$

Figure	Description	Formula

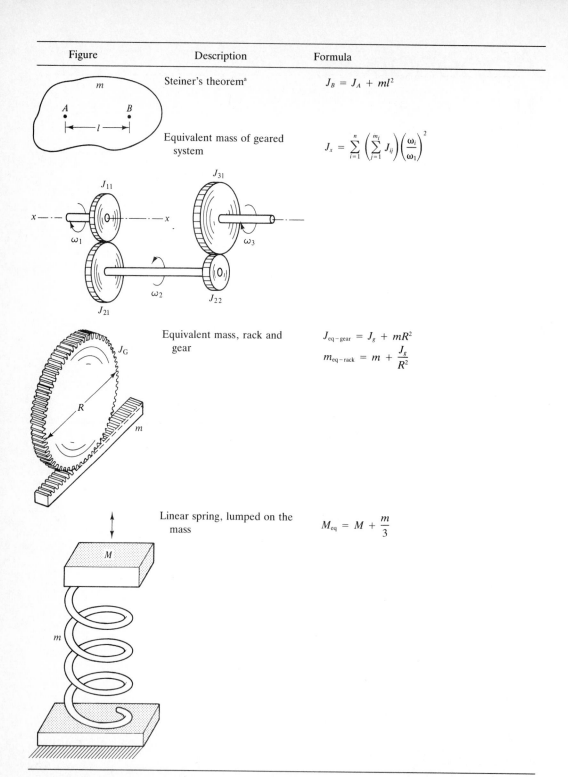

Steiner's theorem[a]

$$J_B = J_A + ml^2$$

Equivalent mass of geared system

$$J_x = \sum_{i=1}^{n} \left(\sum_{j=1}^{m_i} J_{ij} \right) \left(\frac{\omega_i}{\omega_1} \right)^2$$

Equivalent mass, rack and gear

$$J_{\text{eq-gear}} = J_g + mR^2$$

$$m_{\text{eq-rack}} = m + \frac{J_g}{R^2}$$

Linear spring, lumped on the mass

$$M_{\text{eq}} = M + \frac{m}{3}$$

[a]Jacob Steiner, Swiss mathematician (1796–1863), one of the world's greatest synthetic geometers.

Figure	Description	Formula

Beam mass, lumped on the end mass

$$M_{eq} = M + 0.23m$$

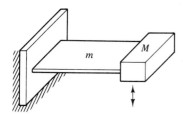

Beam mass, lumped on the midspan

$$M_{eq} = M + 0.5m$$

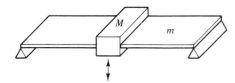

B. Damping Elements

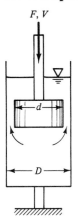

F, V

Dashpot, flow past the piston, viscous damping

$$c = 3\eta\pi\left(\frac{D}{d}\right)^3 \frac{(1 + 2d/D)}{4}$$

d

D

Shear damper, viscous damping

$$c = \frac{\eta A}{h}$$

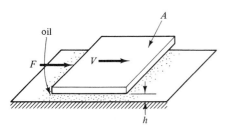

Figure	Description	Formula

Torsional damper, viscous damping

$$c_T = \frac{\pi\mu D^2(l - h)}{2d} \quad \text{(cylinder)} + \frac{\pi\mu D^3}{32h} \quad \text{(face)}$$

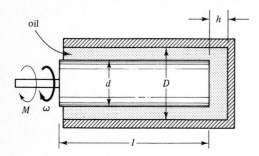

Coulomb damper, dry friction

$$c_{eq} = \frac{4F}{\pi\omega X}$$

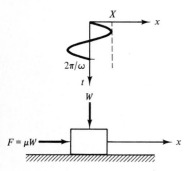

Material damping, viscoelastic damping

$$c_{eq} = \frac{\gamma k}{\omega}$$

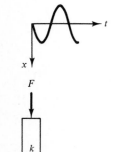

Equivalent dampers, serial

$$\frac{1}{c_{eq}} = \frac{1}{c_1} + \frac{1}{c_2} + \cdots$$

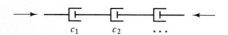

Figure	Description	Formula

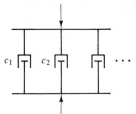

Equivalent dampers, parallel

$$c_{\text{eq}} = c_1 + c_2 + \cdots$$

C. Stiffness Elements
(stiffness $= k$, flexibility $\alpha = 1/k$)

Equivalent springs, serial

$$\frac{1}{k_{\text{eq}}} = \frac{1}{k_1} + \frac{1}{k_2} + \cdots$$

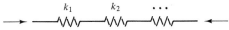

Equivalent springs, parallel, symmetric

$$k_{\text{eq}} = k_1 + k_2 + \cdots$$

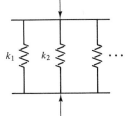

Equivalent springs, parallel, asymmetric

$$k_{\text{eq}} = \frac{(a + b)^2}{b^2/k_1 + a^2/k_2}$$

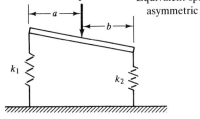

Cantilever cylinder

$$k_{\phi\phi} = \frac{I_P G}{l}$$

$$k_{yy} = \frac{3EI}{l^3}$$

$$k_{xx} = \frac{AE}{l}$$

If end slope $= 0$:

$$k_{yy} = \frac{12EI}{l^3}$$

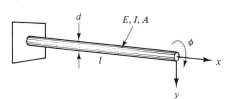

Figure	Description	Formula

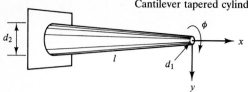

Cantilever tapered cylinder

$$k_{xx} = \frac{Ed_1d_2}{4l}$$

$$k_{yy} = \frac{3E\pi d_2^3 d_1}{64l^3}$$

$$k_{\phi\phi} = \frac{3\pi G(d_2 - d_1)}{32l(T/d_1^3 - 1/d_2^3)}$$

Helical spring

$$k_{xx} = \frac{Gd^4}{64nR^3}$$
$$k_T = Ed^3/(64nR)$$

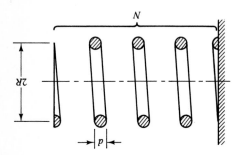

Taut string

$$k_{xx} = \frac{T(a + b)}{ab}$$

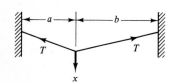

Beam simply supported, $l_i \geq l_j$

$$k_{xx} = \frac{3EI(a + b)^4}{l^3 a^2 b^2}$$

$$\alpha_{ij} = \frac{6EIl}{l_j(l_3 - l_j)^2[2l_i/l_j + l_i/(l - l_j) - l_i^3/l_j(l - l_j)]}$$

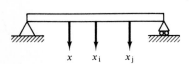

Beam clamped, $l_i \geq l_j$

$$k_{xx} = \frac{3EIl^3}{a^3 b^3}$$

$$\alpha_{ij} = \frac{6EIl^2}{l_j l_i^2(l - l_j)^2[3 - 3l_i/l - l_i l_j/l(l - l_j)]}$$

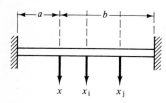

Beam cantilever, $l_i \geq l_j$

$$k_{xx} = \frac{3EI}{l^3}$$

$$\alpha_{ij} = \frac{6EI}{(l - l_j)^3[2 + 3l_i/(l - l_j) - l_i^2/(l - l_j)^3]}$$

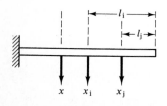

Compound cantilever beam[b]

$$\lambda_{ik} = \frac{l_i}{l_k}, \quad h_i = \frac{l_i^3}{3EI_i}, \quad i, k = 1, 2, 3$$

$$
\begin{bmatrix} \alpha_{11} \\ \alpha_{12} \\ \alpha_{13} \\ \alpha_{22} \\ \alpha_{23} \\ \alpha_{33} \end{bmatrix}
=
\begin{bmatrix}
1 & 0 & 0 \\
1 + 1.5\lambda_{21} & 0 & 0 \\
1 + 1.5(\lambda_{21} + \lambda_{31}) & 0 & 0 \\
1 + 3\lambda_{21}(1 + \lambda_{21}) & 1 & 0 \\
a_{51} & 1 + 1.5\lambda_{32} & 0 \\
a_{61} & 1 + 3\lambda_{32}(1 + \lambda_{32}) & 1
\end{bmatrix}
\cdot
\begin{bmatrix} h_1 \\ h_2 \\ h_3 \end{bmatrix}
$$

$$a_{51} = 1 + 3\lambda_{21}(1 + \lambda_{21} + \lambda_{31}) + 1.5\lambda_{31}$$
$$a_{61} = 1 + 3\lambda_{21}(1 + \lambda_{21} + 2\lambda_{31}) + 3\lambda_{31}(1 + \lambda_{31})$$

Compound cantilever beam

$$\lambda_i = \frac{l_i}{l}, \quad h_i = \frac{l_i^3}{3EI_i}, \quad i = 1, 2, \quad \lambda_{21} = \frac{l_2}{l_1}$$

$$
\begin{bmatrix} \alpha_{11} \\ \alpha_{12} \cdot l \\ \alpha_{13} \\ \alpha_{14} \cdot l \\ \alpha_{22} \cdot l^2 \\ \alpha_{23} \cdot l \\ \alpha_{24} \cdot l^2 \\ \alpha_{33} \\ \alpha_{34} \cdot l \\ \alpha_{44} \cdot l^2 \end{bmatrix}
=
\begin{bmatrix}
1 & 0 \\
1.5/\lambda_1 & 0 \\
1 + 1.5\lambda_{21} & 0 \\
1.5/\lambda_1 & 0 \\
3/\lambda_1^2 & 0 \\
1.5(\lambda_1 + 2\lambda_2)/\lambda_1^2 & 0 \\
3/\lambda_1^2 & 0 \\
(1 + \lambda_2 + \lambda_2^2)/\lambda_1^2 & 1 \\
1.5(\lambda_1 + 2\lambda_2)/\lambda_1^2 & 1.5/\lambda_2 \\
3/\lambda_1^2 & 3/\lambda_2^2
\end{bmatrix}
\cdot
\begin{bmatrix} h_1 \\ h_2 \end{bmatrix}
$$

[b]Compound beam formulas are adapted from: Krämer, E. 1984. *Maschinendynamik*, Berlin: Springer-Verlag. Reprinted by permission of Springer-Verlag.

Figure	Description	Formula

Compound overhang shaft

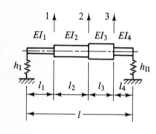

$$\lambda_i = \frac{l_i}{l}, \quad \Lambda_i = \frac{l_i}{l_2 + l_3}, \quad \lambda_{ik} = \frac{l_i}{l_k}, \quad \Lambda_m = \frac{l_2 + l_3 + l_4}{l_2 + l_3}$$

$$h_i = \frac{l_i^3}{3EI_i}, \quad i = 1, 2, 3, 4$$

$$
\begin{bmatrix} \alpha_{11} \\ \alpha_{12} \\ \alpha_{13} \\ \alpha_{22} \\ \alpha_{23} \\ \alpha_{33} \end{bmatrix}
=
\begin{bmatrix}
1 & (1 + \Lambda_3 + \Lambda_3^2)\lambda_{12}^2 & \Lambda_1^2 & 0 & (\Lambda_1/\lambda_1)^2 & \Lambda_1^2 \\
0 & -(0.5 + \Lambda_3)\Lambda_1\lambda_{32} & -\Lambda_1\Lambda_2 & 0 & \Lambda_3\Lambda_1/\lambda_1 & -\Lambda_1\Lambda_2 \\
0 & (0.5 + \Lambda_3)\Lambda_4\lambda_{12} & (0.5 + \Lambda_2)\Lambda_4\lambda_{13} & 0 & -\Lambda_4\Lambda_1/\lambda_1 & -\Lambda_1\Lambda_m \\
0 & \Lambda_3^2 & \Lambda_2^2 & 0 & \Lambda_3^2 & \Lambda_2^2 \\
0 & -\Lambda_3\Lambda_4 & -(0.5 + \Lambda_2)\Lambda_4\lambda_{23} & 0 & -\Lambda_3\Lambda_4 & \Lambda_2\Lambda_m \\
0 & \Lambda_4^2 & (1 + \Lambda_2 + \Lambda_2^2)\lambda_{43}^2 & 1 & \Lambda_4^2 & (\Lambda_2 + \Lambda_3 + \Lambda_4)^2
\end{bmatrix}
\cdot
\begin{bmatrix} h_1 \\ h_2 \\ h_3 \\ h_4 \\ h_1 \\ h_{11} \end{bmatrix}
$$

Compound simply supported shaft

$$a = l_1 + l_2, \quad b = l_3 + l_4, \quad c = l_1 + l_2 + l_3$$
$$d = l_2 + l_3 + l_4$$
$$\lambda_1 = \frac{l_1}{l}, \quad \lambda_4 = \frac{l_4}{l}, \quad \lambda_{12} = \frac{l_1}{l_2}, \quad \lambda_{42} = \frac{l_4}{l_2}, \quad \lambda_{43} = \frac{l_4}{l_3}$$
$$\lambda_a = \frac{a}{l}, \quad \lambda_b = \frac{b}{l}, \quad \lambda_c = \frac{c}{l}, \quad \lambda_d = \frac{d}{l}$$
$$x_2 = \frac{a}{l_2}, \quad \alpha_3 = \frac{a}{l_3}, \quad \beta_2 = \frac{b}{l_2}, \quad \beta_3 = \frac{b}{l_3}$$
$$\gamma = \frac{c}{l_3}, \quad \delta = \frac{d}{l_2}$$
$$h_i = \frac{l_i^3}{3EI_i}$$

$$
\begin{bmatrix} \alpha_{11} \\ \alpha_{12} \\ \alpha_{13} \\ \alpha_{22} \\ \alpha_{23} \\ \alpha_{33} \end{bmatrix}
=
\begin{bmatrix}
\lambda_d^2 & \lambda_1^2 A & \lambda_1^2 D & \lambda_1^2 & \lambda_d^2 & \lambda_1^2 \\
\lambda_b\lambda_d & \lambda_1\lambda_b B & \lambda_1\lambda_a D & \lambda_1\lambda_a & \lambda_b\lambda_d & \lambda_1\lambda_a \\
\lambda_4\lambda_d & \lambda_1\lambda_4 B & \lambda_1\lambda_4 E & \lambda_1\lambda_c & \lambda_4\lambda_d & \lambda_1\lambda_c \\
\lambda_b^2 & \lambda_b^2 C & \lambda_a^2 D & \lambda_a^2 & \lambda_b^2 & \lambda_a^2 \\
\lambda_4\lambda_b & \lambda_4\lambda_b C & \lambda_4\lambda_a E & \lambda_a\lambda_c & \lambda_4\lambda_b & \lambda_a\lambda_c \\
\lambda_4^2 & \lambda_4^2 C & \lambda_4^2 F & \lambda_c^2 & \lambda_4^2 & \lambda_c^2
\end{bmatrix}
\cdot
\begin{bmatrix} h_1 \\ h_2 \\ h_3 \\ h_4 \\ h_I \\ h_{II} \end{bmatrix}
$$

$$A = \beta_2^2 + \beta_2\delta + \delta^2$$
$$B = \lambda_{12}\delta + \alpha_2\beta_2 + 0.5(\alpha_2\delta + \beta_2\lambda_{12})$$
$$C = \alpha_2^2 + \alpha_2\lambda_{12} + \lambda_{12}^2$$
$$D = \beta_3^2 + \beta_3\lambda_{43} + \lambda_{43}^2$$
$$E = \alpha_3\beta_3 + \gamma\lambda_{43} + 0.5(\alpha_3\lambda_{43} + \beta_3\gamma)$$
$$F = \alpha_3^2 + \alpha_3\gamma + \gamma^2$$

Appendix III

Algebraic Identities

$$\sin^2 a + \cos^2 a = 1, \qquad \tan x = \frac{\sin x}{\cos x}$$

$$\sin(a \pm b) = \sin a \cos b \pm \cos a \sin b$$

$$\cos(a \pm b) = \cos a \cos b \mp \sin a \sin b$$

$$\sin a \sin b = \tfrac{1}{2}[\cos(a - b) - \cos(a + b)]$$

$$\cos a \cos b = \tfrac{1}{2}[\cos(a + b) + \cos(a - b)]$$

$$\sin a \cos b = \tfrac{1}{2}[\sin(a + b) + \sin(a - b)]$$

$$a \sin \theta + b \cos \theta = (a^2 + b^2)^{1/2} \cos(\theta - \phi), \qquad \tan \phi = \frac{a}{b}$$

$$= (a^2 + b^2)^{1/2} \sin(\theta + \phi), \qquad \tan \phi = \frac{b}{a}$$

$$\sinh x = \frac{e^x - e^{-x}}{2}, \qquad \cosh x = \frac{e^x + e^{-x}}{2}$$

Triangle cosine law: $a^2 = b^2 + c^2 - 2bc \cos \phi$

Imaginary unit: $i = [-1]^{1/2}, i^2 = -1$

De Moivre's formula[3]:

$$e^{ix} = \cos x + i \sin x, \qquad e^{-ix} = \cos x - i \sin x$$

$$\cos x = \frac{e^{ix} + e^{-ix}}{2}, \qquad \sin x = \frac{e^{ix} - e^{-ix}}{2}$$

$$z = x + iy, \; z_1 + z_2 = (x_1 + x_2) + i(y_1 + y_2), \qquad \bar{z} = x - iy$$

$$z_1 z_2 = (x_1 x_2 - y_1 y_2) + i(x_1 y_1 + x_2 y_2)$$

$$\frac{z_1}{z_2} = \frac{(x_1 x_2 + y_1 y_2) + i(x_2 y_1 - x_1 y_2)}{x_2^2 + y_2^2}$$

$$z = x + iy = R e^{i\phi}, \qquad R = (x^2 + y^2)^{1/2}, \qquad \tan \phi = \frac{y}{x}$$

$$z_1 z_2 = R_1 R_2 e^{i(\phi_1 + \phi_2)}, \qquad \frac{z_1}{z_2} = \frac{R_1}{R_2} e^{i(\phi_1 - \phi_2)}$$

$$z_1 + z_2 = R e^{i\phi}, \qquad R^2 = R_1^2 + R_2^2 - 2R_1 R_2 \cos(\phi_1 - \phi_2)$$

$$\tan \phi = \frac{R_1 \sin \phi_1 + R_2 \sin \phi_2}{R_1 \cos \phi_1 + R_2 \cos \phi_2}$$

$$\sqrt{p + iq} = \pm\lambda \pm \frac{q}{i\lambda}, \qquad \lambda = \sqrt{\frac{p + \sqrt{p^2 + q^2}}{2}}$$

[3]Abraham de Moivre (1997–1754) was born in France but lived most of his life in England. He was a close friend of Newton.

Appendix IV

Elements of Linear Algebra

Matrices

An $n \times m$ *matrix* is an orderly arrangement of $n \times m$ quantities in a rectangular array with n rows and m columns. We use a boldface uppercase letter to represent a matrix. A matrix whose element at the ith row and the jth column is a_{ij} is written as $[a_{ij}]$. Thus

$$\mathbf{A} = \begin{bmatrix} a_{11} & a_{12} & \cdots & a_{1m} \\ a_{21} & a_{22} & \cdots & a_{2m} \\ \vdots & & & \\ a_{n1} & a_{n2} & & a_{nm} \end{bmatrix} = [a_{ij}] \tag{IV.1}$$

The transpose of a matrix $\mathbf{A}$, labeled $\mathbf{A}^T$, is that matrix formed through interchanging of the rows and columns of $\mathbf{A}$. For the $\mathbf{A}$ matrix above, its transpose is a $m \times n$ matrix

$$\mathbf{A}^T = \begin{bmatrix} a_{11} & a_{21} & \cdots & a_{n1} \\ a_{12} & a_{22} & \cdots & a_{n2} \\ \vdots & & & \\ a_{1m} & a_{2m} & \cdots & a_{nm} \end{bmatrix} = [a_{ji}] \qquad \text{(IV.2)}$$

An $n \times 1$ matrix is called a *column vector* or simply a *vector*. The transpose of such a matrix is called a *row vector*. A row vector is written in braces $\{\cdot\}$. A 1×1 matrix is called a *scalar*.

A matrix whose number of rows is equal to the number of columns is called a *square matrix*. An $n \times n$ square matrix is said to be of *order n*. Many matrix operations are defined only relative to a square matrix. We deal here only with square matrices and vectors.

The elements a_{ii}, $i = 1, \ldots, n$ of an $n \times n$ square matrix are called the *diagonal elements;* the remaining elements, characterized by a_{ij} with $i \neq j$ $(i, j = 1, \ldots, n)$, are known as the *off-diagonal elements*.

A square matrix whose off-diagonal elements are all 0 is called a *diagonal matrix*. A diagonal matrix with $a_{ii} = c_i$, $i = 1, \ldots, n$, is sometimes labeled diag$[c_1 \cdots c_n]$.

A diagonal matrix with all of its diagonal elements unity (i.e., the matrix diag$[1, 1, \ldots, 1]$) is known as the *identity* (or *unit*) *matrix*. It is represented by the symbol $\mathbf{I}$.

A matrix with every element 0 is called a *null* or *zero matrix;* it is labeled $\mathbf{0}$.

An $n \times n$ square matrix whose off-diagonal elements are such that $a_{ij} = a_{ji}$ for $i \neq j$ $(i, j = 1, \ldots, n)$ is called a *symmetric matrix*. If such a matrix is given by $\mathbf{A}$, then $\mathbf{A}^T = \mathbf{A}$.

If a square matrix $\mathbf{A}$ has complex numbers as elements, its conjugate transpose matrix, labeled $\mathbf{A}^+$, is the matrix formed by replacing every element of the matrix $\mathbf{A}^T$ by its complex conjugate. A matrix $\mathbf{A}$ with the property that $\mathbf{A}^+ = \mathbf{A}$ is known as a *hermitian matrix*.

A square matrix $\mathbf{A}$ whose determinant $|\mathbf{A}|$ is 0 is called a *singular matrix;* otherwise, it is *nonsingular*.

The determinant of a square matrix $\mathbf{A}$ is a number associated with the matrix and is usually written

$$A = \begin{vmatrix} a_{11} & \cdots & a_{1n} \\ \vdots & & \\ a_{n1} & \cdots & a_{nn} \end{vmatrix}$$

The determinant $|[m_{ij}]|$ of the $(n - 1) \times (n - 1)$ matrix that results when the ith row and the jth column are removed from an $n \times n$ matrix $\mathbf{A}$ is called the *minor* of the element a_{ij} of $\mathbf{A}$. The minor of a_{ij} multiplied by $(-1)^{i+j}$ yields the *cofactor* of the element a_{ij}.

The value of a determinant can be obtained by the so-called *Laplace expansion formula*

$$|[a_{ij}]| = \begin{cases} \sum\limits_{j=1}^{n} (-1)^{i+j} a_{ij} |[m_{ij}]| & \text{for any integer } i \; 1 \le i \le n \\ \\ \sum\limits_{i=1}^{n} (-1)^{i+j} a_{ij} |[m_{ij}]| & \text{for any integer } j \; 1 \le j \le n \end{cases} \quad \text{(IV.3)}$$

Since each determinant $|[m_{ij}]|$ in (IV.3) can be further expanded by means of the Laplace expansion formula, the value of the determinant of any order is uniquely given if we define the value of a 1×1 and 2×2 determinant. These are defined as follows:

$$|a| = a \quad \text{(IV.4)}$$

$$\begin{vmatrix} a_{11} & a_{12} \\ a_{21} & a_{22} \end{vmatrix} = a_{11}a_{22} - a_{21}a_{12} \quad \text{(IV.5)}$$

Basic Matrix Operations

Two matrices $\mathbf{A} = [a_{ij}]$ and $\mathbf{B} = [b_{ij}]$ are equal if and only if (1) their orders are equal and (2) their corresponding elements are equal (i.e., $a_{ij} = b_{ij}$ for all i and i and j). When $\mathbf{A}$ and $\mathbf{B}$ are equal, we write $\mathbf{A} = \mathbf{B}$.

Two matrices $\mathbf{A} = [a_{ij}]$ and $\mathbf{B} = [b_{ij}]$ whose orders are the same can be added or subtracted to result in new matrices $\mathbf{C} = [c_{ij}]$ and $\mathbf{D} = [d_{ij}]$, respectively; the elements of $\mathbf{A}$, $\mathbf{B}$, $\mathbf{C}$, and $\mathbf{D}$ are related by

$$c_{ij} = a_{ij} + b_{ij}, \qquad d_{ij} = a_{ij} - b_{ij} \quad \text{(IV.6)}$$

When a matrix $\mathbf{A} = [a_{ij}]$ is multiplied by a scalar k, a new matrix results, whose element is ka_{ij}.

Two matrices $\mathbf{A}$ and $\mathbf{B}$ can be multiplied to form the product matrix $\mathbf{AB} = \mathbf{C}$ if and only if the number of columns of $\mathbf{A}$ is equal to the number of rows of $\mathbf{B}$. When this holds, $\mathbf{A}$ and $\mathbf{B}$ are said to be *conformable* with regard to multiplication. If $\mathbf{A}$ is $n \times m$ and $\mathbf{B}$ is $m \times l$, the resulting matrix $\mathbf{C} = \mathbf{AB}$ will be $n \times l$ and has elements

$$c_{ij} = \sum_{k=1}^{m} a_{ik}b_{kj} \quad \text{(IV.7)}$$

Note that in the above, if $n \ne l$, the product $\mathbf{BA}$ is not possible. It is also easy to show that even if $n = l$, the matrix multiplication is not commutative, so that in general $\mathbf{AB} \ne \mathbf{BA}$. However, it is readily seen for any square matrix $\mathbf{A}$ of the same dimensions as $\mathbf{I}$ that $\mathbf{AI} = \mathbf{IA} = \mathbf{A}$.

To define an operation that resembles division, the concept of the inverse of a square matrix is necessary. Relative to a square matrix $\mathbf{A}$ that is nonsingular, its inverse $\mathbf{A}^{-1}$ is given by the relation

$$\mathbf{AA}^{-1} = \mathbf{A}^{-1}\mathbf{A} = \mathbf{I} \quad \text{(IV.8)}$$

Eigenvalues and Eigenvectors of a Square Matrix

Occasions often arise where one needs to find vectors $\mathbf{x}$ and values of scalars λ such that the equation

$$\mathbf{Ax} = \lambda\mathbf{x} \tag{IV.9}$$

is satisfied for a given square matrix $\mathbf{A}$. The values of scalar λ's that are found to satisfy (IV.9) are called the *eigenvalues* of the square matrix $\mathbf{A}$.

Writing the form (IV.9)

$$(\mathbf{A} - \lambda\mathbf{I})\mathbf{x} = 0 \tag{IV.10}$$

we can prove that[1]

$$|\mathbf{A} - \lambda\mathbf{I}| = 0 \tag{IV.11}$$

is a necessary and sufficient condition for the existence of nontrivial λ's that satisfy (IV.10).

Expanding the determinant (IV.11), we obtain an nth-degree polynomial in λ which, if set equal to zero, results in an equation called the *characteristic equation* of $\mathbf{A}$. The n values of λ, $\lambda_1, \ldots, \lambda_n$ (which need not be either real or distinct) can then be solved for as the roots of the characteristic equation.

For each value of λ, say λ_i, there is associated a vector $\mathbf{v}_i$ that satisfies (IV.10) (i.e., $\mathbf{Av}_i = \lambda_i\mathbf{v}_i$). Vector $\mathbf{v}_i$ is called the eigenvector corresponding to the eigenvalue λ_i. For example, for the matrix

$$\begin{bmatrix} 0 & 1 & 0 \\ 0 & 0 & 1 \\ -8 & -14 & -7 \end{bmatrix}$$

$$|\mathbf{A} - \lambda I| = \begin{vmatrix} -\lambda & 1 & 0 \\ 0 & -\lambda & 1 \\ -8 & -14 & -7 - \lambda \end{vmatrix} = \lambda^3 + 7\lambda^2 + 14\lambda + 8 = 0 \tag{IV.12}$$

The eigenvalues are then $\lambda_1 = -1$, $\lambda_2 = -2$, and $\lambda_3 = -4$. Corresponding to λ_1, we have $\mathbf{Av}_1 = -\mathbf{v}_1$. This yields the set of equations $u_{12} = -u_{11}$, $u_{13} = -u_{12}$, $8u_{11} - 14u_{12} - 14u_{12} - 7u_{13} = -u_{13}$, where u_{11}, u_{12}, and u_{13} are the components of the vector $\mathbf{v}_1$. This set is homogeneous and hence is determined down to a multiplicative constant. Letting $u_{11} = 1$, we have the solution

$$v_1 = \begin{bmatrix} 1 \\ -1 \\ 1 \end{bmatrix}$$

In a similar manner we can find

$$v_2 = \begin{bmatrix} 1 \\ -1 \\ 4 \end{bmatrix} \quad \text{and} \quad v_3 = \begin{bmatrix} 1 \\ -4 \\ 16 \end{bmatrix}$$

[1]Pontryagin, L. S. 1962. *Ordinary Differential Equations*. Reading, Mass.: Addison-Wesley.

The characteristic equation of a real matrix may turn out to have complex roots appearing in conjugate pairs. Here the eigenvectors may also have complex components. It may be shown, however, that if a matrix is hermitian, the eigenvalues must be real. As a real symmetric matrix is a particular case of hermitian matrix, it therefore has real eigenvalues and eigenvectors.

Quadratic Forms and the Positive Definiteness of a Square Matrix

An expression such as

$$q(x_1, \ldots, x_n) = \sum_{i=1}^{n} \sum_{j=1}^{n} k_{ij} x_i x_j \tag{IV.13}$$

involving terms of second order in x_i and x_j is known as a *quadratic form* of n variables.

It is clear that a quadratic form can be compactly expressed in a vector–matrix form as

$$q(x_1, \ldots, x_n) = \sum_{i=1}^{n} \sum_{j=1}^{n} k_{ij} x_i x_j = \mathbf{x}^T \mathbf{Q} \mathbf{x} \tag{IV.14}$$

where

$$\mathbf{Q} = \begin{bmatrix} k_{11} & k_{12} & \cdots & k_{1n} \\ k_{21} & k_{22} & \cdots & \cdot \\ \vdots & & & \vdots \\ k_{n1} & \cdot & \cdots & k_{nn} \end{bmatrix} = [q_{ij}] = \mathbf{Q}^T \tag{IV.15}$$

A quadratic form is said to be *positive (negative) definite* if $q = 0$ for $x = 0$ and $q > 0$ ($q < 0$) for $x \neq 0$. It is positive semidefinite (negative semidefinite) if $q = 0$ for $x = 0$ and $q \geq 0$ ($q \leq 0$) for $x \neq 0$.

An important test of the positive definiteness of a quadratic form is given by the following theorem: A quadratic $q(x) = \mathbf{x}^T \mathbf{Q} \mathbf{x}$ is positive (negative) definite if and only if the n determinants $Q_1, \ldots, Q_n$ are all positive (negative), where

$$|Q_1| = |q_{11}|, \quad |Q_2| = \begin{vmatrix} q_{11} & q_{12} \\ q_{12} & q_{22} \end{vmatrix}, \quad Q_3 = \begin{vmatrix} q_{11} & q_{12} & q_{13} \\ q_{12} & q_{22} & q_{23} \\ q_{13} & q_{23} & q_{33} \end{vmatrix}$$

$$Q_m = |q_{ij}| \quad j = 1, \ldots, m \leq n, \quad |Q_n| = |Q| \tag{IV.16}$$

Some Useful Relationships

The reader may wish to verify the following useful matrix relations:

1. $[\mathbf{AB} \cdots \mathbf{MN}]^T = \mathbf{N}^T \mathbf{M}^T \cdots \mathbf{B}^T \mathbf{A}^T.$
2. $[\mathbf{AB} \cdots \mathbf{MN}]^{-1} = \mathbf{N}^{-1} \mathbf{M}^{-1} \cdots \mathbf{B}^{-1} \mathbf{A}^{-1}$ if each of the inverse matrices exists.

3. $|\mathbf{AB} \cdots \mathbf{MN}| = |\mathbf{A}| \cdot |\mathbf{B}| \cdots |\mathbf{M}| \cdot |\mathbf{N}|.$

4. For an $n \times n$ matrix partitioned into $\begin{bmatrix} A & \vdots & B \\ -- & \vdots & -- \\ C & \vdots & D \end{bmatrix}$ with $\mathbf{A}$ an $m \times m$ matrix

and $\mathbf{D}$ an $(n - m) \times (n - m)$ matrix,

$$\begin{bmatrix} A & \vdots & B \\ -- & \vdots & -- \\ C & \vdots & D \end{bmatrix}^{-1}$$

$$= \begin{bmatrix} \mathbf{I}_m & -\mathbf{A}^{-1}\mathbf{B} \\ \mathbf{O}_{(n-m) \times m} & \mathbf{I}_m \end{bmatrix} \begin{bmatrix} A^{-1} & \mathbf{O}_{m \times (n-m)} \\ \mathbf{O}_{(n-m) \times m} & (\mathbf{D} - \mathbf{CA}^{-1}\mathbf{B})^{-1} \end{bmatrix} \begin{bmatrix} \mathbf{I}_m & \mathbf{O}_{m \times (n-m)} \\ -\mathbf{CA}^{-1} & \mathbf{I}_m \end{bmatrix}$$

where $\mathbf{O}_{r \times s}$ is an $r \times s$ matrix composed entirely of 0s.

Appendix V

Transfer Matrices

1. Massless beam

$$L_1 = \begin{bmatrix} 1 & l & \dfrac{l^2}{2EI} & \dfrac{l^3}{6EI} & \cdot & 0 \\[2ex] 0 & 1 & \dfrac{l}{EI} & \dfrac{l^2}{2EI} & \cdot & 0 \\[2ex] 0 & 0 & 1 & l & \cdot & 0 \\[1ex] 0 & 0 & 0 & 1 & \cdot & 0 \\[1ex] \cdot & \cdot & \cdot & \cdot & \cdot & \cdot \\[1ex] 0 & 0 & 0 & 0 & & 1 \end{bmatrix}$$

2. Massless beam with axial compressive force, N:

$$L_2 = \begin{bmatrix} 1 & \dfrac{l}{\alpha}\sin\alpha & \dfrac{1-\cos\alpha}{\alpha^2} & \dfrac{l^3}{EI}\dfrac{\alpha-\sin\alpha}{a^3} & \cdot & 0 \\[2ex] 0 & \cos\alpha & \dfrac{l}{EI} & \dfrac{l^2}{EI}\dfrac{1-\cos\alpha}{\alpha^2} & \cdot & 0 \\[2ex] 0 & Nl\dfrac{\sin\alpha}{\alpha} & \cos\alpha & l\dfrac{\sin\alpha}{\alpha} & \cdot & 0 \\[2ex] 0 & 0 & 0 & 1 & \cdot & 0 \\[1ex] \cdot & \cdot & \cdot & \cdot & \cdot & \cdot \\[1ex] 0 & 0 & 0 & 0 & & 1 \end{bmatrix}$$

where $\alpha = (N/EI)^{1/2}l$.

3. Beam with uniform mass:

$$L_3 = \begin{bmatrix} e_1 & e_2 & \dfrac{e_3}{EI} & \dfrac{e_4}{EI} & \cdot & 0 \\[2ex] \beta^4 e_3 & e_1 & \dfrac{e_2}{EI} & \dfrac{e_3}{EI} & \cdot & 0 \\[2ex] \rho\omega^2 e_3 & \rho\omega^2 e_4 & e_1 & e_2 & \cdot & 0 \\[2ex] \rho\omega^2 e_2 & \rho\omega^2 e_3 & \beta^4 e_4 & e_1 & \cdot & 0 \\[1ex] \cdot & \cdot & \cdot & \cdot & \cdot & \\[1ex] 0 & 0 & 0 & 0 & & 1 \end{bmatrix}$$

where

$$\beta = \left(\frac{\rho\omega^2}{EI}\right)^{1-1/4}$$

$$e_1 = \tfrac{1}{2}(\cosh\beta l + \cos\beta l)$$

$$e_2 = \frac{1}{2\beta}(\sinh\beta l + \sin\beta l)$$

$$e_3 = \frac{1}{2\beta^2}(\cosh\beta l - \cos\beta l)$$

$$e_4 = \frac{1}{2\beta^3}(\sinh\beta l - \sin\beta l)$$

4. Concentrated mass with rotatory inertia

$$L_4 = \begin{bmatrix} 1 & 0 & 0 & 0 & \cdot & 0 \\ 0 & 1 & 0 & 0 & \cdot & 0 \\ 0 & Mr^2\omega^2 & 1 & 0 & \cdot & 0 \\ M\omega^2 & 0 & 0 & 1 & \cdot & 0 \\ \cdot & \cdot & \cdot & \cdot & \cdot & \\ 0 & 0 & 0 & 0 & & 1 \end{bmatrix}$$

where M = mass
r = radius of gyration = $(J/M)^{1/2}$

5. Concentrated spring

$$L_5 = \begin{bmatrix} 1 & 0 & 0 & 0 & \cdot & 0 \\ 0 & 1 & 0 & 0 & \cdot & 0 \\ 0 & 0 & 1 & 0 & \cdot & 0 \\ -k & 0 & 0 & 1 & \cdot & 0 \\ \cdot & \cdot & \cdot & \cdot & \cdot & \\ 0 & 0 & 0 & 0 & & 1 \end{bmatrix}$$

6. Concentrated spring K_1 with intermediate mass M and base spring K_2: L_6 is the same as L_5 except that

$$k = \frac{K_2(K_1 - M\omega^2)}{K_1} + K_2 - M\omega^2$$

Appendix
VI

Tables
of
Laplace
Transforms

$$f(s) = \int_0^\infty e^{-st}F(t)\, dt$$

	f(s)	F(t)
1.	$af_1(s) + bf_2(s)$	$aF_1(t) + bF_2(t)$
2.	$f(s/a)$	$aF(at)$
3.	$f(s-a)$	$e^{at}F(t)$
4.	$e^{-as}f(s)$	$u(t-a) = \begin{matrix} F(t-a) & t>a \\ 0 & t<a \end{matrix}$
5.	$sf(s) - F(0)$	$F'(t)$
6.	$s^2 f(s) - sF(0) - F'(0)$	$F''(t)$
7.	$s^n f(s) - s^{n-1}F(0) - s^{n-2}F'(0) - \ldots - F^{(n-1)}(0)$	$F^{(n)}(t)$
8.	$f'(s)$	$-tF(t)$
9.	$f''(s)$	$t^2 F(t)$
10.	$f^{(n)}(s)$	$(-1)^n t^n F(t)$

	$f(s)$	$F(t)$
11.	$\dfrac{f(s)}{s}$	$\displaystyle\int_0^t F(u)\,du$
12.	$f(s)g(s)$	$\displaystyle\int_0^T F(u)G(t-u)\,du$
13.	$\displaystyle\int_0^\infty f(u)\,du$	$\dfrac{F(t)}{t}$
14.	$\dfrac{1}{1-e^{sT}}\displaystyle\int_0^T e^{-su}F(u)\,du$	$F(t) = F(t+T)$
15.	$\dfrac{P(s)}{Q(s)}$ $P(s)$ = Polynomial of degree less than n $Q(s) = (s-a_1)(s-a_2)\cdots(s-a_n)$ where $a_1, a_2, \cdots, a_n$ are all distinct.	$\displaystyle\sum_{k=1}^n \frac{P(a_k)}{Q'(a_k)}e^{a_k t}$

	f(s)	F(t)
1.	$\dfrac{1}{s}$	1
2.	$\dfrac{1}{s^2}$	t
3.	$\dfrac{1}{s^n}$ $n = 1, 2, 3, \ldots$	$\dfrac{t^{n-1}}{(n-1)!}$, $0! = 1$
4.	$\dfrac{1}{s-a}$	e^{at}
5.	$\dfrac{1}{(s-a)^n}$ $n = 1, 2, 3, \ldots$	$\dfrac{t^{n-1}e^{at}}{(n-1)}$, $0! = 1$
6.	$\dfrac{a}{s^2 + a^2}$	$\sin at$
7.	$\dfrac{s}{s^2 + a^2}$	$\cos at$
8.	$\dfrac{a \cos \phi - s \sin \phi}{s^2 + a^2}$	$\sin (at - \phi)$
9.	$\dfrac{1}{s^2 + 2\zeta\omega_o s + \omega_o^2}$	$\dfrac{1}{\omega_o\sqrt{1-\zeta^2}} \exp(-\zeta\omega_0 t) \sin \omega_0 \sqrt{1-\zeta^2}\, t$
10.	$\dfrac{1}{(s-a)(s-b)}$ $a \neq b$	$\dfrac{e^{bt} - e^{at}}{b - a}$
11.	$\dfrac{s}{(s-a)(s-b)}$ $a \neq b$	$\dfrac{be^{bt} - ae^{at}}{b - a}$
12.	$\dfrac{s}{s^4 - a^4}$	$\dfrac{1}{2a^2}$ $(\cosh at - \cos at)$
13.	$\dfrac{s^2}{s^4 - a^4}$	$\dfrac{1}{2a}$ $(\sinh at + \sin at)$
14.	$\dfrac{s^3}{s^4 - a^4}$	$\dfrac{1}{2}$ $(\cosh at + \cos at)$
15.	1	Unit impulse function at t = 0
16.	$\dfrac{1}{as^2}$ $\tanh (as/2)$	Triangular wave function
17.	$\dfrac{1}{s}$ $\tanh (as/2)$	Square wave function

	$f(s)$	$F(t)$
18.	$\dfrac{\pi^a}{(a^2 s^2 + \pi^2)} \coth(as/2)$	Rectified sine wave function
19.	$\dfrac{\pi^a}{(a^2 s^2 + \pi^2)(1 - e^{-as})}$	Half rectified sine wave function
20.	$\dfrac{1}{as^2} - \dfrac{e^{-as}}{s(1 - e^{-as})}$	Saw tooth wave function
21.	$\dfrac{1}{s^2}(1 + e^{-sa})\dfrac{F_0}{a}$	
22.	$\dfrac{e^{-as}}{s}$	Heaviside's unit function $u(t - a)$
23.	$\dfrac{e^{-as}(1 - e^{-\epsilon s})}{s}$	Pulse function
24.	$\dfrac{1}{s(1 - e^{-as})}$	Step function
25.	$\dfrac{e^{-s} + e^{-2s}}{s(1 - e^{-s})^2}$	$F(t) = n^2,\ n \leq t < n + 1,\ n = 0, 1, 2, \ldots$
26.	$\dfrac{1 - e^{-s}}{s(1 - re^{-s})}$	$F(t) = r^n,\ n \leq t < n + 1,\ n = 0, 1, 2, \ldots$
27.	$\dfrac{a(1 + e^{-as})}{a^2 s^2 + \pi^2}$	$F(t) = \begin{cases} \sin(\pi t/a) & 0 \leq t < a \\ 0 & t > a \end{cases}$
28.	$\dfrac{F_0}{a}\left\{\dfrac{1}{s^2} - e^{-as}\left(\dfrac{a}{s} - \dfrac{1}{s^2}\right)\right\}$	

Appendix
VII

Unit Conversion and Nomenclature

General Conversion Formulas

$$N^a m^b s^c = 4.448^a \times 0.3048^b \times lb_f^a ft^b s^c$$

$$N^a mm^b s^c = 4.448^a \times 304.8^b \times lb_f^a ft^b s^c$$

$$N^a m^b s^c = 4.448^a \times 0.0254^b \times lb_f^a in.^b s^c$$

$$N^a mm^b s^c = 4.448^a \times 25.4^b \times lb_f^a in.^b s^c$$

$$lb_f^a ft^b sec^c = 0.2248^a \times 3.2808^b \times N^a m^b s^c$$

$$lb_f^a ft^b sec^c = 0.2248^a \times 0.0032808^b \times N^a mm^b s^c$$

$$lb_f^a in.^b sec^c = 0.2248^a \times 39.37^b \times N^a m^b s^c$$

$$lb_f^a in.^b sec^c = 0.2248^a \times 0.03937^b \times N^a mm^b s^c$$

Symbol	Quantity	B.S. units (lb-force)	Multiply by:	To obtain SI units
a	Acceleration	ft/sec^2	0.3048	m/s^2
ω	Angular velocity	rad/s	1	rad/s
		(rpm	0.1047	rad/s)
A	Area	ft^2	0.09290	m^2
I	Area moment of inertia	ft^4	86.30×10^{-4}	m^4
c	Damping constant, linear	lb-sec/ft	2101	N·s/m
c_T	Damping constant, rotary	lb-ft-sec/rad	0.1360	N·m·s/rad
E, V, T	Energy	ft-lb	1.356	J (joule)
F	Force	lb	4.448	N
n	Frequency	sec^{-1}	1	s^{-1}
l	Length	ft	0.3048	m
m	Mass	lb-sec^2-ft^{-1}	14.59	kg
J	Mass moment of inertia	in.-lb$_f$-sec^2	0.1130	kg·m^2
M	Moment	ft-lb	1.356	N·m
P	Power	lb-ft-sec^{-1}	1.356	W
		(hp	745.7	W)
P	Pressure	psi	6894	Pa (pascal, N/m^2)
k	Stiffness, linear	lb/ft	14.59	N/m
k_T	Stiffness, rotary	lb-ft/rad	1.356	N·m/rad
t	Time	sec	1	s
u	Velocity	ft-sec^{-1}	0.3048	m·s^{-1}
μ	Viscosity	lb-sec-in.$^{-2}$	6894	Pa·s
W	Work	lb-ft	1.356	J

Nomenclature

Symbol	Meaning	Units
A	Vibration amplitude	m
A	Area	m^2
$\mathbf{A}$	Flexibility matrix	m/N
$\mathbf{C}$	Damping matrix	N·s/m
D	Diameter	m
D	Dissipation function	N·m
$\mathbf{D}$	System matrix = $\mathbf{M}^{-1}\mathbf{K}$	s^{-2}
E	Young's modulus of elasticity	N/m^2
$E(x)$	Expected value of x	
F	Force	N
G	Shear modulus	N/m^2
$\mathbf{G}$	Impulse Response Matrix	Depends on problem
H	Angular momentum	kg·m^2/s
$H(\omega)$	Amplification factor	
I	Area moment of inertia	m^4
$\mathbf{I}$	Identity matrix	
J	Mass moment of inertia	kg·m^2
$\mathbf{K}$	Stiffness matrix	N/m
L	Length	m
L	Lagrangian function	N·m
$\mathbf{L}$	Field Transfer Matrix	
M	Bending moment	N·m

Nomenclature (cont.)

Symbol	Meaning	Units
M	Mass matrix	kg
N	Frequency	Hz
P	Force	N
P	Spectral matrix, Point Transfer Matrix	
$P(x)$	Probability distribution function of x	
Q	Quality factor	
Q	Mode Participation Factor	Depends on problem
R	Radius	m
R	Rayleigh's quotient	$1/s^2$
$R(\tau)$	Autocorrelation function	
S	Sommerfeld number $\eta LDN(R/c)^2/W$	
So	Sommerfeld parameter $(W/\eta\omega LD)(c/R)^2$	
$S(\omega)$	Spectral density function	
$S_x(\omega)$	One-sided spectral density function	
T	Tension, axial force	N
T	Vibration Period	s
T	Kinetic energy	N·m
U	Peripheral velocity	m/s
U	Dissipation Function	N/s
V	Shear force	N
V	Potential energy	N·m
V	Liapounov function	
W	Weight	N
W	Work	N·m
X	Vibration amplitude	m
Z	Vibration amplitude, complex	m
Z	Modal matrix	
a	Acceleration	m/s^2
a_i	Fourier sin coefficients, $i = 1, 2, \ldots$	Same as function
a_{ij}	Flexibility coefficient	m/N
b_i	Fourier cos coefficients, $i = 0, 1, 2, \ldots$	Same as function
c	Integration constant	
c	Bearing radial clearance	m
c	Damping constant	N·s/m
c	Sound velocity	m/s
c	Clearance	m
d	Diameter	m
e	Eccentricity	m
f	Frequency	Hz
g	Acceleration of gravity	m/s^2
$g(t)$	Impulse response function	Depends on problem
$h(t)$	Step response function	Depends on problem
i	Imaginary unit $(-1)^{(1/2)}$	
k	Spring constant	N/m
k	Shape Shear Factor	
l	Length	m
m	Mass	kg
p	Force per Unit Length	N/m
p	Complex eigenvalue	s^{-1}
p	Pressure	Pa, N/m^2
$p(x)$	Probability density function of x	

Nomenclature (cont.)

Symbol	Meaning	Units
q	Generalized coordinate	Depends on problem
$\mathbf{q}$	Generalized coordinate vector	Depends on problem
r	Radius	m
r	Frequency ratio ω/ω_n	
r	Radial displacement	m
s	Parameter of Laplace transform	1/s
t	Time	s
u	Linear velocity	m/s
u	Unbalance mass $\times$ eccentricity	kg·m
v	Linear velocity	m/s
w	Lateral deflection	m
x	Displacement	m
y	Displacement	m
z	Displacement, complex	m
$\mathbf{z}$	Vibration Mode, State Vector	Depends on problem
x, y, z	Cartesian coordinates	m

Greek Letters

Symbol	Meaning	Units
Δ	Increment	
Φ	Amplitude of torsional vibration	rad
Φ	Modal Matrix	Depends on problem
Ω	Angular velocity of shaft	rad/s
α	Rayleigh's parameter	
α	Angular acceleration	rad/s²
γ	Material damping ratio	
δ	Logarithmic decrement	
δ	Displacement	m
ϵ	Bearing relative eccentricity	
ϵ	Strain	
ζ	Damping ratio	
ζ	Rotating Cartesian coordinate (complex)	m
η	Fluid viscosity	Pa·s, N·s/m²
η	Rotating Cartesian coordinate	m
θ	Rotation angle	rad
λ	Eigenvalue ω^2	(rad/s)²
μ	Friction coefficient	
μ	$\zeta\omega_n$	s⁻¹
ξ	Rotating Cartesian coordinate	m
π	3.141592653	
ρ	Radius of curvature	m
ρ	Material density	kg/m³
σ	Stress	Pa, N/m²
σ_x	Standard deviation of x	
τ	Elapsed time	s
τ	Shear stress	N/m²
ϕ	Torsional vibration	rad
ϕ	Normal Mode	Depends on problem
χ	Beam slope in the $x - z$ plane	rad
ψ	Beam slope in the $y - z$ plane	rad
ω	Angular velocity	rad/s

Appendix VII

Nomenclature *(cont.)*

Symbol	Meaning	Units

Subscripts

c	Critical
d	Damped
eq	Equivalent
h	Hysteretic
n	Natural
p	Polar
T	Torsional
v	Viscous

Prefixes

Δ	Small increment
$o(\cdot)$	Order of $(\cdot)$
c	centi (/100)
k	kilo ($\times$ 1000)
M	mega ($\times$ 10^6)
m	milli (/1000)
μ	micro ($\times$ 10^{-6})

Operations

$L[x(t)]$	Differential operator on $x(t)$
$L\{x(t)\}$	Laplace transform of $x(t)$
$\text{Im}(\cdot)$	Imaginary part of $(\cdot)$
$\text{Re}(\cdot)$	Real part of $(\cdot)$
$\det(\mathbf{A})$	Determinant of $\mathbf{A}$
$\{\cdot\}$	Transposed column vector
$[\cdot]$	Matrix
$[\cdot]^T$	Transpose of a matrix
$[\cdot]^{-1}$	Inverse of a matrix
$\mathbf{A}$	Matrix
$\mathbf{a}$	Vector
$\cdot$	Time derivative
$|x|$	Absolute value of x
$\text{RMS}(x) = \sqrt{\overline{x^2}}$	

Answers
to
Selected
Problems

(Steel properties used, where applicable: $E = 2.1 \times 10^{11}$ N/m^2, $G = 1.0 \times 10^{11}$ N/m^2, $\rho = 7800$ kg/m^3)

CHAPTER 1

1.1. $A = 5.831$, $\phi = 1.03$ rad.   **1.3.** $x = A \sin(\omega t \cdot \phi)$, $A = 0.1797$ m, $\phi = 0.59$ rad.
1.5. $\phi = -2.30$ rad.   **1.7.** $x = 51 \times 10^{-6} \sin 196.2t$ **1.9.** $A = 0.2$ mm, $A_1 = 0.141$ mm, $A_2 = 0.142$ mm (RMS), $V_{RMS} = 0.707$ V_{max} **1.11.** $A_0 = 0$, $A_1 = A_2 = A_3 = \cdots = 0$, $B_n = 2F_0/\pi n$, $n = 1, 2, 3 \ldots$ **1.13.** $A_0 = F_0$, $A_n = 0$, $B_n = F_0/\pi n$, $n = 1, 2, 3 \ldots$ **1.15.** $A_0 = 4A/\pi$, $A_1 = 0$, $A_2 = 2A/\pi$, $A_3 = 0, \ldots$, $B_n = 0$, $n = 1, 2, 3 \ldots$
1.17. $A_n = 0$, $B_1 = -1.18F_0$, $B_2 = 0$, $B_3 = -0.54F_0$, $B_4 = 0, \ldots$ **1.21.** $C_0 = F_0$, $C_1 = 2F_0/\pi$, $C_2 = 0$, $C_3 = 2F_0/3\pi, \ldots$ **1.23.** $C_1 = 0.318F_0$, $C_2 = 0.159F_0$, $C_3 = 0.106F_0$, $\ldots$ **1.25.** $C_0 = 1.27A$, $C_1 = 0$, $C_2 = 0.212A$, $C_3 = 0, \ldots$ **1.27.** $C_1 = -1.18iF_0$, $C_2 = 0$, $C_3 = -0.54iF_0, \ldots$

CHAPTER 2

2.1. $k = 291{,}666$ N/m   **2.3.** $k_T = 984$ Nm/rad   **2.5.** $k_T = 0.126 \times 10^6$ Nm/rad ($G = E/2$)   **2.6.** $k = 1.61 \times 10^8$ N/m   **2.7.** $h = 77$ mm   **2.10.** $k_1 = 2500$ N/m,

739

$d_2 = 12.03$ mm **2.11.** $\omega_n = 93.5$ rad/sec **2.13.** $\omega_n = 80.5$ rad/sec **2.15.** $d = 34$ mm **2.19.** $\omega_1 = 49.6$ rad/sec, $\omega_2 = 99.2$ rad/sec **2.22.** $\omega_n = 0.66$ rad/sec **2.24.** $\omega_n = 0.8$ rad/sec **2.28.** $\omega_n = \{(kl^2/4 - wa)/[(w/g)(l^2/4 + a^2)]\}^{1/2}$ **2.30.** $\omega_n = \{[\pi d^4 G/16l + kL^2]/[(1/12)mL^2]\}^{1/2}$ **2.31.** $\omega_n = [(k_1 h_1^2)/(mL^2) + (k_2 h_2^2)/(mL^2) \cdot g/L]^{1/2}$ **2.33.** $\omega_n = [(ka^2)/(mL^2)]^{1/2}$ **2.35.** $A = [(mga/kl^2)^2 + (\omega_o/\omega_n^2)^2]^{1/2}$, $\omega_0 = 3m(2gh)^{1/2}/(m_1 l^2 + 3ma^2)$, $\omega_n = (kl^2/J_0)^{1/2}$ **2.37.** $A = [(mg/k)^2 + [(m(2gh)^{1/2}]^2(m/k)]^{1/2}$ **2.41.** $\omega_n = [4g/(15R)]^{1/2}$ **2.43.** $\omega_n = [(3(2)^{1/2}/4)(2(2)^{1/2} + mg)/ma)]^{1/2}$ **2.45.** $\omega_n = 1.02(g/L)^{1/2}$ **2.51.** $A_1 = A \exp(\pi\zeta/2)$, $A_2 = A_1 \exp(\pi\zeta/2)$ **2.53.** $A_1 = A \exp(\pi\zeta/2)$, $A_2 = A_1 \exp(\pi\zeta/2)$

CHAPTER 3

3.1. $X = 6.81 \times 10^{-6}$ m **3.3.** $X = 3.96 \times 10^{-6}$ m **3.5.** $V_{ripple} = 9.48$ Volts **3.8.** $X = 6.35 \times 10^{-3}$ m **3.9.** $F = P_0 A[1 - k/(k - m\omega^2)]\cos \omega t$ **3.11.** $X = 6.81 \times 10^{-6}$ **3.13.** $X = 3.96 \times 10^{-6}$ **3.15.** $V_{ripple} = 0.996$ volts **3.19.** $F = P_0 A[\cos \omega_n t - k \sin \omega t/(c\omega_n)]$, $c = 3\eta\pi(D/d)^3(1 + 2d/D)/4$ **3.21.** $x_s(t) = k_2 e \cos(\omega t + \phi)/[(k_1 + k_2 \cdot m\omega^2)^2 + (c\omega)^2]^{1/2}$ **3.23.** $X_s(t) = -ce\omega \sin(\omega t \cdot \phi)/[(k - m\omega^2)^2 + (c\omega)^2]^{1/2}$ **3.25.** $X_s = k_2 e \cos(\omega t \cdot \phi) - c_2 e\omega \sin (\omega t - \phi)]/[(k_1 + k_2 - m\omega^2)^2 + [(c_1 + c_2)\omega]^2]^{1/2}$ **3.27.** $k = 58500$ N/m **3.29.** $X = 1.55 \times 10^{-3}$

CHAPTER 4

4.3. $\omega_n = (4g/15R)^{1/2}$ **4.5.** $\omega_n = 1.02(g/L)^{1/2}$ **4.7.** $\omega_n = [(l\omega^2 + g)/(l + J/ml)]^{1/2}$ **4.9.** $\omega_n = \{[3m/2 + M(2 + l/r)]/[k_1 + k_2 - Mg(1 + l/r)/r]\}^{1/2}$ **4.12.** $\omega_n = [2mgr \sin \theta/J]^{1/2}$ **4.14.** $\omega_n = (k/J)^{1/2}$ **4.17.** $\omega_n = \{k/[J_1 + 2J_2(\bar{r}_1/r_1)^2]\}^{1/2}$ **4.19.** $\omega_n = [6J(D - h)/L^2]^{1/2}$ **4.22.** $\omega_n = [[3(2)^{1/2}/4][2(2)^{1/2}ka^2 + mg]/ma]^{1/2}$ **4.24.** $\omega_n = 1.02(g/L)^{1/2}$ **4.26.** $\omega_n = [2g/3(R - r)L]^{1/2}$ **4.28.** $\omega_n = [kl^2/(J_+ + ml^2)L]^{1/2}$ **4.30.** $\omega_n = [kL^2/2J_0]^{1/2}$ **4.32.** $\omega_n = [(k_1 + k_2)/J]^{1/2}$ **4.36.** $(3/2)m(R \cdot r)^2 \ddot\theta + mg(R \cdot r)\theta = F_q = -c_v\dot\theta$, $c_v = \delta a(R \cdot r)[(3/2)(R \cdot r)g]^{1/2}/\pi$ **4.40.** $W = 3.9$ kW

CHAPTER 5

5.1. $x_s(t) = F_0/2k + 0.63F_0 \sin \omega_0 t/(k - m\omega^2) + 0.212F_0 \sin 3\omega_0 t/(k - 9m\omega^2) + \cdots$ **5.3.** $a_0 = b_1 = b_2 = b_3 = \cdots = 0$, $a_1 = 0.318F_0/(k - m\omega^2)$, $a_2 = 0.159F_0/(k - 4m\omega^2)$ **5.5.** $a_0 = 1.27A/k$, $a_1 = 0$, $a_2 = 0.212F_0/(k - m\omega^2)$, $a_3 = 0.159F_0 \sin \omega_0 t/(k - 4m\omega^2)$, $\ldots$ **5.7.** $a_0 = a_1 = a_2 = a_3 = \cdots = 0$, $b_1 = 1.18F_0/(k - m\omega^2)$, $b_2 = 0$, $b_3 = -0.54F_0/(k - 4m\omega^2)$, $\ldots$ **5.9.** $a_0 = 5.31 \times 10^{-7} A_0$, $a_1 = 0$, $b_1 = -1.86 \times 10^{-6} F_0$, $a_2 = 0$, $b_2 = 0$, $\ldots$ **5.11.** $a_0 = 0$, $a_1 = -9.41 \times 10^{-7}F_0$, $a_2 = -3.8 \times 10^{-7}F_0$, $\ldots$, $b_1 = b_2 = b_3 = \cdots = 0$ **5.13.** $a_0 = 6.74 \times 10^{-7}A$, $a_1 = b_1 = 0$, $a_2 = -5.07 \times 10^{-8}A$, $b_2 = 0$, $a_3 = 0$, $b_3 = 0$, $\ldots$ **5.15.** $a_0 = 0$, $a_1 = 0$, $b_1 = 3.49 \times 10^{-6}F_0$, $a_2 = 0$, $b_2 = 0$, $a_3 = 0$, $b_3 = 5.103 \times 10^{-8}F_0$, $\ldots$ **5.17.** $a_0 = 0$, $a_1 = -3.49 \times 10^{-6}F_0$, $b_1 = 0$, $a_2 = 0$, $b_2 = 0$, $a_3 = -5.103 \times 10^{-8}F_0$, $b_3 = 0$, $\ldots$ **5.21.** $x(t) = -(F_0/m\omega_d)\{\alpha \exp[-(c/2m)(t - t_0)][\cos \omega_d(t - t_0) \cdot (\beta/\alpha) \sin \omega_d(t - t_0)]/(\alpha^2 + \beta^2)\} + (F_0/m\omega_d) \alpha \exp[-(c/2m)t][\cos \omega_d t - (\beta/\alpha) \sin \omega_d t]/(\alpha^2 + \beta^2)\}$ **5.34.** $x(t) = -(F_0/m\omega_d)\{\alpha \exp[-(c/2m)t - t_0][\cos \omega_d(t - t_0) \cdot (\beta/\alpha) \sin \omega_d(t - t_0)]/(\alpha^2 + \beta^2)\} + (F_0/m\omega_d)\{\alpha \exp[-(c/2m)t][\cos \omega_d t - (\beta/\alpha) \sin \omega_d t]/(\alpha^2 + \beta^2)\}$ **5.54.** $\ddot{x}_{max} = 0.45g$ **5.56.** $F_{max} = 7129$ N **5.57.** Will not derail.

CHAPTER 6

6.1. $\omega_{n1} = 13.66$, $\omega_{n2} = 40.88$ rad/sec **6.3.** $\omega_{n1} = 2.82$ rad/sec **6.5.** $\omega_{n1} = 125.4$ rad/sec **6.7.** $\omega_{n1} = 88.4$ rad/sec **6.9.** $\omega_{n1} = 0$, $\omega_{n2} = 1169$ rad/sec **6.11.** $\omega_{n1} = 14.9$,

ω_{n2} = 16.3 rad/sec **6.13.** ω_{n1} = 8.36, ω_{n2} = 353 rad/sec **6.14.** H = 187 m
6.17. ω_{n1} = 2.17, ω_{n2} = 33.4 rad/sec

CHAPTER 7

7.1. 433.2, 1465.6, 2241.4 rad/sec and their fractions by 2, 12, 16. **7.3.** 155.4, 1249.1, 1745.3 rad/sec. **7.5.** 0, 24, 28.6 rad/sec. **7.8.** 233, 653, 943 rad/sec. **7.9.** 0.7, 2.3, 3.0 rad/sec. **7.11.** 433.2, 1456, 2241 rad/sec **7.13.** 155.4, 1249.1, 1745.3 rad/sec. **7.15.** 0, 24, 28.6 rad/sec. **7.18.** 233, 653, 943 rad/sec. **7.19.** 0.7, 2.3, 3.0 rad/sec. **7.31.** $\ddot{q}_i + 2 \times 0.05\omega_i\dot{q}_i + \omega_i^2 q_i = 0$, ω_i as in problem 7.1. **7.33.** $\ddot{q}_i + 2 \times 0.05\omega_i\dot{q}_i + \omega_i^2 q_i = 0$, ω_i as in problem 7.3. **7.35.** $\ddot{q}_i + 2 \times 0.05\omega_i\dot{q}_i + \omega_i^2 q_i = 0$, ω_i as in problem 7.5. **7.38.** $\ddot{q}_i + 2 \times 0.05\omega_i\dot{q}_i + \omega_i^2 q_i = 0$, ω_i as in problem 7.8. **7.39.** $\ddot{q}_i + 2 \times 0.05\omega_i\dot{q}_i + \omega_i^2 q_i = 0$, ω_i as in problem 7.9.

CHAPTER 8

8.1. $X_1 = 3.86 \times 10^{-6}$, $X_2 = 1.33 \times 10^{-6}$, $X_3 = 2.47 \times 10^{-6}$ m. **8.2.** $X_1 = 9.38 \times 10^{-4}$, $X_2 = 1.33 \times 10^{-5}$, $X_3 = 9.84 \times 10^{-4}$ m. **8.5.** $X_1 = 1.17 \times 10^{-4}$, $X_2 = 3.44 \times 10^{-7}$, $X_3 = 3.44 \times 10^{-7}$ m. **8.7.** cos terms: $X_1 = -8.36 \times 10^{-6}$, $X_2 = -2.99 \times 10^{-14}$, $X_3 = 1.86 \times 10^{-9}$ m. sin terms: $X_1 = 2.94 \times 10^{-14}$, $X_2 = 8.32 \times 10^{-6}$, $X_3 = -6.2 \times 10^{-10}$ m. **8.9.** $X_1 = 4.98 \times 10^{-6}$, $X_2 = 1.18 \times 10^{-5}$, $X_3 = 4.98 \times 10^{-6}$ m. **8.13.** $X_1 = 1.56 \times 10^{-5}$, $X_2 = 3.00 \times 10^{-5}$, $X_3 = 3.29 \times 10^{-5}$ m. **8.15.** $X_1 = 1.56 \times 10^{-5}$, $X_2 = 3.00 \times 10^{-5}$, $X_3 = 3.29 \times 10^{-5}$ m. **8.18.** $X_1 = 2.11 \times 10^{-5}$, $X_2 = 2.61 \times 10^{-5}$, $X_3 = 1.58 \times 10^{-5}$ m. **8.20.** $X_1 = 1.25 \times 10^{-6}$, $X_2 = 1.07 \times 10^{-5}$, $X_3 = 1.25 \times 10^{-6}$ m. **8.22.** $\overline{F} = 1.34 \times 10^6$ N, $\sigma_F = 0.49 \times 10^6$ N. **8.24.** $\overline{\omega}_d$ = 17.3 rad/sec, σ_ω = 8 rad/sec. **8.26.** 4.5% **8.28.** 4.5% **8.30.** 4.5%

CHAPTER 9

9.2. 51.64 m/s **9.4.** 5027 m/s **9.6.** 2808 N **9.8.** 256, 513, 769 rad/sec **9.10.** tan $\omega l/c$ = $1/[\omega EA/c]$ **9.12.** tan $\omega l_1/c + (E_1 A_1/E_2 A_2) \tan \omega l_2/c = 0$ **9.15.** tan $\omega l/c = -AE\omega/kc$ **9.17.** tan $\omega l_1/c + (G_1 J_{p1}/G_2 J_{p2}) \tan \omega l_2/c = 0$ **9.18.** tan $\omega l/c = (\lambda_1 + \lambda_2)/(1 - \lambda_1\lambda_2)$ $\lambda_1 = GI_p/(J_{1p}c\omega)$, $\lambda_2 = GI_p/(J_{2p}c\omega)$ **9.21.** 14.22, 28.44, 42.66 rad/sec. **9.23.** 0, 5185, 16424 rad/sec. **9.24.** 317.18 rad/sec. **9.28.** 26.67, 106.68, 240.05 rad/sec. **9.30.** 94.65, 114.5, 2159 rad/sec.

CHAPTER 10

10.1. $\omega_n = 0.585(EI/ml^3)^{1/2}$ **10.3.** 427% **10.4.** 33/140 **10.6.** $\omega_n = 0.249(IG/JL)^{1/2}$ **10.7.** $\omega_n = 2.74(a^2/l)(Eb/m)^{1/2}$ **10.9.** $\omega_n = [k_1 k_2 l^2/m(k_1 l^2 + k_2 L^2)]^{1/2}$ **10.15.** $\omega_n = [(k_1 + k_2 - lmg/r^2)/[1.5m + m(1 - l/r)^2]]^{1/2}$ **10.21.** $\omega_n = (4g/15R)^{1/2}$ **10.22.** $\omega_n = 1.03(g/L)^{1/2}$ **10.23.** $\omega_n = 1.02(g/L)^{1/2}$ **10.25.** $\omega_n = [12(\pi d^4 G/16l + kL^2)/mL^2]^{1/2}$ **10.28.** $\omega_n = (k/J)^{1/2}$ **10.35.** $\omega_n = [2k(Jr_1^2 + Jr_2^2)/J_1 J_2]^{1/2}$ **10.39.** $\omega_{1n} = (kh^2/J)$, $\omega_{2n} = (3kh^2/J)^{1/2}$, **10.40.** $\omega_{1n} = (k/m)$, $\omega_{2n} = (kh^2/2J)$, **10.71.** $\omega_n = 0.585(EI/ml^3)^{1/2}$ **10.73.** 427% **10.74.** 33/140 **10.76.** $\omega_n = 0.249(IG/JL)^{1/2}$ **10.77.** $\omega_n = 2.74(a^2/l)(Eb/m)^{1/2}$ **10.79.** $\omega_n = [k_1 k_2 l^2/m(k_1 l^2 + k_2 L^2)]^{1/2}$

CHAPTER 11

11.1. $\omega_{cr} = 625$ rad/sec, $\omega_n = 4.21$ rad/sec **11.2.** No gyroscopic effect, $X = 2.36 \times 10^{-6}$ m With gyroscopic effect, $X = 1.32 \times 10^{-6}$ m **11.3.** Amplitude will not change substantially.

CHAPTER 12

12.7. $t = \pi(x_0 k/F - 1)/2\omega_n$ **12.8.** $mu = 0.31$ **12.16.** $T = 4$
$$\int_0^{x_0} dx / \left[2 \int_{x_0}^{x} f(x)dx \right]^{1/2} = 4[(n+1)/2]^{1/2} \int_0^{x_0} dx/(x_o^{n+1} - x_o^{n-1})$$

CHAPTER 13

13.1. Half acceleration, 1.5 hours. **13.2.** Complaint not founded. **13.3.** No need for isolation. **13.4.** $m_u a = 0.45 \times 10^{-3}$ $kg \cdot m$ **13.5.** Very rough. **13.6.** 0.6 **13.7.** 0.18 **13.9.** $k_1 = 4.35 \times 10^9$, $k_2 = 7737$ **13.10.** 11.447 N sec/m **13.11.** 1448 kg

CHAPTER 14

14.1. $a_0 = 0.07$, $a_1 = 0.39$, $a_2 = 0.92$, $a_3 = 2.21$, $a_4 = 6.55$, $a_5 = 6.08$, $a_6 = 6.81$, $a_7 = 3.03$, $a_8 = 1.80$, $a_9 = 1.35$, $a_{10} = 1.09$ **14.2.** $a_0 = 1.00$, $a_1 = 0.50$, $a_2 = 1.00$, $a_3 = 1.00$, $a_4 = 1.00$, $a_5 = 1.00$, $a_6 = 1.00$, $a_7 = 1.00$, $a_8 = 1.00$, $a_9 = 1.00$, $a_{10} = 0.50$ **14.3.** $a_0 = 0.50$, $a_1 = 1.00$, $a_2 = 0.75$, $a_3 = 1.00$, $a_4 = 1.00$, $a_5 = 1.00$, $a_6 = 1.00$, $a_7 = 1.00$, $a_8 = 1.00$, $a_9 = 0.75$, $a_{10} = 0.75$ **14.4.** $a_0 = 1.00$, $a_1 = 0.62$, $a_2 = 1.00$, $a_3 = 0.87$, $a_4 = 1.00$, $a_5 = 1.00$, $a_6 = 1.00$, $a_7 = 1.00$, $a_8 = 0.87$, $a_9 = 0.87$, $a_{10} = 0.50$ **14.5.** $a_0 = 0.62$, $a_1 = 1.00$, $a_2 = 0.75$, $a_3 = 1.00$, $a_4 = 0.94$, $a_5 = 1.00$, $a_6 = 1.00$, $a_7 = 0.94$, $a_8 = 0.94$, $a_9 = 0.69$, $a_{10} = 0.69$ **14.6.** Mean $= 0.11$, Standard deviation $= 0.61266$, Kurtosis $= 6.3957$ **14.7.** Mean $= 0.045236$, Standard deviation $= 0.66206$, Kurtosis $= 5.2034$ **14.8.** SE $= x_{pulse} = X_{pulse} \times \Delta T_{pulse}/T_{pulse} = 1.5 \times .001/0.0166 = 0.09$ **14.9.** SE $= x_{pulse} = X_{pulse} \times \Delta T_{pulse}/T_{pulse} = 1.5 \times .001/0.00033 = 0.45$ **14.10.** $X_{env} = 2|\cos 85t|$ **14.21.** $f_1 = 4.33X$, $f_2 = 5.66X$, $f_3 = 7.36X$, $f_4 = 0.29X$ **14.22.** left: $f_1 = 4.37X$, $f_2 = 5.62X$, $f_3 = 7.87X$, $f_4 = 0.29X$. right: $f_1 = 4.41X$, $f_2 = 5.58X$, $f_3 = 8.45X$, $f_4 = 0.29X$ **14.23.** $f_1 = 0.9X$, $f_2 = 1.03X$ **14.24.** $f_1 = N_s/N_p = 0.9X$, $f_2 = N_p/N_r = 0.34X$ **14.25.** Crack **14.26.** Misalignment. **14.27.** Bearing whirl **14.28.** Looseness, dry friction. **14.29.** Fluid or torque whirl.

Index

Name Index

*Boldface numbers indicate pages with biographical data.

Subject Index